S0-AVN-715

Astrophysics in the Next Decade

Astrophysics and Space Science Proceedings

For other titles published in this series, go to
www.springer.com/series/7395

Astrophysics in the Next Decade

The James Webb Space Telescope and Concurrent Facilities

Edited by

H.A. Thronson

NASA Goddard Space Flight Center, Greenbelt, MD, U.S.A.

M. Stiavelli

Space Telescope Science Institute, Baltimore, MD, U.S.A.

and

A.G.G.M. Tielens

NASA Ames Research Center, Moffett Field, CA, U.S.A.

Editors

Harley A. Thronson
NASA Goddard Space Flight Center (GSFC)
Greenbelt MD 20771
USA
Harley.A.Thronson@nasa.gov

A.G.G.M. Tielens
NASA Ames Research Center
Mail Stop 239-4
Moffett Field CA 94035-1000
USA
Alexander.G.Tielens@nasa.gov

Massimo Stiavelli
Space Telescope Science Institute (STScI)
3700 San Martin Dr.
Baltimore MD 21218
USA
mstiavel@stsci.edu

ISBN: 978-1-4020-9456-9 e-ISBN: 978-1-4020-9457-6

DOI 10.1007/978-1-4020-9457-6

Library of Congress Control Number: 2008939015

Printed on acid-free paper

9 8 7 6 5 4 3 2 1

springer.com

Preface

The conference that eventually became *Astrophysics in the Next Decade: JWST and Concurrent Facilities* was originally proposed by Matthew Greenhouse in 2006. He argued that the astronomical community would be well served by a meeting (and proceedings) that compiled in one place the breadth of scientific and technical information available for the James Webb Space Telescope. As planning for the meeting evolved, several individuals urged that the conference be broadened to include other major astronomical missions of the coming decade. The conference was held September 24–27, 2007 at the Marriott Starr Pass in Tucson, Arizona, and was attended by over 200 astronomers from around the world. The facilities were outstanding and the Sonoran Desert was stunning in early autumn.

Under the general guidance of the local organizers, the Science Organizing Committee selected topics and arranged for speakers to address the key goals of the astronomical community in the next decade. While the conference emphasized astronomical observations from the visible to the sub-millimeter, several sessions dealt with key science goals enabled by dramatic new capabilities in other bands, particularly when these goals were complementary to those of Webb observers. The breadth of the final program and these proceedings underscore the importance of the Webb mission to almost all areas of modern astrophysics research.

Harley A. Thronson — Greenbelt, Maryland
Massimo Stiavelli — Baltimore, Maryland
Alexander G.G.M. Tielens — Moffett Field, California

Acknowledgements

The conference could not have been successful without the participation from a number of individuals and institutions. In addition to the committee membership, Marcia Rieke assisted with logistics in Tucson and Eric Smith provided essential support from the NASA Science Mission Directorate. Mario Livio and other scientists from the Space Telescope Science Institute, the JWST Project, Johns Hopkins University, and NASA Goddard Space Flight Center Flight Programs Directorate contributed institutional and financial support. Of course, the meeting and these proceedings would not be possible without the commitment of the speakers, moderators, and authors, to whom we are indebted.

Science Organizing Committee

Chairperson: Alan Dressler

LOC Liaison: Peter Stockman

Other Members: Crystal Brogan, Dale Cruikshank, Ewine van Dishoeck, Richard Ellis, Rob Kennicutt, Rolf Kudritzki, Avi Loeb, John Mather, Yvonne Pendleton, Massimo Stiavelli, Leonardo Testi, Xander Tielens, Meg Urry, Jeff Valenti.

Local Organizing Committee

Chairperson: Peter Stockman

Other Members: Jonathan Gardner, Matt Greenhouse, Heidi Hammel, John Mather, Neill Reid, Massimo Stiavelli, Harley Thronson.

Photography

Harley Thronson and Dale Cruikshank

Administrative Support Personnel

Karyn Keidel, Tina Schappell, Darlene Spencer, Lee Tinnin
Kirsten Theunissen
Springer Science + Business Media B.V.

Deivanai Loganathan and Team,
Integra Software Services Pvt. Ltd, India.

Contents

1 The James Webb Space Telescope 1
Jonathan P. Gardner and the JWST Science Working Group

2 Beyond the Hubble Space Telescope: Early Development of the Next Generation Space Telescope 31
Robert W. Smith and W. Patrick McCray

3 The Kuiper Belt and Other Debris Disks 53
David Jewitt, Amaya Moro-Martín and Pedro Lacerda

4 The Future of Ultracool Dwarf Science with JWST 101
Mark S. Marley and S.K. Leggett

5 Transiting Exoplanets with JWST 123
S. Seager, D. Deming and J.A. Valenti

6 The Unsolved Problem of Star Formation: Dusty Dense Cores and the Origin of Stellar Masses 147
Charles J. Lada

7 Accretion Disks Before (?) the Main Planet Formation Phase 167
C. Dominik

8 Astrochemistry of Dense Protostellar and Protoplanetary Environments 187
Ewine F. van Dishoeck

9 Extreme Star Formation 215
Jean L. Turner

10 Prospects for Studies of Stellar Evolution and Stellar Death in the *JWST* Era . . . 247
Michael J. Barlow

11 Origin and Evolution of the Interstellar Medium . . . 271
A.G.G.M. Tielens

12 Astrophysics in the Next Decade: The Evolution of Galaxies . . . 309
Alice Shapley

13 The Co-Evolution of Galaxies and Black Holes: Current Status and Future Prospects . . . 335
Timothy M. Heckman

14 The Intergalactic Medium at High Redshifts . . . 357
Steven R. Furlanetto

15 Observing the First Stars and Black Holes . . . 385
Zoltán Haiman

16 Baryons: What, When and Where? . . . 419
Jason X. Prochaska and Jason Tumlinson

17 Observational Constraints of Reionization History in the JWST Era . 457
Xiaohui Fan

18 The Frontier of Reionization: Theory and Forthcoming Observations . . . 481
Abraham Loeb

Index . . . 511

Contributors

Michael J. Barlow Department of Physics and Astronomy, University College London, Gower Street, London WC1E 6BT, UK, mjb@star.ucl.ac.uk

D. Deming NASA Goddard Space Flight Center, Planetary Systems Branch, Code 693 Greenbelt, MD 20771, USA, leo.d.deming@nasa.gov

C. Dominik Sterrenkundig Instituut "Anton Pannekoek", University of Amsterdam, Kruislaan 403, NL-1098 SJ, Amsterdam; and Afdeling Sterrenkunde, Radboud Universiteit Nijmegen, Faculteit NWI, Postbus, 9010, NL-6500 GL Nijmegen, dominik@science.uva.nl

Xiaohui Fan Steward Observatory, University of Arizona, Arizona, USA, fan@as.arizona.edu

Steven R. Furlanetto University of California, Los Angeles, CA, USA, sfurlane@astro.ucla.edu

Jonathan P. Gardner Observational Cosmology Laboratory, Code 665, NASA Goddard Space Flight Center, Greenbelt MD 20771, USA, jonathan.p.gardner@nasa.gov

Zoltán Haiman Department of Astronomy, Columbia University, New York, NY 10027, USA, zoltan@astro.columbia.edu

Timothy M. Heckman Center for Astrophysical Sciences, Department of Physics and Astronomy, Johns Hopkins University, Baltimore, MD, USA, heckman@pha.jhu.edu

David Jewitt Institute for Astronomy, University of Hawaii, 2680 Woodlawn Drive, Honolulu, USA, jewitt@ifa.hawaii.edu

Pedro Lacerda Institute for Astronomy, University of Hawaii, USA, pedro@ifa.hawali.edu

Charles J. Lada Harvard-Smithsonian Center for Astrophysics, Cambridge, MA 02138, USA, clada@cfa.harvard.edu

S.K. Leggett Gemini Observatory, 670 North A'ohoku Place Hilo, HI 96720, USA, skl@gemini.edu

Abraham Loeb Harvard University, CfA, MS 51, 60 Garden Street, Cambridge MA 02138, USA, aloeb@cfa.harvard.edu

Mark S. Marley NASA Ames Research Center, Mail Stop 245-3, Moffett Field CA 94035, USA, Mark.S.Marley@NASA.gov

W. Patrick McCray Department of History, University of California, Santa Barbara CA, USA, pmccray@history.ucsb.edu

Amaya Moro-Martín Department of Astronomy, Princeton University, Princeton, NJ, USA, amaya@astro.princeton.edu

Jason X. Prochaska University of California Observatories - Lick Observatory, University of California, Santa Cruz, CA 95064, USA, xavier@ucolick.org

S. Seager Department of Earth, Atmospheric, and Planetary Sciences, Dept. of Physics, Massachusetts Institute of Technology, 77 Massachusetts Ave, 54-1626, Cambridge, MA, 02139, USA, seager@mit.edu

Alice Shapley Princeton University, Peyton Hall, Ivy Lane, Princeton, NJ 08544, USA, aes@astro.princeton.edu

Robert W. Smith Department of History and Classics, University of Alberta, Edmonton, Alberta, Canada, Robert.Smith@ualberta.ca

A.G.G.M. Tielens MS 245-3, Space Sciences Division, NASA Ames Research Center, Moffett Field, CA 94035, USA, Alexander.G.Tielens@nasa.gov

Jason Tumlinson Yale Center for Astronomy and Astrophysics, Yale University, New Haven, CT, 06520, USA, jason.tumlinson@yale.edu

Jean L. Turner Department of Physics and Astronomy, UCLA, Los Angeles CA 90095-1547 USA, turner@astro.ucla.edu

J.A. Valenti Space Telescope Science Institute, 3700 San Martin Dr., Baltimore, MD 21218, USA, valenti@stsci.edu

Ewine F. van Dishoeck Leiden Observatory, Leiden University, P.O. Box 9513, 2300 RA Leiden, The Netherlands and Max-Planck Institute für Extraterrestrische Physik, Garching, Germany, ewine@strw.leidenuniv.nl

Chapter 1
The James Webb Space Telescope

Jonathan P. Gardner and the JWST Science Working Group[1]

Abstract The James Webb Space Telescope (JWST) is a large (6.6 m), cold (<50 K), infrared (IR)-optimized space observatory that will be launched early in the next decade into orbit around the second Earth–Sun Lagrange point. The observatory will have four instruments: a near-IR camera, a near-IR multi-object spectrograph, and a tunable filter imager that will cover the wavelength range, $0.6 < \lambda < 5.0\,\mu$m, while the mid-IR instrument will do both imaging and spectroscopy from $5.0 < \lambda < 29\,\mu$m. The JWST science goals are divided into four themes. The End of the Dark Ages: First Light and Reionization theme seeks to identify the first luminous sources to form and to determine the ionization history of the early universe. The Assembly of Galaxies theme seeks to determine how galaxies and the dark matter, gas, stars, metals, morphological structures, and active nuclei within them evolved from the epoch of reionization to the present day. The Birth of Stars and Protoplanetary Systems theme seeks to unravel the birth and early evolution of stars, from infall onto dust-enshrouded protostars to the genesis of planetary systems. The Planetary Systems and the Origins of Life theme seeks to determine the physical and chemical properties of planetary systems including our own, and investigate the potential for the origins of life in those systems. To enable these science goals, JWST consists of a telescope, an instrument package, a spacecraft, and a sunshield. The telescope primary mirror is made of 18 beryllium segments, some of which are deployed. The segments will be brought into optical alignment on-orbit through a process of periodic wavefront sensing and control. The instrument package contains the four science instruments and a fine guidance sensor. The spacecraft provides pointing, orbit maintenance, and communications. The sunshield provides passive thermal control. The JWST operations plan is based on that used for previous space

---

J.P. Gardner (✉)
Observational Cosmology Laboratory, Code 665, NASA Goddard Space Flight Center, Greenbelt MD 20771, USA
e-mail: jonathan.p.gardner@nasa.gov

[1]John C. Mather, Mark Clampin, Rene Doyon, Kathryn A. Flanagan, Marijn Franx, Matthew A. Greenhouse, Heidi B. Hammel, John B. Hutchings, Peter Jakobsen, Simon J. Lilly, Jonathan I. Lunine, Mark J. McCaughrean, Matt Mountain, George H. Rieke, Marcia J. Rieke, George Sonneborn, Massimo Stiavelli, Rogier Windhorst and Gillian S. Wright.

H.A. Thronson et al. (eds.), *Astrophysics in the Next Decade,* Astrophysics and Space Science Proceedings, DOI 10.1007/978-1-4020-9457-6_1,

observatories, and the majority of JWST observing time will be allocated to the international astronomical community through annual peer-reviewed proposal opportunities.

1.1 Science

1.1.1 The End of the Dark Ages: First Light and Reionization

The James Webb Space Telescope (JWST; Fig. 1.1; Gardner et al. 2006) seeks to identify the first luminous sources to form and to determine the ionization history of the early universe. The emergence of the first sources of light in the universe marks the end of the "Dark Ages" in cosmic history, a period characterized by the absence of discrete sources of light (Rees 1997). After the appearance of the first sources of light, hydrogen in the intergalactic medium was reionized. Results from the Wilkinson Microwave Anisotropy Probe (Kogut et al. 2003) combined with data on quasars at $z \sim 6$ from the Sloan Digital Sky Survey (Fan et al. 2002) show that this reionization had a complex history (e.g. Gnedin 2004). Although there are indications that galaxies produced the majority of the ultraviolet reionizing radiation, the contribution of quasars could be significant.

What Are the First Galaxies? To identify a sample of high-redshift galaxies, JWST will make an ultra-deep imaging survey using several broad-band filters. The Lyman break technique will identify objects at increasing redshifts up to z = 30 or higher. For dwarf galaxies with 10^6 $M_\odot$ of zero-metallicity massive stars at

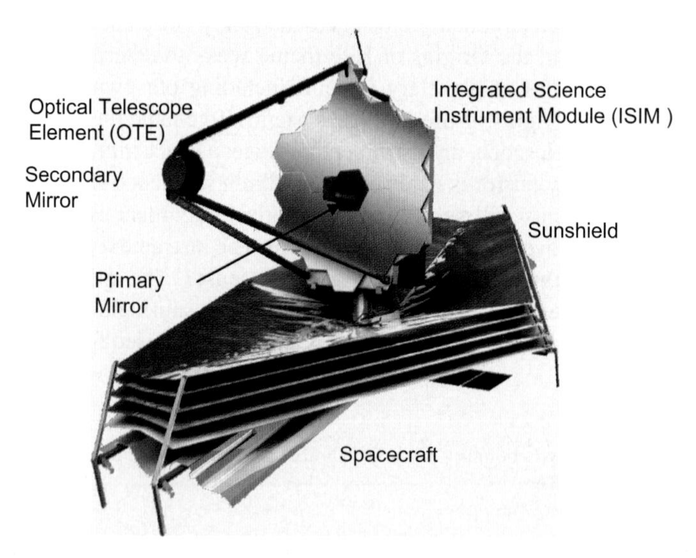

Fig. 1.1 The JWST Observatory. The optical telescope element (OTE) contains the primary and secondary mirrors, the integrated science instrument module (ISIM) element contains the instrumentation and the spacecraft element consists of the spacecraft and the sunshield

$15 < z < 30$, the expected AB magnitude at emitted wavelengths just longward of Lyman α is ∼31 mag. This ultra-deep survey will allow the derivation of galaxy counts as a function of redshift, seeing a drop at redshifts beyond the formation of the first sources. The intensity of the non-ionizing continuum yields the star formation rate as a function of redshift. The deep broad-band imaging in the near infrared will enable follow-up low-resolution spectroscopy and mid-infrared imaging of promising sources. The imaging could be timed to search for supernovae.

When and How Did Reionization Occur? JWST will obtain high signal-to-noise, $R = 1000$, near-infrared spectra of high-redshift QSOs, galaxies or GRBs in order to determine the presence of a Gunn-Peterson trough or of a Lyman α damping wing (Fig. 1.2; X. Fan, private communication). The targets will be the brightest known high-redshift objects, perhaps found in a wide-area Spitzer warm mission survey (Gardner et al. 2007). A damping wing should be present for a few million years, before the ionizing radiation is sufficient to create a large Strömgren sphere around each ionizing source. In order to measure the ionizing continuum of a class of sources, we will measure their hydrogen and helium Balmer lines. Comparison between these lines gives an estimate for the steepness, or hardness of the ionizing continuum, providing a measurement of the rate of production of ionizing photons for any given class of sources under the assumption that the escape fraction is known. Identification of the nature of ionizing sources requires a combination of diagnostics: line shapes, line widths, line ratios, and the shape of the continuum. We expect the intrinsic line shapes and widths of AGN-powered sources to be broader than those of sources ionized by stellar radiation.

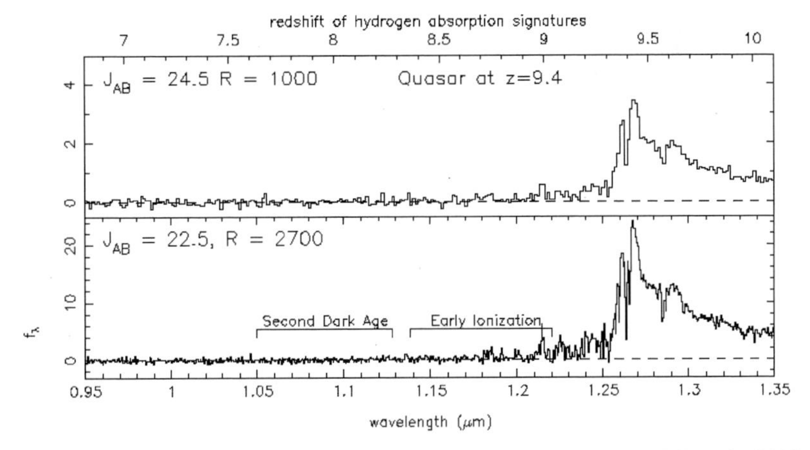

Fig. 1.2 A simulated spectrum of a $z = 9.4$ quasar observed with JWST's NIRSpec for 300 ksec. In the top figure, a faint quasar is observed with spectral resolution $R = 1000$, while in the bottom figure a brighter quasar is observed at higher resolution. The simulation includes an early reionization, followed by a recombination and a second dark age. (X. Fan, private communication.)

1.1.2 The Assembly of Galaxies

JWST seeks to determine how galaxies and the dark matter, gas, stars, metals, morphological structures, and active nuclei within them evolved from the epoch of reionization to the present day. Theory and observation show that galaxies are assembled through a process of the hierarchical merging of dark matter concentrations, accompanied by chemical evolution of the gas and dust, and resulting in the Hubble Sequence of galaxies (e.g. White & Frenk 1991, Barkana & Loeb 2001). However, we do not really know how galaxies are formed, what controls their shapes, and what makes them form stars. We do not know how the chemical elements are generated and redistributed through the galaxies, and whether the central black holes exert great influence over the galaxies. We do not know the global effects of violent events as small and large parts join together in collisions.

How did the Hubble Sequence and Heavy Elements Form? To trace the formation of the Hubble Sequence of galaxies, JWST will determine the morphologies, stellar populations, and star-formation rates in a very large sample of galaxies observed in deep-wide imaging and spectroscopic surveys, as a function of their mass, environment, and cosmic epoch. JWST will also determine when the long-lived stars in a typical galaxy were formed, whether in situ or in smaller galaxies that subsequently merged together to form a large galaxy. Direct characterization of the merging rate of galaxies will provide another angle on this question.

What Physical Processes Determine Galaxy Properties? Despite the variety of galaxy properties observed today, galaxies obey a number of remarkably tight scaling relations between basic properties of luminosity, size, kinematics and metal enrichment. These include the Tully-Fisher relation for disk galaxies and the "fundamental plane", and projections thereof, for spheroids (Tully & Fisher 1977, Bender et al. 1992). More recently, a surprising relationship between the mass of the central black hole and the properties of the surrounding spheroid (e.g., the velocity dispersion) has been established (Magorrian et al. 1998, Gebhardt et al. 2000). JWST will investigate these processes through deep imaging for structural parameters plus high-resolution spectroscopy for kinematical data. The dark matter will be revealed through gravitational weak-lensing surveys reaching $z \sim 2.5$. JWST will also use mid-infrared spectroscopy to diagnose the energy sources in the interior of bright systems, as starbursts have strong PAH features, while AGN have strong [NeVI] 7.66 μm lines (Armus et al. 2004, Soifer et al. 2004).

1.1.3 The Birth of Stars and Protoplanetary Systems

JWST seeks to unravel the birth and early evolution of stars, from infall onto dust-enshrouded protostars, to the genesis of planetary systems. The formation of stars and planets is a complex process, even in the well-developed paradigm for a single, isolated low-mass star (Shu et al. 1987). Things are even more complicated, as stars very rarely form in isolation. The current picture of star formation starts on large scales, as molecular cloud cores cool and fragment to form highly dynamic clusters of protostars, spanning the mass spectrum from O stars to planetary-mass brown dwarfs. Within those clusters, individual young sources are often encircled by disks of warm gas and dust, where material aggregates to form proto-planetary systems. These disks are the source of highly-collimated jets and outflows, which transfer energy and angular momentum from the infalling material into the surrounding medium, and clear away the remainder of the birth core. On larger scales, the intense ultraviolet flux and strong winds of the most massive stars can disperse an entire molecular cloud, while simultaneously ionizing and evaporating the circumstellar disks of the surrounding lower-mass stars.

How do Clouds Collapse and form Protoplanetary Systems? Stars form in small ($\sim$0.1 pc) regions undergoing gravitational collapse within larger molecular clouds. Standard theory predicts that these cores collapse from the inside out (Shu 1977, Terebey et al. 1984), but there are alternative theories which predict different density distributions in the cores. By measuring the extinction to background stars, JWST will map those density distributions in order to understand the relative roles that magnetic fields, turbulence, and rotation play while the clouds collapse to form stars. Once self-gravitating molecular cloud cores have formed, they collapse to form protostellar seeds, which gain material via continuing accretion. These protostars will begin radiating in the 10 μm to 20 μm region, since radiation from the warm central source is scattered off dust grains in the inner envelope into the line-of-sight. Direct imaging of disks around low-mass stars reveal their internal density and temperature structure, and show how they are affected by their environment. Since circumstellar disks are both a product and a mediator of the star-formation process, as well as the progenitors of planetary systems, it is clear that a fuller understanding of the evolution of circumstellar disks will play a key role in our understanding of these central topics. Spectra of circumstellar disks will trace the formation history of planetesimals and other solid bodies, their composition, their processing, and their possible future evolution (Fig. 1.3).

How do Very High and Very Low-Mass Stars Form? The formation of massive stars produces intense winds and ionizing radiation which impacts the surrounding molecular cloud material and the nascent circumstellar disks of adjacent low-mass stars. The mechanism by which massive stars form is not yet known. The standard disk-accretion scenario appropriate for low-mass stars cannot be simply scaled up, since predictions are that the radiation pressure from the growing central source would build up so quickly that no more material could accrete, limiting the mass of the source (Yorke & Sonnhalter 2002). Likewise, the low-mass opacity-limited cutoff for star formation is predicted to be 3 to 10 M_{JUP}, although a more detailed consideration of the role of magnetic fields may bring this down to 1 M_{JUP} (Boss 2001). JWST will conduct imaging and spectroscopic surveys of young star-forming regions to establish the both the high and low-mass limits of the initial mass function, and to determine the processes that operate at the extremes of star formation.

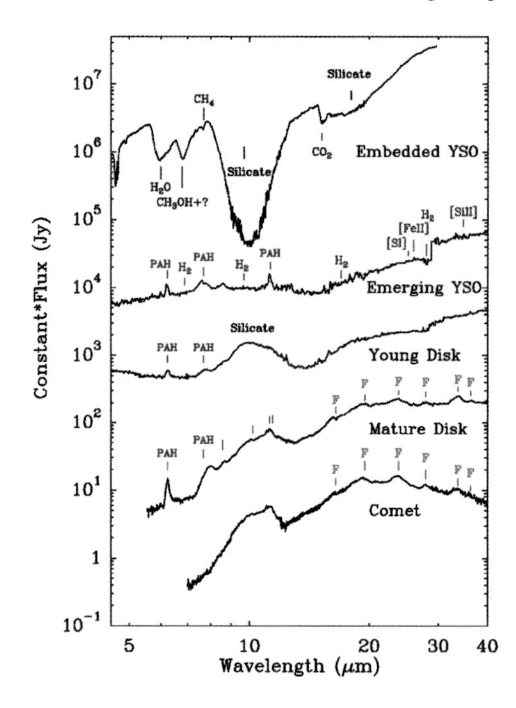

Fig. 1.3 A series of ISO SWS mid-IR spectra of young stars and circumstellar disks at different stages in their evolution (Malfait et al. 1998, Gibb et al., 2000, van den Ancker et al., 2000). From *top* to *bottom*, in a rough evolutionary sequence, the spectra change from being dominated by solid-state absorption features and shocked gas emission lines, to PAH features and photodissociation region lines, to amorphous and crystalline silicates with HI recombination lines. An ISO SWS spectrum of Comet Hale-Bopp is shown for comparison (Crovisier et al. 1997)

1.1.4 Planetary Systems and the Origins of Life

JWST seeks to determine the physical and chemical properties of planetary systems including our own, and to investigate the potential for the origins of life in those systems.

How do Planets Form? The formation of multiple objects is a common outcome of star formation, including binary or higher-order star systems, a central star orbited by brown dwarfs and/or planets, or a star with a remnant disk of particulates. Brown dwarfs and giant planets might arise from two different formation mechanisms, or planets could be the low-mass tail of a single star-formation process (Chabrier et al. 2000, Lunine et al. 2004). The formation of giant planets is a signpost, detectable with JWST, of a process that may also generate terrestrial planets. In contrast, direct collapse formation of brown dwarfs may signal systems in which terrestrial planet formation is rare or impossible, because of the required disk mass, angular momentum and subsequent disk evolution. JWST will conduct coronagraphic surveys of nearby stars to find mature Jovian companions, and of more distant systems to find young planets (Fig. 1.4). Spectroscopy of isolated or widely separated planets or sub-brown dwarfs will reveal basic physical parameters such as gravity, composition, temperature and the effect of clouds. The remnant of the circumstellar disk that formed our solar system is observable today as the smaller planets, moons, asteroids and comets, along with the zodiacal light and interplanetary gas and dust. JWST spectroscopy of comets will reveal how circumstellar disks are like our own solar system.

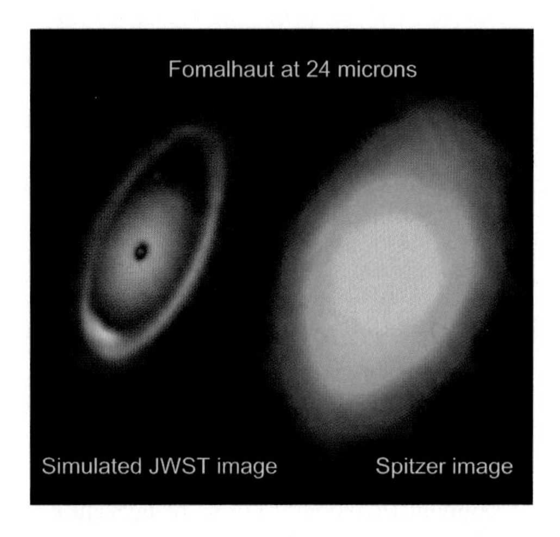

Fig. 1.4 Dusty debris disks. We show the Fomalhaut debris disk as it appears to Spitzer at 24 μm (*right*; Staplefeldt et al. 2004), and in a simulated JWST image. Structure within the disk is resolvable by JWST

How are Habitable Zones Established? Some geochemical evidence suggests that Earth's water did not come from locally formed planetesimals (Morbidelli et al. 2000). The source of water is uncertain; it could have been delivered through impacts of asteroids and comets, although isotopic ratios call the latter source into question. By measuring isotopic ratios in comets and larger Kuiper Belt bodies, JWST can solve this part of the puzzle, removing a major uncertainty in the source of water for our own planet. JWST will also monitor icy and organic-rich bodies in the outer solar system, such as Titan, in order to investigate ongoing organic evolution as well as constrain chemical processes in the early solar system.

1.2 Implementation

To make these scientific measurements, JWST will be a large cold telescope, with a wide field of view (FOV), exceptional angular resolution and sensitivity, and wide wavelength coverage in both imaging and spectroscopy. It will be launched early in the next decade to an orbit around the Earth-Sun second Lagrange point (L2).

The JWST Project is organized into three segments: observatory, ground and launch. The observatory is composed of an optical telescope element, an integrated science instrument module (ISIM) containing the scientific instruments, a spacecraft, and a sunshield (Fig. 1.1). The telescope (Feinberg 2004, Atkinson et al. 2006) is a deployable optical system that provides diffraction-limited performance at 2 μm using active wavefront sensing and control (WFS&C). The ISIM (Greenhouse et al. 2006) contains four science instruments (SIs): a near-infrared camera (NIRCam; Horner & Rieke 2004), a near-infrared multi-object spectrograph (NIRSpec; Zamkotsian & Dohlen 2004), a mid-infrared camera (MIRI; Wright et al. 2004), and a near-infrared tunable filter imager (TFI; Rowlands et al. 2004b). The ISIM also contains a fine guidance sensor (FGS; Rowlands et al. 2004a) to provide active control of pointing. The spacecraft provides the pointing platform and housekeeping functions for the observatory. The wavelength range of JWST and the SIs spans 0.6 to 29 microns, limited at the short end by the gold coatings on the primary mirror and at the long end by the detector technology. The sunshield shades the telescope and ISIM from solar illumination to allow zodiacal-light-limited performance at $\lambda < 10\,\mu m$ and high sensitivity out to 29 μm, and provides a stable thermal environment for the telescope and ISIM.

1.2.1 Observatory

The primary mirror uses semi-rigid primary mirror segments mounted on a stable and rigid backplane composite structure. The architecture is termed "semi-rigid", because it has a modest amount of flexibility that allows for on-orbit compensation of segment-to-segment radius of curvature variations. Semi-rigid segments have a high degree of inherent optical quality and stability, allowing verification of optical performance by 1 g end-to-end ground testing before launch.

The telescope optics are made of beryllium. Spitzer and IRAS had beryllium mirrors and its material properties are known at temperatures as low as 10 K. Beryllium has an extremely small variation in its coefficient of thermal expansion over temperatures of 30 to 80 K, making the telescope optics intrinsically stable to small temperature variations. Beryllium fabrication and figuring procedures were designed using the results from the Advanced Mirror System Demonstrator (AMSD) program (Feinberg 2004, Stahl et al. 2004). An engineering demonstration unit will be used as a pathfinder for the flight mirror processing.

The sunshield enables passive cooling and provides a stable cryogenic environment by minimizing the amount of solar energy incident onto the telescope and ISIM. The observatory will not use active wavefront control during observations.

1.2.2 Observatory Performance (Table 1.1)

Image Quality: The imaging performance of the telescope will be diffraction limited at 2 μm, defined as having a Strehl ratio >0.80 (e.g., Bély 2003). JWST will achieve this image quality using image-based wavefront sensing and control (WFS&C) of the primary mirror. There will also be a fine guidance sensor in the focal plane and a fine steering mirror to maintain pointing during observations. The allocated top-level wavefront error (WFE) is 150 nm root-mean-squared (rms) through to the NIRCam focal plane, and includes both the effect of 7.0 milliarcsec image motion, most of which is line-of-sight jitter, and 51 nm of drift instability.

Sky Coverage and Continuous Visibility: Field of regard (FOR) refers to the fraction of the celestial sphere that the telescope may point towards at any given time. JWST's FOR is limited by the size of the sunshield. A continuous viewing zone within 5° of both the north and south ecliptic poles is available throughout the

Table 1.1 The predicted performance for the JWST Observatory

Parameter	Capability
Wavelength	0.6 to 28.5 μm
Image quality	Strehl ratio of 0.8 at 2 μm
Telescope FOV	Instruments share ~166 arcmin^2 FOV
Orbit	Lissajous orbit about Sun-Earth L2
Celestial sphere coverage	100% annually
	39.7% at any given time
	100% of sphere has >51 contiguous days visibility
	Continuous viewing zone <5° from ecliptic pole
Observing efficiency	Observatory ~80%. Overall efficiency >70%
Mission life	Commissioning at launch + 6 months
	5 yr minimum lifetime after commissioning
	10 yr fuel carried for station keeping

year. Thirty percent of the sky can be viewed continuously for at least 197 continuous days. All regions of the sky have at least 51 days of continuous visibility per year. The architecture provides an instantaneous FOR at any epoch of approximately 40% of the sky. This FOR extends 5° past the ecliptic pole, and provides 100% accessibility of the sky during a one-year period. In addition, a nominal 5° angular safety margin will be maintained when determining the allowable observatory pointing relative to the Sun.

In order to take full advantage of the FOR, JWST will be able to observe any point within it at any allowable roll angle, with a probability of acquiring a guide star of at least 95% under nominal conditions. This will ensure that most of the required targets will be observable without special scheduling or other procedures.

Sensitivity and Stray Light Limitations: Many JWST observations will be background limited. The background is a combination of in-field zodiacal light, scattered thermal emission from the sunshield and telescope, scattered starlight, and scattered zodiacal light. Over most of the sky, the zodiacal light dominates at wavelengths $\lambda < 10\,\mu\text{m}$, so broad-band imaging will be zodiacal-light limited at these wavelengths. The flux sensitivities of the JWST instruments are given in Section 1.2.4.

Observational Efficiency: The observational efficiency is the fraction of the total mission time spent actually "counting photons," rather than occupied by overhead activities such as slewing, stabilizing, calibrating, and adjusting parts of the observatory. The overall efficiency is expected to be 70%, although this depends on the specific observing programs that are implemented. In comparison to missions in low-Earth orbit, there are significant efficiency advantages of the L2 orbit. JWST will use parallel instrument calibration (i.e., darks) to achieve this overall efficiency. The JWST architecture is optimized in the sense that the overheads associated with pointing to targets are the dominant contributors to inefficiency. The spacecraft will carry enough propellant to maintain science operations for a 10-year lifetime.

1.2.3 Observatory Design Description

1.2.3.1 Optical Telescope Element (OTE)

The telescope has 132 degrees of freedom of adjustment mechanisms and a composite structure so that passive figure control and passive disturbance attenuation are possible. Figure 1.5 shows an isometric view of the telescope's primary and secondary mirrors.

The optical telescope has a 25 m^2 collecting area, and is a three-mirror anastigmat, made up of 18 hexagonal segments. The telescope has an effective f/number of 20, and an effective focal length of 131.4 m. Each ISIM instrument reimages the telescope focal plane onto its detectors, allowing for independent selection of detector plate scale for sampling of the optical PSF. A fine steering mirror (FSM) is used for image stabilization.

Figure 1.6 shows the placement of the individual ISIM instrument detectors on the telescope's field of view. The NIRCam detectors are placed in a spatial region with the lowest residual WFE to take full advantage of imaging performance. The telescope mirrors are gold coated, providing a broad spectral bandpass, from 0.6 to 29 μm.

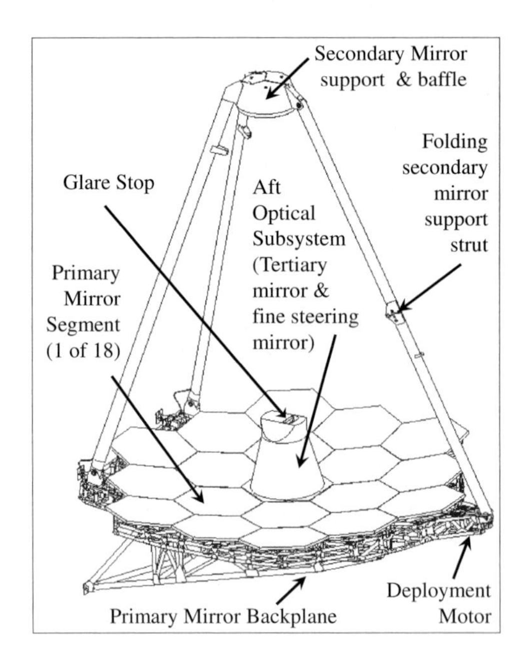

Fig. 1.5 Isometric view of the JWST telescope. The telescope is a three-mirror anastigmat with an f/20 design, made up of 18 hexagonal Beryllium segments totaling 25 m^2 collecting area

1.2.3.2 Mirror Segments

The primary mirror is composed of 18 individual beryllium mirror segments. When properly phased relative to each other, these segments act as a single mirror. Primary

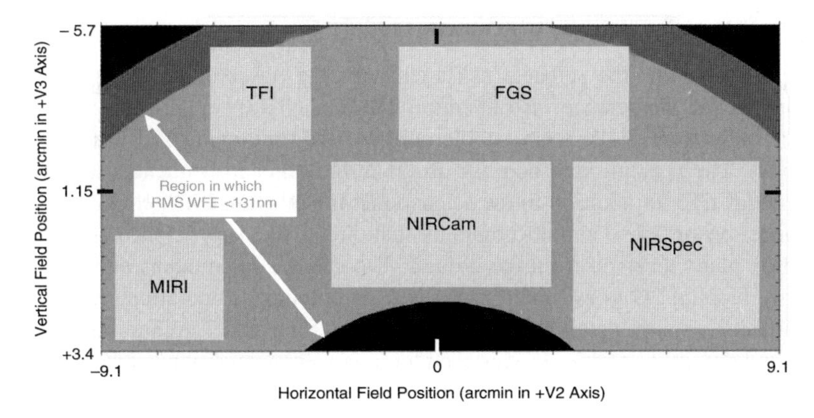

Fig. 1.6 Placement of the science instruments in the telescope field of view

mirror phasing is achieved via six degree-of-freedom rigid body motion of the individual segments, and an additional control for the mirror segment radius of curvature. The six degrees of freedom are decenter (x, y), tip, tilt, piston (z), and clocking. Each mirror segment is hexagonal with a 1.32 m flat-to-flat dimension. There are three separate segment types with slightly different aspheric prescriptions depending on placement. All segments within a single type are completely interchangeable. Figure 1.7 shows the rear portion of the mirror segments and the seven actuators. The secondary mirror has 6 actuators which provide 6 degree-of-freedom rigid body control, although only 5 degree-of-freedom control is actually needed.

Fig. 1.7 A mirror segment with the hexapod support structure and actuators is prepared for acoustic and vibration testing

1.2.3.3 Integrated Science Instrument Module

The ISIM contains the science instruments for the observatory, and the support electronics for the science instruments: NIRCam, NIRSpec, TFI, and MIRI (described in Section 1.2.4). A cryo-cooler will be used for cooling MIRI and its Si:As detectors. The near-infrared detector arrays in the other instruments are passively cooled HgCdTe. In addition to the science instruments, the ISIM contains the fine guidance sensor (FGS) and the computer that directs the daily science observations based on plans received from the ground. The science instruments and FGS have non-overlapping FOVs as shown in Fig. 1.6. Simultaneous operation of all science instruments is possible; this capability will be used for parallel calibration, including darks and possibly sky flats. FGS is used for guide star acquisition and fine pointing. Its FOV and sensitivity are sufficient to provide a greater than 95% probability of acquiring a guide star for any valid pointing direction and roll angle.

1.2.3.4 Wavefront Sensing and Control Subsystem (WFS&C)

The ability to perform on-orbit alignment of the telescope's 18 segments is one of the enabling technologies that allow JWST to be built. The WFS&C subsystem aligns these segments so that their wavefronts match properly, creating a diffraction-limited 6.6 m telescope, rather than overlapping images from 18 individual 1.3 m telescopes.

WFS&C uses a set of deterministic algorithms that control the fundamental and critical initial telescope alignment operation (Acton et al. 2004). Determination of the wavefront error and the necessary telescope mirror commands is done on the ground using the downlinked image data. The algorithm uses least-squares fits to images taken in and out of focus, at different wavelengths, and at multiple field points. The focus is adjusted using weak lenses in the NIRCam filter wheel. This algorithm was used to diagnose and measure the spherical aberration of the HST primary mirror. Mirror adjustment commands are uplinked to the observatory. Figure 1.8 summarizes the WFS&C process.

The initial telescope commissioning process occurs in four phases: (1) image capture and identification, (2) coarse alignment, (3) coarse phasing, and (4) fine phasing. To accomplish commissioning, the telescope is pointed at a celestial region with specific characteristics at the 2 μm NIRCam operation wavelength. A bright (magnitude 10 or brighter) isolated commissioning star is located on the NIRCam detector. A second star (guide star), with magnitude and position restrictions relative to the commissioning star, is used with the fine guiding sensor.

1.2.3.5 Sunshield

The sunshield (Fig. 1.9) enables the passively-cooled thermal design. It is a 5-layer "V" groove radiator design, which is sized (about 19.5 m $\times$ 11.4 m) and shaped to limit solar radiation pressure induced momentum buildup. It reduces the $\sim$200 kW incident solar radiation that impinges on the sunshield to milliwatts incident on the

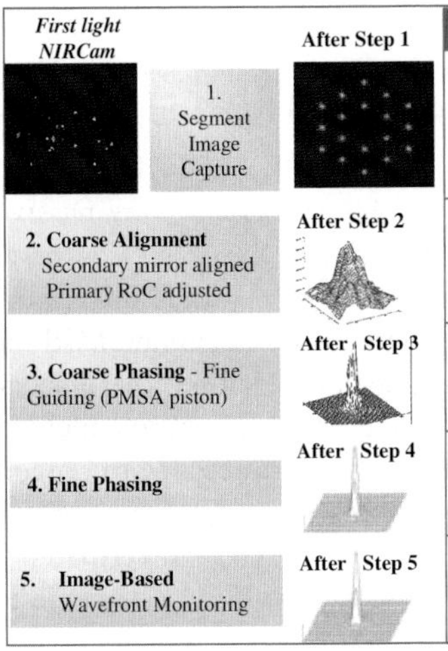

First light NIRCam / Step		Initial Capture	Final Condition
1. Segment Image Capture	After Step 1	18 individual 1.6-m diameter aberrated sub-telescope images PM segments: < 1 mm, < 2 arcmin tilt SM: < 3 mm, < 5 arcmin tilt	PM segments: < 100 μm, < 2 arcsec tilt SM: < 3 mm, < 5 arcmin tilt
2. Coarse Alignment Secondary mirror aligned Primary RoC adjusted	After Step 2	Primary Mirror segments: < 1 mm, < 10 arcsec tilt Secondary Mirror : < 3 mm, < 5 arcmin tilt	WFE < 200 μm (rms)
3. Coarse Phasing - Fine Guiding (PMSA piston)	After Step 3	WFE: < 250 μm rms	WFE < 1 μm (rms)
4. Fine Phasing	After Step 4	WFE: < 5 μm (rms)	WFE < 110 nm (rms)
5. **Image-Based** Wavefront Monitoring	After Step 5	WFE: < 150 nm (rms)	WFE < 110 nm (rms)

Fig. 1.8 A summary of the WFS/C commissioning and maintenance phases

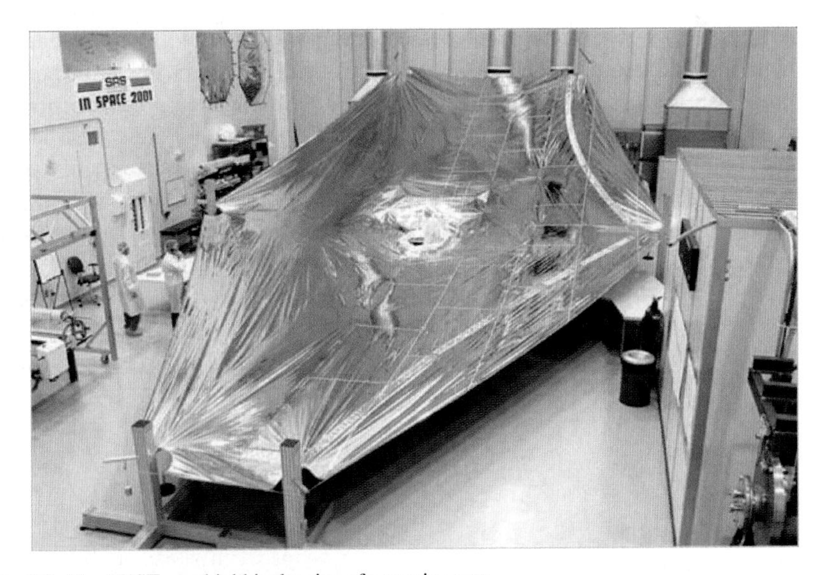

Fig. 1.9 The JWST sunshield is the size of a tennis court

telescope and ISIM. The orientation and angular separation of the individual layers can be seen in the artist's rendition of the JWST observatory in Fig. 1.1.

1.2.3.6 Spacecraft Bus

The spacecraft provides the housekeeping function of the observatory, including data handling, orbit maintenance, solar array power, coarse star tracking, and reaction wheels for fine star tracking. It has a 471 Gbit solid state recorder to hold the science and engineering data collected between and during the daily contacts with the ground station. The downlink operates at Ka band and has a selectable rate of 7, 14 or 28 Mbps. A pair of omni-directional antennas (at S band) provides nearly complete spherical coverage for emergency communications.

1.2.3.7 Integration and Test

All of the JWST flight elements will be qualified to operational environments during their integration and test. Performance verification starts at the sub-system level and continues as the observatory is integrated. Integration and test culminates in an end-to-end cryogenic optical test of the OTE and ISIM, where the telescope's cryogenic performance is verified. The telescope and ISIM combination will be cryogenically tested at Chamber A at NASA's Johnson Space Center (Fig. 1.10) and verified prior to integration to the spacecraft and sunshield. This telescope/ISIM cryogenic test will be the end-to-end optical performance verification for the observatory and will include tests to verify the alignment of the OTE to ISIM, an alignment check of the individual telescope elements (primary, secondary, aft optics system to each other), a test of the full 18-segment phased primary mirror, and model validation of the thermal and structural stability. Crosschecks will be also be performed on the aft-optics system and science instrument wavefront error, and end-to-end wavefront error over portions of the telescope pupil, pupil alignment, and stray light levels. The test chamber is 16.8 m diameter and 35.7 m high, giving adequate room to configure cost-effective assembly, test, and verification equipment. The telescope will be tested with the optical axis vertical to minimize gravity effects. The sunshield deployment will be tested at room temperature, and a scale version of the sunshield will be tested in the deployed configuration with a helium-cooled shroud to validate thermal models.

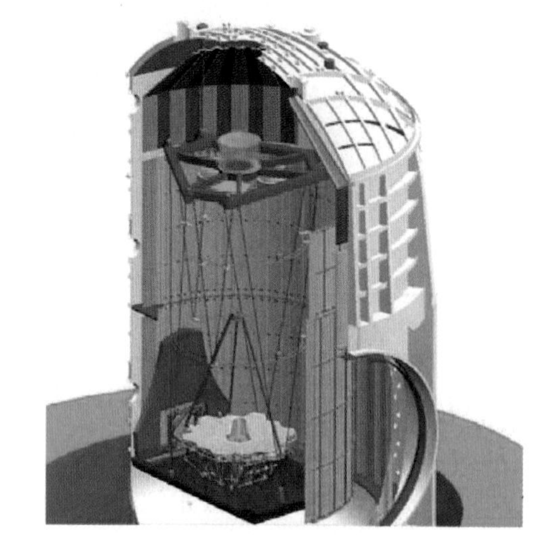

Fig. 1.10 The JWST observatory pictured in the thermal vacuum test chamber at NASA's Johnson Space Center. The test chamber is 16.8 m in diameter and 35.7 m high. The end-to-end optical performance verification for the observatory will test alignment, the full 18 segment phased primary mirror, and validate the thermal and structural stability models

1.2.4 Instrumentation

Table 1.2 lists the main parameters of the science instruments.

11.5

Table 1.3 lists the instrument sensitivities. Sensitivity is defined to be the brightness of a point source detected at 10σ in 10,000 s. Longer or shorter exposures are expected to scale as the square root of the exposure time. Targets at the North Ecliptic Pole are assumed. The sensitivities in this table represent the best estimate at the time of writing and are subject to change.

Although stated as the sensitivity achievable for 10,000 s exposures, it is expected that cosmic ray hits will limit the maximum exposure time for an individual integration to about 1000 s, and that longer total exposure times will be achieved through

Table 1.2 Science instrument characteristics

Instrument	λ (μm)	Detector	mas/pixel	Field of view
NIRCam Short	0.6 to 2.3	Eight 2048 × 2048	32	2.2 × 4.4 arcmin
Long[a]	2.4 to 5.0	Two 2048 × 2048	65	2.2 × 4.4 arcmin
NIRSpec MSA[b]	0.6 to 5.0	Two 2048 × 2048	100	3.4 × 3.1 arcmin
Slits[c]				~0.2 × 4 arcsec
IFU				3.0 × 3.0 arcsec
MIRI Imaging	5.0 to 27.0	1024 × 1024	110	1.4 × 1.9 arcmin
Slit[d]	5.0 to 10.0			0.2 × 5 arcsec
IFU	5.0 to 28.5	Two 1024 × 1024	200 to 470	3.6 × 3.6 to 7.5 × 7.5 arcsec
TFI	1.6 to 4.9[e]	2048 × 2048	65	2.2 × 4.4 arcmin

[a]Use of a dichroic renders the NIRCam long-wavelength FOV cospatial with the short-wavelength channel; the two channels acquire data simultaneously.

[b]NIRSpec includes a microshutter assembly (MSA) with four 365 × 171 microshutter arrays. The individual shutters are each 203 (spectral) × 463 (spatial) milliarcsec clear aperture on a 267 × 528 milliarcsec pitch.

[c]NIRSpec also includes several fixed slits and an integral field unit (IFU)' which provide redundancy and high contrast spectroscopy on individual targets.

[d]MIRI includes a fixed slit for low-resolution (R ~ 100) spectroscopy over the 5 to 10 μm range, and an IFU for R ~ 3000 spectroscopy over the full 5 to 28.5 μm range. The long wavelength cutoff for MIRI spectroscopy is set by the detector performance, which drops longward of 28.0 μm.

[e]The wavelength range for the TFI is 1.6 to 2.6 μm and 3.1 to 4.9 μm. There is no sensitivity between 2.6 and 3.1 μm.

Table 1.3 Instrument sensitivities

Instrument/mode	λ (μm)	Bandwidth	Sensitivity
NIRCam	2.0	R = 4	11.4 nJy, AB = 28.8
TFI	3.5	R = 100	126 nJy, AB = 26.1
NIRSpec/Low Res.	3.0	R = 100	132 nJy, AB = 26.1
NIRSpec/Med. Res.	2.0	R = 1000	1.64×10^{-18} erg s^{-1} cm^{-2}
MIRI/Broadband	10.0	R = 5	700 nJy, AB = 24.3
MIRI/Broadband	21.0	R = 4.2	8.7 μJy, AB = 21.6
MIRI/Spect.	9.2	R = 2400	1.0×10^{-17} erg s^{-1} cm^{-2}
MIRI/Spect.	22.5	R = 1200	5.6×10^{-17} erg s^{-1} cm^{-2}

co-adding. Based on experience with Hubble data, for example in the Hubble Deep Field, we expect the errors to scale as the square root of the exposure time in co-adds as long as 10^5 or even 10^6 s. The absolute photometric accuracy is expected to be 5% for imaging and 10 to 15% for coronagraphy and spectroscopy, based on calibration observations of standard stars.

1.2.4.1 Near-Infrared Camera

NIRCam provides filter imaging in the 0.6 to 5.0 μm range with wavelength multiplexing. It includes the ability to sense the wavefront errors of the observatory. NIRCam consists of an imaging assembly within an enclosure that is mounted in the ISIM. The imaging assembly consists of two fully redundant, identical optical trains mounted on two beryllium benches, one of which is shown in Fig. 1.11. The incoming light initially reflects off the pick-off mirror. Subsequently it passes through the collimator and the dichroic, which is used to split the light into the short (0.6 to 2.3 μm) and long (2.4 to 5.0 μm) wavelength light paths. Each of these two beams then passes through a pupil wheel and filter wheel combination, each beam having its own pair of pupil and filter wheel. After this, the light passes through the camera corrector optics and is imaged (after reflecting off a fold flat in the short wavelength beam) onto the detectors.

The instrument contains a total of ten 2048×2048 detector chips, including those in the identical redundant optical trains. The short wavelength arm in each optical train contains a 2×2 mosaic of these detectors, optimized for the 0.6–2.3 μm wavelength range, with a small gap ($\sim$ 3 mm $=\sim$ 5 arcsec) between adjacent detectors. The long wavelength arm in each optical train contains a single detector

Fig. 1.11 The NIRCam engineering test unit optical bench

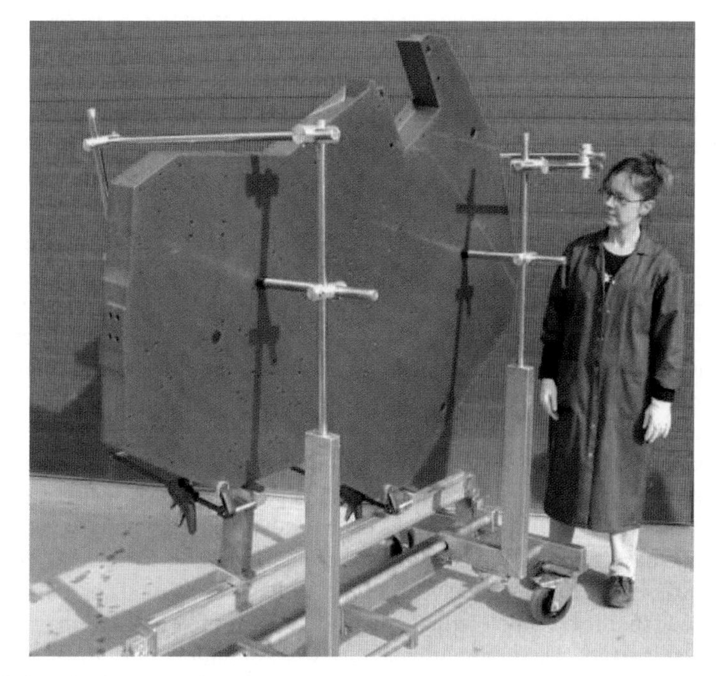

Fig. 1.12 The NIRSpec optical bench

covering 2.4–5.0 μm. The detectors arrays are HgCdTe of HAWAII II heritage built by Rockwell Science Center. The detectors will all have thinned substrates to avoid cosmic ray scintillation issues, as well as to extend their sensitivity below 0.85 μm. Each optical train contains a dual filter and pupil wheel, containing a range of wide-, medium- and narrow-band filters and the WFS&C optics.

Coronagraphy: To enable the coronagraphic imaging, each of the two identical optical trains in the instrument also contains a traditional focal plane coronagraphic mask plate held at a fixed distance from the detectors, so that the coronagraph spots are always in focus at the detector plane. Each coronagraphic plate is transmissive, and contains a series of spots of different sizes, including linear and radial-sinc occulters, to block the light from a bright object. Each coronagraphic plate also includes a neutral density spot to enable centroiding on bright stars, as well as calibration sources at each end that can send light through the optical train of the imager to enable internal alignment checks. Normally these coronagraphic plates are not in the optical path for the instrument, but they are selected by rotating into the beam a mild optical wedge that is mounted in the pupil wheel, which translates the image plane so that the coronagraphic masks are shifted onto the active detector area. Diffraction can also be reduced by apodization at the pupil mask, thus the pupil wheels will be equipped with both a classical and an apodized pupil with integral wedges in each case. Current models predict a contrast of $\sim 10^4$ at 0.5 arcsec, at a wavelength of 4.6 μm.

1.2.4.2 Near-Infrared Spectrograph

NIRSpec (Fig. 1.12) is a near-infrared multi-object dispersive spectrograph capable of simultaneously observing more than 100 sources over a field-of-view (FOV) larger than 3×3 arcmin. In addition to the multi-object capability, it includes fixed slits and an integral field unit for imaging spectroscopy. Six gratings will yield resolving powers of R ~ 1000 and ~2700 in three spectral bands, spanning the range 1.0 to 5.0 μm. A single prism will yield R ~ 100 over 0.6 to 5.0 μm. Figure 1.13 shows a layout of the instrument.

The region of sky to be observed is transferred from the JWST telescope to the spectrograph aperture focal plane by a pick-off mirror and a system of fore-optics that includes a filter wheel for selecting bandpasses and introducing internal calibration sources.

Targets in the FOV are normally selected by opening groups of shutters in a micro-shutter assembly (MSA) to form multiple apertures. The micro-shutter assembly itself consists of a mosaic of 4 subunits producing a final array of 730 (spectral) by 342 (spatial) individually addressable shutters with 203×463 milliarcsec openings and 267×528 milliarcsec pitch. Sweeping a magnet across the surface of the micro shutters opens all of the shutters. Individual shutters may then be addressed and released electronically, and the return path of the magnet closes the released shutters. The minimum aperture size is 1 shutter (spectral) by 1 shutter (spatial) at all wavelengths. Multiple pointings may be required to avoid placing targets near the edge of a shutter and to observe targets with spectra that would overlap if observed simultaneously at the requested roll angle. The nominal slit length is 3 shutters in all wavebands. In the open configuration, a shutter passes light from the fore-optics to the collimator. A slitless mode can be configured by opening all of the micro shutters. As the shutters are individually addressable, long slits, diagonal

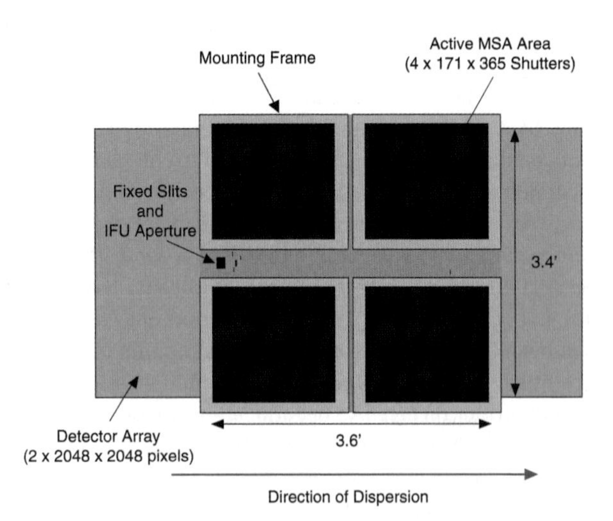

Fig. 1.13 Schematic layout of the NIRSpec microshutter array mask overlaid on the detector array and projected to the same angular scale

slits, Hadamard-transform masks (Riesenberg & Dillner 1999), and other patterns can also be configured with them.

In addition to the slits defined by the micro-shutter assembly, NIRSpec also includes five fixed slits that can be used for high-contrast spectroscopy. They are placed in a central strip of the aperture focal plane between sub-units of the micro-shutter assembly and also provide redundancy in case the micro-shutters fail. Three fixed slits are 3.5 arcsec long and 200 milliarcsec wide. One fixed slit is 4 arcsec long and 400 milliarcsec wide for increased throughput at the expense of spectral resolution. One fixed slit is 2 arcsec long and 100 milliarcsec wide for brighter targets.

The strip between micro-shutter sub-units also contains the 3 by 3 arcsec entrance aperture for an integral field unit (IFU). The IFU has 30 slices, each 100 milliarcsec wide. The IFU relay optics introduce a 2:1 anamorphic magnification of each slicer such that the matching projected virtual slits are properly sampled on the detector by two 50 milliarcsec pixels in the dispersion direction and at the (nominal) 100 milliarcsec per pixel in the spatial direction.

The aperture focal plane is re-imaged onto a mosaic of two NIR detectors by a collimator, a dispersing element (gratings or a double-pass prism) or an imaging mirror, and a camera. Three gratings are used for first-order coverage of the three NIRSpec wavebands at $R \sim 1000$ (1.0 to 1.8 μm; 1.7 to 3.0 μm; 2.9 to 5.0 μm). The same three wavebands are also covered by first-order $R \sim 2700$ gratings for objects in a fixed slit or in the IFU. The prism gives $R \sim 100$ resolution over the entire NIRSpec bandpass (0.6 to 5 μm), but can optionally be blocked below 1 μm with one of the filters. Any of the aperture selection devices (micro-shutter assembly, fixed slits or IFU) can be used at any spectral resolution.

The focal plane array is a mosaic of two detectors (see Fig. 1.13), each 2k $\times$ 2k, forming an array of 2k $\times$ 4k 100 milliarcsec pixels. The detectors will be thinned HgCdTe arrays built by Rockwell Science Center. NIRSpec contains a calibration unit with a number of continuum and line sources.

1.2.4.3 Mid-Infrared Instrument

The Mid-Infrared Instrument (MIRI) on JWST provides imaging and spectroscopic measurements over the wavelength range 5 to 29 μm. MIRI consists of an optical bench assembly (Fig. 1.14) with associated instrument control electronics, actively cooled detector modules with associated focal plane electronics, and a cryo-cooler with associated control electronics. The cryo-cooler electronics interface with the spacecraft command and telemetry processor, while the instrument control electronics interface with the ISIM command and data handling. The optical bench assembly contains two actively cooled subcomponents, an imager and an Integral Field Unit (IFU) spectrograph, plus an on-board calibration unit.

Imaging: The imager module provides broad-band imaging, coronagraphy, and low-resolution ($R \sim 100$, 5 to 10 μm) slit spectroscopy using a single 1024 $\times$ 1024 pixel Si:As detector with 25 μm pixels. Three quarters of the detector is available for imaging, while the remaining quarter is devoted to the coronagraphic masks

Fig. 1.14 MIRI verification model going into thermal vacuum testing

and the slit for the low-resolution spectrometer. The coronagraphic masks include three phase masks for a quadrant-phase coronagraph and one opaque spot for a Lyot coronagraph. The coronagraphic masks each have a square field of view of 26×26 arcsec and are optimized for particular wavelengths. The imager's only moving part is an 18-position filter wheel. Filter positions are allocated as follows: 12 filters for imaging, 4 filter and diaphragm combinations for coronagraphy, 1 ZnS-Ge double prism for the low-resolution spectroscopic mode and 1 dark position. The imager

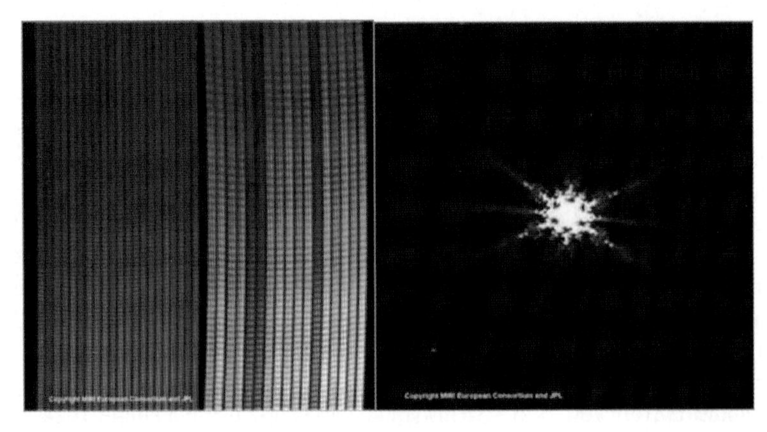

Fig. 1.15 Test data from the MIRI verification model, showing a spectrum (*left*) and point spread function of the imager (*right*)

will have a pixel scale of 0.11 arcsec/pixel and a total field of view of 113×113 arcsec; however, the field of view of its clear aperture is 84×113 arcsec because the coronagraph masks and the low-resolution spectrograph are fixed on one side of the focal plane.

Integral-Field Spectroscopy: The integral-field spectrograph obtains simultaneous spectral and spatial data on a small region of sky. The spectrograph field of view is next to that of the imager so that accurate positioning of targets will be possible by locating the image with the imager channel and off-setting to the spectrograph (Fig. 1.15). The light is divided into four spectral ranges by dichroics, and two of these ranges are imaged onto each of two detector arrays. A full spectrum is obtained by taking exposures at each of three settings of the grating wheel. The spectrograph uses four image slicers to produce dispersed images of the sky on two 1024×1024 detectors, providing $R \sim 3000$ integral-field spectroscopy over the full 5 to 29 μm wavelength range, although the sensitivity of the detectors drops longward of 28 μm. The IFUs provide four simultaneous and concentric fields of view. The slice widths set the angular resolution perpendicular to the slice. The pixel sizes set the angular resolution along the slice.

The spectral window of each IFU channel is covered using three separate gratings (i.e., 12 gratings to cover the four channels). Each grating is fixed in orientation and can be rotated into the optical path using a wheel mechanism (there are two wheel mechanisms which each hold 3 pairs of gratings). The optics system for the four IFUs is split into two identical sections (in terms of optical layout). One section is dedicated to the two short wavelength IFUs and the other handles the two longer wavelength IFUs, with one detector for each section. The two sections share the wheel mechanisms (each mechanism incorporates three gratings for one of the channels in the short wavelength section and three gratings for one of the channels in the long wavelength section). The image slicers in the MIRI IFU dissect the input image plane. The dispersed spectra from two IFU inputs are placed on one detector side-by-side. The spatial information from the IFU is spread out into two adjacent rows with the information from each slice separated by a small gap.

Coronagraphy: MIRI features a coronagraph designed for high contrast imaging in selected mid-infrared bandpasses. Three quadrant phase masks (Boccaletti et al. 2004, Gratadour et al. 2005) provide high contrast imaging to an inner working angle of λ/D, with bandpass of $\lambda/20$, centered at 10.65 μm, 11.4 μm, and 15.5 μm respectively. A fourth, traditional Lyot mask of radius 0.9 arcsec, will provide $R \sim 5$ imaging at a central wavelength of 23 μm. Simulations predict that the quadrant phases masks will achieve a contrast of $\sim 10^4$ at 3 λ/D. The Lyot stop is predicted to deliver a contrast of 2×10^3 at 3 λ/D.

1.2.4.4 Tunable Filter Imager

The tunable filter imager (TFI; Fig. 1.16) provides narrow-band near-infrared imaging over a field of view of 2.2×2.2 arcmin^2 with a spectral resolution $R \sim 100$. The etalon design allows observations at wavelengths of 1.6 μm to 2.6 μm and 3.1 μm

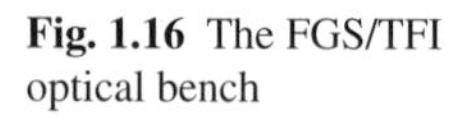

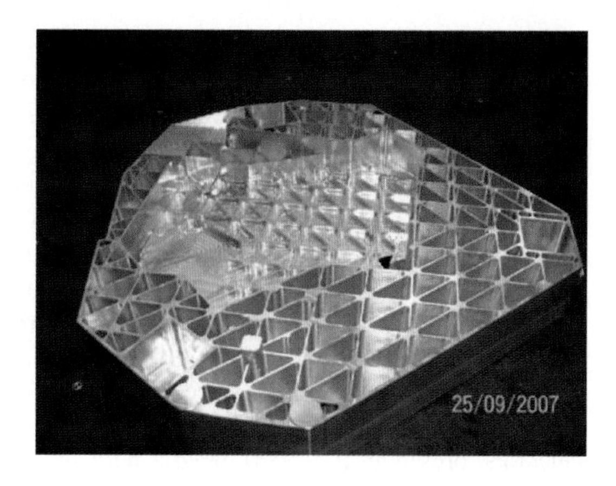

Fig. 1.16 The FGS/TFI optical bench

to 4.9 μm. The gap in wavelength coverage allows one channel to reach more than one octave in wavelength.

The TFI uses dielectric-coated Fabry-Perot etalon plates with a small air (vacuum) gap. The finesse is about 30 and the filters are used in third order. The finesse was chosen to be a compromise between the surface figure requirements and the need to minimize the number of blocking filters, while providing a contrast ratio of about 100. The filters are scanned using piezo-electric actuators, consisting of lead-zirconia-titanite ceramic transducers. The air (vacuum) gap ranges from 2.0 to 8.0 μm plate spacings. The Fabry-Perot operates at a nominal temperature of $\sim$35 K.

To demonstrate the feasibility of a cryogenic Fabry-Perot etalon that meets the requirements, Rowlands et al. (2004b) fabricated a prototype. The etalon surface figure is the most critical requirement, and is influenced by the coatings, as well as structural and cryogenic issues.

The TFI incorporates four coronagraphic occulting spots permanently to one side of the field of view, and occupying a region 20 by 80 arcsec. A set of selectable apodization masks is located at the internal pupil images of each channel by the filter wheels. The coronagraph will deliver a contrast ratio of $\sim 10^4$ (10σ) at 1 arcsec separation. The sensitivity is limited by speckle noise. Contrast ratios of 10^5 may be achievable at sub-arcsec scales using roll or spectral deconvolution techniques (Doyon et al. 2004, Sparks & Ford 2002).

1.2.4.5 Fine Guidance Sensor

The fine guidance sensor (FGS) instrument uses two fields of view in the JWST focal plane to provide fine guidance for the telescope. The field of view locations are chosen to provide optimum lever arms to all instruments for roll about a single guide star. Roll is sensed by star trackers on the spacecraft bus.

The FGS consists of an optical assembly and a set of focal plane and instrument control electronics. The optical assembly of the FGS instrument consists of two channels, imaging separate regions of the sky onto independent 2k $\times$ 2k detectors.

The detectors will be HgCdTe, similar to those in NIRCam. The plate scale is 67 milliarcsec/pixel and the field of view is $2.3 \times 2.3\,\mathrm{arcmin}^2$.

The FGS will provide continuous pointing information to the observatory that is used to stabilize the line of sight, allowing JWST to obtain the required image quality. Each of the independent FGS channels provides $> 90\%$ probability of obtaining a useable guide star for any observatory pointing and roll angle. With both channels, the probability is $>95\%$. The wavelength region and pixel size have been optimized so that in fine-guidance mode the FGS will provide pointing information to a precision of <5 milliarcsec updated at 16 Hz. The guide stars will be chosen from existing catalogs with AB < 19.5 mag in the J band. Fainter stars may be used for location identification. In the event that a suitable guide star is not available for a particular desired pointing and roll angle combination, alternate choices of roll angles can be considered during scheduling.

JWST will be capable of relative pointing offsets with an accuracy of 5 milliarcsec rms, which will enable sub-pixel dithering and coronagraphic acquisition. Absolute astrometric accuracy will be limited to 1 arcsec rms by the accuracy of the guide star catalog.

JWST will be capable of moving object tracking, at linear rates up to 30 milliarcsec s^{-1} relative to the fixed guide stars, although some degradation of the image quality may occur. This will enable observations of Kuiper Belt objects, the moons of the outer planets and comets that come as close to the Sun as Mars. The planets themselves would saturate many of the observing modes.

The FGS is designed to be completely redundant in terms of the guiding function. The loss of any single component would at most result in the loss of one FGS channel. This would reduce the probability of guide star acquisition to $\sim 90\%$ if using current catalogs.

1.2.5 Launch, Orbit and Commissioning

Launch and Orbit: JWST will be launched on an Arianespace Ariane 5 Enhanced Capability-A rocket into orbit about the second Sun-Earth Lagrange (L2) point in the Earth-Sun system, approximately 1.5×10^6 km from the Earth. The orbit (shown in Fig. 1.17) lies in a plane out of, but inclined slightly with respect to, the ecliptic plane. This orbit avoids Earth and Moon eclipses of the Sun, ensuring continuous electrical power. JWST will have a 6-month orbit period about the L2 point in the rotating coordinate system moving with the Earth around the Sun. Station-keeping maneuvers are performed after the end of each orbit determination period of 22 days. The total observatory mass is 6500 kg, including station-keeping propellant sized for a 10-year lifetime.

Deployment: The observatory has the following five deployments: (1) deploy spacecraft bus appendages, including solar arrays and the high-gain antenna, (2) deploy sunshield, (3) extend telescope tower, (4) deploy secondary mirror support structure, and (5) deploy primary mirror wings. The deployment mechanism design

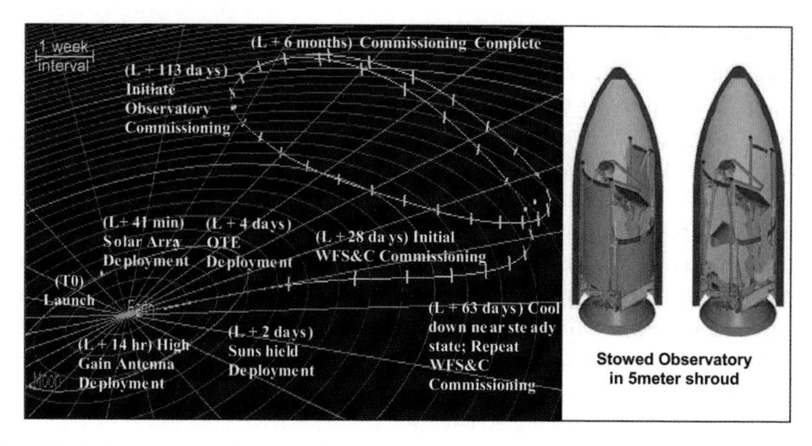

Fig. 1.17 The launch and orbit of JWST

includes heaters and other protections that eliminate the need for time-critical events and allow for unlatching and re-latching to relieve any residual long-term stress in the structure. The secondary-mirror deployment and primary-mirror deployment sequence are shown in Fig. 1.18.

Commissioning: There will be pre-commissioning activities during the transfer to L2. About 28 days after launch (Fig. 1.17), the observatory will cool down to a temperature that permits pre-commissioning activities to begin. Almost continuous ISIM availability for preliminary science observations is provided from this time until 106 days after launch, when a final trajectory burn is required to achieve orbit about L2. The intrinsic passive-thermal stability of the semi-rigid mirror segment and the two-chord-fold primary allows early operation of science instruments, by providing a stable optical image to the ISIM. These pre-commissioning activities will develop an operational experience database that allows formal commissioning and science operations to be conducted efficiently.

A final checkout of all systems is initiated after the L2 orbit is achieved. Commissioning (complete 6 months after launch) includes ISIM, telescope, sunshield, and spacecraft operations that were not feasible during the transfer to L2. It also repeats

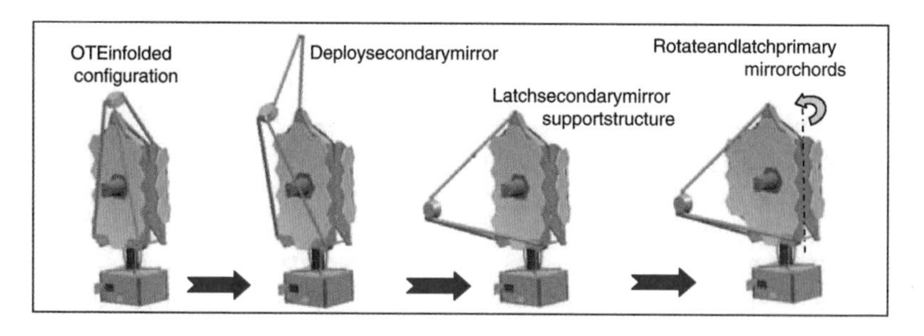

Fig. 1.18 The secondary and primary mirror deployment sequence of JWST

selected operations performed during the transfer, to ensure adequate knowledge of system performance in the final orbital-thermal conditions and allow comparisons with previous measurements. Examples of repeated operations include final optical distortion mappings for the observatory and the characterization of the wavefront sensing and control actuators' transfer function. Allotting 76 days for ISIM pre-commissioning activities allows for operations rehearsal and increases familiarity with observatory operations without reducing operational availability.

Mission Lifetime: JWST will operate with all science instruments for at least five years after completion of commissioning. In order to exploit the full scientific potential of the mission, a lifetime of ten years or longer is desired. Although we will not require mission assurance to guarantee a lifetime greater than five years, JWST will maintain the possibility of a longer mission lifetime, and will carry propellant sized for at least 10 years of operation after launch. There are no other consumables which would limit lifetime.

1.2.6 Operations

JWST will be operated from a Science and Operations Center (S&OC) located at the Space Telescope Science Institute (STScI), the organization that operates HST for NASA. Although the capabilities of JWST are being developed to address science themes discussed in the previous sections, almost all of the observing time on JWST will be competitively selected. Approximately 10% of the observing time for the first 5 years of the mission has already been awarded to the science instrument teams and to other members of the science working group. An additional 5% will be director's discretionary time allocated by the director of the S&OC. The remaining ~85% of the observing time will be awarded through a series of proposal solicitations, which will be open to any astronomer in the world. The proposal solicitations will begin a year or two before the anticipated launch of JWST. The scope of JWST's competitively-selected investigations will range from large legacy-style projects, which last months and address a range of science goals simultaneously, to small programs that target important but very specific science objectives.

1.2.6.1 Operations Concept

JWST will be located at Sun-Earth L2 in order to allow effective cooling of the telescope and ISIM. The distance to the observatory is approximately the same throughout the mission, so the data rate to and from the observatory can be maintained with the same large ground antennae and 1 m class downlink antenna. Operations will be simpler than for HST, as HST's low-Earth orbit means that target visibilities are interrupted every 95 min by Earth occultation. Astronomical targets from an L2 orbit are visible for long periods of time once or twice a year. L2 is a saddle point in the gravitational potential of the Earth-Sun system, and so propellant is required to maintain JWST at this position, and to dump angular momentum that builds up due to solar pressure on the sunshield. Since propellant is limited, its usage will be carefully monitored to meet the design goal of a 10-year operational lifetime.

From an operations perspective, the instruments on JWST are comparable in complexity to those on HST, but more complex than those on Spitzer. A relatively small number of instrument configurations will be used by observers, with a limited selection of readout and dither patterns. Limiting the number of observing configurations also allows better calibrations of the instruments and assures consistency in the archived data. In order to achieve 70% observational efficiency, most calibration exposures will be done in parallel; that is, one instrument will be taking darks while another instrument is doing science observations.

1.2.6.2 Event-Driven Architecture

Commanding HST and most other low-Earth-orbiting satellites is based upon absolute time. In contrast, JWST's observing program will have few absolute time requirements. The commanding concept for JWST is like a command queue; commands in the queue are executed sequentially and the next command starts when an indication is received that the previous command is completed (successfully or unsuccessfully). This approach sacrifices certainty, but simplifies the software system that is needed for operations planning and lowers operations cost, in both the mission's development and operational phases, and improves observational efficiency. The event-driven architecture for JWST will be implemented through a construct of visits and an observation plan. A visit is a logically grouped series of activities and conditions, including a slew to a new target, the steps required to finely point at the target, the setup of the instrument for the observation itself, and the acquisition of all of the science data at that pointing position. The observation plan constitutes the single queue that orders the visits. If a visit ends early because of a failure to identify the appropriate guide stars, for example, the on-board software starts the next visit in the sequence. The observation plan is uploaded to the observatory weekly, and fits within a long-range plan. Orbit maintenance is the only routine activity which will require real-time commanding, because managing its risk requires real-time monitoring.

The S&OC will solicit proposals for JWST observations annually, on behalf of NASA, ESA and CSA. Those proposing observations will use an integrated planning tool, consisting of a graphical user interface, exposure-time calculators and tools for importing sky maps and accurately positioning the JWST apertures on targets. When writing a proposal, the astronomer will complete those portions required for scientific and technical assessment, including any special requirements for timing and/or fixed orientation of the observatory (Phase I). As with Spitzer, time will effectively be allocated in "wall-clock time", including estimates for slew and setup times. Once the proposal is selected, the scientist will fill in the remaining details that would be required to execute the approved observations on JWST (Phase II). Target of opportunity observations, which interrupt the JWST observing plan for time-critical events such as supernovae or gamma-ray burst sources, will be enabled within 2 days of the decision to make these observations.

The ground operations software will be constructed around the largely commercial-off-the-shelf command and telemetry system that will have been used in JWST's integration and testing. The calibration pipeline will include cosmic ray removal, the factor which limits the exposure times for the detectors, and processing to remove the dark current and flat field and any other image artifacts. The S&OC will also manage the wavefront sensing and control observations, which will be implemented using the same system as the science observations. The WFS/C observations will be analyzed and corrections to the mirror positions will be implemented at the next observatory contact.

An archive of all of the data obtained from the observatory will be maintained in the S&OC, using the capabilities of the Multi-Mission Archive at Space Telescope. Processing of the raw data will be done "on-the-fly" as calibrated data are requested over the Internet.

References

Acton, D. S., Atcheson, P. D., Cermak, M., Kingsbury, L. K., Shi, F., & Redding, D. C. 2004, in J. C. Mather, ed., Optical, Infrared, & Millimeter Space Telescopes, Proceedings of SPIE, 5487, (SPIE: Bellingham WA), 887

Armus, L. et al. 2004, ApJS, 154, 178

Atkinson, C., Texter, S., Hellekson, R., Patton, K., Keski-Kua, R., & Feinberg, L. 2006, in J. C. Mather, H. A. MacEwen, & M. W. M. de Graauw, eds., Space Telescopes and Instrumentation I: Optical, Infrared and Millimeter, Proceedings of SPIE, 6265, (SPIE: Bellingham WA), 62650T

Barkana, R., & Loeb, A. 2001, Physics Reports, 349, 125

Bély, P. Y. 2003, The Design and Construction of Large Optical Telescopes, (Springer: New York)

Bender, R., Burstein, D., & Faber, S. M. 1992, ApJ, 399, 462

Boccaletti, A., et al. 2004, PASP, 116, 1061

Boss, A. P. 2001, ApJ, 551, L167

Chabrier, G., Baraffe, I., Allard, F., & Hauschildt, P. 2000, ApJ, 542, L119

Crovisier, J., et al. 1997, Science, 275, 1904

Doyon, R., et al. 2004, in J. C. Mather, ed., Optical, Infrared, and Millimeter Space Telescopes, Proceedings of SPIE, 5487, (SPIE: Bellingham WA), 746

Fan, X., et al. 2002, AJ, 123, 1247

Feinberg, L. 2004, in J. C. Mather, ed., Optical, Infrared, and Millimeter Space Telescopes, Proceedings of SPIE, 5487, (SPIE: Bellingham WA), 814

Gardner, J. P., et al. 2006, Space Sci. Rev., 123, 485

Gardner, J. P., Fan, X., Wilson, G. & Stiavelli, M. 2007, in L. J. Storrie-Lombardi, & N. A. Silberman, eds., The Science Opportunities of the Warm Spitzer Mission Workshop, AIP Conference Proceedings, (Springer-Verlag: New York)

Gebhardt, K., et al. 2000, ApJ, 539, L13

Gibb, E. L., et al. 2000, ApJ, 536, 347

Gnedin, N. Y. 2004, ApJ, 610, 9

Gratadour, D., et al. 2005, A&A, 429, 433

Greenhouse, M. A., et al. 2006, in J. C. Mather, H. A. MacEwen, & M. W. M. de Graauw, eds., Space Telescopes and Instrumentation I: Optical, Infrared and Millimeter, Proceedings of SPIE, 6265, (SPIE: Bellingham WA), 626513

Horner, S. D., & Rieke, M. J., 2004, in J. C. Mather, ed., Optical, Infrared, and Millimeter Space Telescopes, Proceedings of SPIE, 5487, (SPIE: Bellingham WA) 628

Kogut, A., et al. 2003, ApJS, 148, 161

Lunine, J. I., Coradini, A., Gautier, D., Owen, T.C., & Wuchterl, G. 2004, in F. Bagenal, T. Dowling, & W. McKinnon, eds., Jupiter, (Cambridge University Press: Cambridge) 19
Magorrian J., et al. 1998, AJ, 115, 2285
Malfait, K., Waelkens, C., Waters, L. B. F. M., Vandenbussche, B., Huygen, E., & de Graauw, M. S. 1998, A&A, 332, L25
Morbidelli, A., Chambers, J., Lunine, J. I., Petit, J. M., Robert, F., Valsecchi, G. B., Cyr, K. E. 2000, Meteoritics Pl. Sci. 35, 1309
Rees, M. J. 1997, in N. R. Tanvir, A. Aragon-Salamanca, & J. V. Wall eds., The Hubble Space Telescope and the High Redshift Universe, (World Scientific: Singapore), 115
Riesenberg, R., & Dillner, U. 1999, in S. S. Shen ed., Imaging Spectrometry V, Proceedings of SPIE, 3753, (SPIE: Bellingham WA), 203
Rowlands, N., et al. 2004a, in J. C. Mather, ed., Optical, Infrared, and Millimeter Space Telescopes, Proceedings of SPIE, 5487, (SPIE: Bellingham WA), 664
Rowlands, N., et al. 2004b, in J. C. Mather, ed., Optical, Infrared, and Millimeter Space Telescopes, Proceedings of SPIE, 5487, (SPIE: Bellingham WA), 676
Shu, F. H. 1977, ApJ, 214, 488
Shu, F. H., Adams, F. C., & Lizano, S. 1987, ARA&A, 25, 23
Soifer, B. T., et al. 2004, ApJS, 154, 151
Sparks, W. B., & Ford, H. C. 2002, ApJ, 578, 543
Stahl, H. P., Feinberg, L., & Texter, S. 2004, in J. C. Mather, ed., Optical, Infrared, and Millimeter Space Telescopes, Proceedings of SPIE, 5487, (SPIE: Bellingham WA), 818
Terebey, S., Shu, F. H., & Cassen, P. 1984, ApJ, 286, 529
Tully, R. B., & Fisher, J. R., 1977, A&A, 54, 661
van den Ancker, M. E., Tielens, A. G. G. M., & Wesselius, P. R. 2000, A&A, 358, 1035
White, S. D. M., & Frenk, C. S. 1991, ApJ, 379, 52
Wright, G. S., et al. 2004, in J. C. Mather, ed., Optical, Infrared, and Millimeter Space Telescopes, Proceedings of SPIE, 5487, (SPIE: Bellingham WA) 653
Yorke, H. W., & Sonnhalter, C. 2002, ApJ, 569, 846
Zamkotsian, F., & Dohlen, K. 2004, in J. C. Mather, ed., Optical, Infrared, and Millimeter Space Telescopes, Proceedings of SPIE, 5487, (SPIE: Bellingham WA), 635

Xander Tielens, Massimo Stiavelli, and Massimo Robberto pose for the camera

Chapter 2
Beyond the Hubble Space Telescope: Early Development of the Next Generation Space Telescope

Robert W. Smith and W. Patrick McCray

Abstract In this paper we investigate the early history of what was at first called the Next Generation Space Telescope, later to be renamed the James Webb Space Telescope. We argue that the initial ideas for such a Next Generation Space Telescope were developed in the context of the planning for a successor to the Hubble Space Telescope. Much the most important group of astronomers and engineers examining such a successor was based at the Space Telescope Science Institute in Baltimore. By the late 1980s, they had fashioned concepts for a successor that would work in optical, ultraviolet and infrared wavelengths, concepts that would later be regarded as politically unrealistic given the costs associated with them. We also explore how the fortunes of the planned Next Generation Space Telescope were intimately linked to that of its "parent," the Hubble Space Telescope.

2.1 Introduction

Very large-scale machine-centered projects have been a central element in the physical sciences since World War II, especially in North America, Europe, and Japan. Built with the support of national governments, often working together in international partnerships, these endeavors cost hundreds of millions or even billions of dollars and engage the efforts of armies of scientists and engineers. The biggest of these projects have typically taken decades to bring to fruition. For scientists, their construction has been in large part an act of faith that new and powerful new scientific instruments will surely lead to novel and exciting scientific results.[1]

The journey from conception to completion for such endeavors has usually been fraught with assorted challenges and difficulties, and in some cases these have led to a project's demise years after detailed work has begun. A striking example of this is the 1993 cancellation of the Superconducting Super Collider, a high-energy physics accelerator, after the Department of Energy spent over $4 billion on its design and construction.[2] In general, however, scientists and scientific communities have

R.W. Smith (✉)
Department of History and Classics, University of Alberta, Edmonton, Alberta, Canada
e-mail: Robert.Smith@ualberta.ca

H.A. Thronson et al. (eds.), *Astrophysics in the Next Decade,* Astrophysics and Space Science Proceedings, DOI 10.1007/978-1-4020-9457-6_2,

become increasingly adept at enlisting broad involvement in proposed programs, thereby building stronger bases of support that enable advocates to better resist threats of cancellation. Such very large-scale efforts have also resulted in unique and extremely powerful tools that have greatly expanded as well as helped to intellectually reconstitute scientific disciplines. A leading example of this phenomenon is the Hubble Space Telescope.

The space and scientific agencies, as well as the scientific and engineering communities, engaged in such enterprises have often faced a number of critical and sensitive issues. One is when to initiate serious design work on new machines that will replace those in operation, being built, or being planned. Another is the question of when to decommission instruments already doing productive research. Given the very long lead times from conception to operation, engineers and scientists have often wanted (or been forced) to begin planning the next big machine years or decades before securing any scientific results from the one under construction, results that of course might well have the potential to shape or revise design decisions.

In this paper we will examine how a scientific community and its constituent sub-communities took the first steps towards the construction of what was initially called the Next Generation Space Telescope (renamed by NASA in 2002 as the James Webb Space Telescope in honor of NASA's administrator between 1961 and 1968). Scientists and engineers initially conceived the NGST in the mid-1980s as a successor to the Hubble Space Telescope (HST), some years before this observatory began scientific operations in orbit around the Earth in 1990.

There were initially two parallel tracks to the NGST's early history. In one there were a range of developments in infrared astronomy that would prove later to be crucial for NGST planning. In the second track, advocates explored a successor to the Hubble Space Telescope that would operate in ultraviolet, optical, and infrared wavelengths, just as the Hubble was supposed to do. These two tracks would ultimately come together in the mid-1990s and prompt NASA to issue a study contract in October 1995 for feasibility studies for a Next Generation Space Telescope. In this paper, our focus is on the second of these two tracks, and our main narrative thread is provided by the way in which scientists and engineers examined a wide range of design options for the NGST early in its life.

The largest of large-scale scientific tools require not just the enthusiastic endorsement of small groups of scientists and engineers, but the whole-hearted support of entire scientific communities, generally in more than one country because projects of the largest scale typically involve international collaborations. We therefore ask how advocates worked to form a consensus around some basics of the design of the NGST. This effort helped to create a favorable climate of opinion, a key step towards winning broad approval for an NGST by persuading colleagues it might be not just technically feasible but also perhaps politically feasible. As we will see, space astronomy has been in one respect a remarkable adventure of the human spirit, but in the U.S. it has also been pursued in a highly competitive environment with often intense debates and conflicts over resources and priorities. That is, the resultant mission is rarely a consensus design, but rather the "winner" of a contentious process.

2.2 A Successor to the Hubble Space Telescope

In its early years, NASA supported space astronomy in various wavelength regions, although the agency gave the edge to UV and optical astronomy from the start. Studies in the different wavelength regions nevertheless ran a similar course in that research generally started with survey missions, leading in time to very versatile but complex and costly spacecraft. The pace at which a particular wavelength region reached the stage of what would later be called "Great Observatories" or "Flagship" missions differed. As UV and optical investigations had gained an early lead, the large-scale observatory in this region came first in the shape of the Hubble Space Telescope (HST).

From the early 1970s on, the HST was the key space telescope in NASA's planning.[3] By 1974, this was a joint effort of NASA and the European Space Agency, although NASA was the dominant partner. NASA, of course, has been primarily a technical management agency. When it comes to pursuing astronomy, NASA has mostly provided money, facilities and management expertise. The design and construction of hardware and software themselves has come very largely from industrial contractors that NASA oversees and coordinates. NASA, of course, has its own institutional interests, of which astronomers outside the agency must be mindful to get built the tools they desire. NASA, however, is not a monolithic agency. Different groups within the space agency often have somewhat different or even contradictory interests. NASA's history is replete with many examples of the tensions, for example, between its different several field centers as well as between the field centers and NASA Headquarters in Washington D.C.[4]

The idea of building a successor to HST also seemed obvious to some, but by no means all, astronomers almost from the start of serious planning for Hubble. Detailed design and construction of HST began in 1978 following White House and Congressional approval of the project, by which time astronomers had generally accepted the view that "the whole history of science, and particularly of Astronomy in recent years, tells us that progress depends on the development of new instruments which give us new ways of looking at the world."[5] The main advocates in the 1980s of a successor to the Hubble Space Telescope, as we shall see, were based at the AURA-managed Space Telescope Science Institute (STScI) in Baltimore. There was therefore an aspect of "institutional maintenance" too to this effort. Once the HST's mission was over, so was the Institute's, unless new business, perhaps in the form of a successor to the HST, was in the offing. Thus, long range scientific planning meshed nicely with institutional maintenance.

Even before the Hubble Space Telescope was launched, the central question for many astronomers was *not* "Should a successor to the Space Telescope be built?", but rather how much the successor to the Hubble Space Telescope should differ from the HST itself. Of special importance was whether they should simply scale up Hubble to what seemed likely to be a "do-able" size, given whatever technology would be available at that future time. By the mid-1980s, Hubble's mirror, 2.4 m in diameter, was quite small by the standards of state-of-the-art ground-based telescopes, either under construction or in planning. Scaling up to a bigger mirror size and fashioning a generally larger version of Hubble was therefore an attractive option to at least explore. The limit to how big a mirror they might reasonably argue for would likely be set by the perceptions of what cost its patrons in the White House and Congress would support. On the other hand, instead of a simple scaling-up, supporters of the new space telescope could be even bolder and press for radical, and therefore quite probably more risky, technological choices. This could also include a major mission that would operate at different wavelengths than Hubble, thus opening a new "window" in exploring the Universe. Addressing the question of how far to push the technological boundaries as well as simply determining what they were and what the "political system" would support demanded in part that astronomers and engineers integrate various elements from existing state-of-the-art telescope designs into their planning. It also required that eventually they take the expertise of industrial contractors and the military into account as they had very extensive experience in both planning and building complex satellites for a variety of national security purposes.[6]

The various contractors who might be involved in planning for, managing or building such a machine also had self-interest in proposing some sort of successor to the Hubble Space Telescope. For example, in 1980, NASA's Marshall Space Flight Center considered a plan to launch an 8 m space telescope into low-Earth orbit in the external tank of the space shuttle. In the following year, the company responsible for the optical elements of the HST, Perkin-Elmer, published an article on "Space Astronomy." Included here was a concept for an optical-ultraviolet telescope in space with a mirror 8 m in diameter (a very big step up from the 2.4 m diameter mirror for the HST). Malcolm Longair, one of the leading European astronomers working closely on the HST and at the time Astronomer Royal for Scotland, carried this idea further when in 1983 he argued that many "exciting projects are being converted from gleams in the eye of the astronomer into feasible astronomical projects, the only limits being those of the imagination of the astronomer and the more important limits of funding and manpower. Examples of these types of project include a very large space telescope for optical and ultraviolet observations. An aperture of up to 10 m . . . would represent a huge increase in scientific capability over even the Space Telescope."[7]

During the 1980s, the idea of such a successor to the HST was pressed most enthusiastically by a group of staff members at the Space Telescope Science Institute. Key in this respect was the Institute's director, Riccardo Giacconi. Giacconi, arguably the leading figure in the establishment of x-ray astronomy, shared the Nobel Prize for Physics in 2002 for this work. From hard experience of flying and planning x-ray satellites, Giacconi knew that the lead times for large-scale space observatories could be counted in decades. If there was not to be big gap between the end of the life of the Hubble Space Telescope and its successor, it was essential to get planning underway well before HST was launched.

By 1986, a small group at the Institute was thinking hard about such a successor. The group included astronomers Garth Illingworth and Peter Stockman, as well as Pierre Bely, an engineer with experience of large ground-based telescopes as well as space astronomy. In that year, Bely wrote a paper that laid out some of the details for a 10 m optical telescope in space. He considered, among other things, the size, cost, and location of such a telescope. In line with Giacconi's own thinking, Bely contended that although the Hubble Space Telescope was not yet launched, its limited operational life of 15 years and experience that it takes from ten to fifteen years to complete a large astronomical telescope, meant "it soon will be time to start making serious plans for its successor."[8] Bely reported that several designs had already been advanced. He promoted a general purpose observatory in space that would be as "unspecialized"[9] as HST. By "unspecialized," Bely meant that the new observatory, like HST, should have a wide, rather than a narrow, range of capabilities – a general-purpose observatory, which would also have political appeal to a broad range of astronomical communities.

The new telescope, however, should be a much more powerful scientific tool than the HST: among other things, it should have a bigger mirror and greater resolving power, and cover the wavelength range from the ultraviolet through the optical to the infrared. Bely considered costs, where such a telescope should operate from, whether or not it should be an international venture, the options for the size and type of main mirror, and a number of other issues. In the end, he advocated further studies of a telescope with a primary 10 m in diameter placed into geosynchronous orbit; that is, an orbit some 22,000 miles or so above the Earth, so that it would always be above one part of the planet. This location was sufficiently far from the Earth that the Earth would not block out large sections of the sky, a serious handicap for the Hubble Space Telescope designed to orbit only a few hundred miles above the Earth, which was at the time the limit to the altitude to which a larger optical system could be launched by the system – the Space Shuttle – that was available to the astronomers. Bely's initial design criteria proved to have significant longevity as the NGST began to take shape.

2.3 The Space Studies Board

Since the American government established NASA in 1958, the agency has maintained important links with the National Academy of Sciences. Charged, among other things, with providing advice to the government, the National Academy of Sciences had formed the Space Sciences Board (later renamed as the Space Studies Board [SSB]) at the dawn of the Space Age with the specific intent of using the Board to provide advice to NASA. Although the relationship between the SSB and NASA has sometimes been fraught, in the opinion of some critics, the National Academy quickly became a form of "shadow government" whose backing was often reckoned to be critical if a new project was to proceed. The Space Studies Board's recommendations therefore generally carried clout, and its reports can become crucial political resources. If a successor to the HST was to come into being, it would surely need the strong backing of the National Academy.

In 1988, the Space Studies Board released a report that detailed key scientific issues they anticipated that space scientists would tackle from roughly 1995 to 2015.

Space Science in the Twenty-First Century: Imperatives for the Decades 1995–2015, outlined three main areas for astronomy and astrophysics, one being "large area and high-throughput telescopes."[10] The SSB also recommended NASA consider an optical telescope with a mirror 8–16 m in diameter: "A large aperture space telescope for the ultraviolet, optical, and infrared regions has immense scientific potential. The need for such a telescope will be very high after 10 to 20 years use of HST and ground-based 8-to 10 m-class telescopes," the report contended. "Even now we see that some of the most fundamental of all astronomical questions will require the power of a filled-aperture telescope of 8- to 16-m diameter designed to cover a wavelength range of 912 Å to 30 μm [that is, from the ultraviolet to the mid-infrared] with ambient cooling to 100 K to maximize the infrared performance."

The report also noted that both the Hubble Space Telescope and SIRTF (the Space Infrared Telescope Facility, a medium-aperture infrared telescope at this time slated to be carried into space intermittently by the Space Shuttle), which had yet to be launched, with the expected wealth of date yet to be analyzed so "it is difficult but not premature to formulate a detailed concept of such a large-scale telescope for the ultraviolet, optical and infrared regions." The report's authors nevertheless extolled the increased performance of such an instrument compared to the Hubble Space Telescope. In their opinion, there was a wide range of scientific problems that could be tackled only by a telescope of this type. The extra light-gathering power and resolution, combined with advanced instruments and detectors, "would lead to a quantum leap in our understanding of some of the most fundamental questions in astronomy."[11] But they stressed it would "not simply be a scaled-up HST." Unlike the HST, the new telescope's optics could be cooled to "at least the lower limit of passive radiation methods, about 100 K[12]", and so its infrared performance optimized. [The great challenge with infrared observations is to ensure that the thermal emission from the telescope itself does not swamp the radiation the telescope is detecting from astronomical objects.] For an 8 m class telescope, a large launch vehicle could carry it to orbit, while a 16 m telescope would require the segments of its mirror to be lofted into space and then assembled onto a structure. Also, the group reckoned that an 8–16 m telescope was "within closer reach than a simple extrapolation form HST would suggest."[13]

2.4 Beyond the Hubble Space Telescope

In 1988 STScI scientist Garth Illingworth gave a presentation on "The Next Generation: An 8–16 m Space Telescope" as part of the International Astronomical Union's General Assembly in Baltimore in 1988. He posed the basic question "*What is the UV-Visible-Optical Observatory that will follow HST*?" First, he made the case for a successor, what he referred to as "Son of HST" or maybe "Daughter of HST." He presented two arguments in making his case: continuity and discoveries. HST offered broad capabilities and the scientific case for such an observatory would be as true in the future as it had been for HST back in the 1970s, both to carry out major observing programs and to support other missions. New facilities, such as a next generation telescope, would "open up new 'discovery space' by a significant amount." About the same time, a popular book and accompanying articles on *Cosmic Discovery* by Cornell astronomer Martin Harwit was widely quoted among astronomers. According to Harwit, an increase in instrument sensitivity of 2–3 orders of magnitude were typically required to achieve major "discoveries;" that is, an increase in telescope diameter of 3–5 along with new generations of instruments. In line with the 1988 Space Studies Board report, Illingworth described the observatory he would prefer to study, one with a 16 m primary (perhaps with a segmented design made up of four 8 m parts) that would be passively cooled to around 100 K, be in geosynchronous orbit, and operate in the wavelength range from 0.1 μm to longer than 10 μm, thereby providing an extremely powerful telescope that could operate from the ultraviolet well into the infrared. As to timing, he noted that in 1962, the National Academy of Sciences had held a workshop that produced "A Review of Space Science," and recommended a large diffraction-limited space telescope, in effect a recommendation for what became HST. This meant there had been more than 25 years from a major recommendation to launching HST (he was speaking about a year before its launch). Hence, to ensure a successor to the HST in 15–20 years – that is, around 2005–2010 – "now is clearly the time to move."[14] Even more sobering, what would become the Space Infrared Telescope Facility (SIRTF) and today the Spitzer infrared "Great Observatory" was first proposed to NASA in the summer of 1971 and launched slightly more than 30 years later.

A year after Illingworth gave his presentation at the International Astronomical Union, over 130 astronomers, engineers, and science managers met in Baltimore at the Space Telescope Science Institute to plan what was now called "The Next Generation Space Telescope." This gathering proved a key step in moving beyond the Hubble Space Telescope to the 'Next Big Machine' for space astronomy. The Baltimore meeting was of a kind that astronomers frequently held. Such meetings offer a convenient forum for members of the science and engineering communities to acquaint themselves with what their colleagues are doing, exchange ideas, advocate particular choices, impress potential patrons and possible partners, build a base of support among scientists, and generally move a project forward. As one participant explained, "No one goes away from these meetings and says 'I've seen the Holy Grail and so-and-so has it.' They are to inform people and get everyone up to the same level before they can move on to the next level."[15] Attendees at the 1989 workshop learned of several diverse designs already vying to become the Next Generation Space Telescope. Some were relatively new while versions of others had circulated throughout the astronomy community for years, again a common practice in astronomy by this date.

Garth Illingworth chaired the meeting's science committee. He described his colleagues' goals as "very ambitious and challenging, but realistic extrapolations of current technology."[16] The organizers of the Baltimore meeting presented their colleagues with two "straw man" designs to focus their discussion – one was a space telescope with a 10 m mirror orbiting the Earth. The second was even more exotic: a 16 m telescope on the lunar surface, its proposed location reflecting NASA's very short-lived commitment to a new policy announced just weeks earlier by President George H.W. Bush on 20 July 1989, the twentieth anniversary of astronauts landing on the Moon. This proposed program, known as the Space Exploration Initiative (SEI), envisaged a relatively swift return by the United States to the surface of the Moon and the start of planning for a human expedition to Mars.[17]

NASA and the astronomy community responded to Bush's proposal (and the hope of massive funding that might become available) by rapidly developing optimistic plans for Moon-based astronomy facilities. One idea suggested by NASA's Marshall Space Flight Center (at the time the lead NASA center for the development of the Hubble Space Telescope) was a Large Lunar Telescope. These schemes pictured a 16 m telescope (with a segmented mirror) that would be assembled robotically or perhaps by teams of astronauts. In the planning for the Hubble Space Telescope some two decades earlier, the option of a large Moon-based telescope had also been raised. For a time the name of the proposed telescope was the Large Orbital Telescope, but this was changed to the more neutral Large Space Telescope so as to leave open the possibility of basing it on the Moon. Bush's ambitious, but politically unrealistic, plan faded swiftly away and carried off with it the idea of a lunar telescope. Marshall's planning initiatives nevertheless underlined the space center's interest in some sort of role in the NGST project.

Astronomers considered other locations for the proposed Next Generation Space Telescope. One option that engineers advanced was a low-Earth orbit achieved via the Space Shuttle, which is where the HST would be located. Another involved flying the telescope to one of the Lagrangian points. Located about a million miles from Earth, this piece of cosmic real estate was reckoned by many advocates of NGST as having excellent qualities for a space observatory: it is very cold and dark and a long way from the Earth. But placing a telescope at a Lagrangian point would also eliminate the option of having visiting astronauts service and upgrade it, at least so long as the Shuttle was the only means to flying U.S. astronauts into space. The proposed location for an NGST was, as we shall see, a critical element in the efforts to develop a telescope that would be far less expensive than a scaled-up version of the Hubble Space Telescope.

The debate and negotiations that ensued on the possible location of an NGST as well as other issues revealed several critical problems that astronomers and engineers reckoned they needed to resolve if they were to make serious progress towards a Next Generation Space Telescope. One pressing issue was the design and size of the telescope's primary mirror. Astronomers generally regard the primary mirror as the single most important component of any telescope. Its size and quality determine how much light the telescope can collect and the worth of the data it produces. At the Baltimore meeting engineers and industrial contractors touted the optical quality of Hubble's 2.4 m mirror, which at the time was generally reckoned to be superb (which of course was not in fact the case). A 10 m mirror would collect twenty times as much light as HST and, other things being equal, yield much superior images and even more exciting data.

Even a 16 m mirror was discussed. Not widely appreciated at the time was a prescient design by a Swales engineer, Philip J. Tulkoff, that proposed a 10 m space

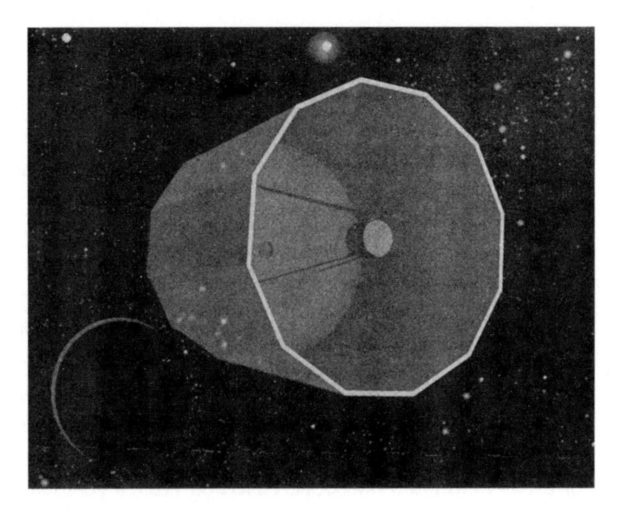

Fig. 2.1 Artist's concept of a 10 m next generation space telescope as envisaged in 1989

aperture telescope that could be passively cooled to between 70 and 100 K, permitting very sensitive observations well into long infrared wavelengths and a modest step in breaking the engineering paradigm of the time that infrared space telescopes required complex and heavy liquid cryogen systems.

As already noted, by 1989 the size of Hubble's mirror was decidedly modest by comparison with ground-based telescope. By the 1980s, the standard size of telescope mirrors for ground based instruments was in the 4 m range while astronomers and engineers were already well-along in building telescopes with mirrors that sported mirrors as large as 10 m (Fig. 2.1). Astronomers reckoned, however, that HST's location far above the turbulent atmosphere would allow it to collect images of exquisite detail while working 24 hours a day and these factors would more than offset the fact that in terms of size its mirror was far from the state-of-the-art in 1989. HST's mirror was also relatively heavy in comparison to the new lightweight technologies NASA and the military desired. Putting objects into orbit in the era of the Space Shuttle was – and remains – extremely expensive, and designers of scientific spacecraft in the 1980s and 1990s, therefore, saw reducing weight wherever possible as a crucial problem. Designers also knew that the heavier a scientific satellite generally the more it would cost. Developing the capability to make very big lightweight mirrors, sometimes called "gossamer optics,"[18] was clearly generally reckoned by the participants at the Baltimore meeting to be the most significant technical obstacle to building the NGST.

The design of NGST's primary mirror certainly became one of the project's so-called tall poles,[19] just as it had been for the Hubble Space Telescope. That is to say, designing and fashioning the primary mirror to the required specifications might entail such difficulties that it would hold up the project (aka, the "tent") and other problems would tend to get lost beneath the canvas – problems that would be present, certainly, but not so pressing or so visible. If astronomers were going to launch any Next Generation Space Telescope, it was already clear to its advocates in 1989 that they needed to reduce radically the weight of the mirror from what they could expect if they simply scaled-up the HST mirror design. Such a procedure would lead to a hugely expensive telescope, and a key concern at the meeting was the overall cost of an NGST. Hanging over the proceedings was the need to "break" the cost curve of the HST as an extrapolation of this curve would mean, unless the political context changed radically, an impossibly expensive telescope that would never be built.

George Field, a prominent astronomer who had been a champion of the Hubble Space Telescope in the 1970s, suggested at the Baltimore meeting that the price of a telescope in low Earth orbit with a mirror of diameter D (meters) would be about $\$3.8(D/10)^{1.7}$ billion in 1986 dollars.[20] If a 16 m telescope were placed on the Moon, the new estimate for the telescope alone would be \$8.4 billion in 1986 dollars. As Garth Illingworth noted in his presentation, one scaling law used for the cost of telescopes reckoned that the cost rose as the 2.7th power of the diameter. HST's cost for design and development was around \$2 billion, so applying "such a factor for a 10–16 m class telescope based on HST's cost leaves one gasping." But Illingworth contended that recent large ground-based telescopes had "broken" this "cost curve," which had been established by telescopes built in the 1950s to the 1970s, by a factor of 4, with more gains in the "pipeline." Critical was savings in weight, which translated into cost savings. Illingworth judged that as "we can see from the discussion at this meeting, this is an area where major improvements in fabrication and polishing techniques are occurring. The combination of improved performance and lower weight for the optical segments will directly and dramatically affect the final cost of the NGST."[21]

In the published report of the Workshop, five statements and recommendations were presented as representing the spirit of the collective opinion of the participants and are worth quoting here at length. There were:

> 1. Scientific Objectives:
>
> There will be a definitive need to continue and extend the observational capability offered by HST beyond its predicted lifetime. A gap of more than 5 years would be a blow to the vitality of forefront astronomical research.
>
> The scientific potential of an HST follow-up mission with enhanced flux collecting power and spatial resolution, and with spectral coverage extended through the near-infrared is enormous.... An observatory providing high sensitivity and high-throughput spectroscopic capability at diffraction-limited spatial resolution from the UV to beyond 10 microns is vital for the study of the most fundamental questions of astrophysics. These include the formation and evolution of galaxies, stars and planets, and the nature of the young universe.
>
> 2. Technological Readiness:
>
> A telescope in the 10–16 m class is not an unrealistically large step beyond the current state of technological development. While development and demonstration programs are clearly needed, many of the core technologies are maturing to the point where the required goals appear to be within a very reasonable extrapolation of the current state-of-the-art. In particular, advances in the fabrication of lightweight optics and new techniques for polishing have the potential for very substantial weight savings and hence cost savings, while offering optical performance beyond what was possible in the past.

3. Siting:
Both the Moon and high Earth orbits are suitable sites for [the] next generation space telescope. Low Earth orbits are undesirable because of high disturbance levels, insufficient passive cooling and low observing efficiency.... Space-based and lunar-based designs should be pursued in parallel for the next few years to clarify the observational, technical, space logistical and cost tradeoffs.

4. Programmatic approach:
A 10–16 m (space-based) to 16 m (lunar-based) aperture is considered a realistic goal. Future workshops should concentrate on further definition of the scientific objectives, review of preliminary studies and the identification of critical technologies. Strawman designs should be prepared to refine the various concepts and ideas and to focus discussion.... In projects of this complexity, efficient design is the result of many compromises that can only be developed by successive iterations and by system-level analyses. The importance of this iterative process involving astronomers, physicists and engineers in the science-engineering tradeoffs and in defining the requirements was emphasized by many participants. The involvement of these different groups needs to occur during all phases of the project, from concept to development, through technology development and fabrication, and finally during system-level testing.

Once clearly identified by the preliminary design process, the development of the key enabling technologies should be integrated with the appropriate long-term program of the national and international Space Agencies....

5. International cooperation:
Like HST, the next generation space telescope project should be carried out cooperatively as an international program. Cost sharing renders such major missions more affordable for each participating country, and international collaboration often enhances quality and performance. Complex and pioneering space missions also benefit from the exchange of ideas and variety of approaches afforded by multicultural associations.[22]

Co-operation between NASA and the European Space Agency was very much "in the air" at the Workshop. Duccio Machetto, an ESA astronomer based at the Space Telescope Science Institute, spoke on "ESA[’s]" Long Term Plans and Status. He noted that "There is a large interest in the astronomical community in Europe for HST and also in a future HST. It is therefore important to include the European astronomical scientific community in a possible joint venture in a 10–16 m next generation space telescope."[23] Further, one page of the published report of the Workshop also carried a section headed "Sage Advice." Here John Bahcall, a leading and very influential astronomer who had played a pivotal role in winning Congressional approval for the Hubble Space Telescope in the 1970s and so making the HST politically feasible, was quoted as arguing that "International cooperation may be critical for such a major project."[24] Ultimately, the James Webb Space Telescope would be a cooperative venture of NASA, the European Space Agency, and the Canadian Space Agency. At this stage, however, none of the astronomers and engineers who attended the workshop was based at Canadian institutions, although there was already strong European interest. Indeed, as we shall see in subsequent work, the European space agencies were in many cases offering the astronomy communities more frequent opportunities at this time to fund advanced post-HST concepts than NASA.

2.5 The Decade of the Infrared

As astronomers and engineers met in Baltimore to discuss an NGST, work was already proceeding on what became known as the Bahcall Committee report, as this very large scale effort at planning was chaired by John Bahcall. We have already discussed the importance of advice from the National Academy of Sciences in shaping NASA's priorities. Hence a significant hurdle came for the NGST when in early 1989 the National Research Council commissioned the Astronomy and Astrophysics Survey Committee to review the field and produce a series of recommendations on new ground-based and space-based programs and observatories. The Bahcall Committee's recommendations were based on studies by fifteen advisory panels that represented various wavelength regions and particular areas of astrophysics. In all, advice came from over 300 astronomers who served on these panels, and another 600 or so wrote letters, essays, or delivered oral presentations at various open meetings.

In line with the usual thinking of astronomers by this point, the Committee argued in its final report that "Progress in astronomy often comes from technological advances that open new windows on the universe or make possible large increases in sensitivity or resolution. During the 1990s, arrays of infrared detectors, the ability to build large optical telescopes, improved angular resolution at a variety of wavelengths, new electronic detectors, and the ability of computers to process large amounts of date will make possible an improved view of the universe."[25] The Committee went on to package their conclusions by proclaiming the 1990s as "The Decade of the Infrared" and expected that the "technological revolution in detectors at infrared wavelengths will increase the power of telescopes by factors of thousands."[26]

The panel on ultraviolet and optical astronomy, which was chaired by Garth Illingworth, who we have seen was an active champion of a successor to the HST, strongly recommended building what the panel termed the "Large Space Telescope." This would be a 6 m telescope operating in the ultraviolet, the infrared, and the optical, and the panel urged that it be flown within a few years of the end of the expected 15 year life of the HST (then assumed to be 2005). The panel therefore proposed starting the Large Space Telescope in 1998 so that it could be completed by 2009. The Large Space Telescope would later be followed by "a telescope of astonishing power," the 16 m Next Generation Space Telescope, which the panel judged should naturally be located at a lunar outpost,[27] again assuming a major NASA program capable of placing and operating complex facilities on the lunar surface at a cost and timeframe acceptable to the astronomy community.

With the Bahcall Committee charged to recommend the most important new initiatives for the decade 1990 to 2000, it in fact advocated neither the Large Space Telescope nor the Next Generation Space Telescope. Even the Large Space Telescope was costed at $2 billion and in 1991 very large scale astronomy projects were standing in the shadow of the Hubble Space Telescope, which had been launched in 1990 but, as we shall see later, its early performance had failed badly to live up to its advance billing. While the Bahcall Committee did not give explicit reasons why it chose not to back the Large Space Telescope, they certainly judged it premature to give it a top priority. In particular, the Bahcall Committee assessed what it reckoned to be the critical technological initiatives for the 1990s so as to "form the basis for frontier science in the decade 2000–2010." $50 million, the Committee argued, should be spent developing technologies for large space telescopes.[28] Contending that "we must begin now the conceptual planning and technological development for the next generation of astronomy missions to follow the Great Observatories," they cited as one example the 6 m Large Space Telescope operating in the ultraviolet to the infrared. They discussed other possible missions too, including a very large x-ray telescope and a submillimeter observatory consisting of a deployable 10 m telescope. It did not, however, make a choice between the different options, judging that the "scientific imperatives and the infrastructure available at the time of selection will influence which missions are chosen."

Among the technical issues, they reckoned, would be the construction and control of lightweight systems, the capability of launch vehicles, advances in robotic constructions techniques, as well as the possible availability of facilities on the Moon. "The technology development programs listed [in this report]," the Bahcall Committee claimed, "will provide part of the factual basis required for decisions about future astronomical missions."[29]

2.6 A Road Not Traveled

By 1991, as Bahcall's committee deliberated its recommendations for astronomy's next decade, the advocates for some sort of Next Generation Space Telescope (what had also been called the Large Space Telescope in the Bahcall Committee deliberations) had made considerable progress in nursing their project along. But they were still a very long way from a go-ahead to start detailed feasibility studies, let alone serious design work or actual construction. The kind of issues that could derail plans for a big space telescope are provided by the story of another of the large space telescopes touched on by the Bahcall Committee, the Large Deployable Reflector (LDR).[30] Far from being a side story, the LDR offers an example of NASA sets about developing new missions and its fate was a warning at the time to NGST advocates.

In the late 1970s, engineers and scientists from two NASA field centers, the Jet Propulsion Laboratory and Ames Research Center, both in California, had proposed the LDR. This mission sprang from JPL's studies that began in 1976 of a space observatory to make observations in the submillimeter wavelength range. These studies led to a series of workshops in 1977 involving American and European astronomers. They recommended pursuing a 10 m space telescope to operate in the infrared and sub-millimeter regions of the spectrum. As the possible designs matured, they suggested one path to bigger yet lighter mirrors. Rather than using a single, massive piece of material for the primary mirror, engineers by the early 1980s proposed a Large Deployable Reflector with a 20 m mirror (Fig. 2.2).

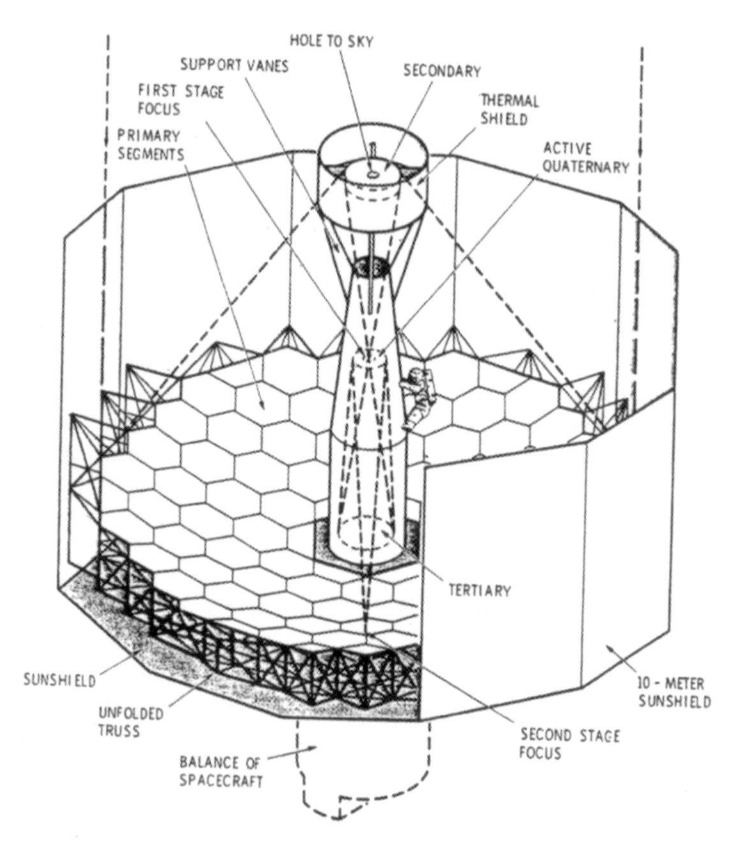

Fig. 2.2 A concept that never was. The planned large deployable reflector with a 20 m primary

Assembled from small lightweight hexagonal segments, this would fit together to make one giant light collector. The mirror could be collapsed to a smaller size to fit inside a rocket or into a spacecraft that would fit into the Space Shuttle's cargo bay (the first flight of which was in 1981), and then, once in space, open like the petals of a flower. In later studies engineers examined assembling the LDR in space using the space station, NASA's flagship program, which had been approved for construction by President Reagan in 1983, illustrating the often *ad hoc* nature of plans for new missions as advocates borrow from existing projects or other planned missions and work within the overall context of NASA's broader institutional goals.

The idea of a segmented primary mirror was attractive to many managers, engineers and scientists, one major reason being the advances astronomers were making in designing ground-based telescopes that exploited segmented mirrors.[31]

Designs for segmented primary mirrors for space projects were based, in part, on new schemes astronomers were proposing for ground-based telescopes. During the 1980s, for example, engineers and astronomers at Caltech and University of California developed segmented mirror technologies for the 10 m Keck Telescope project in Hawaii. The first Keck telescope went into operation in 1991 and a second one soon followed. Other telescope projects were also demonstrating how computers linked with mechanical systems could accurately control lightweight mirrors. This knowledge helped boost the confidence of NASA staffers that segmented mirror technologies could be developed for space telescopes.[32]

The Jet Propulsion Laboratory and Ames sustained the Large Deployable Reflector program with modest funding throughout the 1980s. The two NASA centers held conferences on it every two years which drew dozens of participants. In 1982, for example, a week-long workshop in California, similar in many respects to the Baltimore NGST workshop that was to be held in 1989, attracted around one hundred scientists and engineers to develop the science rationale for the LDR and to formulate what its observational capabilities should be.[33] The National Academy of Sciences twice recommended the Large Deployable Reflector as a high priority.[34] While it existed only on paper, the Large Deployable Reflector was a serious project whose advocates initially had high hopes of advancing it through the NASA bureaucracy and the White House and Congress to the stage where it would indeed be built. Reflecting their seriousness, NASA supported Lockheed Martin and Kodak to perform studies of the Large Deployable Reflector concept.

Although the Bahcall Committee had declared the 1990s as the decade of the infrared, in the end it did not support the LDR. Instead the Committee gave its strong backing to three other infrared projects instead,[35] although none were missions considered capable enough to be considered the successor to Hubble. Hence after several years of funding, NASA's upper management decided not to pursue the Large Deployable Reflector past the initial design stage. Advances in ground-based telescopes, its anticipated very high cost, plus the fact that somewhat similar missions were being pursued by the European Space Agency, kept it limited to the drawing board.

One common pattern in NASA's strategy in the 1980s and 1990s for developing new missions is clear here. The agency sponsored studies of various depths for many possible projects. Relatively few survived to be built, but elements of them lived on in various ways in other programs. In the case of the Large Deployable Reflector, the deployable primary mirror concept became a central feature in NASA's later plans for Next Generation Space Telescope,[36] although the *direct* influence on the designs of the NGST was probably negligible.

2.7 Spherical Aberration

In April 1990, before the results of the deliberations of the Bahcall Committee were complete, the Hubble Space Telescope was launched amid an enormous blaze of publicity. Expectations of the quality of the images it would return were extremely high, among astronomers and the general public. When NASA released the first images to the public, however, its spokesmen had the grim job of announcing that the telescope suffered from an optical defect known as spherical aberration. This meant that its images were not nearly as good as expected and that HST's scientific performance would be crippled, at least until some kind of repair or technological fix could be put into place. Derided by late night comedians and in editorial cartoons, proclaimed to be a "technoturkey" by one U.S. Senator, Hubble swiftly became a national symbol of technological failure.[37]

The space agency put into place a team to locate and review what had gone wrong. It soon concluded that the primary mirror was the culprit. Due to a mistake that had not been caught in assembling a test device, the primary mirror had been polished too flat at its edges.[38] Cast in the now dubious role of successor to the flawed and for a time publicly ridiculed Hubble Space Telescope, the consequences for the Next Generation Space Telescope were severe. Progress on its planning slowed to a crawl. It was not, however, brought to a halt.

In early 1991, there was a two-day workshop with 79 participants at the Jet Propulsion Laboratory on "Technologies for Large Filled-Aperture Telescopes in Space," one result of funding by NASA Headquarters to advance the technologies engineers and scientists reckoned to be needed for new astrophysics missions between 1995 and 2015. A number of the presentations centered on possible successors to the Hubble Space Telescope but, to judge from the conference proceedings, the likely cost of such a Next Big Machine was even more of a concern for the participants than they had they had been two years earlier in Baltimore.

In the executive summary of the two-day workshop, Garth Illingworth described the main conclusions drawn by the participants. As he put it, an 8 m class telescope "in high Earth orbit would have unprecedented power for problems as diverse as planet searches around nearby stars to the way in which galaxies formed in the young universe. It will build upon the discoveries and astronomical understanding of many decades of research with astronomical observatories, and is the natural successor to the Hubble Space Telescope (HST) and the new generation of large 10 m class ground-based telescopes."[39] He reckoned that while the gains with the HST are "impressive," those with the NGST would be "truly astonishing."[40]

Illingworth also claimed that operating an NGST in high Earth orbit would lead to savings in weight, size, power, and the complexity of its operations. These savings would in turn mean an 8 m NGST would be comparable in weight to the HST (12 tons), thereby "breaking away" from the HST cost curve; that is, securing a telescope considerably less expensive than might be anticipated just from the HST experience. But there was also now a key development. In the discussion of one of the papers, Rodger Thompson of the University of Arizona and a Principal Investigator for one of the instruments slated to fly aboard the HST later in the 1990s, challenged the assumptions underlying the planning directed towards a single, multipurpose and multiwavelength telescope that would cost a great deal of money, but not be serviced on orbit. He asked if "maybe a series of identical, let's say 6 m telescopes with individual instrumentation, all sort of lined up to go for specific purposes, might be better. One might be cryogenic, one might be just for spectroscopy or something else. This might be a better way to save money. These are production line types of telescopes."[41]

Illingworth protested that while there was "a significant level of rationality in that argument," politically it would be one that "you'd never be able to sell. Having one of something that was closely similar to the rest of them would essentially kill off the rest, given the cost. In the minds of the folks that are funding these things – and I think Congress in a sense – they're looking at this and saying: Here's astronomers out there wanting the world. We'll give them one and that's it."[42] Thompson replied that HST's fate had changed attitudes.

Other participants also spoke in favor of Thompson's idea. Although various concerns were expressed, in the panel discussion of the chairs of the various working groups, the debate had now shifted the consensus to planning for two telescopes. But the discussants also accepted that no studies had been done on this concept. Basic information was lacking so serious studies were needed. As Illingworth put it, "If it turns out it's do-able, the cost to do two of them, then you know you're not in a position of really selling it. If there are some cost savings to be made, we may be in a much more advantageous position."

One of the possible telescopes now being mentioned was what one speaker referred to as a "super Edison;" that is, a larger and more powerful version of a concept for an ambitious passively cooled "next generation" infrared telescope developed by a team led by University of Wyoming astronomer Harley Thronson in collaboration with a group at the Royal Observatory, Edinburgh.[43] Thronson was in attendance at the workshop. But only later, as we shall discuss elsewhere, would schemes for infrared telescopes mesh with the planning for the NGST.

Hence in the executive summary of the workshop, the conclusion now was that "While we have discussed NGST as being a single all-purpose UV to mid-IR telescope, it has been suggested that it may well be cheaper to design and configure two spacecraft, one for UV-Visible and the other for the Visible-IR. This is not obviously the case. Technical feasibility studies need to be combined with cost trade-off analyses to establish the most cost-effective and timely route to fruition of the program. The current baseline is to consider NGST as a single [high Earth orbit] telescope"[44] operating in the uv-visible and infrared.

2.8 Conclusions

Years before the Hubble Space Telescope was launched in 1990 a number of astronomers and engineers in the US and Europe were thinking hard about a possible successor to the HST as well as working to engage a broad community of researchers in the design of such a new observatory. That the launch of any such successor was likely to be many years away was also widely accepted. However, the fiasco of Hubble's spherical aberration had a serious effect on the pace at which plans were advancing for the Next Generation Space Telescope. Thus crucially for the dynamics of building the "Next Big Machine," the fate of the offspring was intimately tied to that of the parent. In fact, as we will describe in later papers, it was only when in the mid-1990s that the NGST planning was remade by the incorporation of a series of technology developments in infrared astronomy that NASA threw its institutional weight and money behind the development of a Next Generation Space Telescope: until that time, the American space agency had been generally standing on the sidelines as the major astronomical space telescope of the early 21st Century was being debated. The efforts between the mid 1980s and the early 1990s were nevertheless critical to the establishment and success of the later endeavor.

Without a set of committed advocates in these years willing to work away often as individuals or small groups to raise the consciousness of their colleagues about the possibilities of a successor to the HST even before its launch and eventual success, those later efforts would surely have been postponed, if pursued at all.

Acknowledgments We are grateful to Harley Thronson for comments on an earlier draft of this paper. The research for this paper has in part been made possible by the Canadian Space Agency (9F007-060458/001/ST) and NASA (NAG5-13604) and we are most grateful for this support. Robert W. Smith also completed part of his research during his tenure as the Lindbergh Professor of Aerospace History at the Smithsonian Institution in 2007 and he greatly appreciates the Institution's support and the collegial atmosphere provided by the Division of Space History at the National Air and Space Museum.

Notes

1. Smith, R.: Engines of Discovery: Scientific Instruments and the History of Astronomy and Planetary Science in the United States in the Twentieth Century. Journal for the History of Astronomy, **28**, 49–77 (1997).
2. See, for example, Riordan, M.: A Tale of Two Cultures: Building the Superconducting Super Collider, 1988–1993. Historical Studies in the Physical and Biological Sciences, **32**, 125–144 (2001).
3. What eventually became the Hubble Space Telescope in 1983, was also known for lengthy periods as the Large Space Telescope and the Space Telescope.
4. This is an implicit theme, for instance, in Dunar, A.J., and Waring, S.J., Power to Explore. A History of Marshall Space Flight Center 1960–1990. NASA, Washington DC (1999).
5. Brown, R.J.: The Size, Shape and Temperature of Stars. In West, R.M. (ed.) Understanding the Universe. The Impact of Space Astronomy, pp. 73–92. D. Reidel, Dordrecht (1983). On this point, see also Smith, R.W.: Engines of Discovery: Scientific Instruments and the History of Astronomy and Planetary Science in the United States in the Twentieth Century. Journal for the History of Astronomy, **28**, 49–77 (1997).
6. On the development of satellites for national security purposes, see, among others, Day, D.A., Logsdon, J.M., Latell, B. (eds.): Eye in the Sky. The Story of the Corona Spy Satellites. Smithsonian Institution Press, Washington DC and London (1998). Burrows, W.E.: Deep Black. Space Espionage and National Security. Random House, New York (1986).
7. Longair, M.: Space Science and Cosmology. In: West. R.M. (ed.): .The Impact of Space Astronomy, pp. 129–226. D. Reidel, Dordrecht (1983).
8. Bely, P.Y.: A Ten-Meter Optical Telescope in Space. Advanced Technology Optical Telescopes III, in SPIE, **628**, 188–195, (1986).
9. Ibid.

10. Space Science in the Twenty-First Century: Imperatives for the Decade 1995–2015. Washington DC, National Academy Press (1988), in seven volumes, volume 1, Overview, 46. Quotations are from Field, G: Summary of Space Science Board 1995-2015 Study, pp. 11–15. In: Bely, P.Y., Burrows, C.J., Illingworth, G.J. (eds.). The Next Generation Space Telescope. Proceedings of a Workshop Jointly Sponsored by the National Aeronautics and Space Administration and the Space Telescope Science Institute and held at the Space Telescope Science Institute, Baltimore, Maryland, 13–15 September 1989. Space Telescope Science Institute, Baltimore (1990).
11. Ibid.
12. This widely quoted apparent limit to a low temperature achievable via passive (aka, radiative) cooling alone became for some years a major hindrance to even more aggressive use of the natural cold of space in achieving sensitive observations at long infrared wavelengths.
13. Space Science in the Twenty-First Century: Imperatives for the Decade 1995–2015. Washington DC, National Academy Press (1988), in seven volumes, volume 1, Overview, 46. Quotations are from Field, G: Summary of Space Science Board 1995–2015 Study, pp. 11–15. In: Bely, P.Y., Burrows, C.J., Illingworth, G.J. (eds.). The Next Generation Space Telescope. Proceedings of a Workshop Jointly Sponsored by the National Aeronautics and Space Administration and the Space Telescope Science Institute and held at the Space Telescope Science Institute, Baltimore, Maryland, 13–15 September 1989. Space Telescope Science Institute, Baltimore (1990).
14. Illingworth, G.J.: The Next Generation: An 8–16 m Space Telescope. Highlights of Astronomy, **8**, 449–453 (1989).
15. James Breckinridge; July 17, 2001 interview.
16. After HST, What Next?: Sky & Telescope. February 1990, p. 129.
17. On the fate of the SEI, see Hogan, T.: Mars Wars. The Rise and Fall of the Space Exploration Initiative. NASA, Washington DC (2007). For a discussion of some of the policy issues involved, see also Roy, S.A.: The Origin of the Smaller, Faster, Cheaper Approach in NASA's Solar System Exploration Program. Space Policy, **14**, 153–171 (1998).
18. Angel, R.: Future Optical and Infrared Telescopes. Nature, **409**, 427–431 (2001).
19. This is discussed in Smith, R.W. (with contributions by Hanle, P., Kargon, R., Tatarewicz, J.): The Space Telescope: A Study of NASA, Science, Technology and Politics. Expanded paperback edition, Cambridge, University Press, New York (1993), p. 69.
20. Field, G.: Summary of Space Science Board 1995–2015 Study, pp. 11–15. In: Bely, P.Y., Burrows, C.J., Illingworth, G.J. (eds.). The Next Generation Space Telescope. Proceedings of a Workshop Jointly Sponsored by the National Aeronautics and Space Administration and the Space Telescope Science Institute and held at the Space Telescope Science Institute, Baltimore, Maryland, 13–15 September 1989. Space Telescope Science Institute, Baltimore (1990).
21. Illingworth, G.J.: The Next Generation UV-Visible-IR Space Telescope. In: Bely, P.Y., Burrows, C.J., Illingworth, G.J. (eds.). The Next Generation Space Telescope. Proceedings of a Workshop Jointly Sponsored by the National Aeronautics and Space Administration and the Space Telescope Science Institute and held at the Space Telescope Science Institute, Baltimore, Maryland, 13–15 September 1989, pp. 31–36. Space Telescope Science Institute, Baltimore (1990).
22. Conclusions of the Workshop: In: Bely, P.Y., Burrows, C.J., Illingworth, G.J. (eds.). The Next Generation Space Telescope. Proceedings of a Workshop Jointly Sponsored by the National Aeronautics and Space Administration and the Space Telescope Science Institute and held at the Space Telescope Science Institute, Baltimore, Maryland, 13–15 September 1989, pp. 5–6. Space Telescope Science Institute, Baltimore (1990).
23. Macchetto, D.: ESA Long Term Plans and Status. In: Bely, P.Y., Burrows, C.J., Illingworth, G.J. (eds.). The Next Generation Space Telescope. Proceedings of a Workshop Jointly Sponsored by the National Aeronautics and Space Administration and the Space Telescope Science Institute and held at the Space Telescope Science Institute, Baltimore, Maryland, 13–15 September 1989, pp. 27–30. Space Telescope Science Institute, Baltimore.
24. Bahcall, J.: In: Bely, P.Y., Burrows, C.J., Illingworth, G.J. (eds.). The Next Generation Space Telescope. Proceedings of a Workshop Jointly Sponsored by the National Aeronautics and Space Administration and the Space Telescope Science Institute and held at the Space Telescope Science

Institute, Baltimore, Maryland, 13–15 September 1989, p. 7. Space Telescope Science Institute, Baltimore (1990).
25. National Research Council Astronomy and Astrophysics Survey Committee: The Decade of Discovery in Astronomy and Astrophysics. National Academy Press, Washington DC (1991), p. 3.
26. Ibid., p. 10.
27. National Research Council Astronomy and Astrophysics Survey Committee: The Decade of Discovery in Astronomy and Astrophysics. Working Papers...: National Academy Press, Washington DC (1991), pp. IV–1.
28. National Research Council Astronomy and Astrophysics Survey Committee: The Decade of Discovery in Astronomy and Astrophysics. National Academy Press, Washington DC (1991), p. 20.
29. Ibid., p. 27.
30. See, for example, Swanson, P.N., Gulkis, S., Kuiper, T.B.H., Kiya, M.: Large Deployable Reflector (LDR): A Concept for an Orbiting Submillimeter-Infrared Telescope for the 1990s. Optical Engineering, **22**, 725–731 (1983).
31. Pierre Y. Bely; July 18, 2001 interview.
32. Pierre Y. Bely; July 18, 2001 interview.
33. According to a number of sources, in an ironic Spoonerism, long-remembered by scientists who originally had high hopes for LDR, at a presentation on the mission at a workshop at the CalTech at this time, one speaker inadvertently referred to the mission as "the Large *Deplorable* Reflector." The audience winced.
34. Astronomy Survey Committee: Astronomy and Astrophysics for the 1980's. Report of the Astronomy Survey Committee. National Research Council, Washington DC (1982).
35. National Research Council Astronomy and Astrophysics Survey Committee: The Decade of Discovery in Astronomy and Astrophysics. National Academy Press, Washington DC (1991), 75–80.
36. Pierre Y. Bely; July 18, 2001, interview.
37. The finding of spherical aberration and its aftermath is discussed in Smith, R.W. (with contributions by Hanle, P., Kargon, R., Tatarewicz, J.N.): Cambridge University Press, New York (1993), expanded paperback edition, pp. 399–425.
38. The Hubble Space Telescope Optical Systems Failure Report. NASA, Washington DC (1990).
39. Illingworth, G.J.: Executive Summary, in Workshop Proceedings: Technologies for Large Filled-Aperture Telescopes in Space (JPL D-8541, vol. 4, 1991), pp. 3–16.
40. Ibid., p. 5.
41. Ibid., comments in question session, p. 161.
42. Ibid., p. 161.
43. For a brief account of Edison, see Davies, J.K.: It will never work! An idea that changed infrared astronomy from space. The Space Review, 21 August 2006. www.thespacereview.com/article/688/1. Accessed 21 March 2008.
44. Ibid., p. 12.

Michael Shull, Matthew Greenhouse, and Chick Woodward embarrass Harley Thronson, who obviously dressed inappropriately for a meeting in Arizona

Chapter 3
The Kuiper Belt and Other Debris Disks

David Jewitt, Amaya Moro-Martín and Pedro Lacerda

Abstract We discuss the current knowledge of the solar system, focusing on bodies in the outer regions, on the information they provide concerning solar system formation, and on the possible relationships that may exist between our system and the debris disks of other stars. Beyond the domains of the terrestrial and giant planets, the comets in the Kuiper belt and the Oort cloud preserve some of our most pristine materials. The Kuiper belt, in particular, is a collisional dust source and a scientific bridge to the dusty "debris disks" observed around many nearby main-sequence stars. Study of the solar system provides a level of detail that we cannot discern in the distant disks while observations of the disks may help to set the solar system in proper context.

3.1 Introduction

Planetary astronomy is unusual among the astronomical sciences in that the objects of its attention are inexorably transformed by intensive study into the targets of other sciences. For example, the Moon and Mars were studied telescopically by astronomers for nearly four centuries but, in the last few decades, these worlds have been transformed into the playgrounds of geologists, geophysicists, meteorologists, biologists and others. Telescopic studies continue to be of value, but we now learn most about the Moon and Mars from in-situ investigations. This transformation from science at-a-distance to science up-close is a forward step and a tremendous luxury not afforded to the rest of astronomy. Those who study other stars or the galaxies beyond our own will always be forced by distance to do so telescopically.

However, the impression that the solar system is now *only* geology or meteorology or some other science beyond the realm of astronomy is completely incorrect when applied to the outer regions. The outer solar system (OSS) remains firmly entrenched within the domain of astronomy, its contents accessible only to telescopes. Indeed, major components of the OSS, notably the Kuiper belt, were discovered

D. Jewitt (✉)
Institute for Astronomy, University of Hawaii, 2680 Woodlawn Drive, Honolulu, USA
e-mail: jewitt@ifa.hawaii.edu

H.A. Thronson et al. (eds.), *Astrophysics in the Next Decade,* Astrophysics and Space Science Proceedings, DOI 10.1007/978-1-4020-9457-6_3,

(telescopically) less than two decades ago and will continue to be best studied via astronomical techniques for the foreseeable future. It is reasonable to expect that the next generation of telescopes in space and on the ground will play a major role in improving our understanding of the solar system, its origin and its similarity to related systems around other stars.

In this chapter, we present an up-to-date overview of the layout of the solar system and direct attention to the outer regions where our understanding is the least secure but the potential for scientific advance is the greatest. The architecture of the Kuiper belt is discussed as an example of a source-body system that probably lies behind many of the debris disks of other stars. Next, the debris disks are discussed based on the latest observations from the ground and from Spitzer, and on new models of dust transport. Throughout, we use text within grey boxes to highlight areas where the next generation telescopes are expected to have major impact.

3.2 The Architecture of the Solar System

It is useful to divide the solar system into three distinct domains, those of the terrestrial planets, the giant planets, and the comets. Objects within these domains are distinguished by their compositions, by their modes of formation and by the depth and quality of knowledge we possess on each.

3.2.1 Terrestrial Planets

The terrestrial planets (Mercury, Venus, Earth and Moon, Mars and most main-belt asteroids) are found inside 3 AU. They have refractory compositions dominated by iron (~35% by mass), oxygen (~30%), silicon and magnesium (~15% each) and were formed by binary accretion in the protoplanetary disk. About 95% of the terrestrial planet mass is contained within Venus and Earth ($\sim 1\ M_{\oplus} = 6 \times 10^{24}$ kg, each). The rest is found in the small planets Mercury and Mars ($0.1\ M_{\oplus}$ combined) with trace amounts ($\sim 3 \times 10^{-4}\ M_{\oplus}$) in the main-belt asteroids located between Mars and Jupiter. The largest asteroid is Ceres (~900 km diameter). The terrestrial planets grew by binary accretion between solid bodies in the protoplanetary disk of the Sun. While not all details of this process are understood, it is clear that sticking and coagulation of dust grains, perhaps aided at first by hydrodynamic forces exerted from the gaseous component of the disk, produced larger and larger bodies up to the ones we see now in the solar system. The gaseous component, judged mainly by observations of other stars, dissipated on timescales from a few to ~10 Myr. Measurements of inclusions within primitive meteorites show that macroscopic bodies existed within a few Myr of the origin. The overall timescale for growth was determined, ultimately, by the sweeping up of residual mass from the disk, a process thought to have taken perhaps 40 Myr in the case of the Earth. No substantial body is found in the asteroid belt although it is likely that sufficient mass existed there to form an object of planetary class. This is thought to be because growth in this region was interrupted by strong perturbations, caused by the emergence of nearby, massive Jupiter (at ~5 AU).

3.2.2 *Giant Planets*

Orbits of the giant planets (Jupiter, Saturn, Uranus and Neptune) span the range 5–30 AU. The giants are in fact of two compositionally distinct kinds.

3.2.2.1 Gas Giants

Jupiter (310 $M_\oplus$) and Saturn (95 $M_\oplus$) are so-called gas giants because, mass-wise, they are dominated by hydrogen and helium. Throughout the bulk of each planet these gases are compressed, however, into a degenerate (metallic) liquid that supports convection and sustains a magnetic field through dynamo action. The compositional similarity to the Sun suggests to some investigators that the gas giants might form by simple hydrodynamic collapse of the protoplanetary gas nebula (Boss 2001). In hydrodynamic collapse the essential timescale is given by the free-fall time, and this could be astonishingy short (e.g. 1000 yrs). Details of this instability, especially related to the necessarily rapid cooling of the collapsing planet, remain under discussion (Boley et al. 2007, Boss 2007). In fact, measurements of the moment of inertia, coupled with determinations of the equation of state of hydrogen-helium mixtures at relevant pressures and temperatures, show that Saturn (certainly) and Jupiter (probably) have distinct cores containing 5 $M_\oplus$ to 15 $M_\oplus$ of heavy elements (the case of Jupiter is less compelling than Saturn because of its greater mass and central hydrostatic pressure, leading to larger uncertainties concerning the self-compressibility of the gas). The presence of a dense core is the basis for the model of formation through " nucleated instability", in which the core grows by binary accretion in the manner of the terrestrial planets, until the gravitational escape speed from the core becomes comparable to the thermal speed of molecules in the gas nebula. Then, the core traps gas directly from the nebula, leading to the large masses and gas-rich compositions observed in Jupiter and Saturn.

Historically, the main sticking point for nucleated instability models has been that the cores must grow to critical size *before* the surrounding gas nebula dissipates (i.e. 5 $M_\oplus$ to 15 $M_\oplus$ cores must grow in much less than 10 Myr). The increase in the disk surface density due to the freezing of water as ice outside the snow-line is one factor helping to decrease core growth times. Another may be the radial jumping motion of the growing cores, driven by angular momentum and energy exchange with planetesimals (Rice and Armitage 2003). In recent times, a consensus appears to have emerged that Jupiter's core, at least, could have grown by binary accretion from a disk with $\Sigma \sim 50$ to $100\,\mathrm{kg\,m^{-2}}$ on timescales ~1 Myr (Rice and Armitage 2003). The collapse of nebular gas onto the core after this would have been nearly instantaneous.

Recent data show that the heavier elements (at least in Jupiter, the better studied of the gas giants) are enriched relative to hydrogen in the Sun by factors of $\sim$2 to 4 (Owen et al. 1999), so that wholesale nebular collapse cannot be the whole story (and may not even be part of it). The enrichment applies not only to species that are condensible at the $\sim$100 K temperatures appropriate to Jupiter's orbit, but to the noble gases Ar, Kr and Xe, which can only be trapped in ice at much lower temperatures, $<$30 K (e.g. Bar-Nun et al. 1988). Therefore, Owen et al. suggest that the gas giants incorporate substantial mass from a hitherto unsuspected population of ultracold bodies, presumably originating in the outer solar system. In a variant of this model, cold grains from the outer solar system trap volatile gases but drift inwards under the action of gas drag, eventually reaching the inner nebula when they evaporate in the heat of the Sun and enrich the gas (Guillot and Hueso 2006).

3.2.2.2 Ice Giants

Uranus (15 $M_\oplus$) and Neptune (17 $M_\oplus$), in addition to being an order of magnitude less massive than the gas giants are compositionally distinct. These planets contain a few $M_\oplus$ of H and He, and a much larger fraction of the "ices" H_2O, CH_4 and NH_3. For this reason they are known as "ice giants", but the name is misleading because they are certainly not solid bodies but are merely composed of molecules which, if they were much colder, would be simple ices. In terms of their mode of formation, the difference between the ice giants and the gas giants may be largely one of timescale. It is widely thought that the ice giants correspond to the heavy cores of Jupiter and Saturn, but with only vestigial hydrogen/helium envelopes accreted from the rapidly dissipating gaseous component of the protoplanetary disk.

While qualitatively appealing, forming Uranus and (especially) Neptune on the 10 Myr timescale associated with the loss of the gas disk has been a major challenge to those who model planetary growth. The problem is evident from a simple consideration of the collision rate between particles in a disk or surface density $\Sigma(R)\,\mathrm{kg\,m^{-2}}$, where R is the heliocentric distance. The probability of a collision in each orbit varies in proportion to $\Sigma(R)$, while the orbital period varies as $R^{3/2}$. Together, this gives a collision timescale varying as $R^{3/2}/\Sigma(R)$ which, with $\Sigma(R) \propto R^{-3/2}$ gives $t_c \propto R^3$. A giant planet core that takes 1 Myr to form at 5 AU would take $6^3 \sim 200$ Myr to form at 30 AU, and this considerably exceeds the gas disk lifetime. Suggested solutions to the long growth times for the outer planets include augmentation of the disk density, $\Sigma(R)$, perhaps through the action of aerodynamic drag, and formation of the ice giants at smaller distances (and therefore higher Σ) than those at which they reside. The latter possibility ties into the general notion that the orbits of the outermost three planets have expanded in response to the action or torques between the planets and the disk. Still another idea is that Uranus and Neptune are gas giants whose hydrogen and helium envelopes were ablated by ionizing radiation from the Sun or a nearby, hot star (Boss et al. 2002).

3.2.3 Comets

Comets are icy bodies which sublimate in the heat of the Sun, producing observationally diagnostic unbound atmospheres or "comae". For most known comets, the sublimation is sufficiently strong that mass loss cannot be sustained for much longer than $\sim 10^4$ yr, a tiny fraction of the age of the solar system. For this reason, the comets must be continually resupplied to the planetary region from one or more low temperature reservoirs, if their numbers are to remain in steady state. In the last half century, at least three distinct source regions have been identified.

3.2.3.1 Oort Cloud Comets

The orbits of long period comets are highly elliptical, isotropically distributed and typically large (Fig. 3.1), suggesting a gravitationally bound, spheroidal source region of order 100,000 AU in extent (Oort 1950). Comets in the cloud are scattered randomly into the planetary region by the action of passing stars and also perturbed by the asymmetric gravitational potential of the galactic disk (e.g. Higuchi et al. 2007). Unfortunately, the cloud is so large that its residents cannot be directly counted. The population must instead be inferred from the rate of arrival of comets from the cloud and estimates of the external torques. A recent work gives the number larger than $\sim$1 km in radius as 5×10^{11}, with a combined mass in the range $2M_{\oplus}$–$40M_{\oplus}$ (Francis 2005).

Oort cloud comets are thought to have originated in the Sun's protoplanetary disk in the vicinity of the giant planets and were scattered out by interactions with the growing, migrating planetary embryos. Although most were lost, a fraction of the ejected comets, perhaps from 1 to 10%, was subsequently deflected by external perturbations that lifted the perihelia out of the planetary region, effectively decoupling these comets from the rest of the system (Hahn and Malhotra 1999). Over $\sim$1 Gyr, the orbits of trapped comets were randomized, converting their distribution from a flattened one reflecting their disk source to a spherical one, compatible with the random directions of arrival of long period comets. In this model, which is qualitatively unchanged from that proposed by Oort (1950), the scale and population of the Oort cloud are set by the external perturbations from nearby stars and the galactic disk. If the Sun formed in a cluster, then the average perturbations from cluster members would have been bigger than now and a substantial population of more tightly bound, so-called "inner Oort cloud" comets could have been trapped. It is possible that the Halley family comets, whose orbits are predominantly but not exclusively prograde (Fig. 3.2) are delivered from the inner Oort cloud (Levison et al. 2001).

Some 90 to 99% of the comets formed in our system eluded capture and now roam the interstellar medium. If this fraction applies to all stars, then $\sim 10^{23}$ to $\sim 10^{24}$ "interstellar comets" exist in our galaxy, containing several $\times 10^5$ $M_{\odot}$ of metals. Interstellar comets ejected from other stars might be detected with all-sky surveys in the Pan STARRS or Large Synoptic Survey Telescope (LSST) class (Jewitt 2003).

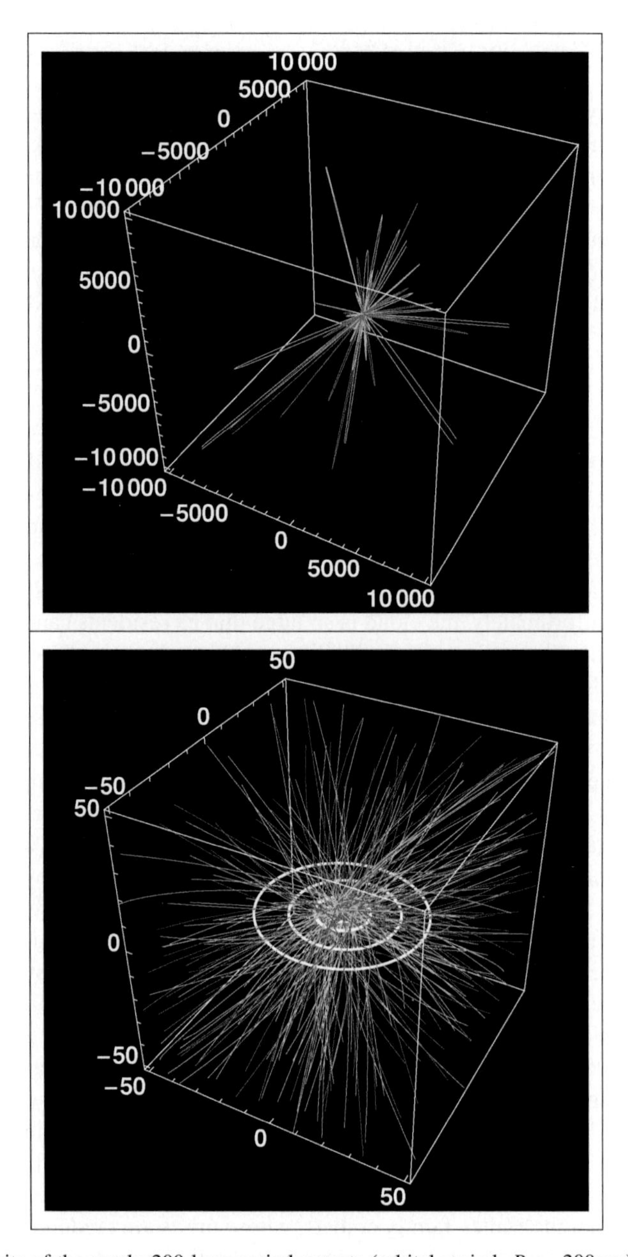

Fig. 3.1 Orbits of the nearly 200 long period comets (orbital period, $P > 200$ yrs) listed in the JPL small-body database. The highly eccentric orbits of many LPCs appear as nearly radial lines on the top (wide angle) view, showing a cube 20,000 AU on a side. The bottom panel shows a narrow angle view of a 100 AU cube. Prograde orbits are shown in cyan while retrograde orbits are in *magenta*. Numbers along the axes are distances in AU from the Sun at (0,0,0). The orbits of the giant planets are shown in *white*

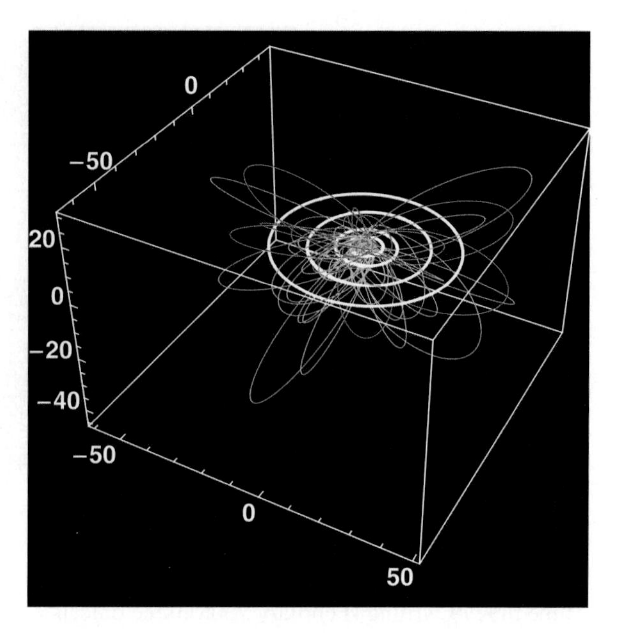

Fig. 3.2 Halley family comets. Their non-isotropic inclination distribution, with more prograde (*cyan*) orbits than retrograde (*magenta*), points to an origin in the inner Oort cloud, where stellar and galactic perturbations have been too small to randomize the orbits. Following the classical definition, we plot the 44 comets in the JPL small-body database having Tisserand parameters with respect to Jupiter ≤ 2 and orbital periods $20 < P(\mathrm{yrs}) < 200$. The orbits of the giant planets are shown in *white*

Such objects would be recognized by their strongly hyperbolic orbits relative to the Sun, quite different from any object yet observed.

3.2.3.2 Kuiper Belt Comets

The orbits of the Jupiter family comets (JFCs) have modest inclinations, with no retrograde examples, and most have eccentricities much less than unity (Fig. 3.3). They are dynamically distinct from the long period Oort comets, and they interact strongly with Jupiter. For many years it was thought that the JFCs were captured from the long period population by Jupiter but increasingly numerical detailed work in the 1980's showed that this was not possible (Fernández 1980, Duncan et al. 1988). The source appears to be the Kuiper belt, although this is not an iron-clad conclusion and the particular region or regions in the Kuiper belt from which the JFCs originate has yet to be identified.

About 200 JFCs are numbered (meaning that their orbits are very well determined) and a further 200 are known. Their survival is limited by a combination of volatile depletion, ejection from the solar system, or impact into a planet or the Sun. Dynamical interactions alone give a median lifetime near 0.5 Myr (Levison and Duncan 1994), which matches the volatile depletion lifetime for bodies smaller than $r \sim 40$ km. The implication is that JFCs smaller than this size will become dormant before they are dynamically removed, ending up as bodies that are asteroidal in appearance but cometary in orbit. Some of these dead comets are suspected to exist among the near-Earth "asteroid" population; indirect evidence for a fraction $\sim$10% comes from albedo measurements (Fernández et al. 2001) and dynamical models (Bottke et al. 2002).

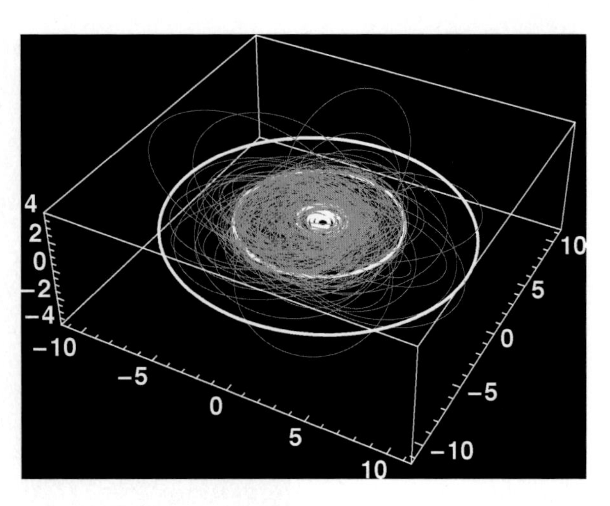

Fig. 3.3 Perspective view of the Jupiter family comets (*salmon*) together with the orbits of the planets out to Saturn. The Sun is at $(x, y, z) = (0, 0, 0)$ and the distances are in AU. Plotted are 166 JFCs, selected from the JPL small-body database based on their Tisserand parameter with respect to Jupiter ($2 < T_J < 3$; see, e.g., Levison and Duncan 1994) and with orbital periods $P < 20$ yr

3.2.3.3 Main Belt Comets

At the time of writing (February 2008) three objects are known to have the dynamical characteristics of asteroids but the physical appearances of comets. They show comae and particle tails indicative of on-going mass loss (Hsieh and Jewitt, 2006, Figs. 3.4 and 3.5). These are the Main-Belt comets (MBCs), most directly interpreted as ice-rich asteroids.

In the modern solar system, the MBCs are dynamically isolated from the Oort cloud and Kuiper belt reservoirs (i.e. they cannot be captured from these other regions given the present-day layout of the solar system, cf. Levison et al. 2006). The MBCs should thus be regarded as a third and independent comet reservoir. Two possibilities for their origin seem plausible. The MBCs could have accreted ice if they grew in place but outside the snow-line. Alternatively, the MBCs might have been captured from elsewhere if the layout of the solar system were very different in the past, providing a dynamical paths from the Kuiper belt that do not now exist. Insufficient evidence exists at present to decide between these possibilities.

These bodies escaped detection until now because their mass loss rates ($\ll 1$ kg s^{-1}) are two to three orders of magnitude smaller than from typical comets.

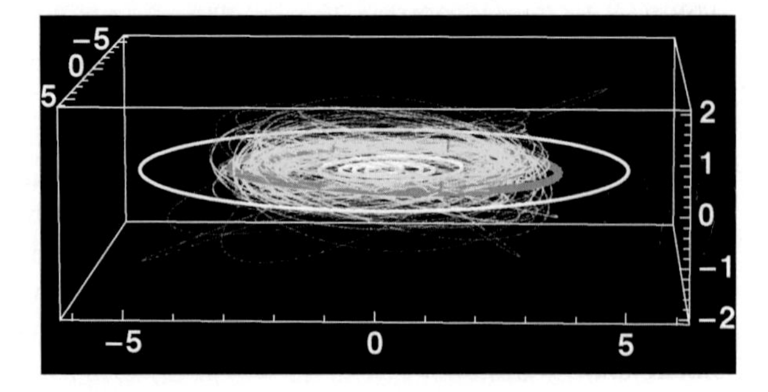

Fig. 3.4 Perspective view of the main-belt comets (*orange*) together with the orbits of 100 asteroids (*thin, yellow lines*) and planets out to Jupiter

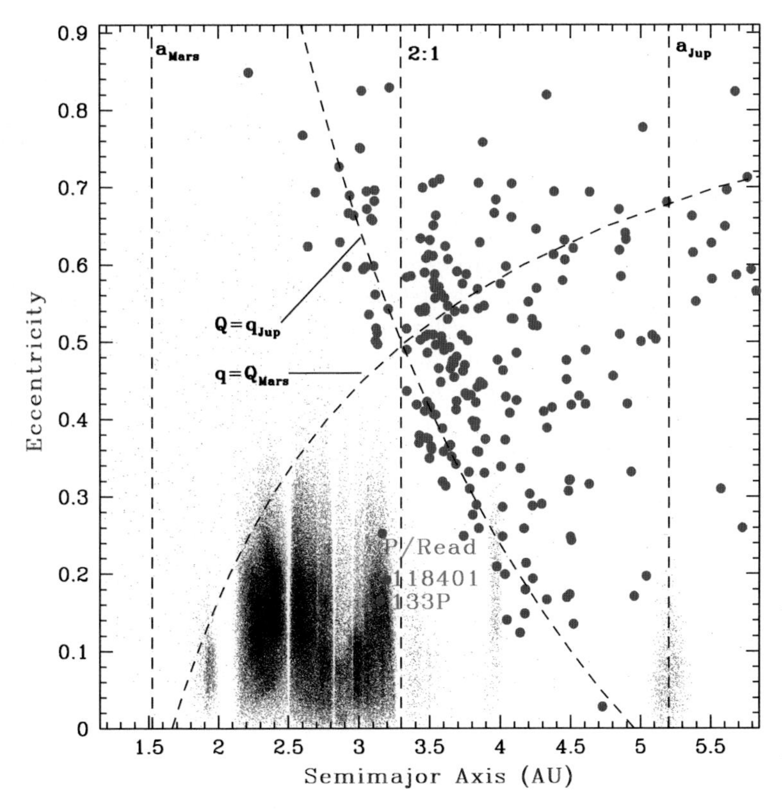

Fig. 3.5 Orbital semimajor axis vs. eccentricity for objects classified as asteroids (*black*) and comets (*blue*), together with the three known main-belt comets 133P/Elst-Pizarro, P/2005 U1 (Read), and 118401 (1999 RE_{70}) (plotted in *red*). Vertical dashed lines mark the semimajor axes of Mars and Jupiter and the 2:1 mean-motion resonance with Jupiter (commonly considered the outer boundary of the classical main belt), as labeled. Curved dashed lines show the loci of orbits with perihelia equal to Mars's aphelion ($q = Q_{Mars}$) and orbits with aphelia equal to Jupiter's perihelion ($Q = q_{Jup}$). Objects plotted above the $q > Q_{Mars}$ line are Mars-crossers. Objects plotted to the right of the $Q < q_{Jup}$ line are Jupiter-crossers. From Hsieh and Jewitt (2006)

The mass loss is believed to be driven by the sublimation of water ice exposed in small surface regions, with effective sublimating areas of $\sim$1000 m^2 and less. Although sublimation at 3 AU is weak, the MBCs are small and mass loss at the observed rates cannot be sustained for the age of the solar system. Instead, a trigger for the activity (perhaps impact excavation of otherwise buried ices) is needed. This raises the possibility that the detected MBCs are a tiny fraction of the total number that will be found by dedicated surveys like *PanSTARRS* and *LSST*. Even more interesting is the possibility that many or even most outer belt asteroids are in fact ice-rich bodies which display only transient activity. Evidence from the petrology and mineralogy of meteorites (e.g. the CI and CM chondrites that are thought to originate in the outer belt) shows the past presence of liquid water in the meteorite parent bodies. The MBCs show that some of this water survived to the present day.

Next generation facilities will offer opportunities to study the physical properties of the satellite systems of the giant planets. The regular satellites occupy prograde orbits of small inclination and eccentricity, resulting from their formation in the accretion disks through which the planets grew. As such, the individual regular satellite systems offer information about the sub-nebulae that must have existed around the growing planets. These sub-nebulae had their own density and temperature structures very different from the local disk of the Sun. Studies of reflected and thermal radiation with *JWST* and *ALMA* will be able to determine compositional differences in even the fainter regular satellites of the ice giants, and may provide constraints on Io-like and Enceladus-like endogenous activity.

The study of comets will be greatly advanced by *JWST* spectroscopy of the nuclei of comets (in order to minimize the effects of scattering from the dust coma, these small objects must be observed when far from the Sun and therefore faint). Simultaneous measurements of reflected and thermal radiation will provide albedo measurements which, with high signal-to-noise ratio spectra, will help determine the nature and evolution of the refractory surface mantles of these bodies.

3.3 Kuiper Belt

More than 1200 Kuiper belt objects (KBOs) have been discovered since the first example, 1992 QB1, was identified in 1992 (Jewitt and Luu 1993). The known objects have typical diameters of 100 km and larger. The total number of such objects, scaled from published surveys, is of order 70,000, showing that there is considerable remaining discovery space to be filled by future surveys. The number of objects larger than 1 km may exceed 10^8.

The KBO orbits occupy a thick disk or sheet outside Neptune's orbit (Fig. 3.6). A major observational result is the finding that the KBOs can be divided on the basis of their orbital elements into several, distinct groups (Jewitt 1998). This is best seen in a (semimajor axis) vs. e (orbital eccentricity) space, shown here in Fig. 3.7. The major Kuiper belt dynamical groups in the figure are:

- Classical Kuiper belt objects (red circles). These orbit between about $a = 42$ AU and the 2:1 mean-motion resonance with Neptune at $a = 47$ AU. Typical orbital eccentricities of the Classicals are $e \sim 0$ to 0.2 while the inclination distribution appears to be bimodal with components near $i = 2°$ (so-called "Cold-Classicals") and $i = 20°$ (the "Hot Classicals", Brown and Trujillo 2001, Elliot et al. 2005). Numerical integrations show that the orbits of the Classical objects are stable on timescales comparable to the age of the solar system largely because their perihelia are always so far from Neptune's orbit that no strong scattering occurs Holman and Wisdom 1993.

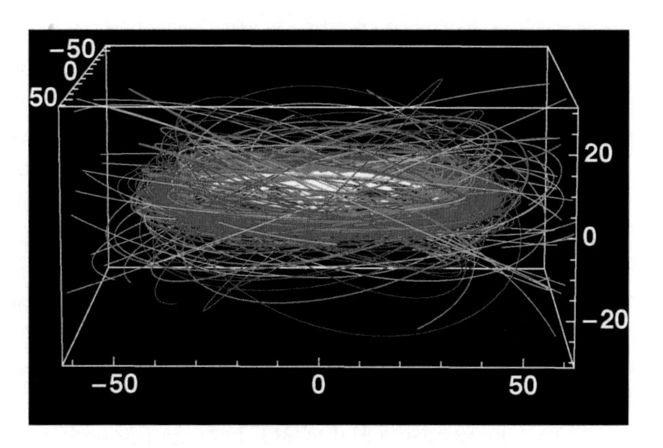

Fig. 3.6 Perspective view of the Kuiper belt. The Sun is at $(x, y, z) = (0, 0, 0)$ and the axes are marked in AU. White ellipses show the orbits of the four giant planets. *Red* orbits mark the classical KBOs. Their ring-like distribution is obvious. Blue orbits show resonant KBOs, while *green* shows the Scattered objects. For clarity, only one fifth of the total number of objects in each dynamical type is plotted

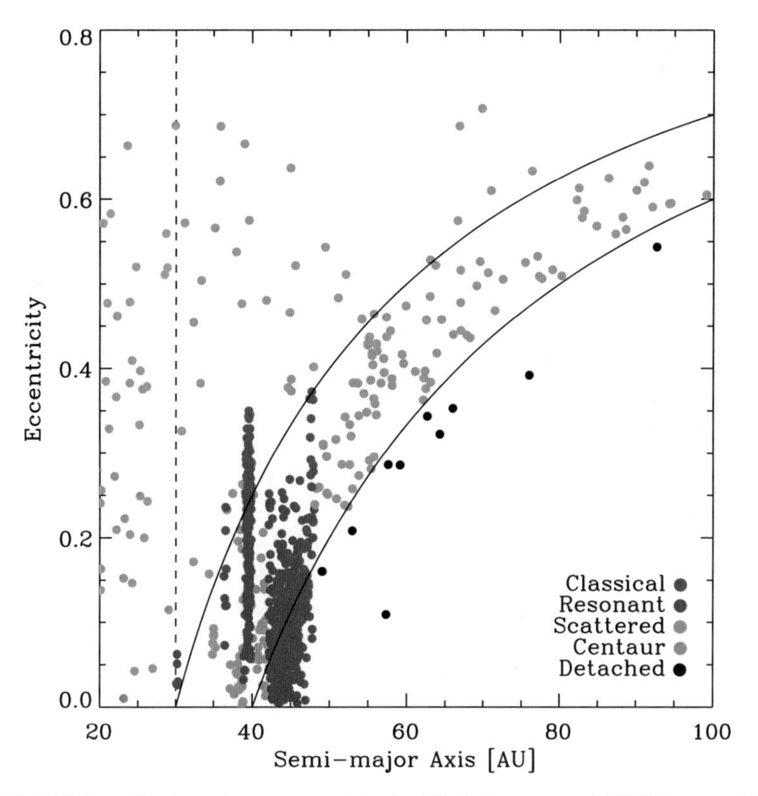

Fig. 3.7 Orbital semimajor axis vs. eccentricity for KBOs known as of 2008 January 30. The objects are color-coded according to their dynamical type, as labeled in the Figure and discussed in the text. A vertical dashed line at $a = 30$ AU marks the orbit of Neptune, the nominal inner-boundary of the Kuiper belt. Upper and lower solid arcs show the loci of orbits having perihelion distances $q = a(1 - e)$ equal to 30 AU and 40 AU, respectively

- Resonant Kuiper belt objects (blue circles). A number of mean-motion resonances (MMRs) with Neptune are populated by KBOs, especially the 3:2 MMR at 39.3 AU and the 2:1 MMR at 47.6 AU (see Fig. 3.7). The 3:2 MMR objects are known as Plutinos, to recognize 134340 Pluto as the first known member of this population. The 2:1 MMR objects are sometimes called "twotinos" while those in 1:1 MMR are Neptune's Trojans (Sheppard and Trujillo 2006). The resonant objects are dynamically stable by virtue of phase protections conferred by the resonances. For example, KBOs in 3:2 MMR can, like Pluto itself, have perihelia inside Neptune's orbit, but their orbits librate under perturbations from that planet in such a way that the distance of closest approach to Neptune is always large. In fact, all objects above the upper solid arc in Fig. 3.7 have perihelia interior to Neptune's orbit.

  The process by which KBOs became trapped in MMRs is thought to be planetary migration (Fernández and Ip 1984). Migration occurs as a result of angular momentum transfer during gravitational interactions between the planets and material in the disk. At late stages in the evolution of the disk (i.e. later than $\sim$10 Myr, after the gaseous component of the disk has dissipated) the interactions are between the planets and individual KBOs or other planetesimals. In a one-planet system, the sling-shot ejection of KBOs to the interstellar medium would result in net shrinkage of the planetary orbit. In the real solar system, however, KBOs can be scattered inwards from planet to planet, carrying energy and angular momentum with them as they go. Massive Jupiter then acts as the source of angular momentum and energy, ejecting KBOs from the system. In the process, its orbit shrinks, while those of the other giant planets expand. The timescale is the same as the timescale for planetary growth, and the distance through which a planet migrates depends on the mass ejected from the system. Outward migration of Neptune carries that planet's MMRs outwards, leading to the sweep-up of KBOs (Malhotra 1995).
- Scattered Kuiper belt objects (green circles), also known as Scattered disk objects. These objects have typically eccentric and inclined orbits with perihelia in the $30 \leq q \leq 40$ AU range (Figs. 3.7 and 3.8). The prototype is 1996 TL66, whose orbit stretches from $\sim$35 AU to $\sim$130 AU (Luu et al. 1997). The brightness of these objects varies strongly around the orbit, such that a majority are visible only when near perihelion. For example, the survey in which 1996 TL66 was discovered lacked the sensitivity to detect the object over 88% of the orbital period. Accordingly, the estimated population is large, probably rivaling the rest of the Kuiper belt (Trujillo et al. 2001).

  Scattered KBOs owe their extreme orbital properties to continued, weak perihelic interactions with Neptune which excite the eccentricity to larger and larger values (Duncan and Levison 1997). The current aphelion record-holder is 87269 (2000 OO67) with $Q = 1123$ AU. As the aphelion grows, so does the dynamical influence of external perturbations from passing stars and the asymmetric gravitational potential of the galactic disk.
- Detached Kuiper belt objects (black circles). The orbits of these bodies resemble those of the Scattered KBOs except that the perihelia are too far from Neptune, $q > 40$ AU, for planetary perturbations to have excited the eccentricities. The

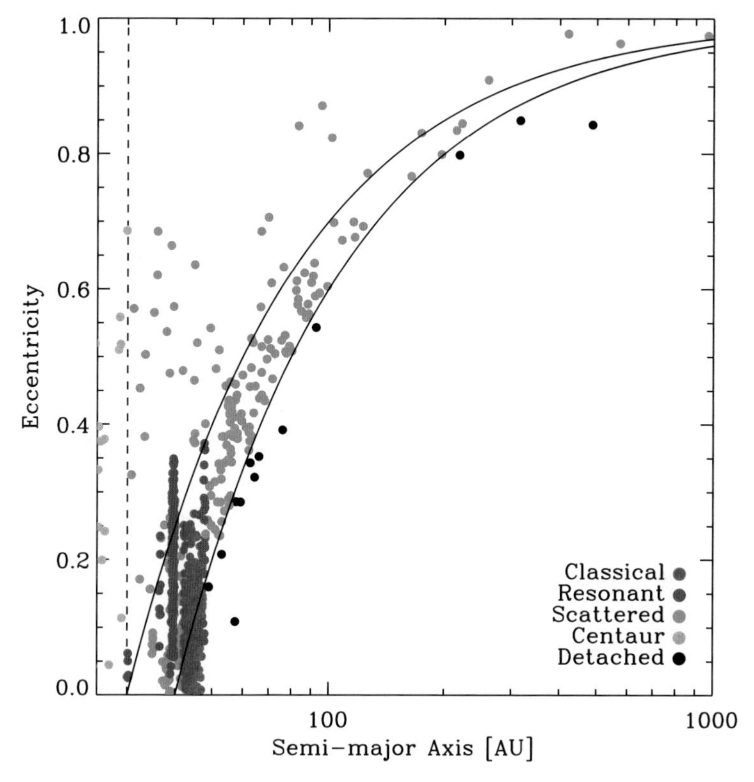

Fig. 3.8 Same as Fig. 3.7 but with a logarithmic x-axis extended to $a = 1000$ AU to better show the extent of the scattered KBOs

prototype is 2000 CR105, with $q = 44$ AU (Gladman et al. 2002) but a more extreme example is the famous Sedna with $q = 74$ AU (Brown et al. 2004).

The mechanism by which the perihelia of the detached objects were lifted away from the influence of Neptune is unknown. The most interesting conjectures include the tidal action of external perturbers, whether they be unseen planets in our own system or unbound passing stars (Morbidelli and Levison 2004).

The Kuiper belt is a thick disk (Fig. 3.9), with an apparent full width at half maximum, $FWHM \sim 10°$ (Jewitt et al. 1996). The apparent width is an underestimate of the true width, however, because most KBOs have been discovered in surveys aimed near the ecliptic, and the sensitivity of such surveys varies inversely with the KBO inclination. Estimates of the unbiased (i.e. true) inclination distribution give $FWHM \sim 25$ to $30°$ (Jewitt et al. 1996, Brown 2001, Elliot et al. 2005 – see especially Fig. 3.20). Moreover, all four components of the Kuiper belt possess broad inclination distributions (Fig. 3.9). The inclination distribution of the Classical KBOs appears to be bimodal, with a narrow core superimposed on a broad halo (Brown and Trujillo 2001, Elliot et al. 2005).

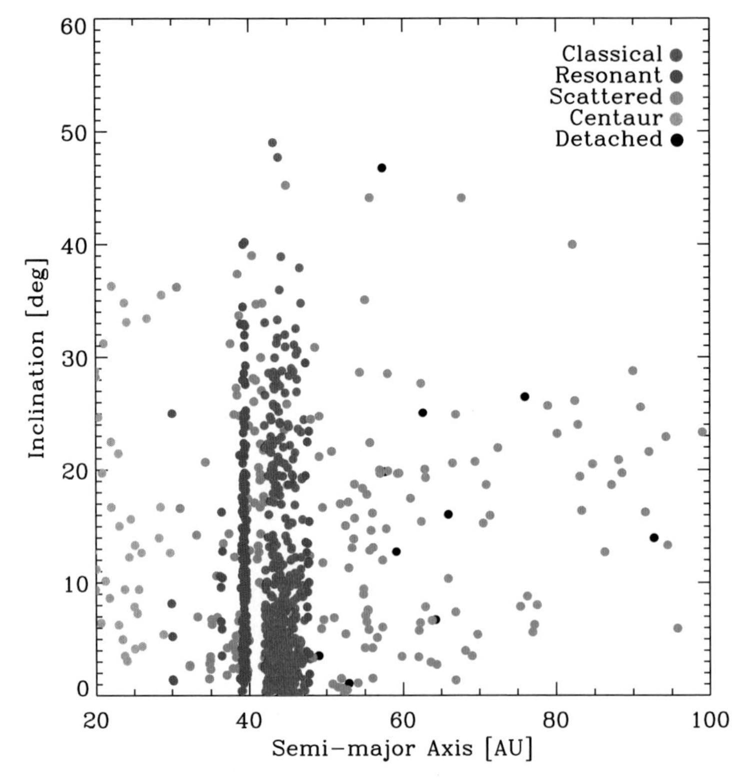

Fig. 3.9 Orbital semimajor axis vs. inclination for KBOs known as of 2008 January 30. The plot shows that the KBOs occupy orbits having a wide range of inclinations. Most of the plotted objects were discovered in surveys directed towards the ecliptic, meaning that an observation bias *against* finding high inclination bodies is imprinted on the sample

The Kuiper belt is not thin like the Sun's original accretion disk and it is clear the the inclinations of the orbits of its members have been excited. The velocity dispersion amongst KBOs is $\Delta V = 1.7\,\mathrm{km\,s^{-1}}$ (Jewitt et al. 1996, Trujillo et al. 2001). At these velocities, impacts between all but the largest KBOs lead to shattering and the production of dust, rather than to accretion and growth.

Next generation facilities should provide unprecedented survey capabilities (through *LSST*) that will provide a deep, all-sky survey of the solar system to 24th visual magnitude, or deeper. The number of objects for which reliable orbits exist will increase from $\sim 10^3$, at present, by one to two orders of magnitude, depending on the detailed strategies employed by the surveys. Large samples are needed to assess the relative populations of the resonances and other dynamical niches that may place limits on the formation and evolution of the OSS. Objects much larger than Pluto, perhaps in the Mars or Earth class, may also be revealed by careful work.

3.4 Interrelation of the Populations

Evidence that the small-body populations are interrelated is provided by dynamical simulations. The interrelations are of two basic types: (a) those that occur through dynamical processes operating in the current solar system and (b) those that might have operated at an earlier epoch when the architecture of the solar system may have been different from now.

As examples of the first kind, it is clear that objects in the Oort cloud and Kuiper belt reservoirs can be perturbed into planet crossing orbits, and that these perturbations drive a cascade from the outer solar system through the Centaurs (bodies, asteroidal or cometary in nature) that are strongly interacting with the giant planets to the Jupiter family comets (orbits small enough for the Sun to initiate sublimation of near-surface ice and strongly interacting with Jupiter) to dead and dormant comets in the near-Earth "asteroid" population. In the current system, simulations show that it is *not* possible to capture the Trojans or the irregular satellites (Fig. 3.11) of the giant planets from the passing armada of small bodies, and there is no known dynamical path linking comets from the Kuiper belt, for example, to comets in the main-belt (the MBCs).

The second kind of interrelation is possible because of planetary migration. The best evidence for the latter is deduced from the resonant Kuiper belt populations which, in one model, require that Neptune's orbit expanded by roughly 10 AU, so pushing its mean-motion resonances outwards through the undisturbed Kuiper belt (Malhotra 1995). The torques driving the migration moved Jupiter inwards (by a few $\times 0.1$ AU, since it is so massive) as the other giants migrated outwards. In one exciting model, the "Nice model", this migration pushed Jupiter and Saturn across the 2:1 mean motion resonance (Gomes et al. 2005). The dynamical consequences of the periodic perturbations induced between the solar system's two most massive planets would have been severe (Fig. 3.10). In published models, these perturbations excite a 30 $M_\oplus$ Kuiper belt, placing large numbers of objects into planet crossing orbits and clearing the Kuiper belt down to its current, puny mass of ~ 0.1 $M_\oplus$. During this clearing phase, numerous opportunities exist for trapping scattered KBOs in dynamically surprising locations. For example, the Trojans of the planets could have been trapped during this phase (Morbidelli et al. 2005). Some of the irregular satellites might likewise have been acquired at this time (Nesvorny et al. 2007; furthermore, any irregular satellites possessed by the planets *before* the mean-motion resonance crossing would have been lost). Other KBOs might have been trapped in the outer regions of the main-asteroid belt, perhaps providing a Kuiper belt source for ice in the MBCs.

Whether or not Jupiter and Saturn actually crossed the 2:1 resonance is unknown. In the Nice model, the 2:1 resonance crossing is contrived to occur at about 3.8 Gyr, so as to coincide with the epoch of the late-heavy bombardment (LHB). The latter is a period of heavy cratering on the Moon, thought by some to result from a sudden shower of impactors some 0.8 Gyr after the formation epoch. However, the interpretation of the crater age data is non-unique and reasonable

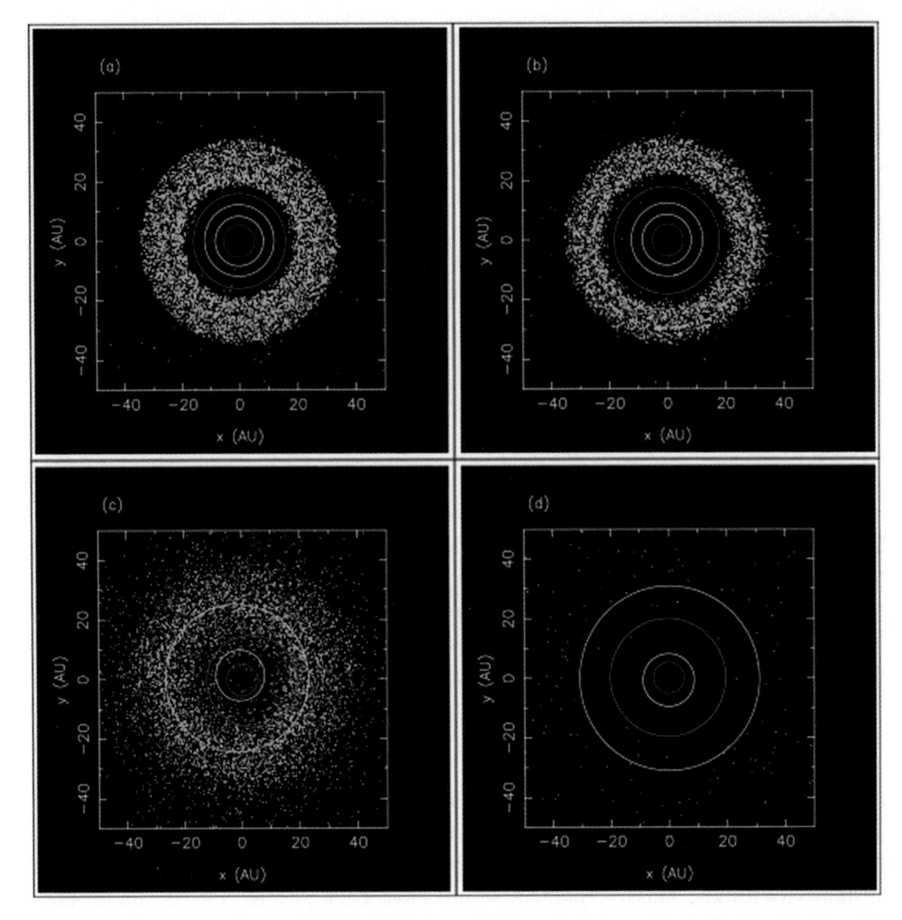

Fig. 3.10 "Nice" model in which the architecture of the solar system is set by the clearing of a massive ($30\,M_\oplus$) Kuiper belt (*stippled green region*) when planets are thrown outwards by strong interactions between Jupiter (*red*) and Saturn (*pink*) at the 2:1 mean-motion resonance. (**a**) The initial configuration with the giant planets at 5.5, 8.2, 11.5 and 14.2 AU (**b**) Just before the 2:1 resonance crossing, timed to occur near 880 Myr from the start (**c**) 3 Myr after resonance crossing (note the large eccentricity of Uranus (*purple*) at this time and the placement of Neptune (*blue*) *in* the Kuiper belt) and (**d**) 200 Myr later, by which time the planetary orbits have assumed nearly their current properties. Adapted with permission from Gomes et al. 2005

arguments exist to interpret the LHB in other ways (Chapman et al. 2007). For example, the cratering rate could merely *appear* to peak at 3.8 Gyr because all earlier (older) surfaces were destroyed by an impact flux even stronger than that at the LHB. (The LHB would then be an analog to the epoch in the expanding early universe when the optical depth first fell below unity and the galaxies became visible).

Fig. 3.11 Saturn's irregular satellite Phoebe, which might be a captured Kuiper belt object. The effective spherical radius is 107±1 km: the largest crater, Jason (*at left*), has a diameter comparable to the radius. The surface contains many ices (water, carbon dioxide) and yet is dark, with geometric albedo 0.08, as a consequence of dust mixed in the ice. Image from Cassini Imaging Team/NASA/JPL/Space Science Institute

3.5 Evidence Concerning the Birth Environment

Several lines of evidence suggest that the Sun was formed in a dense cluster.

The existence of widespread evidence for the decay products of short-lived isotopes in mineral inclusions in meteorites suggests that one or more supernovae exploded in the vicinity of the Sun, shortly before its formation. ^{26}Al (half-life 0.7 Myr) was the first such unstable isotope to be identified (Lee et al. 1977) but others, like ^{60}Fe (half-life 1.5 Myr) are known (and, unlike ^{26}Al, cannot be produced by nuclear spallation reactions; Mostefaoui et al. 2004). It is possible, although not required, that the collapse formation of the protoplanetary nebula was triggered by shock compression from the supernova that supplied the unstable nuclei (Ouellette et al. 2007).

The sharp outer edge to the classical Kuiper belt could be produced by a stellar encounter having an impact parameter ~150 AU to 200 AU (Ida et al. 2000), although this is only one of several possible causes. Such close encounters are highly improbable in the modern epoch but would be more likely in the dense environment of a young cluster. At the same time, the survival and regularity of the orbits of the planets suggests that no very close encounter occurred (Gaidos 1995).

As noted above, the existence of a dense inner Oort cloud, required in some models to explain the source of the Halley family comets (long-period comets which include retrograde examples but which are not isotropically distributed), can only be populated via the stronger mean perturbations exerted between stars in a dense cluster.

Adams and Laughlin (2001) conclude from these and related considerations that the Sun formed in a cluster of 2000 ± 1000 stars.

3.6 Colors and Physical Properties

It has long been recognized that Kuiper belt objects exhibit a diversity in surface colors unparalleled among solar system populations (Luu and Jewitt 1996). The distribution is relatively smooth from neutrally colored to extremely red objects. Indeed, a large fraction of KBOs (and Centaurs) is covered in "ultrared matter" (Jewitt 2002), the reddest material observed on small bodies. This material is absent in other populations, and is thought to be due to irradiated complex organics (Cruikshank et al. 2007). Further, the *UBVRIJ* colors are mutually correlated (Jewitt and Luu 1998, Jewitt et al. 2007), which seems to indicate that the spread is caused by a single reddening agent.

The wide range of colors suggests a broad range of surface compositions. However, such extreme non-uniformity in composition is unlikely to be intrinsic given the uniform and low temperatures across the disk of the Kuiper belt; the compositional spread is probably the result of some evolutionary process. Early theories to explain the color scatter invoke a competition between (the reddening) long-term exposure to cosmic radiation and (the de-reddening) impact resurfacing (Luu and Jewitt 1996, Delsanti et al. 2004), but the implied rotational color variability and correlation between color and collision likelihood are not observed (Jewitt and Luu 2001, Thébault and Doressoundiram 2003).

More recently it has been suggested that the diversity was emplaced when the small body populations were scattered by the outward migration of the ice giant planets (Gomes 2003), as a consequence of the mutual 2:1 resonance crossing by Jupiter and Saturn (see Nice model above). In other words, some of the objects now in the Kuiper belt may have formed much closer to the Sun, in the 10–20 AU region, where the chemistry would have been different and perhaps more diverse. Although appealing, the theory that the KB region was sprinkled with bodies from various heliocentric distances remains a non-unique explanation and the implicit assumption that KBOs formed closer to the Sun should be less red remains ad-hoc. The relative importances of this dynamical mixing and the reddening effect by cosmic irradiation are still poorly understood. Cosmic-ray reddening seems to be important as the Classical KBOs, supposedly formed locally at 40 AU and having passively evolving surfaces only subject to cosmic radiation (their circular orbits protect them from mutual collisions), are on average the reddest KBOs (Tegler and Romanishin 2000).

A powerful way to investigate the physical properties of KBOs is by the analysis of their rotational properties, usually inferred from their lightcurves (Sheppard and Jewitt 2002, Lacerda and Luu 2006). Lightcurves are periodic brightness variations due to rotation: as a non-spherical (and non-azimuthally symmetric) KBO rotates in space, its sky-projected cross-section will vary periodically, and thus modulate the amount of sunlight reflected back to the observer. The period P and range Δm of a KBO's lightcurve provide information on its rotation period and shape, respectively. Multi-wavelength lightcurves may also reveal surface features such as albedo or color patchiness (Buie et al. 1992, Lacerda et al. 2008). These features are usually seen as second order effects superimposed on the principal, shape-regulated lightcurve. By combining the period and the range of a KBO lightcurve it is possible to constrain its density under the assumption that the object's shape is mainly controlled by its self-gravity (Jewitt and Sheppard 2002, Lacerda and Jewitt 2007). Lightcurves can also uncover unresolved, close binary objects (Sheppard and Jewitt 2004, Lacerda and Jewitt 2007).

Interesting KBOs, whose lightcurves have been particularly informative include: (1) 134340 Pluto, whose albedo-controlled light variations have been used to map the distribution of ices of different albedos across the surface (Buie et al. 1992, Young et al. 1999), which is likely controlled by the surface deposition of frosts from Pluto's thin atmosphere (Trafton 1989), (2) 20000 Varuna, whose rapid rotation ($P = 6.34 \pm 0.01$ hr) and elongated shape ($\Delta m = 0.42 \pm 0.03$ mag) indicate a bulk density $\rho \sim 1000\,\mathrm{kg\,m^{-3}}$ and hence require an internally porous structure (Jewitt and Sheppard 2002), (3) 139775 2001 QG298, whose extreme lightcurve (large range $\Delta m = 1.14 \pm 0.04$ mag and slow period $P \sim 13.8$ hr) suggests an extreme interpretation as a contact or near-contact binary (Sheppard and Jewitt 2004, Takahashi et al. 2004, Lacerda and Jewitt 2007), and finally (4) 136198 2003 EL61, exhibiting super-fast rotation (P=3.9 hr) that requires a density $\sim 2500\,\mathrm{kg\,m^{-3}}$, and a recently identified surface feature both redder and darker than the average surface (Lacerda et al. 2008).

Statistically, the rotational properties of KBOs can be used to constrain the distributions of spin periods (Lacerda and Luu 2006) and shapes (Lacerda and Luu 2003), which in turn can be used to infer the importance of collisions in their evolution. For instance, most main-belt asteroids have been significantly affected by mutual collisions, as shown by their quasi-Maxwellian spin rate distribution (Harris 1979, Farinella et al. 1981) and a shape distribution consistent with fragmentation experiments carried out in the laboratory (Catullo et al. 1984). KBOs spin on average more slowly ($\langle P_{KBO} \rangle \sim 8.4$ hr vs. $\langle P_{ast} \rangle \sim 6.0$ hr; Lacerda and Luu 2006) and are more spherical (as derived from the Δm distribution; Luu and Lacerda 2003, Sheppard et al. 2008) than asteroids of the same size, both indicative of a milder collisional history. Typical impact speeds in the current Kuiper belt and main asteroid belt are respectively 2 and 5 $\mathrm{km\,s^{-1}}$. However, three of the four KBOs listed in the previous paragraph show extreme rotations and shapes, likely the result of collision events. Because the current number density of KBOs is too low for these events to occur on relevant timescales, their rotations were probably acquired at an early epoch when the Kuiper belt was more massive and collisions were more frequent (Davis and Farinella 1997, Jewitt and Sheppard 2002).

Next generation telescopes will provide revolutionary new data on the physical properties of KBOs. Their colors and rotational properties can potentially reveal much about these objects' surface and physical natures. *JWST*, in particular, will provide high quality near-infrared spectra that will place the best constraints on the (probably) organic mantles of the KBOs, thought to be some of the most primitive matter in the solar system. Separate measurements of the albedos and diameters, obtained from optical/thermal measurements using *ALMA* and *JWST*, will give the albedos and accurate diameters, needed to fully understand the surface materials.

Survey data will permit the identification of >100 wide binaries, while the high angular resolution afforded by *JWST* will reveal a much larger number (thousands?) of close binaries. Orbital elements for each will lead to the computation of system masses through Kepler's law. Diameters from optical/thermal measurements (using *ALMA* and *JWST*) will then permit the determination of system densities for KBOs over a wide range of diameters and orbital characteristics in the Kuiper belt. Density, as the "first geophysical parameter", provides our best handle on accretion models of the Kuiper belt objects.

3.7 Solar System Dust

Collisions between solids in the early solar system generally resulted in agglomeration due to the prevailing low impact energies. However, at present times, high impact energies dominate and collisions result in fragmentation and the generation of dust from asteroids, comets and KBOs. Due to the effect of radiation forces (see Section 3.7.2) these dust particles spread throughout the solar system forming a dust disk.

3.7.1 Inner Solar System: Asteroidal and Cometary Dust

The existence of dust in the inner solar system (a.k.a. zodiacal cloud) has long been known since the first scientific observations of the zodiacal light by Cassini in 1683, correctly interpreted by de Duiliers in 1684 as produced by sunlight reflected from small particles orbiting the Sun. Other dust-related phenomena that can be observed naked eye are dust cometary tails and "shooting stars".

The sources of dust in the inner solar system are the asteroids, as evident from the observation of dust bands associated with the recent formation of asteroidal families, and comets, as evident from the presence of dust trails and tails. Their relative contribution can be studied from the He content of collected interplanetary dust particles, which is strongly dependent on the velocity of atmospheric entry, expected to be low for asteroidal dust and high for cometary dust. Their present contributions are thought to differ by less than a factor of 10 (Brownlee et al. 1994), but they have likely changed with time. It is thought that due to the depletion of asteroids, the asteroidal dust surface area has slowly declined by a factor of 10 (Grogan et al. 2001) with excursions in the dust production rate by up to an order of magnitude associated with breakup events like those giving rise to the Hirayama asteroid families that resulted in the formation of the dust bands observed by *IRAS* (Sykes and Greenberg 1986). The formation of the Veritas family 8.3 Myr ago still accounts for ∼25% of the zodiacal thermal emission today (Dermott et al. 2002). A major peak of dust production in the inner solar system is expected to have occurred at the time of the LHB (Section 3.4), as a consequence of an increased rate of asteroidal collisions and to the collisions of numerous impactors originating in the main asteroid belt (Strom et al. 2005) with the terrestrial planets.

The thermal emission of the zodiacal cloud dominates the night sky between 5–500 μm and has a fractional luminosity of $L_{dust}/L_{Sun} \sim 10^{-8}$–$10^{-7}$ (Dermott et al. 2002). Studied by *IRAS*, *COBE* and *ISO* space telescopes, it shows a featureless spectrum produced by a dominant population of low albedo (<0.08) rapidly-rotating amorphous forsterite/olivine grains that are 10–100 μm in size and are located near 1 AU. The presence a weak (6% over the continuum) 10 μm silicate emission feature indicates the presence of a small population of ∼1 μm grains of dirty crystalline olivine and hydrous silicate composition (Reach et al. 2003).

Interplanetary dust particles (IDPs) have been best characterized at around 1 AU by in situ satellite measurements, observations of micro-meteorite impact craters on lunar samples, ground radar observations of the ionized trails created as the particles pass through the atmosphere and laboratory analysis of dust particles collected from the Earth's stratosphere, polar ice and deep sea sediments. Laboratory analysis of collected IDPs show that the particles are 1–1000 μm in size, typically black, porous (∼40%) and composed of mineral assemblages of a large number of submicron-size grains with chondritic composition and bulk densities of 1–3 g/cm^3. Their individual origin, whether asteroidal or cometary, is difficult to establish. The cumulative mass distribution of the particles at 1 AU follows a broken power-law such that the dominant contribution to the cross sectional area (and therefore to the zodiacal emission) comes from 10^{-10} kg grains (∼30 μm in radius), while the dominant contribution to the total dust mass comes from $\sim 10^{-8}$ kg grains (Leinert and Grün 1990).

In situ spacecraft detections of zodiacal dust out to 3 AU, carried out by Pioneer 8–11, Helios, Galileo and Ulysses, showed that the particles typically have $i < 30^\circ$ and $e > 0.6$ with a spatial density falling as $r^{-1.3}$ for $r < 1$ AU and $r^{-1.5}$ for $r > 1$ AU, and that there is a population of grains on hyperbolic orbits (a.k.a β-meteoroids), as well as stream of small grains origina jovian system (see review by Grün et al. 2001).

3.7.2 Outer Solar System: Kuiper Belt Dust

As remarked above, collisions in the modern-day Kuiper belt are erosive, not agglomerative, and result in the production of dust. In fact, two components to Kuiper belt dust production are expected: (1) erosion of KBO surfaces by the flux of interstellar meteoroids (e.g. Grun et al. 1994), leading to the steady production of dust at about 10^3 to 10^4 kg s^{-1} (Yamamoto and Mukai 1998); and (2) mutual collisions between KBOs, with estimated dust production of about (0.01–3)$\times 10^8$ kg s^{-1} (Stern 1996). For comparison, the dust production rate in the zodiacal cloud, from comets and asteroids combined, is about 10^3 kg s^{-1} (Leinert et al. 1983). The fractional luminosity of the KB dust is expected to be around $L_{dust}/L_{Sun} \sim 10^{-7}$–$10^{-6}$ (Stern 1996), compared to $L_{dust}/L_{Sun} \sim 10^{-8}$–$10^{-7}$ for the zodiacal cloud (Dermott et al. 2002).

Observationally, detection of Kuiper belt dust at optical wavelengths is confounded by the foreground presence of cometary and asteroidal dust in the zodiacal cloud. Thermally, these near and far dust populations might be distinguished on the basis of their different temperatures (~200 K in the Zodical cloud vs. ~40 K in the Kuiper belt) but, although sought, Kuiper belt dust has not been detected this way (Backman et al. 1995). At infrared thermal wavelengths both foreground zodiacal cloud dust *and* background galactic dust contaminate any possible emission from Kuiper belt dust. The cosmic microwave background radiation provides a very uniform source against which emission from the Kuiper belt might potentially be detected but, again, no detection has been reported (Babich et al. 2007).

Whereas remote detections have yet to be achieved, the circumstances for in-situ detection are much more favorable (Gurnett et al. 1997). The Voyager 1 and 2 plasma wave instruments detected dust particles via the pulses of plasma created by high velocity impacts with the spacecraft. Because the plasma wave detector was not built with impact detection as its primary purpose, the properties and flux of the impacting dust are known only approximately. Still, several important results are available from the Voyager spacecraft. Impacts were recorded continuously as the Voyagers crossed the (then unknown) Kuiper belt region of the solar system. The smallest dust particles capable of generating measurable plasma are thought to be $a_0 \sim 2\mu$m in radius. Measured in the 30 AU to 60 AU region along the Voyager flight paths, the number density of such particles is $N_1 \sim 2 \times 10^{-8}$ m^{-3}. Taking the thickness of the Kuiper belt (measured perpendicular to the midplane) as $H \sim 10$ AU, this corresponds to an optical depth $\tau \sim \pi a_0^2 H N_1 \sim 4 \times 10^{-7}$, roughly 10^3 times smaller than the optical depth of a β-Pictoris class dust circumstellar disk. An upper limit on the density of gravel (*cm*-sized) particles in the Kuiper belt is provided by the survival of a 20-cm propellant tank on the Pioneer 10 spacecraft (Anderson et al. 1998).

3.7.3 Dust Dynamics and Dust Disk Structure

After the dust particles are released from their parent bodies (asteroids, comets and KBOs) they experience the effects of radiation and stellar wind forces. Due to radiation pressure, their orbital elements and specific orbital energy change immediately upon release. If their orbital energy becomes positive ($\beta > 0.5$), the dust particles escape on hyperbolic orbits (known as β-meteoroids – Zook and Berg 1975). If their orbital energy remains negative ($\beta < 0.5$), their semi-major axis increases but they remain on bound orbits. Their new semimajor axis and eccentricity (a', e') in terms of that of their parent bodies (a and e) are $a' = a\frac{1-\beta}{1-2a\beta/r}$ and $e' = |\, 1 - \frac{(1-2a\beta/r)(1-e^2)}{(1-\beta^2)}\, |^{1/2}$ (their inclination does not change), where r is the particle location at release and β is the ratio of the radiation pressure force to the gravitational force. For spherical grains orbiting the Sun, $\beta = 5.7 Q_{pr}/\rho b$, where ρ and b are the density and radius of the grain in MKS units and Q_{pr} is the radiation pressure coefficient, a measure of the fractional amount of energy scattered and/or absorbed by the grain and a function of the physical properties of the grain and the wavelength of the incoming radiation (Burns et al. 1979).

With time, Poynting–Rorbertson (P–R) and solar wind corpuscular drag (which result from the interaction of the dust grains with the stellar photons and solar wind particles, respectively) tend to circularize and decrease the semimajor axis of the orbits, forcing the particles to slowly drift in towards the central star until they are destroyed by sublimation in a time given by $t_{PR} = 0.7\left(\frac{b}{\mu m}\right)\left(\frac{\rho}{kg/m^3}\right)\left(\frac{R}{AU}\right)^2\left(\frac{L_\odot}{L_*}\right)\frac{1}{1+albedo}\, yr$, where R is the starting heliocentric distance of the dust particle and b and ρ are the particle radius and density, respectively (Burns et al. 1979, Backman and Paresce 1993). If the dust is constantly being produced from a planetesimal belt, and because the dust particles inclinations are not affected by radiation forces, this inward drift creates a dust disk of wide radial extent and uniform density. Grains can also be destroyed by mutual grain collisions, with a collisional lifetime of $t_{col} = 1.26 \times 10^4 \left(\frac{R}{AU}\right)^{3/2}\left(\frac{M_\odot}{M_*}\right)^{1/2}\left(\frac{10^{-5}}{L_{dust}/L_*}\right) yr$ (Backman and Paresce 1993).

For dust disks with $M_{dust} > 10^{-3}\, M_\oplus$, $t_{col} < t_{PR}$, i.e. the grains are destroyed by multiple mutual collisions before they migrate far from their parent bodies (in this context, "destruction" means that the collisions break the grains into smaller and smaller pieces until they are sufficiently small to be blown away by radiation pressure). This regime is referred to as collision-dominated. The present solar system, however, is radiation-dominated because it does not contain large quantities of dust and $t_{col} > t_{PR}$, i.e. the grains can migrate far from the location of their parent bodies. This is particularly interesting in systems with planets and outer dust-producing planetesimal belts because in their journey toward the central star the orbits of the dust particles are affected by gravitational perturbations with the planets via the trapping of particles in mean motion resonances (MMRs), the effect of secular resonances and the gravitational scattering of dust. This results in the formation of structure in the dust disk (Fig. 3.12).

Dust particles drifting inward can become entrapped in exterior MMRs because at these locations the particle receives energy from the perturbing planet that can balance the energy loss due to P-R drag, halting the migration. This makes the lifetime of particles trapped in outer MMRs longer than in inner MMRs (Liou and Zook 1997), with the former dominating the disk structure. This results in the formation of resonant rings outside the planet's orbit, as the vast majority of the particles spend most of their lifetimes trapped in exterior MMRs. In some cases, due to the geometry of the resonance, a clumpy structure is created. Figure 3.12 (from Moro-Martín and Malhotra 2002) shows the effect of resonant trapping expected in the (yet to be observed) KB dust disk, where the ring-like structure, the asymmetric

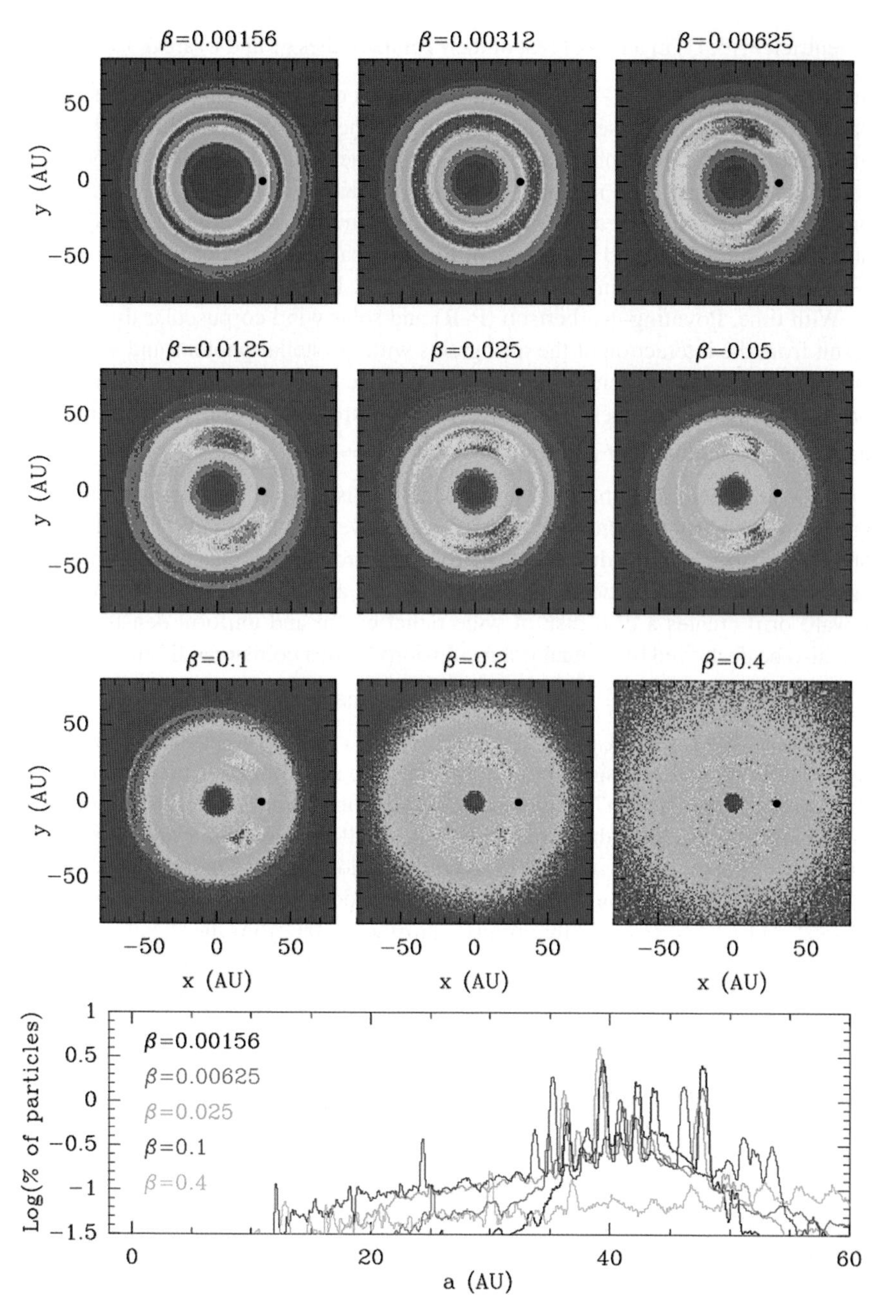

Fig. 3.12 (Caption on next page)

clumps along the orbit of Neptune, and the clearing of dust at Neptune's location are all due to the trapping of particles in MMRs with Neptune (as seen in the histogram of semimajor axis). Neptune plays the leading role in the trapping of dust particles because of its mass and because it is the outermost planet and its exterior resonances are not affected by the interior resonances from the other planets. Trapping is more efficient for larger particles (i.e. smaller β values) because the drag force is weaker and the particles cross the resonance at a slower rate increasing their probability of being captured. The effects of resonance trapping in hypothetical planetary systems each consisting of a single planet on a circular orbit and an outer planetesimal belt similar to the KB are shown in Fig. 3.13. More eccentric planets ($e < 0.6$) can also create clumpy eccentric rings and offset rings with a pair of clumps (Kuchner and Holman 2003). Even though, as mentioned in Section 3.7.2, the KB dust disk has yet to be observed, this is not the case for the Zodical cloud, for which *IRAS* and *COBE* thermal observations show that there is a ring of asteroidal dust particles trapped in exterior resonances with the Earth at around 1 AU, with a 10% number density enhancement on the Earth's wake that results from the resonance geometry (Dermott et al. 1994).

Secular perturbations, the long-term average of the perturbing forces, act on timescales >0.1 Myr. If the planet and the planetesimal disk are not coplanar, the secular perturbations can create a warp in the dust disk as their tendency to align the orbits operates on shorter timescales closer to the star. A warp can also be created in systems with two non-coplanar planets. If the planet is in an eccentric orbit, the secular resonances can force an eccentricity on the dust particles and this creates an offset in the disk center with respect to the star that can result in a brightness asymmetry. Other effects of secular perturbations are spirals and inner gaps (Wyatt et al. 1999). The effect of secular perturbations can be seen in *IRAS* and *COBE* observations on the zodiacal cloud and account for the presence of an inner edge around 2 AU due to a secular resonance with Saturn (that also explains the inner edge of the main asteroid belt), the offset of the cloud center with respect to the

Fig. 3.12 (continued) Expected number density distribution of the KB dust disk for nine different particle sizes (or β values). β is a dimensionless constant equal to the ratio between the radiation pressure force and the gravitational force and depends on the density, radius and optical properties of the dust grains. If we assume that the grains are composed of spherical astronomical silicates ($\rho = 2.5$, Weingartner and Draine 2001), β values of 0.4, 0.2, 0.1, 0.05, 0.025, 0.0125, 0.00625, 0.00312, 0.00156 correspond to grain radii of 0.7, 1.3, 2.3, 4.5, 8.8, 17.0, 33.3, 65.9, 134.7 μm, respectively. The trapping of particles in MMRs with Neptune is responsible for the ring-like structure, the asymmetric clumps along the orbit of Neptune, and the clearing of dust at Neptune's location (indicated with a *black dot*). The disk structure is more prominent for larger particles (smaller β values) because the P-R drift rate is slower and the trapping is more efficient. The disk is more extended in the case of small grains (large β values) because small particles are more strongly affected by radiation pressure. The histogram shows the relative occurrence of the different MMRs for different sized grains, where the large majority of the peaks correspond to MMRs with Neptune. The inner depleted region inside $\sim$10 AU is created by gravitational scattering of dust grains with Jupiter and Saturn. More details on these models can be found in Moro-Martín and Malhotra (2002, 2003)

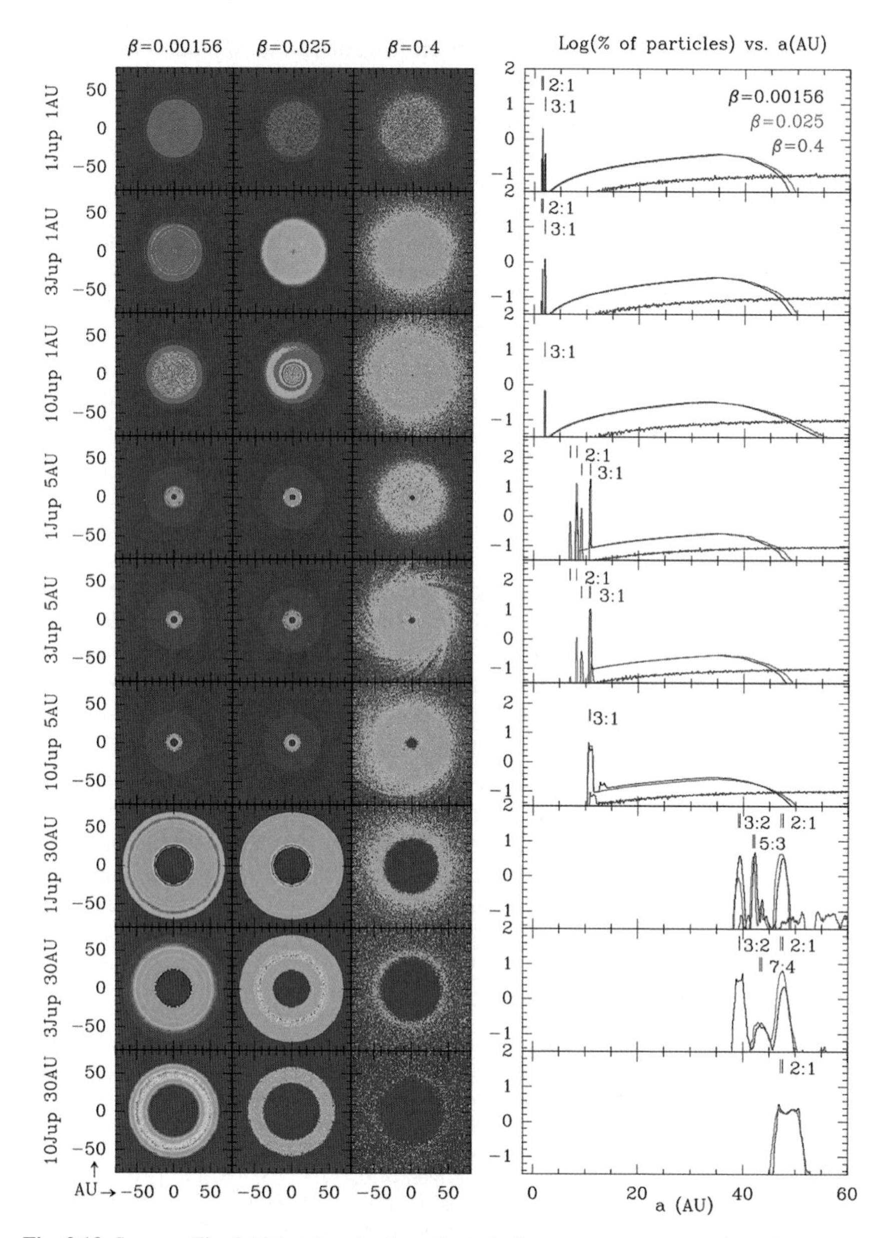

Fig. 3.13 Same as Fig. 3.12 but for nine hypothetical planetary systems around a solar type star consisting of a single planet with a mass of 1, 3 or 10M_{Jup} in a circular orbit at 1, 5 or 30 AU, and a coplanar outer planetesimal belt similar to the KB. The models with 1 M_{Jup} planet at 1 and 5 AU show that the dust particles are preferentially trapped in the 2:1 and 3:1 resonances, but when the mass of the planet is increased to 10 M_{Jup}, and consequently the hill radius of the planet increases, the 3:1 becomes dominant. The resonance structure becomes richer when the planet is further away from the star, and when the mass of the planet decreases (compare the model for a 1M_{Jup} planet at 30 AU with the solar system model in Fig. 3.12, where the structure is dominated by Neptune). From Moro-Martín, Wolf & Malhotra (in preparation)

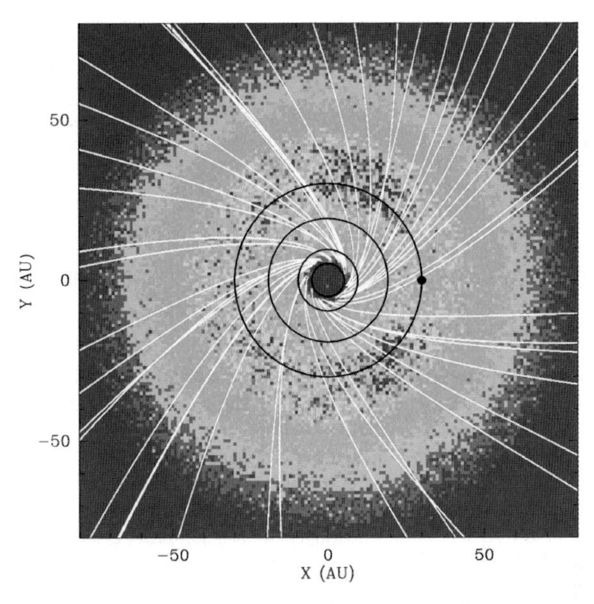

Fig. 3.14 Expected number density distribution of a KB dust disk composed of particles with $\beta = 0.2$ with the trajectories of the particles ejected by Jupiter in white. The black dot indicates the position of Neptune and the circles correspond to the orbits of the Giant planets. In addition to the population of small grains with $\beta > 0.5$ blown-out by radiation pressure, the gravitational scattering by the giant planets (Jupiter and Saturn in the case of the solar system) produces an outflow of large grains ($\beta < 0.5$) that is largely confined to the ecliptic (Moro-Martín and Malhotra 2005b). Interestingly, a stream of dust particles arriving from the direction of β Pictoris has been reported by Baggaley (2000)

Sun, the inclination of the cloud with respect to the ecliptic, and the cloud warp (see review in Dermott et al. 2001).

The efficient ejection of dust grains by gravitational scattering with massive planets as the particles drift inward from an outer belt of planetesimals (Figs. 3.14 and 3.15) can result in the formation of a dust depleted region inside the orbit of the planet (Figs. 3.12 and 3.13), such as the one expected inside 10 AU in the KB dust disk models (due to gravitational scattering by Jupiter and Saturn).

Next generation facilities will probably be unable to detect diffuse emission from Kuiper dust because the optical depth is so low, and the effects of foreground and background confusion so large. However, small Kuiper belt objects are sufficiently numerous that there is a non-negligible chance that the collision clouds of recent impacts will be detected (Stern 1996). Such clouds can potentially be very bright, and evolve on timescales (days and weeks) that are amenable to direct observational investigation using *LSST* and *JWST*. Collision cloud measurements provide our best chance to understand

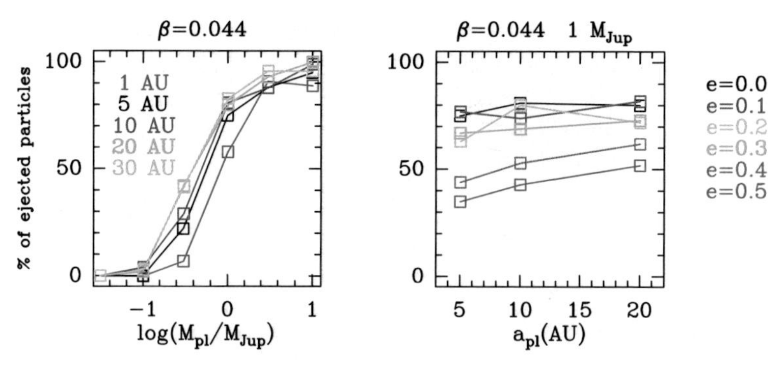

Fig. 3.15 Percentage of dust particles ejected from the system by gravitational scattering with the planet for the single planet models in Fig. 3.13. The particle size is fixed, corresponding to grains with $\beta = 0.044$. Because gravitational scattering is independent of the particle size, the efficiency of ejection is fairly independent of β. *Left*: Dependency of the efficiency of ejection on the planet's mass (x-axis) and the planet's semimajor axis (indicated by the different colors). *Right*: Dependency of the efficiency of ejection on the planet semimajor axis (x-axis) and eccentricity (corresponding to the different colors). The models in the right panel correspond to a 1 M_{Jup} mass planet on a circular orbit around a solar type star. Planets with masses of 3–10 M_{Jup} at 1 AU–30 AU in a circular orbit eject >90% of the dust grains that go past their orbits under P-R drag; a 1 M_{Jup} planet at 30 AU ejects >80% of the grains, and about 50–90% if located at 1 AU, while a 0.3 M_{Jup} planet is not able to open a gap, ejecting <10% of the grains. These results are valid for dust grains sizes in the range 0.7 μm–135 μm. From Moro-Martín and Malhotra 2005

the sub-kilometer population in the Kuiper belt. These objects are too small to be directly detected but are of special relevance as the precursors to the Centaurs and Jupiter family comet nuclei. Measurement of their number is of central importance in understanding the role of the Kuiper belt as the JFC source, and as the source of Kuiper dust.

3.8 Kuiper Belts of Other Stars

Radial velocity studies have revealed that >7% of solar-type stars harbor giant planets with masses <13 M_{Jup} and semimajor axis <5 AU (Marcy et al. 2005). This is a lower limit because the duration of the surveys (6–8 years) limits the ability to detect long-period planets; the expected frequency extrapolated to 20 AU is ~12% (Marcy et al. 2005). As of February 2008, 276 extra-solar planets have been detected with a mass distribution that follows $dN/dM \propto M^{-1.05}$ from 0.3M_{Jup} to 10 M_{Jup} (the surveys are incomplete at smaller masses). A natural question arises whether these planetary systems, some of them harboring multiple planets, also contain planetesimals like the asteroids, comets and KBOs in the solar system. Long before extra-solar planets were discovered we inferred that the answer to this question was yes: colliding planetesimals had to be responsible for the dust disks observed around mature stars. In Section 3.8.1 we will discuss how these dust disks, known as *debris disks*, can help us study indirectly Kuiper belts around other stars; other methods by which extra-solar Kuiper belts might be found and characterized in the future will be discussed in Sections 3.8.2 and 3.8.3.

3.8.1 Debris Disks

3.8.1.1 Evidence of Planetesimals

Theory and observations show that stars form in circumstellar disks composed of gas and dust that had previously collapsed from the densest regions of molecular clouds. For solar type stars, the masses of these disks are $\sim$0.01–0.10 $M_{\odot}$ and extend to 100s of AU, comparable to the minimum mass solar nebula ($\sim$0.015 $M_{\odot}$, which is the mass required to account for the condensed material in the solar system planets). Over time, these primordial or proto-planetary disks, with dust grain properties similar to those found in the interstellar medium, dissipate as the disk material accretes onto the star, is blown away by stellar wind ablation, photo-evaporation or high-energy stellar photons, or is stripped away by passing stars. The primordial gas and dust in these disks dissipate in less than 10^7 years (see e.g. Hartmann 2000 and references therein).

However, it is found that some main sequence stars older than $\sim 10^7$ years show evidence of dust emission. In most cases, this evidence comes from the detection of an infrared flux in excess of that expected from the stellar photosphere, thought to arise from the thermal emission of circumstellar dust. In some nearby stars, like the ones shown in Figs. 3.16, 3.17 and 3.18, direct imaging has confirmed that the emission comes from a dust disk. In Section 3.7.3 we discussed the lifetimes of the dust grains due to radiation pressure, P-R drag and mutual grain collisions. It is found that in most cases these lifetimes are much shorter than the age of the star,[1] and therefore the observed dust cannot be primordial but is more likely produced by a reservoir of undetected dust-producing planetesimals, like the KBOs, asteroids and comets in the solar system (see e.g. Backman and Paresce 1993). This is why these dust disks observed around mature main sequence stars are known as debris disks. Debris disks are evidence of the presence of planetesimals around other main sequence stars. In the core accretion model, these planetesimals formed in the earlier protoplanetary disk phase described above, as the ISM-like dust grains sedimented into the mid-plane of the disk and aggregated into larger and larger bodies (perhaps helped by turbulence) until they became planetesimals, the largest of which could

[1] One needs to be cautious with this argument because the ages of main sequence stars are difficult to determine. For example, the prototype (and best studied) of debris disk β-Pictoris, was dated at $\sim$100 to 200 Myr (Paresce 1991) but later become only $\sim$20 Myr old (Barrado y Navascués et al. 1999).

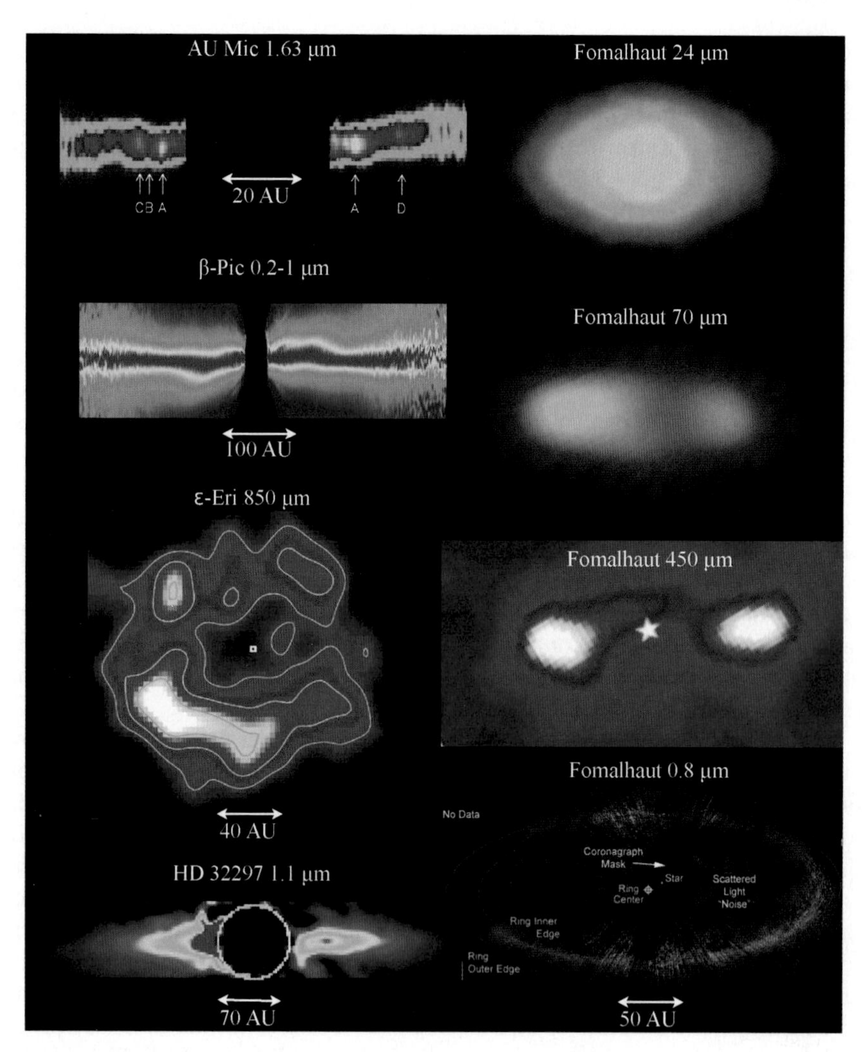

Fig. 3.16 Spatially resolved images of nearby debris disks showing dust emission from 10s to 100s of AU with a wide diversity of complex features including inner gaps, warps, brightness asymmetries, offsets and clumply rings, some of which may be due to the presence of massive planets (7.3). *Left* (from top to bottom): AU Mic (Keck AO at 1.63 μm; Liu 2004), β-Pic (STIS CCD coronography at 0.2–1 μm; Heap et al. 2000), ϵ-Eri (JCMT/SCUBA at 850 μm; Greaves et al. 2005) and HD 32297 (HST/NICMOS coronography at 1.1 μm; Schneider, Silverstone and Hines 2005). *Right:* All images correspond to Fomalhaut (7.7 pc away) and are on the same scale. From top to bottom: Spitzer/MIPS at 24 μm (Stapelfeldt et al. 2004); Spitzer/MIPS at 70 μm (Stapelfeldt et al. 2004); JCMT/SCUBA at 450 μm (Holland et al. 2003) and HST/ACS at 0.69–0.97 μm (Kalas et al. 2005). For this last panel, the annular disk in the scattered light image has an inner radius of ~133 AU and a radial thickness of ~25 AU and its center is offset from the star by about 15 ± 1 AU in the plane, possibly induced by an unseen planet. Its sharp inner edge has also been interpreted as a signature of a planet (Kalas et al. 2005). The dust mass of the Fomalhaut debris disk in millimeter-sized particles is about 10^{23} kg (~0.02 $M_\oplus$; Holland et al. 1998) but, if larger bodies are present, the mass could be 50 to 100 $M_\oplus$, or about 1000 times the mass of our Kuiper belt

Fig. 3.17 Keck Telescope image of the AU Mic debris disk, showing a 100 AU diameter region with central obscuration by a coronagraphic mask (Liu 2004). AU Mic is a 12(+8, −4) Myr old M dwarf (about 0.5 $M_{\odot}$) about 9.9 pc from Earth (Zuckerman et al. 2001) and with a dust mass near 0.01 $M_{\oplus}$ (Liu 2004). In scattered light, it shows a nearly edge-on disk about 200 AU in diameter with evidence for sub-structure (Kalas et al. 2004, Liu 2004)

potentially become the seeds out of which the giant planets form from the accretion of gas onto these planetary cores.

Even though these extra-solar planetesimals remain undetected, the dust they produce has a much larger cumulative surface area that makes the dust detectable in scattered light and in thermal emission. The study of these debris disks can help us learn indirectly about their parent planetesimals, roughly characterizing their frequencies, location and composition, and even the presence of massive planets.

3.8.1.2 Spatially Resolved Observations

Most debris disk observations are spatially unresolved and the debris disks are identified from the excess thermal emission contributed by dust in their spectral energy distributions (SEDs). In a few cases (about two dozen so far), the disks are close enough and the images are spatially resolved. Figures 3.16, 3.17 and 3.18 show the most spectacular examples. These high resolution observations show a rich diversity of morphological features including warps (AU Mic & β-Pic), offsets of the disk center with respect to the central star (ε-Eri and Fomalhaut), brightness asymmetries (HD 32297 and Fomalhaut), clumpy rings (AU Mic, β-Pic, ε-Eri and Fomalhaut) and sharp inner edges (Fomalhaut), features that, as discussed in Section 3.7.3 could be due to gravitational perturbations of massive planets, and that in some cases have been observed in the zodiacal cloud, while in other cases used to belong to the realm of KB disk models. Even though the origin of individual features is still under discussion and the models require further refinements (e.g. in the dust collisional processes and the effects of gas drag), the complexity of these features, in particular the azimuthal asymmetries, indicate that planets likely play a role in the creation of structure in the debris disks. This is of interest because the structure, in particular that created by the trapping of particles in MMRs, is sensitive to the presence of moderately massive planets at large distances (recall the KB dust disk models and the structure created by Neptune). This is a parameter space that cannot be explored with the present planet detection techniques, like the radial velocity and

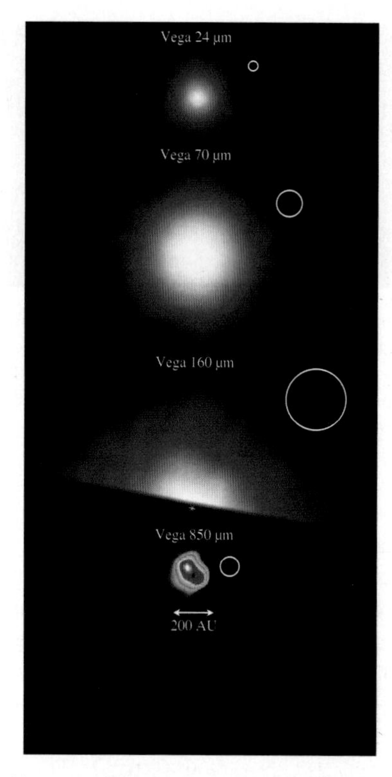

Fig. 3.18 Spatially resolved images of Vega from Spitzer/MIPS at 24, 70, 160 μm (Su et al. 2005) and from JCMT/SCUBA at 850 μm (Holland et al. 1998). All images are in the same scale. The instrument beam sizes (shown in *white circles*) indicate that the wide radial extent of the MIPS disk images compared to the SCUBA disk image is not a consequence of the instrumental PSF but due to a different spatial location of the particles traced by the two instruments. The sub-mm emission is thought to arise from large dust particles originating from a planetesimal belt analogous to the KB, while the MIPS emission is though to correspond to smaller particles with $\beta < 0.5$ (may be due to porosity), produced by collisions in the planetesimal belt traced by the sub-mm observations, and that are blown away by radiation pressure to distances much larger than the location of the parent bodies (Su et al. 2005). This scenario would explain not only the wider extent of the MIPS disk but also its uniform distribution, in contrast with the clumply and more compact sub-mm disk

transient studies, and therefore the study of debris disk structure can help us learn about the diversity of planetary systems. In this context, high resolution debris disk observations over a wide wavelength range are of critical importance, like those to be obtained with *Herschel*, *JWST* and *ALMA*.

3.8.1.3 Spectral Energy Distributions

As we mentioned above, most of the debris disks observations are spatially unresolved and are limited to the study of the SED of the star+disk system. Even assuming that the dust is distributed in a disk (and not, for example, in a spherical shell) there are degeneracies in the SED analysis and the dust distribution cannot be unambiguously determined (e.g. Moro-Moro-Martín, Wolf and Malhotra 2005a). Nevertheless, a wealth of information can be extracted from the SED. *IRAS* and *ISO* made critical discoveries on this front, but the number of known debris disks remained too small for statistical studies. This changed recently with the unprecedented sensitivity of the *Spitzer* instruments, that allowed the detection of hundreds of debris disks in large stellar surveys that searched for dust around 328 single FGK stars (Hillenbrand et al. 2008, Meyer et al. 2008, Carpenter et al. 2008), a different sample of 293 FGK stars (Trilling et al. 2008, Beichman et al. 2006a,b, Bryden et al. 2006), 160 A single stars (Su et al. 2006, Rieke et al. 2005); 69 A3–F8 binary stars (Trilling et al. 2007), and in young stellar clusters (Gorlova et al. 2007, Siegler et al. 2007). As a result, we now possess information concerning their frequencies, their dependency with stellar type and stellar environment, their temporal evolution and the composition of the dust grains (see e.g. Moro-Martín et al. 2008 for a recent review).

Debris Disk Frequencies

The *Spitzer FEPS*[2] survey of 328 FGK stars found that the frequency of 24 μm excess is 14.7% for stars younger than 300 Myr and 2% for older stars, while at 70 μm, the excess rates are 6–10% (Hilllenbrand et al. 2008; Carpenter et al. 2008). These disks show characteristic temperatures of 60–180 K with evidence of a population of colder grains to account for the 70 μm excesses; the implied disk inner radii are > 10 AU and extend over tens of AU (see Fig. 3.19 – Carpenter et al. 2008). Figure 3.20 shows the debris disks incidence rates derived from a combined sample of 350 AFGKM stars from Trilling et al. (2008); for the 225 Sun-like (FG) stars in the sample older than 600 Myr, the frequency of the debris disks are $4.2^{+2}_{-1.1}$% at 24 μm and $16.4^{+2.8}_{-2..9}$% at 70 μm.

The above debris disks incidence rates compare to ~20% of solar-type stars that harbor giant planets inside 20 AU (Marcy et al. 2005). Even though the frequencies seem similar, one should keep in mind that the sensitivity of the *Spitzer* observations is limited to fractional luminosities of $L_{dust}/L_{*} > 10^{-5}$, i.e. >100 times the expected luminosity from the KB dust in our solar system. Assuming a gaussian distribution of debris disk luminosities and extrapolating from *Spitzer* observations (showing that the frequency of dust detection increases steeply with decreasing fractional luminosity), Bryden et al. (2006) found that the luminosity of the solar system dust is consistent with being 10 × brighter or fainter than an average solar-type star, i.e. debris disks at the solar system level could be common. Observations therefore indicate that planetary systems harboring dust-producing KBOs are more common than those with giant planets, which would be in agreement with the core accretion

[2] http://feps.as.arizona.edu/

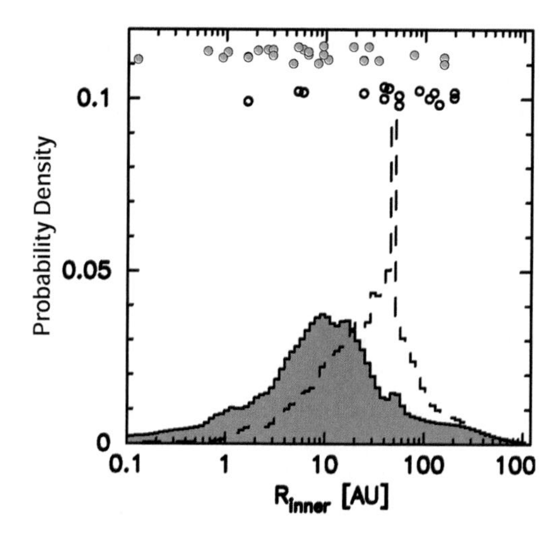

Fig. 3.19 Probability distribution for disk inner radii based on the analysis of the spectra (12–35 μm) of 44 debris disks around FGK stars from the *FEPS* survey. The dashed and grey histograms correspond to sources with and without 70 μm excess, respectively (with best fit parameters are shown as *open* and *grey circles*). Typical disk inner radius are ∼ 40 AU and ∼ 10 AU for disks with and without 70 μm excess, respectively, indicating that most of the debris disks observed are KB-like. Figure adapted with permission from Carpenter et al. 2008

models of planet formation where the planetesimals are the building blocks of planets and the conditions required for to form planetesimals are less restricted than those to form gas giants. Indeed, there is no apparent difference between the incidence rate of debris disks around stars with and without known planetary companions (Moro-Moro-Martín et al. 2007; Bryden et al. in preparation), although planet-bearing stars tend to harbor more dusty disks (Bryden et al. in preparation), which could result from the excitation of the planetesimals' orbits by gravitational perturbations with the planet. Figure 3.20 shows that there is no dependency on stellar type, neither in the frequency of debris disks, nor on the dust mass and location, indicating that planetesimal formation can take place under a wide range of conditions (Trilling et al. 2008).

Debris Disk Evolution

The *FEPS* survey of 328 FGK stars found that at 24 μm, the frequency of excess (> 10.2% over the stellar photosphere) decreases from 14.7% at ages < 300 Myr to 2% for older stars; at 70 μm, there is no apparent dependency of the excess frequency with stellar age, however, the amplitude of the 70 μm excess emission seems to decline from stars 30–200 Myr in age to older stars (Hillenbrand et al. 2008; Carpenter et al. 2008). Figures 3.21 and 3.22 from Trilling et al. (2008) show that for FGK type stars the debris disks incidence and fractional luminosity do not have a strong dependency with stellar age in the 1–10 Gyr time frame, in contrast with the 100–400 Myr evolution timescale of young (0.01–1 Gyr) stars seen in Fig. 3.23. Trilling

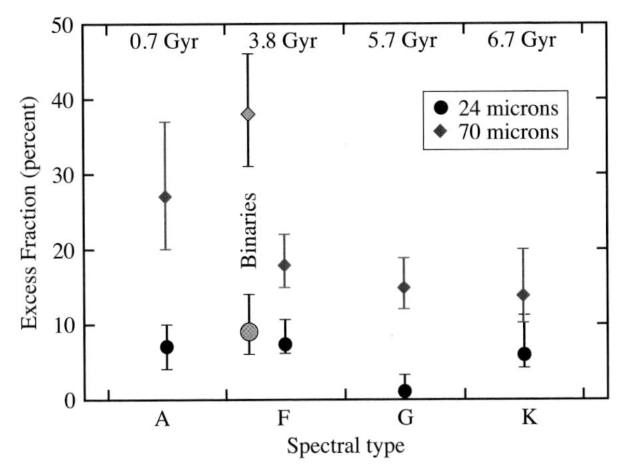

Fig. 3.20 The percentage of stars showing excess dust emission, i.e. with indirect evidence of the presence of dust-producing planetesimal belts, as a function of stellar type for ages >600 Myr (the mean ages within each type are shown at the top). The vertical bars correspond to binomial errors that include 68% of the probability (1 σ for Gaussian errors). Black is for 24 μm excess emission (tracing warmer dust) and red is for 70 μm (tracing colder dust). The data are consistent with no dependence on spectral type. There seems to be a weak decrease with spectral type at 70 μm but so far this is statistically not significant and could be due to an effect of age. Further analysis indicates that percentage of stars showing excess is different in old A stars and in M stars than in FGK stars. The excess rate for old M stars is 0% with upper limits (binomial errors) of 2.9% at 24 μm and 12% at 70 μm (Gautier et al. 2007). The lack of distant disks around K stars may be an observational bias because their peak emission would be at λ >70 μm and therefore remain undetected by *Spitzer*. The upcoming *Herschel* space telescope will provide the sensitivity to explore more distant and fainter debris disks. Figure adapted from Trilling et al. (2008) with data from Su et al. (2006), Trilling et al. (2007), Beichman et al. (2006b) and Gautier et al. (2007)

et al. (2007) argues that this data suggests that the dominant physical processes driving the evolution of the dust disks in young stars might be different from those in more mature stars, and operate on different timescales: while the former might be dominated by the production of dust during transient events like the LHB in the solar system or by individual collisions of large planetesimals (like the one giving rise to the formation of the Moon), the later might be the result of a more steady collisional evolution of a large population of planetesimals. The debris disks evolution observed by *Spitzer* for solar-type (Fig. 3.23 – Siegler et al. 2007) and A-type stars (Rieke et al. 2005, Su et al. 2006) indicate that both transient and more steady state dust production processes play a role; however, their relative importance and the question of how the dust production could be maintained in the oldest disks for billions of years is still under discussion.

An interesting example is HD 69830, one of the outliers in Fig. 3.24, a KOV star (0.8 $M_\odot$, 0.45 $L_\odot$) known to harbor three Neptune-like planets inside 0.63 AU. It shows a strong excess at 24 μm but no emission at 70 μm, indicating that the dust is warm and is located close to the star. The spectrum of the dust excess

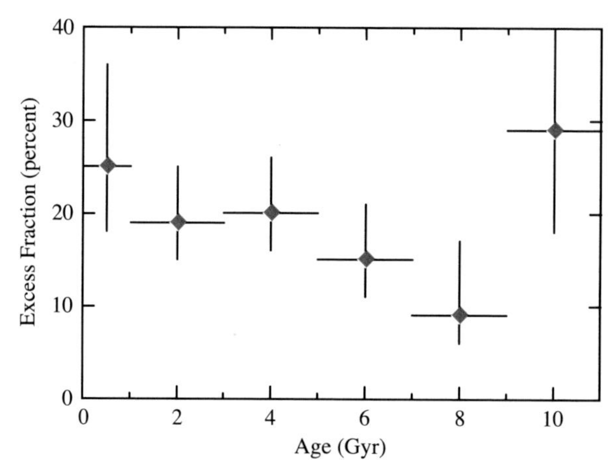

Fig. 3.21 The percentage of stars showing excess dust emission, i.e. with indirect evidence of the presence of dust-producing planetesimal belts, as a function of age for the F0–K5 stars. The horizontal error bars are the age bins (not the age uncertainties). The vertical bars correspond to binomial errors that include 68% of the probability (1 σ for Gaussian errors). The data are consistent with no dependency with age with a rate of ~20%. The data seems to suggest an overall decrease but so far is statistically not significant, and if present may be due to an observational bias (because of the deficiency of excesses around the K stars in the oldest age bin). The number of stars in the bins are (from young to old): 24, 57, 60, 52, 33, and 7 (the high value of the oldest bin may be a small number statistical anomaly). For comparison, A-type stars evolve on timescales of 400 Myr (Su et al. 2006). Figure adapted from Trilling et al. (2008) with data from Beichman et al. (2006b)

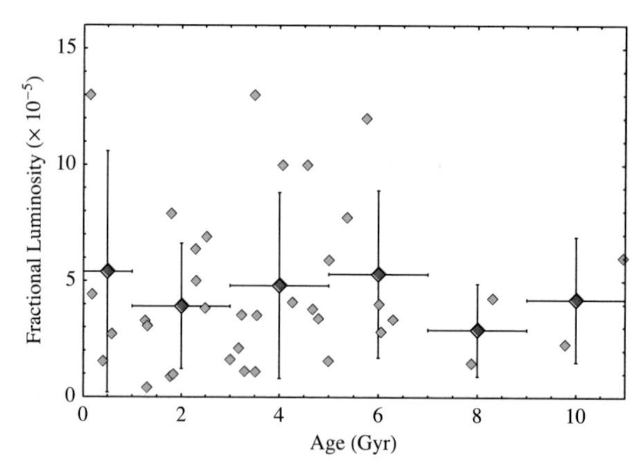

Fig. 3.22 The fractional luminosity of the debris disks, L_{dust}/L_{star}, as a function of age for FGK stars. The open symbols show the means within each age bin; the horizontal and vertical error bars show the bin widths and 1 σ errors, respectively. The data are consistent with no trend of L_{dust}/L_{star} with age, but there seems to be a deficiency of disks with high L_{dust}/L_{star} older than 6 Gyr. For comparison, the solar system is 4.5 Gyr old and is expected to have a dust disk with $L_{dust}/L_{star} \sim 10^{-7}$–$10^{-6}$. Figure adapted from Trilling et al. (2008) with data from Beichman et al. (2006b)

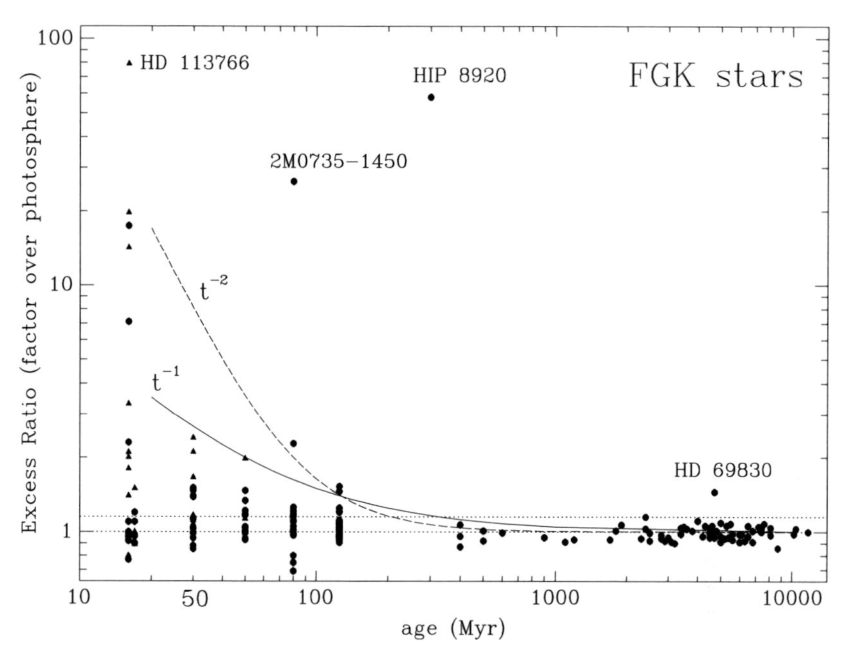

Fig. 3.23 Ratio of the 24 μm Excess Emission over the expected stellar value for FGK stars as a function of stellar age (*triangles* for F0–F4 stars and *circles* for F5–K7 stars). The vertical alignments correspond to stars in clusters or associations. The data agree broadly with collisional cascade models of dust evolution (resulting in a $1/t$ decay for the dust mass) punctuated by peaks of dust production due to individual collisional events. A-type stars show a similar behavior. Figure from Siegler et al. (2007) using data from Gorlova et al. (2004), Hines et al. (2006) and Song et al. (2005)

(Fig. 3.24) shows strong silicate emission lines thought to arise from small grains of highly processed material similar to that of a disrupted P- or D-type asteroid plus small icy grains, likely located outside the outermost planet (Lisse et al. 2007). The observed levels of dust production are too high to be sustained for the entire age of the star, indicating that the dust production processes are transient (Wyatt et al. 2007).

Whether the disks are transient or the result of the steady erosion of planetesimals is of critical importance for the interpretation of the statistics of the incidence rate of 24 μm excesses. For solar type stars, the 24 μm emission traces the 4–6 AU region. Terrestrial planet formation is expected to result in the production of large quantities of dust in this region, due to gravitational perturbations produced by large 1000 km-sized planetesimals that excite the orbits of a swarm of 1–10 km-size planetesimals, increasing their rate of mutual collisions and producing dust (Kenyon and Bromley 2005). This warm dust can therefore serve as a proxy of terrestrial planet formation. Figure 3.25 show the frequency of 24 μm emission for solar type (FGK) stars as a function of stellar age. This rate is <20% inside each age bin and

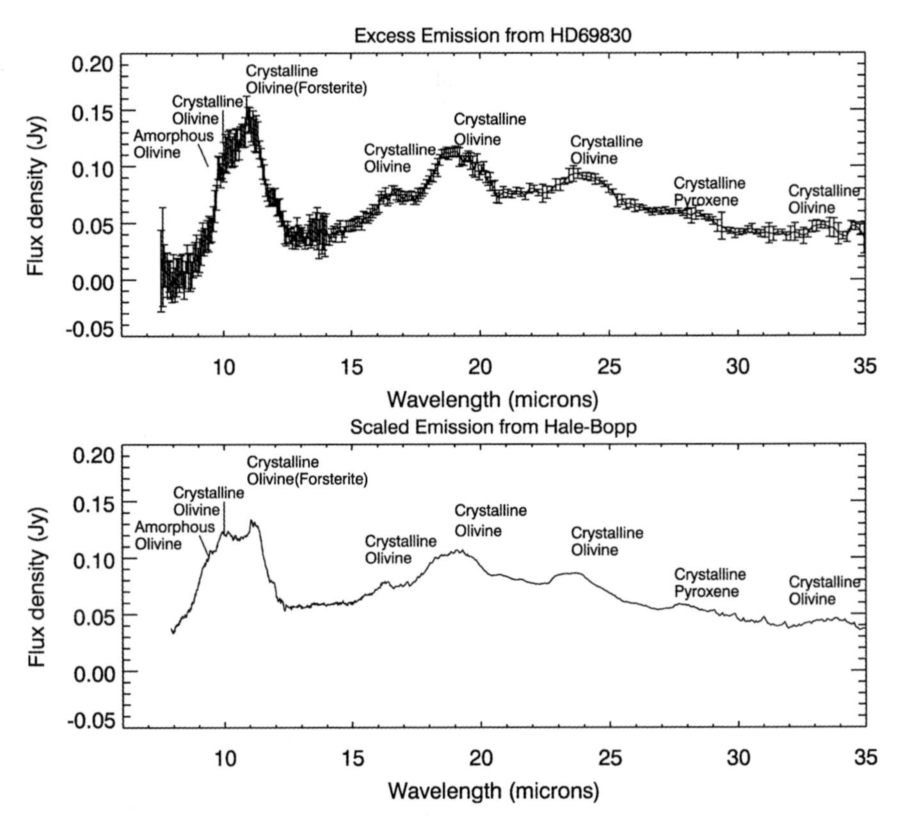

Fig. 3.24 Spectrum of the dust excess emission from HD 69830 (Beichman et al. 2005 – *top*) compared to the spectrum of comet Hale-Bopp normalized to a blackbody temperature of 400 K (Crovisier et al. 1996 – *bottom*). HD 69830 is one of the outliers in Fig. 3.23

decreases with age. If the dust-producing events are very long-lived, the stars that show dust excesses in one age bin will also show dust excesses at later times. In this case the frequency of warm dust (which indirectly traces the frequency of terrestrial planet formation) is $<20\%$. However, if the dust-producing events are short-lived, shorter than the age bins, the stars showing excesses in one age bin are not the same as the stars showing excesses at other age bins, i.e. they can produce dust at different epochs, and in this case the overall frequency of warm dust is obtained from adding all the frequencies in all age bins, which results in $>60\%$ (assuming that each star only has one epoch of high dust production). If this is the case, the frequency of terrestrial planet formation would be high (Meyer et al. 2008). However, the interpretation of the data would change if the observed 24 μm excesses arise from the steady erosion of cold-KB-like disks (Carpenter et al. 2008). Spatially resolved observations able to directly locate the dust would help resolve this issue.

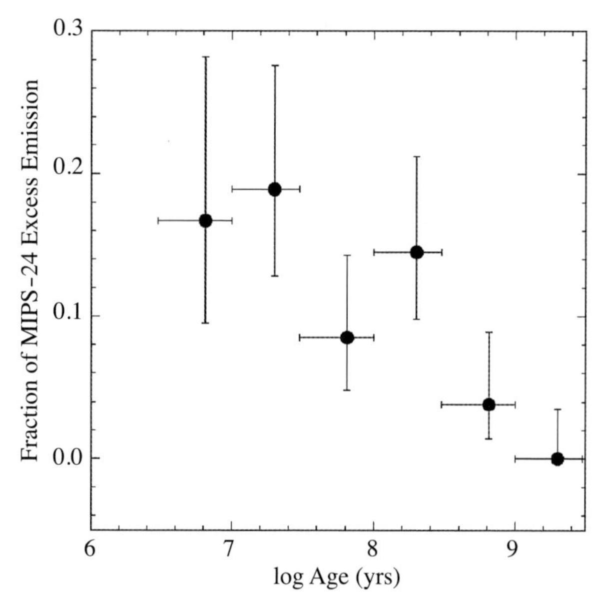

Fig. 3.25 Fraction of FGK type stars with 24 μm excess emission as a function of stellar age from a sample of 328 stars. The data points correspond to average values within a given age bin: (5/30) for stars 3–10 Myr, (9/48) for 10–30 Myr, (5/59) for 30–100 Myr, (9/62) for 100–300 Myr, (2/53) 300–1000 Myr. The widths of the age bins are shown by the *horizontal bars*, while the *vertical bars* show Poisson errors. Figure from Meyer et al. (2008)

Next generation facilities will offer high-sensitivity, high-resolution, multi-wavelength observations that should result in major breakthroughs in the study of debris disks. Debris disks are proxies for the presence of planetesimals around mature stars having a wide diversity of stellar types (A–K), suggesting that planetesimal formation is a robust process. The study of the warm dust can tell us about the frequency of terrestrial planet formation and the presence of asteroid-like bodies, while the study of the cold dust sheds light on the population of small bodies in KB-like regions. In addition, the study of debris disks around stars having a wide diversity of ages can help us learn about the evolution of planetary systems. However, the statistics so far are limited to dust disks 100–1000 × more luminous than that of our solar system and the observations are generally spatially unresolved. High-sensitivity observations with future telescopes like *Herschel*, *JWST* and *ALMA* will be able to detect dust at the solar system level, will help us improve our understanding of the frequency of planetesimals and, together with the result from planet searches, will show the diversity of planetary systems. Multi-wavelength observations are critical to help locate the dust in spatially unresolved disks and fundamental to interpret the debris disks statistics. High-resolution imaging observations are very important to directly locate the dust (circumventing the SED degeneracy), and to study the structure of the debris disks, perhaps serving as a planet-detection method sensitive to long-period Neptune-like planets that otherwise may be undetectable in the foreseeable future. Multi-wavelength observations also play a critical role in the interpretation of the structure because different wavelengths trace different particle sizes which have distinct dynamical characters that affect the disk morphology.

3.8.2 Photospheric Pollution

Dust produced collisionally in a Kuiper belt may spiral to the central star under the action of radiation and/or plasma drag, contaminating the photosphere with metal-enriched material. Separately, gravitational interactions and dynamical instabilities in a Kuiper belt may eject large objects (comets), causing some to impact the central star. Both processes operate in our solar system but neither produces a spectrally distinctive signature, for example a metal enrichment, on the Sun. This is simply because the photosphere of the Sun already contains a large mass of metals and the addition of dust or macroscopic bodies makes only a tiny, fractional contribution.

However, the atmospheres of some white dwarf stars offer much more favorable opportunities for the detection of a photospheric pollution signal. First, many white dwarfs are naturally depleted in metals as a result of sedimentation of heavy elements driven by their strong gravitational fields. Since their atmospheres should be very clean, quite modest masses of heavy-element pollutants can be detected. About one fifth of white dwarfs expected to have pure hydrogen or pure helium atmospheres in fact show evidence for heavier elements, most likely due to pollution from external sources. Second, stellar evolution leading to the white dwarf stage includes the loss of stellar mass through an enhanced wind. As the central star mass decreases, the semimajor axes of orbiting bodies should increase, leading to dynamical instabilities caused by resonance sweeping and other effects (Debes and Sigurdsson 2002). Separately, white dwarf stars with Oort clouds should experience a steady flux of impacts from comets deflected inwards by stellar and galactic perturbations (Alcock et al. 1986).

An observed depletion of carbon relative to iron may suggest that the infalling material is cometary rather than of interstellar origin (Jura 2006).

3.8.3 Thermal Activation

The blackbody temperature, in Kelvin, of an object located at distance, R_{AU}, from a star of luminosity, $L_\star/L_\odot$, is $T_{BB} = 278\ R_{AU}^{-1/2}(L_\star/L_\odot)^{1/4}$. In the Kuiper belt today, at $R_{AU} = 40$, the blackbody temperature is $T_{BB} = 44\,\mathrm{K}$. Water sublimation at this low temperature is negligible. However, the sublimation rate is an exponential function of temperature and, by $T_{BB} = 200$ K, water sublimates rapidly (with a mass flux $\sim 10^{-4}$ kg m^{-2} s^{-1}, corresponding to ice recession at about 3 m yr^{-1} for density 1000 kg m^{-3}). A 1 km scale body would sublimate away in just a few centuries. By the above relation, temperatures of 200 K are reached when $L_{\star}/L_{\odot} = 400$, for the same 40 AU distance.

Stellar evolution into the red giant phase will drive the Sun's luminosity to exceed this value after about 10 Gyr on the main-sequence, with an increase in the luminosity (by up to a factor of $\sim 10^4$), and in the loss of mass through an enhanced stellar wind (from the current value, $\sim 10^{-11}$ M$_{\odot}$ yr^{-1}, to $\sim 10^{-7}$ M$_{\odot}$ yr^{-1}, or more). When this happens, the entire Kuiper belt will light up as surface ices sublimate and dust particles, previously embedded in the KBOs, are ejected into space. Not all KBOs will be destroyed by roasting in the heat of the giant Sun: observations of comets near the Sun show that these bodies can insulate themselves from the heat by the development of refractory mantles, consisting of silicate and organic-rich debris particles that are too large to be ejected by gas drag. Still, the impact of the red giant phase should be dramatic and suggests that the sublimated Kuiper belts of other stars might be detected around red-giants. The key observational signatures would be the thermal excess itself, at temperatures appropriate to Kuiper belt-like distances, and ring-like morphology. Sensitivity at thermal wavelengths combined with high angular resolution will lend *JWST* to this type of observation, although the overwhelming signal from the star itself will present a formidable observational limitation to any imaging studies.

Water vapor has been reported around carbon stars that are not expected to show water and interpreted as produced by sublimated comets (Ford and Neufeld 2001). In IRC +10216, the mass of water is estimated as 3×10^{-5} M$_{\odot}$ (Melnick et al. 2001). This is about 10 M$_{\oplus}$, or 100 times the mass of the modern Kuiper belt but perhaps comparable to the mass of the Kuiper belt when formed. A possible explanation is that the water derives from sublimated comets in an unseen Kuiper belt, but chemical explanations for this large water mass may also be possible (Willacy 2004). On the other hand, a search for (unresolved) thermal emission from dust around 66 first-ascent red giants proved negative, with limits on the Kuiper belt masses of these stars near 0.1 M$_{\oplus}$, the current mass of the Kuiper belt (Jura 2004).

Acknowledgments DJ was supported by a grant from NASA's Origins program, PL by an NSF Planetary Astronomy grant to DJ. A.M.M. is under contract with the Jet Propulsion Laboratory (JPL) funded by NASA through the Michelson Fellowship Program. A.M.M. is also supported by the Lyman Spitzer Fellowship at Princeton University.

References

Adams, F. C., & Laughlin, G. 2001: Constraints on the Birth Aggregate of the Solar System. Icarus, 150, 151

Alcock, C., Fristrom, C. C., & Siegelman, R. 1986: On the Number of Comets Around Other Single Stars. Ap. J., 302, 462

Anderson, J. D., Lau, E. L., Scherer, K., Rosenbaum, D. C., & Teplitz, V. L. 1998: Kuiper Belt Constraint from Pioneer 10. Icarus, 131, 167

Babich, D., Blake, C. H., & Steinhardt, C. L. 2007: What Can the Cosmic Microwave Background Tell Us about the Outer Solar System? Ap. J., 669, 1406

Backman, D. E., Dasgupta, A., & Stencel, R. E. 1995: Model of a Kuiper Belt Small Grain Population and Resulting Far-Infrared Emission. Ap. J. Lett., 450, L35

Backman, D. E., & Paresce, F. 1993: Main-Sequence Stars with Circumstellar Solid Material – The VEGA Phenomenon. Protostars and Planets III, 1253

Baggaley, W. J. 2000: Advanced Meteor Orbit Radar Observations of Interstellar Meteoroids. JGR, 105, 10353

Bar-Nun, A., Kleinfeld, I., & Kochavi, E. 1988: Trapping of Gas Mixtures by Amorphous Water Ice. Phys. Rev. B., 38, 7749

Barrado y Navascués, D., Stauffer, J. R., Song, I., & Caillault, J.-P. 1999: The Age of Beta Pictoris. Ap. J. Lett., 520, L123

Beichman, C. et al. 2005. An Excess Due to Small Grains Around Nearby Køv Star HD 69830. Ap. J., 626, 1061

Beichman, C. A., et al. 2006a: IRS Spectra of Solar-Type Stars: A Search for Asteroid Belt Analogs. Ap. J., 639, 1166

Beichman, C. A., et al. 2006b: New Debris Disks around Nearby Main-Sequence Stars: Impact on the Direct Detection of Planets. Ap. J., 652, 1674

Boley, A. C., Durisen, R. H., Nordlund, Å., & Lord, J. 2007: Three-Dimensional Radiative Hydrodynamics for Disk Stability Simulations: A Proposed Testing Standard and New Results. Ap. J., 665, 1254

Boss, A. P. 2001: Gas Giant Protoplanet Formation: Disk Instability Models with Thermodynamics and Radiative Transfer. Ap. J., 563, 367

Boss, A. P. 2007: Testing Disk Instability Models for Giant Planet Formation. Ap. J. Lett., 661, L73

Boss, A. P., Wetherill, G. W., & Haghighipour, N. 2002: Rapid Formation of Ice Giant Planets. Icarus, 156, 291

Bottke, W. F., Morbidelli, A., Jedicke, R., Petit, J.-M., Levison, H. F., Michel, P., & Metcalfe, T. S. 2002: Debiased Orbital and Absolute Magnitude Distribution of the Near-Earth Objects. Icarus, 156, 399

Brown, M. E. 2001: The Inclination Distribution of the Kuiper Belt. Astron. J., 121, 2804

Brown, M. E., Trujillo, C., & Rabinowitz, D. 2004: Discovery of a Candidate Inner Oort Cloud Planetoid. Ap. J., 617, 645

Brownlee, D. E., Joswiak, D. J., Love, S. G., Bradley, J. P., Nier, A. O., & Schlutter, D. J. 1994: Identification and Analysis of Cometary IDPs. Lunar and Planetary Institute Conference Abstracts, 25, 185

Bryden, G., et al. 2006: Frequency of Debris Disks around Solar-Type Stars: First Results from a Spitzer MIPS Survey. Ap. J., 636, 1098

Buie, M. W., Tholen, D. J., & Horne, K. 1992: Albedo maps of Pluto and Charon – Initial mutual event results. Icarus 97, 21

Burns, J. A., Lamy, P. L., & Soter, S. 1979: Radiation Forces on Small Particles in the Solar System. Icarus, 40, 1

Carpenter, J., et al. 2008: Formation and Evolution of Planetary Systems. Ap. J. Supp, in press.

Catullo, V., Zappala, V., Farinella, P., & Paolicchi, P. 1984: Analysis of the Shape Distribution of Asteroids. Astron. Ap., 138, 464

Chapman, C. R., Cohen, B. A., & Grinspoon, D. H. 2007: What are the Real Constraints on the Existence and Magnitude of the Late Heavy Bombardment? Icarus, 189, 233

Crovisier, J., et al. 1996: The Infrared Spectrum of Comet C/1995 O1 (Hale-Bopp) at 4.6 AU from the Sun.. Astron. Ap., 315, L385

Cruikshank, D. P., Barucci, M. A., Emery, J. P., Fernández, Y. R., Grundy, W. M., Noll, K. S., & Stansberry, J. A. 2007: Physical Properties of Transneptunian Objects. Protostars and Planets V, 879

Davis, D. R., & Farinella, P. 1997: Collisional Evolution of Edgeworth-Kuiper Belt Objects. Icarus 125, 50

Debes, J. H., & Sigurdsson, S. 2002: Are There Unstable Planetary Systems around White Dwarfs? Ap. J., 572, 556

Delsanti, A., Hainaut, O., Jourdeuil, E., Meech, K. J., Boehnhardt, H., & Barrera, L. 2004: Simultaneous Visible-Near IR Photometric Study of Kuiper Belt Object Surfaces with the ESO/Very Large Telescopes. Astron. Ap., 417, 1145

Dermott, S. F., Grogan, K., Durda, D. D., Jayaraman, S., Kehoe, T. J. J., Kortenkamp, S. J., & Wyatt, M. C. 2001: Orbital Evolution of Interplanetary Dust. Interplanetary Dust (E. Grün, B. A. S. Gustafson, S. F. Dermott, H. Fechtig eds.) Springer, A&A Library, pp. 569–639

Dermott, S. F., Jayaraman, S., Xu, Y. L., Gustafson, B. A. S., & Liou, J. C. 1994: A Circumsolar Ring of Asteroidal Dust in Resonant Lock with the Earth. Nature, 369, 719

Dermott, S. F., Kehoe, T. J. J., Durda, D. D., Grogan, K., & Nesvorný, D. 2002: Recent Rubble-pile Origin of Asteroidal Solar System Dust Bands and Asteroidal Interplanetary Dust Particles. Asteroids, Comets, and Meteors: ACM 2002, 500, 319

Duncan, M., Quinn, T., & Tremaine, S. 1988: The origin of short-period comets. Ap. J. Lett., 328, L69

Duncan, M. J., & Levison, H. F. 1997: A Scattered Comet Disk and the Origin of Jupiter Family Comets. Science, 276, 1670

Elliot, J. L., et al. 2005: The Deep Ecliptic Survey: Astron. J., 129, 1117

Farinella, P., Paolicchi, P., & Zappala, V. 1981: Analysis of the Spin Rate Distribution of Asteroids. Astron. Ap., 104, 159

Fernández, J. A. 1980: On the Existence of a Comet Belt Beyond Neptune. MNRAS, 192, 481

Fernández, J. A., & Ip, W.-H. 1984: Some Dynamical Aspects of the Accretion of Uranus and Neptune – The Exchange of Orbital Angular Momentum with Planetesimals. Icarus, 58, 109

Fernández, Y. R., Jewitt, D. C., Sheppard, & S. S. 2001. Low Albedos Among Extinct Comet Candidates. AP. J., 553, L197–L200.

Ford, K. E. S., & Neufeld, D. A. 2001: Water Vapor in Carbon-rich Asymptotic Giant Branch Stars from the Vaporization of Icy Orbiting Bodies. Ap. J. Lett., 557, L113

Francis, P. J. 2005: The Demographics of Long-Period Comets. Ap. J., 635, 1348

Gaidos, E. J. 1995: Paleodynamics: Solar System Formation and the Early Environment of the Sun. Icarus, 114, 258

Gautier, T. N., III, et al. 2007: Far-Infrared Properties of M Dwarfs. Ap. J., 667, 527

Gladman, B., Holman, M., Grav, T., Kavelaars, J., Nicholson, P., Aksnes, K., & Petit, J.-M. 2002: Evidence for an Extended Scattered Disk. Icarus, 157, 269

Gomes, R. S. 2003: The Origin of the Kuiper Belt High-Inclination Population. Icarus 161, 404

Gomes, R., Levison, H., Tsiganis, K., & Morbidelli, A. 2005: Origin of the Late Heavy Bombardment. Nature, 435, 466.

Gorlova, N., et al. 2004: New Debris-Disk Candidates: 24 Micron Stellar Excesses at 100 Million years. Ap. J. Supp., 154, 448

Gorlova, N., Balog, Z., Rieke, G. H., Muzerolle, J., Su, K. Y. L., Ivanov, V. D., & Young, E. T. 2007: Debris Disks in NGC 2547. Ap. J., 670, 516

Greaves, J. S., et al. 2005: Structure in the ϵ Eridani Debris Disk. Ap. J. Lett., 619, L187

Grogan, K., Dermott, S. F., & Durda, D. D. 2001: The Size-Frequency Distribution of the Zodiacal Cloud: Evidence from the Solar System Dust Bands. Icarus, 152, 251

Grün, E., Gustafson, B., Mann, I., Baguhl, M., 1994. Dust in the Heliosphere. Astron. Ap., 286, 915

Grün, E., Baguhl, M., Svedhem, H., & Zook, H. A. 2001: In Situ Measurements of Cosmic Dust. Interplanetary Dust (E. Grün, B. A. S. Gustafson, S. F. Dermott, H. Fechtig, eds.), Springer A&A Library, p. 293. Dust in the Heliosphere. Astron. Ap., 286, 915

Guillot, T., & Hueso, R. 2006: The Composition of Jupiter: Sign of a (relatively) Late Formation in a Chemically Evolved Protosolar Disc. MNRAS, 367, L47

Gurnett, D. A., Ansher, J. A., Kurth, W. S., & Granroth, L. J. 1997: system by the Voyager 1 and 2 plasma wave instruments. Geoph. R. Lett., 24, 3125 24, 3125

Hartmann, L. 2000: Accretion Processes in Star Formation, Cambridge University Press.
Hahn, J. M., & Malhotra, R. 1999: Orbital Evolution of Planets Embedded in a Planetesimal Disk. A. J., 117, 3041
Harris, A. W. 1979: Asteroid Rotation Rates II. A Theory for the Collisional Evolution of Rotation rates. Icarus 40, 145
Heap, S. R., Lindler, D. J., Lanz, T. M., Cornett, R. H., Hubeny, I., Maran, S. P., & Woodgate, B. 2000: Space Telescope Imaging Spectrograph Coronagraphic Observations of β Pictoris. Ap. J., 539, 435
Higuchi, A., Kokubo, E., Kinoshita, H., & Mukai, T. 2007: Orbital Evolution of Planetesimals due to the Galactic Tide: Formation of the Comet Cloud. A. J., 134, 1693
Hillenbrand, L. A., et al. 2008: The Complete Census of 70-um-Bright Debris Disks within the FEPS (Formation and Evolution of Planetary Systems) Spitzer Legacy Survey of Sun-like Stars. ApJ, 677, 630
Hines, D. C., et al. 2006: The Formation and Evolution of Planetary Systems (FEPS): Discovery of an Unusual Debris System Associated with HD 12039. Ap. J., 638, 1070
Holland, W. S., et al. 1998: Submillimetre Images of Dusty Debris Around Nearby Stars. Nature, 392, 788
Holland, W. S., et al. 2003: Submillimeter Observations of an Asymmetric Dust Disk around Fomalhaut. Ap. J., 582, 1141
Holland, W. S., et al. 1998: Submillimetre Images of Dusty Debris Around Nearby Stars. Nature, 392, 788
Holman, M. J., & Wisdom, J. 1993: Dynamical Stability in the Outer Solar System and the Delivery of Short Period Comets. Astron. J., 105, 1987
Hsieh, H. H., & Jewitt, D. 2006: A Population of Comets in the Main Asteroid Belt. Science, 312, 561
Ida, S., Larwood, J., & Burkert, A. 2000: Evidence for Early Stellar Encounters in the Orbital Distribution of Edgeworth-Kuiper Belt Objects. Ap. J., 528, 351
Jewitt, D. C. 2002: From Kuiper Belt Object to Cometary Nucleus: The Missing Ultrared Matter. Astron. J. 123, 1039
Jewitt, D. 2003: Project Pan-STARRS and the Outer Solar System. Earth Moon and Planets, 92, 465
Jewitt, D., & Luu, J. 1993: Discovery of the candidate Kuiper belt object 1992 QB1. Nature, 362, 730
Jewitt, D., Luu, J., & Chen, J. 1996. The MKCT Kuiper Belt and Centaur Survey. Astron J., 112, 1225.
Jewitt, D., Luu, J. 1998: Optical-Infrared Spectral Diversity in the Kuiper Belt. Astron. J., 115, 1667
Jewitt, D. C., Luu, J. X. 2001: Colors and Spectra of Kuiper Belt Objects. Astron. J., 122, 2099
Jewitt, D., Luu, J., & Trujillo, C. 1998: Large Kuiper Belt Objects: The Mauna Kea 8K CCD Survey. Astron. J., 115, 2125
Jewitt, D., Peixinho, N., & Hsieh, H. H. 2007: U-Band Photometry of Kuiper Belt Objects. Astron. J., 134, 2046
Jewitt, D. C., & Sheppard, S. S. 2002: Physical Properties of Trans-Neptunian Object (20000) Varuna. Astron. J., 123, 2110
Jura, M. 2004: Other Kuiper Belts. Ap. J., 603, 729
Jura, M. 2006: Carbon Deficiency in Externally Polluted White Dwarfs: Evidence for Accretion of Asteroids. Ap. J., 653, 613
Kalas, P., Graham, J. R., & Clampin, M. 2005: A Planetary System as the Origin of Structure in Fomalhaut's Dust Belt. Nature, 435, 1067
Kalas, P., Liu, M. C., & Matthews, B. C. 2004: Discovery of a Large Dust Disk Around the Nearby Star AU Microscopii. Science, 303, 1990
Kenyon, S. J., & Bromley, B. C. 2005: Prospects for Detection of Catastrophic Collisions in Debris Disks. Astron. J., 130, 269

Kuchner, M. J., & Holman, M. J. 2003: The Geometry of Resonant Signatures in Debris Disks with Planets. Ap. J., 588, 1110
Lacerda, P., & Jewitt, D. C. 2007: Densities of Solar System Objects from Their Rotational Light Curves. Astron. J., 133, 1393
Lacerda, P., Jewitt, D., & Peixinho, N. 2008: High Precision Photometry of Extreme KBO 2003 EL61. Astron. J., 135, 1749
Lacerda, P., & Luu, J. 2003: On the Detectability of Lightcurves of Kuiper Belt Objects. Icarus 161, 174
Lacerda, P., & Luu, J. 2006: Analysis of the Rotational Properties of Kuiper Belt Objects. Astron. J., 131, 2314
Lee, T., Papanastassiou, D. A., & Wasserburg, G. J. 1977: Aluminum-26 in the Early Solar System - Fossil or Fuel. Ap. J. Lett., 211, L107
Leinert, C., Roser, S., & Buitrago, J. 1983: How to Maintain the Spatial Distribution of Interplanetary Dust. Astron. Ap., 118, 345
Leinert, C., & Grün, E. 1990: Interplanetary Dust. Physics of the Inner Heliosphere I, 207
Levison, H. F., & Duncan, M. J. 1994: The Long-Term Dynamical Behavior of Short-Period Comets. Icarus, 108, 18
Levison, H. F., Dones, L., & Duncan, M. J. 2001: The Origin of Halley-Type Comets: Probing the Inner Oort Cloud. A. J., 121, 2253
Levison, H. F., Terrell, D., Wiegert, P. A., Dones, L., & Duncan, M. J. 2006: On the Origin of the Unusual Orbit of Comet 2P/Encke. Icarus, 182, 161
Liou, J.-C., & Zook, H. A. 1997: Evolution of Interplanetary Dust Particles in Mean Motion Resonances with Planets. Icarus, 128, 354
Lisse C. M., Beichman C. A., Bryden, G., & Wyatt, M. C. 2007: On the Nature of the Dust in the Debris Disk Around HD 69830. ApJ, 658, 584
Liu, M. C. 2004: Substructure in the Circumstellar Disk Around the Young Star AU Microscopii. Science, 305, 1442
Luu, J., & Jewitt, D. 1996: Color Diversity Among the Centaurs and Kuiper Belt Objects. Astron. J., 112, 2310
Luu, J., & Lacerda, P. 2003: The Shape Distribution Of Kuiper Belt Objects. Earth Moon and Planets 92, 221
Luu, J., Marsden, B. G., Jewitt, D., Trujillo, C. A., Hergenrother, C. W., Chen, J., & Offutt, W. B. 1997: A New Dynamical Class in the Trans-Neptunian Solar System. Nature, 387, 573
Marcy, G., Butler, R. P., Fischer, D., Vogt, S., Wright, J. T., Tinney, C. G., & Jones, H. R. A. 2005: Observed Properties of Exoplanets: Masses, Orbits, and Metallicities. Progress of Theoretical Physics Supplement, 158, 24
Malhotra, R. 1995: The Origin of Pluto's Orbit: Implications for the Solar System Beyond Neptune. Astron. J., 110, 420
Melnick, G. J., Neufeld, D. A., Ford, K. E. S., Hollenbach, D. J., & Ashby, M. L. N. 2001: Discovery of Water Vapour Around IRC+10216 as Evidence for Comets Orbiting Another Star. Nature, 412, 160
Meyer, M. R., et al. 2008: Evolution of Mid-Infrared Excess around Sun-like Stars: Constraints on Models of Terrestrial Planet Formation. Ap. J. Lett., 673, L181
Moro-Martín, A., & Malhotra, R. 2002: A Study of the Dynamics of Dust from the Kuiper Belt: Spatial Distribution and Spectral Energy Distribution. Astron. J., 124, 2305
Moro-Martín, A., & Malhotra, R. 2003: Dynamical Models of Kuiper Belt Dust in the Inner and Outer Solar System. Astron. J., 125, 2255
Moro-Martín, A., Wolf, S., & Malhotra, R. 2005a: Signatures of Planets in Spatially Unresolved Debris Disks. Ap. J., 621, 1079
Moro-Martín, A., & Malhotra, R. 2005b: Dust Outflows and Inner Gaps Generated by Massive Planets in Debris Disks. Ap. J., 633, 1150
Moro-Martín, A., et al. 2007: Are Debris Disks and Massive Planets Correlated? Ap. J., 658, 1312

Moro-Martin, A., Wyatt, M. C., Malhotra, R., & Trilling, D. E. 2008: Extra-Solar Kuiper Belt Dust Disks. The Solar System Beyond Neptune (A. Barucci, H. Boehnhardt, D. Cruikshank, A. Morbidelli, eds.) University of Arizona Press, Tucson. arXiv:astro-ph/0703383

Morbidelli, A., & Levison, H. F. 2004: Scenarios for the Origin of the Orbits of the Trans-Neptunian Objects 2000 CR105 and 2003 VB12 (Sedna). Astron. J., 128, 2564

Morbidelli, A., Levison, H., Tsiganis, K., & Gomes, R. 2005: Chaotic Capture of Jupiters Trojans. Nature, 435, 462

Mostefaoui, S., Lugmair, G. W., Hoppe, P., & El Goresy, A. 2004: Evidence for live ^{60}Fe in meteorites. New Astron. Rev., 48, 155

Oort, J. H. 1950: The Structure of the Cloud of Comets Surrounding the Solar System and a Hypothesis Concerning its Origin. Bull. Astron. Inst. Neth., 11, 91

Ouellette, N., Desch, S. J., & Hester, J. J. 2007: Interaction of Supernova Ejecta with Nearby Protoplanetary Disks. Ap. J., 662, 1268

Owen, T., Mahaffy, P., Niemann, H. B., Atreya, S., Donahue, T., Bar-Nun, A., & de Pater, I. 1999: A Low-Temperature Origin for the Planetesimals that Formed Jupiter. Nature, 402, 269

Paresce, F. 1991: On the Evolutionary Status of Beta Pictoris. Astron. Ap., 247, L25

Peixinho, N., Doressoundiram, A., Delsanti, A., Boehnhardt, H., Barucci, M. A., Belskaya, I. 2003: Reopening the TNOs Color Controversy: Centaurs Bimodality and TNOs Unimodality. Astron. AP. 410, L29.

Reach, W. T., Morris, P., Boulanger, F., & Okumura, K. 2003: The mid-infrared spectrum of the zodiacal and exozodiacal light. Icarus, 164, 384

Rice, W. K. M., & Armitage, P. J. 2003: On the Formation Timescale and Core Masses of Gas Giant Planets. Ap. J. Lett., 598, L55

Rieke, G. H., et al. 2005: Decay of Planetary Debris Disks. Ap. J., 620, 1010

Schneider, G., Silverstone, M. D., & Hines, D. C. 2005: Discovery of a Nearly Edge-on Disk around HD 32297. Ap. J. Lett., 629, L117

Siegler, N., Muzerolle, J., Young, E. T., Rieke, G. H., Mamajek, E. E., Trilling, D. E., Gorlova, N., & Su, K. Y. L. 2007: Spitzer 24 μm Observations of Open Cluster IC 2391 and Debris Disk Evolution of FGK Stars. Ap. J., 654, 580

Sheppard, S. S., & Jewitt, D. C. 2002: Time-resolved Photometry of Kuiper Belt Objects: Rotations, Shapes, and Phase Functions. Astron. J., 124, 1757

Sheppard, S. S., & Jewitt, D. 2004: Extreme Kuiper Belt Object 2001 QG298 and the Fraction of Contact Binaries. Astron. J., 127, 3023

Sheppard, S. S., Lacerda, P., & Ortiz, J.-L. 2008: Photometric Lightcurves of Transneptunian Objects and Centaurs: Rotations, Shapes, and Densities. In "The Solar System Beyond Neptune" (Eds. Barucci, A. et al.), University of Arizona, Tucson, p. 129

Sheppard, S. S., & Trujillo, C. A. 2006: A Thick Cloud of Neptune Trojans and Their Colors. Science, 313, 511

Song, I., Zuckerman, B., Weinberger, A. J., & Becklin, E. E. 2005: Extreme collisions between planetesimals as the origin of warm dust around a Sun-like star. Nature, 436, 363

Stapelfeldt, K. R., et al. 2004: First Look at the Fomalhaut Debris Disk with the Spitzer Space Telescope. Ap. J. Supp., 154, 458

Stern, S. A. 1996: Signatures of collisions in the Kuiper Disk. Astron. Ap., 310, 999

Strom, R. G., Malhotra, R., Ito, T., Yoshida, F., & Kring, D. A. 2005: The Origin of Planetary Impactors in the Inner Solar System. Science, 309, 1847

Su, K. Y. L., et al. 2005: The Vega Debris Disk: A Surprise from Spitzer. Ap. J., 628, 487

Su, K. Y. L., et al. 2006: Debris Disk Evolution around A Stars. Ap. J., 653, 675

Sykes, M. V., & Greenberg, R. 1986: The Formation and Origin of the IRAS Zodiacal Dust Bands as a Consequence of Single Collisions Between Asteroids. Icarus, 65, 51

Takahashi, S., and Ip, W. 2004. A Shape and Density Model of EKBO 2001 QG298. Pub. Astr. Soc. Japan, 56, 1099

Tegler, S. C., & Romanishin, W. 2000: Extremely Red Kuiper-Belt Objects in Near-Circular Orbits Beyond 40 AU. Nature 407, 979

Thébault, P., & Doressoundiram, A. 2003: Colors and Collision Rates Within the Kuiper belt: Problems with the Collisional Resurfacing Scenario. Icarus 162, 27

Trafton, L. M. 1989: Pluto's Atmosphere Near Perihelion. Geophy. Res. Lett., 16, 1213

Trilling, D. E., et al. 2007: Debris Disks in Main-Sequence Binary Systems.. Ap. J., 658, 1289

Trilling, D. E., et al. 2008: Debris Disks around Sun-like Stars. Ap. J., 674, 1086

Trujillo, C. A., Jewitt, D. C., & Luu, J. X. 2001: Properties of the Trans-Neptunian Belt. Astron. J. 122, 457

Weingartner, J., & Draine, B. 2001. Dust Grain-Size Distributions and Extinction in the Milky Way and Magellanic Clouds. Ap. J., 548, 296.

Willacy, K. 2004: A Chemical Route to the Formation of Water in the Circumstellar Envelopes around Carbon-rich Asymptotic Giant Branch Stars: Fischer-Tropsch Catalysis. Ap. J. Lett., 600, L87

Wyatt, M. C., Dermott, S. F., Telesco, C. M., Fisher, R. S., Grogan, K., Holmes, E. K., & Piña, R. K. 1999: How Observations of Circumstellar Disk Asymmetries Can Reveal Hidden Planets: Pericenter Glow and Its Application to the HR 4796 Disk. Ap. J., 527, 918

Wyatt, M. C., Smith, R., Greaves, J. S., Beichman, C. A., Bryden, G., & Lisse, C. M. 2007: Transience of Hot Dust around Sun-like Stars. Ap. J., 658, 569

Yamamoto, S., & Mukai, T. 1998: Dust Production by Impacts of Interstellar Dust on Edgeworth-Kuiper Belt Objects. Astron. Ap., 329, 785

Young, E. F., Galdamez, K., Buie, M. W., Binzel, R. P., & Tholen, D. J. 1999: Mapping the Variegated Surface of Pluto. Astron. J., 117, 1063

Zook, H. A., & Berg, O. E. 1975: A Source for Hyperbolic Cosmic Dust Particles. Planetary Space Sci., 23, 183

Zuckerman, B., Song, I., Bessell, M. S., & Webb, R. A. 2001. The Beta Pictoris Moving Group. Ap. J., 562, L87

Yvonne Pendleton graciously allowed a birthday party to be thrown in her honor during the meeting

Chapter 4
The Future of Ultracool Dwarf Science with JWST

Mark S. Marley and S.K. Leggett

Abstract Ultracool dwarfs exhibit a remarkably varied set of characteristics which hint at the complex physical processes acting in their atmospheres and interiors. Spectra of these objects not only depend upon their mass and effective temperature, but also their atmospheric chemistry, weather, and dynamics. As a consequence divining their mass, metallicity and age solely from their spectra has been a challenge. JWST, by illuminating spectral blind spots and observing objects with constrained masses and ages should finally unearth a sufficient number of ultracool dwarf Rosetta Stones to allow us to decipher the processes underlying the complex brown dwarf cooling sequence. In addition the spectra of objects invisible from the ground, including very low mass objects in clusters and nearby cold dwarfs from the disk population, will be seen for the first time. In combination with other ground- and space-based assets and programs, JWST will usher in a new golden era of brown dwarf science and discovery.

4.1 Introduction

The explosive growth of brown dwarf and ultracool dwarf discoveries over the past dozen years has been so extraordinary that it is a rare paper in the field that does not open by remarking upon it. The first undisputed brown dwarf, Gl 229 B, was discovered as a companion to an M dwarf in 1995 (Nakajima et al. 1995). The Two-Micron All Sky Survey (2MASS, Skrutskie et al. 2006) and the Sloan Digital Sky Survey (SDSS, York et al. 2000) subsequently revealed large numbers of ultracool low-mass field dwarfs. These surveys first led to the discovery of the isolated field late-L dwarfs (Kirkpatrick et al. 1999), then the mid-T dwarfs (Burgasser et al. 1999, Strauss et al. 1999) and finally the early-T dwarfs (Leggett et al. 2000). Today over 600 warm ($T_{\mathrm{eff}} \sim 2400$ to 1400 K) L and cool (600 to 1400 K) T dwarfs are known[1] and the quest for the elusive, even cooler "Y" dwarfs is ongoing.

M.S. Marley (✉)
NASA Ames Research Center, Mail Stop 245-3, Moffett Field CA 94035, USA
e-mail: Mark.S.Marley@NASA.gov

[1] See http://www.DwarfArchives.org

H.A. Thronson et al. (eds.), *Astrophysics in the Next Decade,* Astrophysics and Space Science Proceedings, DOI 10.1007/978-1-4020-9457-6_4,

Note that collectively late M and later type dwarfs are often termed 'Ultracool Dwarfs' (or UCDs) to avoid having to distinguish whether particular warm objects in this group are above or below the hydrogen burning minimum mass, the requirement for bestowing the term 'brown dwarf'. Brown dwarfs will continuously cool over time. The more massive UCDs will eventually arrive on the bottom of the hydrogen burning main sequence (Burrows et al. 1997).

Ultracool dwarf science is exciting not only for the rapid pace of discovery, but for a host of other reasons as well. First, since brown dwarfs lack an internal energy source (beyond a brief period of deuterium burning), they cool off over time; they thus reach effective temperatures below those found in stars and enter the realm where chemical equilibrium favors such decidedly 'unstellar' atmospheric species as CH_4 and NH_3. Along with these more typical 'planetary' gasses, silicate and iron clouds are found in their atmospheres, leading to interesting, complex interactions between atmospheric chemical, radiative-transfer, dynamical, and meteorological processes. Ultracool dwarfs thus bridge the domain between the bottom of the stellar main sequence and giant planets. They are a laboratory for understanding processes that will also be important in the characterization of extrasolar giant planets. Second, they occupy the low mass end of the stellar initial mass function. Understanding the IMF requires that we understand the masses of individual field objects, which ultimately requires an understanding of their luminosity evolution as well as the dependence of their spectra on mass, gravity, effective temperature and metallicity. Finally, as *terra incognita*, brown dwarfs (some of our nearest stellar neighbors) have offered a series of surprises that test our ability to understand the universe around us.

The most distinctive features of the UCD spectral sequence are highlighted in Fig. 4.1. At effective temperatures below those of late-M dwarfs, several chemical changes (illustrated in Fig. 4.2) occur that strongly impact the spectral energy distribution. First, major diatomic metal species (particularly TiO and FeH) become incorporated into grains, leading to the gradual departure of hallmarks of the M spectral sequence (Kirkpatrick et al. 1999). Second, the formation of iron and silicate grains produces optically thick clouds that veil gaseous absorption bands and redden the near-IR *JHK* colors of L dwarfs. The atmospheric temperature domain where these clouds are most important is $\sim$1500–2000 K (e.g. Ackerman & Marley 2001). At lower $T_{\rm eff}$, the clouds lie near or below the base of the wavelength-dependent photosphere, and only marginally affect the SEDs of T dwarfs. Finally, CH_4 supplants CO as the dominant carbon-bearing molecule. This transition is first noted in the 3–4 μm spectra of mid-L dwarfs (Noll et al. 2000) and appears in both the *H* and *K* bands of T0 dwarfs (Geballe et al. 2002). Together, increasing CH_4 absorption and sinking cloud decks cause progressively bluer near-IR colors of T dwarfs. For types T5 and later, significant collision-induced H_2 opacity in the *K* band enhances the trend toward bluer near-infrared colors.

These changes in spectral features are used to assign spectral types to UCDs as briefly explained in Section 4.2. Given assigned spectral types, measurement of the bolometric luminosity of individual objects along with their parallaxes connect the spectral sequence to effective temperature. Figure 4.3 illustrates the effective

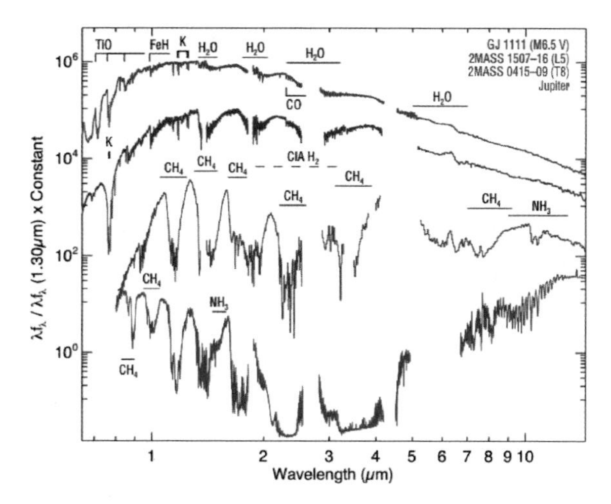

Fig. 4.1 The most prominent signatures of the ultra cool dwarf spectral sequence are seen in these 0.65 to 14.5 μm spectra of a mid-M, L, and T dwarfs as well as Jupiter (adapted from Cushing et al. (2006)). The spectra have been normalized to unity at 1.3 μm and multiplied by constants. Major absorption bands are marked. The collision-induced opacity of H_2 is indicated as a dashed line because it shows no distinct spectral features but rather a broad, smooth absorption. Jupiter's flux shortward of $\sim$ 4 μm is predominantly scattered solar light; thermal emission dominates at longer wavelengths (near- and mid-infrared Jovian spectra from Rayner, Cushing & Vacca (in preparation) and Kunde et al. (2004), respectively)

temperature as a function of spectral type from late M through late T. While the general correlation of increasing spectral type with falling effective temperature is unmistakable, a remarkably rapid set of spectral changes (as expressed in the variation in spectral type) happens over a relatively small span of $T_{\rm eff}$ near 1400 K. As we will discuss, understanding this variation in expressed spectral signatures, the 'L to T transition', is a key subject of current brown dwarf research.

Most of the scientific inquiry into these ultracool dwarfs has focused on their formation and youth, on the resultant initial mass function, and on the determination of their global properties, particularly mass, effective temperature, metallicity, and cloudiness. This review will focus primarily on the latter areas. Burrows et al. (2001) provide a much more in depth review and background to brown dwarf science and is an excellent starting point for those new to the subject. Kirkpatrick (2008) provides a more current look at outstanding issues in the field from an observational perspective. Here we first present a very brief review of the ultracool dwarf spectral types and the nature of the current datasets. We then move on to discuss the role that clouds and atmospheric mixing play in controlling the emitted spectra of these objects and the enigmatic L- to T-type transition. Because clouds control the spectral energy distribution of the L and early T dwarfs, and since clouds are inherently difficult to model, constraining gravity solely by comparison of observations to spectral data is particularly challenging. Finally we will close with a look forward to some of the ultracool dwarf science opportunities that will be enabled by JWST. Because

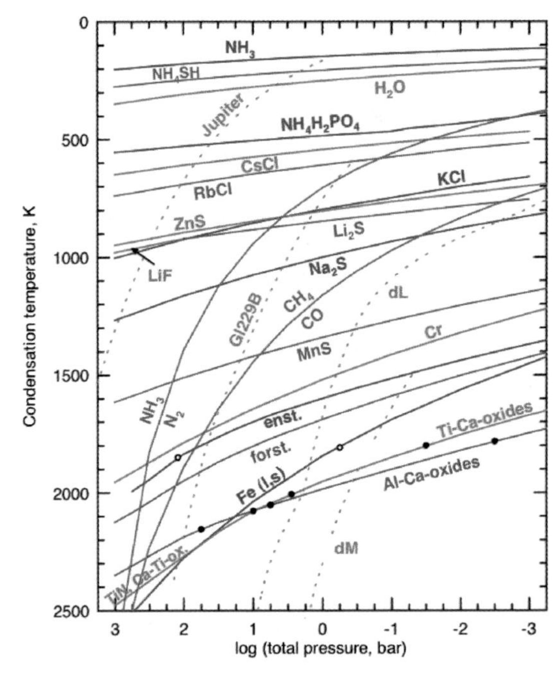

Fig. 4.2 Important chemical equilibrium boundaries for substellar objects (modified from Lodders & Fegley 2006). Green, red, and blue lines denote various condensation boundaries for a solar abundance mixture of gasses in a substellar atmosphere. Light purple lines denote equilibrium boundaries between important gaseous species. Grey dashed lines show model atmospheric temperature-pressure profiles for M, L and T dwarfs (the latter specifically for Gl 229 B) as well as for Jupiter. As one moves upwards in the diagram along the model (T, P) curves, the labeled species will condense at the intersection with the condensation curves and would be expected to be absent from the gas at lower temperatures further up along the model profiles. This figure can be compared with Fig. 4.1 to understand why spectral features for various compounds are present or absent in each observed spectrum

of space limitations we neglect several other important avenues of UCD research, including studies of the IMF, very young objects, and of objects with unusual colors.

4.2 Spectral Type

The currently known ultracool dwarfs span spectral types from late M, L0 through L9, and T0 through T9. The TiO and VO bands, which dominate the optical portions of late-M dwarf spectra, disappear in the L dwarfs (Kirkpatrick et al. 1999), where metallic oxides are replaced by metallic hydrides and where features due to neutral alkali metals are strong. To systematize such objects Kirkpatrick et al. established spectral indices and defined an optical classification scheme for L dwarfs, which is commonly used. The indices measure the strengths of TiO, VO, CrH, Rb and Cs features as well as a red color term, at wavelengths between 0.71 and 0.99 μm.

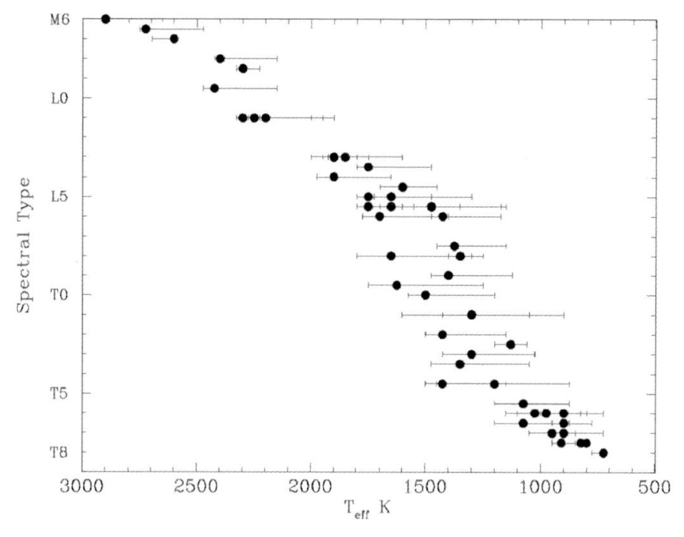

Fig. 4.3 Effective temperature as a function of infrared spectral type for ultracool dwarfs with known parallax (data from Golimowski et al. 2004, Vrba et al. 2004, and Luhman et al. 2007). Note the roughly constant effective temperature for dwarfs of spectral types from late L to early T. See Kirkpatrick (2008) for further discussion

With the discovery of the T dwarfs, which have very little flux in the optical, Geballe et al. (2002) defined a near-infrared classification scheme that encompassed both the L and T dwarfs. The indices measure the strength of the H_2O and CH_4 absorption features between 1.1 and 2 μm, and for the early L dwarfs a red color term is also used which is slightly modified from Kirkpatrick et al. (1999).

Burgasser et al. (2002a) introduced a near-infrared classification scheme for T dwarfs that was very similar to that of Geballe et al. The two schemes were unified in Burgasser et al. (2006) and this near-infrared scheme is the commonly used scheme for typing T dwarfs. For L dwarfs, both the optical Kirkpatrick et al. and the near-infrared Geballe et al. schemes are used; these usually produce the same type (the Geballe et al. scheme was pinned to the Kirkpatrick et al. types), but for L dwarfs with unusual colors they can give significantly different types. Because of this, the classification scheme used for L dwarfs should always be specified.

Leggett et al. (2007a, and other work referenced therein) explore the possible signatures of the next spectral type, for which the letter Y has been suggested (Kirkpatrick 2005). It is likely that NH_3 features will join the familiar water and methane absorption seen in T dwarfs as the effective temperatures approach 600 K. In actuality the situation is likely more complex: it is already known that the atmospheric NH_3 abundance (as seen in *Spitzer* mid-infrared spectra of T dwarfs) is reduced by vertical mixing which drags N_2 up from deeper layers in the atmosphere (e.g. Saumon et al. 2006). If this mechanism continues to act at low effective temperatures (which is not a certainty (Hubeny & Burrows 2007)), the near-infrared ammonia features may be weaker than expected. Another outstanding problem with predicting the signature of the proposed Y dwarfs, is that the linelist for NH_3 is very incomplete at 1.0–1.5 μm. Since the near-infrared flux of Y dwarfs is expected to rapidly 'collapse' with falling $T_{\rm eff}$ (Burrows et al. 2003), it may even be appropriate to ultimately type these objects with mid-, instead of near-infrared, spectra. In this case spectra obtained by *Spitzer* or JWST would be required for spectral typing. Regardless, dwarfs with $T_{\rm eff} \sim 650$ K are now being found (e.g. Warren et al. 2007), and it is likely that soon temperatures where significant spectral changes occur will be reached.

4.3 Ultracool Dwarf Datasets

The L and T dwarfs were discovered primarily as a result of the far-red and near-infrared Sloan Digital Sky Survey and 2 Micron All Sky Survey (e.g. Kirkpatrick et al. 1999, Strauss et al. 1999). This continues with current surveys – the Canada France Hawaii Brown Dwarf Survey and the UKIRT Infrared Deep Sky Survey (e.g., Lodieu et al. 2007) are identifying extreme-T dwarfs by their very red far-red and blue near-infrared colors. Spectral classification is carried out in the far-red or near-infrared. Hence the existing data for L and T dwarfs is primarily far-red and near-infrared imaging and spectroscopy. The spectroscopy has been medium- or low-resolution, both because that is all that is required for the spectral classification, but also because the dwarfs are faint.

Some ground-based imaging and spectroscopy has been carried out at 3.0–5.0 μm (e.g. Noll et al. 2000, Golimowski et al. 2004). Such work is extremely challenging due to the very high and rapidly variable sky background at these wavelengths, and only the brightest dwarfs could be observed from the ground.

This situation changed with the launch of the *Spitzer Space Telescope*. Roellig et al. (2004) and Patten et al. (2006) demonstrated the quality and quantity of mid-infrared imaging and spectroscopy of L and T dwarfs that *Spitzer* could produce, and such work continues through the current, final, cryogenic cycle. IRS spectral data, which span 6 to 15 μm, show a strong 11 μm NH_3 absorption feature in T dwarfs, as well as H_2O and CH_4 absorption features in both L and T dwarfs (Fig. 4.1). IRAC photometry covers the 3 to 8 μm wavelength range, and the 3 to 5 μm range may continue to be available in the warm-*Spitzer* era. These IRAC bandpasses include CH_4, CO and H_2O features, and signatures of vertical transport have been recognized in the photometry (Leggett et al. 2007b).

Warren et al. (2007) further suggest that the $H-$ [4.49] color may be a very good indicator of temperature for dwarfs cooler than 1000 K. For extreme-T dwarfs the near-infrared CH_4 and H_2O bands are so strong that it will be difficult to measure an increase in their strength, hence the mid-infrared may prove to be vital to interpreting the cold objects.

4.4 Ultracool Dwarf Atmospheres

Ultracool dwarf emergent spectra are controlled by the variation in abundances of important atomic, molecular, and grain absorbers both with height in the atmosphere

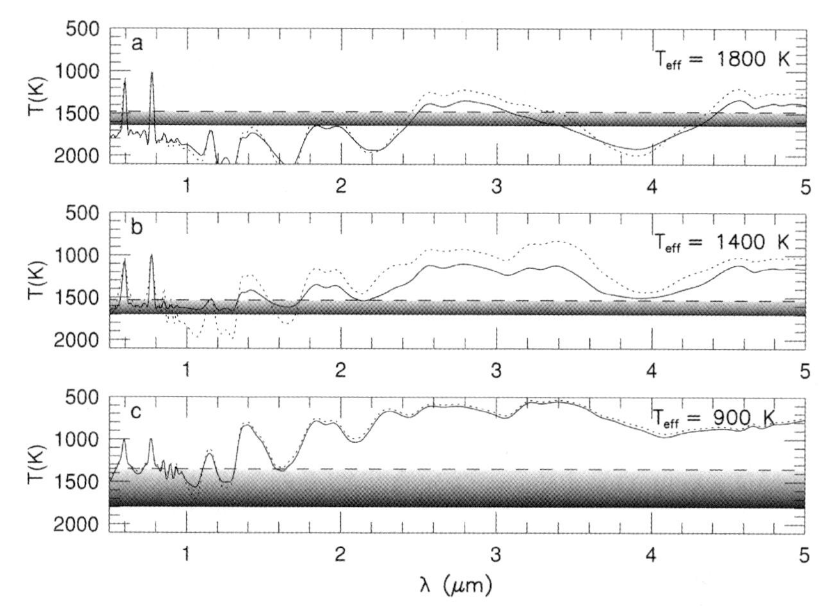

Fig. 4.4 Brightness temperature as a function of wavelength for atmosphere models which include (*solid*) or exclude (*dotted*) silicate and iron clouds (Ackerman & Marley 2001). Brightness temperature increases downward to suggest increasing depth in the atmosphere from which the wavelength-dependent flux emerges. The solid straight line indicates the base of the silicate cloud while the long dashed line denotes the 'top' of the cloud (the level in the atmosphere at which the cloud column extinction reaches 0.1). Shading suggests the decrease in cloud extinction with altitude. Since cloud particle radii exceed 10 μm in these models, the Mie extinction efficiency is not a strong function of wavelength over the range shown. Shown are models characteristic of (**a**) an early-type L dwarf with $T_{\rm eff} = 1800\,$K, (**b**) a late-type L dwarf with $T_{\rm eff} = 1400\,$K, and (**c**) a T dwarf with $T_{\rm eff} = 900\,$K . All of these models are for solar composition and gravity appropriate for a 30 Jupiter-mass brown dwarf. Note that the spectral region just longward of 1 μm is particularly sensitive to the cloud opacity

at a given age and over time as the objects cool. The major atomic and molecular absorption features are imprinted on spectra that have no true continuum.[2] Flux emerges from a many-scale-height-thick range of depths in the atmosphere as a function of wavelength. For example (Fig. 4.4), in an early L dwarf, brightness temperatures[3] range from 1000 K in the depths of alkali absorption lines (Burrows et al. 2000) in the far red, to well over 2000 K in the molecular windows in between strong water absorption bands. By providing a continuum opacity source, clouds can limit the flux emerging in some molecular window regions, but not others. Furthermore the strength of some molecular absorption features, particularly CH_4,

[2] Sharp & Burrows (2007) and Freedman et al. (2008) discuss the atmospheric opacity sources in detail.

[3] Brightness temperature (the temperature that a blackbody that emits radiation of the observed intensity at a given wavelength) is commonly used in planetary atmospheres studies to elucidate the temperature of the emitting level in an atmosphere.

CO, and NH_3, can depend on the strength of mixing in the atmosphere. Thus a full description of an ultracool dwarf atmosphere hinges on the dwarf's gravity, effective temperature, cloud properties, and mixing. In this section we summarize the important unsolved problems related to these atmospheres.

4.4.1 Clouds

At high effective temperatures, the column abundance of condensates (see Marley 2000 for a discussion of influences on cloud optical depth) is low and the difference between models computed with and without cloud opacity is slight (Fig. 4.4). At lower temperatures, however, the cloud substantially alters the temperature profile of the atmosphere and provides a continuum opacity source that limits the depth to which the usual molecular windows probe into the atmosphere. By the effective temperature of the mid-T dwarfs, however, most of the flux emerges from above the cloud level and the clouds are again less important. For the effective temperature range of the mid to late L dwarfs and the early T dwarfs, however, clouds clearly play a very large role in controlling the vertical structure and emergent spectra of brown dwarfs.

Thus, any attempt to model brown dwarf atmospheres must include a treatment of clouds. However, clouds are the leading source of uncertainty in terrestrial atmosphere models and are inherently difficult to model. Their influence depends on the variation of particle size, abundance, and composition with altitude, which in turn depend on the complex interaction of many microphysical processes (e.g. Ackerman & Marley 2001, Helling & Woitke 2006). Fits of model spectra to observational data are highly sensitive to the treatment of clouds in the underlying atmosphere model and the approaches taken by various modeling groups vary widely (see the comparison study in Helling et al. 2008a). For example, Cushing et al. (2008) demonstrate reasonably accurate fits of model spectra to near- and mid-IR spectra of a sample of L and T dwarfs, but the precise values of effective temperature and gravity obtained from the fits depend entirely upon the cloud sedimentation efficiency (Ackerman & Marley 2001) assumed. Since the models are highly dependent on the cloud description, the derived effective temperature and gravities, while plausible, are nevertheless uncertain. A similar conclusion was reached by Helling et al. (2008b). Finding a selection of L dwarfs with known g and $T_{\rm eff}$ that could serve as calibrators of the model spectra would be invaluable. L dwarf companions to main sequence stars with constrained ages, L dwarf binaries with resolved orbits, and L dwarfs in clusters of known ages are all promising targets for such work.

4.4.2 Characterizing Clouds

The spectral range of the InfraRed Spectrometer (IRS) on *Spitzer* includes the 10 μm silicate feature which arises from the Si-O stretching vibration in silicate grains.

The spectral shape and importance of the silicate feature depends on the particle size and composition of the silicate grains. According to phase-equilibrium arguments, in brown dwarf atmospheres the first expected silicate condensate is forsterite Mg_2SiO_4 (Lodders and Fegley 2002), at $T \sim 1700\,K$ ($P = 1$ bar). Since Mg and Si have approximately equal abundances in a solar composition atmosphere, the condensation of forsterite leaves substantial silicon, present as SiO, in the gas phase. In equilibrium, at temperatures about 50 to 100 K cooler than the forsterite condensation temperature, the gaseous SiO reacts with the forsterite to form enstatite, $MgSiO_3$ (Lodders and Fegley 2002). The precise vertical distribution of silicate species depends upon the interplay of the atmospheric dynamics and chemistry and such details have yet to be fully modeled, although efforts to improve the detailed cloud modeling continue (e.g. Cooper et al. 2003, Woitke & Helling 2003, Helling & Woitke 2006, Helling et al. 2008b).

In brown dwarf clouds there is likely a range of particle sizes, ranging from very small, recently condensed grains, to larger grains that have grown by agglomeration. The mean particle size for silicate grains in L dwarf model atmospheres is typically computed to be in the range of several to several tens of microns (Ackerman & Marley 2001, Helling et al. 2008b). Figure 4.5 compares the absorption efficiency of

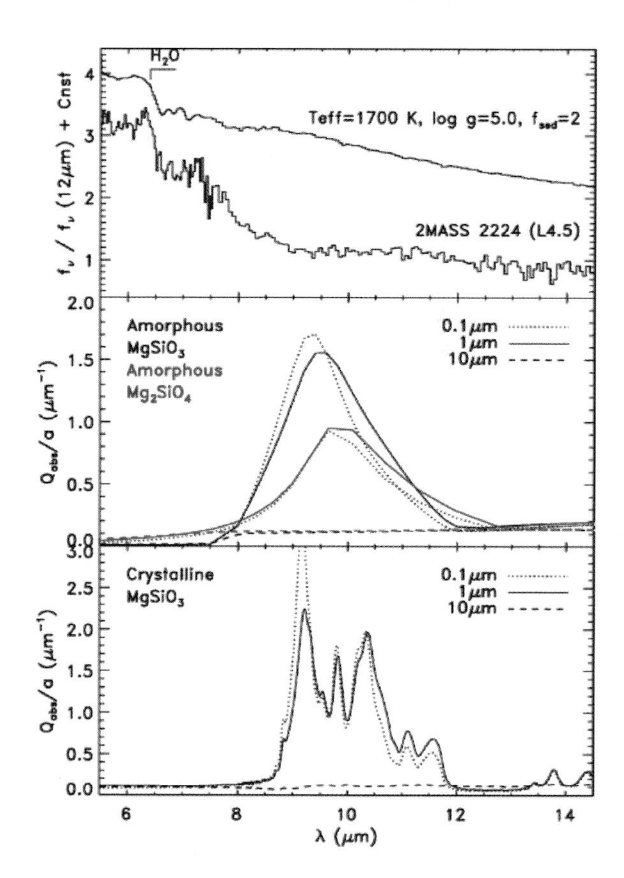

Fig. 4.5 *Top*: *Spitzer* IRS spectrum of 2MASS 2224 (L4.5) and the best fitting model from Cushing et al. (2006). *Middle*: Optical absorption (Q_{abs}/a) for amorphous enstatite ($MgSiO_3$) and forsterite (Mg_2SiO_3) for three different particle sizes, 0.1, 1, and 10 μm. *Bottom*: Optical absorption for crystalline enstatite, also for three different particle sizes. The deviation of the model (shifted vertically) from the data suggests that additional small, and perhaps crystalline, silicate grains are required to adequately account for the observed spectrum

silicate grains of various sizes, composition, and crystal structures to the spectrum of 2MASS J2224-0158 (L4.5). For each species the quantity Q_{abs}/a, or Mie absorption efficiency divided by particle radius, is shown; all else being equal, the total cloud optical depth is proportional to this quantity (Marley 2000). Large grain sizes tend to have a relatively flat absorption spectra (dashed lines) across the IRS spectral range. Only grains smaller than about 3 μm in radius show the classic 10 μm silicate feature (Hanner et al. 1994). Figure 4.5 suggests that the mismatch between the models and data may arise from a population of silicate grains that is not captured by the cloud model used to construct the figure (Ackerman & Marley 2001). The actual silicate cloud may contain both more small particles and a mixture of enstatite and forsterite grains (e.g. Helling & Woitke 2006), although detailed models for this particular dataset have not been attempted. Furthermore the model shown in the figure employs optical properties of amorphous silicate. It is possible, especially at the higher pressures found in brown dwarf atmospheres, that the grains are crystalline, not amorphous. Indeed laboratory solar-composition condensation experiments produce crystalline, not amorphous, silicates (Toppani et al. 2004). Crystalline grains (Fig. 4.5) can have larger and spectrally richer absorption cross sections.

4.4.3 The Transition from L to T

The evolutionary cooling behavior of a given substellar object can be inferred from the field brown dwarf near-infrared color-magnitude diagram (Fig. 4.6a). Over tens to hundreds of millions of years, a given substellar object first moves to redder $J-K$ colors as it cools while falling to fainter J magnitudes. Around $M_J \sim 14-15$ the $J-K$ color turns bluer and the J magnitude slightly (and counter-intuitively) brightens (Dahn et al. 2002, Tinney et al. 2003, Vrba et al. 2004). With further cooling a given dwarf finally falls to fainter J magnitudes and apparently continues to slightly turn somewhat bluer in $J-K$. The behavior with even greater cooling is as yet uncertain until many more objects with $T_{eff} < 700$ K are found.

The 'L to T' transition is the 'horizontal branch' of the color-magnitude diagram as objects move from red to blue in the diagram. From Fig. 4.3 we know that this color (and underlying spectral) change from late-type L dwarfs with $J-K \sim 2.5$ to blue T dwarfs with $J-K \sim -1$ happens rapidly over a small range of effective temperature. No brown dwarf evolution model can currently reproduce the magnitude of the observed color change over such a small range of T_{eff}. Various explanations have been suggested including holes forming in the condensate cloud decks (Ackerman & Marley 2001, Burgasser et al. 2002b), an increase in the efficiency of grain sedimentation (Knapp et al. 2004), or a change in particle size (Burrows et al. 2006). In a series of papers, Tsuji (Tsuji 2002, Tsuji & Nakajima 2003, Tsuji et al. 2004) proposed that a physically very thin cloud could self-consistently explain the rapid L to T transition. These models indeed exhibit a somewhat faster L- to T-like transition, but are still not consistent with the observed rapidity of the color change. More recently Tsuji (2005) has favored a sudden collapse of the global cloud deck at the transition along the lines of the Knapp et al. (2004) suggestion.

Support for a rapid increase in sedimentation efficiency at the L to T transition has come from the model analysis of the 0.8–14.5 μm spectra of four transition dwarfs by Cushing et al. (2008), who find that the cloud sedimentation efficiency (Ackerman & Marley 2001) indeed increases across the transition. The Cushing et al. (2008) sample includes two pairs of mid- to late-L dwarfs with very different near-IR colors. The authors find that the redder L dwarfs have less efficient sedimentation and therefore thicker cloud decks consisting of smaller particles, although gravity also may play a role for one pair. For one of the red L dwarfs, Cushing et al. (2006) identified a broad absorption feature at 9–11 μm which may be due to the presence of small silicate grains (Fig. 4.5).

The difficulty in characterizing the L to T transition arises from our lack of understanding of the masses and effective temperatures of objects at various locations in the ultra-cool dwarf color-magnitude diagram. It is not clear, for example, if the reddest field L dwarfs are more or less massive than bluer objects or if the turn towards the blue in $J - K$ is mass dependent (although there are some indications that it may be (Metchev & Hillenbrand 2006)). There are two ways in which this shortcoming in current understanding could be addressed. First, observing the orbits of binary brown dwarfs allows the total system mass to be measured. Secondly, photometry leading to near-infrared color-magnitude diagrams for many clusters with a variety of ages and metalicities will constrain the nature of transition.

To date, the cluster color magnitude diagram (CMD) has only reached the transition in the Pleiades and perhaps the Sigma Orionis clusters. Figure 4.6b shows the

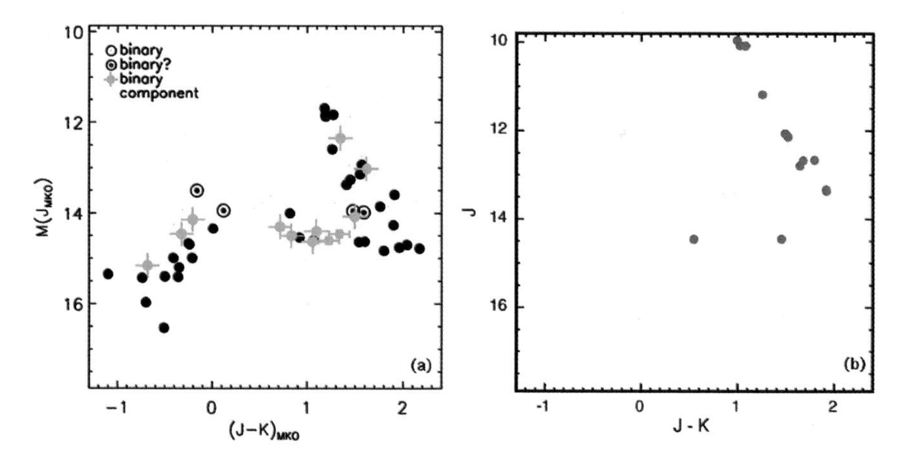

Fig. 4.6 Near-infrared color-magnitude diagrams for field and cluster ultracool dwarfs. (**a**) Black dots show single field L & T dwarfs. Green dots are resolved components of binary systems. Dotted circles are suspected (but unresolved) binaries (figure courtesy M. Liu based on Liu et al. (2006)). (**b**) Candidate ultracool dwarfs in the Pleiades in the most sensitive current survey (Casewell et al. 2007). Faintest objects in this plot have masses of about 11 M_{Jup}. Note that at a fixed magnitude the cluster members tend to be redder than the field objects, which is likely a signature of low gravity. JWST will obtain spectra of quality comparable to Fig. 4.1 for the candidate objects shown on this panel which will help calibrate evolutionary models of the brown dwarf cooling sequence. The detection limit for NIRCam on JWST is at about $J = 22$ for this cluster or $\sim 1\ M_{Jup}$. Model predictions for colors of objects with $J > 15$ are shown in Fig. 4.8

currently best available CMD for the Pleiades. Two objects in this figure can be seen to have turned towards the blue. Deeper searches to fainter magnitudes in this cluster should soon reveal the expected downward turn to the fainter J magnitudes apparent in the field CMD. By constructing evolution models at the age of the Pleiades, it should be possible to constrain the mass at the turnoff from the red L sequence. Given enough clusters of different ages, the turnoff effective temperature and gravity can be constrained, thus illuminating the dependence of the turnoff on gravity and perhaps metallicity.

JWST will be able to obtain moderate resolution ($R \sim 1000$) spectra on the current Pleiades candidates in Fig. 4.6 in about 2.5 hours at $S/N \sim 20$. This spectral resolution should be sufficient to identify, for example, FeH and CH_4 bands as they vary through the spectral sequence. This combination of evolution models and spectra should tightly constrain the empirical cooling sequence. Although this cluster is likely too large on the sky for efficient searching by JWST, NIRCAM could in principle find objects with masses as low as about 1 M_J. Surveys of more compact clusters would not reach to such low masses, but should nevertheless be deep enough to find many young T dwarfs that have already undergone the transition.

4.4.4 The Latest T Dwarfs

At this time only 16 very cool ($T_{eff} < 900$ K) dwarfs with types T7 and later are known, and of these only four are T8 or later. New surveys that go fainter than 2MASS and SDSS have started or are planned, and several groups are attempting to push to later and cooler types (e.g. Warren et al. 2007). All but two of the very late T dwarfs are isolated (the exceptions are Gl 570 D, Burgasser et al. 2000, and HD 3651B, Mugrauer et al. 2007). Since age is unknown for field dwarfs and brown dwarf cool with time, observed spectra must be compared with models or spectra of fiducial objects, to constrain mass and age. Since there are only a few T dwarfs with highly constrained properties, accurate model analysis is crucial for secure determination of gravity and hence mass at the bottom of the T sequence. For field brown dwarfs with ages in the range $\sim$ 1–5 Gyr and masses of 20–50 Jupiter-masses, effective temperature will lie in the range of $\sim$600–800 K. To understand the physical parameters of these elusive, cold, and low-mass dwarfs requires observation of their full spectral energy distribution.

4.4.5 Vertical Mixing and Chemical Disequilibrium

It has long been understood that the abundances of molecules in Jupiter's atmosphere depart from the values predicted purely from equilibrium chemistry (Prinn & Barshay 1977, Barshay & Lewis 1978, Fegley & Prinn 1985, Noll 1988, Fegley

& Lodders 1994, Fegley & Lodders 1996).[4] Rapid upwelling can carry compounds from the deep atmosphere up into the observable regions of the atmosphere on time scales of hours to days. When this convective timescale is shorter than the timescale for chemical reactions to reach equilibrium, then the atmospheric abundances will differ from those that would be found under pure equilibrium conditions. The canonical example of this situation is carbon monoxide in Jupiter's atmosphere. In Jupiter's cold and dense upper troposphere, carbon should almost entirely be found in the form of methane. Deeper into the atmosphere, where temperatures and pressures are higher, CO should be the principal carrier of carbon. Rapid vertical mixing, combined with the strong C–O molecular bond, means that CO molecules can be transported to the observed atmosphere faster than chemical reactions can reduce the CO into CH_4. The observed enhancement of CO, combined with (uncertain) reaction rates places limits on the vigor of convective mixing in the atmosphere.

4.4.5.1 CO

Analyses of the 4.5–5 μm spectra of the T dwarfs Gl 229B and Gl 570 D reveal an abundance of CO that is over 3 orders of magnitude larger than expected from chemical equilibrium calculations (Noll et al. 1997, Oppenheimer et al. 1998, Griffith & Yelle 1999, Saumon et al. 2000), as anticipated by Fegley & Lodders (1996). Photometry in the M band, which overlaps the CO band at 4.6 μm, shows that an excess of CO may be a common feature of T dwarfs. Golimowski et al. (2004) have found that the M band flux is lower than equilibrium models predict based on the K and L' fluxes in all of the T dwarfs in their sample. The low M band flux certainly arises from an excess of CO above that expected by equilibrium models. Since CO is a strong absorber in the M band, a brown dwarf can be much fainter in this band than would be predicted by equilibrium chemistry. Equilibrium models predict that brown dwarfs and cool extrasolar giant planets should be bright at M band, hence any flux decrement at this wavelength would have implications for surveys for cool dwarfs and giant planets. The degree to which this is a concern depends upon how the vigor of mixing declines with effective temperature. Hubeny & Burrows (2007) recently have argued that this will not be a concern at temperatures below about 500 K, however, because the vigor of mixing falls with effective temperature.

4.4.5.2 NH_3

Ammonia forms from N_2 by the reaction $N_2 + 3\,H_2 \Leftrightarrow 2\,NH_3$. $N_2 + 3H_2$ is favored at low pressures and high temperatures because of higher entropy. NH_3 is favored at low temperatures, but since molecular nitrogen is a strongly bound molecule, reactions involving this molecule typically have high reaction energies and proceed very slowly at low temperatures. Like CO, N_2 is favored at the higher temperatures found

[4] In stellar atmospheres, departures from thermochemical equilibrium can arise from interactions of atoms and molecules with the non-thermal radiation field (Hauschildt et al. 1997, Schweitzer 2000). In brown dwarf atmospheres this effect is negligible.

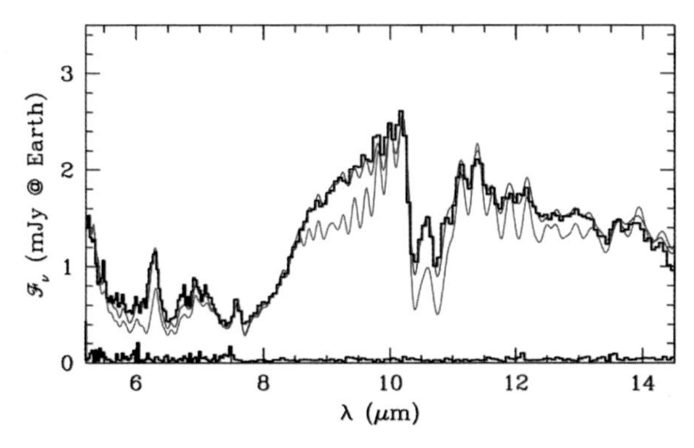

Fig. 4.7 Fits of the IRS spectrum of 2MASS J0415-0935 (Saumon et al. 2007) showing the difference between a model in chemical equilibrium and a model that includes vertical transport that drives the nitrogen and carbon chemistry out of equilibrium. The *red thin curve* is the best-fitting model in chemical equilibrium, and the *blue thin curve* is the best-fitting nonequilibrium model. The data and the noise spectrum are shown by the histograms (*black*). The uncertainty on the flux calibration of the IRS spectrum is ±5%. The model fluxes, which have not been normalized to the data, are shown at the resolving power of the IRS spectrum

deep in brown dwarfs atmospheres. Again, like CO, vigorous vertical transport can bring N_2 in the upper atmosphere faster than it can be converted to NH_3, resulting in an excess of N_2 compared to the values expected from chemical equilibrium. Figure 4.7 illustrates the effect of mixing (Saumon et al. 2007) on the 10 μm ammonia band in the spectra of the T8 dwarf 2MASS0415-0935 (Burgasser et al. 2002a). JWST will obtain higher resolution mid-infrared spectra than the *Spitzer* data analyzed by Saumon et al. (2007). Higher spectral resolution on more targets will allow more in depth studies of atmospheric mixing. Since the vertical profile of mixing also influences cloud particle sizes and optical depths (in L- and L to T transition dwarfs), mapping out the eddy diffusion coefficient (which parameterizes mixing) as a function of mass and effective temperature will help to shed light on cloud dynamics as well as atmospheric chemistry.

4.5 Opportunities for JWST

As we have highlighted in the above sections, there are many unsolved problems in the study of ultracool dwarfs. In this section we will briefly summarize some of the most promising avenues for JWST.

4.5.1 Characterizing Rosetta Stone Dwarfs

The characterization of most field brown dwarfs, particularly the L and early T dwarfs, is hampered by the dependency of model fits on the particulars of the cloud models used to generate model atmospheres and spectra for comparison to data.

Brown dwarfs of known mass, metallicity, and age are thus of particular importance as calibrators for the entire brown dwarf cooling sequence. This section discusses some opportunities for unearthing and deciphering 'Rosetta Stone' ultracool dwarfs with known or easily deduced masses and effective temperatures. Such objects could turn the page to much greater understanding of our library of known L and T dwarfs.

4.5.1.1 Color-Magnitude Diagram for Clusters to Low Masses

By providing cluster color-magnitude diagrams to very low masses (a Jupiter-mass or less) JWST will revolutionize our understanding of brown dwarf cooling in environments controlled for age and metallicity. Comparison of spectra of objects with known properties to models will finally provide insight into the variation in cloud properties with mass and effective temperature. In the Pleiades, Casewell et al. (2007) have detected objects with masses as low as 11 Jupiter masses (Fig. 4.6). In each cluster, a brown dwarf of a given mass will be found either earlier or later on its cooling track, depending on the cluster age. Since the evolutionary cooling of brown dwarfs is well understood, spectra of cluster objects with known masses will definitively connect spectral features with gravity for mid- to late L dwarfs. In the Pleiades such a project should be straightforward for JWST as NIRSPEC will be able to obtain $R \sim 1000$ JHK spectra of the known low-mass cluster members in as little as a few hours. A Jupiter-mass object will be about 6 magnitudes fainter in J band. Assuming JWST could survey a sufficiently large area to find candidates, it would define the brown dwarf cooling curve to a degree still not reached in the disk population. Model photometric predictions (Burrows et al. 2003) for such objects are shown in Fig. 4.8. The actual trajectory in color-magnitude space will ultimately depend upon the interplay of water clouds and atmospheric mixing with the emitted

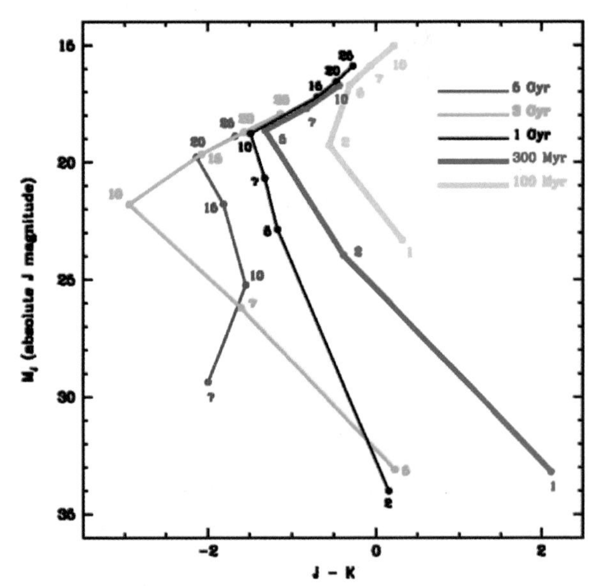

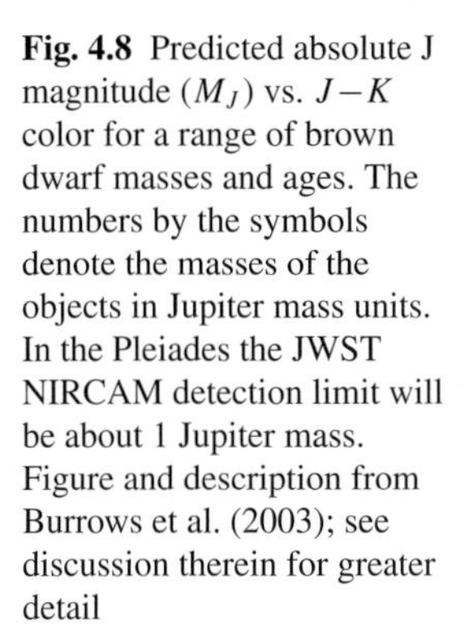

Fig. 4.8 Predicted absolute J magnitude (M_J) vs. $J-K$ color for a range of brown dwarf masses and ages. The numbers by the symbols denote the masses of the objects in Jupiter mass units. In the Pleiades the JWST NIRCAM detection limit will be about 1 Jupiter mass. Figure and description from Burrows et al. (2003); see discussion therein for greater detail

spectra. Comparison of models such as those in the figure with data will test our understanding through this as-yet unexplored range of parameter space.

4.5.1.2 Resolved Spectra of Close Binaries

Binary stars have long served as astronomical workhorses, helping to reveal important details of stellar astrophysics. Likewise binary brown dwarfs are also exceptionally useful. The orbits of close L- and T-dwarf binaries allow the total system mass to be determined as Bouy et al. (2004) have done for the binary 2MASSW J0746425+2000321. Assuming coevality and equal metallicity combined with the total system mass and resolved spectra of the individual dwarfs furthermore allows such systems to elucidate the interpretation of brown dwarf spectra. Many more tight binaries have been found by the combination of HST and ground based adaptive optics imaging (see summary in Bouy et al. 2008). Orbital periods for many of these systems appear to be less than twenty years, so dynamical masses will be available during the JWST mission lifetime. The combination of the ground-based astrometry and photometry and resolved NIRSPEC high R spectra of many of the individual objects, particularly for the tightest binaries, in these systems should provide important constraints on models of brown dwarf evolution, atmospheric structure, and emergent spectra.

4.5.1.3 Cloud Behavior from L to T to Y

Perhaps the greatest single observational result that could drive improvements in understanding of the L to T transition would be a direct measurement of surface gravity (or almost equivalently, mass) and effective temperature of late L and early T dwarfs. This would elucidate the dependence of the initiation of the L to T transition on mass and $T_{\rm eff}$. Constraining the turnoff absolute magnitude in a variety of clusters of known ages and resolving the spectra of close binaries that have measurable dynamical masses would highly constrain the nature of the L to T transition.

Likewise, as brown dwarfs further cool through the T sequence, a number of open issues remain. Although T dwarfs are generally modeled as being entirely cloud free, some models that include very thin cloud decks better reproduce the spectra and color of the T's. As brown dwarfs cool through about 500 K, thin water clouds should appear high in their atmospheres. With falling effective temperature these clouds are expected to thicken and begin to substantially alter the spectra of the Y dwarfs. It is entirely possible that, like the departure of clouds in the late L dwarfs, the arrival of clouds in the early Y dwarfs will produce unexpected and perhaps rapid color and spectral changes. Although we cannot yet identify what these changes might be, the same type of observations noted above will also be invaluable in constraining the Y dwarfs. Since the optical and near-IR flux of the Y dwarfs is expected to rapidly decrease (Burrows et al. 2003), the signatures of the water clouds will be best obtained in the mid-IR by JWST.

4.5.2 Spectra of Very Cool Objects

As of early 2008, the brown dwarf with the lowest estimated effective temperature is ULAS J0034-00 with $T_{\rm eff} \sim 650\,\rm K$ (Warren et al. 2007). A number of ongoing and future searches will certainly find cool objects in the solar neighborhood (e.g., UKIDSS, Pan-STARRS, and the WISE mission). In particular WISE will have sufficient sensitivity to detect a $T_{\rm eff} \sim 200\,\rm K$ brown dwarf ($2\,\rm M_J$ at 1 Gyr) at a distance of about 2 pc. Assuming such nearby targets are found, JWST will produce exquisite spectra that will be unobtainable from the ground. A survey with NIRSPEC in high resolution mode with the YJH and LM gratings would nicely sample the spectra of cool field dwarfs. For a 500K dwarf at 10 pc, we estimate an exposure of about a minute will provide a spectrum with S/N of about 100 in M band. A one hour exposure would be required for the same dwarf at 25 pc. Detection limits and model spectra are shown in Fig. 4.9.

High quality spectra of cool dwarfs will be important for a number of reasons. First, cold disk objects possess effective temperatures comparable to those of middle-aged to old extrasolar giant planets. The disk population of brown dwarfs will thus provide ground truth for the spectral features that such cold objects exhibit (for example NH_3 should appear in the near-IR (Saumon et al. 2000, Burrows et al. 2003, Leggett et al. 2007a)). These objects will also have water clouds. The experience with the challenge of modeling silicate and iron clouds in L dwarfs alluded to above implies that water clouds will be no more tractable. The cold disk population will thus provide a proving ground for exoplanet water cloud modeling, which is undoubtedly needed (see Marley et al. 2007 and references therein).

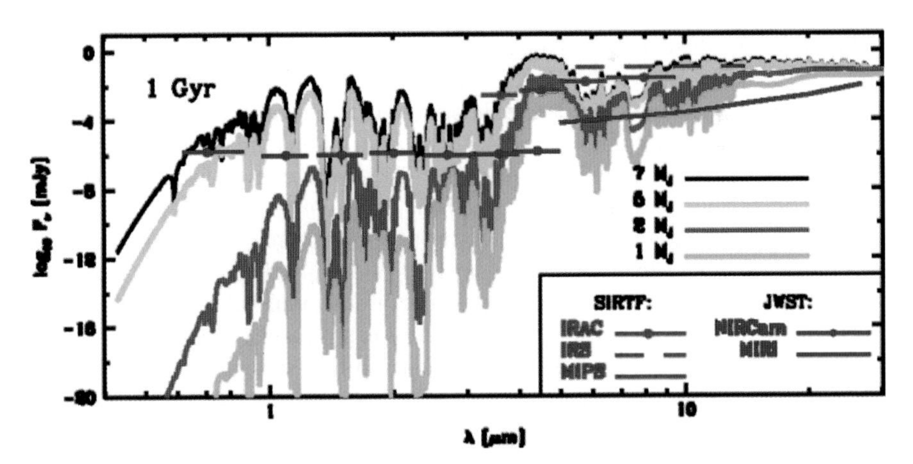

Fig. 4.9 Spectra (flux in millijanskys) vs. wavelength (in microns) for a range of brown dwarf masses at an age of 1 Gyr and a distance of 10 pc. Superposed are the approximate point-source sensitivities for instruments on *Spitzer* (*red*) and *JWST* (*blue*). The JWST/NIRCam sensitivities are $5\,\sigma$ and assume an exposure time of 5×10^4 sec. The JWST/MIRI sensitivity curve from 5.0 to 27 μm is $10\,\sigma$ and assumes an exposure time of 10^4 sec. Figure and description from Burrows et al. (2003); see discussion therein for greater detail

Atmospheric mixing, long recognized in Jupiter's atmosphere, is also important for brown dwarfs (see Section 4.4.5), yet the most diagnostic spectral region for this process (the CO band at 4.6 μm) lies in a blind spot for *Spitzer* spectroscopy and most groundbased observatories, but not JWST. Thus the nearby disk brown dwarfs will elucidate the extent to which *M* band flux of objects in this effective temperature range is impacted by excess atmospheric CO. Vertical mixing could be an important consideration for the direct detection of giant planets around nearby stars (Golimowski et al. 2004, Marley et al. 2007, Hinz et al. 2006). After the discovery of Gl 229B, Marley et al. (1996) suggested that a substantial 4 to 5 μm flux peak should be a universal feature of giant planets and brown dwarfs. This expectation, combined with a favorable planet/star flux ratio, has made the band a favorite for planet detection (Burrows 2005). However, groundbased and IRAC photometry suggests that cool dwarfs are fainter in this region—and the *L* band region is brighter—than predicted by equilibrium chemistry. Given these and other considerations, Leggett et al. (2007b) suggested that the comparative advantage of *M* band for searches for young, bright giant planets, such as the searches planned with the JWST coronagraph, might be somewhat less than currently expected (see also Marley et al. 2007).

Hubeny and Burrows (2007), however, recently predicted that the vigor of atmospheric mixing will decline with effective temperature. If this is indeed the case then *M* band will remain a fruitful hunting ground for extrasolar giant planet coronagraphic imaging. NIRCAM photometry of cold brown dwarfs will certainly illuminate this issue.

4.5.3 Mid-IR Spectra Beyond **Spitzer**

Examples of some of the best available *Spitzer* IRS mid-infrared spectra of L and T dwarfs are shown in Figs. 4.5 and 4.7. The spectral region between about 6 and 15 μm is important for a number of reasons. First, there are several strong molecular bands in this region, including water, methane, and ammonia. As recounted above, methane and ammonia are particularly sensitive to atmospheric mixing. The Si-O vibrational band, seen in the opacity of small silicate grains (Fig. 4.5), also may trace the arrival of silicate clouds. Finally, a number of other molecules, not yet detected in brown dwarf spectra, have absorption features in this range (Mainzer et al. 2007).

MIRI will produce much higher resolution and S/N spectra than the best data from IRS (see the sensitivity curve in Fig. 4.9). Higher quality spectra will allow for more robust detection of silicate features, perhaps including the sort of fine structure in the grain opacity seen in the lower panel of Fig. 4.5, as well as for fine detail on the molecular features (see model prediction in Burrows et al. (2003)). If silicates are indeed detected, high resolution spectra could in principle differentiate between the particular silicate species, including forsterite, enstatite, and even quartz (SiO_2, Helling & Woitke 2006), and whether the grains are in crystalline or amorphous form (Fig. 4.5).

4.6 Conclusions

The next decade holds the potential to substantially improve our fundamental understanding of ultracool dwarfs. Ground and space based surveys for very cool dwarfs, including UKIDSS, Pan-STARRS, and the WISE mission as well as deep surveys of young clusters will provide a host of ultracool dwarf targets for JWST. For cold nearby dwarfs JWST will provide unparalleled near- and especially mid-infrared spectra. These observations will constrain the water clouds expected to be present in objects with $T_{\rm eff} < 500\,{\rm K}$ and measure the degree of atmospheric mixing. Both types of observations are highly relevant to the ultimate direct detection and characterization of extrasolar giant planets by coronagraphy.

In young clusters (e.g. the Pleiades and younger) JWST will provide exceptional quality spectra of many known cluster members and will have the capability of imaging objects down to about one Jupiter mass and below. Such observations will constrain the evolutionary cooling tracks for dwarfs with lower gravities than most field objects and will tightly constrain the nature of the L to T transition by revealing its dependence on gravity.

Many other opportunities, including producing resolved spectra of tight binary dwarfs and searching for spectral signatures of condensates and low abundance gasses are also possible. Combined with the inevitable unexpected discoveries, there is no doubt that JWST will bring brown dwarf astrophysics into the same highly constrained realm as stellar astrophysics. Interpretting these expected datasets will undoubtedly require substantial improvements to atmosphere and evolution modeling, particularly cloud and chemical transport modeling of ultracool dwarf atmospheres.

Acknowledgments The authors thank M. Cushing, M. Liu, and D. Saumon, for helpful conversations on the future of brown dwarf science and A. Burrows, Ch. Helling, X. Tielens and K. Zahnle for thoughtful comments on the manuscript. We thank M. Cushing, M. Liu, and K. Lodders for preparing Figs. 4.1, 4.6a and 4.2, respectively and C. Nixon for kindly providing the Cassini CIRS spectrum of Jupiter for Fig. 4.1.

References

Ackerman, A. S., & Marley, M. S. 2001, ApJ, 556, 872
Barshay, S. S., & Lewis, J. S. 1978, Icarus, 33, 593
Bouy, H., et al. 2004, A&A, 423, 341
Bouy, H., et al. 2008, A&A, 481, 757
Burgasser, A. J., et al. 1999, ApJL, 522, L65
Burgasser, A. J., et al. 2002a, ApJ, 564, 421
Burgasser, A. J., Marley, M. S., Ackerman, A. S., Saumon, D., Lodders, K., Dahn, C. C., Harris, H. C., & Kirkpatrick, J. D. 2002b, ApJL, 571, L151
Burgasser, A. J., Geballe, T. R., Leggett, S. K., Kirkpatrick, J. D., & Golimowski, D. A. 2006, ApJ, 637, 1067
Burrows, A., et al. 1997, ApJ, 491, 856
Burrows, A., Marley, M. S., & Sharp, C. M. 2000, ApJ, 531, 438
Burrows, A., Hubbard, W. B., Lunine, J. I., & Liebert, J. 2001, Reviews of Modern Physics, 73, 719

Burrows, A., Sudarsky, D., & Lunine, J. I. 2003, ApJ, 596, 587
Burrows, A., 2005, Nature, 433, 261
Burrows, A., Sudarsky, D., & Hubeny, I. 2006, ApJ, 640, 1063
Casewell, S. L., Dobbie, P. D., Hodgkin, S. T., Moraux, E., Jameson, R. F., Hambly, N. C., Irwin, J., & Lodieu, N. 2007, MNRAS, 378, 1131
Cooper, C. S., Sudarsky, D., Milsom, J. A., Lunine, J. I., & Burrows, A. 2003, ApJ, 586, 1320
Cushing, M. C., et al. 2006, ApJ, 648, 614
Cushing, M. C., et al. 2008, ApJ, 678, 1372
Dahn, C. C., et al. 2002, AJ, 124, 1170
Fegley, B. J., & Lodders, K. 1994, Icarus, 110, 117
Fegley, B., Jr., & Prinn, R. G. 1985, ApJ, 299, 1067
Fegley, B. J., & Lodders, K. 1996, ApJL, 472, L37
Freedman, R. S., Marley, M. S., & Lodders, K. 2008, ApJSup, 174, 504
Geballe, T. R., et al. 2002, ApJ, 564, 466
Golimowski, D. A., et al. 2004, AJ, 127, 3516
Griffith, C. A., & Yelle, R. V. 1999, ApJL, 519, L85
Hanner, M. S., Lynch, D. K., & Russell, R. W. 1994, ApJ, 425, 274
Hauschildt, P. H., Allard, F., Alexander, D. R., & Baron, E. 1997, ApJ, 488, 428
Helling, Ch., Thi, W.-F., Woitke, P., & Fridlund, M. 2006, A&A, 451, L9
Helling, Ch., & Woitke, P. 2006, Astron. Astroph., 455, 325
Helling, Ch., et al. 2008a, MNRAS, in press, arXiv:0809.3657
Helling, Ch., Dehn, M., Woitke, P., & Hauschildt, P. H. 2008b, ApJL, 675, L105
Hinz, P. M., Heinze, A. N., Sivanandam, S., Miller, D. L., Kenworthy, M. A., Brusa, G., Freed, M., & Angel, J. R. P. 2006, ApJ, 653, 1486
Hubeny, I., & Burrows, A. 2007, ApJ, 669, 1248
Kirkpatrick, J. D., et al. 1999, ApJ, 519, 802
Kirkpatrick, J. D. 2005, Ann. Rev. Astron. & Astrophys., 43, 195
Kirkpatrick, J. D. 2008, 14th Cambridge Workshop on Cool Stars, Stellar Systems, and the Sun, 384, 85
Knapp, G. R., et al. 2004, AJ, 127, 3553
Kunde, V. G., et al. 2004, Science, 305, 1582
Leggett, S. K., et al. 2000, ApJL, 536, L35
Leggett, S. K., Marley, M. S., Freedman, R., Saumon, D., Liu, M. C., Geballe, T. R., Golimowski, D. A., & Stephens, D. C., 2007a, ApJ, 667, 537
Leggett, S. K., Saumon, D., Marley, M. S., Geballe, T. R., Golimowski, D. A., Stephens, D., & Fan, X. 2007b, ApJ, 655, 1079
Liu, M. C., Leggett, S. K., Golimowski, D. A., Chiu, K., Fan, X., Geballe, T. R., Schneider, D. P., & Brinkmann, J. 2006, ApJ, 647, 1393
Lodders, K., & Fegley, B. 2002, Icarus, 155, 393
Lodders, K., & Fegley, B., Jr. 2006, Astrophysics Update 2, 1
Lodieu, N., et al. 2007, MNRAS, 379, 1423
Luhman, K. L., et al. 2007, ApJ, 654, 570
Mainzer, A. K., et al. 2007, ApJ, 662, 1245
Marley, M. S., Saumon, D., Guillot, T., Freedman, R. S., Hubbard, W. B., Burrows, A., & Lunine, J. I. 1996, Science, 272, 1919
Marley, M. 2000, From Giant Planets to Cool Stars, 212, 152
Marley, M. S., Fortney, J., Seager, S., & Barman, T. 2007, Protostars and Planets V, 733
Metchev, S. & Hillenbrand, L. A. 2006, ApJ, 651, 1166
Mugrauer, M., Seifahrt, A., Neuhäuser, R., & Mazeh, T. 2006, MNRAS, 373, L31
Nakajima, T., Oppenheimer, B. R., Kulkarni, S. R., Golimowski, D. A., Matthews, K., & Durrance, S. T. 1995, Nature, 378, 463
Noll, K. S., Knacke, R. F., Geballe, T. R., & Tokunaga, A. T. 1988, ApJL, 324, 1210
Noll, K. S., Geballe, T. R., & Marley, M. S. 1997, ApJL, 489, L87
Noll, K. S., Geballe, T. R., Leggett, S. K., & Marley, M. S. 2000, ApJL, 541, L75

Oppenheimer, B. R., Kulkarni, S. R., Matthews, K., & van Kerkwijk, M. H. 1998, ApJ, 502, 932
Patten, B. M., et al. 2006, ApJ, 651, 502
Prinn, R. G., & Barshay, S. S. 1977, Science, 198, 1031
Roellig, T. L., et al. 2004, ApJS, 154, 418
Saumon, D., Geballe, T. R., Leggett, S. K., Marley, M. S., Freedman, R. S., Lodders, K., Fegley, B., Jr., & Sengupta, S. K. 2000, ApJ, 541, 374
Saumon, D., Marley, M. S., Cushing, M. C., Leggett, S. K., Roellig, T. L., Lodders, K., & Freedman, R. S. 2006, ApJ, 647, 552
Saumon, D., et al. 2007, ApJ, 656, 1136
Schweitzer, A., Hauschildt, P. H., & Baron, E. 2000, ApJ, 541, 1004
Sharp, C. M., & Burrows, A. 2007, ApJSup, 168, 140
Skrutskie, M. F., et al. 2006, AJ, 131, 1163
Strauss, M. A., et al. 1999, ApJL, 522, L61
Tinney, C. G., Burgasser, A. J., & Kirkpatrick, J. D. 2003, AJ, 126, 975
Toppani, A., Libourel, G., Robert, F., Ghanbaja, J., & Zimmermann, L. 2004, Lunar and Planetary Institute Conference Abstracts, 35, 1726
Tsuji, T. 2002, ApJ, 575, 264
Tsuji, T., & Nakajima, T. 2003, ApJL, 585, L151
Tsuji, T., Nakajima, T., & Yanagisawa, K. 2004, ApJ, 607, 511
Tsuji, T. 2005, ApJ, 621, 1033
Vrba, F. J., et al. 2004, AJ, 127, 2948
Warren, S. J., et al. 2007, MNRAS, 381, 1400
Woitke, P., & Helling, Ch. 2003, A & A, 399, 297
York, D. G., et al. 2000, AJ, 120, 1579

Xander Tielens, Yvonne Pendleton, and Jean Turner smile for the cameraman, Yvonne's husband

Chapter 5
Transiting Exoplanets with JWST

S. Seager, D. Deming and J.A. Valenti

Abstract The era of exoplanet characterization is upon us. For a subset of exoplanets—the transiting planets—physical properties can be measured, including mass, radius, and atmosphere characteristics. Indeed, measuring the atmospheres of a further subset of transiting planets, the hot Jupiters, is now routine with the *Spitzer Space Telescope*. The *James Webb Space Telescope* (*JWST*) will continue *Spitzer's* legacy with its large mirror size and precise thermal stability. *JWST* is poised for the significant achievement of identifying habitable planets around bright M through G stars—rocky planets lacking extensive gas envelopes, with water vapor and signs of chemical disequilibrium in their atmospheres. Favorable transiting planet systems, are, however, anticipated to be rare and their atmosphere observations will require tens to hundreds of hours of *JWST* time per planet. We review what is known about the physical characteristics of transiting planets, summarize lessons learned from *Spitzer* high-contrast exoplanet measurements, and give several examples of potential *JWST* observations.

5.1 Introduction

The existence of exoplanets is firmly established with over 300 known to orbit nearby, sun-like stars. Figure 5.1 shows the known exoplanets as of July 2008 with symbols indicating their discovery techniques (http://exoplanet.eu/). The majority of the known exoplanets have been discovered by the Doppler technique which measures the star's line-of-sight motion as the star orbits the planet-star common center of mass (Butler et al. 2006; Udry et al. 2007). While most planets discovered with the Doppler technique are giant planets, the new frontier is discovery of super Earths (loosely defined as planets with masses between 1 and $10\,M_\oplus$). About a dozen radial velocity planets with $M < 10\,M_\oplus$ and another dozen with $10\,M_\oplus < M < 30\,M_\oplus$ have been reported. The transit technique finds planets by

S. Seager (✉)
Department of Earth, Atmospheric, and Planetary Sciences, Dept. of Physics, Massachusetts Institute of Technology, 77 Massachusetts Ave., 54-1626, Cambridge, MA, 02139, USA
e-mail: seager@mit.edu

H.A. Thronson et al. (eds.), *Astrophysics in the Next Decade,* Astrophysics and Space Science Proceedings, DOI 10.1007/978-1-4020-9457-6_5,

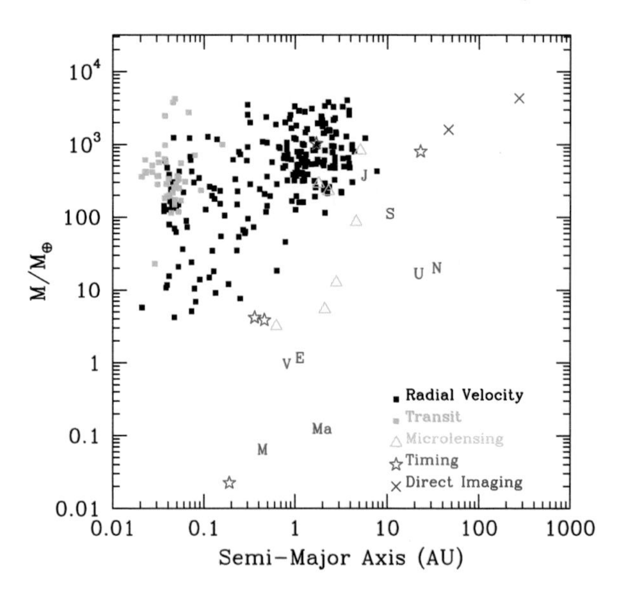

Fig. 5.1 Known planets as of July 2008. We have defined planet to have a maximum mass of a 13 Jupiter mass. The symbols indicate the discovery technique; see text for details. Data from (http://exoplanet.eu/)

monitoring thousands of stars, looking for the small drop in brightness of the parent star that is caused by a planet crossing in front of the star. At the time of writing this article, around 50 transiting planets are known. Due to selection effects, transiting planets found from ground-based searches are limited to small semi-major axes (Charbonneau et al. 2007). Gravitational microlensing has recently emerged as a powerful planet-finding technique, discovering 6 planets, two belonging to a scaled down version of our own solar system (Gaudi et al. 2008). Direct imaging is able to find young or massive planets with very large semi-major axes. The mass of directly imaged planets (e.g., (Chauvin et al. 2004) and references therein) is inferred from the measured flux based on evolution models, and is hence uncertain. The timing discovery method includes both pulsar planets (Wolszczan & Frail 1992) and planets orbiting stars with stable oscillation periods (Silvotti et al. 2007).

Many fascinating properties of exoplanets have been uncovered by the initial data set of hundreds of exoplanets. A glance at Fig. 5.1 shows one of the most surprising features: that exoplanets exist in an almost continuous range of mass and semi-major axis. Not shown in Fig. 5.1 are the equally wide range of eccentricities; several different theories for the origin of planet eccentricities have been proposed. Because there is not enough solid material close to the star in a protoplanetary disk, the giant planets are believed to have formed further out in the disk and migrated inwards. The migration stopping mechanisms, and even the details of planet migration are not fully understood. Out of the ~50 known transiting exoplanets, several have very large radii, and are too big for their mass and age according to planet evolution models (see Fig. 5.5). These "puffed-up" planets must have extra energy in their core that prevents cooling and contraction, but no satisfactory explanation yet exists.

The next step beyond discovery is to characterize the physical properties of exoplanets by measuring densities, atmospheric composition, and atmospheric temperatures.

There are two paths to exoplanet characterization. The first is direct imaging where the planet and star are spatially separated on the sky. Direct imaging has been successful for discovering hot or massive planetary candidates with large ($\sim$50–100 AU) orbital and projected spatial separation (Chauvin et al. 2004, 2005; Neuhäuser et al. 2005). Although *JWST* will incorporate several coronagraphic modes, neither the telescope nor the instruments were optimized for coronagraphy. A relatively large inner working angle and limited planet-star contrast restrict *JWST* to studying young or massive Jupiters with large semi-major axes. The case for *JWST* coronagraphic observations of exoplanets is presented in Gardner et al. (2006).

Solar-system aged small planets are not observable via direct imaging with current technology, even though an Earth at 10 pc is brighter than the faintest galaxies observed by the *Hubble Space Telescope* (*HST*). The major impediment to direct observations is instead the adjacent host star; the Sun is 10 million to 10 billion times brighter than Earth (for mid-infrared and visible wavelengths, respectively). No existing or planned telescope is capable of achieving this contrast ratio at 1 AU separations.

The second path to exoplanet characterization is via the transit technique. A subset of exoplanets cross in front of their stars as seen from Earth ("primary eclipse" or "transit"). Planets that cross in front of their star, also pass behind the star (secondary eclipse), provided that the transiting planet is on a circular orbit. The probability to transit is $\sim R_*/a$, where R_* is the stellar radius and a the semi-major axis. Transits are therefore most easily found for planets orbiting close to the star. Indeed, all but one of the $\sim$50 known transiting exoplanets have $a < 0.09$ AU (See (http://exoplanet.eu/) and references therein).

Observations of transiting planets exploit separation of photons in time, rather than in space (see Fig. 5.2). That is, observations are made in the combined light of the planet-star system. When the planet transits the star as seen from Earth the starlight gets dimmer by the planet-to-star area ratio. If the size of the star is known, the planet size can be derived. During the planet transit, some of the starlight passes through the optically thin part of the planet atmosphere, picking up spectral features from the planet. A planetary transmission spectrum can be obtained by dividing the spectrum of the star and planet during transit by the spectrum of the star alone (the latter taken before or after transit).

When the planet disappears behind the star, the total flux from the planet-star system drops. The drop is related to both relative sizes of the planet and star and their relative brightnesses (at a given wavelength). The flux spectrum of the planet can be derived by subtracting the flux spectrum of the star alone (during secondary eclipse) from the flux spectrum of both the star and planet (just before and after secondary eclipse). The planet's flux gives information on the planet composition and temperature gradient (at infrared wavelengths) or albedo (at visible wavelengths). Finally, non-transiting exoplanets can in principle also be observed in the combined planet-star flux as the planet goes through illumination phases as seen from Earth.

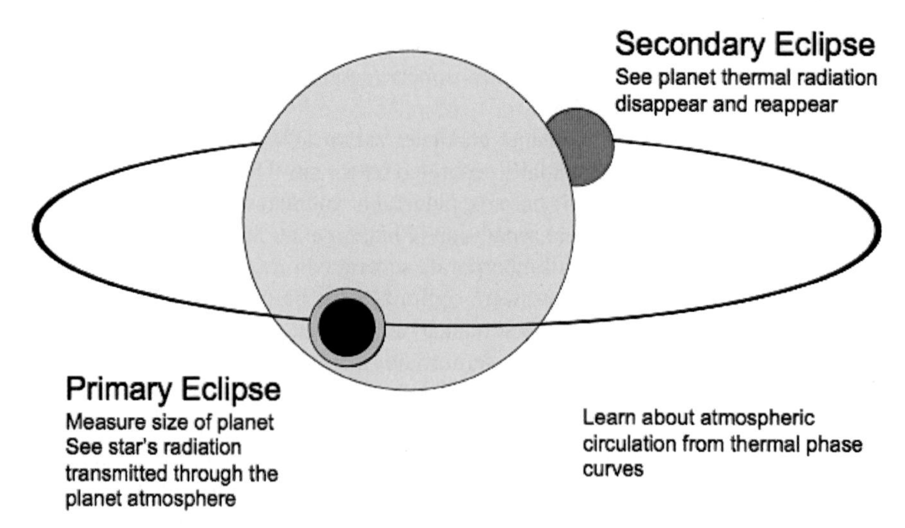

Fig. 5.2 Schematic of a planet transit. Not to scale

In the thermal infrared the phase observations provide information on energy redistribution of absorbed stellar radiation (see Section 5.2.3) and at visible wavelengths the phase observations give information on scattering particles (gas or clouds).

Primary and secondary eclipses enable high-contrast measurements because the precise on/off nature of the transit and secondary eclipse events provide an intrinsic calibration reference. This is one reason why the *Hubble Space Telescope* and the *Spitzer Space Telescope* have been so successful in measuring high-contrast transit signals that were not considered in their designs.

5.2 *Spitzer's* Legacy

5.2.1 Background

The *Spitzer* Space Telescope is a cryogenically cooled 85 cm diameter telescope launched into an Earth-trailing orbit in 2003. All three of *Spitzer*'s science instruments (IRAC (Fazio et al. 2004), IRS (Houck et al. 2004), and MIPS (Rieke et al. 2004)) have been used to study exoplanets. *Spitzer* has revolutionized the field of exoplanets by making measurements of hot Jupiter atmospheres routine. The *Spitzer* exoplanet studies are directly relevant to *JWST* because *JWST* will make similar measurements, but with much higher S/N, for much smaller exoplanets, or for planets with semi-major axes beyond 0.05 AU.

We describe one reason why *Spitzer* has been so successful in detecting photons from exoplanets during secondary eclipse. Figure 5.3 shows the relative fluxes for the Sun, Jupiter, Earth, Venus, and Mars, approximating each as a black body. The planets also reflect light from the Sun at visible wavelengths, giving them two flux

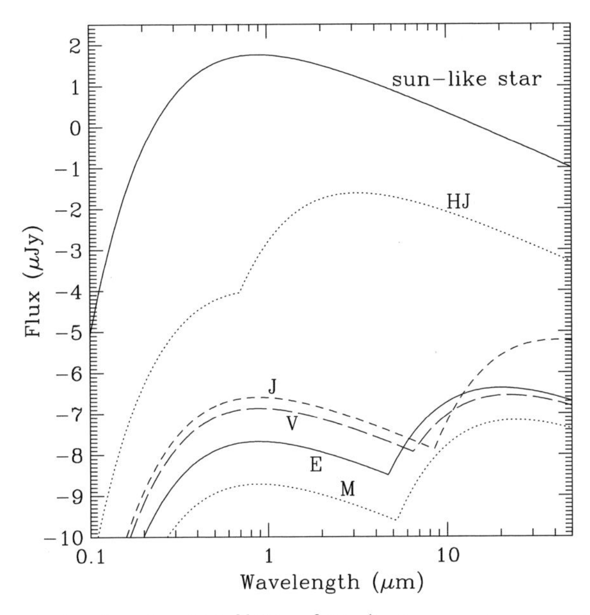

Fig. 5.3 Black body flux (in units of 10^{-26} W m^{-2} Hz^{-1}) of some solar system bodies as "seen" from 10 pc. The Sun is represented by a 5750 K black body. The planets Jupiter, Venus, Earth, and Mars are shown and are labeled with their first initial. A putative hot Jupiter is labeled with "HJ". The planets have two peaks in their spectra. The short-wavelength peak is due to sunlight scattered from the planet atmosphere and is computed using the planet's geometric albedo. The long-wavelength peak is from the planet's thermal emission and is estimated by a black body of the planet's effective temperature. Data from (Cox 2000). The Hot Jupiter albedo was assumed to be 0.05 and the equilibrium temperature to be 1600 K

peaks in their schematic spectrum. We see from Fig. 5.3 that at infrared wavelengths ($<$10 μm) the solar system planets are more than 7 orders of magnitude fainter than the Sun. A generic hot Jupiter, with assumed geometric albedo of 0.05, equilibrium temperature of 1600 K, and a radius of 1.2 R_J is also shown on the same figure. This representative hot Jupiter is less than 3 orders of magnitude fainter than the Sun at some wavelengths. Equally important is that the planet-to-star flux ratio is favorable where the star and planet flux are high, i.e. plenty of photons are available to reach the telescope. The ∼8 μm region is therefore a sweet spot for *Spitzer* observations of hot Jupiter exoplanets.

5.2.2 Exoplanet Radii

A precise planet radius together with planet mass enables a study of the planet's density and interior bulk composition. We have shown that infrared wavelengths are ideal for deriving a precise planet radius from the transit light curve, due to the miniscule amount of stellar limb darkening at infrared wavelengths (Richardson et al. 2006). In contrast, at visible wavelengths limb darkening affects the shape of the transit light curve. Because limb darkening is imperfectly known for stars other than the Sun, limb darkening must be solved for from planet transit light curves at visible wavelengths in order to fit the planet's radius. *Spitzer* measurements of the HD 209458 primary eclipse with MIPS at 24 μm yielded a planetary radius of $1.26 \pm 0.08\ R_J$ (Richardson et al. 2006). At shorter infrared wavelengths with more photons from the star and where limb darkening is still negligible, and for host stars of later spectral type than HD 209458, transit light curves will enable even more precise planet radii to be derived.

Transit observations at infrared wavelengths are especially useful for planets transiting M stars. M stars are faint at visible wavelengths with peak flux output at near-IR wavelengths. *Spitzer* is arguably the best existing telescope for determining precise radii of planets transiting M stars. The Neptune-mass planet GJ 436b (Butler et al. 2004; Gillon et al. 2007) was observed by *Spitzer* to have a Neptune-like radius of $4.33 \pm 0.18 R_\oplus$ (Deming et al. 2007; Gillon et al. 2007).

5.2.3 Exoplanet Atmosphere Summary

Several different exoplanets have published *Spitzer* secondary eclipse atmosphere measurements. These secondary eclipse measurements have detection significances ranging from 60σ (Knutson et al. 2007) down to 5σ. Rather than describe each planet individually, we present a summary based on an important question related to atmospheric circulation.

Hot Jupiters are expected to be tidally locked to their parent stars—presenting the same face to the star at all times. This causes a permanent day and night side. A long standing question has been about the temperature difference from the day to night side. Are the hot Jupiters scorchingly hot on the day side and exceedingly cold on the night side? Or, does atmospheric circulation efficiently redistribute the absorbed stellar radiation from the day side to the night side?

Surprisingly, *Spitzer* has found that both scenarios are possible. *Spitzer* has measured the flux of the planet and star system as a function of orbital phase for several hot Jupiter systems (Harrington et al. 2006; Cowan et al. 2007; Knutson et al. 2007). Assuming that the star is constant in flux, the resulting brightness change is due to the planet alone. The HD 189733 star and planet shows some variation at 8 μm during the 30 hour continuous observation of half an orbital phase (Knutson et al. 2007). This variation corresponds to about 20% variation in planet temperature (from a brightness temperature of 1212 to 973 K). In contrast, the non-transiting exoplanet Ups And shows a marked day-night contrast suggesting that the day and night side temperatures differ by over 1000 K (Harrington et al. 2006).

Once the stellar radiation is absorbed on the planet's day side, there is a competition between reradiation and advection. If the radiation is absorbed high in the atmosphere, the reradiation timescale is short and reradiation dominates over advection. In this case the absorbed stellar radiation is reradiated before it has a chance to be advected around the planet, resulting in a very hot planet day side and a correspondingly very cold night side. If the radiation penetrates deep into the planet atmosphere where it is finally absorbed, the advective timescale dominates and the absorbed stellar radiation is efficiently circulated around the planet. This case would generate a planet with a small temperature variation around the planet. See also (Seager et al. 2005; Harrington et al. 2007; Burrows et al. 2008; Fortney et al. 2007). See (Showman et al. 2007) and references therein for a review discussion of atmospheric circulation models.

In Fig. 5.4, we plot measured brightness temperatures of seven hot exoplanets together with two equilibrium temperature (T_{eq}) curves. One of the T_{eq} curves is for a planet with evenly redistributed absorbed stellar radiation ($f = 1/4$ below), corresponding to a planet with little temperature difference between the day and night sides. The other T_{eq} curve is for a planet with instantaneous reradiation of absorbed stellar radiation ($f = 2/3$ below), corresponding to a planet with a strong day-night temperature difference. The cooler of the hot exoplanets in Fig. 5.4 lie along the evenly redistributed energy curve, while the hotter exoplanets lie nearer to the instantaneous reradiation T_{eq} curve.

Physically, T_{eq} is the effective temperature attained by an isothermal planet after it has reached complete equilibrium with the radiation from its parent star. T_{eq} is described by

$$T_{eq} = T_* \left(\frac{R_*}{a}\right)^{1/2} \left[f(1 - A_{\mathrm{B}})\right]^{1/4}, \tag{5.1}$$

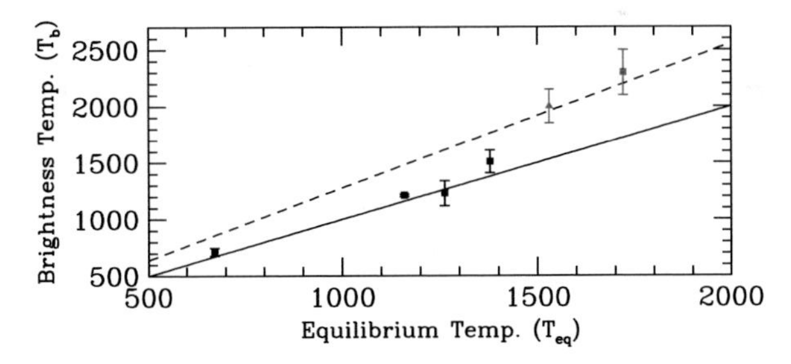

Fig. 5.4 Brightness temperature (8 μm) as a function of the day side equilibrium temperature for six hot Jupiters. Brightness temperature is the measured flux converted to a temperature. The day side equilibrium temperature T_{eq}, is defined in equation (5.1) and the accompanying text. The parameter f is used to approximate atmospheric circulation: in this formulation of equation (5.1), $f = 1/4$ corresponds to a uniform temperature around the planet (*solid line*), whereas $f = 2/3$ corresponds to instantaneous reradiation on the planet's dayside hemisphere (*dashed line*). The cooler planets lie near the uniform temperature line whereas the hotter planets lie near the hot day side line. From left to right, the brightness temperatures are GJ 436b (Deming et al. 2007), HD 189733 b (Knutson et al. 2007), TrES-1 (Charbonneau et al. 2005), HD 209458b (Knutson et al. 2008), Ups And (Harrington et al. 2006), HD 149026 (Harrington et al. 2007). (Note that the Ups And day side temperature is estimated from its thermal phase curve)

where T_* and R_* are the effective temperature and the radius of the star, a is the planet semi-major axis, and f and A_B are the re-radiation factor and the Bond albedo of the planet.

Here we explain how hot Jupiters can exist both with and without large day-night temperature variations. (See also, Hubeny et al. 2003; Harrington et al. 2007; Burrows et al. 2008; Fortney et al. 2007). Hot planets such as Ups And are on one side of a temperature-driven chemical composition boundary, while cooler planets such as HD 209458b are on the cooler side. Specifically, if the hot Jupiter planet atmosphere is relatively cool, TiO is locked into solid particles that have little absorbing power in the atmosphere. In the hotter atmosphere, TiO is a "deadly" gas that absorbs so strongly it puts the planet in the reradiation regime leading to a large day-night contrast. At the temperature of these hot day side exoplanets, some elements will be in atomic (instead of molecular) form and atomic line opacities may also play a significant absorbing role.

What evidence do we have for the temperature-induced two-atmosphere hypothesis? Cool stars (M stars) have visible-wavelength spectra that are dominated—and indeed dramatically suppressed—by TiO gas. Hot brown dwarfs also have spectra with TiO absorption features wheras cooler brown dwarfs do not, implying that, for cooler brown dwarfs, Ti is sequestered in solid particles. Temperature and pressures in hot Jupiter atmospheres are similar to brown dwarfs (although the temperature gradient is different) so that we expect a similar temperature-induced chemical composition.

Other notable exoplanet atmosphere discoveries by *Spitzer* come from spectrophotometry and include discovery of a temperature inversion on the day side of HD 209458b (Burrows et al. 2007; Knutson et al. 2008) and a tentative detection of water vapor in transmission spectra during primary transit (Tinetti et al. 2007; Ehrenreich et al. 2007).

This interesting "two types of hot Jupiter atmospheres" hypothesis shows just how complex hot Jupiters are. The results also imply that 3D coupled radiative transfer-atmospheric circulation models are needed to fully understand hot Jupiters. Next generation data with *JWST* (Section 5.4) in terms of high SNR low-resolution spectra as a function of orbital phase will lead to a deeper understanding of hot Jupiter atmospheres.

5.2.4 Lessons Learned from Spitzer

Spitzer and its instruments were not designed for high contrast observations. Here we describe some of the instrumental effects that become important at the part-per-thousand level and consequently affect exoplanet observations. These may be useful to consider when planning *JWST* exoplanet transit observations.

The most notable instrumental effect is the "ramp": a gradual detector-induced rise of up to 10% in the signal measured in individual pixels over time. This rise is illumination-dependent; pixels with high levels of illumination converge to a constant value within the first two hours of observations and lower-flux pixels increase linearly over time. Knutson et al. (2007) have attributed this to chargetrapping.

The ramp is present at the long wavelength detectors (8 and 16 μm and possibly also at 5 μm). The charge trapping is likely caused by the ionized impurities in the arsenic-doped silicon detector. The first electrons that are released by photons get trapped by the ions and therefore they are not immediately read out.

The ramp can be removed from a data set by a fitting a linear plus logarithmic function. A method to avoid the ramp is to "preflash" the detector before an observation, by observing a bright star. We note that at 24 μm no ramp effect is observed (Deming et al. 2005); this may be because the detector is always illuminated by the zodiacal background. The ramp effect may be important for *JWST* observations because similar detector materials are being used on MIRI. We recommend preflashing before transit observations.

A second instrumental effect that is significant for exoplanet transit observations is the IRAC intrapixel sensitivity variation for the short-wavelength channels (3.5 and 4.5 μm) (Morales-Calderón et al. 2006). This intrapixel sensitivity variation is a property of the detector, is spatially asymmetric, and does not change with time. One possible explanation of this detector effect is that there are physical gaps between the pixels which are unresponsive to light. When an image is centered on a pixel the highest sensitivity arises. A second possibility is related to the bump-bond contact between the detector and the underlying multiplexer. Each pixel makes electrical contact with the multiplexer at pixel center. In this scenario, the electrons generated close to the contact point may be collected more efficiently than electrons generated at pixel edges. The IRAC intrapixel sensitivity variation can be corrected for by a natural mapping of the pixels, by exploiting the telescope pointing jitter (Deming and Seager submitted).

A third significant instrumental effect found from exoplanet observations is a telescope pointing oscillation with a period of roughly one hour. This pointing oscillation affects IRS slit spectra. As the star becomes uncentered and recentered on the slit the number of photons going through the slit changes. This change is wavelength-dependent, because of the wavelength-dependent PSF, and therefore can generate erroneous features in an exoplanetary spectrum. The cause of the pointing oscillation is still a matter of debate.

A fourth effect is a longer-term telescope drift in telescope pointing. This telescope drift is less certain than the well-documented 1 hour oscillation. This affects the IRS spectral flux in terms of the broad distribution of energy, because the slit width is comparable to the diffraction width of the telescope PSF. This is also a wavelength-dependent effect.

Despite not being designed for high-contrast observations, *Spitzer* has revolutionized the study of exoplanet atmospheres by routinely detecting secondary eclipses of hot Jupiters and measuring their brightness temperatures at different wavelengths. Because of *JWST*'s larger mirror diameter and higher spectral resolution, and by taking care to understand and remove instrumental effects, transit observations with *JWST* hold even more promise.

5.3 Transiting Planet Radii with JWST

JWST will continue to study the physical characteristics of transiting exoplanets in the tradition of *Spitzer*. See Chapter 1 of this volume for the *JWST* telescope, instrument, and performance. *JWST* has 40 times the collecting area of *Spitzer*, enabling studies of smaller exoplanets—pushing down to potentially habitable exoplanets.

The overall goal of planet transit observations is to combine the radius with the planet mass to infer an interior bulk composition. Figure 5.5 shows transiting exoplanets on a mass-radius diagram with curves for planets of homogeneous composition. For example, we would like to know if planets in the mass range 5 to 20 $M_\oplus$ have significant gas envelopes like Neptune (∼10% by mass), or instead consist

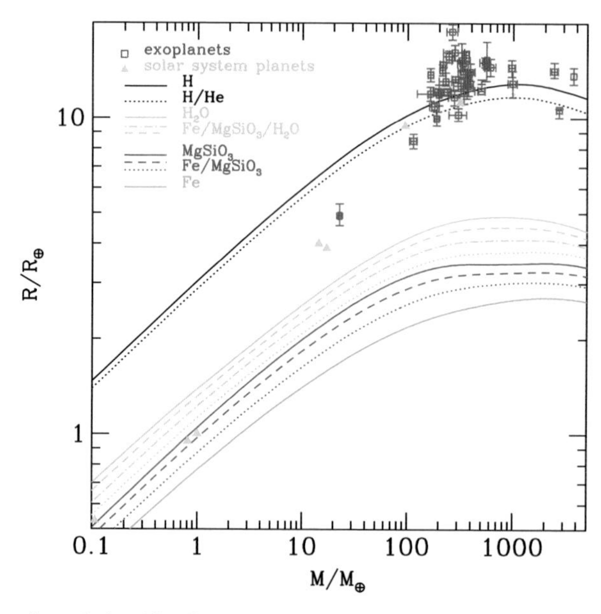

Fig. 5.5 Mass-radius relationships for solid planets. The solid lines are models of homogeneous planets. From top to bottom the homogeneous planets are: hydrogen (*cyan solid line*); a hydrogen-helium mixture with 25% helium by mass (*cyan dotted line*); water ice (*blue solid line*); silicate ($MgSiO_3$ perovskite; *red solid line*); and iron (Fe (ε); *green solid line*). The *non-solid lines* are models of differentiated planets. The *red dashed line* is for silicate planets with 32.5% by mass iron cores and 67.5% silicate mantles (similar to Earth) and the red dotted line is for silicate planets with 70% by mass iron core and 30% silicate mantles (similar to Mercury). The blue dashed line is for water planets with 75% water ice, a 22% silicate shell and a 3% iron core; the blue dot-dashed line is for water planets with 45% water ice, a 48.5% silicate shell and a 6.5% iron core (similar to Ganymede); the blue dotted line is for water planets with 25% water ice, a 52.5% silicate shell and a 22.5% iron core. The blue triangles are solar system planets: from left to right Mars, Venus, Earth, Uranus, Neptune, Saturn, and Jupiter. The magenta squares denote the transiting exoplanets, including HD 149026b at 8.14 $R_\oplus$ and GJ 436b at 3.95 $R_\oplus$. Note that electron degeneracy pressure becomes important at high mass, causing the planet radius to become constant and even decrease for increasing mass. From (Seager et al. 2007)

almost entirely of rock/iron. Owing to high temperatures at a deeply submerged surface, the former are not habitable, while the latter are.

To illustrate why precise radii are needed to constrain a planet's mass we show the range of interior compositions possible for a 10 $M_\oplus$, 2 $R_\oplus$ planet on a ternary diagram (Fig. 5.6). For an explanation of ternary diagrams in this context see (Valencia et al. 2007) and (Zeng and Seager in press). There is a degeneracy in interior composition for a solid exoplanet made of the three typical planetary materials: an iron core, silicate mantle, and water outer layer. This is because of the very different densities of the three components. For example, a planet of a given mass and composition could have the same radius if some of the silicate were exchanged for a combination of water and iron. By showing contour curves of growing observational uncertainties, Figure 5.6 emphasizes how a precise radius reduces the interior composition uncertainty.

JWST should be able to measure precise transit light curves for a wide range of star brightnesses. With its spectral dispersion and high cadence observing, NIRSpec will be capable of high-precision spectrophotometry on bright stars. NIRSpec data

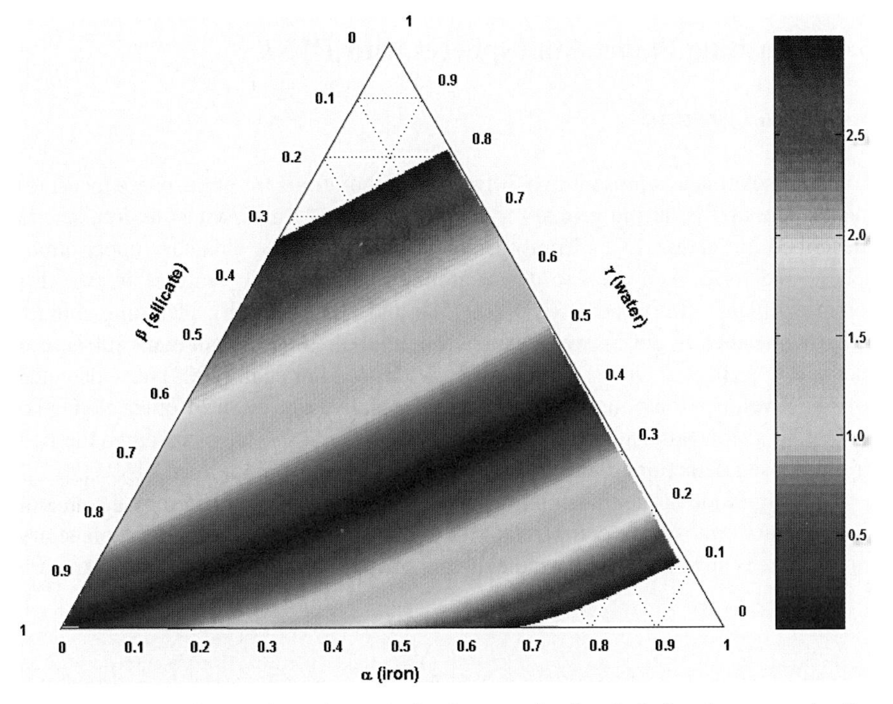

Fig. 5.6 A ternary diagram for a planet of a fixed mass and radius, including the mass and radius uncertainties. This example is for a planet with $M_p = 10 \pm 0.5 M_\oplus$ and $R_p = 2 \pm 0.1 M_\oplus$, showing the 1-, 2- and 3-σ uncertainty curves as indicated by the color bar. Notice that considering the 3-σ uncertainties almost the entire ternary diagram is covered—in other words there is little constraint on the planet internal composition. See (Zeng and Seager in press) for a discussion of the direction and spacing of the curves, as well as for other examples

can then be used in the same way that the *HST* STIS spectra data for HD209458 was rebinned into "photometry" (Brown et al. 2001). For fainter M stars, NIRCam should be sufficient. For all stars, the near infrared is ideal because of negligible stellar limb darkening, removing one of the uncertainties in deriving an accurate planet radius from a planet transit light curve.

By scaling from the (Brown et al. 2001) *HST* data, we provide two interesting examples of *JWST* precision radii (Seager and Lunine 2007; Beichman et al. 2007) with NIRSpec at 0.7 μm. For the first example we consider *Kepler* Earth-size planet candidates orbiting Sun-like stars in 1 AU orbits. *Kepler* stars are about 300 pc distant and Earth-analogs have a transit duration of about 8 hours. With high-cadence observing, *JWST* will be able to obtain a 35-σ transit detection for *Kepler* Earth-analog planet candidates. This *JWST* confirmation would be very significant, because *Kepler*'s SNR detection is 7 over 4 binned transits (Basri et al. 2005).

A second example is that of an Earth-sized moon orbiting the transiting giant planet HD 209458b. At 47 pc and with a 3 hour transit time (and 6 hour observation), *JWST* will also be capable of a moon transit detection at 35-σ SNR.

5.4 Transiting Planet Atmospheres with *JWST*

5.4.1 Background

Transiting exoplanets present two different configurations for atmosphere measurements. The first is during primary transit and is called transit transmission spectra (described in Section 15.1). Transmission spectra probe the planetary upper atmosphere and have been used to detect atomic and molecular features in two different exoplanet atmospheres (HD 209458b and HD 189733b), including sodium (Charbonneau et al. 2002), water vapor (Tinetti et al. 2007; Swain et al. 2008), and methane (Swain et al. 2008). Additionally, Vidal-Madjar et al. (2003) have detected a large envelope of atomic hydrogen (and a tentative detection of other elements) indicating a slow atmospheric escape. Redfield et al. (2008) have presented the first ground-based detection of an exoplanet atmosphere via sodium in HD 189733b.

The magnitude of the transmission spectra signal can be estimated by the area of the planetary atmosphere compared to the area of the star. The area of the planetary atmosphere is an annulus with a radial height of about $5 \times H$. Here H is the planetary scale height

$$H = \frac{kT}{\mu_m g}, \tag{5.2}$$

where k is Boltzmann's constant, T is temperature, μ_m is the mean molecular mass, and g is the planet's surface gravity. The magnitude of the planet transmission spectra signature is approximately

$$5 \times \frac{2R_p H}{R_*^2}, \tag{5.3}$$

where R_p is the planet's radius and R_* is the star's radius. A hot Jupiter's transmission spectra signature is approximately 10^{-4}. We further note that, from the planet's equilibrium effective temperature (5.1) $T \sim 1/\sqrt{a}$ so that

$$Transmission \sim \frac{1}{\sqrt{a}}. \tag{5.4}$$

We emphasize a very critical difference between planets with hydrogen-rich atmospheres (including both the upper layers of giant planet envelopes and thinner atmospheres of super Earths) and terrestrial planets (including Earths or super Earths) with relatively thin N_2 or CO_2 atmospheres. The factor μ_m is 2 for an H_2 atmosphere, but 44 for a CO_2 atmosphere! Hence, the difference in transmission spectra between hydrogen-rich atmospheres and terrestrial-like planet atmospheres is a factor of 20 (with T and g being equal; see (Miller-Ricci submitted) for further discussion). This implies that, while smaller space telescopes can study hydrogen-rich exoplanet atmospheres, *JWST*'s 6.5 m effective mirror diameter is needed to study CO_2- or N_2-dominated atmospheres similar to terrestrial planet atmospheres in our solar system.

The second configuration available for transiting atmosphere studies with *JWST* is secondary eclipse. Observations during secondary eclipse measure the planet's thermal emission. This contains information about the planet's temperature and temperature gradient. Spectral features can also be detected with the planet's thermal emission flux. If absorption lines are detected, the planet has a temperature that is decreasing towards the top of the atmosphere. If emission lines are detected, the planet has a temperature that is increasing towards the top of the atmosphere.

Estimating the magnitude of the planet's thermal emission (in the form of a planet-to-star flux ratio) is not easy because planet model atmospheres are usually needed. That said, we can bracket an estimate with two extremes. One extreme is the case where the thermal emission spectra could be observed over a broad infrared wavelength range to estimate the "bolometric" planet flux. In this case, we can estimate the planet-star flux ratios by a ratio of black bodies, and considering the Stefan-Boltzmann law $F = \sigma_R T^4$,

$$\text{Emission} \sim \frac{R_p^2 T_p^4}{R_*^2 T_*^4}, \tag{5.5}$$

where we have written $T_p = T_{eq}$. Again using the scaling relation $T_p \sim 1/\sqrt{a}$, we find

$$\text{Emission} \sim \frac{1}{a^2}. \tag{5.6}$$

As a separate extreme to estimate the thermal emission planet-star contrast ratio, we can take the Rayleigh-Jeans tail of the black body spectrum $h\nu \ll kT$ to get

$$\text{Emission} = \frac{R_p^2 T_p}{R_*^2 T_*}, \tag{5.7}$$

and again using the $T_{eq} \sim 1/\sqrt{a}$ scaling,

$$\text{Emission} \sim \frac{1}{\sqrt{a}}. \tag{5.8}$$

More reasonably, we can assume that at the peak of the planet's output, which is neither represented by the bolometric flux nor is it in the Rayleigh-Jeans tail (see Figure 5.3), the dependence with a falls between that of $\sim 1/a^2$ (5.6) and $\sim 1/\sqrt{a}$ (5.8).

We have gone through these estimates to make a single main point: a comparison between the semi-major axis (a) dependence of transmission and emission spectra. Emission spectra have a stronger signal than transmission spectra for planets orbiting close to their parent stars (and indeed emission spectra are only possible for planets close to their stars). While transmission spectra are weaker than emission spectra for planets close to their stars, they are still attainable for planets orbiting far from their stars.

5.4.2 Giant Planet Atmospheres

The *JWST* thermal IR detection capability can be explored by scaling the *Spitzer* results (this discussion is from Beichman et al. (2007)). The 5–8 μm range is ideal for solar-type stars because the planet-star contrast is high and the exo-zodiacal background is low. For an estimate we can take the TrES-1 5σ detection at 4.5 μm, taking into account that *JWST* has 40 times the collecting area of *Spitzer* and assuming that the overall efficiency of *JWST* is almost 2x lower, giving an effective collecting area improvement of ~25 times. *JWST* will therefore be able to detect hot Jupiter thermal emission at an SNR of 25 around stars at TrES-1's distance (~150 pc; a distance that includes most stars from shallow ground-based transit surveys). Similarly, *JWST* can detect a hot planet 5 times smaller than TrES-2, or down to 2 Earth radii, for the same set of stars, assuming that instrument systematics are not a limiting factor. Scaling with distance, *JWST* can detect hot Jupiters around stars 5 times more distant than TrES-1 to SNR of 5, which includes all of the *Kepler* and *COROT* target stars. Beyond photometry, *JWST* can obtain thermal emission spectra (albeit at a lower SNR than for photometry of the same planet). Rebinning the R = 3,000 NIRSpec data to low-resolution spectra will enable detection of H_2O, CO, CH_4, and CO_2.

For hot Jupiter transmission spectra, we turn to simulations of NIRSpec transmission spectra (Valenti et al. 2005). NIRSpec will have three "high-resolution"

(2200 $< R < 4400$) gratings (G140H, G235H, G395H) that span the wavelength range 1–5 μm. Observations of planet host stars will be made with the largest fixed aperture (4 × 0.4), rather than microshutters. A 2048 × 64 subarray on each of the two detectors will be nondestructively read "up-the-ramp" every 0.3 s for 2.4 s (G140H, G235H) or 3.6 s (G395H). The detector is then reset to avoid charge saturation (>60,000 e^-), and the process is repeated thousands of times to build up 2 hours of total exposure time. The total spacecraft time required to achieve this exposure time depends on the idle time between individual subarray exposures. *JWST* has the potential to characterize the atmospheres of dozens of transiting planets, if overheads and calibration errors can be controlled.

One of us (Valenti et al. 2005) has simulated NIRSpec observations with *JWST*. They estimated NIRSpec performance based on observatory requirements and some lab data. For reference, *JWST* has 25 m^2 clear aperture and a 33% peak efficiency for the NIRSpec/G140H mode considered here. The simulated observations include those in and out of transit, based on NextGen models (Hauschildt et al. 1999) and planetary absorption spectra from (Brown 2001). Simulated noise takes into account the details of how the detector will be read and how spectra will be extracted. (This software is available upon request from valenti@stsci.edu.) Valenti et al. (2005) do not attempt to estimate the impact of systematic errors that will undoubtedly affect actual observations.

Figure 5.7 shows a simulation of a $K = 12$, $V = 13.4$ star with a 6 hour observation using NIRSpec/G140H, centered on a 2h planetary transit. Both the theoretical spectra and the simulated observations are shown. In the simulated observation, the water vapor absorption features are obvious and their detection significance is high. This is in contrast to the 3 to 4 SNR transmission spectra detections measured with *HST*, *Spitzer*, and from the ground.

A magnitude of $V = 13.4$ encompasses most stars surveyed in the shallow ground-based transit surveys, which may discover well over 100 hot Jupiters by the time of *JWST*'s launch. *Kepler*'s star magnitude range is $V = 9$ to 16 (most transits will be found around stars $V = 14$ to 16), and therefore the simulated spectrum in Figure 5.7 applies to *Kepler*'s hot Jupiters as well (although with a slightly lower SNR for the fainter-end stars). By binning NIRspec data in wavelength, Jupiter

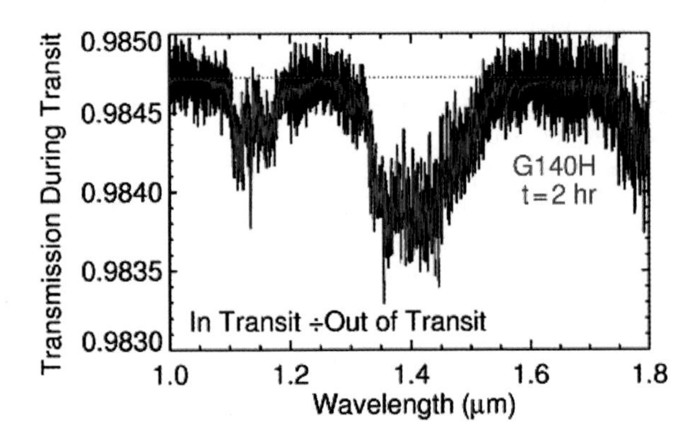

Fig. 5.7 Simulated NIRSpec measurements of transmission spectra of a hot Jupiter exoplanet. The spectrum is the ratio of the simulated spectrum in and out of transit. The blue spectrum shows the Brown et al. (2001) model used to generate the simulated observations (*black curve*). The dominant absorption features are water vapor

transiting planets orbiting out to 1 AU are accessible for NIRSpec transmission spectra. Such observations may help resolve the puzzling question on the origin of the "puffed-up" hot Jupiters that are too large for their mass and age (upper right corner in Fig. 5.5).

Other outstanding questions for hot Jupiters that *JWST* can address include the atmospheric circulation. Tidally-locked to their parent stars, hot Jupiters have a permanent day and night side. By studying the thermal emission spectra as a function of phase we can get a handle on how the temperature of different layers of the planet is changing. Intriguing are the eccentric hot Jupiters whose atmospheres are intensely heated for a brief period of time. Spectra as a function of phase will help us to determine the radiation time constant, a fundamental factor in understanding atmospheric circulation.

By binning NIRSpec data in wavelength, transiting planets smaller than Jupiter can be studied as well. Between Jupiters and super Earths, we expect many hot transiting Neptune-sized exoplanets to be known and accessible to study by *JWST*.

5.4.3 Terrestrial Planet Atmospheres

One of the most interesting exoplanet questions *JWST* can address is whether a planet is habitable. By habitable, we mean, in the conventional sense, one with surface liquid water. Atmospheric water vapor is a good indication of surface liquid water. On Earth, O_2 is considered the most significant biosignature, given that it is a highly reactive gas with a very short lifetime, and only produced in large quantities by biological processes. For signs of life, one would ideally want to observe signatures of molecules that are highly out of redox equilibrium (such as methane and oxygen) (Lederberg 1965; Lovelock 1965), but in reality it will be difficult enough to observe any molecular signature robustly from an Earth-temperature, near Earth-size exoplanet. Finally, CO_2, while not a biosignature, indicates an terrestrial-planet atmosphere. See (Des Marais et al. 2002) for a discussion of Earth's biosignatures.

We choose six examples to illustrate *JWST*'s capabilities for studying terrestrial exoplanets orbiting in their star's habitable zone.

The first is for a large Earth-like exoplanet. In Seager and Lunine (2007), Gilliland considered a 1.5 Earth-radius planet orbiting a Sun-like star at 1 AU at 20 pc distance. *JWST* could achieve a 500-σ detection with 0.25 sec exposures, 0.5 sec down time, and 3×10^8 photons per integration (i.e., 10^5 pixels for 50k electrons in the brightest pixel). This kind of detection can distinguish between Earth- and Venus-like atmospheres. The observations required for such a 4-σ discrimination in this case would span 30 hours centered on a 10-hour transit. This measurement relies on a *JWST* capability of efficiently recording high photon fluxes. More significantly, this hypothetical planet-star system is optimistic; transiting Earths around $V = 6$ dwarfs will be rare and currently no transit survey is capable of finding them.

The second and third examples are for Earth-size planets orbiting M stars. There is new excitement in finding transiting planets orbiting in the habitable zones of M stars. Recall that the habitable zone is defined as the location around a star where a planet's surface temperature will permit liquid water. Owing to their low luminosity, M stars have habitable zones much closer (3 to 40 day period orbits) than Sun-like stars (1 year orbits). In comparison to the above example of a 1.5 Earth-radius planet in a 1 AU orbit transiting a Sun-like star: the probability to transit is high (transit probability is R_*/a); transits are deep; the radial velocity signature is higher and mass measurements are possible; the large planet-star contrast may permit thermal emission measurements. A ground-based targeted transit search for bright M stars is underway (Nutzman & Charbonneau 2007) and expects to yield a few potentially habitable planets, as are ongoing radial velocity surveys of M stars.

One of us (Valenti et al. 2008) has simulated NIRSpec spectra for transiting Earth-like spectra (Ehrenreich et al. 2006) orbiting bright M stars (Fig. 5.8). Orbiting in the habitable zone (P = 29 days) of an M3V, J = 8 star, an ocean planet will have a strong atmospheric water vapor signature at the level of 40 to 60 ppm. CO_2 features may also be detectable. With a 1.4 hour transit duration, 10 transits and 28 hours of observing time are needed for suitable SNR. This ocean planet example is an 0.5 $M_\oplus$, 1 $R_\oplus$ planet. Its lower density makes its scale height more than 2 times higher than Earth's (Section 5.4.1) making a transmission spectrum twice as easy to detect (see Section 5.4.1). For a "small Earth" (defined as 0.1 Earth masses and 0.5 Earth radii) also orbiting in the habitable zone of M3V star (but now for a slightly brighter, J = 6 star), the water vapor signatures are several ppm. Fifty four hours of observing time would be needed for a significant detection. With the period of 29 days in both of these examples, scheduling to observe 10 or more transits is critical. We emphasize that more work both in modeling the planet atmospheres and in *JWST* simulations needs to be done. Regardless, M3V or later stars as bright as J = 6 to J = 8 are rare, making transiting planets even more rare.

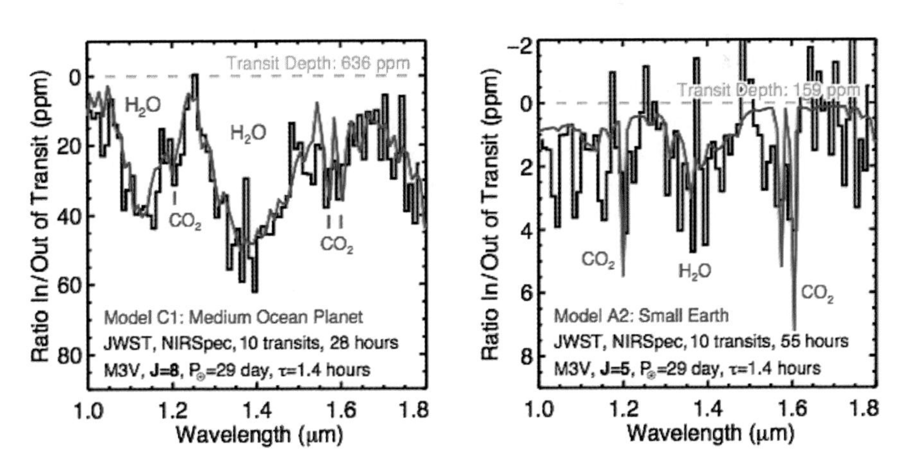

Fig. 5.8 Simulated NIRSpec measurements of transmission spectra of habitable exoplanets. The spectra are the ratio of simulated spectra in and out of transit. Details of the planet and star scenarios are indicated on the Figure. The blue spectrum shows the Ehrenreich et al. (2006) models used to generate the simulated observations (*black curve*). The dominant absorption features are indicated

The fourth example is for a super Earth with a hydrogen-rich atmosphere instead of a hydrogen-poor atmosphere. A hydrogen-rich atmosphere would be created by outgassing on a super Earth with a surface gravity high enough to prevent loss of all of the hydrogen. We have described in Section 5.4.1 how the scale height is inversely proportional to surface gravity and mean molecular weight. We take a 5 $M_{\oplus}$, 1.5 $R_{\oplus}$ planet (Seager et al. 2007) with a corresponding surface gravity 2.2 times higher than Earth's and a mean molecular weight 44 times lower. The transmission spectra signal will be 10 times stronger than the medium ocean planet described in example 2 and Fig. 5.8. Such a hydrogen-rich kind of planet would require $\sim$ three times less observing time than the Earth-like planet atmosphere in example 2 above, or approximately 10 hours. For further discussion on hydrogen-rich atmospheres on super Earths see Miller-Ricci (submitted).

For our final examples, we consider the possibility of mid-IR thermal emission detection during secondary eclipse of an Earth-temperature planet with MIRI. MIRI spectra, in general, may be very useful because with no slit, no detrimental effect from pointing errors will occur (see Section 5.2.4). Regarding Earth-type planets, detection of the ozone (O_3) and CO_2 spectral features are key (Fig. 5.9).

One of us (Deming) has simulated MIRI observations (also presented in Charbonneau & Deming (2007) but with minor corrections here). The planet is modeled as a black body in thermal equilibrium with its star, using a Kurucz model atmosphere for the star (Kurucz 1992). Since the star is bright, its statistical photon fluctuations are a dominant noise source. Noise from the thermal background of the telescope and Sun shield, and background noise from zodiacal emission in our solar system are also included. The efficiency of the telescope/MIRI optical system (electrons out/photons in) was taken to be 0.3.

Figure 5.10 shows the SNR for MIRI R = 20 spectroscopy of a 2 $R_{\oplus}$ super Earth orbiting at 0.03 AU around a 20-pc-distant M5V host star as a function of wavelength. Figure 5.10 aims to show that the SNR would be high enough to detect the O_3 or CO_2 features shown in Fig. 5.9. With a 100 hour observation (for $\sim$40 transits and with the total time divided between in-eclipse and out-of eclipse) a SNR of 10 to 15 is possible. Figure 5.11 shows a similar example, but for a 200 hour observation of a 1 $R_{\oplus}$ planet orbiting in the habitable zone of a 10-pc-distant M8V star with R = 50. For comparison, we note that 100 hours is a bit less than half of the Hubble Deep Field observing time.

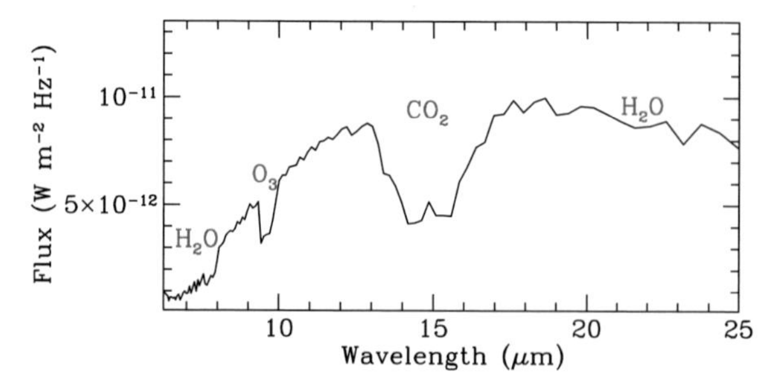

Fig. 5.9 Earth's mid-infrared spectrum as observed by Mars Global Surveyor enroute to Mars (Christensen & Pearl 1997). Major molecular absorption features are noted

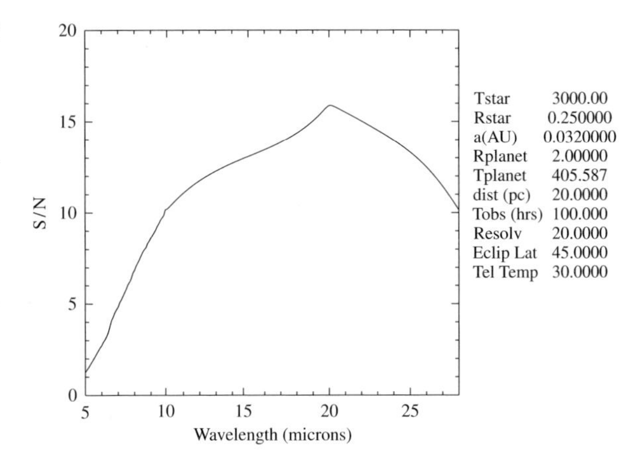

Fig. 5.10 MIRI simulation of the SNR as a function of wavelength for R = 20 spectroscopy of a 2 $R_{\oplus}$ super Earth orbiting at 0.03 AU around an 20-pc-distant M5V host star (where the planet's $T_{eq} = 406$ K). The SNR would be high enough to detect the O_3 or CO_2 features shown in Fig. 5.9

For both SNR vs. wavelength examples in Figs. 5.10 and 5.11, the SNR increases at the blue wavelength range due to the increasing number of photons from the planet. At longer wavelengths, the SNR decreases because of greater thermal background noise from the telescope and the Sun shield. Like the above transmission spectra examples, transiting exoplanets suitable for *JWST*/MIRI followup observations of secondary eclipses are anticipated to be rare.

The only *JWST* instrument we have not discussed is NIRCam, simply due to the current lack of available exoplanet simulations. A brief discussion of transiting exoplanet science with the NIRCam grisms is presented in Greene et al. (2008). NIRCam will be suitable for both observing primary transit and secondary eclipse. Due to the lack of slit (see Section 5.2.4), the grisms, and the near-infrared wavelengths where many molecules have absorption features, NIRCam holds huge promise for transiting exoplanet atmosphere studies.

In summary, *JWST* has the capability to study spectral features of Earth-size or larger planets in the habitable zones of main sequence stars. We need to first find

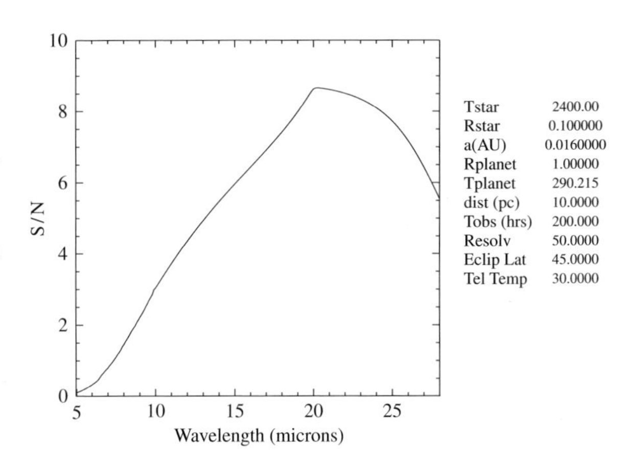

Fig. 5.11 MIRI simulation of the SNR as a function of wavelength for R = 50 spectroscopy of a 1 $R_{\oplus}$ planet orbiting at 0.016 AU around an 10-pc-distant M8V host star (where the planet's $T_{eq} = 290$ K). The SNR would be high enough to detect the O_3 or CO_2 features shown in Fig. 5.9

these rare transiting planets and second be patient with the tens to hundreds of hours of *JWST* time with the concomitant complex scheduling to cover periodic transits.

5.5 Discussion

The *Hubble Space Telescope* and the *Spitzer Space Telescope* (Section 5.2.4) were not designed to achieve the very high SNRs necessary to study transiting exoplanets, but they have succeeded nonetheless. We have learned from these observations that with enough photons, systematics that were unknown in advance can often be corrected for (as long as the errors are uncorrelated). Given the expected thermal stability of *JWST* and the differential nature of transit observations, we are optimistic that *JWST*, too, will succeed in high-contrast transit observations.

We have described a few examples where *JWST* will have a significant impact, including measuring precise exoplanet radii for all sizes of exoplanets and spectra for giant planets at a variety of semi-major axes. Individually, these observations will be "cheap" in terms of telescope time, with single transit measurements being sufficient.

The most significant exoplanet observations *JWST* is poised to make are those for potentially habitable exoplanets. In the most optimistic case, *JWST* is able to identify planets with atmospheric water vapor, or even chemical disequilibrium indicative of biological origin. *JWST* has the capability to do so if the tens to hundreds of hours per target are allocated, and if such rare transiting planets can be discovered in sufficient numbers.

Acknowledgments We thank Mark Clampin, George Ricki, Dave Charbonneau, and Heather Knutson for useful discussions.

References

http://exoplanet.eu/
Basri, G., Borucki, W. J., & Koch, D. 2005, New Astronomy Review, 49, 478
Beichman, C. A., Fridlund, M., Traub, W. A., Stapelfeldt, K. R., Quirrenbach, A., & Seager, S. 2007, Protostars and Planets V, 915
Brown, T. M. 2001, ApJ, 553, 1006
Brown, T. M., Charbonneau, D., Gilliland, R. L., Noyes, R. W., & Burrows, A. 2001, ApJ, 552, 699
Burrows, A., Budaj, J., & Hubeny, I. 2008, ApJ, 678, 1436
Burrows, A., Hubeny, I., Budaj, J., Knutson, H. A., & Charbonneau, D. 2007, ApJ, 668, L171
Butler, R. P., et al. 2006, ApJ, 646, 505
Butler, R. P., Vogt, S. S., Marcy, G. W., Fischer, D. A., Wright, J. T., Henry, G. W., Laughlin, G., & Lissauer, J. J. 2004, ApJ, 617, 580
Charbonneau, D., & Deming, D. 2007, ArXiv e-prints, 706, arXiv:0706.1047
Charbonneau, D., Brown, T. M., Burrows, A., & Laughlin, G. 2007, Protostars and Planets V, 701
Charbonneau, D., Brown, T. M., Noyes, R. W., & Gilliland, R. L. 2002, ApJ, 568, 377
Charbonneau, D., et al. 2005, ApJ, 626, 523

Chauvin, G., et al. 2005, A&A, 438, L29
Chauvin, G., Lagrange, A.-M., Dumas, C., Zuckerman, B., Mouillet, D., Song, I., Beuzit, J.-L., & Lowrance, P. 2004, A&A, 425, L29
Christensen, P. R., & Pearl, J. C. 1997, JGR, 102, 10875
Cowan, N. B., Agol, E., & Charbonneau, D. 2007, MNRAS, 379, 641
Cox, A. N. 2000, Allen's astrophysical quantities, 4th ed. Publisher: New York: AIP Press; Springer, 2000. Edited by Arthur N. Cox. ISBN: 0387987460
Deming, D., & Seager, S., submitted to ApJ
Deming, D., Harrington, J., Laughlin, G., Seager, S., Navarro, S. B., Bowman, W. C., & Horning, K. 2007, ApJ, 667, L199
Deming, D., Seager, S., Richardson, L. J., & Harrington, J. 2005, Nature, 434, 740
Des Marais, D. J.,. et al. 2002, Astrobiology, 2, 153
Ehrenreich, D., Hébrard, G., Lecavelier des Etangs, A., Sing, D. K., Désert, J.-M., Bouchy, F., Ferlet, R., & Vidal-Madjar, A. 2007, ApJ, 668, L179
Ehrenreich, D., Tinetti, G., Lecavelier Des Etangs, A., Vidal-Madjar, A., & Selsis, F. 2006, A&A, 448, 379
Fazio, G. G., et al. 2004, ApJS, 154, 10
Fortney, J. J., Lodders, K., Marley, M. S., & Freedman, R. S. 2007, ArXiv e-prints, 710, arXiv:0710.2558
Gardner, J. P., et al. 2006, Space Science Reviews, 123, 485
Gaudi, B. S., et al. 2008, Science, 319, 927
Gillon, M., et al. 2007, A&A, 471, L51
Gillon, M., et al. 2007, A&A, 472, L13
Greene, T., et al. 2008, SPIE, in press
Harrington, J., Hansen, B. M., Luszcz, S. H., Seager, S., Deming, D., Menou, K., Cho, J. Y.-K., & Richardson, L. J. 2006, Science, 314, 623
Harrington, J., Luszcz, S., Seager, S., Deming, D., & Richardson, L. J. 2007, Nature, 447, 691
Hauschildt, P. H., Allard, F., & Baron, E. 1999, ApJ, 512, 377
Houck, J. R., et al. 2004, ApJS, 154, 18
Hubeny, I., Burrows, A., & Sudarsky, D. 2003, ApJ, 594, 1011
Knutson, H. A., Charbonneau, D., Allen, L. E., Burrows, A., & Megeath, S. T. 2008, ApJ, 673, 526
Knutson, H. A., et al. 2007, Nature, 447, 183
Kurucz, R. L. 1992, The Stellar Populations of Galaxies, 149, 225
Lederberg, J. 1965, Nature, 207, 9
Lovelock, J. E. 1965, Nature, 207, 568
Miller-Ricci, Seager, Sasselov, ArXiv archive astrophysical preprints ArXiv0808.1902
Morales-Calderón, M., et al. 2006, ApJ, 653, 1454
Neuhäuser, R., Guenther, E. W., Wuchterl, G., Mugrauer, M., Bedalov, A., & Hauschildt, P. H. 2005, A&A, 435, L13
Nutzman, P., & Charbonneau, D. 2007, ArXiv e-prints, 709, arXiv:0709.2879
Redfield, S., Endl, M., Cochran, W. D., & Koesterke, L. 2008, ApJ, 673, L87
Richardson, L. J., Harrington, J., Seager, S., & Deming, D. 2006, ApJ, 649, 1043
Rieke, G. H., et al. 2004, ApJS, 154, 25
Seager and Lunine, eds. *JWST* and Astrobiology white paper, NASA Astrobiology Institute
Seager, S., Kuchner, M., Hier-Majumder, C. A., & Militzer, B. 2007, ApJ, 669, 1279
Seager, S., Richardson, L. J., Hansen, B. M. S., Menou, K., Cho, J. Y.-K., & Deming, D. 2005, ApJ, 632, 1122
Showman, A. P., Menou, K., & Y-K. Cho, J. 2007, ArXiv e-prints, 710, arXiv:0710.2930
Silvotti, R., et al. 2007, Nature, 449, 189
Swain, M. R., Vasisht, G., & Tinetti, G. 2008, Nature, 452, 329
Tinetti, G., et al. 2007, Nature, 448, 169
Udry, S., Fischer, D., & Queloz, D. 2007, Protostars and Planets V, 685
Valencia, D., Sasselov, D. D., & O'Connell, R. J. 2007, ApJ, 665, 1413

Valenti, J. A., et al. 2005, Bulletin of the American Astronomical Society, 37, 1350
Valenti, J. A., Turbull, M., McCullough, P., & Gilliland, R. 2008, in prep.
Vidal-Madjar, A., Lecavelier des Etangs, A., Désert, J.-M., Ballester, G. E., Ferlet, R., Hébrard, G., & Mayor, M. 2003, Nature, 422, 143
Wolszczan, A., & Frail, D. A. 1992, Nature, 355, 145
Zeng, L., Seager, S., 2008, PASP, 120, 983

The availability of wireless internet access during the presentations was put to good use by many in the audience

Chapter 6
The Unsolved Problem of Star Formation: Dusty Dense Cores and the Origin of Stellar Masses

Charles J. Lada

Abstract The development of a predictive theory of star formation is one of the prime goals of modern astrophysics. The problem of star formation is a very complex one and its solution depends critically on empirical data. In this contribution, I review some of the progress that has been achieved in identifying and measuring the fundamental boundary and initial conditions that constrain the solution and must be met by any complete theory of star formation. In particular, I discuss what is known about the stellar IMF and the nature of dense cores on the verge of star formation. I then discuss how these two topics contribute to our general understanding of the origin of stellar mass, the key mystery in the star formation process.

6.1 Introduction

The development of a theoretical understanding of the physical process of star formation remains one of the most fundamental unsolved problems of modern astrophysics. This problem is a very complex one and its solution necessarily requires a close synergy between observation and theory. In particular, similar to the process of solving a system of time-dependent differential equations, developing a predictive theory of star formation requires the specification of both the appropriate boundary and initial conditions. These conditions are empirically specified through observations.

6.2 Boundary Conditions

The ultimate goal of a star formation theory is, of course, to produce stars. The fundamental boundary conditions, therefore, are those which describe the basic properties of the objects shown in Fig. 6.1, and those properties are: the compositions, luminosities, temperatures, radii and masses of stars. Because stars form

C.J. Lada (✉)
Harvard-Smithsonian Center for Astrophysics, Cambridge, MA, USA
e-mail: clada@cfa.harvard.edu

H.A. Thronson et al. (eds.), *Astrophysics in the Next Decade,* Astrophysics and Space Science Proceedings, DOI 10.1007/978-1-4020-9457-6_6,

from interstellar gas and dust whose composition is well known, the first boundary condition is trivially satisfied. Stellar luminosities, temperatures and radii are not only known from observation, but also exactly understood and predicted by an existing theory. The theory of stellar structure and evolution is one of the great theories of science and explains the nature of stars as thermonuclear furnaces that through their evolution convert hydrogen, the primary product of the Big Bang, into essentially all the heavier elements of the periodic table. An extremely powerful and elegant property of this theory is that, given only a composition and mass of a star, the stellar luminosity, temperature and radius can be exactly predicted over the entire span of a star's life, from shortly after its birth until its death. Thus, these three boundary conditions are pre-determined and satisfied by the theory of stellar structure and evolution, once the mass and composition of a star are specified.

Given our knowledge of the composition of interstellar material, stellar mass is the single most fundamental boundary condition that remains to be observationally specified. More specifically, it is the distribution of stellar masses at birth, the so-called Initial Mass Function (or IMF), that is the most fundamental boundary condition that needs to be observationally determined to enable a complete and predictive theory of star formation to be constructed. With knowledge of the IMF and how this function varies in space in time, one can, in principle, predict the future evolution of all stellar systems, from galactic clusters to massive galaxies.

Figure 6.1 also illustrates another important stellar property that is also a boundary condition that needs to be considered in the development of a theory of star formation, the tendency of stars to be born in groups and clusters (e.g., Lada & Lada 2003). Perhaps an even more significant boundary condition is the related tendency of stars to be formed in binary or multiple stars systems. As shown in Fig. 6.2, modern observations clearly demonstrate that stellar binarity or multiplicity is a function of spectral type or equivalently of stellar mass. Massive OB stars have binary frequencies that approach unity, whereas the considerably more numerous low-mass M stars have binary frequencies no greater $\approx$ 25–30%. Both the magnitude and variation of stellar multiplicity are boundary conditions that also need to be ultimately satisfied by any complete star formation theory. However, the

Fig. 6.1 Hubble Space Telescope image of the rich, young galactic star cluster NGC 3603 and its surroundings

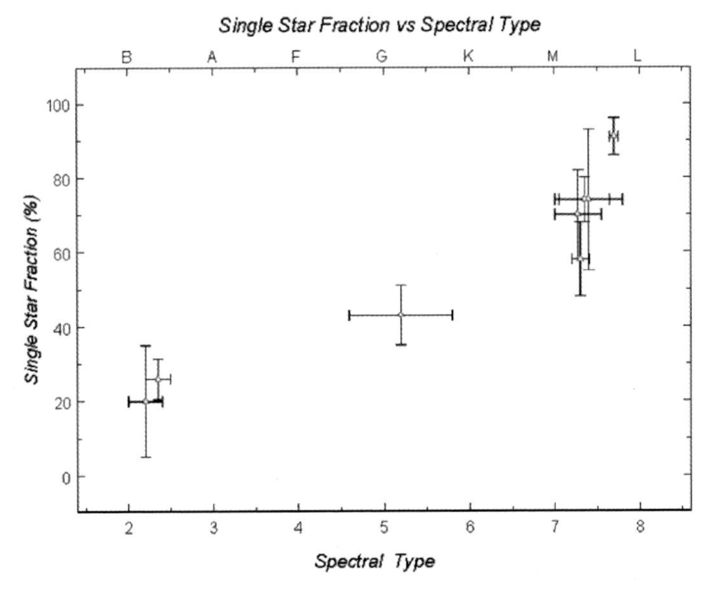

Fig. 6.2 Single star fraction as a function of spectral type. The plot shows that stellar multiplicity is a clear function of spectral type with the single star fraction increasing with increasing spectral types. Early type, massive stars, have the lowest single star fraction and highest multiplicity fraction while late-type, low mass, stars have the highest single star fraction and lowest multiplicity fraction. Adapted from Lada (2006)

remainder of this discussion will focus on the empirical determination of the stellar initial mass function, the most fundamental boundary condition for star formation theory.

6.2.1 The Stellar Initial Mass Function

The origin of stellar masses is the critical problem of star formation. The theory of stellar structure and evolution does not predict or in any way account for stellar mass. The stellar IMF must therefore be empirically determined in order to develop a complete theory of star formation. Observational determination of the IMF has proved to be a difficult task. This is because stellar mass is not an observable quantity. Stellar radiant flux or luminosity is the observable quantity. Thus, the starting point for the determination of an IMF is the construction of the appropriate luminosity function. This turns out to be very difficult since it requires compilation of a volume complete sample of stars both with known distances and with brightnesses and/or luminosities that span a range of at least 5–6 orders of magnitude. The next step in the process is to transform the observed luminosities into masses and this requires the existence and application of a mass-luminosity relation. In mathematical terms:

$$\Psi(M_\lambda) = \phi(\log m)\frac{d\log m}{dM_\lambda} \tag{6.1}$$

Here $\Psi(M_\lambda)$, is the observed luminosity function, by convention, the frequency distribution of absolute magnitudes at some wavelength, λ, of the stellar sample. And $\phi(\log m)$ is the desired mass function of stellar masses, m, and $\frac{d\log m}{dM_\lambda}$ is the mass-luminosity relation. With knowledge of the mass-luminosity relation and the observed luminosity function, (6.1) can be inverted to derive the desired mass function.

The first published attempt to derive an IMF was the pioneering study of Edwin Salpeter (1955) who derived the IMF from the luminosity function of near-by main-sequence field stars. Because this sample consisted of stars formed over the entire 10–13 billion year history of the Milky Way, it had to be corrected for the effects of stellar evolution which significantly de-populated its high mass end. This correction is, however, straightforward if one assumes a constant star formation rate for the Milky Way and employs the theory of stellar structure and evolution to specify the main-sequence lifetimes of the stars as a function of mass. This theory also conveniently provides the stellar mass-luminosity relation for main-sequence stars. This mass-luminosity relation is unique and can be expressed as a power-law, $L \sim m^p$ and since stellar luminosities are measured in terms of magnitudes, $M \sim \log(L) \sim p\log(m)$. Thus, by convention the IMF is expressed in terms of $\log(m)$. To first approximation $p \approx 3.3$ for main-sequence stars. Salpeter found that the IMF could be expressed as a power-law function over the mass range of approximately 1–10 solar masses. That is, $\phi(\log m) \sim m^\gamma$ where $\gamma \approx -1.6$. The number of stars born at any one time was thus found to increase non-linearly with *decreasing* stellar mass.

Thirty years later using more extensive data Scalo (1978) re-evaluated the field-star IMF and showed that the IMF departed from a simple power-law below 1 $M_\odot$. He proposed a form for the IMF that was log-normal in shape with a peak near the hydrogen-burning limit at about 0.1 $M_\odot$. More recent consensus favors a multi (3) power-law form for the field star IMF (i.e., Kroupa 2002) with breaks near 0.6 and 0.1 $M_\odot$ and an overall maximum near the hydrogen-burning limit (HBL). The difficulties inherent in deriving an IMF from local field stars, such as incompleteness, stellar evolution corrections, limited range of stellar masses, etc., prompted the use of open clusters, rather than local field stars, to define the stellar samples. In particular, young, embedded clusters offered a number of advantages for IMF determinations. These included: (1) low mass pre-main sequence stars that are brighter than any other time in their main-sequence lifetimes, enabling statistically significant sampling of stars across a wide range of stellar mass, and (2) no stars lost to stellar evolution, so that the present-day cluster mass function is essentially identical to the initial mass function.

Figure 6.3 displays the IMF for the well-studied, young, embedded Trapezium cluster in Orion derived from infrared imaging observations (Muench et al. 2002). This IMF is representative of IMFs derived for embedded clusters in general and is also indistinguishable from the field star IMF within the uncertainties inherent in both determinations. Therefore, the basic properties of this IMF can be taken as

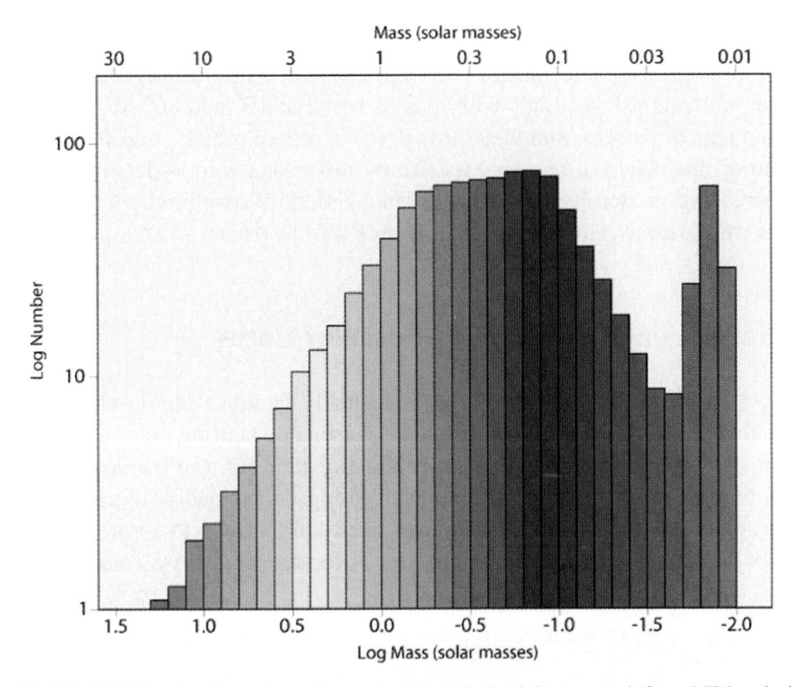

Fig. 6.3 The IMF for the Trapezium cluster in Orion derived from near-infrared K-band observations by Muench et al. (2002). In this figure mass decreases toward the right. The IMF is characterized by a clear but broad peak between 0.1 and 0.5 $M_{\odot}$. On either side of this peak it falls off rapidly. This suggests a characteristic mass for star formation of about 0.25–0.30 $M_{\odot}$. The sharp secondary peak at the low mass end of the IMF does correspond to a similar peak in the K luminosity function, but is likely an artifact of an inadequate mass-luminosity relation for brown dwarfs for this age and mass range

representative of the IMF produced by the star formation process in the Galactic disk. The similarity of the Trapezium cluster IMF, derived for one million year old stars contained in a volume less than a parsec in extent and 400 pc from the sun, to that of the field, which is a collection of stars formed continuously over the last 10-13 billion years over a considerable volume of the Galactic disk, speaks to the robust nature of the IMF and this boundary condition for star formation theory.

The fundamental feature of the IMF is the presence of a broad peak in the distribution of initial stellar masses. The IMF is not a scale-free function. There is a definite mass scale that characterizes the star formation process in the Galaxy. Nature preferentially forms low mass M stars, (i.e., $0.1 \leq m \leq 0.6\ M_{\odot}$).

One consequence of this is that the Milky Way probably contains 400 billion stars of which 336 billion are M stars. It is also interesting that the peak of the IMF is very close to the hydrogen burning limit ($\approx 0.1\ M_\odot$). It is remarkable that the physical process of star formation somehow manages to start with diffuse, cold interstellar gas that is contained in clouds with masses between 10^4 and $10^6\ M_\odot$ and extents reaching tens of parsecs, and then through the action of gravity, transforms a small fraction of this material, reducing its size by more than eight orders of magnitude and increasing its density more than twenty orders of magnitude, into numerous objects with just the right mass ($\sim 0.1\ M_\odot$) to fuse hydrogen.

6.3 Initial Conditions: Dense Molecular Cores

Setting appropriate starting conditions can greatly facilitate the development of a theory for the formation of stars. The most reasonable starting point would appear to be a starless, dense cloud core. In the Galaxy stars form in the dark molecular clouds and an example of an active star forming dark cloud is nicely illustrated by Fig. 6.4. More specifically, stars have been long known to form in the dense cores of such clouds. However, once a star is formed in a dense core it alters the

Fig. 6.4 Infrared ESO VLT image of the Serpens cloud, a site of very active star formation

physical conditions within the core. The physical conditions, that is, mass, density, temperature, pressure, dynamical state, that characterize *starless* dense cores therefore best represent the initial conditions for star formation. Although most of what is known about dense cores derives from studies of cores containing protostars or more evolved young stellar objects, such as PMS T Tauri stars (e.g., Myers 1999), tremendous inroads in determining the properties of starless cores are beginning to be made (Di Francesco et al. 2007).

In astronomy, considerable information about a class of objects can be derived from systematic surveys. To date, most starless cores have been identified in combined surveys of heterogenous samples of star forming clouds. However, complete surveys of individual cloud complexes for dense cores, starless or otherwise are rare but yet extremely valuable because in such samples all the cores are at the same distance, facilitating inter-comparison of their relative properties. To obtain adequate angular resolution, nearby clouds are the preferred targets. But because of their proximity, such clouds subtend large angles on the sky requiring extensive observational campaigns to completely survey. Recently, two nearby relatively large molecular clouds have been surveyed for dense cores using observations of dust within the clouds: the rich star forming Perseus cloud (Kirk et al. 2006, 2007, Enoch et al. 2006) and the Pipe Nebula (Lombardi et al. 2006). The Pipe Nebula is of particular interest because of the almost complete lack of star formation within it. This cloud is an extremely rare example of a relatively massive (10^4 $M_\odot$) molecular cloud that may be in a very early state of evolution, prior to the onset of significant star-forming activity. With a well determined distance of 130 ($\pm$ 20) pc (Lombardi et al. 2006) it is also the closest molecular cloud of its size and mass to the sun, yet until recently, it has received very little attention in the published literature.

6.3.1 The Core Mass Function and the Origin of the IMF

Figure 6.5 shows the wide-field extinction map of the Pipe Nebula constructed from infrared imaging observations by Lombardi et al. (2006). Measurements of the individual infrared extinctions to more than 3 million stars were combined with a gaussian smoothing kernal one arc min in size to construct this map. Applying a wavelet transform to this map, Alves et al. (2007) were able to identify and extract about 160 small scale dust structures or cores. With only one known exception (B59), all these cores appear to be starless. The wavelet transformation permitted a relatively accurate removal of background, intercore, extinction enabling robust measurements of the individual masses (after correction for a standard gas-to-dust ratio of 100) and sizes of the entire population of dust cores in the cloud. The total mass of the core population is 250 $M_\odot$. From this database, Alves et al. constructed the mass spectrum of the core population. The Pipe core mass function (CMF) is shown Fig. 6.6 and compared to stellar IMFs derived for field stars and the Trapezium cluster. The similarity in shapes of the CMF to the IMFs is notable. As mentioned earlier, the similarity in shapes of the field star IMF and that of the young Trapezium cluster suggests a robust nature for the IMF produced by star formation in the Galaxy.

Fig. 6.5 Wide-field extinction map of the Pipe Nebula constructed from 2MASS infrared imaging observations by Lombardi et al. (2006). Nearly one-hundred-sixty individual extinction (dense) cores have been resolved and identified in this grey-scale map (Alves et al. 2007). The width of this image corresponds to about 20 parsecs and the measured extinction range to $A_V \approx$ 0.5–25 magnitudes

The similarity in shapes between the Pipe CMF and the stellar IMFs is equally remarkable and suggests that the IMF originates directly from the CMF after modification of the CMF by a mass independent star formation efficiency (SFE) of about 30%.

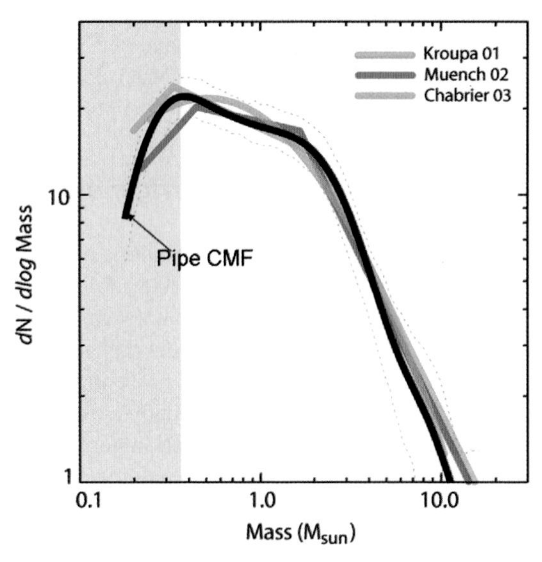

Fig. 6.6 The core mass function (CMF) of the Pipe Nebula (*solid black line*) in the form of a probability density function. Also plotted are the stellar IMF for the Trapezium cluster and two versions of the field star IMF all shifted in mass scale by a factor of ≈ 3. The observations of the CMF are incomplete in the shaded gray region. The shapes of the CMF and the IMFs are remarkably similar suggesting that the IMF may directly originate from the CMF once modified by a constant star formation efficiency of about 30%. This figure has been adapted from Alves et al. (2007)

Studies of a number of other clouds using millimeter-wave observations of dust emission have also suggested a similarity in shapes between CMFs and the IMF (e.g., Testi 1998; Motte et al. 1998; Enoch et al. 2006). However, none of these other CMF measurements were sensitive enough to conclusively detect the break in the CMF and measure the characteristic core mass and SFE.

6.3.2 Physical Properties of the Dusty Cores in the Pipe Nebula

With robust knowledge of the dust column densities and sizes of the cores in the Pipe Nebula, basic physical properties such as the core densities are readily obtained (Lada et al. 2008). Figure 6.7 shows the derived distribution of mean core densities for the entire core population of the Pipe Nebula. The distribution is very narrowly peaked near a density of 7×10^3 cm^{-3}, although there is a significant tail on the high density side. The identified dust cores in the Pipe are dense cores. The breadth of this narrow peak is less than a factor of 2 in density and, as we discuss later, this provides an important insight relating to the fundamental physical nature of the population of dense cores in this cloud.

Observations of the dust are not sufficient to describe the complete set of physical conditions or properties that characterize these starless dense cores. To do this requires direct observation of the gaseous component of the cores which, of course, accounts for most all the mass. To obtain a more complete understanding of the physical conditions characterizing the cores, Muench et al. (2007) and Rathborne et al. (2008) obtained molecular-line emission surveys of $C^{18}O$ and NH_3 lines, respectively, of a significant sample of the core population in the Pipe. These surveys

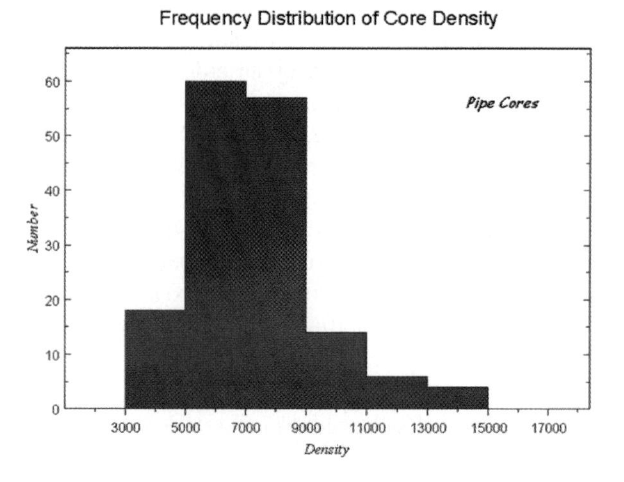

Fig. 6.7 The distribution of mean core density for the dust core population of the Pipe Nebula. The density distribution is strongly peaked around a density of 7000 cm^{-3}and shows that the dust cores are dense cores. This figure has been adapted from Lada et al. (2008)

provided crucial information concerning the kinematics and dynamical states of the cores, which I will now briefly summarize.

Linewidth-size relation: The first notable finding was an absence of a relation between linewidth and core size or mass over the entire two order of magnitude range in core mass. Linewidth-size relations typically characterize gas within molecular clouds and are widely believed to indicate that the physical state of molecular gas is best described by a common hierarchy of supersonic turbulent motions under the action of gravity (Larson 1981). Apparently the gas contained in the dense cores in the Pipe is not part of the turbulent hierarchy that describes the bulk of the gas in a typical massive molecular cloud. Note here that the cores account for only about 2.5% of the total cloud mass and are likely not representative of the cloud as a whole.

Thermally dominated cores: The non-thermal component of both the CO and NH_3 linewidths was found to be subsonic for the large majority ($\sim 70\%$) of the cores, suggesting that thermal motions and pressure may be significant components of the physical state of the gas. The measured non-thermal motions could be due to turbulence, or systematic motions such as rotation, collapse, expansion, oscillation, etc. If due to turbulence, then the turbulence is of a very different nature than the turbulence that generally characterizes molecular clouds. The turbulence in the cores is subsonic rather than supersonic and likely microturbulent rather than macroturbulent in nature.

Figure 6.8 shows the ratio of thermal to turbulent pressure as a function of mass for the entire core population derived from CO (circles) and NH_3 observations. Thermal pressure dominates the internal pressures of the cores in this cloud, independent of mass. In some cases, the lines are so narrow that a non-thermal component is undetectable and the thermal pressures exceed the turbulent pressures by two orders of magnitude, although more typically the difference is on the order of a factor of 2.

Pressure-confined cores: The run of internal core pressure with mass for the cores in the Pipe Nebula is plotted in Fig. 6.9. To first order the total internal core pressure is not a function of core mass over the entire (two order of magnitude) range of measured core masses. That the internal pressures of these cores agree to within a factor of 2–3, independent of core mass, strongly suggests that the cores are in pressure equilibrium with a surrounding external pressure source. The pressure from the general ISM falls short by an order of magnitude to account for this external pressure. However, the pressure due to the weight of the Pipe cloud (i.e., $P/k \sim \pi G\Sigma^2 \approx 10^5\,\mathrm{K\,cm^{-3}}$; e.g., Bertoldi & McKee 1992)[1] is sufficient in magnitude to pressure-confine these cores (Lada et al. 2008). Indeed, in the absence of such pressure, the large majority of the cores would be gravitationally unbound.

Pressure-confined balls of gas can be described by a well-known theoretical model, that of a pressure-truncated isothermal sphere, the Bonnor-Ebert sphere. The NH_3 observations (Rathborne et al. 2008) indicate that the cores in the Pipe cloud all have temperatures near 10–12 K and are likely isothermal to a fair degree of approx-

[1] Here Σ is the mass surface density of the cloud, k is Boltzman's constant and G the gravitational constant.

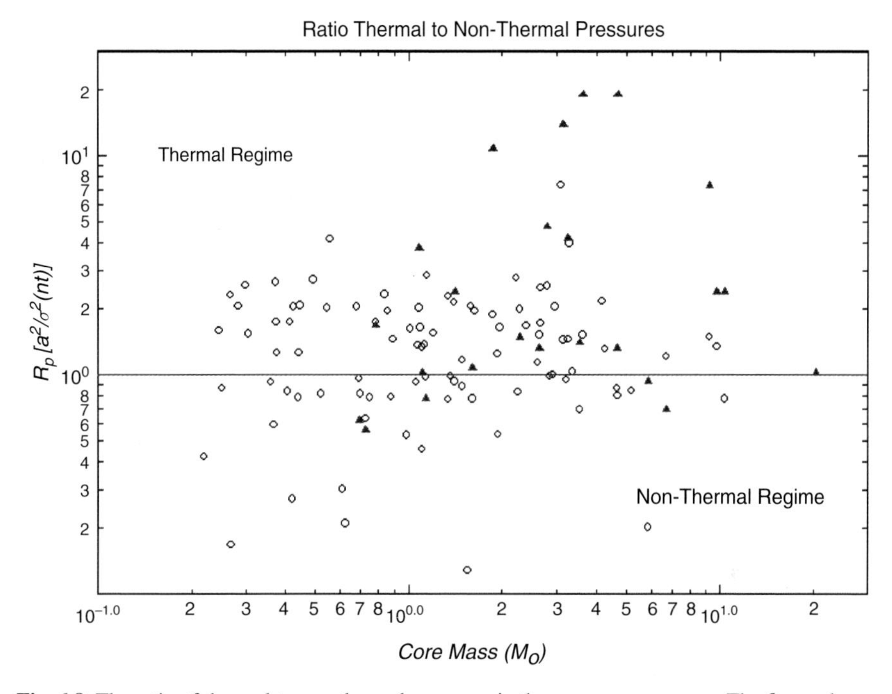

Fig. 6.8 The ratio of thermal to non-thermal pressures in the cores vs core mass. The figure shows that the internal pressure in the large majority ($\sim$ 70%) of the cores is dominated by thermal pressure. For another 20% of the cores thermal pressure is within a factor of two of the turbulent pressure. After Lada et al. (2008)

imation. In support of that idea are observations of the internal structure of specific cores in this cloud. The radial column density profile for one of these (B 68) is shown in Fig. 6.10 along with the predicted column density profile of a pressure-truncated isothermal sphere derived from the Lane-Emden equation of stellar structure (Alves et al. 2001). The data closely matches the theoretical prediction suggesting that B 68 is indeed a pressure-truncated isothermal spheroid in hydrostatic equilibrium. This is confirmed by millimeter-wave molecular-line observations that reveal that the velocity dispersion of gas in this cloud is dominated by thermal motions and show that thermal pressure exceeds non-thermal pressure by a factor of 10–14 (Hotzel et al. 2002; Lada et al. 2003).

That the cores in the Pipe Nebula appear to be pressurized by an external pressure source has a potentially profound implication regarding their very appearance and nature. The physical structure of dense cores is dictated by a single requirement: pressure equilibrium with a surrounding source of external pressure and this source of external pressure is likely produced by the weight of the surrounding cloud itself.

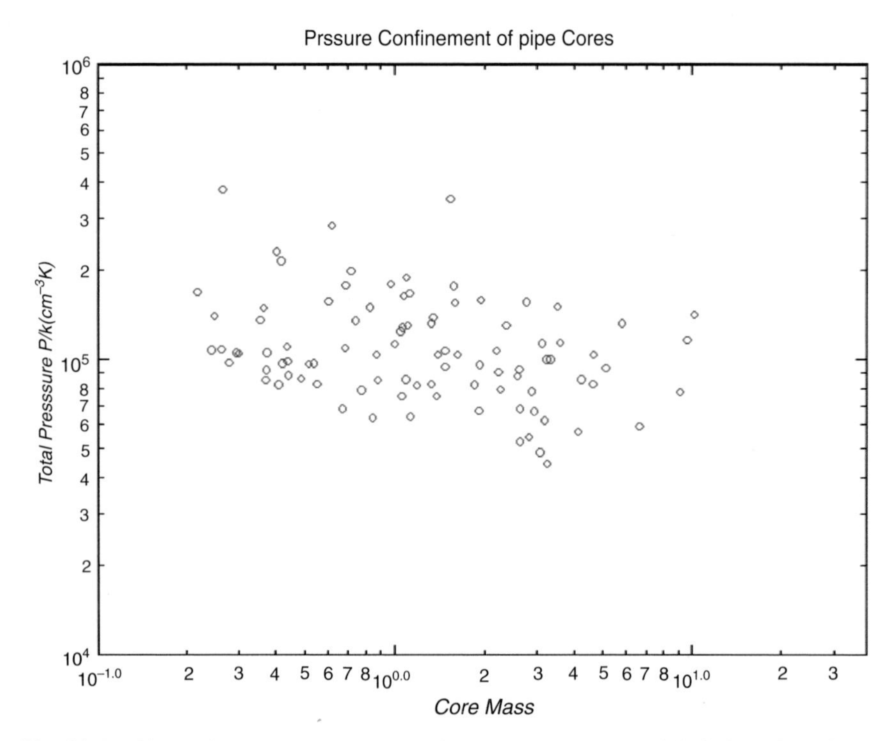

Fig. 6.9 Total internal pressure vs. core mass. These pressures are essentially independent of core mass over entire range of core mass. That the pressures of all the cores agree to within a factor of 2–3 independent of mass indicates that the cores must be in pressure equilibrium with a source of external pressure. After Lada et al. (2008)

Thus, for example, the radial stratification and density structure of the cores (e.g., Fig. 6.10) are a direct consequence of dynamical, perhaps even hydrostatic, balance in a pressurized medium. The narrow peak in the frequency distribution of mean core densities (Fig. 6.7) is likewise a direct result of this requirement. That is, given that the cores are isothermal, it follows from the perfect gas law that they must all have similar densities, at least to within the range of variation inferred for the surrounding pressure (i.e. a factor of 2–3).

6.4 Pressure and the Origin of Core Masses

As mentioned above, pressurized dense cores can be theoretically understood as Bonnor-Ebert (BE) spheres. A BE sphere is a pressure-truncated isothermal ball of gas within which internal pressure everywhere precisely balances the inward push of self-gravity and external surface pressure. The fluid equation that describes such a self-gravitating, isothermal sphere in hydrostatic equilibrium is the following well known variant of the Lane-Emden equation:

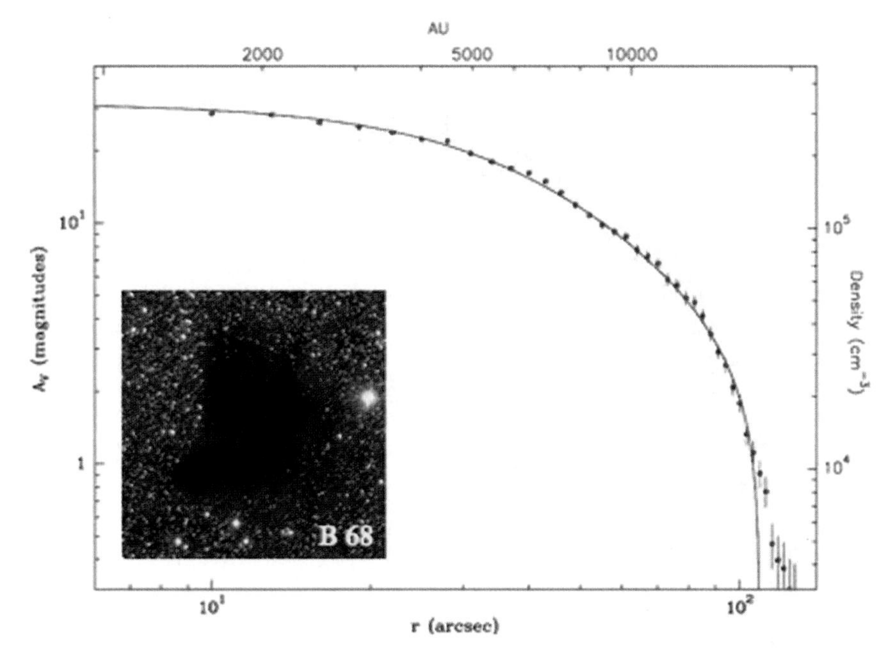

Fig. 6.10 The radial dust extinction or column density profile for B 68 (insert) one of the dark cores in the Pipe complex. The solid trace is the expected density profile of a critically stable Bonnor-Ebert sphere. The close agreement of the theory and empirical data (points) suggests that B 68 is in hydrostatic equilibrium with internal pressure balancing both gravity and external pressure. After Alves et al. (2001)

$$\frac{1}{\xi^2}\frac{d}{d\xi}(\xi^2\frac{d\psi}{d\xi}) = e^{-\psi} \tag{6.2}$$

where ξ is the dimensionless radius:

$$\xi = \mathrm{r}/r_c \tag{6.3}$$

and r_c, is the characteristic or scale radius,

$$r_c = c_s/(4\pi G\rho_0)^{1/2}, \tag{6.4}$$

where c_s is the sound speed in the cloud and ρ_0 is the density at the origin. Equation (6.2) is Possion's equation in dimensionless form where $\psi(\xi)$ is the dimensionless potential and is set by the requirement of hydrostatic equilibrium to be $\psi(\xi)$ = -ln(ρ/ρ_0). The equation can be solved using the boundary conditions that the function ψ and its first derivative are zero at the origin. Equation (6.2) has an infinite family of solutions that are characterized by a single parameter, the dimensionless radius at outer edge (R) of the sphere:

$$\xi_{max} = \mathrm{R}/r_c. \tag{6.5}$$

Each solution thus corresponds to a truncation of the infinite isothermal sphere at a different outer radius, R. The external pressure at a given R must then be equal to that which would be produced by the weight of material that otherwise would extend from R to infinity in an infinite isothermal sphere. The shape of the BE density profile for a pressure truncated isothermal cloud therefore depends on the single parameter $\xi_{\max}$. As it turns out, the higher the value of $\xi_{\max}$ the more centrally concentrated the cloud is. Although there is an infinite family of solutions for the Lame-Emden equation, not all solutions represent stable equilibria. The stability of such pressure truncated clouds was investigated by Bonnor (1956) and Ebert (1955) who showed that those solutions with $\xi_{\max} > 6.5$, correspond to states of unstable equilibria, which are susceptible to gravitational collapse.

The maximum stable or critical BE mass (i.e., for $\xi_{\max} = 6.5$) is given by:

$$M_{BE} = 1.15 \left(\frac{n_s}{10^4}\right)^{-0.5} \left(\frac{\mathrm{T}}{10}\right)^{1.5} M_{\odot} \tag{6.6}$$

Here, n_s is the volume density (i.e., cm^{-3}) at the core surface and T the core's gas temperature (K). The functional dependence of the BE critical mass on density and temperature is identical to that of the Jeans mass. However, the scaling constant is about a factor of 5 higher for the Jeans mass, i.e., $M_J \approx 4.7 M_{BE}$. The critical BE mass corresponds to the Jeans criteria in a pressurized medium. It can also be informative to express the critical BE mass in terms of the external pressure and internal sound speed, a :

$$M_{BE} = 1.15 \left(\frac{a}{0.2\ \mathrm{kms}^{-1}}\right)^4 \left(\frac{P_{ext}/k}{10^5\ \mathrm{K\ cm}^{-3}}\right)^{-0.5} \tag{6.7}$$

Figure 6.11 plots the individual ratios of core mass to critical BE mass as a function of core mass. The ratios form a well-behaved locus of points as a function of mass, which intersects the line $M/M_{BE} = 1$ at a more or less single value of core mass, which is $M \approx 2$–$3\ M_{\odot}$. This indicates that the entire population of cores is characterized by a single critical BE mass. This has an interesting implication when one considers the nature of the CMF for this core population (Fig. 6.6). The CMF has a characteristic mass of about 2–3 $M_{\odot}$, nearly identical to the critical BE mass for the core population. The fact that the characteristic mass of the CMF is very close to the BE critical mass of the core population could be a chance coincidence. On the other hand, this correspondence of the two mass scales may provide an important clue concerning the very origin of the CMF (and ultimately the IMF). The physical interpretation of this coincidence is straightforward: the characteristic mass of the CMF (and the CMF itself) is the result of thermal (i.e. Jeans) fragmentation in a pressurized medium. Indeed, the frequency distribution of nearest neighbor distances for the cores in the Pipe Nebula is narrowly peaked at a spatial scale

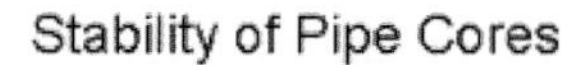

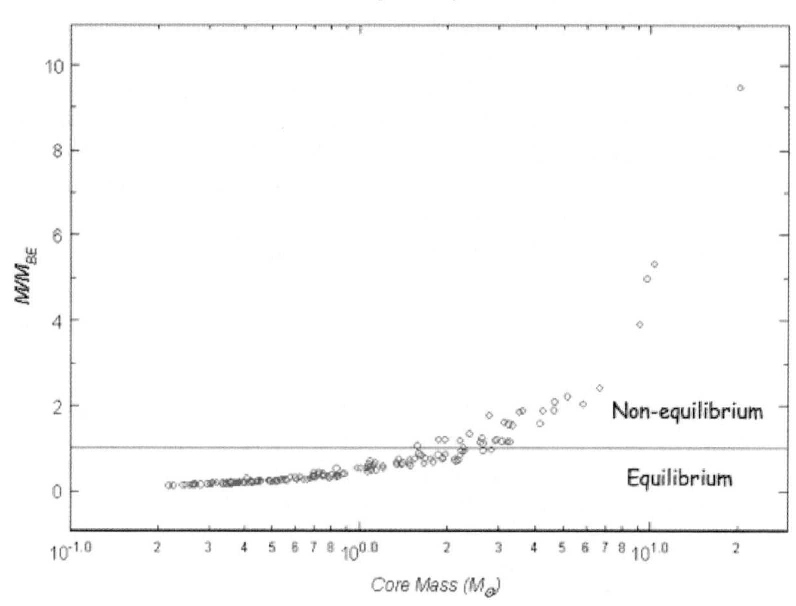

Fig. 6.11 The ratio of core mass to Bonnor-Ebert critical mass for each individual core in the Pipe Nebula vs the log of core mass. The entire core population appears to be characterized by a single critical mass of ≈ 2–$3\ M_{\odot}$. Cores with masses in excess of this mass may be non-equilibrium structures and likely to collapse to form stars. Cores with masses below this mass appear to be mostly stable. After Lada et al. (2008)

($\sim$ 0.2 pc) that corresponds almost exactly to the predicted Jeans length for the observed densities and temperatures that characterize the Pipe cores.

The CMF can be expressed as follows:

$$CMF(\log m) = b_0 \Psi(\log(m/m_c), s_i) \tag{6.8}$$

Here b_0 is a scaling constant which is related to the birthrate of cores (i.e., $b_0 = b(t)\Delta t$) and m_c is the characteristic mass of the CMF, which is given by $m_c = M_{BE}$. The term s_i is the shape parameter, a term that is meant to account for the overall shape and width of the function with respect to the characteristic mass. In a perfect process of thermal fragmentation, one would expect all cores to form with the (same) critical BE mass and s_i to represent a delta function. However, nature is never perfect and in a realistic fragmentation process one would expect a spread of core masses around the critical BE mass (e.g., Adams & Fatuzzo 1996). In this case, the shape parameter is likely determined by the interplay of complex, even non-linear, physical processes. However, statistical considerations suggest that in such a situation the shape parameter could be represented by a lognormal form (e.g., Adams & Fatuzzo 1996) which may or may not be invariant. If, as suggested in Section 6.3.1, the stellar IMF descends directly from the CMF, then the close agreement of shapes of the CMF and the IMFs of both embedded clusters and the field (Fig. 6.6) would seem to further suggest that the shape parameter for the CMF is invariant in most star-forming events.

For a fixed shape parameter, the functional form of the CMF is then essentially determined by equation (6.7), the critical BE mass. Thus, the characteristic mass of the CMF may be controlled by only two basic physical parameters, the internal gas temperature, T, (via the sound speed, a) and the external pressure of the medium from which the cores form. This mass could vary depending on the physical conditions in a given region of star formation.

6.5 From Cores to Stars

The similarity of the Pipe CMF and the IMFs (Fig. 6.6) suggests that star formation in the Pipe cloud could produce a standard IMF by having every dense core form a star. Thus, the IMF would be the direct descendant of the CMF. If the IMF derives directly from the CMF, then the IMF can be expressed as:

$$\phi(\log m) = b_1 \Psi(\log(m/m_c), s_j) \tag{6.9}$$

with

$$m_c = SFE \cdot M_{BE}$$

Here, b_1 is the corresponding stellar birthrate scaling factor which may or may not be equal to b_0, the birthrate scaling factor for cores (6.8). Although, s_j and s_i (6.8) may or may not be the same, Figure 6.6 suggests that they could be very similar.

Figure 6.11, however, indicates that not all cores in the Pipe CMF will necessarily form stars since a significant number of cores have masses below the critical value and, if nothing else changes, are unlikely to collapse to form stars. In this case, the resulting IMF would not resemble the standard IMF, but instead have a peak nearer to 1 $M_\odot$ rather than to 0.1 $M_\odot$.

Figure 6.12 displays the IMFs derived for two star-forming regions. The Taurus clouds are in many ways similar to the Pipe Nebula cloud. However, the Taurus clouds are apparently more evolved since they have produced a large number of dispersed low-mass stars. The Taurus clouds also represent a rare example of a star-forming region characterized by an IMF that appears to differ from that of the field or embedded clusters. The Taurus IMF has its maximum near 1 $M_\odot$ rather than near the HBL (Luhman 2004). IC 348 is a relatively rich cluster formed in the Perseus Cloud and is representative of the perhaps more common cluster-forming physical star-forming environment. Its IMF is similar to that of the field and embedded clusters such as the Trapezium (Fig. 6.3). It has its maximum near 0.1 $M_\odot$ and the HBL. If only the cores that are presently out of equilibrium were to form stars in the Pipe, its IMF might end up resembling that of the Taurus population rather than the standard field/cluster IMF.

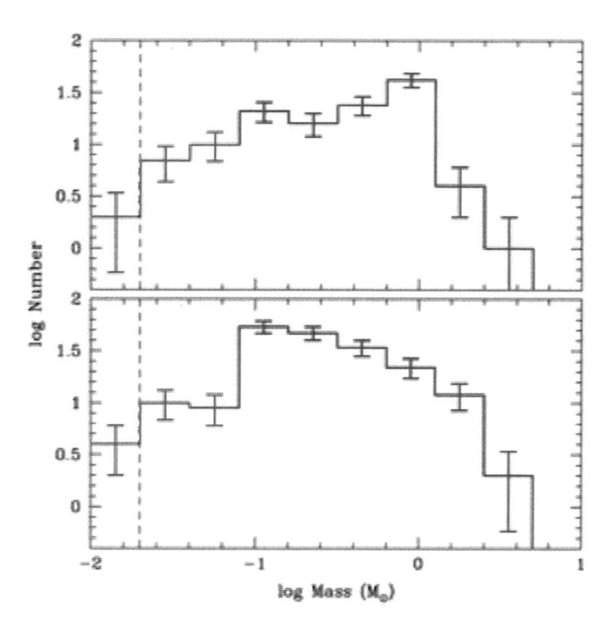

Fig. 6.12 Derived IMFs for the Taurus cloud (Luhman 2004) and the young cluster IC 348 (Luhman et al. 2003). The Taurus IMF peaks near 1 $M_{\odot}$ and is a rare example of an IMF that differs from the IMFs of the field and most embedded clusters. The IC 348 IMF peaks near 0.1 $M_{\odot}$ and is similar to the IMFs of the field and rich clusters such as the Trapezuim (Fig. 6.3). The Taurus observations provide the strongest evidence yet for possible variations in the IMFs of differing star forming regions

In order for the Pipe cores to produce a standard IMF the critical BE mass needs to be lowered. Inspection of equation (6.7) shows that this could be accomplished in two ways, by either increasing the external pressure or lowering the internal pressure support via a decrease in the sound speed, a (= $(\mathrm{kT}/\mu m_H)^{0.5}$). The external pressure would have to be increased by about an order of magnitude (to $P_{ext}/k \approx 10^6$ K cm^{-3}) in order for the critical BE mass to be reduced sufficiently to enable cores that will spawn 0.1–0.2 $M_{\odot}$ stars to cross the equilibrium threshold. This may be difficult to do but could happen if the cloud globally contracted by a factor of about two.

We note here that massive cloud cores that form clusters (e.g., Orion, ρ Oph) have pressures between 10^{6-7} K cm^{-3} and corresponding critical BE masses that would permit the formation of the lowest mass stars and the full sampling of a standard IMF if the sound speed was the same as in the Pipe cores. The fact that most stars form in such clusters likely explains why the field star IMF and the cluster IMF are so similar to each other and dissimilar to the Taurus IMF.

The critical BE mass is very sensitive to the sound speed, varying as a^4. A drop in the internal pressure by a factor of 2 would be enough to lower the critical mass to enable the lower mass cores to form stars down to the hydrogen burning limit. However, the measured temperatures of the Pipe cores are already close to 10 K, the equilibrium value for gas heated by cosmic rays and interstellar light and cooled by line radiation. It is unlikely that these cores could further cool to decrease their temperatures by the necessary factor of 2. Dissipation of any additional sources of internal pressure support such as static magnetic fields and turbulence in the cores could help, but it may be in the end that star formation in the Pipe will produce an IMF more similar to that of Taurus than that of the field.

The above considerations suggest that most star formation occurs in high pressure environments, ($P_{ext}/k > 10^6$ K cm^{-3}) such as the massive dense cores that produce rich young clusters similar to the one shown in Fig. 6.1. One might expect that in very high pressure regions, the critical BE mass would be so small that a shift in the peak of the resulting IMF to lower masses would result. However, such regions are also characterized by higher internal gas temperatures ($\geq$ 20 K), which could also counteract the higher external pressures (6.7) to still produce an initial critical mass similar to that of the Pipe region. Unlike the case of the Pipe, the higher internal temperatures and pressures expected for the cores formed in these regions allow considerably more room for the possibility of cooling and internal pressure reduction as, for example, could occur with gas cooling to the general molecular equilibrium temperature of 10 K. This could produce an IMF with the same mass scale as the field star IMF and simultaneously allow very low mass stars, even brown dwarfs, to form in abundance.

In closing, it is interesting to note that Fig. 6.11 may also provide a clue concerning the origin of stellar multiplicity, the other remaining important boundary condition for a theory of star formation to satisfy. Consider that as the mass of a core increases above the critical value, the core becomes increasingly distant from an equilibrium state and becomes more likely to fragment and form multiple stellar systems. Cores at or near the critical mass are very close to hydrostatic/mechanical equilibrium and are very unlikely to fragment further. These cores will tend to form single stars. Thus, stellar multiplicity will be expected to increase with stellar mass, similar, at least qualitatively, to what is observed (Fig. 6.2).

In any event, the results described in this review suggest that stellar masses may ultimately originate in a relatively simple physical process of pressurized thermal (Jeans) fragmentation. This process in turn is prescribed by only a few measurable physical parameters, including internal gas temperature and surrounding pressure. The former being set by the general process of heating and cooling in dense cores and the latter by the basic properties of the surrounding environment. In principle, these properties can be empirically determined over a wide range in scale, from star-forming regions in the solar neighborhood, to the center of the Galaxy, to distant external galaxies. New instrumentation and facilities such JWST, ALMA and giant ground-based optical and infrared telescopes will provide the essential capabilities for such observations.

Acknowledgments Much of the research reviewed here is based on the product of joint research conducted and published over the past decade with a research team that includes João Alves, Elizabeth Lada, Marco Lombardi, August Muench and Jill Rathborne. Without their dedicated efforts these results and this article largely based on them would not have been possible. I thank Kevin Luhman for providing material used to construct Fig. 6.12.

References

Adams, F. C., & Fatuzzo, M. 1996, *ApJ*, 464, 256.
Alves, J.F., Lada, C.J., & Lada, E.A. 2001, Nature, **409**, 159.

Alves, J., Lombardi, M. & Lada, C.J. 2007, *A&A*, 462, L17.
Bertoldi, F. & McKee, C.F. 1992, *ApJ*, 395, 140.
Bonnor, W.B. 1956, *MNRAS*, 116, 351.
Di Francesco, J., Evans, N.J. II, Caselli, P. Myers, P.C., Shirley, Y., Aikawa, Y., & Tafalla, M. 2007 in *Protostars & Planets V*, eds. B. Reipurth, D. Jewitt & K. Keil (University of Arizona Press, Tucson), p. 17.
Ebert, R. 1955, *Zeitschrift fur Astrophysik*, 36, 222.
Enoch, M. L., et al. 2006, *ApJ*, 638, 293.
Hotzel, S., Harju, J., & Juvela, M. 2002, *A&A*, 395, L5.
Kirk, H., Johnstone, D., & Tafalla, M. 2007, *ApJ*, 668, 1042.
Kirk, H., Johnstone, D., & Di Francesco, J. 2006, *ApJ*, 646, 1009.
Kroupa, P. 2002, *Science*, 295, 82.
Lada, C.J. 2006, *ApJ*, 640, L63.
Lada, C.J. & Lada, E.A, 2003, *ARAA*, 41, 57.
Lada, C.J., Bergin, E.A., Alves, J.F., & Huard, T.L. 2003, *ApJ*, 586, 286.
Lada, C.J., Muench, A.A., Rathborne, J., Alves, J., & Lombardi, M. 2008, *ApJ*, 672, 410.
Larson, R.B. 1981, *MNRAS*, 194, 809.
Lombardi, M., Alves, J., & Lada, C.J. 2006, Astronomy & Astrophysics, **454**, 781.
Luhman, K.L. 2004, *ApJ*, 617, 1216.
Luhman, K.L., Stauffer, J.R., Muench, A.A., Rieke, G.H., Lada, E.A., Bouvier, J., & Lada, C.J. 2003, *ApJ*, 593, 1093.
Motte, F., Andre, P., & Neri, R. 1998, *A&A*, 336, 150.
Muench, A.A., Lada, E.A., Lada, C.J. & Alves, J.F. 2002, *ApJ*, 573, 366.
Muench, A.A., Lada, C.J., Rathborne, J.M., Alves, J.F., & Lombardi, M. 2007, *ApJ*, 671, 1820.
Myers, P.C. 1999 in *The Origin of Stars and Planetary Systems*, eds. C.J. Lada & N.D. Kylafis, (Kluwer, Dodrecht), p. 67.
Rathborne, J.M., Lada, C.J., Muench, A.A., Alves, J.F., & Lombardi, M. 2008, *ApJS*, 174, 396.
Salpeter, E. E. 1955, *ApJ*, 121, 161.
Scalo, J.M. 1978, in *Protostars and Planets*, ed. T. Gehrels, (University of Arizona Press, Tucson), p. 265.
Testi, L., & Sargent, A.I. 1998, *ApJ*, 508, L91.

Ray Villard and Mary Estacion discuss the intricacies of astronomical discoveries and the 24-hour news cycle

Chapter 7
Accretion Disks Before (?) the Main Planet Formation Phase

C. Dominik

Abstract Protoplanetary disks are the sites of planet formation and therefore one of the foremost targets of future facilities in astronomy. In this review, I will discuss the main options for using JWST and concurrent facilities to study the early, gas-rich, massive phases of protoplanetary disks. We discuss the opportunities to study gas and dust in such disks and to derive constraints on physical and chemical processes taking place there.

7.1 Protoplanetary Disks and Planet Formation

One of the most intriguing questions in astronomy today is the question how planets form. After knowing the existence of our on solar system for hundreds of years, only in the last decade have we been able to demonstrate that other stars are surrounded by more or less similar planetary systems that harbor both Jupiter-like giant planets, and lighter, rocky planets possibly similar to the terrestrial planets in our solar system. These discoveries have shown that the basic processes that lead to the formation of planets are universal processes, and that studying the formation of planets in extrasolar planetary systems can tell us about which processes lead to the formation of our own planetary system, including the Earth on which life could arise. The main question we would like to address are how planets form, what processes play a role, and how do these processes depend on the environment (cluster versus isolated star formation), density, temperature, mass and rotation of the prestellar core, and metal content. What are the requirements a system has to fulfil in order to produce the different classes of planets and physical and orbital structure of observed systems? Ultimately we wish to understand how our own solar system is similar to other system, and how it might still be a special place in the universe.

C. Dominik (✉)
Sterrenkundig Instituut "Anton Pannekoek", University of Amsterdam, Kruislaan 403, NL-1098 SJ Amsterdam; and Afdeling Sterrenkunde, Radboud Universiteit Nijmegen, Faculteit NWI, Postbus 9010, NL-6500 GL Nijmegen
e-mail: dominik@science.uva.nl

H.A. Thronson et al. (eds.), *Astrophysics in the Next Decade,* Astrophysics and Space Science Proceedings, DOI 10.1007/978-1-4020-9457-6_7,

It is clear now that planet formation takes place in the circumstellar disks surrounding almost all young stars. The last two decades of astronomical research have seen an explosion of information about protoplanetary disks. The presence of protoplanetary disks around young pre-main-sequence stars was first derived from spectral information showing infrared and excess radiation over a broad range of range of wavelengths (e.g. Adams et al. 1987). In the mean time, imaging and detailed spectroscopy of disks at many wavelengths is the norm and has lead to increasing understanding of these objects. Upcoming facilities are bound to and have in part been constructed to take these observations to the next level.

Massive, optically thick circumstellar disks with strong excess signature in the near infrared initially surround all low and intermediate stars, but disappear on timescales of the order of 5–10 Myr (Haisch et al. 2001). Much more tenuous disks, the so-called debris disks consist of secondary dust created in collisions between leftover comets and planetesimals can last much longer (Habing et al. 1999) or flare up again (Rieke et al. 2005). Debris disks will be the subject of the next chapter (contribution by M. Meyer, this volume).

In order to understand specific information about structure, composition, and dynamics of protoplanetary disks, it is useful to have a schematic picture of the planet formation process in mind. From a combination of observations and theoretical modelling, the following basic picture is currently used as the working hypothesis to study the formation of planets.

1. At the beginning of star formation stands the collapse of a high density molecular core with a typical mass of a solar mass, and a typical size of 0.1 pc. Even a small value for the initial rotation of the star will force the collapsing matter into a disk shape, with the location of infalling matter moving from the center outward. In addition to this, viscous spreading leads to an increase of the disk size with time (Hueso & Guillot 2005).
2. During the active infall period, the disk mass increases until it reaches a value of about 10^{-2} to a few 10^{-1} solar masses at the end of the infall phase. During that phase matter is streaming through the disk toward the star at a rate of order $10^{-6}\,M_{\odot}$/yr. This accretion flow is driven by viscous coupling in the disk, requiring an internal viscosity of the disk much larger than the molecular viscosity. It is widely agreed that the mechanism generating this viscosity is the magneto-rotational instability which relies on the presence of a weak magnetic field and a level ionization in the disk material. This phase lasts for at most one Myr. The temperatures in the disk range from several thousand degrees close to the star to a mere 10-20 degrees further out in the disk. Not much is know about the density and surface density distribution in the disk. At a distance of 1AU from the star, the surface density in the disk can be estimated from the minimum solar mass nebula to be at least of order 1000 g/cm^2, corresponding to a density of 1 few times 10^{14} cm^{-3}. In the outer disk, hundreds of AU away from the star, densities of 10^8 cm^{-3} are typical.

3. In the subsequent phase with duration of a few Myr, the disk is becoming more quiescent as the accretion rate and therefore the energy production in the disk drops by several orders of magnitude. The disk will disappear after about 5 Myr, due to accretion and possibly planet formation in the inner disk and photo evaporation in the outer disk (e.g. Dullemond et al., 2007).
4. It is not entirely clear when and how the process of planet formation really gets started. During the main accretion phase, the entire disk mass is accreted several and possibly many times onto the star, and it is likely that most larger dust grains and planetary seeds get lost by evaporation and eventually by accretion onto the star. However, a part of the matter processed in this phase can be preserved, either by growing large, asteroid-like bodies quickly that may not follow the gas into the star, or by the diffusive and turbulent mixing processes counter acting the radial accretion flow.
5. At the start of the planet formation process stands the slow growth of dust particles immersed in the disk gas (see Fig. 7.1). The relative motion of these particles is initially driven by Brownian motion, and it is well understood now that this growth process produces very porous aggregates, with fractal dimensions of typically 1.5 (Krause & Blum 2004, Paszun & Dominik 2006). Spectroscopically, such aggregates would be difficult to distinguish from small grains (e.g. Min et al. 2006). As particles continue to grow, they start coupling to the turbulent velocity field. The aggregation speed increases (Dullemond & Dominik 2005), and higher velocity collisions start to compact particles (Ormel et al. 2007). The more compact particles settle toward the midplane of the disk, but turbulent motions keep smaller and more porous grains high in the disk atmosphere. High speed collisions between aggregates lead to fragmentation and replenish the small particle population in the disk.
6. As the accretion rate goes down, the disk becomes more quiescent, and evolutionary timescales for the disk structure increase to a Myr. If the strength of the turbulence decreases as well, particles can settle toward the disk midplane more efficiently, and the growth to larger and larger bodies somehow continues. It is unclear so far whether the driving process is aggregation in slow 2-body collisions, gravitational instabilities in a dust-enriched midplane, or the formation of gravitationally bound clusters of intermediate (meter?) sized particles in turbulent eddies (Johansen et al. 2007).
7. Continues growth leads eventually to planetary embryos. Embryos with masses of typically 10 Earth masses can start to accrete gas while there is still gas in the disk, i.e. during at most 10 Myrs. Rocky planet formation may take longer, the estimated formation time for the Earth is 30 Myr. Growth processes in the outer disk can take much longer, up to a Gyr. As large planets form the gravitational interaction between planets and the remaining disk material becomes the dominant dynamical process. While the disk is still massive and gas-rich, large-scale planet migration may cause the creation of hot Jupiters or even the loss of planets into the star. Later, when the disk mass is dominated by the planetesimal population, the dynamical interaction between this population and the planets may reshape the system (e.g. Levison & Morbidelli 2003, Gomes et al. 2005)

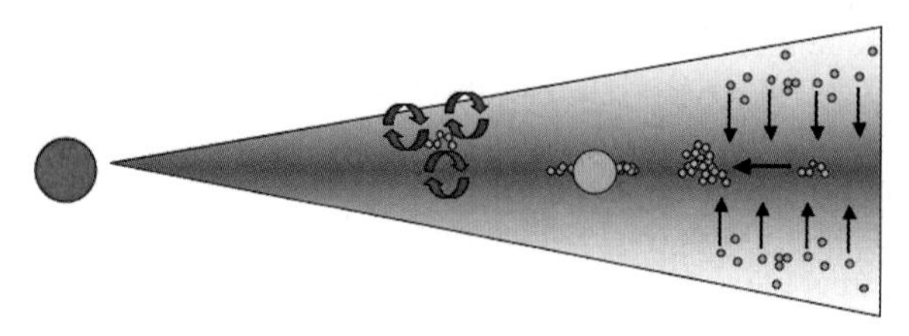

Fig. 7.1 Basic processes that lead to the formation of large particles in disks. Initially, dust grains are suspended in the disk gas at all heights. Relative motions, caused by Brownian motion, differential settling and radial drift, and turbulence, cause these grains to collide and to grow. Finally, accumulations of solids in the disk midplane might lead to gravitationally bound objects that may lead to the formation of planetesimals

7.1.1 What Would We Like to Know

All the processes sketched in the previous section will depend on the composition and distribution of matter in the protoplanetary disk. In order to advance the understanding of the formation and evolution of protoplanetary disks, we can write down a not very moderate list of things we would like to know:

- Full spectral energy distributions of disks, ranging from optical to radio wavelengths.
- Dust grain sizes, internal structures, composition and mineralogical state as a function of distance from the star, height in the disk, and time.
- Gas mass, ionization state, and molecular abundances as a function of distance from the star, height in the disk, and time.
- Age of the object.
- Kinematic and dynamic data, i.e. rotation and inflow, disk winds, viscosity, turbulence, magnetic fields, accretion and instabilities.

Of course, this ideal list of quantities we will never know in full. The rest of this article is about how to use the different observations that will become possible in the next decade to approach this goal and to derive ever tighter constraints for dynamical and radiative transfer models of protoplanetary disks.

Figure 7.2 shows an overview over the expected structure of a protoplanetary disk, and what kinds of components we can expect in different parts of the disk. The presence and state of different components is in part governed and generated by the components of the radiation field that can reach a particular location in a disk - therefore we will describe the radiation field as well.

In the innermost region, close to the star, temperatures are so high that no solid particles can exist. All ices and refractory materials have been evaporated and will form a molecular gas consisting mostly of H_2, along with stable molecules like CO, and H_2O. It is important to know that this gas has no strong continuum opacities and therefore may be optically thin to stellar radiation Calvet et al. 2002.

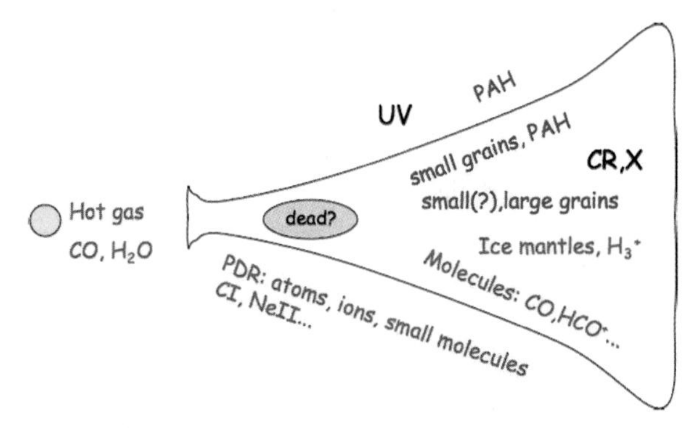

Fig. 7.2 Sketch of the main components of disk material and radiation field present in protoplanetary disks. The figure shows as an outline the structure of the region where the opacity is dominated by dust grains. In the inner disk, that region starts with the puffed-up inner rim that is caused by the opacity jump between dust-free and dust-containing matter (Dullemond et al. 2001a). The bole-shape of the disk surface is what is commonly referred to as a flaring disk, with the disk thickness increasing faster than the distance from the star, a consequence of hydrostatic equilibrium in an irradiated disk (Chiang & Goldreich 1997). Due to this shape, much of the surface received stellar light directly. The figure shows in blue the different grain components, and in red the gas components expected in the disk environment

Further away from the star, when the temperature decreases to about 1500 K, iron, oxide and silicate dust particles can exist and form an inner opacity rim in the disk. On a radial ray outward from the star, the optical depth becomes high very quickly, making the material completely opaque to optical and near-IR radiation. Even rays reaching over the inner rim or scattered toward the disk will not reach deep into the disk. Therefore, the disk naturally develops a number of distinct regions with different chemistry.

1. A surface region which is still penetrated by stellar UV and optical light. In this region, a PDR plasma develops in which most molecules are dissociated, and some atoms become ionized. Because of dust settling, the only grains that will exist in this region are very small grains and polycyclic aromatic hydrocarbons (PAH's).
2. Deeper into the disk, at a layer where the optical depth for optical and UV photons, reaches unity, the photo dissociation of molecules stops. This region is dominated by small grains, PAH's, and molecules like CO, HCO^+.
3. Even deeper into the disk, the material will become colder as the heating through radiation from the disk surface decreases. In this region, most molecules will be frozen out onto dust grains. Outside of a few AU, the disk material will still be weakly ionized due to the effects of X rays and cosmic rays that can penetrate here. However, in the inner disk regions at column densities deeper that $100\,g/cm^2$, the ionisation is expected to be very low, possibly leading to dead zones because of decoupling of the gas from magnetic fields in the disk (Gammie 1996).

7.1.2 Emission Locations

If we would like to use observations to discover properties of protoplanetary disks, it is very important that we keep in mind how vastly different the regions of the disk are that are probed by observations at different wavelengths (see Figs. 7.2 and 7.4):

UV lines and continuum
: UV radiation from the disk may originate mostly as recombination lines in the photon-dominated regions (PDR's) at the inner disk and the disk surface.

Optical lines
: Optical emission lines originate in the inner disk, in the disk surface PDR, and also possibly in shocks.

Optical and NIR imaging: Scattered light
: Scattered light is originating from dust grains in the entire disk surface. Scattered light images can trace the disk out to large distances and show structure in the disk surface as well as changing dust properties.

MIR imaging: Warm dust
: The dust emitting the MIR radiation in disks is the warm dust in the disk surface. The inner regions of the disk are still optically thick at 10 μm, so the midplane cannot contribute here. The outer disk is optically thin at these wavelengths, but the dust quickly becomes too cold to contribute. It is very important to remember that these observations probe the surface dust only and that using them to make statements over the bulk dust composition assumes efficient vertical mixing.

MIR spectroscopy: warm dust and PAH's
: As far as the solid state features from dust grains are concerned, the same constraints as for MIR imaging apply: only the dust at the disk surface contributes. PAH emission features on the other hand can originate from the entire disk surface because of the non-thermal nature of the excitation and emission process. Therefore PAHs trace dust in the entire disk surface, much like the scattered light does.

Absorption studies in edge-on disks
: Absorption studies in edge-on disks provide information about material along a pencil beam toward the star at optical wavelength. At MIR wavelength, the continuum emission originates from the warm dust in the inner disk, and absorption features originate in the outer disk. This offers a unique technique to study cold mid-plane dust, including icy mantles made of CO and/or water.

Submillimeter and millimeter line emission
: Rotational lines are a very powerful method to study gas in disks. The high spectral resolution that can be reached in this wavelength regime allows detailed dynamical studies for disk to be made. However, in the coldest parts of the disk they suffer from the freeze-out of the most important tracer molecules like CO.

Sub-millimeter and millimeter continuum imaging

This is the technique to study the dust in the entire disk. Submm radiation is emitted by all dust grains that are not much larger than the wavelengths of observation. Both warm dust in the inner disk regions and cold dust in the outer regions will contribute to this emission, and submm photometry is a good way to determine disk masses.

7.2 Measuring Disks

In this chapter, I will go through the main disk properties we would like to measure, and the kind of measurements that can achieve this goal. I will be focusing on the observations facilitated by 4 major facilities coming online in the next decade: JWST, HERSCHEL, ALMA, and SOFIA. Section 7.3 will contain a table summarizing the main conclusions from this discussion.

7.2.1 Infall Onto the Star/Disk System

One of the important parameters governing disk evolution is the actual infall rate of new matter onto the disk. While the accretion luminosity and line veiling have been used frequently to observe accretion onto the star itself, thereby tracing the accretion flow in the inner most disk regions, infall from the collapsing core is much more difficult to observe (Zhou et al. 1993, Di Francesco et al. 2001). The successful searches have used molecular tracers CS, H_2CO, and N_2H^+. Such observations require high velocity resolution because the infall velocities are only a km/s or less. They also require high spatial resolution in order to reduce confusion with other velocity fields in the inner disk region like molecular cloud turbulence and outflows. It has also been argued that such observations are only possible during a short window in the evolution of protostars (Rawlings 1996). An important contribution in this field will come from JWST, allowing high resolution imaging in the mid-IR region in order to constrain the warm inner structure of collapsing clouds. ALMA on the other hand with its superior spatial resolution will allow to isolate the infall regions and to provide direct observations of the asymmetric line profiles resulting from infall without confusion.

7.2.2 SED's and Disk Geometry

As imaging with spatial resolution is only available for a limited number of protoplanetary disks, the spectral energy distribution of protoplanetary disks remains an important tool to study large numbers of disks and to derive basic geometrical properties of these disks. It is clear that this kind of work is challenging because fitting an SED with a radiative transfer model suffers from non-uniqueness and degeneracies (e.g. Thamm et al. 1994, Miroshnichenko et al. 2003). Nevertheless it is possible to use energy conservation arguments in order to derive basic geometrical properties of disks (Dullemond et al. 2001b). Two distinct features of SEDs that do reflect geometrical aspects are the onset and energy content of the near-infrared excess that reflects the temperature and luminosity, and therefore the location and height of the inner disk boundary (e.g. Dullemond et al. 2001a, Calvet et al. 2002, 2005, Espaillat et al. 2007), and the total fraction of reprocessed light that portrays the extent of flaring of the outer parts of the disk (Dullemond & Dominik 2004, Dominik et al. 2003). The size of inner holes in disks has received lots of attention recently as it serves as a strong tracer of disk evolution and in particular disk clearing by possible planets in the disk. The recent discovery of a planet in the inner disk of TW Hya (Setiawan et al. 2008) re-enforces the connection between planets and inner holes. The near-IR emission and direct measurement of disk hole sizes is feasible with current ground-based telescopes and optical/IR interferometers (Eisner et al. 2005, 2006). However, the geometrical structure of the outer disk, flaring versus non-flaring disk influences the SED mostly in the far infrared. Measuring SEDs at these wavelengths will be possible for large samples using in particular HERSCHEL. As an example, Figure 7.3 show a number of model spectral energy distributions of a Herbig Ae star. These are SEDs from theoretical models, but observed SEDs look very much like these curves. In the model, different amounts of dust have been driven into large grains and settled to the midplane. The corresponding loss of opacity in the upper layers of the disk changes the disk geometry from flaring to self-shadowed, and the differences are most clearly seen between 20 and 400 μm. Much of this wavelength region is hidden for ground-based instrumentation, but will be easily observed with PACS and SPIRE on board of the HERSCHEL space observatory. Other tracers of

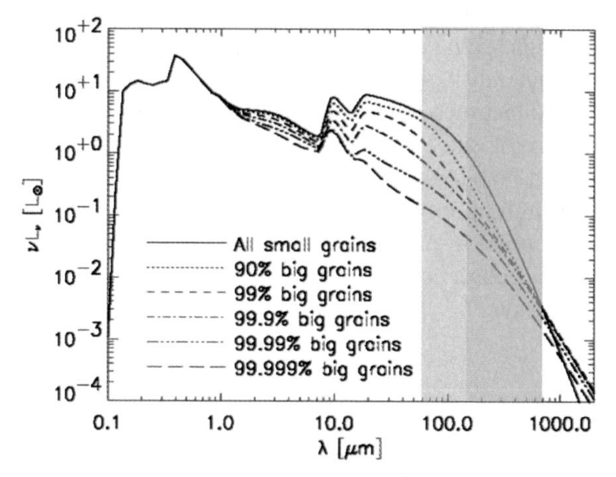

Fig. 7.3 The spectral energy distribution of circumstellar disks around a Herbig Ae star. The different lines show models with the specified fraction of the dust mass driven into mm-sized grains and moved to the midplane, thereby changing the SED group from I to II in the Meeus et al. (2001) classification. The shaded areas shows the HERSCHEL spectral range for photometric observations with PACS and SPIRE. Figure from Dullemond & Dominik (2004)

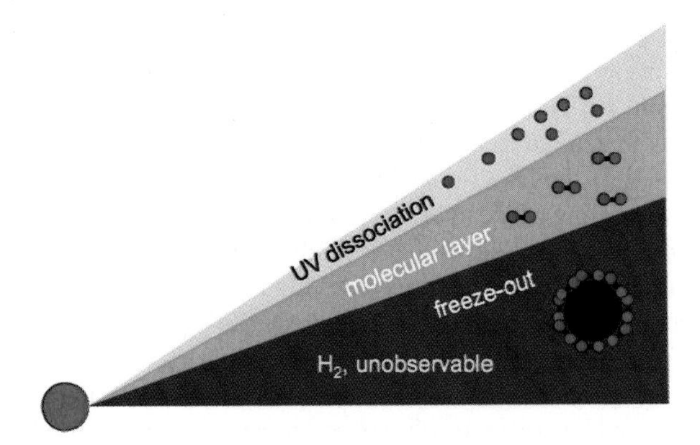

Fig. 7.4 The chemical structure of protoplanetary disks with the different zones: UV dominated PDR on top which is consisting mainly of atoms and ions; the molecular layer in the middle which is shielded from UV irradiation and therefore contains most of the molecules like CO in gas form; ice-mantle regime in the disk midplane where all molecules made of 'metals' are frozen onto the grain surface

the disk scale height and structure may be lines originating from the PDR on top of the disk (Dullemond et al. 2007), for example the CI and CII lines (Kamp et al. 2004).

7.2.3 Gas Chemistry and Mixing

The study of disks using molecular tracers is still in its beginning, mostly because of the limitations of currently available instrumentation for the study of protoplanetary disks. The most important tracer so far has been, of course, CO and its isotopologues, which have been used to trace the velocity field and the size of disks in a number of pioneering studies (Koerner & Sargent 1995, Dutrey et al. 1997, Mannings & Sargent 1997). Subsequent studies found HCO^+, HCN, CN, HNC, H_2CO, CS, and C_2H in the most important disks in Taurus (GG Tau, DM Tau, Dutrey et al. (1997)) and in TW Hya. Further studies found rare molecules, such as DCO^+ (van Dishoeck et al. 2003, Guilloteau et al. 2006), N_2H^+ (Dutrey et al. 2007) and more recently DCN (Qi et al. 2008). Detailed mapping of such lines can reveal information about the vertical temperature structure in the disk (Dartois et al. 2003). All these studies show that most molecules are severely depleted with respect to their ISM abundances, as the temperatures in the outer disks are so low that water, CO, and all other molecules containing 'metals' do indeed freeze out (see Fig. 7.4). In fact, simple models predict that freezing-out should be so severe that even the currently observed abundances of molecules like CO are difficult to explain. A number of different processes have been invoked for keeping detectable amounts of these molecules in the gas, including cosmic-ray-desorption (e.g.Roberts et al. 2007), photo-desorption (Willacy & Langer 2000, Dominik et al. 2005, Willacy 2007) and turbulent mixing (Willacy et al. 2006, Semenov et al. 2006). Photodesorption of

CO may have also been severely underestimated (Öberg et al. 2007). Also in the inner disk regions, mixing processes can lead to transport of molecules into regions where they would not be expected from a local chemical model (e.g. Ilgner et al. 2004). As far as freezing out of molecules is concerned, the recent discovery of an emission line of H_2D^+ in DM Tau and possibly in TW Hya (Ceccarelli et al. 2004), and successful modeling of the chemistry capable to produce this molecule (Ceccarelli & Dominik 2005) has shown that it might be possible to access the midplane gas component in disks with this molecule. First attempts to image the chemistry in protoplanetary disk have been made (e.g. Qi 2001, Kessler-Silacci 2004) and have shown the potential revolution that ALMA will bring to this field. ALMA will allow detailed imaging of the rotational transitions of a large number of molecules with high spatial and kinematic resolution for nearby protoplanetary disks and bring chemical models to the test. ALMA has the potential to study all key molecular ions (HCO^+, N_2H^+ and their deuterated counterparts, but also H_2D^+ and D_2H^+), and thus will be an essential instrument for the study of the ionization state of disks. Simulations indicate that the ALMA interferometer will allow to distinguish the effects of temperature gradients and chemical stratification in disks through molecular line observations, in particular for highly inclined objects (Semenov et al. 2008). JWST will give access to atomic and molecular gas emission lines from, for example, H_2, CO, H_2O, CH_4, SiII, SI, FeII, and NeII. An important diagnostic comes from absorption lines in nearly edge-on disks where the radiation from the star or from the warm inner disk passes through the disk. If the inclination of the disk is favorable, this makes it possible to detect transitions of, for example, C_2H_2, HCN, CO, and CO_2 (Rettig et al. 2003, Lahuis et al. 2006).

HERSCHEL has the potential to detect new lines in disks, but the large beam of HERSCHEL does make this a difficult program. SOFIA can contribute to chemical studies by providing high resolution spectra to study the inner, warm regions of disks, and it may detect water lines.

7.2.4 Grain Sizes and Spatial Distribution

Probing the early stages of planet formation is one of the central goals in the study of young protoplanetary disk. It is clear that the formation of terrestrial planets and very likely also the formation of the cores of giant planets proceeds by gradual growth of dust particles, starting from the unprocessed, 0.1 μm-sized grains. Demonstrating the presence of grains larger than these initial sizes therefore indicates the presence of processes that pave the way to the formation of planets. The presence of large grains has indeed been demonstrated in number of different ways, and most of these methods will be taken to a new level with the upcoming new facilities.

1. The shadow images of proplyds in Orion provide a way to measure the optical depth of the disk as a function of distance from the star. Executing such observations at different wavelengths allows to probe the mean absorption cross section of the dust as a function of wavelength, and therefore the grain size. So

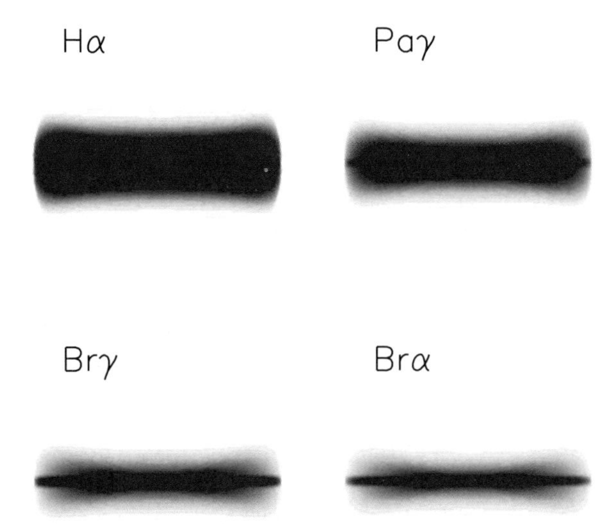

Fig. 7.5 Model for the change in shadow darkness and shape for a protoplanetary disk with a mid-plane layer of mm-sized grains. The model disk has a radius of 100 AU and an inner temperature of 1500 K. The disk surrounds a 0.5 $M_{\odot}$, 4000 K T Tauri star, and has a powerlaw surface density with \$$\Sigma \propto r^{-1}$. The vertical structure is computed from hydrostatic equilibrium according to the model of Chiang & Goldreich (1997). In addition, the disk contains a mid-plane layer of large grains, containing 10% of the mass of the small grain component. Four plot are shown, with the shadow depth plotted for 4 different wavelengths (0.66, 1.1, 2.63, and 4.05 μm). The figure shows how the small particle part of the disk becomes more transparent and the midplane layer becomes more and more visible. Figure courtesy from C.P. Dullemond, private communication

far these have been limited to a maximum wavelength of 4.05 μm using the Brα line (Throop et al. 2001, Shuping et al. 2003). Figure 7.5 shows an illustration for the changes in shadow shape for an edge-on disk with a midplane layer made of mm-sized grains. JWST will allow to extend such observations to longer wavelengths, probing larger grains.

2. Observations of scattered light from disk can give indications of large grains. McCabe et al. (2003) observed the disk of HK Tau B at a wavelength of 11.8 μm and showed that there are clear indications for dust grains several microns in size. Differences of scattering images at a series of wavelengths may also trace the spatial distribution and settling of dust (Duchêne et al. 2004).
3. A big industry has developed around studying the 10 μm and 20 μm silicate emission features for signs of grains growth. This feature becomes weaker and broader when the grains grow from sub-micron to about 3μm sizes, an effect seen for Herbig AeBe stars (van Boekel et al. 2003) and T Tauri stars (Przygodda et al. 2003, Kessler-Silacci et al. 2006, 2007). The 10 and 20 μm features as well as several ice bands can also be seen in absorption toward edge-on disks (e.g. Pontoppidan et al. 2007 Fig. 7.6). Also here, the feature shape can be interpreted in terms of grain size and shape. JWST will allow these studies to be expanded to much weaker sources at larger distances. In particular for absorption studies of edge-on disks, JWST will be the ideal instrument.

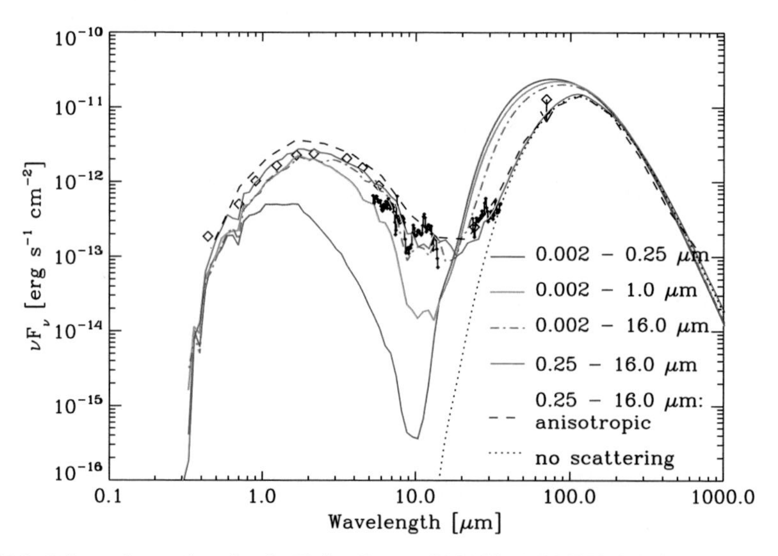

Fig. 7.6 Spitzer observations for the Flying Saucer (2MASS J16281370-2431391) edge-on protoplanetary disk which demonstrate the diagnostic value of absorption studies in protoplanetary disks. Grain size distributions weighted toward increasing grain sizes move the mid-infrared minimum in the SED to progressively longer wavelengths. The fit shows a significant contribution of grains in the 5-10 μm size range. Figure from Pontoppidan et al. (2007), with permission

4. Submm and mm measurements of the dust continuum emission can be used to measure much larger grains. In particular, spatially resolved observations in combination with modelling has proved to be a powerful method (Testi et al. 2003, Natta et al. 2004, Rodmann et al. 2006). Measurements of the scale height of the distribution of large grains responsible for the mm continuum have been performed in HH 30 (Guilloteau et al. 2008). The interpretation of different scale heights as function of observing wavelength is ambiguous, as vertical temperature gradients have the same apparent effect as dust settling. Such measurements will be much more accurate with ALMA, and it should be possible to derive dust sizes as a function of distance from the star.

7.2.5 Dust Mineralogy

Besides grain growth, the other interesting effect on dust in disks is thermal processing. Thermal processing turns amorphous dust into crystalline dust. Observing the distribution of crystalline dust in a disk therefore carries information about the thermal history of the material. A basic indication about the overall crystallinity of small dust grains can be obtained by looking at the emission features of dust grains (van Boekel et al. 2005, Kessler-Silacci et al. 2006). However, the interpretation of this data is difficult, because dust particles from different regions in the disk contribute. Interferometric observations of disks with MIDI-VLTI have made it possible to extract the correlated spectra of a number of Herbig stars, which for the typical distance of such stars corresponds to the emission of the innermost few AU, i.e.

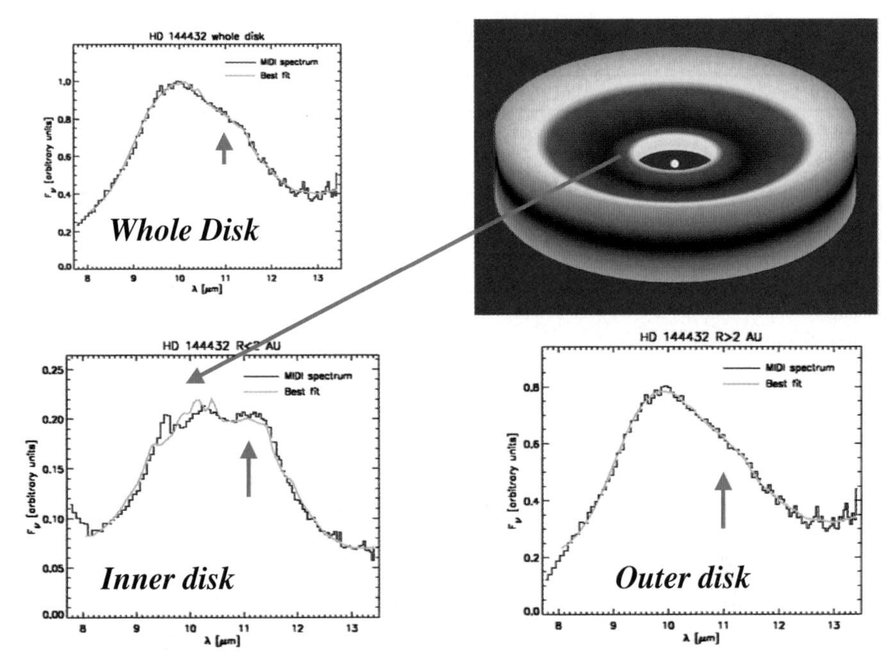

Fig. 7.7 10 μm feature of The Herbig Ae star HD 144432, with separate spectra for whole disk (as measured by single-telescope spectra), the inner 2 AU of the disk (as measured with the VLTI/MIDI interferometric setup at ESO), and outer disk (computed as difference between whole disk and inner disk). The inner disk shows a flat spectrum with a weak peak/over continuum ratio indicative for grains grown to about 3 μm. It also shows several sharp peaks indicating the presence of enstatite and forsterite in crystalline form. The outer disk spectrum shows the typical shape and strength for a feature produced my small ($\sim$ 0.1 μm) amorphous silicate grains. The whole-disk spectrum mostly resembles the ISM feature as well, but with a clearly visible feature at 11.3 μm (forsterite). van Boekel et al. (2004)

a region comparable to the formation region of the terrestrial planets in our solar system (see Fig. 7.7). JWST will provide spectra of extremely weak sources, without giving spatial resolution on most of these sources, so the interpretation will largely hinge on theoretical modelling. A recent attempt (Meijer et al. (2008), submitted to A&A) shows that the typical observations of disks with ISO and Spitzer require high crystallinity in the innermost region, steep decrease to a low level of a few percent, which has to be sustained out to large distances from the star.

While the contribution of JWST to this field will be limited, SOFIA and in particular HERSCHEL offer the possibility to detect new dust features in the far infrared. Table 7.1 contains possible features that are expected in this region.

7.2.6 Dynamics: Rotation, Radial Flows and Turbulence

Finally, we turn to the dynamical information that can be obtained from disk observations with the new facilities. An important measurement is that of rotational velocities as a function of distance from the star. If the disk is in Keplerian rotation,

Table 7.1 Far-IR features for a number of dust species. From Posch et al. (2005), Koike et al. (2000, 2006)

Material	Formula	Features at [μm]
Forsterite	Mg_2SiO_4	69–70
Fayalite	Fe_2SiO_4	93–94,110
Diopside	$CaMgSi_2O_6$	65-66
Calcite	$CaCO_3$	92
Graphite	C	50-70
Water ice	H_2O	62
Methanol ice	a-CH_3OH	68,88.5
Dry ice	CO_2	85
Montmorillionite	$(Na,Ca)_{0.33}(Al,Mg)_2(Si_4O_{10})(OH)_2$+n H_2O	80–100
Serpentine	$((Mg,Fe)_3Si_2O_5(OH)_4$	47
Chlorite	$Mg(ClO_2)_2$	69,86,268

the dynamical mass of the star can be determined (e.g. Simon et al. 2000, Isella et al. 2007). However, also deviations from Keplerian velocity can be detected, showing either inflow (Hogerheijde 2001) or outflow (Piétu et al. 2005). These are both very important indications for the evolution of the disk. Inflow shows accretion of an initially large disk, while outflow may indicate viscous spreading (Hueso & Guillot 2005), even though the velocities will be extremely slow and difficult to measure.

Another important measurement is the determination of turbulence in disks. Turbulence is thought to be the carrier of the viscosity that drives accretion and spreading in disks, as well as being responsible for chemical mixing processes and the stirring up of dust that is trying to settle to the midplane of the disk. Direct measurements of turbulence have been reported using hot water (Carr et al. 2004) and first overtone emission of CO (Hartmann et al. 2004) and the lowest rotational transitions of CO and isotopologues (Dartois et al. 2003, Piétu et al. 2007).

7.2.7 Tracing Proto-Planets

Although these dense disks are called protoplanetary disks, planet formation may already have occurred in some or even most of them, as shown by the detection of a planet around TW Hya (Setiawan et al. 2008). Planets will create gaps or inner cavities which can only be revealed at the longest wavelengths, where the dust is optically thin. Such cavities are at the limit of the possibilities of current instruments (see LkCa15, Piétu et al. (2006)) . Confusion with binary objects (e.g. HH30, Guilloteau et al. (2008)) may however happen. Cavities and gaps will be easily accessible to ALMA (Wolf et al. 2002). JWST will lack the spatial resolution to do so, but could play a role in revealing the excess radiation due to the accretion on the (proto-)planets in the latest stages.

7.3 Summary

The following table summarizes the main opportunities for studying protoplanetary disks with the upcoming observatories JWST, HERSCHEL, ALMA, and SOFIA.

Task	JWST	ALMA	HERSCHEL	SOFIA
Infall	High resolution imaging in MIR	Isolate infall region, observe absorption		High resolution [OI] 63 μm
SED's and disk geometry	NIR, MIR for weak/distant source, go beyond Spitzer	Measure weak submm fluxes	Fill SED holes in MIDIR, critical PDR lines like CI, C^+	Full SED from 3…500 μm, for bright sources, sensitivity similar to ISO
Chemistry and mixing	Atomic/molecular gas lines like H_2, CO, H_2O, CH_4, SiII, SI, FeII, NeII…Absorption studies high inclination disks.	Many new species, imaging to few AU level, huge impact	Potential to see new lines, but beam/sensitivity issues	High resolution line spectra. H_2O, C_2H_2, CH_4 in ro-vibrational lines.
Grain sizes and spatial distribution	Coronographic imaging of scattered light, MIR spectra from weak sources, absorption from edge-on disks, shadow imaging of proplyds	Measure mm-sized grains as function of distance	by way of SED measurements	Constraints on grain sizes through full SED's.
Dust Mineralogy	Mineralogy of weak sources, absorption in edge-on disks ($\lambda = 3..28\mu m$)		FIR features with PACS ($\lambda = 60 \ldots 200\ \mu m$)	FIR features, absorption studies ($\lambda = 3 \ldots 500\ \mu m$)
Dynamics		Line profiles as function of distance, measure turbulence. Face-on disks, different molecules for different heights		High resolution line profiles from inner disk, e.g. H_2 (28 μm), S I (25 μm), Fe II (26 μm), also H_2O, CH_4, CO, HCN and C_2H_2.

References

Adams, F. C., Lada, C. J., & Shu, F. H. 1987, ApJ, 312, 788
Calvet, N., D'Alessio, P., Hartmann, L., Wilner, D., Walsh, A., & Sitko, M. 2002, ApJ, 568, 1008
Calvet, N., D'Alessio, P., Watson, D. M., Franco-Hernández, R., Furlan, E., Green, J., Sutter, P. M., Forrest, W. J., Hartmann, L., Uchida, K. I., Keller, L. D., Sargent, B., Najita, J., Herter, T. L., Barry, D. J., & Hall, P. 2005, ApJ, 630, L185
Carr, J. S., Tokunaga, A. T., & Najita, J. 2004, ApJ, 603, 213
Ceccarelli, C., & Dominik, C. 2005, A & A, 440, 583
Ceccarelli, C., Dominik, C., Lefloch, B., Caselli, P., & Caux, E. 2004, ApJ, 607, L51
Chiang, E. I., & Goldreich, P. 1997, ApJ, 490, 368
Dartois, E., Dutrey, A., & Guilloteau, S. 2003, A & A, 399, 773
Di Francesco, J., Myers, P. C., Wilner, D. J., Ohashi, N., & Mardones, D. 2001, ApJ, 562, 770
Dominik, C., Ceccarelli, C., Hollenbach, D., & Kaufman, M. 2005, ApJ, 635, L85
Dominik, C., Dullemond, C. P., Waters, L. B. F. M., & Walch, S. 2003, A & A, 398, 607
Duchêne, G., McCabe, C., Ghez, A. M., & Macintosh, B. A. 2004, ApJ, 606, 969
Dullemond, C. P., & Dominik, C. 2004, A & A, 417, 159
Dullemond, C. P., & Dominik, C. 2005, A & A, 434, 971
Dullemond, C. P., Dominik, C., & Natta, A. 2001a, ApJ, 560, 957
Dullemond, C. P., Dominik, C., & Natta, A. 2001b, in Astronomische Gesellschaft Meeting Abstracts, Vol. 18, Astronomische Gesellschaft Meeting Abstracts, ed. E. R. Schielicke, 45–+
Dullemond, C. P., Hollenbach, D., Kamp, I., & D'Alessio, P. 2007, in Protostars and Planets V, ed. B. Reipurth, D. Jewitt, & K. Keil, 555–572
Dutrey, A., Guilloteau, S., & Guelin, M. 1997, A & A, 317, L55
Dutrey, A., Henning, T., Guilloteau, S., Semenov, D., Piétu, V., Schreyer, K., Bacmann, A., Launhardt, R., Pety, J., & Gueth, F. 2007, A & A, 464, 615
Eisner, J. A., Chiang, E. I., & Hillenbrand, L. A. 2006, ApJ, 637, L133
Eisner, J. A., Hillenbrand, L. A., White, R. J., Akeson, R. L., & Sargent, A. I. 2005, ApJ, 623, 952
Espaillat, C., Calvet, N., D'Alessio, P., Hernández, J., Qi, C., Hartmann, L., Furlan, E., & Watson, D. M. 2007, ApJ, 670, L135
Gammie, C. F. 1996, ApJ, 457, 355
Gomes, R., Levison, H. F., Tsiganis, K., & Morbidelli, A. 2005, Nature, 435, 466
Guilloteau, S., Dutrey, A., Pety, J., & Gueth, F. 2008, A & A, 478, L31
Guilloteau, S., Piétu, V., Dutrey, A., & Guélin, M. 2006, A & A, 448, L5
Habing, H. J., Dominik, C., Jourdain de Muizon, M., Kessler, M. F., Laureijs, R. J., Leech, K., Metcalfe, L., Salama, A., Siebenmorgen, R., & Trams, N. 1999, Nature, 401, 456
Haisch, K. E., Jr., Lada, E. A., & Lada, C. J. 2001, ApJ, 553, L153
Hartmann, L., Hinkle, K., & Calvet, N. 2004, ApJ, 609, 906
Hogerheijde, M. R. 2001, ApJ, 553, 618
Hueso, R., & Guillot, T. 2005, A & A, 442, 703
Ilgner, M., Henning, T., Markwick, A. J., & Millar, T. J. 2004, A & A, 415, 643
Isella, A., Testi, L., Natta, A., Neri, R., Wilner, D., & Qi, C. 2007, A & A, 469, 213
Johansen, A., Oishi, J. S., Low, M.-M. M., Klahr, H., Henning, T., & Youdin, A. 2007, Nature, 448, 1022
Kamp, I., van Zadelhoff, G.-D., & van Dishoeck, E. 2004, in Astronomical Society of the Pacific Conference Series, Vol. 324, Debris Disks and the Formation of Planets, ed. L. Caroff, L. J. Moon, D. Backman, & E. Praton, 265–+
Kessler-Silacci, J. 2004, PhD thesis, (California Institute of Technology)
Kessler-Silacci, J., Augereau, J.-C., Dullemond, C. P., Geers, V., Lahuis, F., Evans, II, N. J., van Dishoeck, E. F., Blake, G. A., Boogert, A. C. A., Brown, J., Jørgensen, J. K., Knez, C., & Pontoppidan, K. M. 2006, ApJ, 639, 275
Kessler-Silacci, J. E., Dullemond, C. P., Augereau, J.-C., Merín, B., Geers, V. C., van Dishoeck, E. F., Evans, II, N. J., Blake, G. A., & Brown, J. 2007, ApJ, 659, 680

Koerner, D. W., & Sargent, A. I. 1995, Astron. J. , 109, 2138
Koike, C., Mutschke, H., Suto, H., Naoi, T., Chihara, H., Henning, T., Jäger, C., Tsuchiyama, A., Dorschner, J., & Okuda, H. 2006, A & A, 449, 583
Koike, C., Tsuchiyama, A., Shibai, H., Suto, H., Tanabé, T., Chihara, H., Sogawa, H., Mouri, H., & Okada, K. 2000, A & A, 363, 1115
Krause, M., & Blum, J. 2004, Phys Rev Let, 93, 021103
Lahuis, F., van Dishoeck, E. F., Boogert, A. C. A., Pontoppidan, K. M., Blake, G. A., Dullemond, C. P., Evans, II, N. J., Hogerheijde, M. R., Jørgensen, J. K., Kessler-Silacci, J. E., & Knez, C. 2006, ApJ, 636, L145
Levison, H. F., & Morbidelli, A. 2003, Nature, 426, 419
Mannings, V., & Sargent, A. I. 1997, ApJ, 490, 792
McCabe, C., Duchêne, G., & Ghez, A. M. 2003, ApJ, 588, L113
Meeus, G., Waters, L. B. F. M., Bouwman, J., van den Ancker, M. E., Waelkens, C., & Malfait, K. 2001, A & A, 365, 476
Min, M., Dominik, C., Hovenier, J. W., de Koter, A., & Waters, L. B. F. M. 2006, A & A, 445, 1005
Miroshnichenko, A. S., Vinković, D., Ivezić, Ž., & Elitzur, M. 2003, in Astrophysics and Space Science Library, Vol. 299, Open Issues in Local Star Formation, ed. J. Lépine & J. Gregorio-Hetem, 339–+
Natta, A., Testi, L., Neri, R., Shepherd, D. S., & Wilner, D. J. 2004, A & A, 416, 179
Öberg, K. I., Fuchs, G. W., Awad, Z., Fraser, H. J., Schlemmer, S., van Dishoeck, E. F., & Linnartz, H. 2007, ApJ, 662, L23
Ormel, C. W., Spaans, M., & Tielens, A. G. G. M. 2007, A & A, 461, 215
Paszun, D., & Dominik, C. 2006, Icarus, 182, 274
Piétu, V., Dutrey, A., & Guilloteau, S. 2007, A & A, 467, 163
Piétu, V., Dutrey, A., Guilloteau, S., Chapillon, E., & Pety, J. 2006, A & A, 460, L43
Piétu, V., Guilloteau, S., & Dutrey, A. 2005, A & A, 443, 945
Pontoppidan, K. M., Stapelfeldt, K. R., Blake, G. A., van Dishoeck, E. F., & Dullemond, C. P. 2007, ApJ, 658, L111
Posch, T., Kerschbaum, F., Richter, H., & Mutschke, H. 2005, in ESA Special Publication, Vol. 577, ESA Special Publication, ed. A. Wilson, 257–260
Przygodda, F., van Boekel, R., Àbrahàm, P., Melnikov, S. Y., Waters, L. B. F. M., & Leinert, C. 2003, A & A, 412, L43
Qi, C. 2001, PhD thesis, AA(California Institute of Technology)
Qi, C., Wilner, D. J., Aikawa, Y., Blake, G. A., & Hogerheijde, M. R. 2008, ArXiv e-prints, 803
Rawlings, J. M. C. 1996, Ap&SS, 237, 299
Rettig, T. W., Brittain, S. D., Simon, T., & Kulesa, C. 2003, in Bulletin of the American Astronomical Society, Vol. 35, Bulletin of the American Astronomical Society, 1227–+
Rieke, G. H., Su, K. Y. L., Stansberry, J. A., Trilling, D., Bryden, G., Muzerolle, J., White, B., Gorlova, N., Young, E. T., Beichman, C. A., Stapelfeldt, K. R., & Hines, D. C. 2005, ApJ, 620, 1010
Roberts, J. F., Rawlings, J. M. C., Viti, S., & Williams, D. A. 2007, Mon. Not. R. Astron. Soc. , 382, 733
Rodmann, J., Henning, T., Chandler, C. J., Mundy, L. G., & Wilner, D. J. 2006, A & A, 446, 211
Semenov, D., Pavlyuchenkov, Y., Henning, T., Wolf, S., & Launhardt, R. 2008, ApJ, 673, L195
Semenov, D., Wiebe, D., & Henning, T. 2006, ApJ, 647, L57
Setiawan, J., Henning, T., Launhardt, R., Müller, A., Weise, P., & Kürster, M. 2008, Nature, 451, 38
Shuping, R. Y., Bally, J., Morris, M., & Throop, H. 2003, ApJ, 587, L109
Simon, M., Dutrey, A., & Guilloteau, S. 2000, ApJ, 545, 1034
Testi, L., Natta, A., Shepherd, D. S., & Wilner, D. J. 2003, A & A, 403, 323
Thamm, E., Steinacker, J., & Henning, T. 1994, A & A, 287, 493
Throop, H. B., Bally, J., Esposito, L. W., & McCaughrean, M. J. 2001, Science, 292, 1686

van Boekel, R., Min, M., Leinert, C., Waters, L. B. F. M., Richichi, A., Chesneau, O., Dominik, C., Jaffe, W., Dutrey, A., Graser, U., Henning, T., de Jong, J., Köhler, R., de Koter, A., Lopez, B., Malbet, F., Morel, S., Paresce, F., Perrin, G., Preibisch, T., Przygodda, F., Schöller, M., & Wittkowski, M. 2004, Nature, 432, 479

van Boekel, R., Min, M., Waters, L. B. F. M., de Koter, A., Dominik, C., van den Ancker, M. E., & Bouwman, J. 2005, A & A, 437, 189

van Boekel, R., Waters, L. B. F. M., Dominik, C., Bouwman, J., de Koter, A., Dullemond, C. P., & Paresce, F. 2003, A & A, 400, L21

van Dishoeck, E. F., Thi, W.-F., & van Zadelhoff, G.-J. 2003, A & A, 400, L1

Willacy, K. 2007, ApJ, 660, 441

Willacy, K., Langer, W., Allen, M., & Bryden, G. 2006, ApJ, 644, 1202

Willacy, K., & Langer, W. D. 2000, ApJ, 544, 903

Wolf, S., Gueth, F., Henning, T., & Kley, W. 2002, ApJ, 566, L97

Zhou, S., Evans, II, N. J., Koempe, C., & Walmsley, C. M. 1993, ApJ, 404, 232

Blake Bullock and Rolf Danner discuss the approximate dimensions of something

Heidi Hammel and Harley Thronson apparently show no signs of embarrassment at Yvonne Pendleton's birthday celebration

Chapter 8
Astrochemistry of Dense Protostellar and Protoplanetary Environments

Ewine F. van Dishoeck

Abstract Dense molecular clouds contain a remarkably rich chemistry, as revealed by combined submillimeter and infrared observations. Simple and complex (organic) gases, polycyclic aromatic hydrocarbons, ices and silicates have been unambiguously detected in both low- and high-mass star forming regions. During star- and planet formation, these molecules undergo large abundance changes, with most of the heavy species frozen out as icy mantles on grains in the cold pre-stellar phase. As the protostars heat up their immediate surroundings, the warming and evaporation of the ices triggers the formation of more complex molecules, perhaps even of pre-biotic origin. Water, a key ingredient in the chemistry of life, is boosted in abundance in hot gas. Some of these molecules enter the protoplanetary disk where they are exposed to UV radiation or X-rays and modified further. The enhanced resolution and sensitivity of ALMA, Herschel, SOFIA, JWST and ELTs across the full range of wavelengths from cm to μm will be essential to trace this lifecycle of gas and dust from clouds to planets. The continued need for basic molecular data on gaseous and solid-state material coupled with powerful radiative transfer tools is emphasized to reap the full scientific benefits from these new facilities.

8.1 Introduction and Overview

The study of exo-planetary systems and their formation is one of the fastest growing topics in astronomy, with far-reaching implications for our place in the universe. Intimately related to this topic is the study of the chemical composition of the gaseous and solid state material out of which new planets form and how it is modified in the dense protostellar and protoplanetary environments. Interstellar clouds are known to have a rich chemical composition, with more than 130 different species identified in the gas, ranging from the simplest diatomic molecules such as CO to long carbon chains such as HC_9N and complex organic molecules like CH_3OCH_3. Polycyclic

E.F. van Dishoeck (✉)
Leiden Observatory, Leiden University, P.O. Box 9513, 2300 RA Leiden, The Netherlands and Max-Planck Institute für Extraterrestrische Physik, Garching, Germany
email: ewine@strw.leidenuniv.nl

H.A. Thronson et al. (eds.), *Astrophysics in the Next Decade,* Astrophysics and Space Science Proceedings, DOI 10.1007/978-1-4020-9457-6_8,

aromatic hydrocarbons (PAHs) are a ubiquitous component of star-forming regions throughout the universe, and ices have been found to be a major reservoir of heavy elements in the coldest and densest clouds. Indeed, the gas, ice and dust undergo a rich variety of chemical transformations in response to the large changes in physical conditions during star- and planet formation, where temperatures vary from <10 K to more than 1000 K and densities from $\sim 10^4$ to $>10^{10}$ H_2 molecules cm^{-3}. Intense ultraviolet (UV) radiation and X-rays from the young star impact the envelopes and disks causing further chemical changes. At the same time, these chemical changes can also be used as diagnotics of the physical processes. Tracing this chemical and physical evolution in its diversity requires a combination of observations from infrared (IR) to millimeter (mm) wavelengths, where gaseous and solid material have their principal spectroscopic features and where the extinction is small enough to penetrate the dusty regions.

The different evolutionary stages in star- and planet formation are traditionally linked to their Spectral Energy Distributions (SEDs) (Lada 1999), which illustrate how the bulk of the luminosity shifts from far- to near-infrared wavelengths as matter moves from envelope to disk to star. In the standard scenario for the formation of an isolated low-mass star, the earliest stage is represented by a cold pre-stellar core which contracts as magnetic and turbulent support are lost. This phase is characterized by heavy freeze out of molecules onto the grains to form ices. Subsequent inside-out collapse forms a protostellar object which derives most of its luminosity from accretion and heats its immediate surroundings, resulting in evaporation of ices back into the gas phase and in an active high-temperature chemistry. Because the initial core has some angular momentum, much of the infalling material ends up in a dense rotating disk from which further accretion onto the star occurs. Soon after formation, a stellar wind breaks out along the rotational axis of the system and drives a bipolar outflow entraining surrounding cloud material. The main chemical characteristics of outflows are high temperature shock chemistry and sputtering of grains returning heavy elements to the gas. Low-mass young stellar objects (YSOs) in this embedded phase of star formation are often denoted as Class 0 and I objects.

Once the protostellar envelope has been dispersed, an optically visible pre-main sequence star with a disk is revealed, a so-called Class II object. Inside this disk, grains collide and stick owing to the high densities, leading to pebbles, rocks and eventually planetesimals which settle to the midplane and interact to form planets. High densities lead to freeze-out in the cold mid-plane, but UV radiation heats the surface layers leading to an active gas-phase chemistry. Changes in the mineralogical structure of the dust indicate thermal processing in the hot inner disk. Ices survive in the outer parts of the disk, providing the volatiles for atmospheres, oceans and possibly life on rocky planets. The remaining gas and dust which do not make it into planets or smaller (icy) bodies are gradually lost from the disk through a combination of processes, including photoevaporation and stellar winds.

The above scenario holds largely for low- and intermediate mass stars, with luminosities up to $\sim 10^3$ $L_{\odot}$. The formation of massive O and B stars is much more poorly understood due to larger distances, shorter timescales and heavy extinction. Recent observations show that the embedded phase of massive star formation can be divided into several stages. Massive pre-stellar cores with local temperature minima and density maxima represent the initial conditions. The next phase are the high-mass protostellar objects (HMPO) which contain a central young star surrounded by a massive envelope with a centrally peaked temperature and density distribution. HMPOs show signs of active star formation through outflows and/or masers. The next, related phase are the hot molecular cores which have large compact masses of warm and dense dust and gas, and high abundances of complex organic molecules. Finally, hyper-compact HII regions with pockets of ionized gas develop, but they have rather weak free-free emission since they are still confined to the star. The embedded phase ends when the ionized gas expands hydrodynamically and disrupts the parental molecular cloud, producing a classical compact HII region.

In this chapter, an overview will be given of the different chemical characteristics and their modifications during the various phases of star- and planet formation, for both low- and high mass. Also, the use of molecules as probes of physical structure (temperature, density, radiation field, ionization fraction, geometry, . . .) is discussed. Finally, the importance of chemistry and spectroscopic features in tracing evolutionary processes (timescales, grain growth, mixing, gap formation, . . .) is emphasized. Note that our current knowledge is almost entirely based on spatially unresolved observations which encompass the entire star + disk + envelope system in a single beam. A key aspect of new facilities like ALMA, JWST, Herschel, SOFIA and future ELTs is that they will have the combined spatial and spectral resolution as well as the sensitivity to resolve the individual physical components and image the key chemical processes on all relevant scales. Other recent reviews on this topic include (van Dishoeck 2006, Ceccarelli et al. 2007, Bergin et al. 2007, Bergin & Tafalla 2007).

Astrochemistry is a highly interdisciplinary subject and many of the basic chemical processes that occur under the exotic conditions in space are not yet well understood. Continued interaction of astronomers with chemists and physicists to determine the basic atomic and molecular data is essential to make progress.

8.2 Infrared and Submillimeter Observations

The two main observational techniques for studying gaseous and solid-state material in protostellar regions are infrared and (sub)millimeter spectroscopy (Table 8.1). Both techniques have their strengths and weaknesses, and a combination of them is essential to obtain a full inventory and probe the different physical processes involved (Table 8.2).

8.2.1 Infrared Spectroscopy

At infrared wavelengths, the vibrational bands of both gas-phase and solid-state material can be detected, including those of symmetric molecules like CH_4 and C_2H_2. The most abundant molecule in the universe, H_2, also has its principle rotational

Table 8.1 Need for broad wavelength coverage to study astrochemistry and star formation

Radiation	Mid-IR 3–28 μm	Far-IR 28–350 μm	Submm 0.3–7 mm	Radio ≥1 cm
Continuum	Warm dust	Cooler dust, SED peak	Cold dust	Large grains (pebbles), ionized gas
Broad features: solids	Silicates, ices, oxides, PAHs	(Hydrous) silicates, ices, carbonates, PAHs	...	...
Narrow lines: gas	Simple (symmetric) molecules	H_2O, OH, O I hot CO, C II	CO, myriad of simple + complex molecules, O_2	Heavy complex mol. H I, masers

transitions at mid-infrared wavelengths. Resolving powers $R = \lambda/\Delta\lambda$ range from a few $\times 10^2$ up to 10^5, which is sufficient to resolve intrinsically broad solid-state features but not the narrower profiles of quiescent gas. In protostellar sources, bands of silicates, cold ices and gases are seen in *absorption* superposed on the infrared continuum due to hot dust close to the young star (Fig. 8.1). Thus, these observations probe only a pencil-beam line-of-sight and are more sensitive to the warm inner part

Table 8.2 Chemical characteristics of YSOs

Component	Size[a] (″)	Chemical characteristics	Submillimeter diagnostics	Infrared diagnostics
Pre-stellar cores	60	Low-T chemistry, Heavy freeze-out	Ions, deuterated, simple mol. (N_2H^+, H_2D^+)	Simple ices (H_2O, CO_2, CO)
Outer cold envelope	20	Low-T chemistry, Heavy freeze-out	Ions, deuterated, simple mol. (N_2H^+, HCO^+, H_2CO...)	Ices (H_2O, CO_2, CO, CH_3OH?)
Inner warm envelope	1–2	Evaporation, X-rays (CO^+, OH^+,...)	High T_{ex} (H_2CO, CH_3OH, CO, CO_2)	High gas/solid, High T_{ex}, Heated ices
Hot core	≤1	Evaporation, High-T chemistry	H_2O, Complex organics (CH_3OCH_3, CH_3CN)	Hot gas (HCN, NH_3, HNCO)
Outflow: direct impact	<1–20	Shock chemistry, Sputtering cores	H_2O, OH, Si- and S-species (SiO, SO_2)	H_2, Atomic lines ([O I], [Si II], [S I])
Outflow: gentle impact	<1–20	Sputtering ices	Ice products (CH_3OH, H_2O, ..)	H_2
Outer disk: surface	∼2	UV irradiation, Ion-molecule	Ions, radicals (CN/HCN, HCO^+)	PAHs, heated ices (NH_4^+)
Outer disk: midplane	∼2	Heavy freeze-out,	Deuterated mol. (DCO^+, H_2D^+)	...
Inner disk	≤0.02	Evaporation, High-T chemistry, X-rays	...	Silicates, PAHs, [Ne II], CO, H_2, hot gas (H_2O,C_2H_2, HCN)

[a] Typical angular size for a low-mass YSO at 150 pc, or a high-mass YSO at 3 kpc.

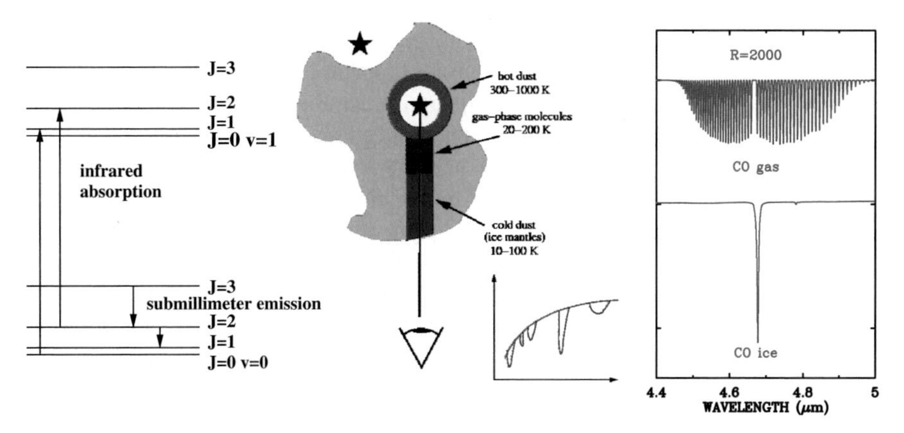

Fig. 8.1 **Left**: Energy level diagram showing typical (sub)millimeter and infrared lines. **Middle**: Cartoon illustrating infrared absorption line observations toward an embedded young stellar object. **Right**: Laboratory spectra of gas-phase and solid CO. Note that gas and ice can readily be distinguished at a resolving power $R = 2000$ and higher. The gas-phase CO spectrum gives direct information on the excitation temperature of the molecule since the individual lines originate from levels with different energies (see *left panel*)

than the submillimeter data. Toward pre-main sequence stars with disks, features of silicates, PAHs and hot gases are usually found in *emission*, originating in the warm surface layers and the hot inner disk. In near edge-on geometries, the bands occur in absorption. Because of the limited spectral resolution and sensitivity, abundances down to $10^{-7}-10^{-8}$ with respect to H_2 can be probed, sufficient to detect the major O, C and N-bearing species but not the minor ones.

Only a small fraction of the infrared wavelength range can be observed from Earth. The *Infrared Space Observatory* (ISO, 1995–1998) provided the first opportunity to obtain complete 2.5–200 μm mid-infrared spectra above the Earth's atmosphere at $R = 200-10^4$ but was limited in sensitivity to YSOs 10^3-10^5 times more luminous than the Sun (for review, see van Dishoeck 2004). The *Spitzer Space Telescope*, launched in 2003, has orders of magnitude higher sensitivity and can obtain 5–40 μm spectra of YSOs down to the brown dwarf limit, albeit only at low resolution $R = 50-600$. The slits of both *ISO* and *Spitzer* are large, typically $>10''$, so that most emission is spatially unresolved.

Complementary ground-based spectra can be obtained from 8–10 m class telescopes in atmospheric windows at 3–20 μm. These facilities have two advantages: R up to 10^5 can be achieved and emission can be spatially resolved on subarcsec scales, comparable to those of disks (Table 8.2).

8.2.2 Submillimeter Spectroscopy

At (sub)millimeter wavelengths, the pure rotational lines of molecules are seen in emission. Only gaseous molecules with a permanent electric dipole moment can be observed, but not symmetric molecules like H_2, N_2, CH_4, and C_2H_2 nor solid-state species. O_2 only has weak magnetic dipole transitions. The level spacing scales with $BJ(J + 1)$, with J the rotational quantum number and B the rotational constant, which is inversely proportional to the reduced mass of the system. Thus, light hydrides such as H_2O, OH, CH, ... have their lowest transitions at submillimeter/far-infrared wavelengths, whereas those of heavy molecules like CO occur at longer wavelengths. A big advantage is that heterodyne receivers naturally have very high resolution $R \geq 10^6$, so that line profiles are fully resolved, providing information on dynamical processes occurring in the protostellar environment. Thus, infall, outflow and rotation can in principle be kinematically distinguished. Also, the sensitivity is very high: molecules with abundances as low as a few$\times 10^{-12}$ can be seen. Detection of linear molecules is favored compared with heavy asymmetric rotors because their population is spread over fewer energy levels.

Because the lines are in emission and can be readily excited by collisions even in cold gas, the spatial distribution of the molecule can be mapped. However, the angular resolution of the most advanced single-dish telescopes is only $\sim$15–20″ which corresponds to a linear scale of a few thousand AU at the distances of the nearest star-forming regions of $\sim$150 pc. Since typical sizes of protostellar envelopes are about 3000 AU, the single-dish beams encompass the entire protostellar envelope and are much larger than protoplanetary disks which have radii of order 100 AU. This means that the different chemical and physical regimes are 'blurred' together in a single spatial pixel, and that the data are generally more biased toward the cold, outer part.

Direct information on smaller spatial scales can be obtained by (sub)millimeter interferometry, which can reach resolutions down to $\sim$1″ or $\sim$50 AU. However, sensitivities of existing arrays are still limited and chemical surveys have not yet been feasible. Fortunately, the fact that molecules have so many different lines and transitions can be put to good use: single-dish observations of lines with a range of excitation conditions combined with detailed radiative transfer modeling can provide constraints on physical and chemical gradients on scales much smaller than the beam, down to a few ″ (see below).

8.2.3 From Intensities to Physical Conditions and Abundances

The rotational and vibrational levels are excited by a combination of collisional and radiative processes. In the simplest case of a 2 level system, the critical density $n_{cr} = A_{u\ell}/q_{u\ell}$, where $A_{u\ell}$ is the Einstein A constant for spontaneous emission and $q_{u\ell}$ the rate coefficient for collisional de-excitation. Since $A_{u\ell}$ is proportional to $\mu^2 \nu^3$ with μ the dipole moment of the molecule, n_{cr} increases with frequency ν. Also, the rotational energy level separation of (linear) molecules is proportional to the quantum number J. Thus, the higher frequency, higher J transitions probe regions with higher densities and temperatures. For example, the critical density of the CS $J = 2 - 1$ transition at 97 GHz is 8×10^4 cm^{-3} ($E_u = 7$ K), whereas that of the 7−6 transition at 346 GHz is 3×10^6 cm^{-3} ($E_u = 66$ K). For optically thick lines, line trapping will lower the critical densities. CO has a particularly small dipole moment, at least 20 times smaller than that of CS and other commonly observed species. Combined with its high abundance of $\sim10^{-4}$ with respect to H_2, this implies that CO is readily excited and detected in molecular clouds, even at densities as low as 10^3 cm^{-3}.

The translation of the observed intensities to chemical abundances, temperatures and densities requires a determination of the excitation of the molecule and an understanding of how the photon is produced in the cloud and how it makes its way from the cloud to the telescope. Significant progress has been made in recent years in the *quantitative* analysis of molecular line data through the development of rapid, accurate and well-tested radiative transfer codes which can be applied to multi-dimensional geometries (for overview, see van Zadelhoff et al. 2002).

8.2.4 Importance of Laboratory Astrophysics

The computation of the excitation of gas-phase molecules requires availability of accurate frequencies, radiative rates and collisional rate coefficients with the main collision partner, usually H_2. For atoms, large data bases are available thanks to many decades of dedicated work by physicists in laboratories across the globe, allowing astronomers to readily interpret their optical and X-ray spectra. For molecules, such information is much more limited. In molecular clouds, the main collision partner is H_2 and state-to-state collisional rate coefficients are in principle required for both para-H_2, $J = 0$ (cold clouds) and ortho-H_2, $J = 1$ (warm regions). In practice, such data are available for only a few molecules and H_2 $J = 0$ is usually taken as the main collision partner. Much of the molecular data gathered over the last 30 years from laboratory experiments and quantum-chemical calculations has been summarized by (Schöier et al. 2005) and made publicly available through the Web[1] together with a simple on-line radiative excitation code. New results, especially those coordinated through the EU 'Molecular Universe' network program, [2] are being added. Nevertheless, accurate collision rates are still lacking for many astrochemically important molecules, especially the larger ones. The analysis of observational data on these species is often limited to the simple rotation diagram approach, in which it is assumed that the level populations can be characterized by a single excitation temperature and that the lines are optically thin (see Section 8.4.3 for example).

PAHs are a special class of very large molecules. Huge efforts have been made in the last decade to determine the basic spectroscopy and intensities of neutral and ionic PAHs by a variety of experimental and theoretical methods. (e.g., Hudgins & Allamandola 1999[3], see review by Tielens 2008). One problem is that most studies

[1] http://www.strw.leidenuniv.nl/~moldata
[2] http://molecular-universe.obspm.fr/
[3] http://www.astrochem.org/databases.htm

are still limited to the smaller PAHs with up to about 30 carbon atoms, whereas the PAHs seen in interstellar and circumstellar regions are likely somewhat larger, containing up to 100 carbon atoms. Another problem is that many of the experiments have been done using matrix isolation techniques rather than in the gas phase, introducing shifts which are a priori unknown.

For solid-state species, information on infrared band positions, profiles and strengths are needed for analysis. Because these quantities depend on specific composition, e.g., whether the material is in pure or mixed form, large series of experiments are needed. For ices, systematic studies have been performed by a variety of groups (e.g., Hudgins et al. 1993, Gerakines et al. 1996, Öberg et al. 2007), with data available through the Web.[4] For silicates and oxides, the Jena database[5] (e.g., Jaeger et al. 1998) is an invaluable resource for interpreting mid-infrared spectra.

Besides basic spectroscopy, there continues to be a great need for basic reaction rates of gas-phase processes from 10 to 1000 K (ion-molecule, neutral-neutral, radiative association, dissociative recombination), of reactions under the influence of UV radiation or X-rays (photodissociation, photoionization) and of reactions between the gas and the grains (e.g., H_2 formation on silicates; molecule formation on and in ices, binding energies). Summaries of rate coefficients are publicly available from UMIST[6] and Ohio state.[7] Ultimately, the science return from the new billion dollar/Euro/Yen investments will be limited by our poor knowledge of these basic processes.

8.3 Cold Clouds and Pre-Stellar Cores

8.3.1 Dense Starless Clouds

Catalogs of isolated dark clouds and globules have been available for nearly a century. Some of them have become favorite targets of astrochemical studies, in particular the starless core TMC-1. Its conditions, $T \approx 10\,\mathrm{K}$ and $n \approx 2 \times 10^4\,\mathrm{cm}^{-3}$, are average of those of other cold dense clouds, but this core shows a remarkable variety of molecules. In particular, the unsaturated long carbon chains—e.g., HC_5N, HC_7N, HC_9N, C_4H, C_6H—have all been detected with relatively high abundances. An exciting recent discovery is that negative ions such as C_6H^- and C_8H^-, which have long been neglected in astrochemical models, are present at abundances of 1–5% of their neutral counterparts (McCarthy et al. 2006, Brünken et al. 2007) (Fig. 8.2).

Clouds such as TMC-1 are traditionally used as benchmarks for pure gas-phase astrochemical models. Reasonable agreement with the observed abundances (within

---

[4] http://www.strw.leidenuniv.nl/~lab

[5] http://www.astro.uni-jena.de/Laboratory/Database/databases.html

[6] http://www.udfa.net

[7] http://www.physics.ohiostate.edu/~eric/research.html

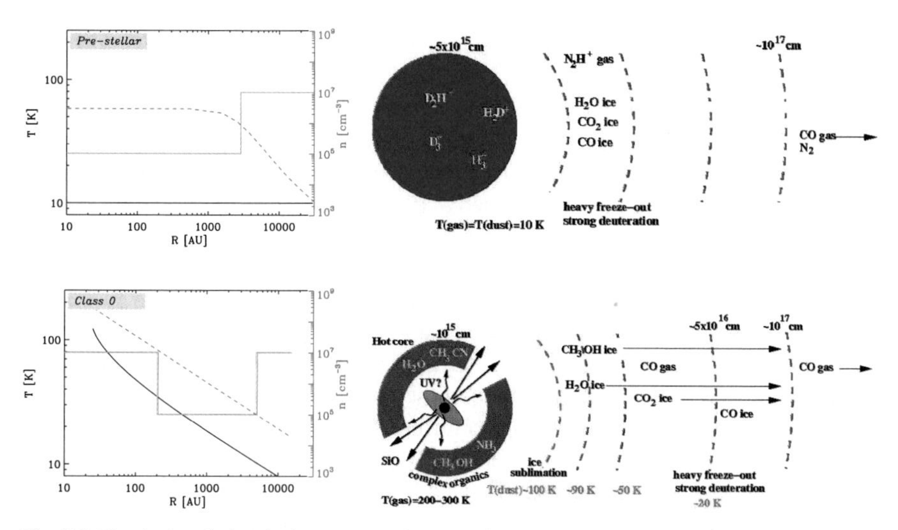

Fig. 8.2 Physical and chemical structure of a prestellar and a protostellar (Class 0) object. **Left**: density (*red*), temperature (*black*) and typical abundance (*green*) profiles. **Right**: cartoons illustrating the main chemical characteristics of each stage (based on Jørgensen et al. 2005, van Dishoeck & Blake 1998)

factors of 3) is obtained for the majority of species if TMC-1 is assumed to be 'chemically young', i.e., if the time since most of the carbon was in atomic form is only $\sim 10^5$ yr (e.g., Wakelam et al. 2006). This time most likely refers to the period elapsed since the dense cloud formed out of the more diffuse gas, but any other event that may have replenished atomic carbon in the cloud could also have reset the 'clock'. However, the inclusion of gas-grain interactions and use of different elemental abundance ratios may affect these conclusions. Also, the recent discovery of the saturated propene molecule, CH_2CHCH_3, which was not included or predicted in any models, casts doubt on their completeness (Marcelino et al. 2007). Indeed, the role of unobservable (at submm) molecules such as the bare carbon-chains C_2 and C_3 and saturated hydrocarbons like CH_4 and C_2H_6 is still poorly understood in dark clouds.

8.3.2 Pre-Stellar Cores

A subset of dark clouds with a clear central density concentration have been identified recently. These so-called pre-stellar cores are believed to be on the verge of collapse and thus represent the initial conditions for low-mass star formation (for review, see Bergin & Tafalla 2007). The physical and chemical state of these clouds is well established on scales of few thousand AU by single dish millimeter observations combined with extinction maps. The cores are cold, with temperatures varying from 10 to 15 K at the edge to as low as 7–8 K at the center, and have density profiles that are well described by Bonnor-Ebert profiles. It is now widely accepted that most molecules are highly depleted in the inner dense parts of these cores (Caselli et al. 1999, Bergin et al. 2002): images of clouds such as B68 show only a ring of $C^{18}O$ emission, indicating that more than 90% of the CO is frozen out toward the center.

This 'catastrophic' freeze-out of CO and other molecules has also been probed directly through observations of ices as functions of position in the cloud. Owing to the increased sensitivity of ground-based 8-10 m IR telescopes, *Spitzer* and *Akari*, it is now possible to make maps of ice abundances on the same spatial scale ($\sim$15″) as those of gas-phase molecules, by taking spectra toward closely spaced background or embedded stars (Pontoppidan et al. 2004). The main features seen in mid-IR spectra are the 9.7 and 18 μm Si-O stretch and bending modes of the silicate grain cores, together with ice bands of relatively simple species (Fig. 8.3). H_2O ice, seen through its 3 and 6 μm bands, has the largest abundance, $\sim(0.5-1.5) \times 10^{-4}$ with respect to H_2, whereas the combined contribution of CO and CO_2 may be as high as that of H_2O. Thus, the ices are a major reservoir of the heavy elements in cold clouds, exceeding even that of gaseous CO. Indeed, toward the center of the ρ Oph F core at least 60% of CO is observed to be frozen out (Pontoppidan 2006).

Another chemical characteristic of cold cores is their extreme deuterium fractionation: doubly and even triply deuterated molecules like D_2CO, NHD_2, ND_3, and CD_3OH have been found with abundances that are enhanced by up to 13 orders of magnitude compared with the overall [D]/[H] ratio (e.g., Lis et al. 2002, Parise et al. 2004). The origin of this strong deuteration is now understood to be directly linked to the observed heavy freeze-out: the basic ions H_3^+ and H_2D^+, which initiate

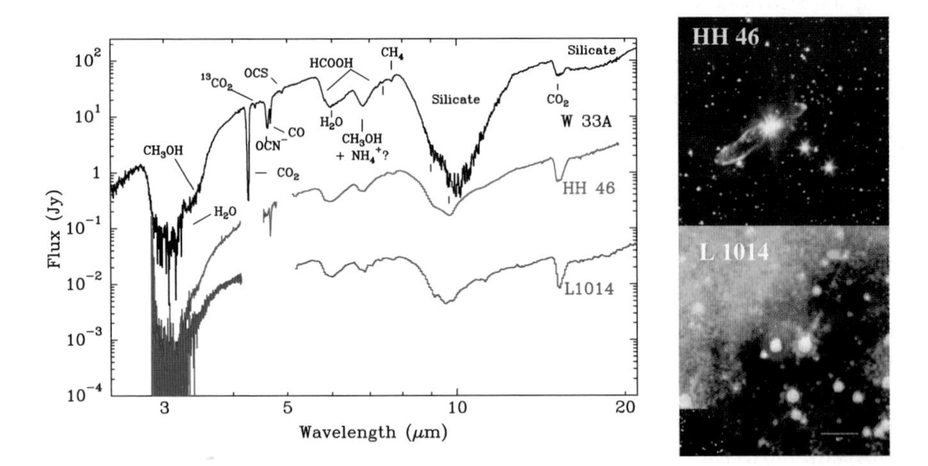

Fig. 8.3 Ices observed toward protostars with a large range of luminosities. **Top**: W 33A ($\sim 10^5$ $L_\odot$) from Gibb et al. (2000b); **Middle**: HH 46 ($\sim$10 $L_\odot$) from Boogert et al. (2004); **Bottom**: L 1014 ($\sim$0.1 $L_\odot$) from Boogert et al. (2008). Note the similarity in major ice features between high-mass and substellar mass YSOs. **Right**: Spitzer image of HH 46 IRS and its bipolar outflow (top; Noriega-Crespo et al. 2004) and of L 1014 (bottom; Young et al. 2004)

much of the low-temperature ion-molecule chemistry and deuteration, are enhanced significantly when their main destroyer, CO, is depleted onto grains. Grain surface chemistry may also play a role (Tielens 1983). Models predict that D_3^+ can become as abundant as H_3^+ (Roberts et al. 2003) and lines of both H_2D^+ and D_2H^+ have indeed been detected in cold clouds at submillimeter wavelengths (Caselli et al. 2003, Vastel et al. 2004). These ions may well be the best probes of the ionization fraction and kinematics in regions where all heavy molecules are depleted.

The freeze-out of CO is reflected in the abundances of many other molecules, either through correlations or anti-correlations (Jørgensen et al. 2004b). A striking example of the latter case is N_2H^+, which, like H_3^+ and H_2D^+, is mainly destroyed by reactions with CO. The binding energy of its precursor, N_2, to ice is comparable to that of CO (Öberg et al. 2005), but the lack of a rapid gas-phase destruction channel together with the boost in H_3^+ keeps the N_2H^+ abundance high even if some N_2 is frozen out. Thus, N_2H^+ is an excellent and easy to observe tracer of the densest and coldest parts of cores.

The above discussion focussed on low-mass cores. Massive cold cores which likely represent the initial conditions for high mass star formation are only just starting to be identified, for example as infrared dark clouds. Their chemical characteristics are likely to be similar to those of the low-mass pre-stellar clouds, in particular heavy freeze-out and strong deuteration toward the center (Pillai et al. 2007).

8.4 Embedded Protostars

8.4.1 Cold Outer Envelope: Ices

Ices are a prominent component of protostellar systems. With *Spitzer*, infrared spectra have now been observed toward several dozen low-mass YSOs including some of the most deeply embedded Class 0 YSOs and sources of substellar luminosity (Boogert et al. 2008) (Fig. 8.3). At first sight, all spectra look remarkably similar, not only to each other, but also to those toward much more luminous YSOs (HMPOs) and background stars. This implies that most ices are formed prior to star formation, and that orders of magnitude higher luminosities do not result in major changes in the ice composition. The detected species are consistent with a simple theory of grain surface chemistry put forward 25 years ago (Tielens & Hagen 1982), in which the major ice components are those formed by hydrogenation of the main atoms and molecules arriving from the gas on the grains. Thus, O, C, and N are hydrogenated to H_2O, CH_4, NH_3 and CO is hydrogenated to H_2CO and CH_3OH, a reaction confirmed in the laboratory (Watanabe et al. 2004). The origin of CO_2 ice is not yet fully clear, but likely involves reactions of CO with either O or OH. Reactions of CO with both H, O, C and N may lead to more complex organics (Fig. 8.4), although not all reactions in this scheme have yet been confirmed in the laboratory (Bisschop et al. 2007a).

On closer inspection, differences in ice composition are found, both between different types of sources and within one class of objects. In particular, abundances

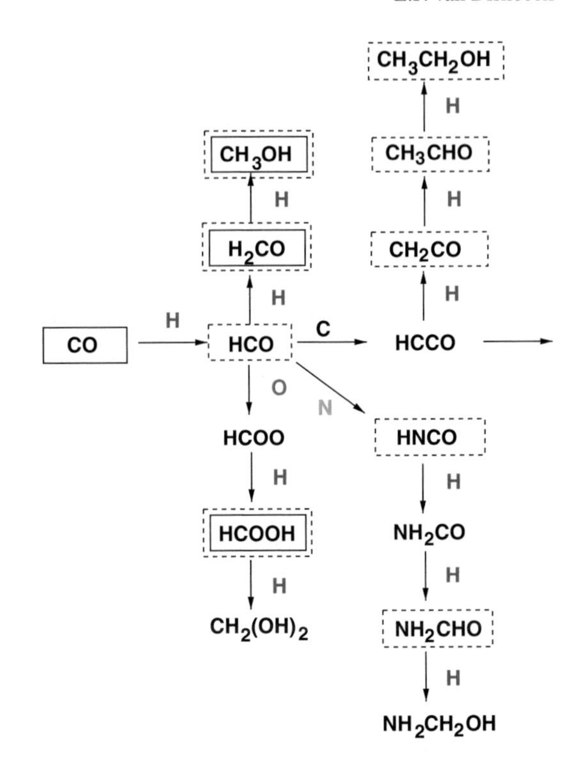

Fig. 8.4 Grain surface chemistry scheme leading to more complex organics proposed by Tielens & Charnley (1997). Solid rectangular boxes contain molecules which have been detected in interstellar ices, whereas dashed boxes indicate molecules that have been detected in the gas phase

of minor species like CH_3OH, OCN^- and NH_3 vary by more than an order of magnitude among the low-mass YSOs and even on spatial scales as small as 1000 AU (Pontoppidan et al. 2004, van Broekhuizen et al. 2005). Since these molecules are the precursors of more complex organic species formed in the gas, one of the main future challenges is to understand the origin of these large abundance variations. Is this due to passive heating of the ices, or are UV radiation or cosmic ray bombardment (i.e., different forms of 'energetic processing') involved? The line profiles of solid CO, CO_2 and the carrier of the 6.8 μm feature (likely NH_4^+, (Schutte & Khanna 2003)) show subtle changes which are related to heating of the ices up to at least 50 K (Keane et al. 2001, Boogert et al. 2008). At such temperatures, various atoms and radicals stored in the ices become mobile and lead to a rich chemistry (Garrod & Herbst 2006). An open question is to what extent complex organic molecules are formed on the grains during this heating phase rather than in the 'hot core' gas (see Section 8.4.3).

8.4.2 Warm Inner Envelopes: Ice Evaporation

Once the protostars have formed, they heat the envelopes from the inside, setting up a strong temperature gradient (see Fig. 8.2). As a result, molecules will evaporate from the grains back to the gas in a sequence according to their sublimation temperatures. The most volatile species like CO and N_2 will start to evaporate around

20 K in the outer envelope, whereas more strongly bound molecules like H_2O only evaporate around 100 K in the inner part. For mixed molecular ices (e.g., CO mixed with H_2O), the evaporation temperature depends on the type of mixed ice. Abundant but volatile species will rapidly sublimate as the mixed ice is heated up. However, a small abundance of these volatile species can remain trapped in the water ice in a clathrate-like structure. These minor species will not come off until the water ice itself starts to sublimate around 100 K. As evolution progresses, the envelope mass decreases and its overall temperature becomes higher.

Observational evidence for ice evaporation is found in both infrared and submillimeter data. Mid-IR spectra of CO and H_2O toward high-mass YSOs directly reveal changes in gas/solid ratios with increasing source temperature (van Dishoeck et al. 1996, Pontoppidan et al. 2003). Also, more complex molecules like HCN, C_2H_2, HNCO and NH_3 freshly evaporated off the grains are found toward a few massive protostars, with orders of magnitude enhanced abundances compared with cold clouds (Evans et al. 1991, Lahuis & van Dishoeck 2000, Knez et al. 2008) (Fig. 8.5).

Independent evidence for freeze-out and evaporation comes from analysis of submillimeter data. The best fit to multi-transition CO data is obtained with a so-called 'drop' abundance profile (Fig. 8.2) in which the CO abundance is normal ($\sim 2 \times 10^{-4}$) in the warm inner part and in the outermost parts where the density is too low for significant freeze-out within the lifetime of the core. In the cold, dense intermediate zone, however, the abundance is at least an order of magnitude lower,

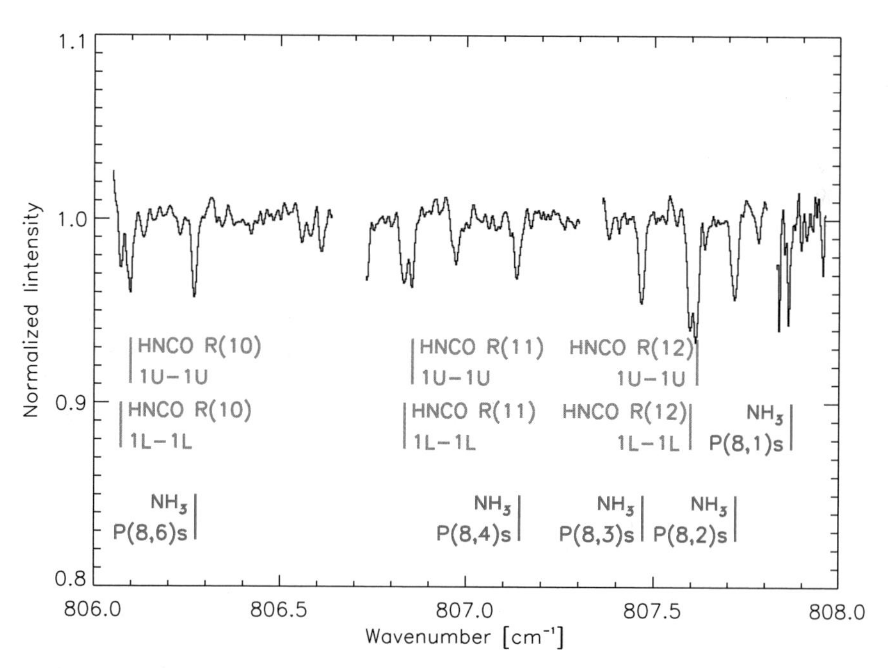

Fig. 8.5 Mid-IR spectrum toward the massive YSO NGC 7538 IRS1 showing high abundances of NH_3 and HNCO in the hot core, likely reflecting freshly evaporated ices (Knez et al. 2008)

since the timescale for freeze-out is short, $\sim 2 \times 10^9/n$ yr. This chemical structure, inferred from spatially unresolved data, has been confirmed by millimeter interferometer studies at higher angular resolution ($\sim$500 AU) for a selected set of objects where the freeze-out zone is directly imaged (e.g., Jørgensen 2004). It is also found for other molecules, especially those directly chemically related to CO.

8.4.3 Hot Cores: Complex Organics and Prebiotic Molecules

In regions where the dust temperature reaches 90–100 K, even the most strongly-bound ices like H_2O start to evaporate, resulting in a 'jump' in the gas-phase abundances of molecules trapped in H_2O ice. For high-mass YSOs, this 100 K radius typically lies at 1000 AU, whereas for low-mass YSOs it is around 100 AU. Thus, taking typical source distances into account, these 'hot cores' have angular sizes of less than 1″ (Table 8.2), i.e., their emission is severely diluted in the $>10''$ single-dish submillimeter beams. Other effects such as holes, cavities and disks also start to become important on these scales. Nevetheless, if the abundance enhancements are sufficiently large (typically more than a factor of 100), they can be detected even in unresolved data.

Hot core regions have been a prime focus of submillimeter observations for more than 30 years with line surveys focussing tranditionally on the prototypical Orion and SgrB2 hot cores (e.g., Blake et al. 1987, Schilke et al. 2001, Nummelin et al. 2000) but now moving to more general high-mass regions (e.g., Helmich & van Dishoeck 1997, Gibb et al. 2000a) and even low-mass YSOs (e.g., Cazaux et al. 2003). Spectra of these objects are littered with lines and often confusion limited (Fig. 8.6). Indeed, such complex spectra are now used as signposts of sources in the earliest stages of massive star formation, since this stage even preceeds the ultracompact H II region stage. Most of the lines can be ascribed to large, saturated organic molecules including CH_3OH (methanol), CH_3OCH_3 (di-methyl ether), $HCOOCH_3$ (methyl formate) and CH_3CN (methyl cyanide). With increasing frequency and sensitivity the laboratory line lists become more and more incomplete resulting in a large fraction of unidentified lines in observed data. For example, a recent 80–280 GHz IRAM-30 m line survey of Orion-KL by Tercero & Cernicharo (in prep.) reveals 16000 lines of which 8000 were unidentified in 2005. Two years later, thanks to new laboratory data on just two molecules –CH_3CH_2CN and CH_2CHCN – together with their isotopes and vibrationally excited states, the number of U-lines has been reduced to 6000. Significantly more laboratory work is needed to speed up this process. Also, the spectra of known molecules (including their isotopes and vibrationally excited states) need to be fully characterized and 'weeded out' before searches for new, more complex and pre-biotic species can be properly undertaken.

One of the main future questions is how far this chemical complexity goes. Molecules as complex as acetamide (CH_3CONH_2, the largest interstellar molecule with a peptide bond) and glycol-aldehyde (CH_2OHCHO, the first interstellar 'sugar')

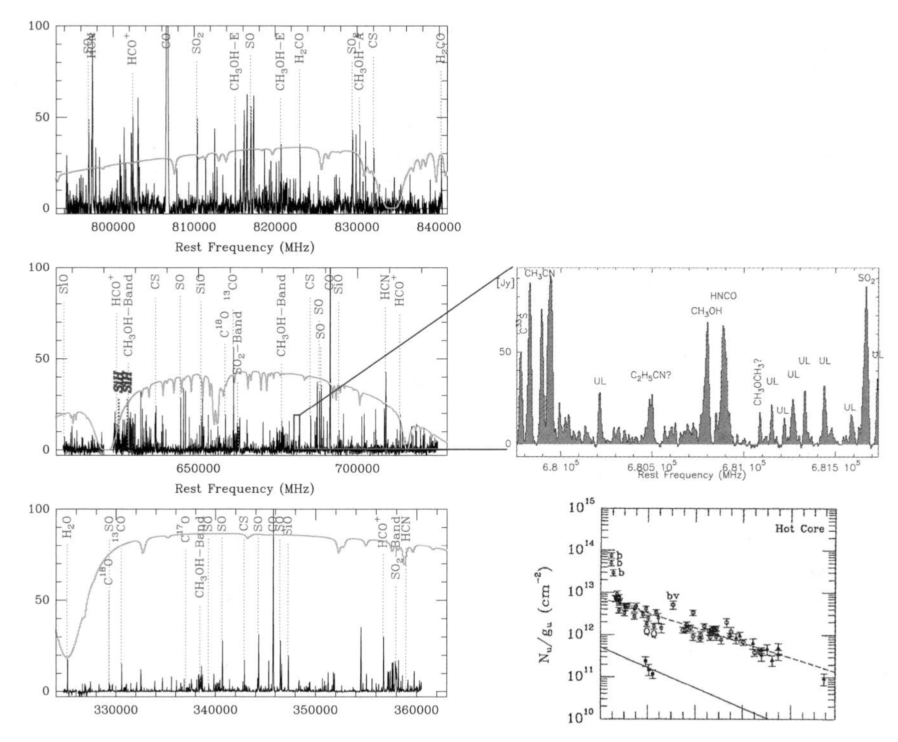

Fig. 8.6 Submillimeter line surveys of the Orion-KL region. **Top**: CSO 794-840 GHz survey by Comito et al. (2005); **Middle**: CSO 600-720 GHz survey by Schilke et al. (2001); **Bottom**: CSO 325-360 GHz line survey by Schilke et al. (1997). The green line in each panel indicates the typical atmospheric transmission. **Right, middle**: SMA interferometric survey around 680 GHz by Beuther et al. (2006). UL denotes unidentified lines. **Right, bottom**: rotation diagram of CH_3OH in the Orion hot core using JCMT data in the 345 GHz window by Sutton et al. (1995), giving $T_{rot}\approx 250$ K (*dashed line*). The optically thin $^{13}CH_3OH$ lines give a somewhat lower temperature of $\sim$180 K (*full line*)

have been found (e.g., Hollis et al. 2000, Hollis et al. 2006). However, in spite of literature claims, the simplest amino-acid glycine (NH_2CH_2COOH) has not yet been convincingly detected, although the chemically related amino acetonitrile (NH_2CH_2CN) has (Belloche et al. 2008). Other prebiotic species like aziridine and pyrimidine have also not yet been seen (Kuan et al. 2004). ALMA will be able to push the searches for larger, perhaps prebiotic molecules two orders of magnitude deeper to abundances of $<10^{-13}$ with respect to H_2, because it will have much higher sensitivity to compact emission and it will resolve the hot cores so that spatial information can be used to aid in the identification of lines. Indeed, chemical differentiation between, for example, O- and N-rich complex organics is seen on small spatial scales in both high- and low-mass YSOs from direct imaging (e.g., Wyrowski et al. 1999, Bottinelli et al. 2004) and is also inferred from (lack of) abundance correlations (Bisschop et al.

2007b, but see Fontani et al. 2007). Abundance ratios such as C_2H_5OH/CH_3OH are remarkably constant in high-mass YSOs, even in very diverse regions including the Galactic Center clouds (Requena-Torres et al. 2006), pointing toward a common origin for these O-rich organics. However, other O-containing molecules such as CH_2CO and CH_3CHO apparently avoid the warm gas, illustrating that not all complex organics co-exist (e.g., Ikeda et al. 2001, Bisschop et al. 2007b).

The origin of these complex organic molecules is still under debate, in particular whether they are first generation molecules directly evaporated from the ices or whether they are second generation products of a 'hot core chemistry' involving high-temperature gas-phase reactions between evaporated molecules (e.g., Charnley et al. 1992). A related problem is that the timescales for crossing the hot core region are very short for low-mass YSOs in a pure infall scenario, only a few hundred yr, much shorter than the timescales of $\sim 10^4 - 10^5$ yr needed for the hot core chemistry (Schöier et al. 2002). Thus, unless some mechanism has slowed down the infall or unless the molecules are in a dynamically stable region such as a disk, there is insufficient time for the hot core gas-phase chemistry to proceed.

8.4.4 Outflows and Shocks

Bipolar outflows and jets are known to be associated with all embedded YSOs, both low- and high-mass. They impact the quiescent envelope and create shocks which can sputter ices and, if sufficiently powerful, even the silicate grain cores themselves (e.g., Blake et al. 1995, Bachiller & Pérez-Gutiérrez 1997). The SiO molecule is therefore a good tracer of shocks: its abundance is greatly enhanced when Si atoms are liberated from the grains by the direct impact. Ices containing species like CH_3OH can be released in the less violent, turbulent shear zones. Indeed, in outflow lobes offset from the central protostar, CH_3OH is observed to be enhanced by more than two orders of magnitude (e.g., Jørgensen et al. 2004a, Benedettini et al. 2007). This could also provide an alternative explanation for the origin of the complex organic molecules, especially if most of them are formed on grains.

Owing to the high temperatures in shocks (up to a few 1000 K), reactions with energy barriers become very important in this hot gas compared with the colder cloud material. The most important example is the reaction of $O + H_2 \rightarrow OH + H$, followed by $OH + H_2 \rightarrow H_2O + H$ (reaction barriers $\sim$2000 K), driving most of the atomic oxygen into water (e.g., Kaufman & Neufeld 1996). Thus, H_2O, OH and [O I] far-infrared lines are predicted to dominate the shock emission, and this has been confirmed observationally for the Orion shock with ISO (Harwit et al. 1998). The precise balance between these three species depends on the H/H_2 ratio of the pre-shock gas, since reactions with H drive H_2O back to OH and O. Since lines of these species, together with those of CO and H_2, are the main coolants of the gas, their observations also provide direct information on the energetics of the flows.

8.4.5 Water and the Oxygen Budget

Water is one of the most important molecules to study in protostellar and protoplanetary environments, because it is a dominant form of oxygen, the third most abundant element in the universe. Like CO, it thus controls the chemistry of many other species. Water also plays an important role in the energy balance as a strong gas coolant, allowing clouds to collapse up to higher temperatures. It can serve as a heating agent as well, if its levels are pumped by infrared radiation followed by collisional de-excitation. As discussed in Section 8.4.1, water is the most abundant molecule in icy mantles, and its presence may help the coagulation process that ultimately produces planets. Asteroids and comets containing ice have likely delivered most of the water to our oceans on Earth, where water is directly associated with the emergence of life. Thus, the distribution of water vapor and ice during the entire star and planet formation process is a fundamental problem relevant to our own origins.

Because water observations are limited from Earth, most information to date has come from satellites (see review by Cernicharo & Crovisier 2005). SWAS and ODIN observed the 557 GHz ground-state line of ortho-H_2O at poor spatial resolution, $\sim 3'$. The emission was found to be surprisingly weak, implying that most water is frozen out on grains in cold clouds, consistent with the direct ice observations. In contrast, ISO found hot water near massive protostars in mid-IR absorption line data (e.g., van Dishoeck et al. 1996, Boonman et al. 2003), as well as strong far-IR water emission lines near low-mass YSOs (e.g., Liseau et al. 1996, Ceccarelli et al. 1998, Nisini et al. 2002). The implication is that water undergoes orders of magnitude changes in abundance between cold and warm regions, from $\sim 10^{-8}$ up to 2×10^{-4} (Boonman & van Dishoeck 2003). Thus, water acts like a 'switch' that turns on whenever energy is deposited in molecular clouds, and it is a natural filter for warm gas.

An important question is the origin of the warm water seen near protostars. One option is shocks, as discussed in Section 8.4.4. ISO has detected strong H_2O and related OH and [O I] emission lines not only from Orion but also from low-mass YSO outflow lobes (e.g., Nisini et al. 1999). Another option is the quiescent inner warm envelope or hot core, where temperatures above ~ 230 K should be high enough to produce copious water both through ice evaporation and the above mentioned gas-phase reactions (Ceccarelli et al. 1996, Charnley 1997). Finally, an intriguing option is hot water arising from the accretion shock as material falls onto the disk. Indeed, strong mid-IR water lines have recently been detected with *Spitzer* with excitation conditions consistent with a disk accretion shock (Fig. 8.7, Section 8.5.1) (Watson et al. 2007).

A related puzzle is that of the total oxygen budget in clouds. Table 8.3 contains a summary of the various forms of oxygen for a cold dense cloud without star formation (Whittet et al. 2007, Pontoppidan et al. 2004). In low density diffuse clouds, optical and UV absorption line data have established that about 30% of the total (gas + solid) elemental O abundance of 4.6×10^{-4} is contained in refractory material

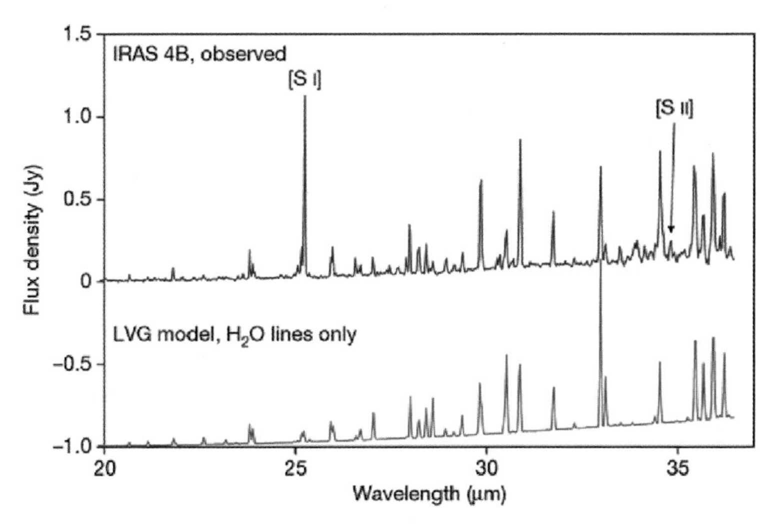

Fig. 8.7 Detection of hot water lines toward the deeply embedded Class 0 protostar NGC 1333 IRAS4B using the *Spitzer Space Telescope*. These lines are thought to originate from the accretion shock onto the growing disk (Watson et al. 2007)

Table 8.3 Typical oxygen budget in a quiescent cold cloud

Component	Material	Fraction of oxygen[a] (%)	Observations
Refractory	Silicates	30	Mid-IR, UV
Ices	H_2O, CO_2, CO, ...	26	Mid-IR
Gas-phase	CO	9	Submm, Mid-IR
Remainder	O? O_2? H_2O?	35	UV, Far-IR, Submm, Mid-IR

[a] Values from Whittet et al. (2007) for several lines of sight in Taurus.

such as silicates, with the remainder in gaseous atomic oxygen (Meyer et al. 1998). Assuming that the refractory budget stays the same in dense clouds, one can make an inventory of the remaining volatile components. As Table 8.3 shows, ices contain about 25–30% and gaseous CO up to 10% of the oxygen, with precise values varying from cloud to cloud. That leaves a significant fraction of the budget, about 1/3, unaccounted for. SWAS and ODIN have shown that the O_2 abundance in dense clouds is surprisingly low, $\leq 10^{-7}$ (Goldsmith et al. 2000, Larsson et al. 2007), as is H_2O in the cold gas (see above discussion). Thus, gaseous atomic O is the most likely reservoir. Limited data on [O I] 63 and 145 μm emission or absorption from cold clouds exist (e.g., Caux et al. 1999, Vastel et al. 2002), but are difficult to interpret because of the high optical depth of the lines coupled with the fact that they are spectrally unresolved.

8.5 Protoplanetary Disks

8.5.1 From Envelope to Disk

In Section 8.4, it has been shown that the inner envelopes around protostars contain a wealth of simple and complex molecules, but it is not yet clear whether and where these molecules end up in the planet-forming zones of disks since the dynamics of gas in the inner few hundred AU are poorly understood. In the standard theory for inside-out collapse and disk formation (e.g., Terebey et al. 1984, Cassen & Moosman 1981), most of the material enters the disk close to the centrifugal radius, which grows with time as t^3. Also, the disk spreads as some material moves inward to accrete onto the star and some moves outward to conserve angular momentum. Thus, in the earliest phase of the collapse, gas falls in very close to the star where it will experience such a strong accretion shock that all ices evaporate and all molecules dissociate (Neufeld & Hollenbach 1994). However, at the later stages, once molecules enter the disks beyond a few AU, all species except the most volatile ices (e.g., CO ice) survive. Some observational evidence for accretion shocks has been presented through mid-IR CO observations (Pontoppidan et al. 2003) and H_2O data (Watson et al. 2007) (see Fig. 8.7).

8.5.2 Disk Chemistry

Once in the disk, the chemistry of the outer part is governed by similar gas-phase and gas-grain interactions as in envelopes, but at higher densities. From analyses of SEDs, it is now well established that disks have temperature gradients not only in the radial direction but also vertically. If disks are flared, they intercept a large fraction of the stellar UV and X-rays which heat the dust and gas in the optically thin surface layers, with the IR emission from this layer subsequently warming the lower parts of the disk. Moreover, this radiation dissociates molecules and ionizes atoms, thus modifying the chemistry in the upper layers. The result is a layered chemical structure (Aikawa et al. 2002), with a cold mid-plane where most molecules are frozen out, a top layer consisting mostly of atoms, and an intermediate layer where the dust grains are warm enough to prevent complete freeze-out and where molecules are sufficiently shielded from radiation to survive (for review, see Bergin et al. 2007).

8.5.2.1 Outer Disk

Indirect evidence for freeze-out in disks comes from the inferred low gas-phase abundances of various molecules (e.g., Dutrey et al. 1997, van Zadelhoff et al. 2001). Direct detection of ices in disks has been possible for a few sources with a favorable near edge-on geometry where the line of sight passes through the outer part (e.g., Pontoppidan 2006, Terada et al. 2007). Because foreground clouds can also contribute to the observed ice absorptions, it has not yet been possible to make an ice inventory for disks. At longer wavelengths, the ice bands can be in emission, and therefore do not require a special geometry. Bands of crystalline water ice at 45 and 62 μm have been seen in a few disks with ISO (e.g., Creech-Eakman et al. 2002).

Most molecules other than CO have only been detected in spatially unresolved single-dish submillimeter spectra from which disk-averaged abundances can be derived (e.g., Dutrey et al. 1997, Thi et al. 2004). Chemical 'images' of disks have so far been limited to just a few pixels across a handful of disks in a few lines (e.g., Qi et al. 2003, Dutrey et al. 2007). The data obtained so far support the layered structure picture but obviously ALMA will throw this field wide open. Besides being interesting in its own right, chemistry and molecular lines can also constrain important physical processes in disks, such as the level of turbulence and the amount of vertical mixing (e.g., Semenov et al. 2006).

8.5.2.2 Inner Disk

The hot gas in the inner disk is readily detected in the mid-IR lines of CO (e.g., Najita et al. 2003, Blake & Boogert 2004), as well as the UV and IR lines of H_2 (e.g., Herczeg et al. 2002, Martin-Zaïdi et al. 2007). PAH emission has also been spatially resolved, with both the inner and outer disk contributing to the various features (e.g., Habart et al. 2004, Geers et al. 2007). The chemistry in the inner disk differs in several aspects from that in the outer parts: the densities are so high that three-body reactions become important; X-rays from the young star may be significant; gas columns may be so high that cosmic rays can no longer penetrate to the mid-plane, thus stopping ion-molecule reactions; and Fischer-Tropsch catalysis can take place on hot metallic grains. A particularly exciting topic is the chemistry inside the 'snow-line' where all molecules evaporate and the chemistry approaches that at LTE (e.g., Markwick et al. 2002). *Spitzer* absorption line data have revealed highly abundant and hot (300–700 K) HCN and C_2H_2 in the inner few AU of two young near edge-on disks consistent with these models (Lahuis et al. 2006, Gibb et al. 2007). More recently, *Spitzer* and ground-based Keck and VLT data have revealed surprisingly strong mid-IR emission lines of hot H_2O ($\sim$800 K), together with OH, HCN, C_2H_2 and/or CO_2, toward a number of disks originating from the inner AU (Carr & Najita 2008, Salyk et al. 2008).

8.5.3 *Disk Chemical Evolution*

Near- and mid-infrared surveys have shown that inner dust disks ($<$10 AU) disappear on timescales of a few Myr (e.g., Cieza et al. 2007). A major question is how the gas and dust dissipate from the disk, and whether they do so at the same time. Examination of hundreds of SEDs of stars with disks in *Spitzer* surveys show that there may be multiple evolutionary paths from the massive gas-rich disks to the tenuous gas-poor debris disks, involving both grain growth and gap opening, either by photoevaporation or planet formation (e.g., Alexander et al. 2006, Varnière et al.

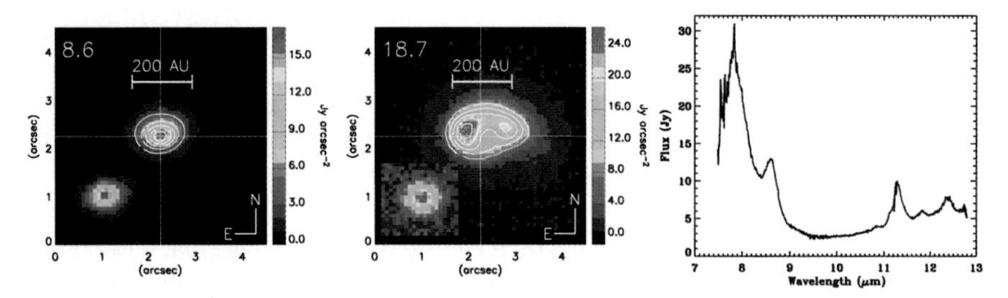

Fig. 8.8 VLT-VISIR mid-infrared images of the disk around the young T Tauri star IRS 48, showing strong centrally peaked PAH emission at 11.3 μm as well as a 60 AU diameter gap devoid of large grains emitting at 19 μm. The inserts show the PSF of a standard star. The 8–13 μm spectrum with the strong PAH features is included (Geers et al. 2007)

2006). Some transitional disks show evidence that molecular gas and PAHs are still present inside the dust gaps (Goto et al. 2006, Jonkheid et al. 2006, Geers et al. 2007, Pontoppidan et al. 2008) (see Fig. 8.8). [O I] is the dominant coolant of dense gas, and as such may also be a particularly powerful gas mass probe of transitional disks with future facilities (Meijerink et al. 2008). The chemistry with such high gas/dust ratios and large grains (which inhibit H_2 formation) is very different from that under normal cloud conditions (e.g., Aikawa & Nomura 2006, Jonkheid et al. 2007). Moreover, the shape of the radiation field plays a critical role, in particular whether there are far-UV photons with energies $>$11 eV which can dissociate H_2 and CO and ionize C (van Zadelhoff et al. 2003, van Dishoeck et al. 2006). Also, young stars are known to have UV flares resulting from discrete accretion events, which temporarily increase the disk temperature and affect the chemistry.

8.6 Prospects for Future Facilities

The major new facilities at infrared and submillimeter wavelengths will be crucial to answer many of the questions raised in the previous sections. The principle limitation at millimeter wavelengths is the low spatial resolution and sensitivity: current single-dish telescopes and interferometers can hardly resolve the envelopes, hot cores and disks. ALMA will be a tremendous leap forward in this field, with the capability and sensitivity to image molecules down to scales of tens of AU where planet formation takes place. In the coldest pre-stellar cores, ALMA can detect the earliest signs of collapse through line profiles of species like N_2H^+ and ortho-H_2D^+ on 100 AU scales. Once the protostar has formed, ALMA is the instrument of choice to unravel the chemistry in hot cores and search for the most complex prebiotic molecules, by spatially imaging the organics and 'weeding out' the more common species. Low-frequency cm data, such as could be provided by one version of the proposed Square Kilometer Array, are well suited to probe the heaviest rotors. ALMA can also resolve individual bowshocks and thus distinguish the shock chemistry from that of the other protostellar components. For example, are the complex organics located in the passively heated hot core region, in the disk, or in the region where the outflow impacts the dense envelope?

Studies of the chemistry in the outer disk will be opened up completely by ALMA, which will provide chemical images in many different species. Since the brightness temperatures of the submillimeter lines from the inner disk can be as high as several hundred K, ALMA can image lines in the nearest disks down to $\sim$10 AU. Because of lower optical depth, ALMA has the advantage that it can probe deeper into the disk than infrared telescopes. ALMA will also be critical to study transitional objects by imaging the holes or gaps in their dust disks down to a few AU and by measuring the remaining gas mass through tracers like CO and [C I] on scales of more than 10 AU.

The strength of JWST-MIRI and NIRSPEC lies in their raw sensitivity coupled with moderate spectral resolution. They will be particularly powerful to probe the ices in the densest, most obscured parts of the cores and determine when and where ices are formed through ice mapping on $\sim 10''$ scale. The high sensitivity will also allow searches for minor ice species toward highly obscured Class 0 protostars to address the question, together with ALMA, which (complex) molecules are formed on the grains as first generation species and which in the gas as second generation. Mid-IR imaging of shocked H_2, [S I] and [Fe II] with JWST will trace the physical structure of shocks in the deeply embedded phase, necessary for understanding the chemistry.

JWST will also allow observations of ices toward a much larger fraction of edge-on disks, and can perhaps even spatially resolve the absorption against the extended mid-IR continuum. Moreover, with its medium resolution mode $R \approx 3000$, JWST can search for mid-infrared vibration-rotation lines of water and organic building blocks like HCN, C_2H_2 and CH_4 in absorption or emission in protostars and disks, although analysis will be hampered by the fact that the lines are spectrally unresolved.

The unique power of ground-based ELTs lies in their very high spectral resolving power up to 10^5 combined with high sensitivity and spatial resolution, allowing quantitative studies of gas-phase lines. Toward protostellar objects, gas/dust ratios can be measured directly in the coldest regions, and some of the complex organics freshly evaporated off the grains can be probed in absorption even toward solar-mass sources. Emission lines of hot water and organics from disks should have booming feature-to-continuum ratios and such spectra will also allow searches for less common species. Moreover, ELTs can image the emission and kinematics at AU resolution and thus determine the chemistry *distribution* in the inner ($<$10 AU) disk, which ALMA cannot probe. This includes the distribution of PAHs, the most complex organics known to date.

Herschel's strength lies in its ability to observe cold and warm H_2O in protostellar environments through the myriad of pure rotational lines. Indeed, the combination of Herschel and SOFIA will be uniquely suited to observe H_2O, OH and O far-infrared lines with orders of magnitude higher spatial and/or spectral resolution and sensitivity compared with previous missions. With no other far-infrared space missions on the horizon, this will be the only chance for decades to follow the water trail through star formation and determine the oxygen budget in a variety of sources. SOFIA is a key complement to Herschel, since it is the only mission that can provide spectrally resolved data on the [O I] lines at 63 and 145 μm.

Herschel and SOFIA will also fully open up the far-infrared wavelength range through line surveys, with ample opportunities for unexpected chemical surprises. Higher frequency THz data can help in the identification of more complex species (including PAHs) since their low-frequency vibrational modes may be more easily recognized than their pure rotational spectra observed with ALMA, where the intensity is spread over many lines. SOFIA will also be the only instrument available to probe the lowest 1370 GHz transition of para-H_2D^+ in the coldest clouds whereas both SOFIA and Herschel can observe the lowest ortho-D_2H^+ transition at 1476 GHz. Finally, Herschel and SOFIA can search for the far-infrared bands of ices and (hydrous) silicates in emission from protostars and disks. This is the only option to probe ices in disks which do not require any special viewing geometry.

8.7 Conclusions

Substantial progress has been made in determining the inventories of the gases and solids in protostellar and protoplanetary regions, not only for bright high-mass YSOs but also for weaker sources thought to be representative of our own early solar system. Each evolutionary phase has its own chemical characteristics related to the changing physical conditions, with freeze-out and ice evaporation playing a major role. The combination of observations from near-infrared to millimeter wavelengths has been essential to probe all chemical components. Basic laboratory data and sophisticated radiative transfer tools remain crucial for quantitative analysis. With the orders of magnitude enhancements in sensitivity, spatial and spectral resolution of future instruments, there is no doubt that astrochemistry will continue to blossom in the next decade of JWST and its concurrent facilities.

Acknowledgments I am grateful to the editor A.G.G.M. Tielens for comments, and to moderator of this session, J. Cernicharo, for stimulating the discussion, the content of which has been included in this chapter. This work is supported by a Spinoza grant from the Netherlands Organization for Scientific Research (NWO).

References

Aikawa, Y., & Nomura, H. 2006, ApJ, 642, 1152
Aikawa, Y., van Zadelhoff, G.J., van Dishoeck, E.F., & Herbst, E.F. 2002, A&A, 386, 622
Alexander, R.D., Clarke, C.J., & Pringle, J.E. 2006, MNRAS, 369, 229
Bachiller, R., & Pérez-Gutiérrez, M. 1997, ApJ, 487, L93
Belloche, A., Menten, K.M., Comito, C., Mueller, H.S.P., Schilke, P., Ott, J., Thorwirth, S., & Hieret, C. 2008, A&A, 482, 179
Benedettini, M., Viti, S., Codella, C., et al. 2007, MNRAS, 381, 1127

Bergin, E.A., Aikawa, Y., Blake, G. A., & van Dishoeck, E. F. 2007, in Protostars and Planets V, eds. B. Reipurth, D. Jewitt, & K. Keil, 751–766
Bergin, E.A., & Tafalla, M. 2007, ARA&A 45, 339
Bergin, E.A., Alves, J., Huard, T., & Lada, C. J. 2002, ApJ, 570, L101
Beuther, H., Zhang, Q., Reid, M.J., et al. 2006, ApJ, 636, 323
Bisschop, S.E., Fuchs, G.W., van Dishoeck, E.F., & Linnartz, H. 2007a, A&A, 474, 1061
Bisschop, S.E., Jørgensen, J.K., van Dishoeck, E.F., & de Wachter, E.B.M. 2007b, A&A, 465, 913
Bisschop et al. 2008, A & A, 488, 959
Blake, G.A., & Boogert, A.C.A. 2004, ApJ, 606, L73
Blake, G.A., Sutton, E.C., Masson, C.R., & Phillips, T.G. 1987, ApJ, 315, 621
Blake, G.A., Sandell, G., van Dishoeck, E.F., Aspin, C., Groesbeck, T., & Mundy, L.G. 1995, ApJ, 441, 689
Boogert, A., Pontoppidan, K., Knez, C., et al. 2008, ApJ, 678, 985
Boogert, A., Pontoppidan, K., Lahuis, F., et al. 2004, ApJS, 154, 359
Boonman, A.M.S., Doty, S.D., van Dishoeck, E.F., Bergin, E.A., Melnick, G.J., Wright, C.M., & Stark, R. 2003, A&A, 406, 937
Boonman, A.M.S. & van Dishoeck, E.F. 2003, A&A, 403, 1003
Bottinelli, S., Ceccarelli, C., Neri, R., et al. 2004, ApJ, 617, L69
Brünken, S., Gupta, H., Gottlieb, C.A., McCarthy, M.C., & Thaddeus, P. 2007, ApJ, 664, L43
Carr, J.S., & Najita, J. 2008, Science, 319, 1504
Caselli, P., Walmsley, C. M., Tafalla, M., Dore, L., & Myers, P. C. 1999, ApJ, 523, L165
Caselli, P., van der Tak, F.F.S., Ceccarelli, C., & Bacmann, A. 2003, A&A, 403, L37
Cassen, P., & Moosman, A. 1981, Icarus, 48, 353
Caux, E., Ceccarelli, C., Castets, A., Vastel, C., Liseau, R., Molinari, S., Nisini, B., Saraceno, P., & White, G.J. 1999, A&A, 347, L1
Cazaux, S., Tielens, A.G.G.M., Ceccarelli, C., et al. 2003, ApJ, 593, L51
Ceccarelli, C., Hollenbach, D.J., & Tielens, A.G.G.M. 1996, ApJ, 471, 400
Ceccarelli, C., Caux, E., White, G.J., et al. 1998, A&A, 331, 372
Ceccarelli, C., Caselli, P., Herbst, E., Tielens, A.G.G.M., & Caux, E. 2007, in Protostars and Planets V, eds. B. Reipurth et al., 47–62
Cernicharo, J., Crovisier, J. 2005, Space Science Rev., 119, 29
Chandler, C.J., Brogan, C.L., Shirley, Y.L., & Loinard, L. 2005, ApJ, 632, 371
Charnley, S.B. 1997, ApJ, 481, 396
Charnley, S.B., Tielens, A.G.G.M., & Millar, T.J. 1992, ApJ, 399, L71
Cieza, L., Padgett, D.L., Stapelfeldt, K.R., et al. 2007, ApJ, 667, 308
Comito, C., Schilke, P., Phillips, T.G., Lis, D.C., Motte, F., & Mehringer, D. 2005, ApJS, 156, 127
Creech-Eakman, M.J., Chiang, E.I., Joung, R.M.K., Blake, G.A., & van Dishoeck, E.F. 2002, A&A, 385, 546
Dutrey, A., Guilloteau, S., & Guelin, M. 1997, A&A, 317, L55
Dutrey, A., Henning, Th., Guilloteau, S., et al. 2007, A&A, 464, 615
Evans, N.J., Lacy, J.H., & Carr, J.S. 1991, ApJ, 383, 674
Fontani, F., Pascucci, I., Caselli, P., Wyrowski, F., Cesaroni, R., & Walmsley, C.M. 2007, A&A, 470, 639
Garrod, R.T., & Herbst, E. 2006, A&A, 457, 927
Gerakines, P.A., Schutte, W.A., & Ehrenfreund, P. 1996, A&A, 312, 289
Geers, V.C., Pontoppidan, K.M., van Dishoeck, E.F., et al. 2007, A&A, 469, L35
Gibb, E.L., Nummelin, A., Irvine, W.M., Whittet, D.C.B., & Bergman, P. 2000a, ApJ, 545, 309
Gibb, E.L., Whittet, D.C.B., Schutte, W.A. et al. 2000b, ApJ, 536, 346
Gibb, E.L., van Brunt, K.A., Brittain, S.D., & Rettig, T.W. 2007, ApJ, 660, 1572
Goldsmith, P.F., Melnick, G.J., Bergin, E.A., et al. 2000, ApJ, 539, L123
Goto, M., Stecklum, B., Linz, H., Feldt, M., Henning, Th., Pascucci, I., & Usuda, T. 2006, ApJ, 649, 299
Habart, E., Natta, A., Krügel, E. 2004, A&A, 427, 179

Harwit, M., Neufeld, D.A., Melnick, G.J., & Kaufman, M.J. 1998, ApJ, 497, L105
Helmich, F.P, & van Dishoeck, E.F. 1997, A&AS, 124, 205
Herczeg, G.J., Linsky, J.L., Valenti, J.A., Johns-Krull, C.M., & Wood, B.E. 2002, ApJ, 572, 310
Hollis, J.M., Lovas, F.J., & Jewell, P.R. 2000, ApJ, 540, L107
Hollis, J.M., Lovas, F.J., Remijan, A.J., et al. 2006, ApJ, 643, L25
Hudgins, D.M., Sandford, S.A., Allamandola, L.J., & Tielens, A.G.G.M. 1993, ApJS, 86, 713
Hudgins, D.M., & Allamandola, L.J. 1999, ApJ, 516, L41
Ikeda, M., Ohishi, M., Nummelin, A., Dickens, J.E., Bergman, P., Hjalmarson, Å, & Irvine, W.M. 2001, ApJ, 560, 792
Jaeger, C., Molster, F.J., Dorschner, J., Henning, Th., Mutschke, H., & Waters, L.B.F.M. 1998, A&A, 339, 904
Jonkheid, B., Kamp, I., Augereau, J.-C., & van Dishoeck, E.F. 2006, A&A, 453, 163
Jonkheid, B., Dullemond, C.P., Hogerheijde, M.R., & van Dishoeck, E.F. 2007, A&A, 463, 203
Jørgensen, J.K. 2004, A&A424, 589
Jørgensen, J.K., Bourke, T. L., Myers, P. C., et al. 2005, ApJ, 632, 973
Jørgensen, J.K., Hogerheijde, M.R., Blake, G.A., van Dishoeck, E.F., Mundy, L.G., & Schöier, F.L. 2004a, A&A, 415, 1021
Jørgensen, J.K., Schöier, F.L., & van Dishoeck, E.F. 2004b, A&A, 416, 603
Kaufman, M.J., & Neufeld, D.A. 1996, ApJ, 456, 611
Keane, J.V., Tielens, A.G.G.M., Boogert, A.C.A., Schutte, W.A., & Whittet, D.C.B. 2001, A&A, 376, 254
Knez, C., Lacy, J., Evans, N.J., II, et al. 2008, ApJ, submitted
Kuan, Y.-J., Charnley, S.B., Huang, H.-C., Kisiel, Z., Ehrenfreund, P., Tseng, W.-L., & Yan, C.-H. 2004, Adv Space Res, 33, 31
Lada, C.J. 1999, in The Origin of Stars and Planetary Systems, eds. C.J. Lada & N.D. Kylafis (Dordrecht: Kluwer), p. 143
Lahuis, F., & van Dishoeck, E.F. 2000, A&A, 355, 699
Lahuis, F., van Dishoeck, E.F., Boogert, A.C.A., et al. 2006, ApJ, 636, L145
Larsson, B., Liseau, R., Pagani, L., et al. 2007, A&A, 466, 999
Lis, D.C, Roueff, E., Gerin, M., Phillips, T.G., Coudert, L.H., van der Tak, F.F.S., & Schilke, P. 2002, ApJ 571, L55
Liseau, R., Ceccarelli, C., Larsson, B., et al. 1996, A&A, 315, L181
Marcelino, N., Cernicharo, J., Agúndez, M., et al. 2007, ApJ, 665, L127
Markwick, A.J., Ilgner, M., Millar, T.J., & Henning, Th. 2002, A&A, 385, 632
Martin-Zaïdi, C., Lagage, P.-O., Pantin, E., & Habart, E. 2007, ApJ, 666, L117
McCarthy, M.C., Gottlieb, C.A., Gupta, H., & Thaddeus, P. 2006, ApJ, 652, L141
Meijerink, R., Glassgold, A.E., & Najita, J.R. 2008, ApJ, 676, 518
Meyer, D.M., Jura, M., & Cardelli, J.A. 1998, ApJ, 493, 222
Najita, J., Carr, J.S., & Mathieu, R.D. 2003, ApJ, 589, 931
Neufeld, D.A. & Hollenbach, D.J. 1994, ApJ, 428, 170
Nisini, B., Benedettini, M., Giannini, T., et al. 1999, A&A, 350, 529
Nisini, B., Giannini, T., & Lorenzetti, D. 2002, ApJ, 574, 246
Noriega-Crespo, A., Morris, P., Marleau, F.R., et al. 2004, ApJS, 154, 352
Nummelin, A., Bergman, P., Hjalmarson, Å, et al. 2000, ApJS, 128, 213
Öberg, K.I., van Broekhuizen, F., Fraser, H.J., Bisschop, S.E., van Dishoeck, E.F., & Schlemmer, S. 2005, ApJ, 621, L33
Öberg, K.I., Fraser, H.J., Boogert, A.C.A., Bisschop, S.E., Fuchs, G.W., van Dishoeck, E.F., & Linnartz, H. 2007, A&A, 462, 1187
Parise, B., Castets, A., Herbst, E., Caux, E., Ceccarelli, C., Mukhopadhyay, I., & Tielens, A.G.G.M. 2004, A&A, 416, 159
Pillai, T., Wyrowski, F., Hatchell, J., Gibb, A.G., & Thompson, M.A. 2007, A&A, 467, 207
Pontoppidan, K.M., Blake, G.A., van Dishoeck, E.F., Smette, A., Ireland, M.I., & Brown, J.M. 2008, ApJ, in press

Pontoppidan, K.M., Fraser, H.J., Dartois, E., et al. 2003, A&A, 408, 981
Pontoppidan, K.M., van Dishoeck, E.F., & Dartois, E. 2004, A&A, 426, 925
Pontoppidan, K.M., Dullemond, C.P., van Dishoeck, E.F., et al. 2006, ApJ, 622, 463
Pontoppidan, K.M. 2006, A&A, 453, L47
Qi, C., Kessler, J.E., Koerner, D.W., Sargent, A.I., & Blake, G.A. 2003, ApJ, 597, 986
Requena-Torres, M.A., Martín-Pintado, J., Rodríguez-Franco, A., Martín, S., Rodríguez-Fernández, N.J., & de Vicente, P. 2006, A&A, 455, 971
Roberts, H., Herbst, E., & Millar, T.J. 2003, ApJ, 591, L41
Salyk, C., Pontoppidan, K.M., Blake, G.A., Lahuis, F., van Dishoeck, E.F., & Evans, N.J. 2008, ApJ, 676, L49
Schilke, P., Benford, D.J., Hunter, T.R., Lis, D.C., & Phillips, T.G. 2001, ApJS, 132, 281
Schilke, P., Groesbeck, T.D., Blake, G.A., & Phillips, T.G. 1997, ApJS, 108, 301
Schöier, F.L., Jørgensen, J.K., van Dishoeck, E.F., & Blake, G.A. 2002, A&A, 390, 1001
Schöier, F.L., van der Tak, F.F.S., van Dishoeck, E.F., & Black, J.H. 2005, A&A, 432, 369
Schutte, W.A., & Khanna, R.K. 2003, A&A, 398, 1049
Semenov, D., Wiebe, D., & Henning, Th. 2006, ApJ, 647, L57
Sutton, E.C., Peng, R., Danchi, W.C., Jaminet, P.A., & Sandell, G., Russell, A.P.G. 1995, ApJS, 97, 455
Terada, H., Tokunaga, A.T., Kobayashi, N., Takato, N., Hayano, Y., & Takami, H. 2007, ApJ, 667, 303
Terebey, S., Shu, F.H., & Cassen, P. 1984, ApJ, 286, 529
Thi, W.-F., van Zadelhoff, G.-J., & van Dishoeck, E.F. 2004, A&A, 425, 955
Tielens, A.G.G.M. 1983, A&A, 119, 177
Tielens, A.G.G.M. 2008, ARA&A, in press
Tielens, A.G.G.M., & Charnley, S.B. 1997, Origins Life Evol. B., 27, 23
Tielens, A.G.G.M., & Hagen, W. 1982, A&A, 114, 245
van Broekhuizen, F.A., Pontoppidan, K.M., Fraser, H.J., & van Dishoeck, E.F. 2005, A&A, 441, 249
van Dishoeck, E.F. 2004, ARA&A, 42, 119
van Dishoeck, E.F. 2006, Proc Nat Ac Sci, 103, 12249
van Dishoeck, E.F., & Blake, G.A. 1998, ARA&A, 36, 317
van Dishoeck, E.F., van Hemert, M.C., & Jonkheid, B.J. 2006, Faraday Discussions, 133, 231
van Dishoeck, E.F., Helmich, F.P., de Graauw, T., et al. 1996, A&A, 315, L349
van Zadelhoff, G.-J., van Dishoeck, E.F., Thi, W.-F., & Blake, G.A. 2001, A&A, 377, 566
van Zadelhoff, G.-J., Dullemond, C.P., van der Tak, F.F.S., et al. 2002, A&A 395, 373
van Zadelhoff, G.-J., Aikawa, Y., Hogerheijde, M.R., & van Dishoeck, E.F. 2003, A&A, 397, 789
Varnière, P., Blackman, E.G., Frank, A., & Quillen, A.C. 2006, ApJ, 640, 1110
Vastel, C., Phillips, T.G., & Yoshida, H. 2004, ApJ, 606, L127
Vastel, C., Polehampton, E.T., Baluteau, J.-P., Swinyard, B.M., Caux, E., & Cox, P. 2002, ApJ, 581, 315
Wakelam, V., Herbst, E., & Selsis, F. 2006, A&A, 451, 551
Watanabe, N., Nagaoka, A., Shiraki, T., & Kouchi, A. 2004, ApJ, 616, 638
Watson, D.M., Bohac, C.J., Hull, C., et al. 2007, Nature, 448, 1026
Whittet, D.C.B., Shenoy, S.S., Bergin, E.A., et al. 2007, ApJ, 655, 332
Wyrowski, F., Schilke, P., Walmsley, C.M., & Menten, K.M. 1999, ApJ, 514, L43
Young, C.H., Jørgensen, J.K., Shirley, Y.L., et al. 2004, ApJS, 154, 352

John Mather explains how JWST will make all the important astronomical discoveries of the first half of the 21st Century

Chapter 9
Extreme Star Formation

Jean L. Turner

Abstract Extreme star formation includes star formation in starbursts and regions forming super star clusters. We survey the current problems in our understanding of the star formation process in starbursts and super star clusters—initial mass functions, cluster mass functions, star formation efficiencies, and radiative feedback into molecular clouds—that are critical to our understanding of the formation and survival of large star clusters, topics that will be the drivers of the observations of the next decade.

9.1 Extreme Star Formation in the Local Universe

Extreme star formation is the violent, luminous star formation that occurs in starbursts and luminous infrared galaxies. It is the formation of super star clusters that may eventually become globular clusters. It is the source of galactic winds and metal enrichment in galaxies. It is probably what most star formation in the universe was like several gigayears ago.

The process of star formation and its associated microphysics is most easily studied in the local universe where we can examine the process of star formation in detail. While there are regions in the Galaxy that may qualify as extreme star formation, most extreme systems are extragalactic. Advances in the study of extragalactic star formation during the next decade are likely to come from improvements in spatial resolution and sensitivity, particularly in the infrared and submillimeter parts of the spectrum. The refurbished HST, forthcoming JWST, and ground-based adaptive optics systems will make fundamental contributions to our understanding of the stellar content of extreme star forming regions. Herschel, ALMA, CARMA, Plateau de Bure, SMA, and SOFIA are the far-infrared, submillimeter, and millimeter telescopes that will deliver images and spectra of molecular gas in galaxies, enabling the study of the earliest stages of star formation, and the regulation of star formation by feedback into molecular clouds. With the subarcsecond and

J.L. Turner (✉)
Department of Physics and Astronomy, UCLA, Los Angeles CA 90095-1547 USA
e-mail: turner@astro.ucla.edu

H.A. Thronson et al. (eds.), *Astrophysics in the Next Decade,* Astrophysics and Space Science Proceedings, DOI 10.1007/978-1-4020-9457-6_9,

milliarcsecond resolutions now possible we can study the star formation process in other galaxies on the parsec spatial scales of molecular cores, young clusters, and Stromgren spheres.

This review is an attempt to distill a very active area of research on extreme star formation, covering both the stellar content and studies of the star-forming gas, and to project this research into the observations of the next decade. The field is a remarkably broad one, because in the process of star birth and cluster evolution, stars and gas are physically interrelated. The observations discussed here cover the range from ultraviolet spectroscopy of hot stars to millimeter line imaging of cold molecular clouds. The focus will be on star formation in the local universe where individual star-forming regions can be resolved, and the star formation process itself can be studied. Star formation in the early universe, where extreme star formation may have been more the norm than the exception, is covered elsewhere in this volume in contributions by Tom Abel and Alice Shapley.

9.2 What Constitutes Extreme Star Formation in the Local Universe?

Advances in instrumentation have shaped and refined our current view of "extreme star formation" (ESF). Many of the features of ESF in the local universe were known half a century ago: giant HII regions (Burbidge & Burbidge 1962), galactic winds (Burbidge et al. 1964, Lynds & Sandage 1963), young "populous clusters" (Gascoigne & Kron 1952, Hodge 1961), O-star dominated compact dwarf galaxies (Sargent & Searle 1970), and bright extragalactic radio sources (Weedman et al. 1981). The recognition that there were individual star formation events that could energetically dominate the evolution of a galaxy came with the development of infrared and high resolution radio capabilities (Gehrz et al. 1983, Rieke & Lebofsky 1978, 1979, Rieke & Low 1975). However, it was the IRAS all-sky survey in the mid and far-infrared that established the universality and energetic importance of the "starburst" to galaxy evolution. IRAS demonstrated that the luminous output of galaxies can be dominated by infrared emission and recent star formation (de Jong et al. 1984, Soifer et al. 1984, 1986, 1987a,b) and that extreme star formation may even be linked to the development of nuclear activity in galaxies (Sanders et al. 1988). Many of the early IRAS results have been followed up with the subsequent ISO (Genzel & Cesarsky 2000) and Spitzer (Lonsdale et al. 2003) infrared space observatories.

The definition of starburst has evolved since the time of IRAS. In early incarnations it described systems that would deplete their gas in substantially less than a Hubble time. However, this definition can exclude galaxies with substantial reservoirs of gas that are forming stars at prodigious rates. Infrared luminosity can be used to classify these extreme star-forming systems: "Ultraluminous infrared galaxies" (ULIRGs) have luminosities of $L_{IR} > 10^{12}\ L_{\odot}$ (Soifer et al. 1987b) and "luminous" infrared galaxies have $L_{IR} >\sim 10^{11}\ L_{\odot}$. These luminous galaxies owe most of their energetic output to star formation (Genzel et al. 1998, Sanders et al. 1986). Another definition captures the localized intensity of starbursts: (Kennicutt 1998a) defines starburst in terms of a star-forming surface density, 100 $M_{\odot}$ pc^{-2} Gyr^{-1}, or in terms of luminosity, $10^{38.4}$–$10^{39.4}$ erg s^{-1} kpc^{-2}.

High spatial resolution has also modified our view of starbursts, revealing that they often, and perhaps nearly always, consist of the formation of large numbers of extremely large clusters, "super star clusters." For the purposes of this review, we will use the term super star cluster (SSC) to denote massive young clusters of less than ~100 Myr in age, and globular clusters to be those systems more than 7–10 Gyr in age. Both starbursts and SSCs comprise "extreme star formation."

The Hubble Space Telescope (HST) is largely responsible for the recognition of the ubiquity of SSCs and their importance in starbursts. The idea that massive clusters similar to globular clusters are actually forming in large numbers at the present time was slow to germinate, probably due to lack of spatial resolution and our inability to resolve them, although the possibility was recognized early on in the large clusters of the Magellanic clouds (Hodge 1961). The cluster R136, the star cluster responsible for the lovely 30 Doradus Nebula in the Large Magellanic Clouds, was believed by many to be a single supermassive star before it was resolved with speckle observations from the ground (Weigelt & Baier 1985). Other mysterious objects included two bright sources in the nearby dwarf galaxy NGC 1569, which were difficult to classify due to the uncertainty in distance to this nearby galaxy. Regarding the two "super star cluster" candidates, Arp & Sandage (1985) stated:

> A definite resolution of the present problem in NGC 1569, and for the same problem with the bright object in NGC 1705, lies in the spatial resolution into stars of these three high-luminosity blue objects using the imaging instrument of the wide-field camera of Space Telescope.

HST did indeed resolve the objects in NGC 1569, revealing that they were large and luminous star clusters. In Fig. 9.1 is shown the HST image of NGC 1569, with objects A and B referred to by Arp and Sandage. Object A consists of two superimposed clusters; crowded conditions and confusion complicate the study of SSCs, even in the closest galaxies and with the angular resolution of HST! R136 was resolved into a compact and rich cluster of stars by HST (de Marchi et al. 1993, Hunter et al. 1996). HST imaging also revealed super star clusters in NGC 1275, M82, NGC 1705, the Antennae, and numerous other starburst galaxies (Holtzman et al. 1992, Maoz et al. 1996, Meurer et al. 1995, O'Connell et al. 1994, Whitmore & Schweizer 1995, Whitmore et al. 1993) The discovery of young, blue star clusters with luminosities consistent with those expected for young globular clusters in local starburst galaxies meant not only that conditions favorable to the formation of protoglobular clusters exist in the present universe, but also that this extreme form of star formation is close enough for the star formation process itself to be studied.

Super star clusters appear to be sufficiently massive and rich to be globular clusters, differing from them only in age. Table 9.1 lists the general characteristics of different classes of Galactic star clusters and SSCs. The brightest young SSCs have $M_V \sim -14$. They are brighter than globular clusters because of their youth. It is

Fig. 9.1 HST revealed that the bright sources NGC 1569-A and NGC 1569-B were large clusters of stars

convenient to take the lower bound for SSCs to be $M_V \sim -11$ (Billett et al. 2002), approximately the magnitude of R136 in the LMC, which is also comparable in size to the most massive young Galactic clusters. However, R136 is considered by some to be on the small side for a globular cluster. The older LMC cluster NGC 1866, at $L \sim 10^6\ L_\odot$ and an intermediate age of 100 Myr, is closer to a genuine globular cluster (Meylan 1993). The upper limit to the ages of SSCs is also somewhat arbitrary; while there is evidence that typical cluster dissolution timescales are about 10 Myr, there are also intermediate age clusters with ages of ~1 Gyr even within the Local Group.

Table 9.1 Super star clusters in context

Type	N_*	Mass ($M_\odot$)	r_h (pc)	ρ_h^a ($M_\odot\ pc^{-3}$)	M_V	Age (yrs)
globular cluster	$>10^5$	$10^{3.5}$–10^6	0.3–4	10^{-1}–$10^{4.5}$	−3 to −10	$>10^{10}$
open cluster	20–2000	350–7000	2.5–4.5	1–100	−4.5 to −10	10^6–$10^{9.8}$
embedded cluster	35–2000	350–1100	0.3–1	1–5	...	10^6–10^7
SSC	$>10^5$	10^5–10^6	3–5	...	−11 to −14	10^6–10^7

[a]Half mass mean density. Number of members, N_*, is not as well-defined for the larger clusters as it is for the local open and embedded clusters. References: Battinelli & Capuzzo-Dolcetta 1991, Billett et al. 2002, Harris 1996, Harris & Harris 2000, Lada & Lada 2003, Mackey & Gilmore 2003, McLaughlin & Fall 2008, McLaughlin & van der Marel 2005, Noyola & Gebhardt 2007.

9.3 Extreme Star-Forming Regions of the Local Universe

The most luminous young SSCs of the Galactic neighborhood are the testbeds for study of the star formation process in large clusters and in starburst systems. What are the Orions of the SSC world? In Table 9.2 we have compiled from the literature properties of some well-studied and spatially resolved SSCs in the local universe. Included are Galactic center clusters and large Galactic star-forming regions. While smaller than many extragalactic SSCs, these Galactic clusters are close and more easily studied and should share many of the star-forming properties. The super star clusters of Table 9.2 reflect a wide range of environments and evolutionary stages in the formation and evolution of SSCs.

The Galactic massive young clusters are readily resolved into stars and contain a wealth of information on young massive cluster evolution. However, even for these nearby clusters, confusion, contamination, and rapid dynamical evolution introduce great complexity into observational interpretations. Westerlund 1 is the closest of these large clusters, located in the Carina arm. NGC 3603 is a large southern cluster somewhat more distant. The Arches, Quintuplet, and Galactic Center nuclear clusters have formed in the immediate vicinity of a supermassive black hole, and may differ in structure and evolution from large clusters in more benign environments. The study of the Galactic Center clusters has been made possible by high resolution infrared observations. Included in Table 9.2 are luminous embedded star-forming regions SgrB2 and W49A; their relation to the massive, unembedded star clusters is unclear, although they have similar total luminosities. The other SSCs listed are in galaxies within $\sim$20 Mpc, in which clusters can be spatially resolved. Many of these SSCs have been identified by their location within "Wolf-Rayet" galaxies, those galaxies with a strong HeII 4686 line indicating the presence of significant numbers of Wolf-Rayet stars of age $\sim$3–4 Myr (Conti 1991, Schaerer et al. 1999b). The Wolf-Rayet feature is a relatively easy way to identify in systems with large clusters of young stars, and these are often found in SSCs. Many of the clusters in Table 9.2 have dwarf galaxy hosts. This may be a selection effect due to the difficulty of isolating clusters amid the higher confusion and extinctions in large spirals, since SSCs are definitely present in many local spirals, such as NGC 253, NGC 6946, and Maffei 2 (Condon 1992, Maoz et al. 1996, 2001, Rodríguez-Rico et al. 2006, Roy et al. 2008, Tsai et al. 2006, Turner & Ho 1994, Watson et al. 1996). The dominance of dwarf galaxy hosts may also be due to "downsizing," the tendency for star formation to occur in smaller systems at later times.

It is evident from Table 9.2 that it can be difficult to compare these clusters because embedded and visible clusters are characterized in different ways. Embedded clusters are often characterized by photons that have been absorbed by gas or dust, with well-defined Lyman continuum fluxes and infrared luminosities. Visible clusters have star counts, cluster magnitudes, colors, and stellar velocity dispersions; these clusters can have good masses and ages. Putting together an evolutionary sequence of objects thus requires multiwavelength observations at high spectral resolution. Extinction is observed to decrease with increasing cluster age (Mengel et al.

Table 9.2 Massive young star clusters in the local universe

Host	Cluster	D (Mpc)	R[a] (pc)	log $\frac{L_*}{L_\odot}$	log $\frac{M_*}{M_\odot}$	log N_{Lyc}	N_O	M_V	Age (Myr)
Galaxy	Arches	0.008	>0.5	8.0	4.1	51.0	160	...	2–2.5
Galaxy	Quintuplet	0.008	1.0	7.5	3–3.8	50.9	100	...	3–6
Galaxy	Center	0.008	0.23	7.3	3–4	50.5	100	...	3–7
Galaxy	Sgr B2	0.008	0.8	7.2	...	50.3	(100)	...	...
Galaxy	NGC 3603	0.0076	4.5	7.0	3.4	50.1	>50	...	1–4
Galaxy	Westerlund 1	0.0045	1	...	4.7	51.3	120	...	3–4
Galaxy	W49A	0.014	5	7.4	...	50.1	80	...	...
LMC	R136[c]	0.05	2.6	...	4.8	51.7	>65	−11	1–3
NGC1569	NGC1569-A1	2.2	1.6–1.8	...	6.11	...	...	−13.6[b]	...
NGC1569	NGC1569-A2	2.2	1.6–1.8	...	5.53	...	...	...	...
NGC1569	NGC1569-B	2.2	3.1	...	5.6	...	...	−12.7	15–25
NGC 1705	NGC1705-1	5.3	1.6	...	5.68	<51	...	−14.0	12
He 2–10	He 2-10-1	3.8	1.5	...	5.7	...	1300	−14.3	5.2
He 2–10	He 2-10-A-4	3.8	3.9	...	...	52.4	...	...	...
He 2–10	He 2-10-A-5	3.8	1.7	...	...	51.9	...	...	...
He 2–10	He 2-10-B-1	3.8	1.8	...	...	51.9	...	...	...
He 2–10	He 2-10-B-2	3.8	1.8	...	...	52.0	...	...	...
M82	M82-A1	3.6	3.0	7.9	6.0	50.9	100	−14.8	6.4
M82	M82-F	3.6	2.8	7.73	5.8	...	...	−14.5	50–60
M82	M82-L	3.6	...	...	7.6	...	...	...	...
NGC3125	NGC3125-A1	11.5	...	...	...	52.4[d]	250–3000[e]	...	3–4
NGC3125	NGC3125-A2	11.5	...	...	...	...	550–3000	...	3–4

Table 9.2 (continued)

Host	Cluster	D (Mpc)	R[a] (pc)	$\log \frac{L_*}{L_\odot}$	$\log \frac{M_*}{M_\odot}$	$\log N_{Lyc}$	N_O	M_V	Age (Myr)
NGC3125	NGC3125-B1,2	11.5	...	...	...	52.2	450	...	3–4
Antennae	Antennae-IR	13.3	<32	...	6.48	52.6	120	−17[f]	4
NGC4214	NGC4214-1	4.1	<2.5	...	...	...	280	−13.1	...
NGC5253	NGC5253-5	3.8	...	5.8	...	51.9	155	∼−14	2
NGC5253	NGC5253-IR	3.8	0.7	9.0	...	52.5	1200–6000[g]	...	2–3

[a]Cluster radii are half-light radii. [b]Cluster A is two clusters, de Marchi et al. 1997. [c]Bright core of a larger, complex cluster, NGC 2070. [d]A1 and A2. [e]Range in O stars is due to differences in reddening. [f]M_K. [g]Resolved source; lesser number for $r < 1$ pc. References: Arches, Quintuplet, Galactic nuclear center clusters: Figer et al. 1999, Lang et al. 2001, Figer et al. 2005, Stolte et al. 2003, Stolte et al. 2002, 2005, 2007, Najarro et al. 2004, Figer 2003, 2004, 2008. Kim et al. 2000, 2004, 2007, Kim & Morris 2003. Sgr B2: Dowell 1997, Gaume et al. 1995. NGC 3603: de Pree, Nysewander, & Goss 1999, Eisenhauer et al. 1998, Pandey et al. 2000, Drissen et al. 2002, Nürnberger & Petr-Gotzens 2002, Stolte et al. 2006, Harayama et al. 2008, Melena et al. 2008. W49: Smith et al. 1978, Welch et al. 1987, Conti & Blum 2002, Homeier & Alves 2005. Westerlund 1: Clark et al. 1998, 2005, Nürnberger et al. 2002, Nürnberger 2004, Crowther et al. 2006, Mengel & Tacconi-Garman 2007, 2008, Brandner et al. 2008. R136: Mills et al. 1978, Meylan 1993, Hunter et al. 1995, Massey & Hunter 1998, Noyola & Gebhardt 2007. M82-A1: Smith et al. 2006. M82-F: Smith & Gallagher 2001, O'Connell et al. 1995, McCrady et al. 2003. M82-L: McCrady & Graham 2007. NGC1569 A and B: O'Connell et al. 1994, Sternberg 1998, Hunter et al. 2000, Ho & Filippenko 1996a, Smith & Gallagher 2001, Origlia et al. 2001, Gilbert 2002, Larsen et al. 2008. NGC1705-1: Ho & Filippenko 1996b, Sternberg 1998, Smith & Gallagher 2001, Johnson et al. 2003, Vázquez et al. 2004. He 2-10-1: Chandar et al. 2003. One of five clusters within He 2-10A. He 2-10-A, B: Vacca & Conti 1992, Johnson & Kobulnicky 2003. NGC3125: Vacca & Conti 1992, Schaerer et al. 1999ab, Schaerer et al. 1999b, Stevens et al. 2002, Chandar et al. 2004a, Hadfield & Crowther 2006. Region A has $\log Q_0 = 52.39$, for 4000 O stars, region B $\log Q_0 = 52.19$, for 3200 O stars, Hadfield & Crowther. Antennae: Gilbert et al. 2000, for 13.3 Mpc. NGC5253-5: Gorjian 1996, Calzetti et al. 1997, Schaerer et al. 1997, Tremonti et al. 2001, Chandar et al. 2004b, Vanzi & Sauvage 2004, Cresci et al. 2005. NGC5253-IR: Obscured IR/radio source offset by $\sim 0.5''$ from NGC5253-5. Beck et al. 1996, Turner et al. 1998, 2000, 2003, Mohan et al. 2001, Alonso-Herrero et al. 2004, Turner & Beck 2004, Martín-Hernández et al. 2005, Rodríguez-Rico et al. 2007.

2005), as one might expect from Galactic star-forming regions, so the embedded clusters are likely to also be the youngest clusters.

High angular resolution is key to the study of even the closest super star clusters, which are often found forming in large numbers. One of the first known super star clusters, NGC 1569-A, consists of two superimposed clusters (de Marchi et al. 1997), which is not immediately obvious even in the HST image (Fig. 9.1). The embedded IR/radio SSC in the center of NGC 5253 was found to be offset by a fraction of an arcsecond from the brightest optical cluster, NGC 5253-5, (Calzetti et al. 1997) only 5–10 pc away (Alonso-Herrero et al. 2004, Turner et al. 2003).

Extinction can also be extreme in bright IR-identified starburst regions, and can obscure even the brightest clusters through the near-IR. Observed differential extinctions between the infrared Brackett lines at 2 and 4 μm indicate $A_V > 1$ and even $A_K > 1$ in many starbursts (Ho et al. 1990, Kawara et al. 1989). IR-derived extinctions are often higher than those derived from Balmer recombination lines toward the same regions (Simon et al. 1979) because of extinctions internal to the HII regions themselves. In M82, near-IR and mid-infrared spectroscopy indicates extinctions of $A_V \sim 25$ (Simon et al. 1979, Willner et al. 1977) to $A_v \sim 50$ (Förster Schreiber et al. 2001), similar to values observed in Galactic compact HII regions, but over much larger areas. The clusters in Arp 220 are heavily obscured, with estimated $A_V \sim$ 10–45 mag (Genzel et al. 1998, Shioya et al. 2001); regions behind the molecular clouds can reach $A_V \sim 1000$ (Downes & Solomon 1998).

What does the next decade hold for the clusters of Table 9.2 and other nearby clusters like them? First, there are more SSCs to discover in the local universe, particularly embedded ones. IRAS is still a valuable tool for discovering young SSCs, but with arcminute resolution, it is not sensitive to bright, subarcsecond sources. There undoubtedly remain many compact, young ESF events to be found in the local universe. The WISE mission, an all sky mid-IR survey, will provide an extremely valuable dataset for discovering young and embedded SSCs. The enhanced sensitivities and high spatial resolution of the next generation of telescopes (JWST, EVLA, ALMA, SOFIA), will redefine our concept of "local" SSC formation, extending this list to more distant systems and to young SSCs within large, gas-rich spirals. The near-infrared, in particular, is an valuable link between visible and embedded SSCs. Subarcsecond resolutions are necessary to resolve individual clusters in regions of high and patchy extinction, and the closest galaxies are even now being pursued with adaptive optics. JWST and future extremely large ground-based telescopes will play an important role in connecting embedded clusters to their older, visible siblings to enable a longitudinal study of SSC evolution.

9.4 Initial Mass Functions and the Most Massive Stars in SSCs

The initial mass functions (IMFs) of SSCs have important consequences for cluster masses, their long-term survival, and potentially, for the eventual remnants left by their dissolution. Are the IMFs of clusters power law? Is there evidence for top-heavy IMFs? How do the IMFs of SSCs compare to Galactic IMFs? Is there evidence for initial mass segregation in young clusters? The most massive star in the universe is likely to be in an SSC; is there a fundamental limit to the mass of stars? There is a review of the outpouring of recent IMF work on massive young clusters by Elmegreen (2008).

Mass functions have been determined for the large star clusters of the Local Group. Many appear to be Salpeter. The Salpeter mass functions are defined as $\xi(M) \sim M^{-\alpha}$, where $\alpha = 2.35$, or, expressed in logarithmic mass intervals as $\xi_L\, d\, M \sim M^{\Gamma} d \log M$, with $\Gamma = -1.35$ (Scalo 1986, 1998). The Kroupa IMF is Salpeter at higher masses, and flattens to $\alpha = 1.3$ for stars below 0.5 $M_\odot$ (Kroupa 2001). Kroupa IMFs often cannot be distinguished from Salpeter in extragalactic SSCs. We will adopt the Γ convention here and note that while in many cases these are referred to as IMFs, what is observed is actually a present day mass function (PDMF), from which an IMF may be inferred.

The PDMF of R136 in the LMC has been measured down to 0.6 $M_\odot$ (Massey & Hunter 1998), where it is consistent with Salpeter, $\Gamma \sim -1.3$, and then flattens below 2 $M_\odot$ to $\Gamma \sim -0.3$ (Sirianni et al. 2000). For a Salpeter or Kroupa power law IMF, flattening or turnover of the power law corresponds to an effective "characteristic mass" for the cluster (Lada & Lada 2003). R136 thus appears to have a characteristic mass of 1–2 $M_\odot$, as compared to ~0.5–1 $M_\odot$ for the much smaller nearby Galactic embedded clusters (Lada & Lada 2003). The southern cluster NGC 3603 has a somewhat flatter-than-Salpeter power law slope of $\Gamma = -0.7$ to -0.9 from 0.4 to 20 $M_\odot$ (Harayama et al. 2008, Stolte et al. 2006, Sung & Bessell 2004). Arches has a very similar power law PDMF with $\Gamma = -0.8$ down to 1.3 $M_\odot$ (Stolte et al. 2002), which may correspond to an IMF that is close to Salpeter $\Gamma = -1.0$–1.1 (Kim et al. 2007). The Arches cluster is mass segregrated, with a flatter slope in the center, $\Gamma \sim 0$, than in the outer parts of the cluster, and the mass function may turn over at 6–7 $M_\odot$ in the core (Stolte et al. 2005); however, the MF at larger radii does not show this effect (Kim et al. 2007). Trends appear to be toward flatter MF power laws and higher characteristic masses for the largest clusters in the Local Group, but the numbers of clusters are very small, and dynamical effects are poorly understood as yet (Stolte et al. 2002).

Beyond the Local Group, it is more difficult to determine IMFs. Observational quantities for more distant SSCs are integral properties such as total luminosity and dynamical masses from cluster velocity dispersions. Constraints on IMFs from these integral properties require assumptions about mass cutoffs, cluster ages, and cluster structure. There is accumulating evidence, however, that IMFs in starbursts and the IMFs in SSCs, if Salpeter, may have higher characteristic cutoffs than typical Galactic clusters. Rieke et al. (1980) modeled the starburst in M82 from its IR emission properties, and concluded that the IMF of the starburst must have a low mass cutoff of 3–8 $M_\odot$. Sternberg (1998) used visible mass-to-light ratios to argue for a cutoff of 1 $M_\odot$ for NGC1705-1, although he does not find a cutoff for NGC 1569A. If the IMF in the M82 SSCs is Salpeter, then McCrady et al. (2003) find that the individual M82 clusters must have truncated mass functions, although some clusters show strong evidence for mass segregation and possible dynamical evolution, which complicates this interpretation (Boily et al. 2005, McCrady & Graham 2007, McCrady et al. 2005, McMillan & Portegies Zwart 2003).

What the low mass cutoffs are for SSCs, whether this varies with environment and how, and what is the likely cause of the low mass cutoffs, effective characteristic masses, and the effects of mass segregation on cluster masses and evolution will be fertile ground for SSC research in the next decade.

At the other end of the IMF, there is the question of what is the limiting mass of a star. If the cluster IMFs bear any resemblance to the Galactic power law Salpeter function, then newly-formed clusters consisting of hundreds of thousands to millions of stars are the place to find the most massive and rare O stars. In addition, in dense star clusters there is the possibility of the formation of very massive stars through stellar collisions, which becomes a viable mechanism in dense environments (Bonnell et al. 1998, Clarke & Bonnell 2008). What is the limiting mass of stars in the local universe? Where are the most massive stars found?

Spectroscopy of individual stars is the most reliable way to identify the most massive stars, but is only possible within the Local Group. Ultraviolet spectroscopy using the STIS instrument on HST has allowed the classification of 45 known stars of spectral class O2 and O3; of these, 35 are found in the LMC cluster R136 (Walborn et al. 2002). R136 alone contains more than 65 O stars (Massey & Hunter 1998). In addition to these visible O stars, there is also a significant population of infrared O stars and Wolf-Rayet stars in the Galactic Center, including the three large star clusters, Arches, Quintuplet, and Galactic Center. These infrared clusters, each about an order of magnitude less massive than R136, contain an estimated 360 O stars among them, nearly 60 Wolf-Rayet stars, and 2-3 LBVs (Figer 2008).

The highest inferred stellar mass in the Local Group is $\sim$170 $M_{\odot}$ in the LMC, and $\sim$200 $M_{\odot}$ in the Galaxy; however the latter could eventually turn out to be a binary system; the largest dynamical mass measured is 90 $M_{\odot}$ (Walborn et al. 2002). Weidner & Kroupa (2004) estimate that based on its IMF, observed to be Salpeter, there should be one 750 $M_{\odot}$ star in the R136 cluster, and instead the upper mass limit appears to be 200 $M_{\odot}$ (Koen 2006). Oey & Clarke (2005) extend this to a larger sample of OB associations, obtaining a cutoff of 120–200 $M_{\odot}$, although they caution that this cutoff is sensitive to the IMF slope for stars $>$10 $M_{\odot}$. Figer et al. (2005) finds that there should be one star of $M \sim 500$ $M_{\odot}$ in the Galactic center, where the current upper mass limit instead seems to be 130 $M_{\odot}$. These observations argue for a stellar upper limit close to the observed 150–200 $M_{\odot}$. However, given the small statistics, it may be that the absence of extremely massive stars is simply an evolutionary effect (Elmegreen 2008). For example, modeling suggests that the Pistol Star in the Galactic Center may have had an initial mass of 200–250 $M_{\odot}$ (Figer et al. 1998, Najarro & Figer 1998). Stars with masses of several hundred times solar would evolve rapidly (Yungelson et al. 2008), will lose significant fractions of their initial masses, and in addition, might be dust-enshrouded for most of their lives. Moreover small number statistics at the upper mass end mean that for most clusters, the power law slopes of IMFs are uncertain to a few tenths for clusters of the size of NGC 3603 (Elmegreen 2008). Supermassive stars are elusive by nature.

If extremely high mass stars do exist and we have not seen them simply because of their rapid evolution, the best place to look for them is in the youngest and largest clusters. These will probably be found in starbursts. There is a good chance that the youngest regions containing the most massive stars will be deeply embedded, perhaps extinguished even in the near-infrared. In these cases, nebular diagnostics provide another way of gauging the upper end of the mass function. The mid-infrared has a number of fine structure lines that can be observed in large HII regions in external galaxies (Dale et al. 2006, Helou et al. 2000, Thornley et al. 2000), including fine structure lines of Ar, Ne, S, and O, that can provide valuable nebular diagnostics of the input stellar radiation field even in embedded sources.

Line ratios of the mid-IR nebular fine structure lines in starburst galaxies measured with the SWS spectrometer on ISO revealed unexpectedly low excitation HII regions. Nebular models of the ISO lines indicate upper mass cutoffs of $M_{upper} \sim 30\ M_{\odot}$ for the IMFs in these starbursts (Thornley et al. 2000). Given the luminosities and inferred stellar masses, the low upper mass cutoffs are surprising. However, there are a number of possible explanations for low excitation that would allow for the presence of more massive stars (Rigby & Rieke 2004). One of these is lack of spatial resolution. Starburst regions often have extended regions of ionized gas which, when combined with the comparatively hard spectra of the compact SSC HII regions, will tend to wash out the high excitation lines. That the lines of [OIV]25.9 μm, [NeIII]15.55 μm, and [SIV]10.51 μm have been detected in dwarf starbursts suggests that hot stars are indeed present in some SSCs (Beck et al. 1996, 2007, Crowther et al. 1999, 2006, Lutz et al. 1998). Strong dependences on metallicity and differences in the input radiation fields from existing stellar models that must be considered in these interpretations (Crowther et al. 1999, Martín-Hernández et al. 2002, Rigby & Rieke 2004).

High spatial resolution and spectroscopy from space will allow great improvements in our knowledge of the most massive stars and the upper mass cutoffs of SSC IMFs in external galaxies in the next decade. Optical and ultraviolet emission lines including Wolf-Rayet features will continue to be important diagnostics of the high mass stellar content and ages of SSCs in an enlarged sample of sources. Access to mid-IR fine structure lines from space, via JWST, will give valuable information on the most massive stars in embedded young SSC nebulae, allowing observation of homonuclear line ratios of Ar and Ne that are not possible from the ground. The improvement in spatial resolution is also extremely important for isolating the spectral signatures of compact SSC nebulae. The MIRI IFU on JWST, with order-of-magnitude improvements in sensitivity and spatial resolution over previous instruments, will be an extremely powerful tool for isolating compact SSC nebulae from more diffuse and potentially lower excitation ionized gas within the galaxies. The form of the cluster IMFs is key input to the question of the long term survivability of clusters.

9.5 What are the Initial Cluster Mass Functions for SSCs?

How are SSCs related to globular clusters? Are young SSCs in the local universe precursors to globular clusters? One characteristic that could link globular clusters to SSCs is the cluster mass function. Globular clusters in the Galaxy have a characteristic mass of a few $\times 10^5$ $M_\odot$, reflecting what appears to be a near-universal globular cluster mass function (GCMF) (Harris 1998). Were globular clusters born with this mass function? Or is the present globular cluster mass function the result of evolution due to dynamical forces such as tidal stripping and disk shocking over billions of years (de Grijs 2007, Fall & Rees 1977, Gnedin & Ostriker 1997, McLaughlin & Fall 2008, Parmentier & Gilmore 2007)? SSCs, as potential precursors to globular clusters, could give us valuable information about the nature vs. nurture question for the GCMF. Is the initial cluster mass function (ICMF) for SSCs universal, and if so, what is it? How does it evolve?

The ICMF has been extensively studied in the Antennae galaxies, where large numbers of SSCs facilitate the statistics. The Antennae are a nearby (13.3 Mpc, Saviane et al., 2008) IR-bright pair of gas-rich interacting galaxies. HST images revealed a system of thousands of young SSCs spread across the galaxy pair (Whitmore & Schweizer 1995, Whitmore et al. 1999). ISO imaging determined that the youngest clusters and brightest infrared emission occur in a dusty and embedded region between the two galaxies (Mirabel et al. 1998). Cluster masses and ages for the visible SSCs of the Antennae have been obtained from multicolor (UBVRI and Hα) HST photometry combined with modeling of the cluster colors and luminosities with STARBURST99 (Leitherer et al. 1999, Vázquez & Leitherer 2005) The cluster mass function (CMF) in the Antennae is power law with $\alpha \sim -2$ over the range of masses 10^4–10^6 $M_\odot$ (Fall et al. 2005). The median cluster mass in the Antennae appears to be about an order of magnitude less than the median globular cluster mass of giant ellipticals such as M87 (Harris 2002). Ages determined from the SSC colors in the Antennae range from 10^6–10^8 yr (Zhang & Fall 1999), with a median of 10-20 Myr (Fall et al. 2005, Mengel et al. 2005), but with a small population extending to more than 10^9 yrs. If the SSC system of the Antennae is typical, one could infer that the characteristic mass of globular clusters is due to selective evolution of clusters of different mass (Fall & Zhang 2001). However, the fact that 70% of the clusters in the Antennae are less than ~10–20 Myr of age could indicate that very few if any of the Antennae clusters will survive to become globular clusters.

Cluster mass functions have been determined for a number of other nearby galaxies, mostly inferred from cluster luminosity functions. A list of nearby starburst systems is given in de Grijs et al. (2003); many cluster systems appear to have power law slopes of $\alpha \sim -2$.

Direct detection of cluster masses via their stellar velocity dispersions and cluster sizes can be done for the closest systems, thereby giving direct determinations of cluster mass functions. Even in smaller SSC systems, dynamical estimates of cluster masses are an important check on mass functions obtained from cluster luminosity functions. Masses at the high ends are consistent with globular cluster masses. In M82, McCrady & Graham (2007) find a power law mass distribution with slope of $\alpha = -1.91$ for fifteen SSCs, nearly identical to the Antennae.

An intriguing indication of ICMF evolution is the cluster system in NGC 5253. There are several hundred clusters in this galaxy (Caldwell & Phillips 1989), many of intermediate age, $\sim$1 Gyr. There is also a younger population of visible and embedded SSCs with ages of $\sim$2–50 Myr (Alonso-Herrero et al. 2002, Calzetti et al. 1997, Gorjian et al. 2001). The cluster mass function for the intermediate age clusters appears to turn over at a mass of 5×10^4 $M_\odot$, while the younger SSCs have a power law mass function (Cresci et al. 2005). Parmentier et al. (2008) suggest that the evolution of the ICMF is due to star formation efficiency (see Section 9.7).

Future work on stellar cluster mass functions in the coming decade will be done from space with HST and JWST and from the ground with IR AO observations. High resolution and sensitivity are important for these studies, since fields where SSCs are found are usually very crowded. Infrared observations are critical for getting the mass functions of the youngest clusters, the best reflection of the ICMF, since they are likely to be embedded in regions of high extinction.

9.6 What Environments Lead to Extreme Star Formation?

In Fig. 9.2 is shown the nearby M81-M82-NGC 3077 interacting group of galaxies (Yun et al. 1994). On the left is the Palomar Sky survey image, showing the stellar luminosity; on the right, is the VLA mosaic of the group in the 21 cm line of HI.

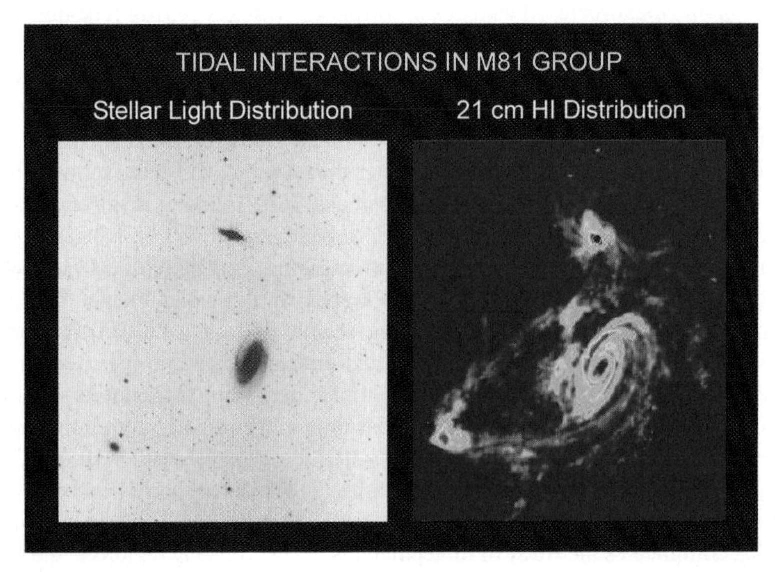

Fig. 9.2 The M81-M82-NGC 3077 group. (*Left*) Palomar Sky Survey Image. (*Right*) VLA image of the 21 cm line of HI. Yun et al. (1994)

The ties that bind this group are obvious in the 21 cm line emission. M82 is one of the best-known starburst galaxies, with $L_{IR} \sim 6 \times 10^{10}$ $L_{\odot}$, and an estimated 10^4–10^5 O stars ($N_{Lyc} \sim 8 \times 10^{54}$ s^{-1}). NGC 3077 also has a modest starburst, of $L_{IR} \sim 3 \times 10^8$ $L_{\odot}$ ($N_{Lyc} \sim 2 \times 10^{52}$ s^{-1}: Sanders et al. 2003), and M81 has a mildly active nucleus. Clearly the conditions for "extreme" star formation are favorable in this group. The atomic hydrogen of this interacting group has its own history, which is different from that of the stars within the galaxies; there is evidence that some of the starburst activity is caused by a delayed "raining down" of orbiting gas onto the galaxies several Myr after their closest encounters (Meier et al. 2001).

The most luminous infrared galaxies in the universe, with $L_{IR} > 10^{11}$ $L_{\odot}$, are merging and interacting systems, and these tend to be systems dominated by star formation. Our knowledge of the stellar distributions is greater than our knowledge of the gas: potentially many groups of galaxies have the connected appearance of the M81 group with tidal loops in atomic hydrogen, and perhaps even in molecular gas, since starburst galaxies are especially rich in molecular gas (Mirabel & Sanders 1989). Arrays with wide-field capability, such as the Allen Telescope Array, the Green Bank Telescope, and the SKA, are well-suited to the mapping of large HI fields in nearby galaxy groups.

Star formation efficiency (SFE) is an important characteristic of the star formation process, since it is a measure of how efficiently molecular clouds are turned into stars. SFE is closely tied to the "infant mortality" of SSCs described in the next section. There are numerous measures of the efficiency of star formation, broadly defined as the proportion of star formation per unit gas. The Schmidt law, in which stars follow a power law correlation with density (Schmidt 1959), or the corresponding Kennicutt law, in terms of gas surface density (Kennicutt 1998b) show that star formation on global scales in galaxies is correlated with gas density. The observable $L_{IR}/M(H_2)$ is also used as an indication of SFE.

While from a global perspective the total gas content, HI + H_2, of galaxies appears to be well correlated with star formation tracers (Crosthwaite & Turner 2007, Wong & Blitz 2002), observations of star-forming regions in the Galaxy indicate that stars form from molecular gas clouds, rather than atomic. CO observations show a good correlation of L_{CO} with L_{IR} (Sanders et al. 1986, Young et al. 1996, Young & Scoville 1991, Young et al. 1986b).

The atomic and molecular gas distribution in the spiral galaxy M83 is shown in Fig. 9.3 along with an overlay of the GALEX ultraviolet image. This figure illustrates the general result that while the atomic hydrogen disk can far exceed the visible disk of a spiral galaxy, the optical portion of a spiral galaxy is primarily molecular gas. Gas that forms stars is molecular. While a good correlation of star formation tracers is seen with CO emission, an even tighter correlation is seen between star formation and the dense ($n > 10^{5-6}$ cm^{-3}) gas tracer HCN (Gao & Solomon 2004). This should not be surprising, since denser gas is more likely to form stars.

What complicates the study of molecular gas in star-forming regions is the necessity of using proxies, generally CO, to map out the distribution of H_2. A conversion

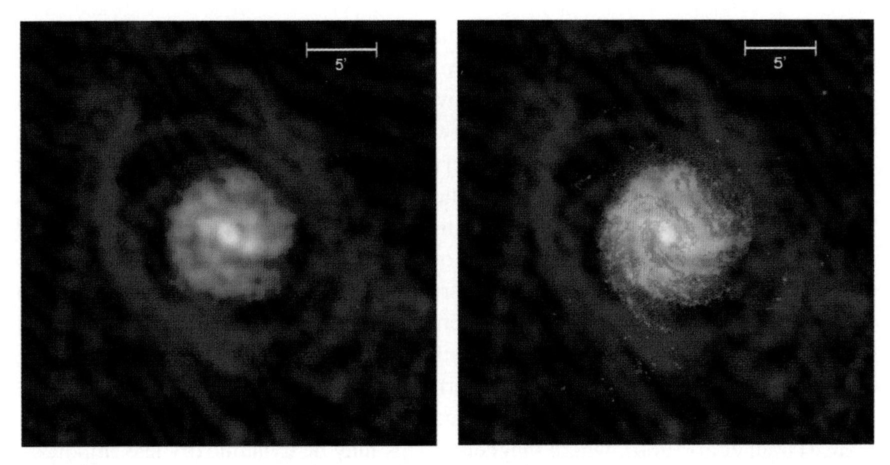

Fig. 9.3 (*Left*) The barred spiral galaxy M83. Red is a VLA image of HI 21 cm line emission, Tilanus & Allen (1993), and yellow is CO emission mapped with the NRAO 12 Meter Telescope, Crosthwaite et al. (2002). (*Right*) Neutral gas with GALEX image overlay; GALEX image, Thilker et al. (2005), Gil de Paz et al. (2007)

factor between CO line intensity and H_2 mass seems to work well in the Galaxy, but will it do as well in ESF environments?

Because of high energies of its first excited states, H_2 tends to be in the ground state for temperatures less than 100 K. Most Galactic giant molecular clouds (GMCs) have temperatures of 7–12 K (Sanders et al. 1985, Scoville et al. 1987), although clouds in starbursts can be warmer. By contrast, CO is relatively abundant, easily excited and thermalized, with a lowest energy level equivalent temperature $E/k \sim 5.5\,\mathrm{K}$. For these reasons, CO lines are generally very optically thick. Yet CO is observed to be a good tracer of mass (Solomon et al. 1987). This is because Galactic disk GMCs appear to be turbulently supported against gravity, and near virial equilibrium (Myers 1983). GMCs, which consist of optically thick clumps with turbulent motions larger than systematic motions, have line profiles that are Gaussian in spite of high optical depths, and "Large Velocity Gradient" (Sobolev approximation) radiative transfer holds (Wolfire et al. 1993). The empirically-determined Galactic conversion factor, $X_{CO} = N_{H2}/I_{CO}$, is thus a dynamical mass tracer (Dickman et al. 1986, Solomon et al. 1987), in effect a Tully-Fisher relation for molecular clouds. Gamma ray observations indicate that a conversion factor of $X_{CO} = 1.9 \times 10^{20}\,\mathrm{cm^{-2}\,(K\,km\,s^{-1})^{-1}}$ predicts H_2 mass to within a factor of two within the Galaxy, with some radial variation (Strong et al. 1988, 2004). As an indicator of dynamical mass, X_{CO} may actually be more robust than optically thin gas tracers in extreme environments, since mass estimates based on optically thin tracers depend upon temperature and relative abundance (Dame et al. 2001, Maloney & Black 1988).

While the CO conversion factor seems to work well in the Galaxy, and as a dynamical mass tracer may be more robust than tracers that are abundance-dependent, the association of CO and H_2 has not been extensively tested in extreme environments. There are clearly some situations in which X_{CO} fails to work well. The Galactic value of X_{CO} does not give good masses for the gas-rich centers of ultraluminous galaxies. In Arp 220, it overestimates the mass by a factor of ~5 due to a gas-rich nucleus, consisting of two counterrotating disks (Sakamoto et al. 1999), which are dominated by warm, pervasive molecular gas in which systematic motions dominate over turbulence (Downes & Solomon 1998, Downes et al. 1993, Solomon et al. 1997). X_{CO} also appears to overpredict H_2 masses in the centers of local gas-rich spiral galaxies, including our own (Dahmen et al. 1998) by factors of 3–4. CO appears to systematically misrepresent H_2 mass in spiral galaxies when the internal cloud dynamics may be different from Galactic disk clouds, such as in the nuclear regions where tidal shear visibly elongates clouds, causing systematic cloud motions to dominate (Meier & Turner 2004, Meier et al. 2008). It may also fail where cloud structure may fundamentally differ from Galactic clouds, as in the LMC (Israel et al. 1986), where magnetic fields may be dynamically less important (Bot et al. 2007).

For an understanding of the links between star formation and molecular gas in ESF regions, we require improved confidence in molecular gas masses in environments that are atypical of the Galaxy. Systematic studies of molecular clouds in different tracers of molecular gas, including dust, in ESF galaxies at high resolution in the millimeter and submillimeter with ALMA, CARMA, Plateau de Bure, and SMA will shed light on when we can confidently use CO as a tracer of molecular gas mass, and under what conditions it ceases to be a good tracer.

Environmental factors other than gas mass are also important in the fostering of star formation, but these are not as yet well understood. Tacconi & Young (1990) concluded that the efficiency of massive star formation is higher in spiral arms than between the arms, consistent with the "strings of pearls along the spiral arms" description of nebulae by Baade (1957). Clearly star formation is enhanced by spiral arms, but how their large-scale influence trickles down in a turbulent GMC to a parsec-scale core is not at all clear (Padoan et al. 2007). Starburst rings also appear to facilitate star formation, particularly young SSCs (Barth et al. 1995, Maoz et al. 2001, 1996). SFE appears to steadily increase with the ferocity of the star formation. While $L_{IR}/M_{H_2} \sim 4\ L_{\odot}/M_{\odot}$ for the Galaxy, it is ~10–20 $L_{\odot}/M_{\odot}$ in regions of active star formation, ~20–100 in ULIRGs (Sanders et al. 1986). These studies rely on L_{IR}, which may have contributions from older stellar populations. Studies of star formation efficiency using tracers of recent star formation are ongoing.

Extreme star formation should require extreme amounts of molecular gas, and luminous infrared galaxies have plenty of it (Downes et al. 1993, Sanders & Mirabel 1985, 1996, Young et al. 1986a). The Antennae interacting system, with its thousands of young SSCs, contains exceptional amounts of gas. A CO image made with the Owens Valley Millimeter Array by Wilson et al. (2000, 2003) is shown on the HST image of (Whitmore et al. 1999) in Fig. 9.4. The greatest concentration of molecular gas is in the dusty, obscured region between the two galaxies where the brightest infrared emission is found (Mirabel et al. 1998). There is an estimated $10^9\ M_{\odot}$ of molecular gas in the Antennae, supporting a current

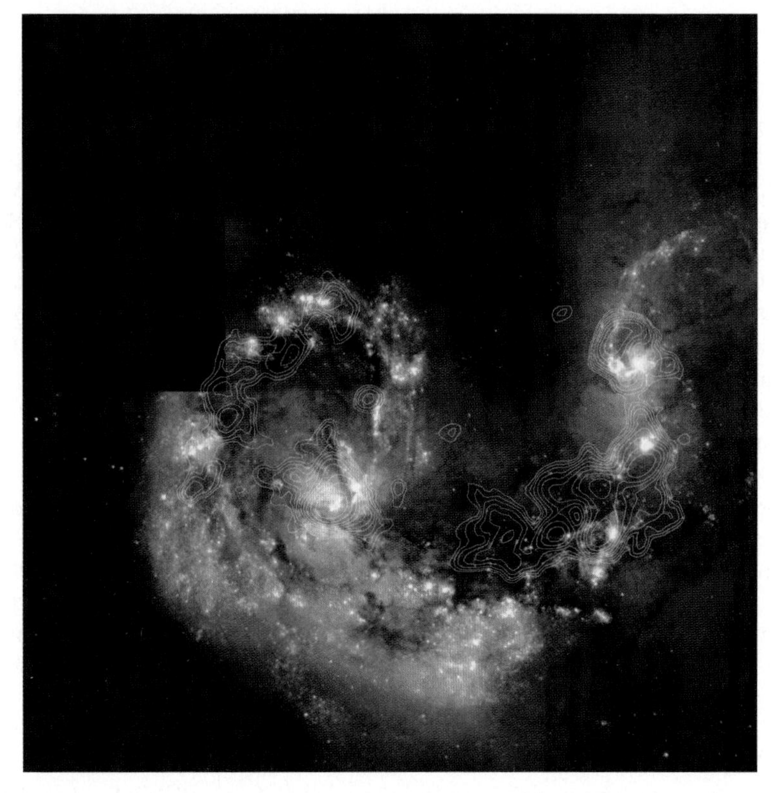

Fig. 9.4 CO in the Antennae. Contours are emission in the J=1–0 line of CO at 3 mm, imaged with the Owens Valley Millimeter Array. In color is the HST image. Wilson et al. (2000), Whitmore et al. (1999)

star formation rate of $N_{Lyc} \sim 10^{54}\,s^{-1}$ (Stanford et al. 1990, for D=13 Mpc, and $X_{CO} \sim 2 \times 10^{20}\,cm^{-2}\,K\,km\,s^{-1}$). Based on the total mass of K-band selected clusters, the SFE is about 3–6% (Mengel et al. 2005), slightly higher than Galactic efficiencies on these scales (Lada et al. 1984) but not by much. H_2 emission observed using Spitzer in the dusty collision region suggests that "pre-starburst" shocks may trigger the star formation there (Haas et al. 2005) as seen in other interacting ULIRGS (Higdon et al. 2006). This is a good illustration of the "Super Giant Molecular Clouds" posited for interacting systems by Harris (2002), Harris & Pudritz (1994). Like the stellar cluster mass function, which is power law with $\alpha \sim -2$, the mass function of giant molecular clouds in the Antennae is also power law, but with a slightly different slope, $\alpha \sim -1.4$ (Wilson et al. 2003).

A counterexample to the "lots of gas, lots of stars" theory is the case of NGC 5253. In this starburst dwarf galaxy, the star formation efficiency appears to be extremely high, SFE ~75% on 100 pc scales, with H_2 masses based on both CO (Meier et al. 2002, Turner et al. 1997) and dust emission (Turner et al. 2008, in prep.) This SFE is nearly two orders of magnitude higher than generally seen on GMC spatial scales in the Galaxy. NGC 5253 has several hundred young clusters, including several SSCs (Calzetti et al. 1997, Gorjian 1996). Why is this galaxy so parsimonious in its usage of gas compared to the Antennae? One difference between this starburst and that of the Antennae is that NGC 5253 is a dwarf galaxy, with an estimated mass of $\sim 10^9$ $M_\odot$, which was probably originally a gas-poor dwarf spheroidal galaxy that has accreted some gas (Caldwell & Phillips 1989). Unlike the Antennae, which are in the midst of a full-blown major merger, NGC 5253 is relatively isolated, although part of the Cen A–M83 group (Karachentsev et al. 2007). The prominent dust lane entering the minor axis is the probable cause of the starburst. Molecular gas is present in the dust lane, and this gas is observed to be falling into the galaxy near the current starburst (Kobulnicky & Skillman 2008, Meier et al. 2002). Radio recombination line emission imaged at high resolution with the VLA of the central "supernebula" shows a velocity gradient in the same direction as that of the infalling streamer (Rodríguez-Rico et al. 2007). High star efficiency is a necessary condition that the SSCs can evolve into globular clusters (Goodwin 1997). NGC 5253 may be the best case yet for a galaxy in which the clusters might survive to become globular clusters.

The high resolution and sensitivity of ALMA, the new CARMA array, Plateau de Bure and SMA will allow many more such starburst systems to be imaged in molecular lines for study of the variation in SFE with environment. Star formation efficiency is a key parameter in the formation of long-lived clusters.

9.7 Super Star Cluster Mortality

Birth and death go hand in hand in young SSCs, since O stars barely stop accreting before they die (Zinnecker & Yorke 2007). During their short lifetimes, O stars find many different ways to lose mass. Windy and explosive by nature, O stars emit copious numbers of destructive ultraviolet photons and are responsible for much of the mass loss and mechanical feedback within a star cluster. Outflows, wind bubbles, LBV mass ejections, and SNR from a single O star can influence a region the size of a young SSC; imagine what an SSC consisting of thousands of O stars can do to a parsec-scale volume! Figure 9.5 illustrates a few of the many ways that O stars can destructively interact with their environments: CO outflows in adolescence (1 Myr); LBV outflows similar to that responsible for the Homunculus Nebula in early adulthood (2–3 Myr); Wolf-Rayet wind bubbles at retirement (3–4 Myr); death by supernova (5–10 Myr). If the young cluster manages to survive through the paroxysms of its riotous O star siblings, then there is most likely a molecular cloud nearby to unbind it.

There is strong evidence that most SSCs in the local universe will not survive to old age. The odds of survival even in the relatively benign environments of the Galactic disk and halo are slim. Only an estimated $\sim 7\%$ of embedded young clusters in the solar neighborhood will live to the age of the Pleiades (Lada & Lada 2003). Dynamical models of the Galactic globular cluster system suggest that as much as 75% of the Galactic stellar halo may have originally been in the form of globular

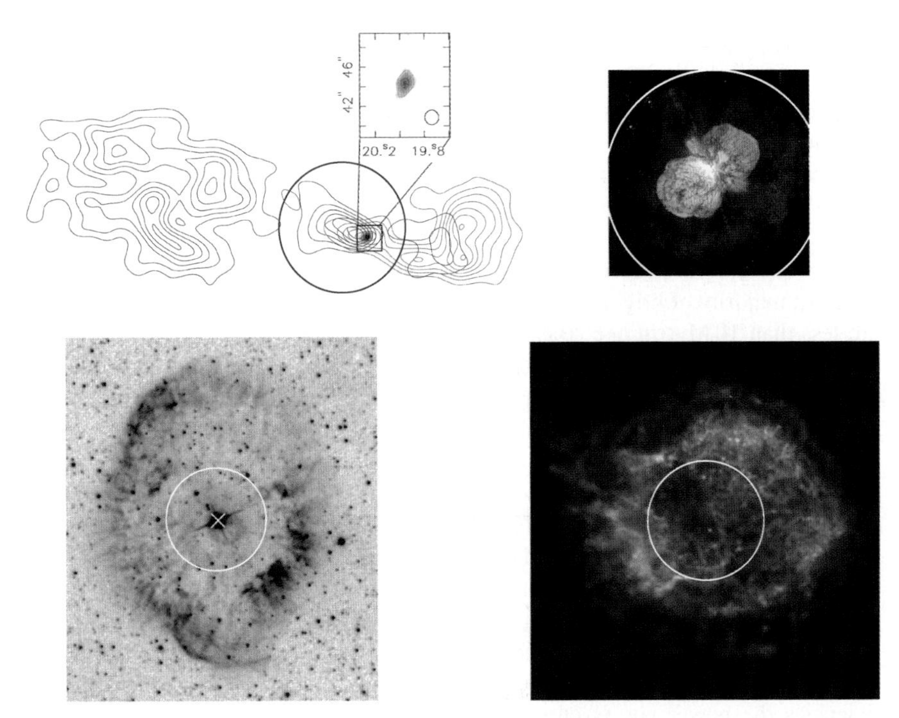

Fig. 9.5 The many ways that O stars can be destructive. Circles represent a region 1 pc across, the size of the core of an SSC. (*top left*) Owens Valley Millimeter Array image of the CO outflow source around the massive protostar G192.16-3.82, Shepherd & Kurtz (1999). (*top right*) The Homunculus Nebula in Eta Carina imaged by HST, Morse et al. (1998). (*lower left*) Wolf-Rayet bubble RCW58 in Hα, Gruendl et al. (2000). (*lower right*) Chandra image of the 1000-yr-old SNR Cas A, Hughes et al. (2000)

clusters (Gnedin & Ostriker 1997, Shin et al. 2008), which now account for only 1% of visible halo stars (Harris 1998).

The first hurdle that a young SSC must overcome is star formation efficiency, defined here as $\eta = M_{\rm stars}/(M_{\rm stars} + M_{\rm gas})$. The gas mass contribution includes contributions from ionized gas and atomic and molecular gas from the natal clouds. Nominally $\eta > 50\%$ is required to leave a bound cluster (Hills 1980, Lada et al. 1984, Mathieu 1983). Lower values of $\eta \sim 30\%$ can be accommodated if the cluster loses stars, but retains a smaller bound core, leading to smaller clusters. The cluster may also survive if gas is lost quasistatically, so that the cluster adjusts to the new equilibrium (Kroupa & Boily 2002); it may also survive, although with a significantly reduced stellar mass, if it suffers rapid mass loss ("infant weight-loss") early on (Bastian & Goodwin 2006). Models suggest that clusters with masses less than 10^5 M$_\odot$ lose their residual gas quickly, and that 95% of all clusters are so destroyed within a few tens of Myr, and that rapid gas expulsion may give a natural explanation for the lognormal PDMF for globular clusters (Baumgardt et al. 2008).

Based on the SFE/η as observed in the Galaxy, the picture looks bleak for young SSCs. On the sizescales of giant molecular clouds, η is at most a few percent (Lada & Lada 2003), a number that appears to be determined by the turbulent dynamics of clouds (Krumholz et al. 2006, Padoan & Nordlund 2002, Padoan et al. 2007). On smaller scales in the Galaxy, $\eta \sim$ 10–30% (NGC 3603; Nürnberger et al. 2002), which is still rather small to preserve a significant bound cluster on globular cluster scales.

There is both fossil and structural observational evidence that clusters dissolve. The vast majority of bright SSCs with masses over 10^5 $M_\odot$ in known SSC systems are less than 10 Myr in age (Bastian et al. 2005, Mengel et al. 2005). From STIS spectroscopy, Tremonti et al. (2001) and Chandar et al. (2005) find that the field stars in the nearby starburst galaxy NGC 5253 can be modeled by dispersed cluster stars, with cluster dissolution timescales of 7–10 Myr. In the Antennae system, Fall et al. (2005) find that the number of clusters falls sharply with age, with ~50% of the stars in clusters having dispersed after 10 Myrs. They estimate that at least 20% and perhaps all, of the disk stars in the Antennae have formed within clusters. On the other hand, there is fossil evidence, in the form of globular clusters, that large clusters can and do survive for many Gyr.

Star formation efficiency and cluster disruption may imprint upon cluster mass functions. Parmentier et al. (2008) suggest that at $\eta \sim$ 20%, a power-law core mass function turns into a bell-shaped cluster mass function, while at higher efficiencies the power law is preserved. Gieles & Bastian (2008) suggest that the maximum cluster mass and age is a diagnostic of cluster disruption, and they see evidence in cluster mass function, that formation/disruption does vary among galaxies.

The question remains of how globular clusters have managed to live to such a ripe old age. Can we identify SSCs forming today that might indeed live to become 10 billion years old? What initial conditions favor SSC longevity? Going to deeper limits in the cluster luminosity function could illuminate the connection between today's SSCs and the older globular cluster population (Chandar et al. 2004b). This is an extremely active area of research, but there are currently few examples of high resolution studies of the efficiency of star formation on GMC sizescales in starbursts. ALMA will have the sensitivity and resolution to enable these studies in many nearby galaxies.

9.8 Radiative Feedback: Effects on Molecular Clouds and Chemistry

The starburst galaxy, M82, has one of the earliest known and best studied examples of a galactic wind. Both mechanical luminosity in the form of stellar winds and supernovae and radiative luminosity from starbursts are feedback mechanisms that can potentially disrupt star formation and end the starburst phase. Yet there are galaxies, such as the Antennae, observed to have thousands of SSCs spread over regions of hundreds of pc extent, with cluster ages spanning many tens of Myr during tidal interactions lasting $\sim$100 Myr. Evidently, episodes of intense star formation can take place over periods of time far longer than the lifetimes of individual massive stars in spite of feedback. The subject of galactic winds and feedback is a large and active one, and has been recently reviewed by Veilleux et al. (2005). The effects of feedback on denser molecular gas in ESF regions is not yet well characterized, and it is the molecular gas from which the future generations of stars will form.

Starburst feedback and star formation occur on different spatial scales. Giant molecular clouds consist of clumps that are governed by turbulence; only a small fraction of these clumps contain cores, which are the regions that collapse to form stars or star clusters (McKee & Ostriker 2007). Current computational models of turbulent clouds can explain the canonical star formation efficiencies of a few percent as that fraction of turbulent clumps that become dense enough for gravity to dominate (Krumholz et al. 2006, Padoan & Nordlund 2002, Padoan et al. 2007). Once a core begins to collapse, free fall is rapid and there is little time for feedback to operate. It is more likely that feedback operates on longer-lived and lower density molecular cloud envelopes dominated by turbulence, but how this large scale effect communicates down to the small and rapidly collapsing star forming cores is unclear (Elmegreen 2007, Krumholz & Tan 2007, Padoan et al. 2007). If rich star clusters form stars for several dynamical times, the energetic input feedback could become important (Tan et al. 2006).

One surprising characteristic of the interstellar medium in regions of ESF is the ubiquity of dense ($n_e \sim 10^{4-5}\,cm^{-3}$), "compact" HII regions, a stage of star formation that should be fleeting and relatively rare. First detected spectroscopically in dwarf galaxies, such as NGC 5253 and He 2–10 (Beck et al. 2000, Kobulnicky & Johnson 1999, Turner et al. 1998), these nebulae are detected by their high free-free optical depths at cm wavelengths. These are the ESF analogs of dense Galactic "compact" HII regions, only much larger in size because of the high Lyman continuum rates from these large clusters. If the expansion of HII regions is governed by classical wind bubble theory, then the dynamical ages implied by the sizes of these HII regions are extremely short, tens of thousands of years. M82-A1 is a young SSC with both a visible HII region and a cluster, in which the dynamical age of the HII region is too small for the cluster age (Smith et al. 2006). The HII region around the Galactic cluster NGC 3603 is also too small for the cluster age (Drissen et al. 1995). These may be scaled-up versions of the classic Galactic ultracompact HII problem, in which there are "too many" compact HII regions given their inferred dynamical ages (Dreher et al. 1984, Wood & Churchwell 1989). A possible explanation for the long lifetimes of these HII regions is confinement by the high pressure environment of nearby dense ($n_{H_2} > 10^5$ cm^{-3}) and warm molecular clouds (e.g., de Pree et al. 1995, Dopita et al. 2005, 2006). Radiative cooling may provide an important energy sink for the output of young super star clusters (Silich et al. 2007). However, the standard wind-blown bubble theory that is generally applied to the development of Galactic HII regions (Castor et al. 1975, Chevalier & Clegg 1985) may not always apply to the HII regions surrounding SSCs, which are massive enough to exert a non-negligible gravitational pull on their HII regions (Kroupa & Boily 2002). The "supernebula" in NGC 5253 appears to be gravity-bound, if not in equilibrium (Turner et al. 2003). Gravity could either stall the expansion of an SSC HII region's expansion or create a cluster wind akin to a stellar wind, depending on conditions. It is clear that the high interstellar pressures in starburst regions are critical to their development, and to the evolution of the nearby molecular clouds.

Molecular gas has a tremendous ability to absorb energy and radiate it away. This could explain the ability of galaxies to sustain starburst events of extended duration such as the one that has produced the thousands of SSCs in the Antennae. Irradiation causes heating, ionization, dissociation, and pronounced chemical changes in molecular clouds. It also provides us with a rich spectrum of potential diagnostics of radiative feedback.

In Fig. 9.6 is shown a schematic of the ionization structure of a photodissociation region (PDR), adapted from Tielens & Hollenbach (1985). (For a full description of PDRs, see Tielens 2005 or Hollenbach & Tielens 1999 and also Bertoldi & Draine 1996, Draine & Bertoldi 1996, Kaufman et al. 1999, 2006, Tielens & Hollenbach 1985.) Molecular clouds tend to form H_2 at $A_V < \sim 1$. The translucent edge to the molecular cloud can be quite warm, a few hundred K, warm enough for excited H_2 to be visible. Between $A_V = 1$ and $A_V \sim 5$, while the clouds are molecular, they also have significant abundances of heavy ions such as C^+ and S^+. The presence of ions can drive a rich gas-phase chemistry through ion-molecule reactions. The high temperatures can also liberate molecules that have formed on the surfaces of grains in the form of ices in the coldest clouds. Warming the grains in either shocks or radiatively in PDRs brings these molecules into the gas phase. PAH emission is also bright from these regions, and dominates Spitzer images in the 8 μm IRAC band where it shows a close association with star-forming regions (Peeters et al. 2004, Galliano et al. 2008). These chemical diagnostics have been used to great effect in modeling the effects of protostars on their surrounding protostellar disks, and

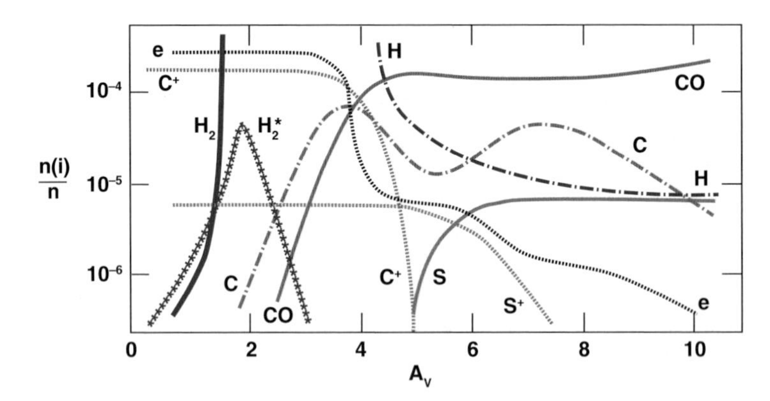

Fig. 9.6 Schematic of the ionization structure of the Orion photodissociation region (PDR), with relative elemental abundance plotted versus visual extinction. Adapted from Tielens & Hollenbach (1985) and Tielens (2005)

in determining the shapes and orientations of disks (van Dishoeck & Blake 1998). Clouds near sources of high X-ray radiation are subject to a similar phenomenon, but with chemistry that is driven by a hard radiation field; these regions are called "XDRs" (Lepp & Dalgarno 2006, Maloney et al. 1996, Meijerink & Spaans 2005, Meijerink et al. 2006).

It might seem that processes that occur on scales of $A_v \sim 1-5$ would be difficult to detect in other galaxies, on the spatial scales of GMCs, but this is not the case. Molecular clouds are porous, and there are many surfaces within clouds of relatively low A_v individually; an estimated 90% of molecular gas is in PDRs (Hollenbach & Tielens 1999). Some of the first indications of the importance of PDR chemistry were the detections of warm CO and the tracers of warmed, potentially shocked, gas such as the CII 158 μm line in starburst galaxies (Mauersberger & Henkel 1991, Stacey et al. 1991). Temperatures as high as 400–900K have been inferred from lines of interstellar ammonia (Mauersberger et al. 2003).

Lines of numerous heavy molecules have been detected from other galaxies, and the brightest sources are the star-forming galaxies. Molecules such as formaldehyde (H_2CO), methanol (CH_3OH), and cyanoacetylene (HC_3N) have been detected in nearby starburst galaxies (Mangum et al. 2008, Martín et al. 2006, Meier & Turner 2008). Many of these models can provide discriminants between PDR and XDR-heated gas (Aalto et al. 2007, Meijerink et al. 2006). A spectral line survey of NGC 253 in the 2 mm atmospheric window reveals 111 identifiable spectral features from 25 different molecular species; the spectrum suggests that the molecular abundances in NGC 253 are similar to those of the Galactic Center, with a chemistry dominated by low-velocity shocks (Martín et al. 2006, 2008).

Imaging adds another dimension to the molecular line spectra. In Fig. 9.7 are shown interferometric images in several molecules of the nuclear "minispiral" of the nearby Scd galaxy, IC 342. The lines shown are all at $\lambda = 3$ mm, and have comparable excitation energies and similar critical densities. The images show a clear variation in cloud chemistry and molecular abundances with galactic location. A principal component analysis (Meier & Turner 2005) shows that the molecules N_2H^+, HNC, and HCN have similar spatial distributions to CO and its isotopologues, and are good tracers of the overall molecular cloud distribution. The molecule CH_3OH (methanol) is well known from Galactic studies to be a "grain-chemistry" molecule, which forms on grain surfaces and is liberated by shocks or warm cloud conditions; here, methanol and HNCO follow the arms of the minispiral. Methanol and HNCO appear to be tracing the gentle shocks of the gas passing through the spiral arms. The final group of molecules, $C^{34}S$ and C_2H are found in the immediate vicinity (50 pc) of the nuclear cluster; these molecules presumably reflect the intense radiation fields of the nuclear star cluster.

The next decade will see a blossoming of molecular line studies of ESF. The current state of molecular line work in galaxies has been recently reviewed by Omont (2007), and also is represented by contributions in the volume by Bachiller & Cernicharo (2008). Spectroscopy of ESF gas in galaxies with the APEX, ASTE, CARMA, IRAM 30 m, Plateau de Bure, CARMA, SMA, Spitzer, and the VLA telescopes will continue to probe the star-forming ISM in nearby galaxies through the

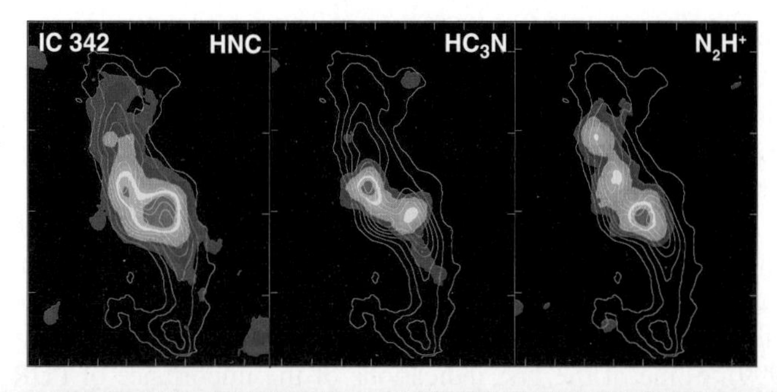

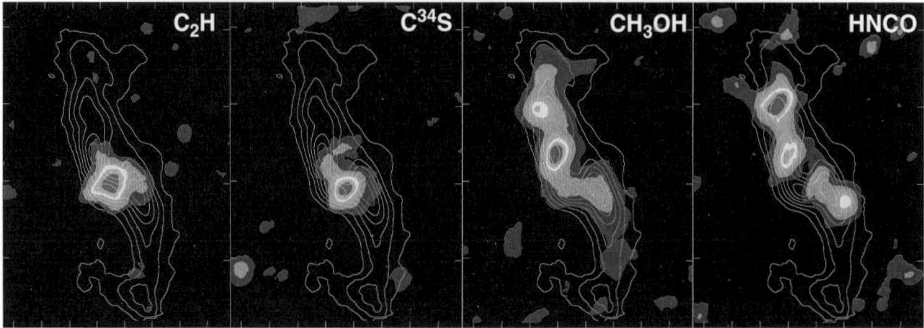

Fig. 9.7 Spatially-resolved (5″ = 75 pc) chemistry of the central 300 pc of the Scd spiral galaxy IC 342. Meier & Turner (2005). Panels at top show molecules that are overall molecular gas tracers. *Bottom, left*: C_2H and $C^{34}S$ trace clouds in high radiation fields (PDRs); *right*: CH_3OH (methanol) and HNCO trace grain-chemistry along the arms of the minispiral

next decade. Herschel and SOFIA will soon provide spectroscopy of the important 158μm line of CII from PDRs in ESF galaxies. ALMA will add tremendous sensitivity, milliarcsecond resolution, km/s velocity resolution, access to the southern hemisphere, and submillimeter capability, allowing us to study extreme star formation and extreme star-forming gas and its effects on galaxies in exquisite detail.

Acknowledgments I am grateful to Nate McCrady, David S. Meier, and Andrea Stolte for their helpful discussions and comments, and to Xander Tielens for his good humor and patience.

References

Aalto, S., Spaans, M., Wiedner, M.C., Hüttemeister, S., Astron. Astrophys., **464**, 193–200 (2007)

Alonso-Herrero, A., Rieke, G.H., Rieke, M.J., Scoville, N.Z., Astron. J., **124**, 166–182 (2002)

Alonso-Herrero, A., Takagi, T., Baker, A.J., Rieke, G.H., Rieke, M.J., Imanishi, M., Scoville, N.Z., Astrophys. J., **612**, 222–237 (2004)

Arp, H., Sandage, A., Astron. J., **90**, 1163–1171 (1985)

Baade, W., The Observatory, **77**, 165–171 (1957)

Bachiller, R., Cernicharo, J., Science with the Atacama Large Array (New York: Springer) (2008)

Barth, A.J., Ho, L.C., Filippenko, A.V., Sargent, W.L., Astron. J., **110**, 1009–1018 (1995)
Bastian, N., Goodwin, S.P., Mon. Not. R. Astron. Soc., **369**, L9–L13 (2006)
Bastian, N., Gieles, M., Efremov, Y.N., Lamers, H.J.G.L.M., Astron. Astrophys., **443**, 79–90 (2005)
Battinelli, P., Capuzzo-Dolcetta, R., Mon. Not. R. Astron. Soc., **249**, 76–83 (1991)
Baumgardt, H., Kroupa, P., Parmentier, G., Mon. Not. R. Astron. Soc., **384**, 1231–1241 (2008)
Beck, S.C., Turner, J.L., Ho, P.T.P., Lacy, J.H., Kelly, D.M., Astrophys. J., **457**, 610–615 (1996)
Beck, S.C., Turner, J.L., Kloosterman, J., Astron. J., **134**, 1237–1244 (2007)
Beck, S.C., Turner, J.L., Kovo, O., Astron. J., **120**, 244–259 (2000)
Bertoldi, F., Draine, B.T., Astrophys. J., **458**, 222–232 (1996)
Billett, O.H., Hunter, D.A., Elmegreen, B.G., Astron. J., **123**, 1454–1475 (2002)
Boily, C.M., Lançon, A., Deiters, S., Heggie, D.C., Astrophys. J., **620**, L27–L30 (2005)
Bonnell, I.A., Bate, M.R., Zinnecker, H., Mon. Not. R. Astron. Soc., **298**, 93–102 (1998)
Bot, C., Boulanger, F., Rubio, M., Rantakyro, F., Astron. Astrophys., **471**, 103–112 (2007)
Brandner, W., Clark, J.S., Stolte, A., Waters, R., Negueruela, I., Goodwin, S.P., Astron. Astrophys., **478**, 137–149 (2008)
Burbidge, E.M., Burbidge, G.R., Astrophys. J., **135**, 694–710 (1962)
Burbidge, E.M., Burbidge, G.R., Rubin, V.C., Astron. J., **69**, 535–535 (1964)
Caldwell, N., Phillips, M.M., Astrophys. J., **338**, 789–803 (1989)
Calzetti, D., Meurer, G.R., Bohlin, R.C., Garnett, D.R., Kinney, A.L., Leitherer, C., Storchi-Bergmann, T., Astron. J., **114**, 1834–1849 (1997)
Castor, J., McCray, R., Weaver, R., Astrophys. J., **200**, L107–L110 (1975)
Chandar, R., Leitherer, C., Tremonti, C., Calzetti, D., Astrophys. J., **586**, 939–958 (2003)
Chandar, R., Leitherer, C., Tremonti, C.A. 2004, Astrophys. J., **604**, 153–166 (2004a)
Chandar, R., Leitherer, C., Tremonti, C.A., Calzetti, D., Aloisi, A., Meurer, G.R., de Mello, D., Astrophys. J., **628**, 210–230 (2005)
Chandar, R., Whitmore, B., Lee, M.G., Astrophys. J., **611**, 220–244 (2004b)
Chevalier, R.A., Clegg, A.W., Nature, **317**, 44–45 (1985)
Clark, J.S., Fender, R.P., Waters, L.B.F.M., Dougherty, S.M., Koornneef, J., Steele, I.A., van Blokland, A., Mon. Not. R. Astron. Soc., **299**, L43–L47 (1998)
Clark, J.S., Negueruela, I., Crowther, P.A., Goodwin, S.P., Astron. Astrophys., **434**, 949–969 (2005)
Clarke, C.J., Bonnell, I. A., Mon. Not. R. Astron. Soc., 388, 1171 (2008)
Condon, J.J., Annu. Rev. Astron. Astrophys., **30**, 575 (1992)
Conti, P.S. 1991, Astrophys. J., **377**, 115–125
Conti, P.S., Blum, R.D., Astrophys. J., **564**, 827–833 (2002)
Cresci, G., Vanzi, L., Sauvage, M., Astron. Astrophys., **433**, 447–454 (2005)
Crosthwaite, L.P., Turner, J.L., Buchholz, L., Ho, P.T.P., Martin, R.N., Astron. J., **123**, 1892–1912 (2002)
Crosthwaite, L.P., Turner, J.L., Astron. J., **134**, 1827–1842 (2007)
Crowther, P.A., Beck, S.C., Willis, A.J., Conti, P.S., Morris, P.W., and Sutherland, R.S., Mon. Not. R. Astron. Soc., **304**, 654–668 (1999)
Crowther, P.A., Hadfield, L.J., Clark, J.S., Negueruela, I., Vacca, W.D., Mon. Not. R. Astron. Soc., **372**, 1407–1424 (2006)
Dahmen, G., Huttemeister, S., Wilson, T.L., Mauersberger, R. 1998, Astron. Astrophys., **331**, 959–976
Dale, D.A., et al., Astrophys. J., **646**, 161–173 (2006)
Dame, T.M., Hartmann, D., Thaddeus, P. 2001, Astrophys. J., **547**, 792–813
de Grijs, R. to appear in "Young massive star clusters—Initial conditions and environments", E. Perez, R. de Grijs, R. M. Gonzalez Delgado, eds. (Springer: Dordrecht) e-prints, 711, arXiv:0711.3540 (2007)
de Grijs, R., Anders, P., Bastian, N., Lynds, R., Lamers, H.J.G.L.M., O'Neil, E.J., Mon. Not. R. Astron. Soc., **343**, 1285–1300 (2003)

de Jong, T., Clegg, P.E., Rowan-Robinson, M., Soifer, B.T., Habing, H.J., Houck, J.R., Aumann, H.H., Raimond, E., Astrophys. J., **278**, L67–L70 (1984)
de Marchi, G., Nota, A., Leitherer, C., Ragazzoni, R., Barbieri, C., Astrophys. J., **419**, 658–669 (1993)
de Marchi, G., Clampin, M., Greggio, L., Leitherer, C., Nota, A., Tosi, M., Astrophys. J., **479**, L27–L30 (1997)
de Pree, C.G., Nysewander, M.C., Goss, W.M., Astron. J., **117**, 2902–2918 (1999)
de Pree, C.G., Rodriguez, L.F., and Goss, W.M., Revista Mexicana de Astronomia y Astrofisica, **31**, 39–44 (1995)
Dickman, R.L., Snell, R.L., Schloerb, F.P., Astrophys. J., **309**, 326–330 (1986)
Dopita, M.A., et al., Astrophys. J., **619**, 755–778 (2005)
Dopita, M.A., et al., ApJS, **167**, 177–200 (2006)
Dowell, C.D., Astrophys. J., **487**, 237–247 (1997)
Downes, D., Solomon, P.M., Astrophys. J., **507**, 615–654 (1998)
Downes, D., Solomon, P.M., Radford, S.J.E., Astrophys. J., **414**, L13–L16 (1993)
Draine, B.T., Bertoldi, F., Astrophys. J., **468**, 269–289 (1996)
Dreher, J.W., Johnston, K.J., Welch, W.J., Walker, R.C., Astrophys. J., **283**, 632–639 (1984)
Drissen, L., Moffat, A.F.J., Walborn, N.R., Shara, M.M., Astron. J., **110**, 2235–2241 (1995)
Eisenhauer, F., Quirrenbach, A., Zinnecker, H., Genzel, R., Astrophys. J., **498**, 278–292 (1998)
Elmegreen, B.G., Astrophys. J., **668**, 1064–1082 (2007)
Elmegreen, B.G., ArXiv e-prints, 803, arXiv:0803.3154 (2008)
Fall, S.M., Rees, M.J., Mon. Not. R. Astron. Soc., **181**, 37P–42P (1977)
Fall, S.M., Zhang, Q., Astrophys. J., **561**, 751–765 (2001)
Fall, S.M., Chandar, R., Whitmore, B.C., Astrophys. J., **631**, L133–L136 (2005)
Figer, D.F., Astronomische Nachrichten Supplement, **324**, 255–261 (2003)
Figer, D.F., The Formation and Evolution of Massive Young Star Clusters, **322**, 49 (2004)
Figer, D.F., Nature, **434**, 192–194 (2005)
Figer, D.F., ArXiv e-prints, 803, arXiv:0803.1619 (2008)
Figer, D.F., in Massive Stars as Cosmic Engines, IAU Symposium, **250**, 247–256 (2008)
Figer, D.F., Kim, S.S., Morris, M., Serabyn, E., Rich, R.M., McLean, I.S., Astrophys. J., **525**, 750–758 (1999)
Figer, D.F., Najarro, F., Morris, M., McLean, I.S., Geballe, T.R., Ghez, A.M., Langer, N., Astrophys. J., **506**, 384–404 (1998)
Förster Schreiber, N.M., Genzel, R., Lutz, D., Kunze, D., Sternberg, A., Astrophys. J., **552**, 544–571 (2001)
Galliano, F., Madden, S.C., Tielens, A.G.G.M., Peeters, E., Jones, A.P., Astrophys. J., **679**, 310–345 (2008)
Gao, Y., Solomon, P.M., Astrophys. J., **606**, 271–290 (2004)
Gascoigne, S.C.B., Kron, G.E., Publ. Astron. Soc. Pac., **64**, 196–200 (1952)
Gaume, R.A., Claussen, M.J., de Pree, C.G., Goss, W.M., Mehringer, D.M., Astrophys. J., **449**, 663–673 (1995)
Gehrz, R.D., Sramek, R.A., Weedman, D.W., Astrophys. J., **267**, 551–562 (1983)
Genzel, R., Cesarsky, C.J. Annu. Rev. Astron. Astrophys., **38**, 761–814 (2000)
Genzel, R., et al., Astrophys. J., **498**, 579–605 (1998)
Gieles, M., Bastian, N., Astron. Astrophys., **482**, 165–171 (2008)
Gil de Paz, A., et al., ApJS, **173**, 185–255 (2007)
Gilbert, A.M., Ph.D. Dissertation, University of California, Berkeley (2002)
Gilbert, A.M., et al., Astrophys. J., **533**, L57–L60 (2000)
Goodwin, S.P. 1997, Mon. Not. R. Astron. Soc., **286**, 669–680 (1997)
Gorjian, V., Astron. J., **112**, 1886–1893 (1996)
Gorjian, V., Turner, J.L., Beck, S.C., Astrophys. J., **554**, L29–L33 (2001)
Gnedin, O.Y., Ostriker, J.P., Astrophys. J., **474**, 223–255 (1997)
Gruendl, R.A., Chu, Y.-H., Dunne, B.C., Points, S.D., Astron. J., **120**, 2670–2678 (2000)
Haas, M., Chini, R., Klaas, U., Astron. Astrophys., **433**, L17–L20 (2005)

Hadfield, L.J., Crowther, P.A., Mon. Not. R. Astron. Soc., **368**, 1822–1832 (2006)
Harayama, Y., Eisenhauer, F., Martins, F., Astrophys. J., **675**, 1319–1342 (2008)
Harris, W.E., Extragalactic Star Clusters, **207**, 545 (2002)
Harris, W.E., Astron. J., **112**, 1487–1488 (1996)
Harris, W.E., in Galactic Halos, ed. D. Zaritsky, (San Francisco:A. S. P.) **136**, 33–41 (1998)
Harris, W.E., Harris, H.C., in Astrophysical Quantities, A.N. Cox ed. (Springer, London), 545–568 (2000)
Harris, W.E., Pudritz, R.E., Astrophys. J., **429**, 177–191 (1994)
Helou, G., Lu, N.Y., Werner, M.W., Malhotra, S., Silbermann, N., Astrophys. J., **532**, L21–L24 (2000)
Higdon, S.J.U., Armus, L., Higdon, J.L., Soifer, B.T., Spoon, H.W.W., Astrophys. J., **648**, 323–339 (2006)
Hills, J.G., Astrophys. J., **235**, 986–991 (1980)
Ho, P.T.P., Beck, S.C., Turner, J.L., Astrophys. J., **349**, 57–66 (1990)
Ho, L.C., Filippenko, A.V., Astrophys. J., **466**, L83–L86 (1996a)
Ho, L.C., Filippenko, A.V., Astrophys. J., **472**, 600–610 (1996b)
Hodge, P.W., Astrophys. J., **133**, 413–419 (1961)
Hollenbach, D.J., Tielens, A.G.G.M., Reviews of Modern Physics, **71**, 173–230 (1999)
Holtzman, J.A., et al., Astron. J., **103**, 691–702 (1992)
Homeier, N.L., Alves, J., Astron. Astrophys., **430**, 481–489 (2005)
Hughes, J.P., Rakowski, C.E., Burrows, D.N., Slane, P.O., Astrophys. J., **528**, L109–L113 (2000)
Hunter, D.A., Shaya, E.J., Holtzman, J.A., Light, R.M., O'Neil, E.J., Jr., Lynds, R., Astrophys. J., **448**, 179–194 (1995)
Hunter, D.A., O'Neil, E.J., Jr., Lynds, R., Shaya, E.J., Groth, E.J., Holtzman, J.A., Astrophys. J., **459**, L27–L30 (1996)
Hunter, D.A., O'Connell, R.W., Gallagher, J.S., Smecker-Hane, T.A., Astron. J., **120**, 2383–2401 (2000)
Israel, F.P., de Graauw, T., van de Stadt, H., de Vries, C.P., Astrophys. J., **303**, 186–197 (1986)
Johnson, K.E., Kobulnicky, H.A., Astrophys. J., **597**, 923–928 (2003)
Johnson, K.E., Indebetouw, R., Pisano, D.J., Astron. J., **126**, 101–112 (2003)
Karachentsev, I.D., et al. 2007, Astron. J., **133**, 504–517 (2007)
Kaufman, M.J., Wolfire, M.G., Hollenbach, D.J., Astrophys. J., **644**, 283–299 (2006)
Kaufman, M.J., Wolfire, M.G., Hollenbach, D.J., Luhman, M.L., Astrophys. J., **527**, 795–813 (1999)
Kawara, K., Nishida, M., Phillips, M.M., Astrophys. J., **337**, 230–235 (1989)
Kennicutt, R.C., Jr., Annu. Rev. Astron. Astrophys., **36**, 189–232 (1998a)
Kennicutt, R.C., Jr., Astrophys. J., **498**, 541–552 (1998b)
Kim, S.S., Figer, D.F., Lee, H.M., Morris, M., Astrophys. J., **545**, 301–308 (2000)
Kim, S.S., Figer, D.F., Kudritzki, R.P., and Najarro, F., J. Korean Astronomical Soc., **40**, 153–155 (2007)
Kim, S.S., Figer, D.F., Morris, M., Astrophys. J., **607**, L123–L126 (2004)
Kobulnicky, H.A., Johnson, K.E., Astrophys. J., **527**, 154–166 (1999)
Kobulnicky, H.A., Skillman, E.D., Astron. J., **135**, 527–537 (2008)
Koen, C., Mon. Not. R. Astron. Soc., **365**, 590–594 (2006)
Kroupa, P., Mon. Not. R. Astron. Soc., **322**, 231–246 (2001)
Kroupa, P., Boily, C.M., Mon. Not. R. Astron. Soc., **336**, 1188–1194 (2002)
Krumholz, M.R., Matzner, C.D., McKee, C.F., Astrophys. J., **653**, 361–382 (2006)
Krumholz, M.R., Tan, J.C., Astrophys. J., **654**, 304–315 (2007)
Lada, C.J., Margulis, M., Dearborn, D., Astrophys. J., **285**, 141–152 (1984)
Lada, C.J., Lada, E.A., Annu. Rev. Astron. Astrophys., **41**, 57–115 (2003)
Lang, C.C., Goss, W.M., Morris, M., Astron. J., **121**, 2681–2705 (2001)
Larsen, S.S., Origlia, L., Brodie, J., Gallagher, J.S., Mon. Not. R. Astron. Soc., **383**, 263–276 (2008)
Leitherer, C., et al., ApJS, **123**, 3–40 (1999)

Lepp, S., Dalgarno, A., Astron. Astrophys., **306**, L21–L24 (2006)
Lonsdale, C.J., et al., Publ. Astron. Soc. Pac., **115**, 897–927 (2003)
Lutz, D., Kunze, D., Spoon, H.W.W., Thornley, M.D., Astron. Astrophys., **333**, L75–L78 (1998)
Lynds, C.R., Sandage, A.R., Astron. J., **68**, 285 (1963)
Mackey, A.D., Gilmore, G.F., Mon. Not. R. Astron. Soc., **338**, 85–119 (2003)
Maloney, P., Black, J.H., Astrophys. J., **325**, 389–401 (1988)
Maloney, P.R., Hollenbach, D.J., Tielens, A.G.G.M., Astrophys. J., **466**, 561–584 (1996)
Mangum, J.G., Darling, J., Menten, K.M., Henkel, C., Astrophys. J., **673**, 832–846 (2008)
Maoz, D., Barth, A.J., Ho, L.C., Sternberg, A., Filippenko, A.V., Astron. J., **121**, 3048–3074 (2001)
Maoz, D., Barth, A.J., Sternberg, A., Filippenko, A.V., Ho, L.C., Macchetto, F.D., Rix, H.-W., Schneider, D.P., Astron. J., **111**, 2248–2264 (1996)
Maoz, D., Filippenko, A.V., Ho, L.C., Macchetto, F.D., Rix, H.-W., Schneider, D.P., ApJS, **107**, 215–226 (1996)
Martín, S., Mauersberger, R., Martín-Pintado, J., Henkel, C., García-Burillo, S., ApJS, **164**, 450–476 (2006)
Martín, S., Requena-Torres, M.A., Martín-Pintado, J., Mauersberger, R. 2008, ApSS, **313**, 303–306 (2008)
Martín-Hernández, N.L., Schaerer, D., Sauvage, M., Astron. Astrophys., **429**, 449–467 (2005)
Martín-Hernández, N.L., Vermeij, R., Tielens, A.G.G.M., van der Hulst, J.M., Peeters, E., Astron. Astrophys., **389**, 286–294 (2002)
Massey, P., Hunter, D.A., Astrophys. J., **493**, 180 (1998)
Mathieu, R.D., Astrophys. J., **267**, L97–L101 (1983)
Mauersberger, R., Henkel, C., Astron. Astrophys., **245**, 457–466 (1991)
Mauersberger, R., Henkel, C., Weiß, A., Peck, A.B., Hagiwara, Y., Astron. Astrophys., **403**, 561 (2003)
McCrady, N., Graham, J.R., Astrophys. J., **663**, 844–856 (2007)
McCrady, N., Gilbert, A.M., Graham, J.R., Astrophys. J., **596**, 240–252 (2003)
McCrady, N., Graham, J.R., Vacca, W.D., Astrophys. J., **621**, 278–284 (2005)
McKee, C.F., Ostriker, E.C., Annu. Rev. Astron. Astrophys., **45**, 565–687 (2007)
McLaughlin, D.E., Fall, S.M., Astrophys. J., **679**, 1272–1287 (2008)
McLaughlin, D.E., van der Marel, R.P., ApJS, **161**, 304–360 (2005)
McMillan, S.L.W., Portegies Zwart, S.F., Astrophys. J., **596**, 314–322 (2003)
Meier, D.S., Turner, J.L., Beck, S.C., Astron. J., **122**, 1770–1781 (2001)
Meier, D.S., Turner, J.L., Beck, S.C., Astron. J., **124**, 877–855 (2002)
Meier, D.S., Turner, J.L., Astron. J., **127**, 2069–2084 (2004)
Meier, D.S., Turner, J.L., Astrophys. J., **618**, 259–280 (2005)
Meier, D.S., Turner, J.L., submitted (2008)
Meier, D.S., Turner, J.L., Hurt, R.L., Astrophys. J., **675**, 281–302 (2008)
Meijerink, R., Spaans, M., Astron. Astrophys., **436**, 397–409 (2005)
Meijerink, R., Spaans, M., Israel, F.P., Astrophys. J., **650**, L103–L106 (2006)
Melena, N.W., Massey, P., Morrell, N.I., Zangari, A.M., Astron. J., **135**, 878–891 (2008)
Mengel, S., Lehnert, M.D., Thatte, N., Genzel, R., Astron. Astrophys., **443**, 41–60 (2005)
Mengel, S., Tacconi-Garman, L.E., Astron. Astrophys., **466**, 151–155 (2007)
Mengel, S., Tacconi-Garman, L.E., ArXiv e-prints, 803, arXiv:0803.4471 (2008)
Meurer, G.R., Heckman, T.M., Leitherer, C., Kinney, A., Robert, C., Garnett, D.R., Astron. J., **110**, 2665–2691 (1995)
Meylan, G., The Globular Cluster-Galaxy Connection, **48**, 588–600 (1993)
Mills, B.Y., Turtle, A.J., Watkinson, A., Mon. Not. R. Astron. Soc., **185**, 263–276 (1978)
Mirabel, I.F., Sanders, D.B., Astrophys. J., **340**, L53–L56 (1989)
Mirabel, I.F., et al., Astron. Astrophys., **333**, L1–L4 (1998)
Mohan, N.R., Anantharamaiah, K.R., Goss, W.M., Astrophys. J., **557**, 659–670 (2001)
Morse, J.A., Davidson, K., Bally, J., Ebbets, D., Balick, B., Frank, A., Astron. J., **116**, 2443–2461 (1998)

Myers, P.C., Astrophys. J., **270**, 105–118 (1983)
Najarro, F., Figer, D.F., Ap&SS, **263**, 251–254 (1998)
Noyola, E., Gebhardt, K., Astron. J., **134**, 912–925 (2007)
Nürnberger, D.E.A. 2004, Ph.D.Dissertation, University Würzburg
Nürnberger, D.E.A., Bronfman, L., Yorke, H.W., Zinnecker, H., Astron. Astrophys., **394**, 253–269 (2002)
Nürnberger, D.E.A., Petr-Gotzens, M.G., Astron. Astrophys., **382**, 537–553 (2002)
O'Connell, R.W., Gallagher, J.S., III, and Hunter, D.A., Astrophys. J., **433**, 65–69 (1994)
O'Connell, R.W., Gallagher, J.S., III, Hunter, D.A., Colley, W.N., Astrophys. J., **446**, L1–L4 (1995)
Oey, M.S., Clarke, C.J., Astrophys. J., **620**, L43–L46 (2005)
Omont, A., Reports of Progress in Physics, **70**, 1099–1176 (2007)
Origlia, L., Leitherer, C., Aloisi, A., Greggio, L., Tosi, M., Astron. J., **122**, 815–824 (2001)
Padoan, P., Nordlund, Å., Astrophys. J., **576**, 870–879 (2002)
Padoan, P., Nordlund, Å., Kritsuk, A.G., Norman, M.L., Li, P.S., Astrophys. J., **661**, 972–981 (2007)
Pandey, A.K., Ogura, K., Sekiguchi, K., PASJ, **52**, 847–865 (2000)
Parmentier, G., Gilmore, G., Mon. Not. R. Astron. Soc., **377**, 352–372 (2007)
Parmentier, G., Goodwin, S.P., Kroupa, P., Baumgardt, H., Astrophys. J., **678**, 347–352 (2008)
Peeters, E., Spoon, H.W.W., Tielens, A.G.G.M., Astrophys. J., **613**, 986–1003 (2004)
Rieke, G.H., Lebofsky, M.J., Astrophys. J., **220**, L37–L41 (1978)
Rieke, G.H., Lebofsky, M.J., Annu. Rev. Astron. Astrophys., **17**, 477–511 (1979)
Rieke, G.H., Lebofsky, M.J., Thompson, R.I., Low, F.J., Tokunaga, A.T., Astrophys. J., **238**, 24–40 (1980)
Rieke, G.H., Low, F.J., Astrophys. J., **197**, 17–23 (1975)
Rigby, J.R., Rieke, G.H., Astrophys. J., **606**, 237–257 (2004)
Rodríguez-Rico, C.A., Goss, W.M., Turner, J.L., and Gómez, Y., Astrophys. J., **670**, 295–300 (2007)
Rodríguez-Rico, C.A., Goss, W.M., Zhao, J.-H., Gómez, Y., Anantharamaiah, K.R., Astrophys. J., **644**, 914–923 (2006)
Roy, A.L., Goss, W.M., Anantharamaiah, K.R., Astron. Astrophys., **483**, 79–88 (2008)
Sakamoto, K., Scoville, N.Z., Yun, M.S., Crosas, M., Genzel, R., Tacconi, L.J., Astrophys. J., **514**, 68–76 (1999)
Sanders, D.B., Mirabel, I.F., Astrophys. J., **298**, L31–L35 (1985)
Sanders, D.B., Mirabel, I.F., Annu. Rev. Astron. Astrophys., **34**, 749–792 (1996)
Sanders, D.B., Mazzarella, J.M., Kim, D.-C., Surace, J.A., Soifer, B.T., Astron. J., **126**, 1607–1664 (2003)
Sanders, D.B., Scoville, N.Z., Solomon, P.M., Astrophys. J., **289**, 373–387 (1985)
Sanders, D.B., Scoville, N.Z., Young, J.S., Soifer, B.T., Schloerb, F.P., Rice, W.L., and Danielson, G.E., Astrophys. J., **305**, L45–L49 (1986)
Sanders, D.B., Soifer, B.T., Elias, J.H., Neugebauer, G., Matthews, K., Astrophys. J., **328**, L35–L39 (1988)
Sargent, W.L.W., Searle, L., Astrophys. J., **162**, L155–L160 (1970)
Saviane, I., Momany, Y., da Costa, G.S., Rich, R.M., Hibbard, J.E., Astrophys. J., **678**, 179–186 (2008)
Scalo, J.M., Fundamentals of Cosmic Physics, **11**, 1–278 (1986)
Scalo, J., The Stellar Initial Mass Function (38th Herstmonceux Conference), **142**, 201 (1998)
Schaerer, D., Contini, T., Kunth, D., Meynet, G. 1997, Astrophys. J., **481**, L75–L78
Schaerer, D., Contini, T., Kunth, D., Astron. Astrophys., **341**, 399–417 (1999a)
Schaerer, D., Contini, T., Pindao, M., A&AS, **136**, 35–52 (1999b)
Schmidt, M., Astrophys. J., **129**, 243–258 (1959)
Scoville, N.Z., Yun, M.S., Sanders, D.B., Clemens, D.P., and Waller, W.H., ApJS, **63**, 821–915 (1987)
Shepherd, D.S., Kurtz, S.E., Astrophys. J., **523**, 690–700 (1999)
Shin, J., Kim, S.S., Takahashi, K., Mon. Not. R. Astron. Soc., **386**, L67–L71 (2008)

Shioya, Y., Taniguchi, Y., Trentham, N., Mon. Not. R. Astron. Soc., **321**, 11–17 (2001)
Silich, S., Tenorio-Tagle, G., Muñoz-Tuñón, C., Astrophys. J., **669**, 952–958 (2007)
Simon, M., Simon, T., Joyce, R.R., Astrophys. J., **227**, 64–66 (1979)
Sirianni, M., Nota, A., Leitherer, C., De Marchi, G., Clampin, M., Astrophys. J., **533**, 203–214 (2000)
Smith, L.F., Mezger, P.G., Biermann, P., Astron. Astrophys., **66**, 65–76 (1978)
Smith, L.J., Gallagher, J.S., Mon. Not. R. Astron. Soc., **326**, 1027–1040 (2001)
Smith, L.J., Westmoquette, M.S., Gallagher, J.S., O'Connell, R.W., Rosario, D.J., de Grijs, R., Mon. Not. R. Astron. Soc., **370**, 513–527 (2006)
Soifer, B.T., et al., Astrophys. J., **283**, L1-L4 (1984)
Soifer, B.T., Sanders, D.B., Neugebauer, G., Danielson, G.E., Lonsdale, C.J., Madore, B.F., Persson, S.E., Astrophys. J., **303**, L41-L44 (1986)
Soifer, B.T., Neugebauer, G., Houck, J.R., Annu. Rev. Astron. Astrophys., **25**, 187–230 (1987a)
Soifer, B.T., Sanders, D.B., Madore, B.F., Neugebauer, G., Danielson, G.E., Elias, J.H., Lonsdale, C.J., Rice, W.L., Astrophys. J., **320**, 238–257 (1987b)
Solomon, P.M., Downes, D., Radford, S.J.E., Barrett, J.W., Astrophys. J., **478**, 144–161 (1997)
Solomon, P.M., Rivolo, A.R., Barrett, J., Yahil, A., Astrophys. J., **319**, 730–741 (1987)
Stacey, G.J., Geis, N., Genzel, R., Lugten, J.B., Poglitsch, A., Sternberg, A., Townes, C.H., Astrophys. J., **373**, 423–444 (1991)
Stanford, S.A., Sargent, A.I., Sanders, D.B., Scoville, N.Z., Astrophys. J., **349**, 492–496 (1990)
Sternberg, A., Astrophys. J., **506**, 721–726 (1998)
Stevens, I.R., Forbes, D.A., Norris, R.P., Mon. Not. R. Astron. Soc., **335**, 1079–1084 (2002)
Stolte, A., Ph.D.Thesis, Combined Faculties for the Natural Sciences and for Mathematics of the Ruperto-Carola University of Heidelberg (2003)
Stolte, A., Brandner, W., Grebel, E.K., Lenzen, R., Lagrange, A.-M., Astrophys. J., **628**, L113–L116 (2005)
Stolte, A., Grebel, E.K., Brandner, W., Figer, D.F., Astron. Astrophys., **394**, 459–478 (2002)
Stolte, A., Brandner, W., Brandl, B., Zinnecker, H., Astron. J., **132**, 253–270 (2006)
Strong, A.W., et al., Astron. Astrophys., **207**, 1–15 (1988)
Strong, A.W., Moskalenko, I.V., Reimer, O., Digel, S., and Diehl, R., Astron. Astrophys., **422**, L47–L50 (2004)
Sung, H., Bessell, M.S., Astron. J., **127**, 1014–1028 (2004)
Tacconi, L.J., Young, J.S., Astrophys. J., **352**, 595–604 (1990)
Tan, J.C., Krumholz, M.R., McKee, C.F., Astrophys. J., **641**, L121–L124 (2006)
Thilker, D.A., et al., Astrophys. J., **619**, L79–L82 (2005)
Thornley, M.D., Schreiber, N.M.F., Lutz, D., Genzel, R., Spoon, H.W.W., Kunze, D., Sternberg, A., Astrophys. J., **539**, 641–657 (2000)
Tielens, A.G.G.M., The Physics and Chemistry of the Interstellar Medium, by A.G.G.M. Tielens, pp. ISBN 0521826349. Cambridge, UK: Cambridge University Press (2005)
Tielens, A.G.G.M., Hollenbach, D., Astrophys. J., **291**, 722–754 (1985)
Tielens, A.G.G.M., Hollenbach, D., Astrophys. J., **291**, 747–754 (1985)
Tilanus, R.P.J., Allen, R.J., Astron. Astrophys., **274**, 707–729 (1993)
Tremonti, C.A., Calzetti, D., Leitherer, C., Heckman, T.M., Astrophys. J., **555**, 322–337 (2001)
Tsai, C.-W., Turner, J.L., Beck, S.C., Crosthwaite, L.P., Ho, P.T.P., Meier, D.S., Astron. J., **132**, 2383–2397 (2006)
Turner, J.L., Beck, S.C., Astrophys. J., **602**, L85–L88 (2004)
Turner, J.L., Ho, P.T.P., Astrophys. J., **421**, 122 (1994)
Turner, J.L., Beck, S.C., Crosthwaite, L.P., Larkin, J.E., McLean, I.S., Meier, D.S., Nature, **423**, 621–623 (2003)
Turner, J.L., Beck, S.C., Ho, P.T.P., Astrophys. J., **532**, L109–L112 (2000)
Turner, J.L., Beck, S.C., Hurt, R.L., Astrophys. J., **474**, L11–L14 (1997)
Turner, J.L., Ho, P.T.P., Beck, S.C., Astron. J., **116**, 1212–1220 (1998)
Turner, J.L., et al., in preparation (2008)
Vacca, W.D., Conti, P.S., Astrophys. J., **401**, 543–558 (1992)

van Dishoeck, E.F., Blake, G.A., Annu. Rev. Astron. Astrophys., **36**, 317–368 (1998)
Vanzi, L., Sauvage, M., Astron. Astrophys., **415**, 509–520 (2004)
Vázquez, G.A., Leitherer, C., Astrophys. J., **621**, 695–717 (2005)
Vázquez, G.A., Leitherer, C., Heckman, T.M., Lennon, D.J., de Mello, D.F., Meurer, G.R., Martin, C.L., Astrophys. J., **600**, 162–181 (2004)
Veilleux, S., Cecil, G., Bland-Hawthorn, J., Annu. Rev. Astron. Astrophys., **43**, 769–826 (2005)
Walborn, N.R., et al., Astron. J., **123**, 2754–2771 (2002)
Watson, A.M., et al., Astron. J., **112**, 534–544 (1996)
Weedman, D.W., Feldman, F.R., Balzano, V.A., Ramsey, L.W., Sramek, R.A., Wuu, C.-C., Astrophys. J., **248**, 105–112 (1981)
Weidner, C., Kroupa, P., Mon. Not. R. Astron. Soc., **348**, 187–191 (2004)
Weigelt, G., Baier, G., Astron. Astrophys., **150**, L18–L20 (1985)
Welch, W.J., Dreher, J.W., Jackson, J.M., Terebey, S., Vogel, S.N., Science, **238**, 1550–1555 (1987)
Whitmore, B.C., Schweizer, F., Astron. J., **109**, 960–980 (1995)
Whitmore, B.C., Schweizer, F., Leitherer, C., Borne, K., Robert, C., Astron. J., **106**, 1354–1370 (1993)
Whitmore, B.C., Zhang, Q., Leitherer, C., Fall, S.M., Schweizer, F., Miller, B.W., Astron. J., **118**, 1551–1576 (1999)
Willner, S.P., Soifer, B.T., Russell, R.W., Joyce, R.R., Gillett, F.C., Astrophys. J., **217**, L121–L124 (1977)
Wilson, C.D., Scoville, N., Madden, S.C., Charmandaris, V., Astrophys. J., **542**, 120–127 (2000)
Wilson, C.D., Scoville, N., Madden, S.C., Charmandaris, V., Astrophys. J., **599**, 1049–1066 (2003)
Wolfire, M.G., Hollenbach, D., Tielens, A.G.G.M., Astrophys. J., **402**, 195–215 (1993)
Wong, T., Blitz, L., Astrophys. J., **569**, 157–183 (2002)
Wood, D.O.S., Churchwell, E. 1989, ApJS, **69**, 831–895 (1989)
Young, J.S., Allen, L., Kenney, J.D.P., Lesser, A., Rownd, B. 1996, Astron. J., **112**, 1903–1927 (1996)
Young, J.S., Kenney, J.D., Tacconi, L., Claussen, M.J., Huang, Y.-L., Tacconi-Garman, L., Xie, S., Schloerb, F.P., Astrophys. J., **311**, L17–L21 (1986a)
Young, J.S., Scoville, N.Z., Annu. Rev. Astron. Astrophys., **29**, 581–625 (1991)
Young, J.S., Schloerb, F.P., Kenney, J.D., Lord, S.D., Astrophys. J., **304**, 443–458 (1986b)
Yun, M.S., Ho, P.T.P., Lo, K.Y., Nature, **372**, 530–532 (1994)
Yungelson, L.R., van den Heuvel, E.P.J., Vink, J.S., Portegies Zwart, S.F., de Koter, A., Astron. Astrophys., **477**, 223–237 (2008)
Zhang, Q., Fall, S.M., Astrophys. J., **527**, L81–L84 (1999)
Zinnecker, H., Yorke, H.W., Annu. Rev. Astron. Astrophys., **45**, 481–463 (2007)

Michael Shull enjoys a healthy breakfast to start the day

Chapter 10
Prospects for Studies of Stellar Evolution and Stellar Death in the *JWST* Era

Michael J. Barlow

Abstract I review the prospects for studies of the advanced evolutionary stages of low-, intermediate- and high-mass stars by the *JWST* and concurrent facilities, with particular emphasis on how they may help elucidate the dominant contributors to the interstellar dust component of galaxies. Observations extending from the mid-infrared to the submillimeter can help quantify the heavy element and dust species inputs to galaxies from AGB stars. *JWST*'s MIRI mid-infrared instrument will be so sensitive that observations of the dust emission from individual intergalactic AGB stars and planetary nebulae in the Virgo Cluster will be feasible. The *Herschel Space Observatory* will enable the last largely unexplored spectral region, from the far-IR to the submm, to be surveyed for new lines and dust features, while *SOFIA* will cover the wavelength gap between *JWST* and *Herschel*, a spectral region containing important fine structure lines, together with key water-ice and crystalline silicate bands. *Spitzer* has significantly increased the number of Type II supernovae that have been surveyed for early-epoch dust formation but reliable quantification of the dust contributions from massive star supernovae of Type II, Type Ib and Type Ic to low- and high-redshift galaxies should come from *JWST* MIRI observations, which will be able to probe a volume over 1000 times larger than *Spitzer*.

10.1 Introduction

The bulk of the heavy element enrichment of the interstellar media of galaxies is a result of mass loss during the final evolutionary stages of stars, yet these final stages are currently the least well understood parts of their lives. This is because (a) relatively short evolutionary timescales often lead to comparatively small numbers of objects in different advanced evolutionary stages; (b) strong mass loss with dust formation during some late phases can lead to self-obscuration at optical wavelengths, requiring infrared surveys for their detection; (c) the mass loss process itself can have a complex dependence on pulsational properties, metallicity and the physics

M.J. Barlow (✉)
Department of Physics and Astronomy, University College London, Gower Street,
London WC1E 6BT, UK
e-mail: mjb@star.ucl.ac.uk

H.A. Thronson et al. (eds.), *Astrophysics in the Next Decade,* Astrophysics and Space Science Proceedings, DOI 10.1007/978-1-4020-9457-6_10,

of dust formation. For low and intermediate mass stars ($\leq 8\ M_{\odot}$), strong mass loss during the asymptotic giant branch (AGB) phase (Iben & Renzini 1983) exposes the dredged-up products of nucleosynthesis (nitrogen from CNO-cycle H-shell burning and carbon from helium-shell burning). For higher mass stars, mass loss stripping eventually exposes the products of H- and He-burning. During the late stages of massive stars these products can dominate their spectra, as luminous blue variables (LBVs) or as WN, WC or WO Wolf-Rayet (WR) stars.

Current or planned optical and infrared surveys of the Milky Way Galaxy (MWG) and of other Local Group galaxies will yield much larger samples of single and binary stars in the various brief evolutionary phases that occur near the end-points of their lives. Ground-based optical surveys include the nearly complete INT Photometric Hα Survey of the northern galactic plane (IPHAS; Drew et al. 2005) and the complementary VST Photometric Hα Survey (VPHAS+) of the southern galactic plane that will begin in 2009, while ground-based near-IR surveys include the UKIRT Infrared Deep Sky Survey (UKIDSS; Lawrence et al. 2007) in the north, and the VISTA sky surveys in the south. From space, we have the recent *Spitzer* GLIMPSE (Benjamin et al. 2003) and MIPSGAL (Carey et al. 2005) mid-IR surveys of the MWG between $l = \pm 60^{\circ}$, which will be joined by the *Herschel* 70–500 μm Hi-GAL Survey of the same MWG regions, while the *Spitzer* SAGE survey (Meixner et al. 2006) and the *Herschel* Heritage survey will map the contents of the lower metallicity LMC and SMC dwarf galaxies from 3.6 to 500 μm. These surveys will enable many stars in different advanced evolutionary stages to be identified, from their characteristic optical/IR spectral energy distributions (SEDs).

The *JWST* is an infrared observatory and a significant number of its concurrent facilities will operate at infrared and submillimeter wavelengths. The gaseous component of mass loss outflows can be studied in detail at these wavelengths using high resolution spectrometers. In addition, towards the end-points of their evolution many stars become extremely luminous at infrared wavelengths due to re-emission by dust particles formed in their outflows. This dust emission can be used as a tool for studying stellar populations, for studying the mass loss process and its history and for investigating the effects of stellar evolution on the enrichment of galaxies. Carbon and silicate dust particles are thought to be ubiquitous throughout our Galaxy but where are these particles formed? Evolved stars (AGB stars, M supergiants, WR stars, supernovae, etc.) are known to be significant sources of dust particles but what are the dominant sources of dust in our own and other galaxies? Tielens et al. (2005) summarised current estimates for the gas and dust inputs to the ISM of our galaxy from various classes of evolved stars; their review pinpoints the fact that the integrated contributions by some of these stellar types have large uncertainties at present.

As a result of ongoing measurements of mass loss rates by the *Spitzer* SAGE Survey, the dominant stellar dust contributors to the metal-poor LMC and SMC dwarf galaxies, at known distances, will soon be known. For the much more massive Milky Way Galaxy, the optical and IR surveys discussed above will lead to much larger samples of hitherto rare evolved star types. Once accurate parallaxes become available for one billion MWG stars down to 20th magnitude from ESA's GAIA

Mission (2012-20), then we will have reliable distances, population numbers and mass loss rates for a wide range of evolutionary types, allowing accurate gas and dust enrichment rates for the MWG to be established. In this review I will focus on future observations by the *JWST* and other new facilities of low-, intermediate- and high-mass stars in the advanced evolutionary phases that are likely to provide the dominant contributions to the gas and dust enrichment of galaxies.

10.2 AGB Stars, Post-AGB Objects and Planetary Nebulae

10.2.1 The Infrared Spectra of Evolved Stars

Figure 10.1 illustrates the mid-infrared spectra of eight oxygen-rich stars that are at different stages in their evolution up the AGB (Sylvester et al. 1999).

Mira (Omicron Ceti) exhibits pure emission silicate bands at 10 and 18 μm, while CRL 2199 and WX Psc (=IRC+10011) show self-absorbed emission features and the remaining five sources show absorption features. This sequence is interpreted as one of increasing mass loss rate and increasing circumstellar self-absorption. Although the silicate-absorption objects are more luminous, the time spent at such high luminosities and mass loss rates is much shorter than the time spent lower down

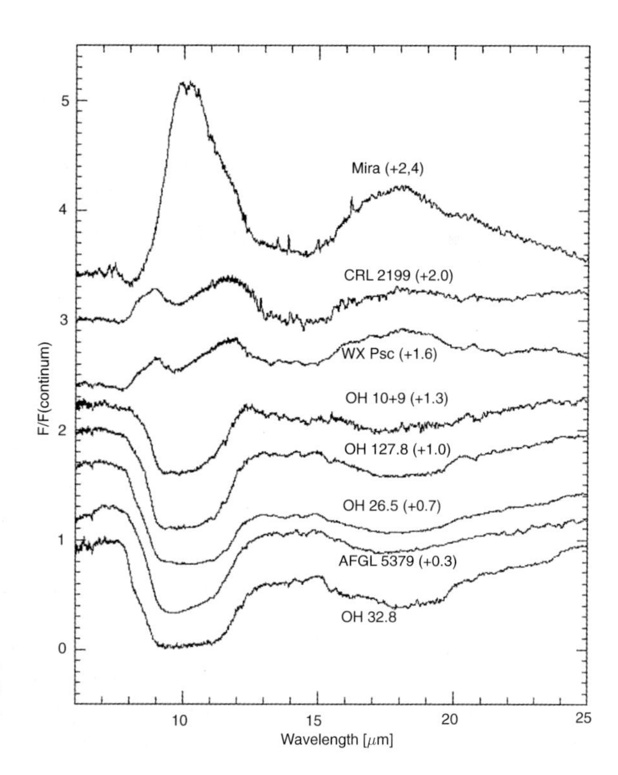

Fig. 10.1 The 6-25-μm *ISO* SWS spectra of eight oxygen-rich AGB stars, in order of increasing mass loss rate, from top to bottom. The broad amorphous silicate features at 9.7 μm and 18 μm are increasingly self-absorbed with increasing mass loss rate (from Sylvester et al. 1999)

the AGB. The mid-IR spectra of ellliptical galaxies show 10-μm silicate features in emission (Bregman et al. 2006), interpreted as resulting from the combined spectra of many AGB stars having silicate emission features.

For the objects shown in Fig. 10.1, Sylvester et al. (1999) found that after division by a smooth continuum fit, a number of sharp features were evident longwards of 20 μm that corresponded very well to crystalline silicate emission features that had previously been discovered in the *ISO* spectra of several classes of objects, including comets, very young stars and very evolved objects. An example of the latter is NGC 6302, a likely descendant of an extreme OH/IR star. It has a massive edge-on dust disk which completely obscures the central star, whose effective temperature is estimated from photoionization modelling to exceed 200,000 K. An *HST* image of NGC 6302 is shown in Fig. 10.2. Its *ISO* spectrum (Figs. 10.3 and 10.4) exhibits many sharp crystalline silicate emission features, a well as features due to other species, such as crystalline water ice.

The crystalline particles in NGC 6302 and other sources are thought to have been produced by the annealing of amorphous grains following upward temperature excursions. The peak wavelengths and widths of some of the features have been found to be good indicators of the temperature of the emitting particles. Figure 10.5 (left panel) shows laboratory spectra of the 69-μm band of forsterite (crystalline olivine), obtained at three different temperatures (Bowey et al. 2002). As the temperature changes from 295 K to 3.5 K, the peak wavelength of the band moves shortwards by nearly a micron, while the FWHM of the feature becomes narrower. The tem-

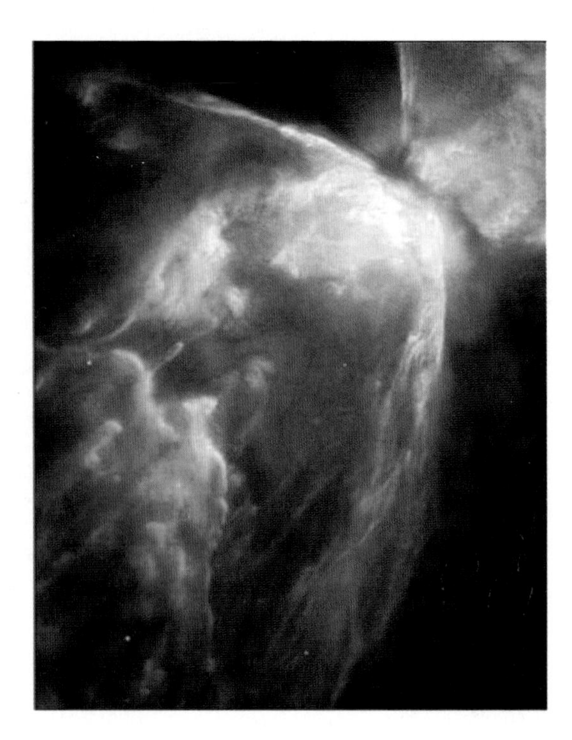

Fig. 10.2 *HST* WFPC2 F656N image of the central parts of NGC 6302, showing the obscured region (top right) attributed to an edge-on dust disk that is collimating the bipolar outflow geometry. See Matsuura et al. (2005)

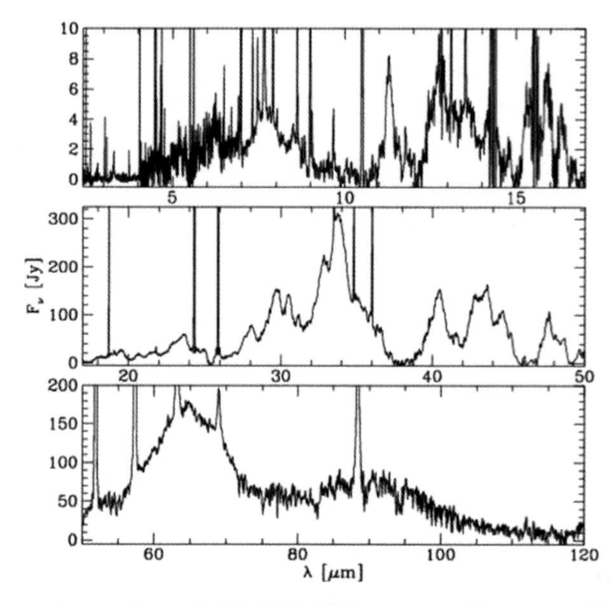

Fig. 10.3 The continuum-subtracted *ISO* SWS+LWS spectrum of the extreme bipolar planetary nebula NGC 6302, which has a massive edge-on dust disk, illustrating the presence of PAH features (top panel), as well as many crystalline silicate and water-ice features (bottom two panels). From Molster et al. (2001)

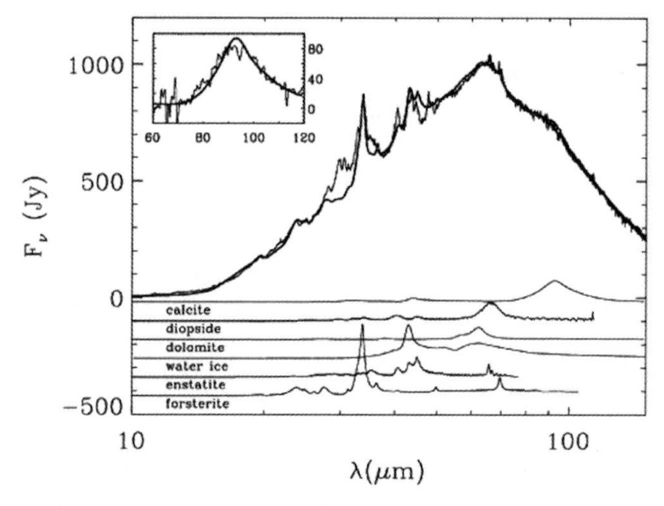

Fig. 10.4 A multi-component continuum and dust feature fit to the *ISO* far-infrared spectrum of NGC 6302 (Kemper et al. 2002). *JWST*-MIRI stops at 28 μm. *Herschel*-PACS starts at 57 μm. In between, there are many crystalline silicate bands, plus the crucial 44-μm crystalline ice band and the [O III] 52-μm line. In the next ten years, only *SOFIA* will be able to observe the 28-57-μm wavelength range

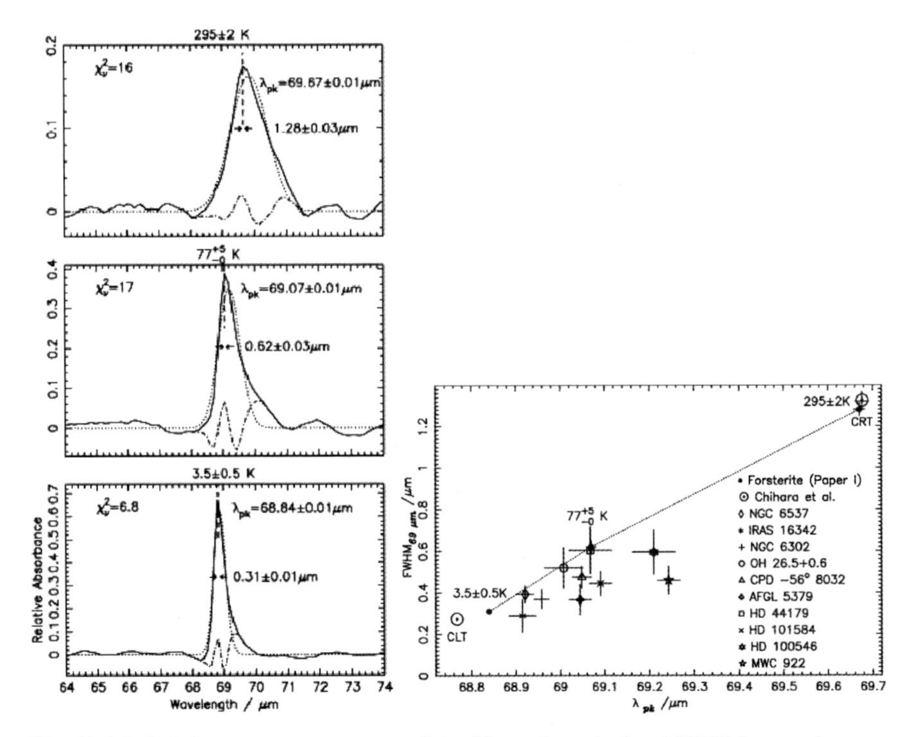

Fig. 10.5 *Left*: Laboratory measurements of the 69-μm forsterite band FWHM vs. peak wavelength, for three different temperatures. *Right*: The FWHM versus observed peak wavelength of the 69-μm feature for ten astronomical sources (Bowey et al. 2002)

perature derived from the peak wavelength of the 69-μm forsterite band (Fig. 10.5; right panel) is well-correlated with the continuum dust temperature and thus is an excellent dust thermometer that can be used by *Herschel*-PACS and by *SOFIA* for studies of post-MS and pre-MS objects.

As well as exhibiting strong crystalline silicate emission features longwards of 20 μm, NGC 6302 also exhibits a number of emission bands in the 5-15 μm region of its spectrum that are usually attributed to C-rich PAH particles (Fig. 10.3, panel 1). This 'dual dust chemistry' phenomenon has also been encountered in the infrared spectra of a number of other post-AGB objects and planetary nebulae, particularly objects having late WC-type central stars (Cohen et al. 2002). An archetypal example of the dual dust chemistry phenomenon is the Red Rectangle bipolar nebula around the post-AGB star HD 44179 (Waters et al. 1998; Fig. 10.6). Longwards of 20 μm, its spectrum is dominated by strong crystalline silicate emission features, whereas shortwards of 15 μm it is dominated by strong PAH emission bands. The current most favoured interpretation of dual dust chemistry sources is that the crystalline silicates reside in a cool shielded dust disk around a binary system, having been captured there following an earlier phase of AGB mass loss; later on, the AGB star evolved from an O-rich to a C-rich chemistry, following

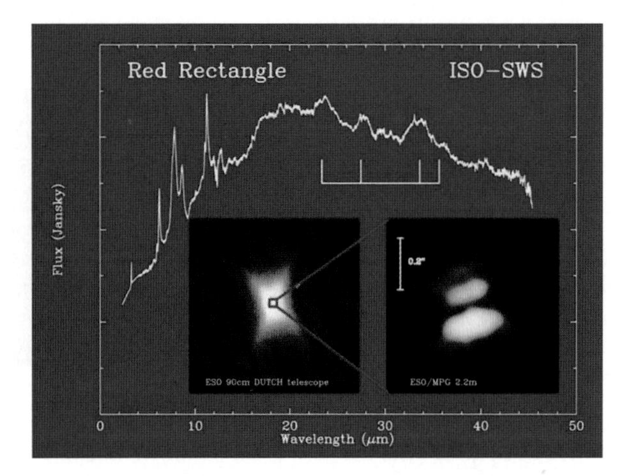

Fig. 10.6 The *ISO* SWS 2.4-45-μm spectrum of HD 44179, the Red Rectangle, illustrating its strong PAH emission features shortwards of 15 μm and its prominent crystalline silicate features longwards of 20 μm. The insets show (*left*) a ground-based optical image of the Red Rectangle, and (*right*) an interferometric optical image of the central region, showing a dark dust lane cutting across the center. Figure courtesy of Frank Molster and Rens Waters

several episodes of the 3rd dredge-up, with newly produced C-rich particles being channeled into polar outflows by the O-rich dust disk.

AFGL 618 and AFGL 2688 are Galactic carbon-rich bipolar post-AGB objects which, unlike the Red Rectangle, do not exhibit dual dust chemistries, with only carbon-rich molecular or dust features having been detected in their spectra. Cernicharo et al. (2001) identified absorption features due to benzene and several other complex hydrocarbon molecules in the *ISO* SWS spectrum of AFGL 618, while Bernard-Salas et al. (2006) have detected the same absorption features in an R = 600 *Spitzer* IRS spectrum of SMP LMC11 (Fig. 10.7), an LMC object that has been classified in the past as a planetary nebula, but whose IR spectrum and *HST* image (Fig. 10.7) indicate it as likely to be transiting between the post-AGB and ionized PN phases. As can be seen in Fig. 10.7, the 14.85-μm benzene absorption feature is even stronger in the spectrum of SMP LMC11 than in the spectrum of AFGL 618.

10.2.2 Potential JWST Studies of Extragalactic Evolved Stars and Planetary Nebulae

SMP LMC11 is an example of a potential target for higher spectral resolution infrared spectroscopy by *SOFIA* (apart from the 13.5–16.5 μm region, which is inaccessible even from airborne altitudes) or by *JWST*-MIRI. However, it is worth bearing in mind that even at the distance of the LMC, SMP LMC11, with a peak flux of ∼0.5 Jy (Fig. 10.7), will only just be observable using MIRI's R = 3000 integral field spectroscopy mode, given the expected 0.5 Jy saturation limit of this

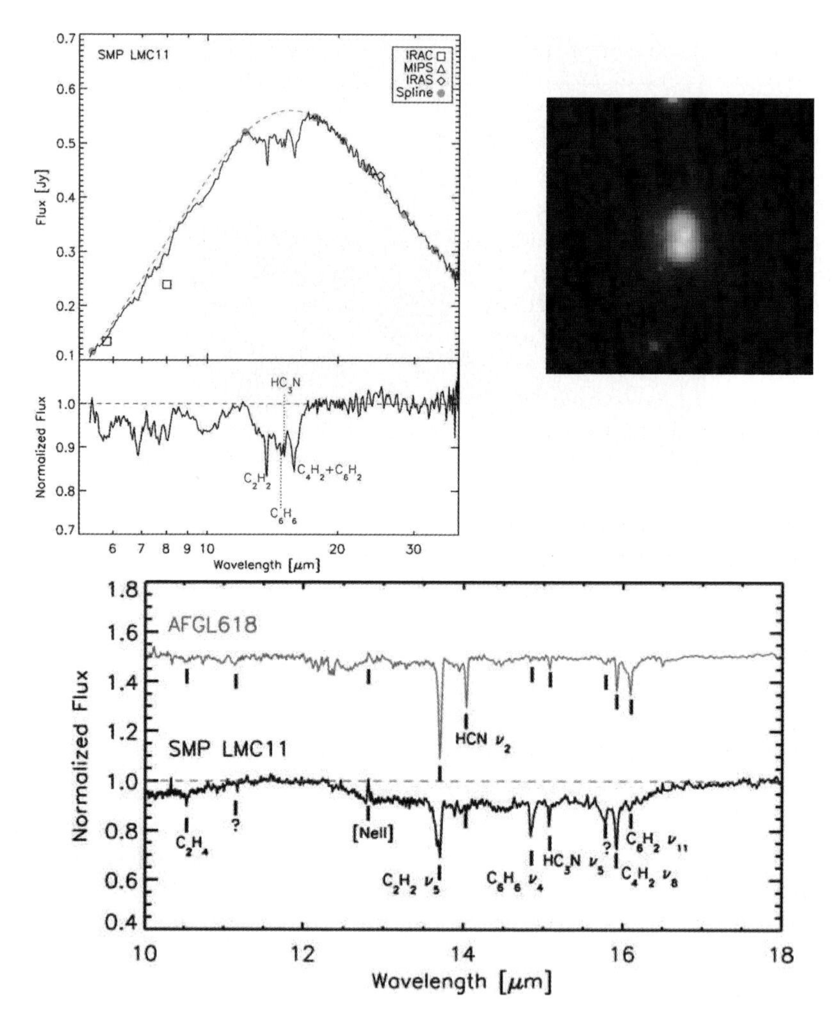

Fig. 10.7 *Top-left*: The *Spitzer*-IRS spectrum of the proto-planetary nebula SMP LMC11 (Bernard-Salas et al. 2006) showing both fluxed and normalized versions, with several hydrocarbon absorption features identified. *Bottom*: Normalized 10-18-μm spectra of AFGL 618 (Cernicharo et al. 2001) and SMP LMC11 (Bernard-Salas et al. 2006), showing absorption features due to benzene and other complex hydrocarbons. *Top-right*: *HST* V-band image of SMP LMC11, from Shaw et al. (2006), showing its bipolar structure

mode. This starkly illustrates the huge sensitivity gains that MIRI will confer relative to previous mid-infrared instrumentation. Photometric studies with MIRI of AGB stars and post-AGB nebulae will generally only be feasible for targets lying at much greater distances than the Magellanic Clouds.

Using *JWST*'s MIRI in its $1.3' \times 1.7'$ imaging mode, SMP LMC11 would be detectable at 10σ in 10^4 sec out to a distance of 18 Mpc at 18 μm; out to a distance of 38 Mpc at 10 μm; and out to a distance of 47 Mpc at 7.5 μm. Objects discovered via imaging photometry could be followed up from 5 to 10 μm with MIRI's long-slit R = 100 spectrometer. SMP LMC11 would yield 10σ per spectral resolution element in 10^4 sec at a distance of 14.5 Mpc. These numbers indicate that in the above modes MIRI will be comfortably capable of studying similar objects out to the Virgo Cluster (D $\sim$ 14 Mpc) and beyond.

A case study: the intergalactic stellar population of the Virgo Cluster In recent years planetary nebulae have become important probes of extragalactic stellar systems. Up to 10% of the total luminosity of a planetary nebula, $\sim$500 $L_\odot$, can be emitted in the dominant cooling line, [O III] λ5007. This, coupled with the narrowness of the line ($\sim$15–25 km s^{-1}), makes it extremely easy to detect PNe in external galaxies using a narrow-band filter tuned to the galaxy redshift.

Arnaboldi et al. (1996) measured velocities for 19 PNe in the outer regions of the giant elliptical galaxy NGC 4406, in the southern Virgo extension region. Although this galaxy has a peculiar radial velocity of -227 km s^{-1}, three of the PNe had velocities close to 1400 km s^{-1}, the mean radial velocity of the Virgo cluster. It was concluded that they were intracluster PNe. Theuns & Warren (1997) discovered ten PN candidates in the Fornax cluster, in fields well away from any Fornax galaxy - consistent with tidal stripping of cluster galaxies. They estimated that intracluster stars could account for up to 40% of all the stars in the Fornax cluster. Méndez et al. (1997) surveyed a 50 arcmin^2 area near the centre of the Virgo cluster, detecting 11 PN candidates. From this, they estimated a total stellar mass of about 4×10^9 $M_\odot$ in their survey area and that such a population could account for up to 50% of the total stellar mass in the Virgo cluster.

Follow-up observations have confirmed large numbers of intergalactic PN candidates in Virgo. However, they represent a tiny tip of a very large iceberg. How to sample more of the iceberg? The discussion above of *JWST* MIRI capabilities indicates that mid-infrared observations of dust-emitting AGB stars and planetary nebulae located in intergalactic regions of the Virgo Cluster are feasible.

Figure 10.8 shows the *Spitzer* mid-infrared spectra of two LMC planetary nebulae. The fluxes measured for the SiC-emitting nebula SMP LMC8 indicate that in 10^4 sec MIRI imaging would yield 10σ detections for a distance of D = 19 Mpc with the 7.7- and 11.3-μm filters and for a distance of 13 Mpc with the 18-μm filter. For the strongly PAH-emitting PN SMP LMC36, in 10^4 sec MIRI imaging would yield a 10σ detection for a distance of 44 Mpc with the 7.7-μm filter and for a distance of 23 Mpc with the 11.3-μm filter. The carbon star MSX SMC 159, whose *Spitzer* IRS spectrum is shown in Fig. 10.9 (from Sloan et al. 2006), would yield 10σ detections in 10^4 sec with MIRI out to a distance of D =29 Mpc with the 5.6-μm filter, to 23 Mpc with the 7.7-μm filter and to 12 Mpc with the 11.3-μm filter. Similar detection limits apply for oxygen-rich AGB stars. As shown by Groenewegen et al. (2007; see Fig. 10.9), mid-infrared colour indices can be used to estimate mass loss rates and thereby mass inputs from stellar populations hosting AGB stars.

As well as MIRI, the other *JWST* instruments will also be useful for such studies, e.g. with the JWST Tunable Filter Imager (R = 100), a carbon star similar to

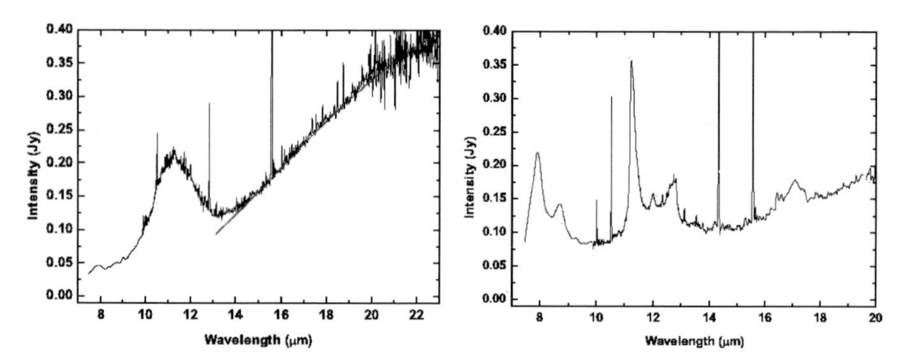

Fig. 10.8 *Spitzer*-IRS spectra of (*left*) SMP LMC8, showing a strong SiC 11.2-μm emission feature and (*right*) SMP LMC36, showing very strong PAH emission bands. Both planetary nebulae also exhibit narrow fine structure ionic emission lines in their spectra

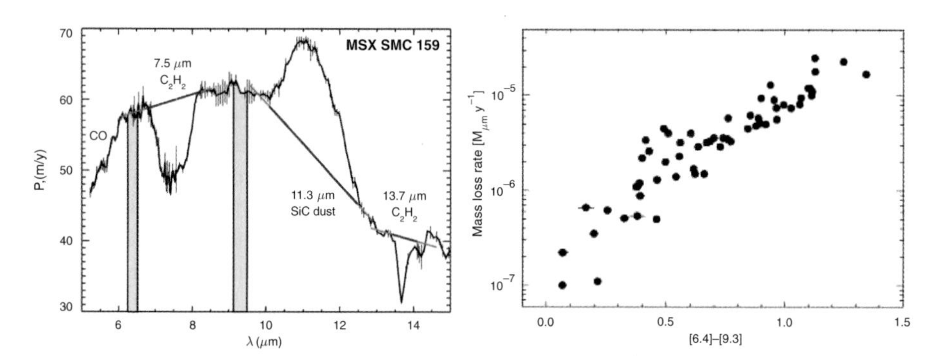

Fig. 10.9 *Left*: the *Spitzer* IRS spectrum of the carbon star MSX SMC 159, illustrating the 11.2-μm SiC dust emission feature, as well as several important molecular absorption features and the narrow 'continuum' regions defined at 6.4 μm and 9.3 μm by Sloan et al. (2006). *Right*: a plot from Groenewegen et al. (2007) of the correlation between the *Spitzer* [6.4]-[9.3] colors of SMC carbon stars and their derived mass loss rates

MSX SMC 159 would give 10σ per resolution element in 10^4 sec for distances out to D = 24-29 Mpc, for wavelengths from 2 to 4 μm. There are at least 20–30 AGB stars for every PN, so there should be many detectable AGB stars in Virgo Cluster fields, allowing the total stellar population to be determined, as well as their gas and dust mass inputs into the intracluster medium.

10.2.3 Exploring New Wavelength Regions at High Spectral Resolution

The 200–650 μm range is the last largely unexplored region of the astronomical spectrum. Many spectral lines and features were observed for the first time in the 45–200 μm region by *ISO*'s Long Wavelength Spectrometer, e.g. Fig. 10.10 shows

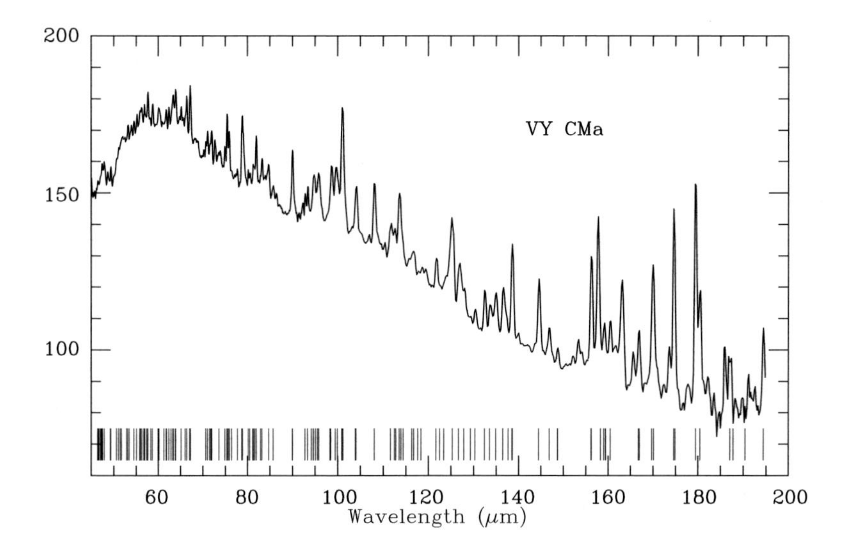

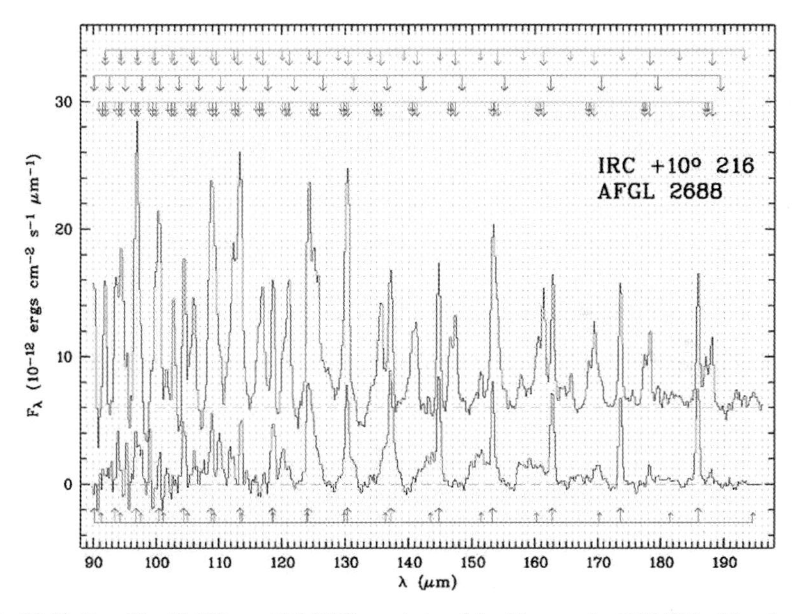

Fig. 10.10 *Top:* The 43-197 μm *ISO* LWS spectrum of the M supergiant VY CMa. The observed flux (F_λ) has been multiplied by λ^4 to show this Rayleigh-Jeans region of the spectrum more clearly. The tick marks at the bottom indicate the wavelengths of some of the ortho- and para-H_2O rotational lines in this spectral region. Over 100 water lines are detected. *Bottom:* The continuum-subtracted 90–197 μm *ISO* LWS spectra of the carbon star IRC+10°216 (*upper spectrum*) and of the C-rich post-AGB object AFGL 2688 (*lower spectrum*). The upward pointing arrows indicate the positions of v=0 and v=1 rotational lines of CO. The downward pointing arrows indicate the positions of rotational lines of HCN and $H^{13}CN$, as well as of vibrationally excited HCN rotational lines. See Cernicharo et al. (1996) and Cox et al. (1996) for more details on the carbon-rich source spectra

the LWS spectra of the oxygen-rich M supergiant VY CMa and the carbon star IRC+10°216. Similar numbers of new lines and features can be expected to be found in the 200–650 μm spectral region. The HIFI, PACS and SPIRE spectrometers onboard ESA's *Herschel Space Observatory* will observe a large number of targets in this wavelength range. HIFI is capable of obtaining complete spectral scans from 157–625 μm at a resolving power of R = 10^6 for a range of archetypal O-rich and C-rich sources, which would allow an unprecedented line inventory to be built up. By measuring line fluxes and profiles with high spectral resolution, HIFI and later ALMA will be able to probe the dynamics of stellar wind outflows and to use a wide range of atomic and molecular species to study the wind chemistry and thermal structure as a function of distance from the central source, with the strong cooling lines of CO, HCN and H_2O being particularly well-suited to probing the thermal and density structures of stellar winds. The superb sensitivity of ALMA will allow detailed studies to be made of objects located throughout the Milky Way and in the Magellanic Clouds. In addition, its high angular resolution will make it very well-suited to probe the dynamical and physical conditions in the complex nebulae found around some AGB stars, post-AGB objects and planetary nebulae.

The *Herschel* SPIRE FTS will obtain spectra at up to R = 1000 for a large number of O-rich and C-rich sources from 200–650 μm, searching particularly for new dust features that may be present. These features may also occur in the spectra of star forming regions and galaxies, but the best place to isolate and identify them is in the spectra of objects with known chemistries, around which they have formed. In addition, the continuum spectral properties of different dust species, particularly their emissivity laws, have yet to be fully characterised in this spectral region.

10.2.4 The 'Missing Mass' Problem for Intermediate Mass Stars

An 'average' planetary nebula has a central star mass of ~0.6 $M_\odot$ and a nebular mass of ~0.3 $M_\odot$. Population modelling predicts a typical main sequence progenitor mass of 1.3 $M_\odot$, so about 0.4 $M_\odot$ appears to have been lost during earlier stages of evolution. A much greater discrepancy exists for intermediate mass progenitors. Populations of white dwarfs have been found in open clusters which have main sequence turn-off masses of 6–8 $M_\odot$ (e.g. in NGC 2516; see Weidemann 2000). So, 5–7 $M_\odot$ must have been lost in order to allow such stars to get below the Chandrasekhar limit, yet the most massive PNe (e.g. NGC 6302, NGC 7027) contain no more than 2 $M_\odot$ of nebular material. So when was the rest of the mass lost (and how do stars that develop degenerate cores know about the Chandrasekhar limit?) There is an obvious need for a comprehensive survey of the mass loss histories of evolved stars.

The most sensitive method to search for ejected material around evolved stars is to image the FIR/submm emission from dust particles in the ejecta that are being heated by the interstellar radiation field to temperatures of 20–30 K, peaking at

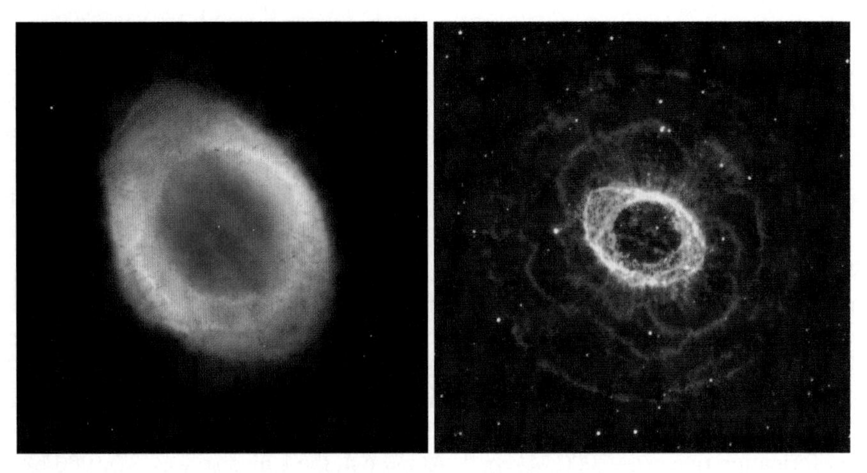

Fig. 10.11 *Left*: An *HST* WFPC2 publicity image of the Ring Nebula, NGC 6720, taken by combining exposures in each of three filters. *Red*: F658N ([N II]), *green*: F501N ([O III]), *Blue*: F469N (He II). The FoV is 2×2 arcmin. *Right*: a publicity image of the Ring Nebula seen observed in the H_2 v=1-0 S(1) 2.122-μm line with the MPA Omega near-IR camera on the Calar Alto 3.5-m telescope. The FoV is 5×5 arcmin. The bright inner emission corresponds to the optical nebula; the outer rosette-shaped filaments correspond to H_2 emission from material that is situated well beyond the optically emitting nebula

far-IR wavelengths. Ionized gas can only be detected following the onset of the planetary nebula phase, while cool rarefied atomic or molecular gas can be extremely difficult to detect except in favorable circumstances (Fig. 10.11 shows the H_2-emitting halo detected around the Ring Nebula). On the other hand, spatially extended dust emission from ejected material is much easier to detect, during any phase of evolution. Very extended dust shells have been detected around a number of AGB stars, in *IRAS* and *ISO* far-infrared images (see e.g. Izumiura et al. 1996, and references therein).

Compared to earlier facilities, the *Herschel Space Observatory* will have a greatly enhanced sensitivity to extended emission at far-infrared and submillimeter wavelengths. Following its launch in 2008/9, one of its Guaranteed Time programmes will carry out mapping observations aimed at detecting and determining the masses of extended dust shells around a wide range of evolved star classes, in order to trace their mass loss histories. Shells produced by past mass loss events over periods of up to 40,000 years are potentially detectable and can yield information on the mass loss process itself, e.g. whether it has been continuous or episodic. Multi-wavelength photometric imaging can yield fluxes, dust temperatures and dust shell masses. The PACS and SPIRE instruments will obtain scanned maps of up to 30×30 arcmin at 70, 110, 250, 350 and 520 μm for a large sample of targets, including AGB stars (O-rich and C-rich), post-AGB objects and PNe. High galactic latitude targets will be favoured, to minimise background confusion. High spectral resolution follow-up observations of the [C II] 158-μm line with *Herschel*-HIFI can provide kinematic information on extended circumstellar shells detected via imaging. Overall, these observations can lead to a better understanding and quantification of the mass loss histories of low- and intermediate-mass stars.

10.3 Dust Production by Massive Stars

Where did the large quantities of dust detected in many high redshift galaxies originate from? Bertoldi et al. (2003) detected redshifted warm dust emission at millimeter wavelengths from three QSOs with $z>6$, i.e. dust had formed less than 1 Gyr after the Big Bang. Dwek et al. (2008) considered the case of the ultraluminous galaxy SDSS J1148+5251, at $z = 6.4$. Its IR luminosity and dust mass were estimated to be 2×10^{13} $L_{\odot}$ and 2×10^{8} $M_{\odot}$, respectively, with its luminosity implying a current star formation rate of ~ 3000 $M_{\odot}$ yr^{-1}. At $z = 6.4$, the Universe was only 900 Myr old; if the galaxy formed at $z = 10$, then it is only 400 Myr old. In fact, given its estimated dynamical mass of 5×10^{10} $M_{\odot}$, its current star formation rate would give an age of only 20 Myr. An age of 20–400 Myr would be insufficient for AGB stars to appear – only massive stars would have had sufficient time to evolve and produce dust. Elvis, Marengo & Karovska (2002) suggested that QSO winds could reach temperatures and pressures similar to those found around cool dust-forming stars and that up to 10^{7} $M_{\odot}$ of dust could be formed. Markwick-Kemper et al. (2007) observed the $z = 0.466$ broad absorption line QSO PG 2112+059 with the *Spitzer* IRS and detected mid-IR emission features which they attributed to amorphous and crystalline formed in the quasar wind. It is not yet clear whether all dust-emitting high-z galaxies possess such AGN central engines. If massive stars should turn out to be the dominant sources of dust in high-z galaxies, which of their evolutionary phases is the dominant dust producer? Is it the late-type supergiant phase, the Luminous Blue Variable phase, the Wolf-Rayet phase, or the final core-collapse supernova event?

10.3.1 Late-Type Supergiants and Hypergiants

The M2 Iab supergiant α Orionis, with a luminosity of $\sim 2.5 \times 10^{5}$ $L_{\odot}$, has been estimated to have a gas mass loss rate of $\sim 1.5 \times 10^{-5}$ $M_{\odot}$ yr^{-1} (Jura & Morris 1981). More luminous late type supergiants (sometimes dubbed ‘hypergiants’) can have even higher mass loss rates and are often self-obscured by their own circumstellar dust at optical wavelengths, e.g. the red hypergiants VY CMa (Fig. 10.10), VX Sgr and NML Cyg. Similarly high mass loss rates can be exhibited by yellow hypergiants, e.g. ρ Cas, IRC+10 420 and HR 8752. Such objects could potentially make a very significant contribution to the dust enrichment of galaxies but currently we do not know the duration of the yellow/red hypergiant phase, nor the total population of such objects in our galaxy. Better statistics from current optical and near-IR surveys, together with distances from Gaia and more precise mass loss rate determinations using improved wind modelling techniques, should lead to a much improved understanding of the contribution of these objects to the dust and gas evolution of galaxies.

10.3.2 Luminous Blue Variables and Wolf-Rayet Stars

The ejecta nebulae around Luminous Blue Variables (LBVs) can contain large masses of dust, as in the cases of η Car and AG Car (though not in the case of P Cygni); the 1840's outburst of η Car has been estimated to have produced 0.2 $M_{\odot}$ of dust (Morris et al. 1999). The M 1-67 ejecta nebula around the WN8 Wolf-Rayet star WR124 (Fig. 10.12) also contains large quantities of dust and is thought to have originated from the outburst of an LBV precursor.

The dust in most LBV nebulae appears to be dominated by oxygen-rich silicate grains. Massive carbon-rich WC9 Wolf-Rayet stars often show hot (~900 K) featureless dust emission that has been attributed to carbon grains. How does dust form in outflows from stars whose effective temperatures are in the region of 30,000 K? The answer was found by Tuthill et al. whose masked-aperture Keck imaging at 2.27-μm revealed a rotating pinwheel plume of dust emission in the WC9+OB binary system WR104 (Fig. 10.13). The dust appears to be formed in the compressed shock interaction region (hotspot) between the stellar winds of the WC9 primary and the OB secondary, the relative motions of the two stars creating an Archimedean spiral. Similar pinwheel structures have since been found around several more WC9

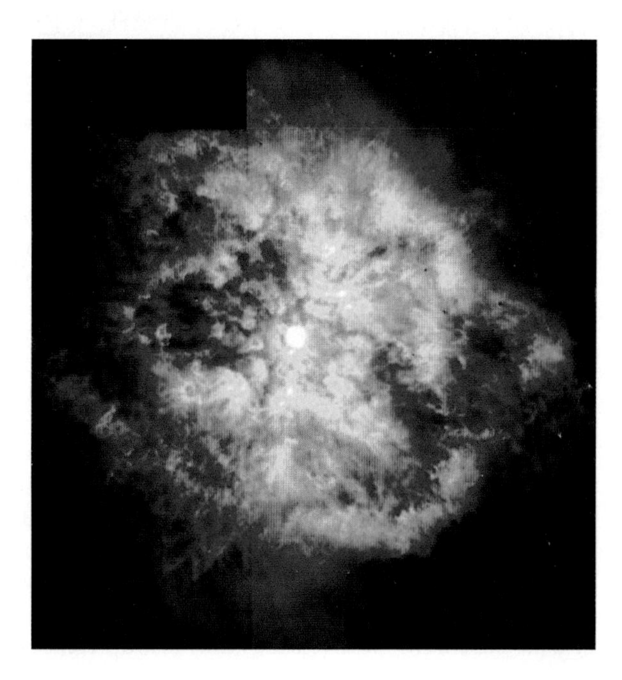

Fig. 10.12 *HST* WFPC2 F656N image of the nebula M 1-67 around the WN8 Wolf-Rayet star WR124 (=BAC 209). For further details, see Grosdidier et al. (1998)

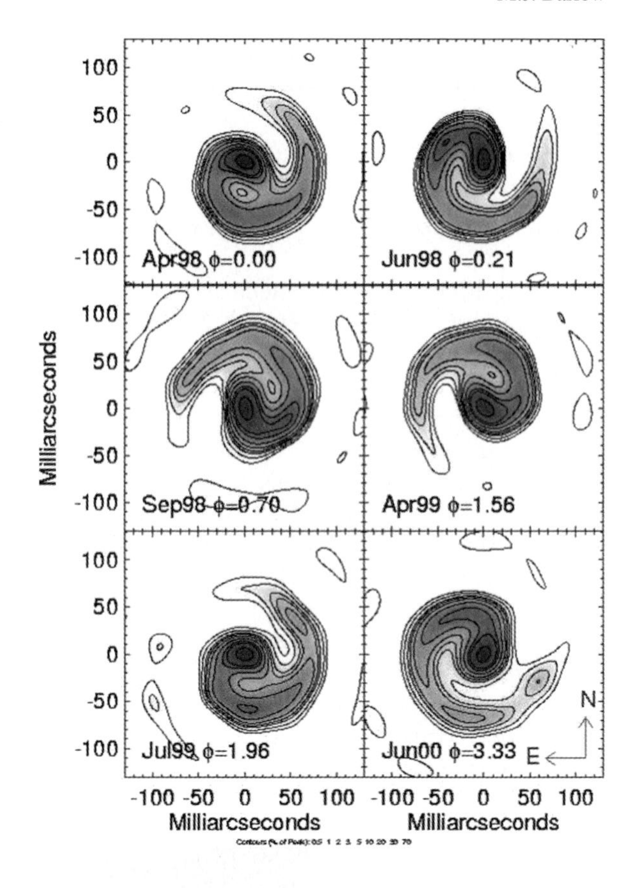

Fig. 10.13 Keck aperture-mask 2.27-μm imaging of a rotating pinwheel plume of dust emission from the WC9+OB binary WR 104. The dust is formed in the compressed shock interaction region (hotspot) between the two stellar winds. The relative motions of the two stars creates an Archimedean spiral. From Tuthill et al. (2002)

systems, including two located in the Galactic Center Quintuplet Cluster (Tuthill et al. 2006). The overall contribution of late WC-type Wolf-Rayet stars to the dust enrichment of galaxies is currently extremely uncertain. Future submm observations by ALMA can help to quantify their overall dust production rates, plus search for molecular emission from the hot-spot shocked wind compression region, while the GAIA parallax survey of the Galaxy should enable their total numbers to be more accurately estimated.

10.3.3 Core Collapse Supernovae

There is plenty of evidence that supernovae can synthesise dust particles that are able to survive the shock buffeting that must take place following their formation. The isotopic analysis of pre-solar meteoritic grain inclusions (those that have non-solar ratios) has found many examples that are dominated by r-process isotopes, indicating a supernova origin (e.g. Clayton et al. 1997). The onion-skin abundance structure of a pre-supernova massive star means that different layers of the ejecta can have

C/O<1 and C/O>1, allowing the formation of O-rich grains and C-rich grains in the respective zones. Infrared photometry and spectrophotometry of SN 1987A demonstrated the onset of thermal dust emission by day 615 (Bouchet & Danziger 1993; Wooden et al. 1993), as shown in Fig. 10.14, the emission being attributed to newly formed dust particles in the ejecta.

From dust nucleation modelling, Todini & Ferrara (2001) predicted that 0.08-1.0 $M_{\odot}$ of dust could condense in the ejecta of a typical high-redshift core collapse supernova within a few years of outburst, corresponding to a condensation efficiency for the available refractory elements of > 0.2. Similarly high condensation efficiencies appear to be required to explain the $\sim 10^8$ solar masses of dust deduced to exist in high redshift QSOs (Morgan & Edmunds 2003; Dwek et al. 2008). However, prior to the launch of *Spitzer*, for the handful of recent core-collapse SNe for which dust formation had been inferred, the derived masses of newly formed dust were $\leq 10^{-3}$ $M_{\odot}$ (e.g. the examples shown in Figs. 10.14 and 10.15)

A different approach to estimating how much dust can be formed in the ejecta of a core-collapse SN was taken Dunne et al. (2003) and Morgan (2003), who used SCUBA submillimeter maps to deduce that $1 - 2$ $M_{\odot}$ of dust were present in the Cas A and Kepler supernova remnants (SNRs), both of which were less than 400 years old. However, Krause et al. (2004) argued that most of the observed submm emission observed from Cas A originated from a foreground molecular cloud that could be seen in CO maps. Deeper observations of more young SNRs across a wide wavelength range are planned in *Herschel Space Observatory* Guaranteed Time; 57–650 μm PACS and SPIRE photometric and spectroscopic maps will be obtained of five galactic SNRs having ages of less than 1000 yrs (Cas A, Kepler, Tycho, Crab and 3C58)

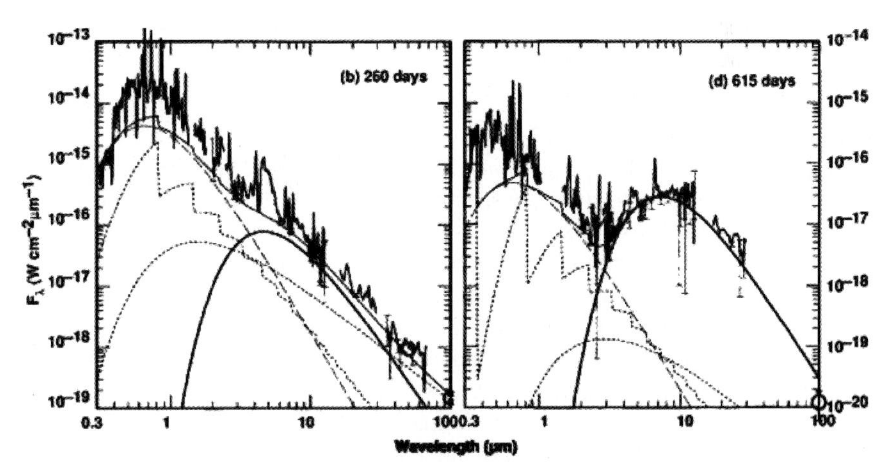

Fig. 10.14 Optical and KAO IR spectrophotometry of SN 1987A, illustrating the definite onset of thermal dust emission by day 615. From Wooden et al. (1993), who derived a lower limit of 10^{-4} $M_{\odot}$ to the mass of dust that had formed by day 775

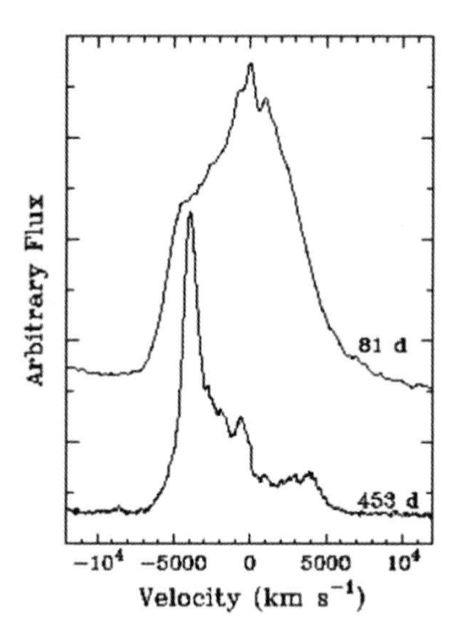

Fig. 10.15 The Hα line profile of SN 1998S at days 81 and 453 (Leonard et al. 2000), illustrating the loss of flux on the red side of the profile, attributed to absorption by newly formed dust preferentially removing photons from the far side of the ejecta. Pozzo et al. (2004) estimated that 10^{-3} $M_{\odot}$ of dust had formed

Returning to observations of the dust formation phase itself (t <1000 days), there are currently three methods for inferring the formation of dust in supernova ejecta:

1. via the detection of thermal IR emission from the newly formed dust. However, the use of this method alone can be compromised by pre-existing nearby dust (e.g. circumstellar dust), which can be heated by the supernova light flash.
2. via the detection of a dip in the SN light curve that can be attributed to extinction by newly formed dust. Pre-existing dust cannot produce such a dip.
3. via the detection of the development of a red-blue asymmetry in the SN emission line profiles, attributable to the removal by newly formed dust of some of the redshifted emission from the far side of the SN ejecta (Lucy et al. 1989; see Fig. 10.15).

Method (1) is normally required if dust masses are to be quantified but ideally it ought to be supported by one or both of (2) and (3).

Since the launch of the *Spitzer Space Telescope*, two teams have been conducting observing programmes to study the spectral evolution of young supernovae. Examples of the spectra of SNe less than 250 days after outburst are shown in Fig. 10.16. The day 135 IRS spectrum of SN 2005df (Gerardy et al. 2007) shows numerous ionic fine structure lines, including lines from radioactive cobalt and nickel, whose decay is the main heating source for the ejecta. The day 214 IRS spectrum of SN 2005af (Kotak et al. 2006) is dominated by the $v = 1 - 0$ band-head of gaseous SiO and is very similar to that of SN 1987A at a similar epoch.

Early in the *Spitzer* mission, dust emission from the Type IIP supernova SN 2002hh was detected in SINGS IRAC images and confirmed by higher angular resolution Gemini Michelle imaging (Barlow et al. 2005). The day 600 spectral energy distribution (SED) could be fitted by a 290 K blackbody (Fig. 10.17, upper), yielding a minimum emitting radius of $R_{min} \sim 10^{17}$ cm and a luminosity of $L = 1.6 \times 10^7$ $L_{\odot}$. A

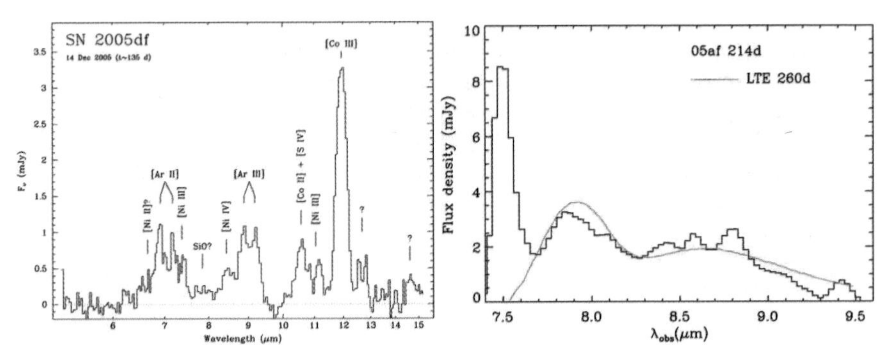

Fig. 10.16 *Left*: The *Spitzer* IRS spectrum of SN 2005df at day 135, showing ionic fine structure emission lines from radioactive cobalt and nickel (Gerardy et al. 2007). *Right*: The *Spitzer* IRS spectrum of SN 2005af at day 214, showing the 8-μm v=1-0 bandhead of gaseous SiO (Kotak et al. 2006). By analogy to SN 1987A, this was suggested to be a precursor to dust formation

more realistic λ^{-1} grain emissivity gave $R_{min} = 5\times10^{17}$ cm. This was far too large for the emitting dust to have formed in the ejecta (it would have taken $>$ 10 yrs for material in the ejecta to reach this radius). It was therefore inferred that the emitting dust must have been pre-existing. All three of the model fits shown in Fig. 10.17 yielded total emitting dust masses in the range 0.10–0.15 $M_\odot$. SN 2002hh is a Type IIP (plateau) supernova, whose very extended optical light curve (Welch et al. 2007) appears explicable in terms of a just-resolved light echo that has been revealed by *HST* ACS/HRC images (Sugerman 2005). Preliminary analysis indicates that the echo has occurred from a thick dust distribution that is located about 2–8 light years ($2 - 8 \times 10^{18}$ cm) in front of the supernova.

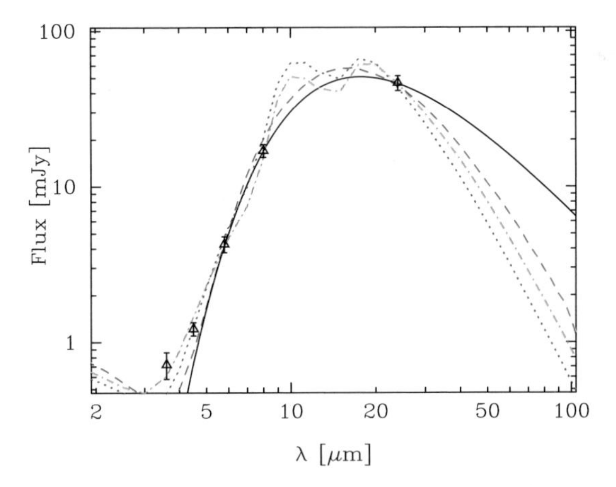

Fig. 10.17 The measured day-600 *Spitzer* 3.6 – 24 μm fluxes for SN 2002hh, in NGC 6946, are shown as open triangles, with vertical bars indicating the flux uncertainties. The *solid black line* is a 290-K blackbody normalised to the 8.0-μm flux point. The *dashed, dotted* and *dash-dotted lines* correspond to radiative transfer models with differing amounts of silicates and amorphous carbon; see Barlow et al. (2005)

Fig. 10.18 A *Spitzer* SINGS multi-band IRAC image of NGC 628. There is a clear detection of SN 2003gd on day-499 (*upper inset*) relative to day-670 (*lower inset*). See Sugerman et al. (2006)

Sugerman et al. (2006) detected the onset of dust emission from SN 2003gd in NGC 628 (Messier 74; see Fig. 10.18); unlike SN 2002hh, the emitting dust was inferred to have formed inside the supernova ejecta. Between days 157 and 493 the Hα feature developed an asymmetric profile, with a reduction in flux on the red side. This was attributed to dust forming in the ejecta preferentially extinguishing emission from receding (red-shifted) gas. There was also an increase in optical extinction after day 500, as evidenced by a dip in its light curve from that date, similar to the behaviour of SN 1987A at the same epochs. Additional extinction by dust was inferred to have occurred after day 500 for both SNe, and for SN 2003gd corresponded to 0.25–0.5 magnitudes in the R-band on day 500, and to 0.8–1.9 magnitudes on day 678.

Both smooth and clumped SN ejecta model fits to the SN 2003gd observations were presented by Sugerman et al. (2006), using a 3D Monte Carlo radiative transfer code with a mother-grid of 61^3 cells (the mother cells that contained clumps were resolved by a subgrid of 5^3 cells). To match both the optical-IR SEDs and the derived R-band extinction estimates for the SN ejecta, their day 499 data could be fitted with smoothly distributed dust having a dust mass of 2×10^{-4} $M_{\odot}$, whereas up to 2×10^{-3} $M_{\odot}$ of dust could be accommodated by a clumped dust model. For day 678, their best-fit smooth dust model required 3×10^{-3} $M_{\odot}$ of dust, while up to 2×10^{-2} $M_{\odot}$ could be accommodated by a clumpy model, the latter implying a heavy element condensation efficiency of about 10%. The *Spitzer* observations of SN 2003gd have also been studied by Meikle et al. (2007).

Table 10.1 Thermal infrared studies of core collapse supernovae – results so far

Name	D(Mpc)	Progenitor mass, $M_{\odot}$	Dust emission?
SN 1987A	0.05	16–22	Yes
SN 1999bw	14.5	unknown	Yes, but very late
SN 2002hh	5.6	8–14	Yes, but pre-existing
SN 2003gd	9.3	6–12	Yes
SN 2004dj	3.3	12–15	Maybe; pre-existing?
SN 2004et	5.6	13–20	Yes
SN 2005cs	8.0	7–12	No

Table 10.1 summarises the results to date of mid-infrared searches for dust emission from young supernovae. Apart from SN 1987A, all are based on *Spitzer* observations, supplemented in two cases by Gemini North Michelle observations. Although *Spitzer* has delivered a very large increase in mid-infrared sensitivity relative to prior facilities, its distance limit for the detection of dust forming around young SNe is effectively 10-15 Mpc, corresponding to a volume within which relatively few new SNe occur each year. All of the SNe listed in Table 10.1 are of Type II, from progenitors with masses $< 20\ M_{\odot}$. No Type Ib or Ic SNe, whose immediate precursors are believed to be H-deficient Wolf-Rayet stars descended from much more massive stars, have so far been close enough to be detected by *Spitzer*, although the Type Ib SN 2006jc has shown evidence for dust formation via the development of red-blue emission line asymmetries and a transient far-red and near-IR continuum excess, interpreted by Smith et al. (2008) as due to dust formation in the dense region created by the impact between the SN ejecta and slower moving material from an LBV-type eruption that was discovered two years before the supernova outburst.

For point source imaging, *Spitzer*'s IRAC is 135× more sensitive at 8 μm than mid-IR instruments on ground-based 8-m telescopes (due to *Spitzer*'s vastly lower thermal backgrounds). For point source imaging, *JWST*-MIRI is expected to be $\sim$ 40× more sensitive at 8 μm than IRAC (see www.stsci.edu/jwst/science/sensitivity/). In addition to these sensitivity gains, the 8× higher angular resolution of *JWST*-MIRI compared to *Spitzer*-IRAC will greatly reduce point source confusion effects in dense starfields, such as encountered when observing galaxies.

The angular resolution advantages of a >6-m class telescope compared to a 0.85-m telescope are illustrated in Fig. 10.19, where an IRAC 8-μm image of the field of SN 2002h, in NGC 6946, is compared to a 8× higher resolution Gemini Michelle 11-μm image of the same field. The five sources that are blended together in the IRAC image are completely resolved from each other in the Gemini image. MIRI's much greater sensitivity and angular resolution should enable SNe out to nearly 200 Mpc to be detected at mid-IR wavelengths, corresponding to a volume $\sim$1000 times larger than for *Spitzer*. So MIRI will be able to quickly observe large numbers of new SNe of all classes, both photometrically and spectroscopically, enabling the dust contribution by each class to be accurately assessed.

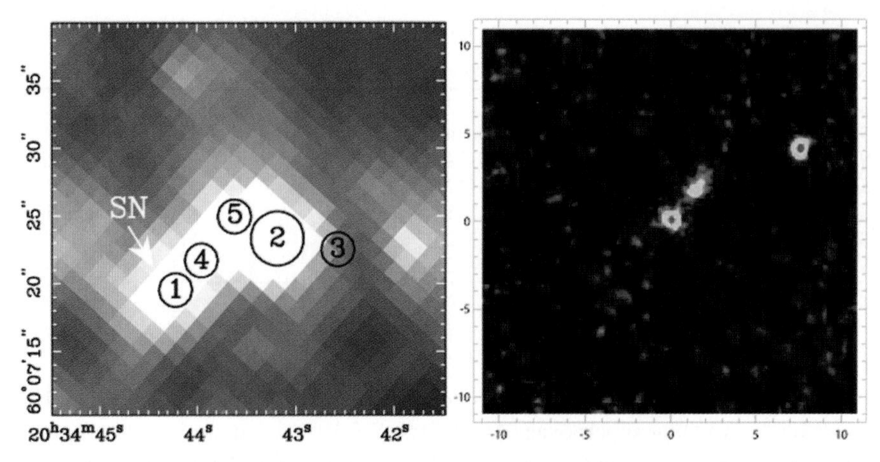

Fig. 10.19 *Left*: Day-590 *Spitzer* IRAC 8-μm image of the region around SN 2002hh. With *Spitzer's* 2.4 arsec angular resolution, five different sources are blended together. Right: the 0.3 arsec angular resolution of this day-698 Gemini-N Michelle 11-μm image completely resolves the sources (the SN is at the center), illustrating how in the case of crowded-field galaxy observations the high angular resolution of JWST-MIRI will supplement its raw sensitivity

Acknowledgments I would like to thank Christoffel Waelkens and Xander Tielens for their comments and suggestions.

References

Arnaboldi, M., et al.: ApJ, **472**, 145 (1996)
Barlow, M. J., et al.: ApJ, **627**, L113 (2005)
Benjamin, R. A., Churchwell, E. B., et al.: PASP, **115**, 953 (2003)
Bernard-Salas, J., Peeters, E., Sloan, G. C., Cami, J., Guiles, S., Houck, J. R.: ApJ, **652**, L29 (2006)
Bertoldi, F., Carilli, C. L., Cox, P., Fan, X., Strauss, M. A., Beelen, A., Omont, A., Zylka, R.: A&A, **406**, L55 (2003)
Bouchet, P., Danziger, I. J.: A&A, **273**, 451 (1993)
Bowey, J. E., et al.: MNRAS, **331**, L1 (2002)
Bregman, J. N., Temi, P., Bregman, J. D.: ApJ, **647**, 265 (2006)
Carey, S. J.; Noriega-Crespo, A.; Price, S. D., et al.: BAAS, **207** 63.33 (2005)
Cernicharo, J., et al.: A&A, **315**, L201 (1996)
Cernicharo, J., Heras, A. M., Tielens, A. G. G. M.., Pardo, J. R., Herpin, F., Gulin, M., Waters, L. B. F. M.: ApJ, **546**, L123 (2001)
Clayton, D. D., Amari, S., Zinner, E.: Ap&SS, **251**, 355 (1997)
Cohen, M., Barlow, M. J., Liu, X.-W., Jones, A. F.: MNRAS, **332**, 879 (2002)
Cox, P., et al.: A&A, **315**, L265 (1996)
Drew, J. E., et al.: MNRAS, **362**, 753 (2005)
Dunne, L., Eales, S., Ivison, R., Morgan, H., Edmunds, M.: Nature, **424**, 285 (2003)
Dwek, E., Galliano, F., Jones, A.P.: in A Century of Cosmology: Past, Present and Future, II Nuovo Cimento, 122, 959 (2008)
Elvis, M., Marengo, M., Karovska, M.: ApJ, **567**, L107 (2002)
Gerardy, C. L., et al.: ApJ, **661**, 995 (2007)
Groenewegen, M. A. T., et al.: MNRAS, **376**, 313 (2007)

Grosdidier, Y., Moffat, A. F. J., Joncas, G., Acker, A.: ApJ, **506**, L127 (1998)
Iben, I. Jr., Renzini, A.: ARAA, **21**, 271 (1983)
Izumiura, H., Hashimoto, O., Kawara, K., Yamamura, I., Waters, L. B. F. M.: A&A, **315**, L221 (1996)
Jura, M., Morris, M.: ApJ, **251**, 181 (1981)
Kemper, F., Jaeger, C., Waters, L. B. F. M., Henning, Th., Molster, F. J., Barlow, M. J., Lim, T., de Koter, A.: Nature, **415**, 295 (2002)
Kotak, R., et al.: 2006, ApJ, **651**, L117 (2006)
Krause, O., Birkmann, S. M., Rieke, G. H., Lemke, D., Klaas, U., Hines, D. C., Gordon K. D.: Nature, **432**, 596 (2004)
Lawrence, A., Warren, S. J., et al.: MNRAS, **379**, 1599 (2007)
Leonard, D. C., Filippenko, A. V., Barth, A. J., Matheson, T.: ApJ, **536**, 239 (2000)
Lucy, L. B., Danziger, I. J., Gouiffes, C., Bouchet, P.: in IAU Colloq. 120, Structure and Dynamics of the Interstellar Medium, ed. G. Tenorio-Tagle, M. Moles, J. Melnick, Springer-Verlag, 164 (1989)
Markwick-Kemper, F., Gallagher, S. C., Hines, D. C., Bouwman, J., ApJ, **668**, L107 (2007)
Matsuura, M., Zijlstra, A. A., Molster, F. J., Waters, L. B. F. M., Nomura, H., Sahai, R., Hoare, M. G.: MNRAS, **359**, 383 (2005)
Meikle, W. P. S., et al.: ApJ, **665**, 608 (2007)
Meixner, M., et al.: AJ, **132**, 2268 (2006)
Méndez, R. H., et al.: ApJ, **491**, L23 (1997)
Molster, F. J., Lim, T. L., Sylvester, R. J., Waters, L. B. F. M., Barlow, M. J., Beintema, D. A., Cohen, M., Cox, P., Schmitt, B. L.: A&A, **372**, 165 (2001)
Morgan, H. L., Edmunds, M. G.: MNRAS, **343**, 427 (2003)
Morgan, H. L., Dunne, L., Eales, S. A., Ivison, R. J., Edmunds, M. G.: 2004, ApJ, **597**, L33 (2003)
Morris, P. W., et al.: Nature, **402**, 502 (1999)
Pozzo, M., Meikle, W. P. S., Fassia, A., Geballe, T., Lundqvist, P., Chugai, N. N., Sollerman, J.: MNRAS, **352**, 457 (2004)
Shaw, R.A., Stanghellini, L., Villaver, E., Mutchler, M., 2006, ApJS, 167, 201
Sloan, G. C., Kraemer, K. E., Matsuura, M., Wood, P. R., Price, S. D., Egan, M. P.: ApJ, **645**, 1118 (2006)
Smith, N., Foley, R. J., Filippenko, A. V.: ApJ, **680**, 568 (2008)
Sugerman, B. E. K., ApJ, **632**, L17 (2005)
Sugerman, B. E. K., et al.: Science, **313**, 196 (2006)
Sylvester, R. J., Kemper, F., Barlow, M. J., de Jong, T., Waters, L. B. F. M., Tielens, A. G. G. M., Omont, A.: A&A, **352**, 587 (1999)
Theuns, T., Warren, S. J.: MNRAS, **284**, L11 (1997)
Tielens, A. G. G. M., Waters, L. B. F. M., Bernatowicz, T. J.: in 'Chondrites and the Protoplanetary Disk', ASP Conference Series, vol. 341, 605 (2005)
Todini P., Ferrara A.: MNRAS, **325**, 726 (2001)
Tuthill, P. G., Monnier, J. D., Danchi, W. C.: in 'Interacting winds from massive stars', ASP Conf. Series, **260**, 321 (2002)
Tuthill, P., Monnier, J., Tanner, A., Figer, D., Ghez, A., Danchi, W.: Science, **313**, 935 (2006)
Waters, L. B. F. M., et al.: Nature, **391**, 868 (1998)
Weidemann, V.: A&A, **363**, 647 (2000)
Welch, D. L., Clayton, G. C., Campbell, A., Barlow, M. J., Sugerman, B. E. K., Meixner, M., Bank, S. H. R.: ApJ, **669**, 525 (2007)
Wooden, D. H., Rank, D. M., Bregman, J. D., Witteborn, F. C., Tielens, A. G. G. M., Cohen, M., Pinto, P. A., Axelrod, T. S.: ApJS, **88**, 477 (1993)

After a long day in the conference, Eric Smith and Rowin Meijerink start the evening with a healthy dinner

Chapter 11
Origin and Evolution of the Interstellar Medium

A.G.G.M. Tielens

Abstract The evolution of the interstellar medium is driven by a number of complex processes which are deeply interwoven, including mass accretion from nearby (dwarf) systems and the intergalactic medium, mass ejection into the halo and interglactic medium, stellar mass injection into the interstellar medium, star formation, mechanical energy input by stellar winds and supernova explosions, and radiative energy input. These processes are mediated by dust and molecules in an only partially understood way. This complex feedback between stars and the medium they are formed in drives the evolution of galaxies and their observational characteristics. This review describes our understanding of the synergetic interaction of these processes and culminates in a set of key questions.

JWST is set to expand studies of the global interstellar medium to the far reaches of the universe and the earliest times. Yet, our understanding of what these observations tell us about what really happens at those epochs will depend very much on our understanding of the microscopic physical and chemical processes and their dependence on the local conditions. These are best studied in the local universe. In order to reap the full benefits of JWST, a concerted program of key observations is required involving not only JWST but also, and in particular, SOFIA and Herschel. This is illustrated by a personal selection of key program.

11.1 Introduction

The origin and evolution of galaxies are closely tied to the cyclic processes in which stars eject gas and dust into the ISM, while at the same time gas and dust clouds in the ISM collapse gravitationally to form stars. The ISM is the birthplace of stars, but stars regulate the structure of the gas, and therefore influence the star formation rate. Winds from low mass stars – hence, the past star formation rate – control the total mass balance of interstellar gas and contribute substantially to the injection of

A.G.G.M. Tielens (✉)
MS 245-3, Space Sciences Division, NASA Ames Research Center, Moffett Field, CA 94035, USA
e-mail: Alexander.G.Tielens@nasa.gov

H.A. Thronson et al. (eds.), *Astrophysics in the Next Decade,* Astrophysics and Space Science Proceedings, DOI 10.1007/978-1-4020-9457-6_11,

dust, an important opacity source, and Polycyclic Aromatic Hydrocarbon molecules (PAHs), an important heating agent of interstellar gas. High mass stars (i.e., the present star formation rate) dominate the mechanical energy injection into the ISM, through stellar winds and supernova explosions, and thus the turbulent pressure which helps support clouds against galactic- and self-gravity. Through the formation of the hot coronal phase, massive stars regulate the thermal pressure as well. Massive stars also control the FUV photon energy budget and the cosmic ray flux, which are important heating, ionization, and dissociation sources of the interstellar gas, and they are also the source of intermediate mass elements which play an important role in interstellar dust. Eventually, it is the dust opacity which allows molecule formation and survival. The enhanced cooling by molecules is crucial in the onset of gravitational instability of molecular clouds.

Clearly, therefore, there is a complex feedback between stars and the ISM. And it is this feedback that determines the structure, composition, chemical evolution, and observational characteristics of the interstellar medium in the Milky Way and in other galaxies all the way back to the first stars and galaxies that formed at $z > 5$. If we want to understand this interaction, we have to understand the fundamental physical processes that link interstellar gas to the mechanical and FUV photon energy inputs from stars.

11.2 The ISM

11.2.1 Interstellar Gas

Figure 11.1 illustrates the circulation of mass in a galaxy. Stars form from gas in the inner star forming disk of the galaxy. Winds from massive stars as well as supernova explosions continuously stir up the interstellar medium. This (localized) mechanical energy input into the ISM heats the gas to high temperatures creating bubbles of hot gas. The concerted efforts of the many massive stars in an OB association will create superbubbles which may blow out of the disk and vent their hot material into the halo. Slow cooling of this gas then leads to precipitation and rain out of the halo gas onto the disk. This 'weather' pattern – known as the galactic fountain – puffs up the gas disk and transports material, including dust and PAHs, to great heights above the plane (Bregman 1980, Norman & Ikeuchi 1989). Gas also slowly flows inwards in the disk, fuelling starformation activity. On cosmological timescales, much of this inward flow may however be driven by gravitational interaction with nearby galaxies funneling material into the inner regions and leading to intense starburst activity. Mergers are more extreme examples of this interaction where the actual collision of clouds in the interacting gas disks trigger widespread star formation as well as fuel the central engines. These extreme starburst and black hole activity may eject material out into intergalactic space, enriching it with metals and dust. Conversely, galaxies accrete gas from intergalactic space. Figure 11.2 shows a local, modest version of this interaction process and ilustrates how gas, PAHs, and dust are

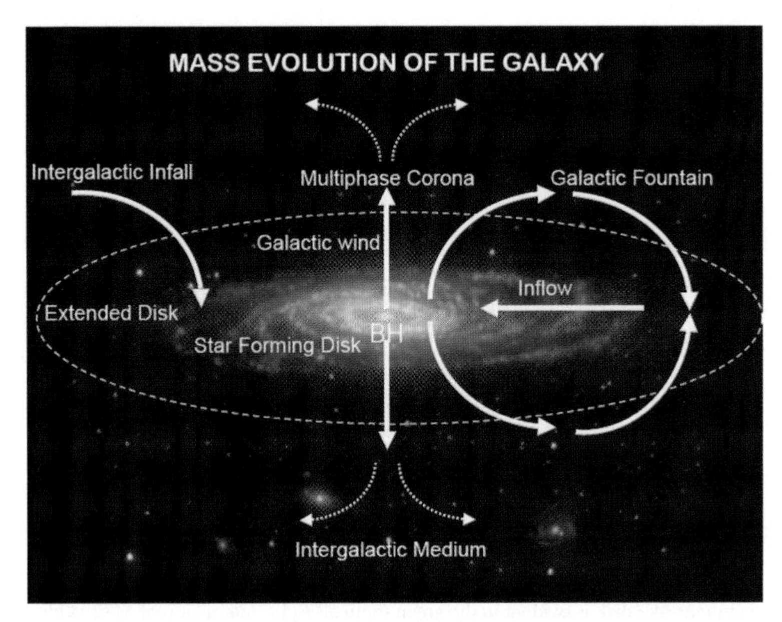

Fig. 11.1 The mass cycles driving the formation and evolution of galaxies. The concerted efforts of stellar winds and supernova activity will shock and heat interstellar gas and create superbubbles in the ISM which may vent much of their hot material into the halo. Extreme starbursts as well as black hole activity can even eject material from the disk into the intergalactic medium. Conversely, the gas disk of the galaxy is fed by infall from intergalactic space and by gas raining down from the halo. Gas will also slowly flow inwards from an extended disk and fuel star formation as well as the activity of the central black hole. These circulation patterns are schematically outlined by arrows

transported to great heights above the plane by the starburst activity in the nucleus of the irregular galaxy, M82 (Engelbracht et al. 2005). Such high lattitude dust and PAHs are very common (Irwin et al. 2007, Thompson et al. 2004, Howk & Savage 1997, Burgdorf et al. 2007).

The mass budget of the ISM is summarized in Fig. 11.3 (Tielens 2005). Stars pollute the Milky Way with about 2 $M_{\odot}$/yr of gas and associated dust and molecules. The gas return is dominated by the numerous low mass stars during their Asymptotic Giant Branch phase. Supernovae from massive stellar progenitors contribute only some 10% of the gas mass but a comparable amount of heavy elements, reflecting their factor 10 enrichment by nucleosynthesis. The timescale for stars to replenish the local interstellar gas mass is $\simeq 5 \times 10^9$ years. This is very comparable to the star formation rate which converts the available gas mass into stars on a timescale of $\simeq 3 \times 10^9$ years. The galactic fountain, setting up the circulation between the halo and the disk, involves more gas (e.g., $\simeq 5$ $M_{\odot}$/yr). The Milky Way grows due to accretion of nearby dwarf galaxies – the Magellanic stream provides a prime example – but this cannibalism is only of minor importance for the overall mass balance. The G-star problem indicates that the Milky Way may have been accreting some 1 $M_{\odot}$/yr of metal-poor intergalactic gas over much of its history. Finally, the gas disk

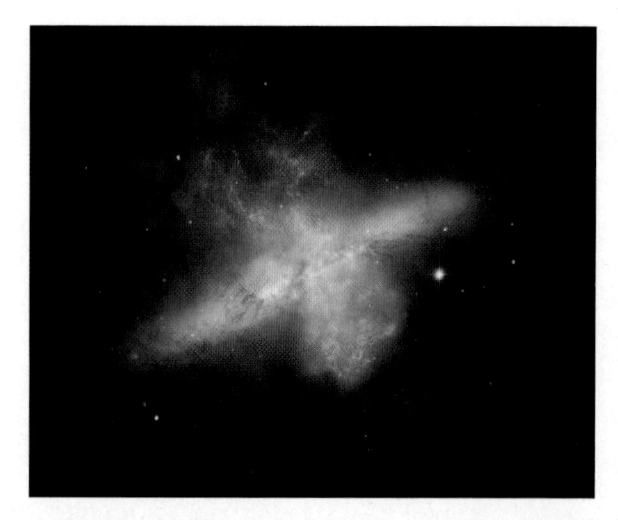

Fig. 11.2 A false color image of the irregular galaxy, M82, and its superwind. The hot gas traced by X-ray data obtained by Chandra is shown in blue. The filamentary ionized gas is traced by Hα and is shown in orange. PAH emission at 8 μm observed by IRAC on Spitzer is shown in red. The stellar, cigar-like disk is evident in the green (optical) light. Interaction of M82, with its more massive neighbor, M81, has ignited a moderate starburst in its center converting about 1 $M_\odot$/yr in the nucleus into stars. The many supernovae explosions associated with this nuclear starburst have created a superwind which has transported material to some 8 kpc above the plane

is very extended and this gas may slowly flow inwards replenishing gas lost to star formation in the inner disk. Unfortunately, existing evidence for radial motion may actually reflect non-axisymmetric distribution of gas in space or velocity and limits on inflow are very forgiving ($\simeq$5 km/s; Blitz private communication). Figure 11.3 adopts 1 km/s. It should be recognized that the magnitude of the various contributions to this mass budget are uncertain. Nevertheless, all of the processes involved are relatively rapid compared to the lifetime of the Milky Way and hence a quasi-steady state has been established, balancing stellar mass injection and intergalactic mass accretion with star formation and mass loss. It should be recognized, though, that some of these processes are highly punctuated, driven by temporal interaction with nearby systems, and steady state may only apply averaged over long timespans.

11.2.2 Interstellar Dust

The characteristics of circumstellar dust and the contribution of different stardust birthsites are reviewed elsewhere in this volume (Barlow 2008). Table 11.1 summarizes the various compound identified either through IR spectroscopy of circumstellar or interstellar dust or through studies of stardust recovered from meteorites. Astronomical identifications which are particularly ambiguous are labelled with (?) in this table. Each of the entries in this table is a story in itself, but space does not allow a detailed discussion. Instead the interested reader is referred to various

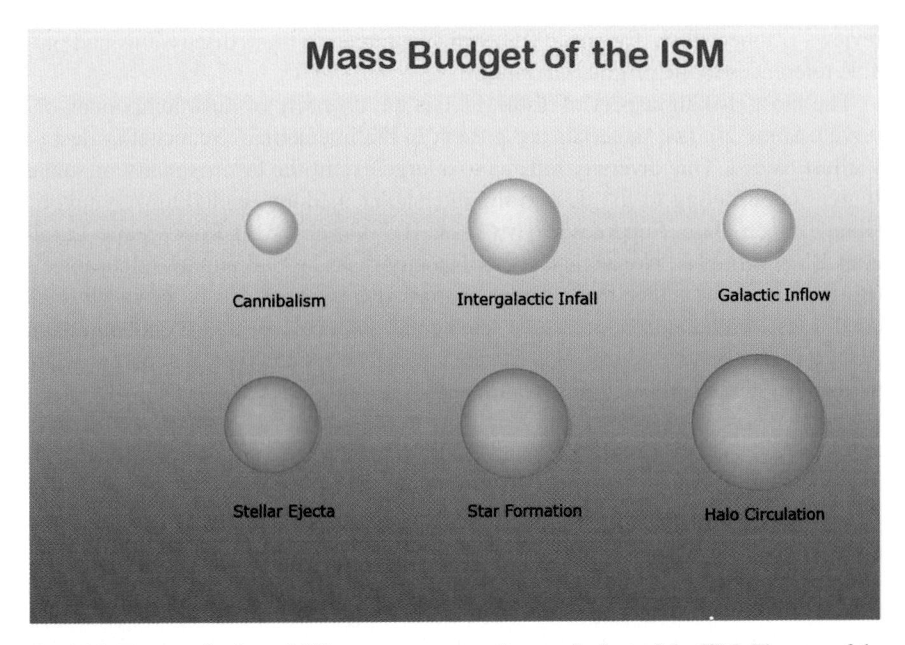

Fig. 11.3 The contribution of different processes to the mass budget of the ISM. The area of the circles indicates the relative importance of each process. Processes indicated by grey circles are more uncertain than those associated with blue circles. See text for details

Table 11.1 Composition of interstellar dust

Class	Compounds	Class	Compounds
Silicates:		Carbonaceous compounds:	
	amorphous FeMg silicates		graphite
	forsterite		diamonds
	enstatite		hydrogenated amorphous carbon
	montmorillonite ?		polycyclic aromatic hydrocarbons
Oxides:		Carbides:	
	corundum		silicon carbide
	spinel		titanium carbide
	wuestite		other metal carbides
	hibonite	Sulfides:	
	rutile		magnesium sulfide
Ices:			iron sulfide ?
	simple molecules	Others:	
	such as		silicon nitride
	H_2O, CH_3OH, CO, CO_2		metallic iron ?
			silicon dioxide ?

reviews (Zinner 2003, Tielens 2001), which summarize these discussions and provide references to the original literature.

The most striking aspect of Table 11.1 is the diversity of dust compounds observed. Some 20 dust materials are present in the interstellar/circumstellar dust or stardust record. This diversity reflects to a large extent the heterogeneity of stellar sources contributing to the dust in the interstellar medium, including Asymptotic Giant Branch stars, Supernovae (type I & II), (C-rich) Wolf-Rayet stars, Luminous Blue Variables, Novae, and protoplanetary disks around young stellar objects (Barlow 2008). Together these stellar sources represent a wide range in physical conditions (temperatures, pressures, elemental abundances) and, concommittant, dust compounds. Asymptotic Giant Branch (AGB) stars and type II supernovae are particularly rich in detected dust compounds.

11.2.3 Interstellar PAHs

The mid-IR spectra of almost all objects are dominated by a very rich set of emission features with major bands at 3.3, 6.2, 7.7, 8.6, 11.2, 12.7, and 16.4 μm (Fig. 11.4). In addition, there are weaker features at 3.4, 3.5, 5.25, 5.75, 6.0, 6.9, 7.5, 10.5, 11.0, 13.5, 14.2, 17.4 and 18.9 μm. These features are perched on broad emission plateaus from 3.2–3.6, 6–9, 11–14, and 15–19 μm. Moreover, many of the well known features shift in peak position, vary in width, and/or show substructure, revealing a sensitivity to the local physical conditions. These IR emission features are ubiquitous and are present in almost all objects, including HII regions, reflection nebulae, young stellar objects, planetary nebulae (PNe), post-AGB objects, nuclei of galaxies, and UltraLuminous InfraRed Galaxies (ULIRGs) (Verstraete et al. 2001, Moutou et al. 1999, Hony et al. 2001, Peeters et al. 2002, Genzel et al. 1998, Acke & van den Ancker 2004, Armus et al. 2007, Brandl et al. 2006, Geers et al. 2006). In addition, the IR cirrus, the surfaces of dark clouds, and indeed the general interstellar medium of galaxies are set aglow in these IR emission features (Mattila et al. 1996, Burton et al. 2000, Abergel et al. 2002, Smith et al. 2007, Regan et al. 2004). These IR emission features are (almost) universally attributed to IR-fluorescence of FUV-pumped Polycyclic Aromatic Hydrocarbon molecules (PAHs) containing some 50 C-atoms (Sellgren 1984, Puget & Leger 1989, Allamandola et al. 1989). This assignment is based upon an energetics argument: small species have a low heat capacity and will get so hot upon absorption of a single photon that they emit in the mid-IR even far from the illuminating star. In addition, the high feature-to-continuum ratio implicates a molecular rather than solid state origin. PAHs are very abundant in the ISM locking up about 4% of the elemental carbon and an abundance of 3×10^{-7} relative to H nuclei.

Observations of the IR emission features and the spectroscopic, physical and chemical characteristics of interstellar PAHs have recently been reviewed elsewhere (Tielens 2008). Here, I highlight one aspect: the observed variations in the relative strength of the CH (3.3 and 11.2 μm features) to the CC modes (6.2 and

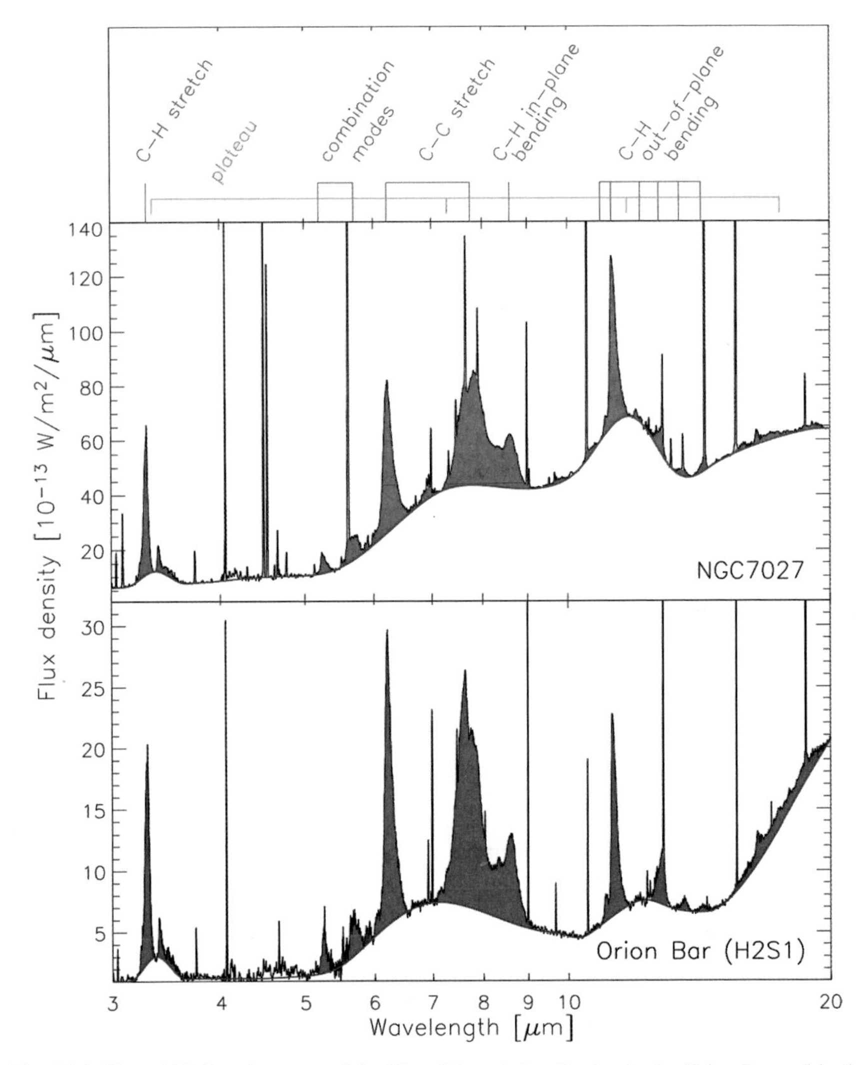

Fig. 11.4 The mid-infrared spectra of the PhotoDissociation Region in the Orion Bar and in the Planetary Nebulae, NGC 7027 are dominated by a rich set of emission features. Assignments of these features with vibrational modes of PAH molecules are labeled at the top. Figure adapted from Peeters et al. (2002). Note the relative intensity variations between the CC (6.2 and 7.7 μm) and CH (3.3 & 11.2 μm) modes and among the CH out-of-plane bending modes (11.2 versus 12.7 μm). The narrow features are atomic or ionic lines originating in the HII region or PDR

7.7 μm features). The two spectra in Figure 11.4 illustrate these variations. An example of the observed large varations in this ratio within one source is shown in Fig. 11.5. These types of variations between sources and also within sources are common (Hony et al. 2001, Galliano et al. 2008b, Rapacioli et al. 2005). These variations are attributed to the effect of charge state on the intrinsic strength of

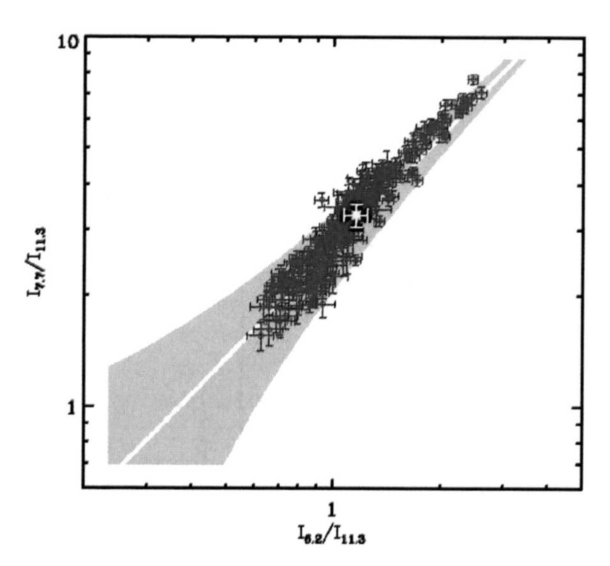

Fig. 11.5 Variations in the relative strength of the CH to the CC modes in the starburst galaxy, M82. The gray stripe denotes the correlation obtained for the data $\pm$ 1σ. Figure taken from Galliano et al. (2008b)

the vibrational modes. Extensive laboratory studies and quantum chemical caculations have demonstrated that, while peak positions can shift somewhat, the intrinsic strength of modes involving CC stretching vibrations can increase manifold upon ionization while the CH stretching and to a lesser extent the out-of-plane bending vibrations decrease in strength (Allamandola et al. 1999, Langhoff 1996). This is illustrated in Fig. 11.6 which compares the calculated/measured IR absorption cross

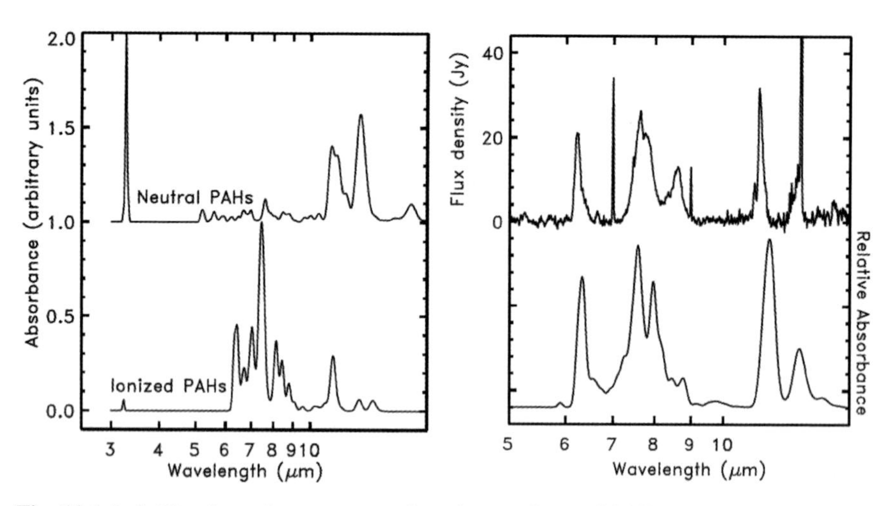

Fig. 11.6 *Left*: The absorption spectrum of a mixture of neutral PAHs (*top*) compared to the spectrum of the same species in their cationic states (*bottom*). The strength of the CC modes has increased considerably in strength relative to the CH modes in the 3 and 11-15 μm region. Figure taken from (Allamandola et al. 1999). *Right*: The combined spectrum of a well chosen mixture of neutral and cationic PAHs (*bottom*) is compared to the observed spectrum (*top*) of the Orion Bar (Peeters et al. 2002)

sections for a family of PAHs in neutral with that of the same species in their cationic state. A cursory glance at the observed spectra shows then that the IR emission features are carried by a mixture of neutral and ionic PAHs (cf., Fig. 11.6). Thus, the observed ratio of the 6.2 to the 11.2 μm bands is a direct measure of the ionized fraction of PAHs in the emission region with important ramifications for our understanding of the photo-electric effect (cf., Section 11.4.1) as well as for the ionization balance of interstellar gas. In addition, observations of the PAH emission features can provide a handle on the physical conditions of the gas (cf., Section 11.6.2).

11.2.4 Lifecycle of the ISM

Star formation – e.g., the timescale for stellar astration – as well as the injection of material by stars and accretion from the nearby universe are very slow processes operating on timescales measured in billions of years. In contrast, mixing between the different phases of the ISM is very rapid (tens to hundreds of millions of years), reflecting the continuous stirring up of the ISM by stellar winds and supernova explosions. As a result, the ISM is well mixed and an average batch of the interstellar medium will contain the ashes of $\sim$100 million AGB stars plus $\sim$1 million SNe. Nevertheless, the action of a local stellar ejection event may well leave a recognizable signature over some hundred million years (de Avillez & Mac Low 2002).

In addition, the stellar birthsite heterogeneity and specificity coupled with ISM processing will lead to variations in for example the dust composition. A number of examples have been well documented. In particular, the silicate abundance is enhanced in the galactic center region by about a factor 2 (Roche & Aitken 1985). Likely, this reflects the increase in metallicity with galactocentric radius and the concommittant decrease in the C-rich versus O-rich AGB stars, which represent major factories of interstellar dust. Likewise, the UV extinction curve observed in the SMC is very different from that of the Milky Way, steadily increasing towards shorter wavelengths while showing no evidence for the well known 2200 Å bump generally ascribed to graphitic carbon dust (Cartledge et al. 2005). This difference is somewhat puzzling since – given the low metallicity of the SMC – C-rich AGB stars should dominate the dust balance in this environment. Likewise, the decreased dust/metallicity ratio as well as the decreased PAH/dust abundance ratio in low metallicity environments is intriguing (Galliano et al. 2008a). The nuclei of ULIRGs provide another example of localized differences in dust characteristics (Fig. 11.18). The IR absorption spectra of a wide variety of ULIRGs show evidence for crystalline silicate material as well as pronounced absorption due to Hydrogenated Amorphous Carbon grains (Spoon, et al. 2006, Armus et al. 2007). These differences may reflect the increased importance of massive stars as dust factories (in their LBV, Red Supergiant, or supernova phase) immediately ($<$100 million years) after an intense starburst phase before low mass stars kick in their contribution or supernova shocks and cosmic rays have had the chance to process the grains. The IR emission spectra of the quasar, PG 2112+059, also show evidence for crystalline silicates (Fig. 11.17) either because the material in the inner accretion disk was recently enriched by a massive circumnuclear starburst associated with the merging event driving the AGN activity or because dust forms in the winds associated with the nuclear jets (Markwick-Kemper et al. 2007). Finally, Hydrogenated Amorphous Carbon grains are an important component of dust in the diffuse ISM but not in dense clouds as evidenced by the strength of the 3.4 μm band in the former and the absence in the latter environment (Pendleton et al. 1994). In contrast, IR spectra of dense cloud material often show a 4.62 μm band which is generally attributed to energetic processing of ice materials, forming organic refractory mantles (Greenberg & Li 1999) which is absent in spectra of dust in the diffuse ISM. These differences in the composition of carbonaceous dust in the diffuse and dense ISM may reflect the growth of ice mantles in the latter phase which inhibit the conversion of graphitic into aliphatic grains by attacking H atoms (Mennella et al. 2003). The (limited) processing of these ices in dense clouds may then create the carrier of the 4.62 μm band while, in the more intense radiation field in the diffuse ISM, this carrier is perhaps rapidly destroyed.

It is clear that the composition of interstellar dust varies considerably with galactic environment, reflecting the local balance between stellar dust injection and dust processing and/or destruction. On the one hand, given that the transfer of radiation in the ISM, the formation of molecules, the gas heating rate, and the gas ionization balance depend strongly on the characteristics of the dust, this may well have important ramification for the large scale structure of the ISM and its observational characteristics. On the other hand, these variations provide a tool to study the characteristics of distant sources, at least, if we understand the processes driving them.

11.3 Phases of the ISM

11.3.1 Thermal and Pressure Equilibrium

The ISM contains a number of different phases each characterized by distinct physical conditions (Table 11.2). They range from cold, dense clouds to low density, warm intercloud gas, to tenuous, hot gas. Quite generally, each of these different phases reflects the importance of a distinct cooling or heating process (Shull 1987).

As originally discussed by (Field 1965, Field et al. 1969), the coexistence of cold clouds and a warm intercloud medium directly reflects the cooling curve of interstellar gas: at low temperatures and high densities, cooling is dominated by emission in the [CII] 158 μm line while at high temperatures and low densities, the [OI] 6300 Å and HI Lyα line take over. These cooling processes act as thermostats coupling the temperature of the medium to (a fraction of) the energy level separation of the relevant processes. The original models (Field et al. 1969) for the phase structure of the ISM balanced these cooling processes by cosmic ray heating, adopting a cosmic ray ionization rate which is an order of magnitude higher than current molecular observations allow (Tielens 2005). Presently, it is generally accepted that heating of

Table 11.2 Characteristics of the phases of the interstellar medium

Phase	n_o^a [cm^{-3}]	T^b [K]	ϕ_v^c [%]	M^d [10^9 M$_\odot$]	$<n_o>^e$ cm^{-3}	H^f pc	Σ^g M$_\odot$ pc^{-2}
Hot intercloud	0.003	10^6	~50	–	0.0015	3000	0.3
Warm Neutral Medium	0.5	8000	30	2.8	0.1^h	220^h	1.5
					0.06^h	400^h	1.4
Warm Ionized Medium	0.1	8000	25	1	0.025^i	900^i	1.1
Cold Neutral Medium[j]	50	80	1	2.2	0.4	94	2.3
Molecular clouds	>200	10	0.05	1.3	0.12	75	1
HII regions	$1–10^5$	10^4	–	0.05	0.015^k	70^k	0.05

[a] Typical gas density for each phase.
[b] Typical gas temperature for each phase.
[c] Volume filling factor (very uncertain and controversial !) of each phase.
[d] Total mass.
[e] Average mid-plane density.
[f] Gaussian scale height, $\sim exp[-(z/H)^2/2]$, unless otherwise indicated.
[g] Surface density in the solar neighborhood.
[h] Best represented by a gaussian and an exponential.
[i] WIM represented by an exponential.
[j] Diffuse clouds.
[k] HII regions represented by an exponential.

the neutral ISM is dominated by the photo-electric effect on large molecules and small dust grains (Section 11.4.1).

Figure 11.7 shows the results of detailed model calculations taking into account all the relevant heating and cooling processes of neutral gas in the ISM (Wolfire et al. 1995). Assuming pressure equilibrium, the intersection of horizontal lines in Fig. 11.7a with the calculated curve will give the phases that can coexist in thermal equilibrium. Over a limited range of pressures ($P/k \simeq 1000 - 4000\,\mathrm{K\,cm^{-3}}$), two stable phases coexist (the third intersection is unstable). For a typical pressure of $3000\,\mathrm{K\,cm^{-3}}$ in the local ISM, these consist of a cold dense phase ($\simeq 60\,\mathrm{cm^{-3}}$ and 50 K) – known as the Cold Neutral Medium (CNM) or HI clouds or Spitzer clouds – and a warm phase ($\simeq 0.4\,\mathrm{cm^{-3}}$ and 7500 K) – known as the Warm Neutral Medium (WNM). Perusing the model results in Fig. 11.7 shows that indeed heating in the ISM is due to the photo-electric effect while cooling is dominated by the CII [158] μm line in the CNM and by [OI] 6300 Å and the HI Lyα lines in the WNM. We note that besides the WNM, there is also a Warm Ionized Medium (WIM), which differs from the WNM in the degree of ionization (Table 11.2). The high ionization of this latter phase implies the widespread penetration of a (weak) ionizing radiation field ($h\nu > 13.6\,\mathrm{eV}$) in the ISM.

Through the detailed microphysical processes of heating and cooling, the characteristics of the phases of the ISM will depend on, for example, the metallicity, abundance of PAHs and small grains, the local FUV radiation field, the penetration of EUV photons, the cosmic ray ionizing flux, and the X-ray flux. Many of these factors vary in tandem because they are controlled by the local star formation rate or the integrated star formation history. Nevertheless, these characteristics will vary with location in the galaxy as well as from galaxy to galaxy and throughout the

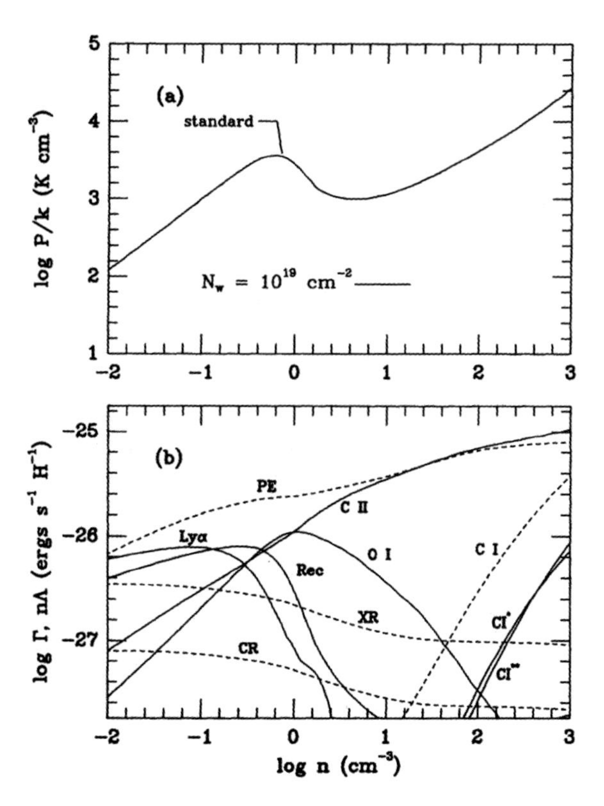

Fig. 11.7 Detailed two phase models for the ISM (Wolfire et al. 1995). Panel (**a**) shows the pressure-density relation for interstellar gas in thermal equilibrium. Panel (**b**) shows the individual heating (*dashed*) and cooling processes (*solid*) at different densities in thermal equilibrium (i.e., for the pressure-density relation of panel (a)). PE, XR and CR are photo-electric, X-ray and cosmic ray heating rate. The solid lines are the obvious cooling lines labeled by the cooling species

history of the universe. This is illustrated through model calculations appropriate for our galaxy in Fig. 11.8 (Wolfire et al. 2003). The calculations for this particular model illustrate that the pressure range over which two phases can coexist vary by an order of magnitude in our galaxy, mainly due to variations in the local FUV radiation field. Comparing these models to the pressure for a thermally or turbulently supported ISM illustrates that two phases will coexist over much of the (starforming) HI disk of the galaxy.

11.3.2 SNe and the Coronal Phase of the ISM

The mechanical energy input through expanding supernova ejecta and – to a lesser extent – by stellar winds has a profound influence on the structure of the ISM by generating turbulent motions of the HI and by sweeping up and shocking the surrounding interstellar gas to very high temperatures. Thus, while the total mechanical energy injected by massive stars into the ISM is only a small fraction of the radiative energy budget ($L_{\rm mech} \simeq 2 \times 10^8$ L$_\odot$ versus $L_{\rm rad} \simeq 4 \times 10^{10}$ L$_\odot$), the interstellar medium reacts to this mechanical energy by becoming very turbulent, the gas disk puffs up, and part of the volume becomes filled with hot gas. The latter may, even

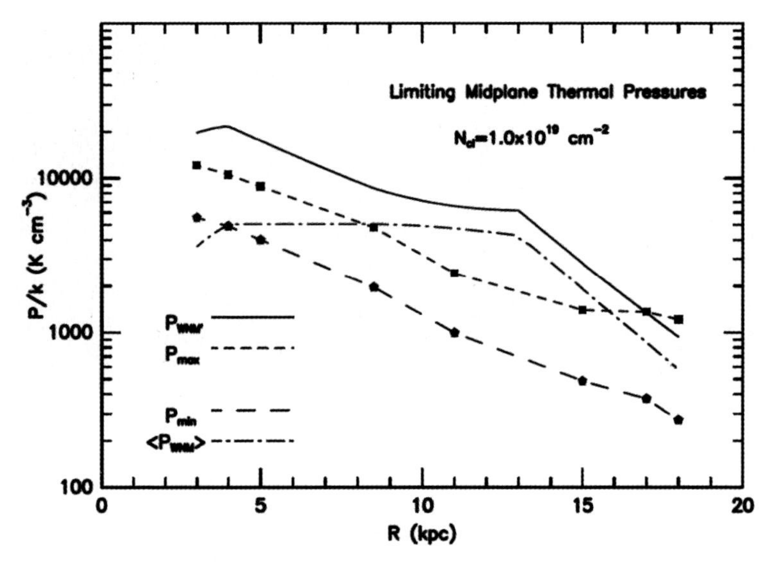

Fig. 11.8 Limiting thermal pressures as a function of galacto-centric radius. P_{max} (*short dashed line with squares*) and P_{min} (*long dashed line with pentagons*) are the maximum and minimum pressures over which a two phase medium exists in pressure and thermal equilibrium (cf., Fig. 11.7). $< P_{WNM'} >$ (*solid line*) is the pressure if all the HI gas were in the WNM phase in hydrostatic equilibrium supported by the thermal pressure. $< P_{WNM} >$ (*dashed-dotted line*) is the pressure if all the HI gas were in the WNM phase at the measured mid-plane, average HI density. The additional pressure support for this has to be provided by turbulent pressure. Figure adapted from Wolfire et al. (2003)

in the plane, become a dominant, separate phase of the ISM, the Hot Ionized (intercloud) Medium (HIM) in which the WNM/WIM and CNM are embedded.

The expansion of supernova remnants and hence the temperature, density, and filling factor of the HIM depend very much on physical processes which are only partly understood, on the concerted effects of multiple supernova explosions in OB associations, and on venting of the hot gas into the coronal halo of the Milky Way. Specifically, at high temperatures ($\sim 10^7$ K), radiative cooling is very slow and the hot gas cools down through expansion and/or heat conduction/mass loading with the surrounding medium. The earliest models for the interaction of stellar winds from O stars envisioned the formation of a hot, tenuous stellar bubble surrounded by a "cold", dense shell of swept-up interstellar medium gas photo-ionized by the central star (Weaver et al. 1977). In these models, the gas in the bubble mainly cools through evaporation of material from the cold, dense shell. However, it is now well understood that champagne flows are an important aspect of the evolution of HII regions. Basically, HII regions formed near molecular cloud surfaces will expand and eventually break through into the surrounding lower density intercloud material, forming a blister HII region (Tenorio-Tagle 1979). Combined observations using Chandra, Spitzer, and ground-based observatories (Fig. 11.9) illustrate the resulting geometry and characteristics of such a region and the importance of

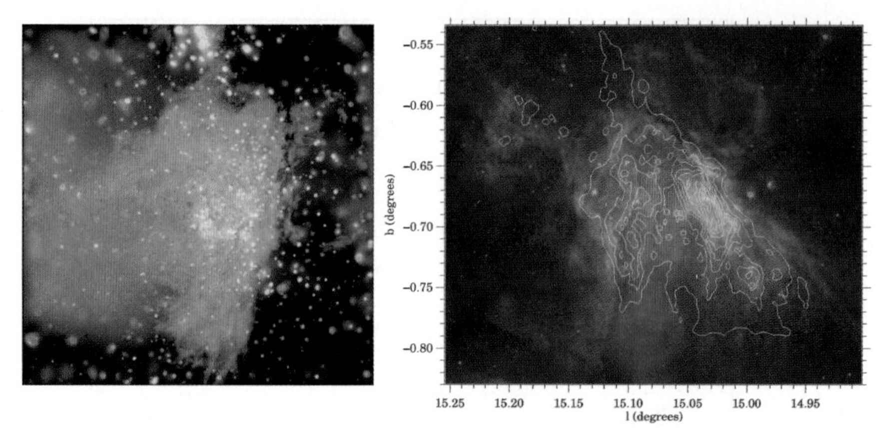

Fig. 11.9 *Right*: A combined Spitzer-chandra view of the region of massive star formation, M17 (Povich et al. 2007). X ray emission from a $\sim 7 \times 10^6$ K plasma (*blue*; (Townsley et al. 2003)) fills a central cavity surrounding the O stars (*Cyan circles*). The wite contours trace the radio emission from the ionized gas. Warm dust in the ionized gas is shown in red (21 μm, MSX). PAH emission from the PDR surfaces of the surrounding molecular clouds is traced by green (5.8 μm IRAC-Spitzer). The image is about 10 pc in size. *Left*: A blow up revealing the intricate interaction between hot gas and cool PDR material (private communication Townsley 2008). Note the change in orientation to RA-Dec and the change in color palet (Blue is soft X-rays, Green is hard X-rays, red is 5.8 μm PAH emission

adiabatic expansion in a blister HII region environment. In particular, the hot gas emits bright X-ray emission but only a small portion of the wind energy and mass appears in the observed diffuse X-ray plasma. In blister HII regions, most of the energy and mass flows without cooling into the surrounding low-density interstellar medium (Townsley et al. 2003). In addition, the filamentary nature of the PDR gas as revealed by IRAC-Spitzer demonstrates the importance of entrainment of cold material (Fig. 11.9). The resulting denser gas structures will continue to load the hot gas as it is swept along outwards but this whole process is poorly understood (e.g., in terms of physics, how much energy is transferred into the cold gas by energetic electrons as well as in "fluid" characteristics like exposed surface area). Dust and PAH emission may provide a dye with which the mixing and geometry of the phases can be followed but only if we understand the destruction of these species/compounds in the hot gas. Thus, the concerted efforts of awide range of observatories may well provide an observational handle on some of these questions.

M17 is a very young OB association containing over 100 O stars and eventually these massive stars will go supernova. The coalescence and rejuvenation of the hot gas in this bubble by each successive SN will drive the formation and expansion of a superbubble. When this superbubble breaks out of the (cloud) disk, the energy will be vented into the halo (cf., Section 11.2.1) – the galactic fountain – mixing material to great height and over large distances within the disk. Figure 11.2 shows a more extreme example of such a break-out associated with the moderate starburst in the nucleus of M82. Studies of IR emission from edge on galaxies have revealed that this levitation of dust is a common characteristic (Irwin et al. 2007, Thompson et al. 2004, Howk & Savage 1997, Burgdorf et al. 2007).

11.3.3 Three Phase Models of the ISM

These theoretical concepts have given much insight in the phases observed in the ISM with 21 cm HI studies and optical and FUV absorption lines as well as X-ray emission studies (McKee & Ostriker 1977). Cold HI clouds and the warm intercloud medium result from the increased importance of [CII] cooling at higher densities and Lyα and [OI] 6300 Å cooling at higher temperatures, respectively. The hot phase reflects the recent input of supernova energy. These models predict the right physical conditions of these phases as well as reproduce the observed [CII] cooling rate per H-nuclues of the CNM (Wolfire et al. 1995). There are however also a number of challenges. These include the filling factor of the hot gas in the plane. Also, the assumption of thermal equilibrium for the WNM is not fully justified since the cooling time is relatively long for this low density gas. Observationally, some $\sim$50% of the WNM is in the unstable phase, exemplifying the importance of turbulence and slow shocks in driving the ISM from thermal equilibrium (Heiles & Troland 2003). In addition, HI observations reveal the presence of very cold ($\simeq$20 K) HI clouds. These must imply either the absence of photo-electric heating in these clouds – perhaps because the PAHs have been destroyed – or the presence of additional cooling – perhaps due to efficient molecule formation. The widespread presence of the [OVI] FUV absorption line is another indicator of the importance of rapidly cooling gas or locally heated gas; in this case in the range 1–5 $\times$ 10^5 K (Savage et al. 2003). These may well be tracing the conduction fronts between the hot and the warm intercloud phases (Heckman et al. 2002). Absorption line studies provide only "spot"-access to the interaction between the phases. In that regard, IR dust and PAH studies might be more amenable to fully inventorize this interaction.

11.4 Physics of the ISM

11.4.1 PAHs and the Energy Balance of the ISM

Photo-electric heating is the dominant process that couples the energy balance of the gas to the non-ionizing radiation field of stars in HI regions. As such, photo-electric heating ultimately controls the phase structure of the ISM and the physical conditions in PDRs – which includes most of the HI gas mass of the ISM – and therefore the evolution of the ISM of galaxies (Hollenbach & Tielens 1999). It has long been recognized that photoelectric heating is dominated by the smallest grains present in the ISM (Watson 1973) and PAHs and very small grains dominate the heating of interstellar gas (Bakes & Tielens 1994). Essentially, absorption of an FUV photon creates an electron-hole pair in the material. The electron diffuses towards the surface, losing its excess kinetic energy along the way through "collisions". Because the FUV absorption depth (~100 Å) can be much larger than the mean free path of low energy electrons in solid materials, the yield is very low for large grains. Because of the Coulomb attraction, the photo-electric heating efficiency is sensitive to the grain charge (e.g., ionization potential). The charge of a species is set by the ratio of the ionization rate over the recombination rate which is proportional to the charging parameter, $G_o T^{1/2}/n_e$ (Tielens 2005).

Extensive theoretical calculations on the heating by an interstellar grain size distribution of PAHs and small grains, including the effects of charge, have been performed by (Bakes & Tielens 1994, Weingartner & Draine 2001) and the resulting efficiencies (ratio of gas heating to FUV absorption rate of grains and PAHs) have been fitted to a simple analytical formula. Figure 11.10 shows the calculated photo-electric heating rate as a function of grains size, illustrating that only species less than ~100 Å contribute effectively to the photo-electric heating rate (Bakes & Tielens 1994). Figure 11.11 compares the calculated photo-electric efficiency as a function of the charging parameter, γ, to observations of the heating in diffuse sightlines and in well-known PDRs. For the diffuse sightlines, the photo-electric heating rate is determined from the upper level populations of the CII finestructure transition measured through FUV absorption lines (Pottasch et al. 1979, Gry et al. 1992) where the physical conditions have been taken from the definitive study by van Dishoeck & Black (1986). For PDRs, the photo-electric efficiency has been determined directly by comparing the intensity of the dominant cooling lines ([OI], [CII]) to the far-IR dust emission. Physical conditions in these regions were determined from atomic and molecular line ratios (Hollenbach & Tielens 1999) (cf., Section 11.4.2). The

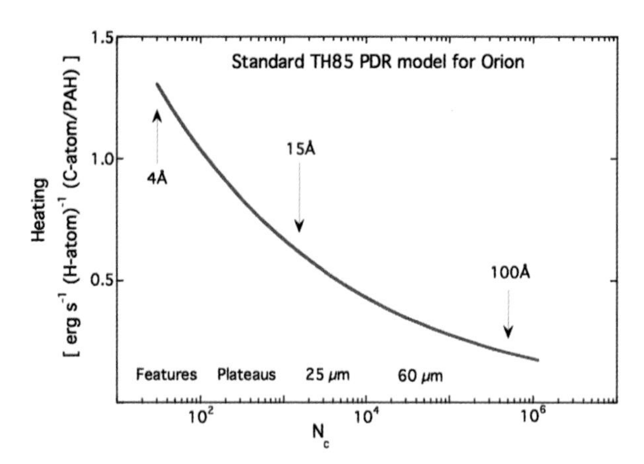

Fig. 11.10 The contribution to the photoelectric heating of interstellar gas by species of different sizes (Bakes & Tielens 1994), here traced by the number of carbon atoms, N_C. The results of these calculations are presented in such a way that equal areas under the curve correspond to equal contributions to the heating. Typically, about half of the heating originates from PAH and PAH clusters ($<10^3$ C-atoms). The other half is contributed by very small grains ($15 < a < 100$ Å). Classical grains do not contribute noticably to the heating. The typical IR emission characteristics as a function of size are indicated at the bottom of the figure

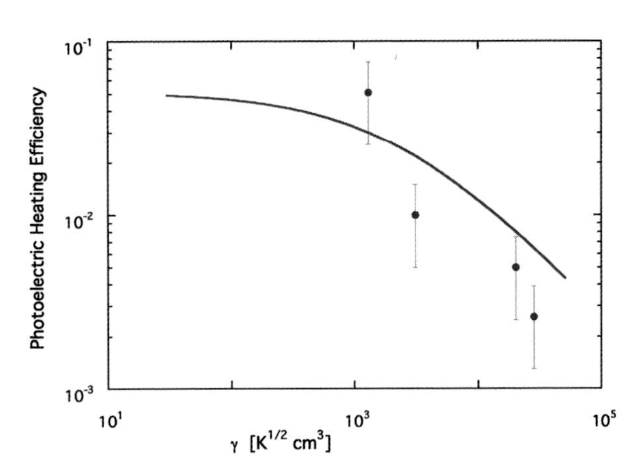

Fig. 11.11 The photo-electric efficiency as a function of the charging parameter, $\gamma = G_o T^{1/2}/n_e$ (proportional to the ionization rate over the recombination rate). Neutral species are located to the left in this figure while to the right the charge of species increases. The data points indicate the measured heating efficiency for the diffuse ISM sightlines, ζ Oph and ξ Per, and the well-studied PDRs, NGC 2023 and the Orion Bar

calculated efficiency compares reasonably well to these observations. It can be expected that SOFIA and Herschel – in combination with ALMA – can substantially increase the sample of sources for which this comparison can be made. More importantly, since the emission spectra of PAHs is a sensitive function of grain charge (cf., Tielens (2008)), combining mid-IR PAH spectra studies and far-IR gas cooling line and dust emission studies can directly test the photo-electric heating model and observationally determine its dependence on such important parameters as the local radiation field, gas temperature, electron density, and metallicity.

11.4.2 PhotoDissociation Regions

Photodissociation regions (PDRs) are neutral atomic/molecular regions where penetrating FUV photons dominate the energy balance and chemical composition of the gas (Hollenbach & Tielens 1999). Heating of the gas is dominated by the photoelectric effect on large PAH molecules and very small grains (cf., Section 11.4.1), while cooling occurs through the atomic fine-structure lines of [CII] and [OI] and the rotational lines of H_2 and CO. While traditionally, the name PDRs is beholden to bright regions associated with compact HII regions (e.g., Orion Bar, M17) and reflection nebulae (e.g., NGC 2023 and 7023), the processes that control the characteristics of diffuse clouds in the general ISM are very similar to those in PDRs. From that perspective, much of the ISM is essentially a large PDR. PDRs are bright in the IR dust continuum, the IR emission features due to PAHs, atomic and ionic finestructure lines, the UV pumped ro-vibrational lines of H_2 (e.g., the 1-0 S(1) line at 2.1218 μm), the pure rotational lines of H_2 (e.g., the 0-0 S(0) and S(1) lines at 28.221 and 17.035 μm), the rotational lines of CO (up to $J = 17 - 16$ have been seen in PDRs), and the rotational lines of reactive species which are produced in the radical zone near the PDR surface (e.g., CN).

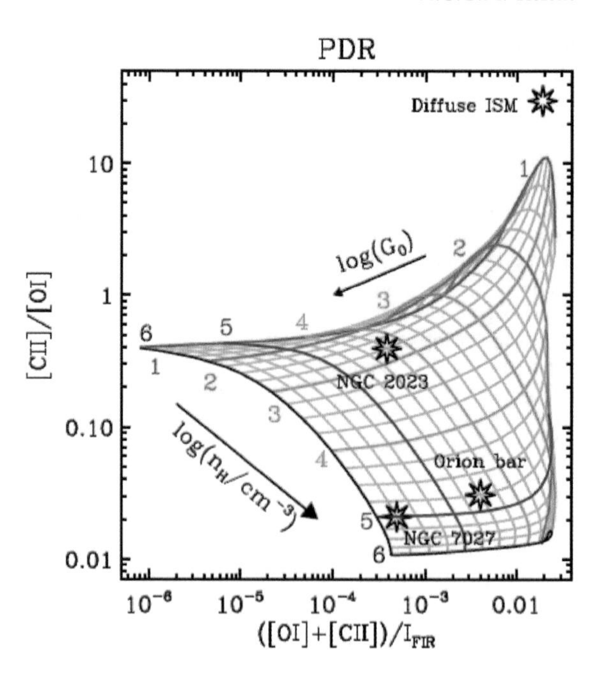

Fig. 11.12 Diagnostic diagram for PDRs (Kaufman et al. 2006). Line ratios of atomic fine-structure lines and molecular rotational lines are sensitive to the incident FUV radiation field, G_o, and density n. Observed values for a few characteristic PDRs are also indicated

Because of their high surface brightness, PDRs lend themselves very well to a detailed study of the effects of the interaction of FUV photons with neutral atomic and molecular gas. Because of differences in excitation energy and critical density, the relative intensities of various lines are sensitive to the local physical conditions (e.g., n and T; cf., (Tielens 2005)). This is illustrated in Fig. 11.12 for the atomic fine-structure lines of [OI] and [CII] using the results of detailed PDR models. Actually, because these two lines are the dominant cooling lines of the gas, this figure plots the ratio of these two lines versus the photo-electric heating efficiency (e.g., the ratio of the gas cooling to the total FUV energy (mostly absorbed by dust and reradiated in the IR)). Thus, the model results displayed in this figure does "entail" a buy-in of the photo-electric heating model. However, there are many similar line combinations possible of a multitude of species whose ratios directly reflect the difference in critical density and excitation energy and hence the physical conditions in PDRs can be determined largely independent of model assumptions.

11.4.3 Dust and Interstellar Shocks

11.4.3.1 Dust Destruction

Supernova eject material into the ISM at velocities of ~10,000 km/s. This high velocity gas energetically processes the swept up material by a strong shock wave. As the supernova remnant expands, the strength of the shock decreases until it merges with the general ISM. Such interstellar shocks are an important destruction agent of interstellar dust (Draine & Salpeter 1979, Seab & Shull 1983, Jones et al. 1994, 1996). While high velocity shocks are more destructive, low velocity shocks process a larger volume. Very low velocity shocks lack the energy to damage grains. Shocks will propagate fastest into the lowest density medium ($v_s \sim n^{-1/2}$). However, the HIM contains very little mass and is from this point of view not important. Hence, processing of interstellar dust is dominated by ~100 km/s shocks propagating into the WNM or WIM which allow strong shocks and yet contain ~50% of the interstellar mass.

Gas is "instantaneously" stopped in a shock front but, because of their inertia, dust grains will keep moving at 3/4 of the shock speed relative to the gas. Since the grains are charged, they will gyrate around the magnetic field and acquire velocities with respect to each other. Moreover, given that the Larmor radius associated with this circular motion is typically very small, interstellar grains are position coupled to the gas. Upon compression, the magnetic field strength will increase and, because of conservation of magnetic moment, the grains will spin up (betatron acceleration). This acceleration is counteracted by collisions with the gas. This drag is more effective for small grains which are therefore less susceptible to betatron acceleration. In all, large grains will move at considerable speeds relative to the gas and to each other over much of the shock structure. These relative motions leads to destruction due to sputtering and due to grain-grain collisions. At grain-gas velocities of 100 km/s, impinging H-atoms have some 50 eV of energy and He atoms some 200 eV. At these energies, this will, typically, result in sputtering yields of about 0.01 per impinging H-atom. Likewise, these velocities are well above the threshold for cratering, melting, and vaporization in collisions between two grains.

Over the years, a number of theoretical studies have appeared that detail the destructive effects of interstellar shocks (Draine & Salpeter 1979, Seab & Shull 1983, Jones et al. 1994, 1996). These studies calculate the shock structure, grain velocities, and sputtering rate and grain-grain collision rate as a function of position behind a shock of a given velocity. For a given grain size (distribution), this allows then the calculation of the fraction of a grain destroyed for shocks of different velocities. The results of these studies show that sputtering is the dominant process returning solid material to the gas phase. This is a slow 'chipping' away of large grains where a 100 km/s shock will sputter a ~30 Å layer, returning some 10% of the grain mass to the gas phase. Grain-grain collisions are important in redistributing the grain mass from large grains to small grains. Indeed, calculations show that a single 100 km/s shock will lead to the complete disruption of essentially all large (~1000 Å) grains, providing a source of very small grains and PAHs (Jones et al. 1996).

While a single shock does little damage, the cumulative effect of many shocks can slowly erode grains away. Convolving such shock calculations with models for the frequency of shocks of different velocities, results then in the lifetime against destruction of interstellar dust grains. Ignoring reaccretion, all calculations agree that, in the local ISM, the overall lifetime of grains against the cumulative effect of strong shocks is some 500 million years, which is much less than the timescale (2.5 billion years) at which grains are replenished by stellar sources (Jones et al.

1994, 1996, Dwek & Scalo 1980). However, reaccretion is now known to be very important (cf., Section 11.4.3.2) and this thin veneer protects the underlying grain material from further destructive action of shocks.

11.4.3.2 Shock Destruction and Elemental Depletions

Observations show that the gas phase abundance in the interstellar medium of major dust forming elements (e.g., Si, Fe, Mg) are much less than Solar or stellar abundances (Fig. 11.13; (Savage & Sembach 1996, Cartledge et al. 2006)). This depletion is generally taken as evidence that a substantial fraction ($\sim$0.9) of these elements is locked up in dust grains (Field 1974). However, gas phase abundances are much higher in high velocity clouds (Routly & Spitzer 1952). Because these high velocities reflect the recent passage of a shock, this provides qualitative support for the importance of dust destruction by shock waves. A more quantitative comparison between models and observations is hampered by our lack of knowledge on the detailed velocity history of the observed gas (Cowie 1978).

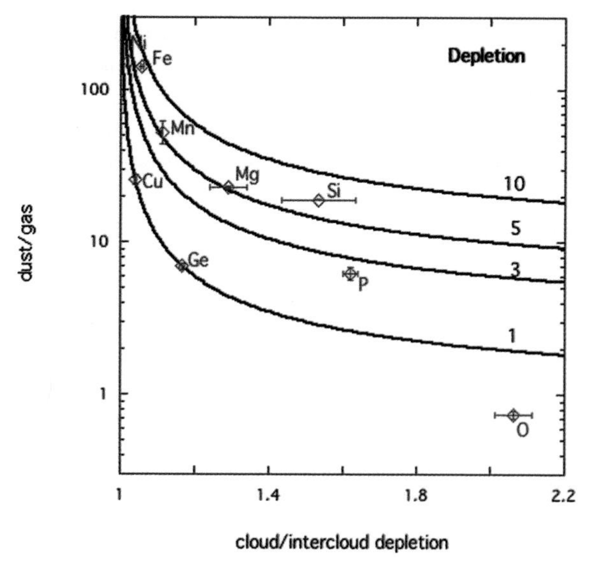

Fig. 11.13 Observed depletions of elements in the ISM (Savage & Sembach 1996). The y-axis is the ratio of the abundance of an element in the dust phase to its abundance in the gas phase, both measured in diffuse clouds. The x-axis is the ratio of the depletion of these elements in the diffuse cloud medium to that in the intercloud medium. The *solid lines* are the results of a simple model balancing destruction in the intercloud medium with accretion in the cloud medium (Tielens 1998). The labels indicate the adopted values for the accretion rate relative to the cloud-to-intercloud mixing ratio. As these observations demonstrate, the rates for destruction in the intercloud phase, accretion in the cloud phase, and mixing between these phases have to be within a factor of a few of each other and are thus very rapid compared to the injection rate of dust by stars into the ISM

Observations have also revealed a large and systematic difference in the elemental depletions in the different phases of the interstellar medium (Fig. 11.13; (Savage & Sembach 1996, Cartledge et al. 2006)). Thus, gas phase depletions are systematically much less in the intercloud medium than in diffuse clouds. The depletion pattern reflects a balance between shock destruction in the intercloud medium and reaccretion in the denser environment of HI clouds. The observed large difference in the depletion between these two phases of the interstellar medium demonstrates then directly that the processes involved – shock destruction and accretion – operate on a timescale similar to the timescale at which material is mixed from the cloud to the intercloud medium and back (Tielens 1998). This mixing timescale is much less ($\simeq$30 million years) than the timescale at which new dust is injected into the ISM. Thus, specifically, some 10% of the iron, 15% of the magnesium, and some 30% of the silicon is returned to the gas phase upon each sojourn into the intercloud medium and then rapidly reaccreted once the material is cycled back to the (diffuse) cloud phase. This leads to the formation of a thin, outer skin that protects underlying grain material against destructive processing in shocks.

For silicon and magnesium, these differences in depletion between cloud and intercloud phase are slightly larger than expected based upon the material properties (e.g., sputtering yields) of silicates, indicating that the re-accreted material is in the form of mantle with somewhat lower binding energy than silicates. The depletion pattern of oxygen indicates that it participates in this shock-sputtering and reaccretion cycle. However, carbon does not show a difference in depletion between the cloud and intercloud medium. So, likely this thin outer layer has an oxide rather than a carbide structure. This difference in chemical behavior is not understood: perhaps, accreted carbon is rapidly cycled to volatile compounds (e.g., CH_4, CO) that are readily photodesorbed rather than become integrated into a silicate or oxide network.

11.4.3.3 Dust and the Origin of Cosmic Rays

The Larmor radius of large grains can become comparable to the relevant shock-size scales (Slavin et al. 2004) and their position decoupled from the gas in the shock. At that point, the fate of the grain depends very much on the specifics (grain size, charge, composition, velocity). Globally, three different regimes can be discerned. Small ($<0.3\ \mu$m) grains are position coupled and their behavior, including destruction, is well described by 'classical' grain shock calculations (Fig. 11.14). For intermediate sized grains and intermediate shock velocities, the grain is reflected many times back and forth across the shock front, and every time it is accelerated to higher velocities (Fig. 11.14). This Fermi acceleration process leads to very high velocities ($\sim$3000 km/s) before the grain is completely sputtered away. At the largest sizes ($>1\ \mu$m), the Larmor radius of the grains is so large that they traverse the shock without noticing its presence. The location of this transition – from fully destroyed to largely unharmed – is very sudden and its location (e.g., critical grain size) depends on the details (e.g., charge, size, material) but that large grains decouple follows directly from simple considerations. We further note that this also implies that

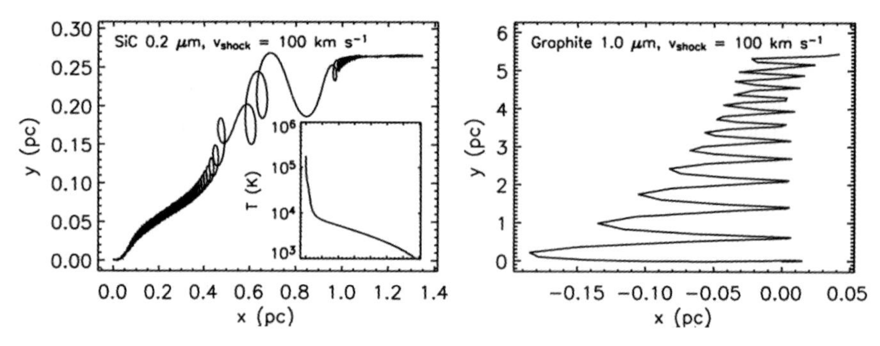

Fig. 11.14 Calculated trajectories for individual dust grains in interstellar shocks (Slavin et al. 2004). The magnetic field direction is perpendicular to the $x - y$ plane. These calculations are for plane-parallel shocks where the material enters from the left. Inserts show the calculated gas temperature as a function of the distance behind the shock front. The x-scale is the same as that in the trajectory panel. *Left*: A typical interstellar grain ($a = 0.2\,\mu$m) is position coupled in a 100 km/s shock and partly sputtered as it slows down. The effects of betatron acceleration and charge reversal behind the shock are apparent. *Right*: a 1 μm-sized grain is rapidly Fermi accelerated across the shock front and reaches velocities exceeding 1000 km/s before complete destruction. The atoms injected into the gas phase will be further accelerated by the shock and are likely the origin of galactic cosmic rays. Note the difference in x- and y-scale

young supernova remnants will selectively sweep up large interstellar grains in their interiors, while the smaller ones remain localized in the outer shells of swept-up interstellar material. Given that the Sun is presently traversing such a young SNR – the local bubble, the Ulysses stream of interstellar grains (Grün et al. 1993) may well be sampling this collection process (Slavin et al. 2004).

Measured cosmic ray abundances show a distinctive non-solar pattern (Fig. 11.15). Highly refractory elements which are largely locked up in dust grains in the ISM have considerably enhanced abundances in cosmic rays. It is, therefore, likely that interstellar dust grains not only participate but are actually prime movers in the cosmic ray production process. Based upon energetics, SNR are thought to be the prime source of cosmic rays. The acceleration of micron-sized dust grains to high velocities through multiple Fermi reflections across supernova shock fronts may be the first step in the production of cosmic rays since the fast moving ions produced by sputtering can be further Fermi-accelerated to cosmic ray energies (Ellison et al. 1997, Slavin et al. 2004).

11.4.4 Cosmic Rays and the Amorphous Structure of Interstellar Dust

Crystalline silicates are an important component of the dust in circumstellar environments. Specifically, AGB stars which are one of the major 'dust factories' inject some 15% of their silicates in crystalline form (Sylvester et al. 1999, Kemper et al. 2004). Likewise, protoplanetary disks surrounding low and intermediate mass

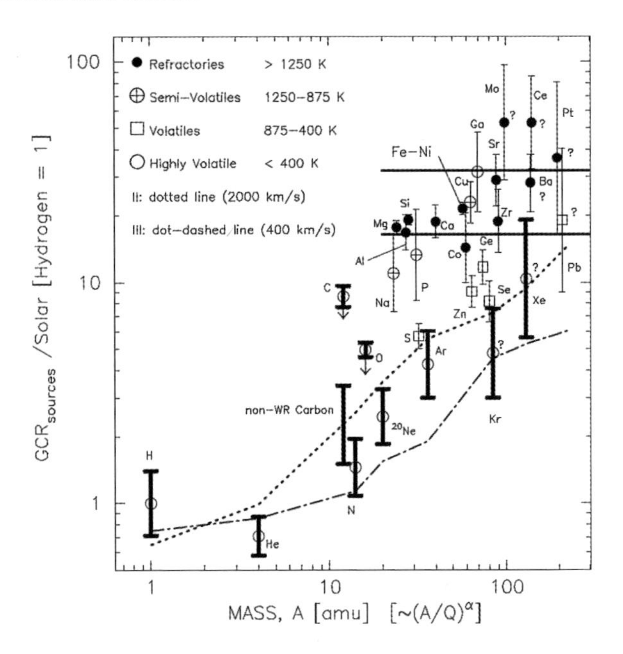

Fig. 11.15 Galactic cosmic-ray source abundance relative to solar abundance versus atomic mass number (Ellison et al. 2004). All values are measured relative to cosmic-ray hydrogen at a given energy per nucleon. The dust-forming elements are enhanced by a factor 10–50 in the cosmic ray spectrum relative to Solar abundances

protostars show ample evidence for crystalline silicates in their mid-IR spectra (Malfait et al. 1998). It is therefore disconcerting that the interstellar medium shows no evidence for crystalline silicates; after all, in the lifecycle of the dust, the ISM separates the AGB dust formation sites from the end stages – newly formed stars and planets. Upper limits on the fraction of crystalline relative to amorphous silicates in the ISM are very strict (<0.01) compared to the fraction injected (~0.05; Kemper et al. 2004). Apparently, crystalline silicates are rapidly converted into amorphous silicates in the ISM and back to crystalline silicates in protoplanetary disks.

Energetic ions impacting on crystalline silicates can rapidly amorphize their internal structure (Demyk et al. 2004, Jäger et al. 2003, Brucato et al. 2004, Bringa et al. 2007). Penetrating high energy ions will create defects (vacancies, interstitials, dislocations) in the crystal structure of a material. Experimentally, changes in the structure of the material can be followed using a variety of techniques, including IR spectroscopy and Rutherford Back Scattering. At high enough doses, the material will be completely amorphized. This is illustrated in Fig. 11.16 which shows the results of laboratory experiments of 10 MeV Xe ions impacting on crystalline forsterite (Bringa et al. 2007). At low energies, the interaction is dominated by nuclear interaction; at high energies, electronic interaction takes over. The latter is actually more effective in driving the crystalline-amorphous transition both because only a very small dose is required (10^{14} ions/cm^2 versus 10^{18} ions/cm^2) and because the penetration depth increases rapidly with increasing energy. In particular, 10 keV H and He ions are required to amorphize the full depth of 0.1 μm silicate grains.

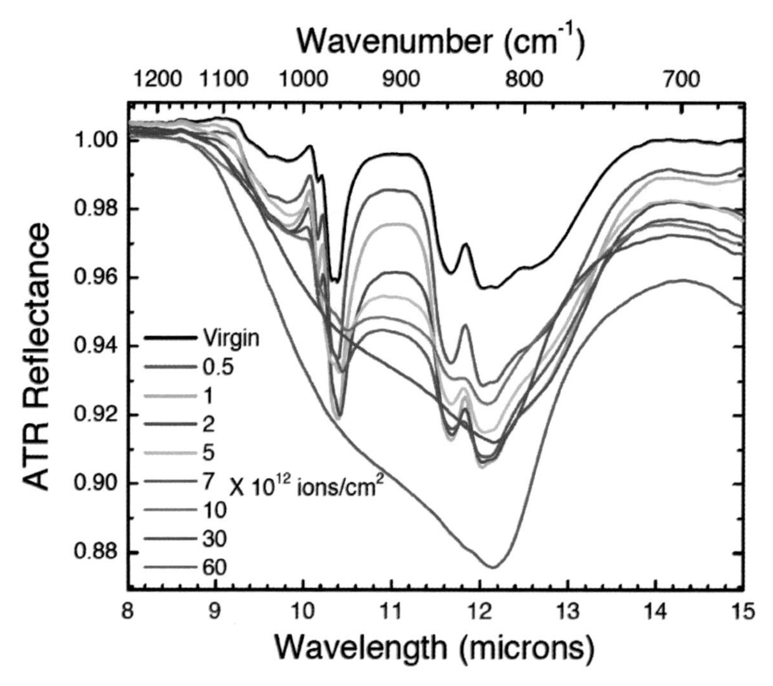

Fig. 11.16 Changes in the IR spectra of crystalline forsterite (Mg_2SiO_4) with increasing exposure to 10 MeV Xe ions (Bringa et al. 2007). The IR spectra are measured in reflection and the narrow features "sticking" downward reflect vibrational modes of the material. Initially, the spectrum is characterized by a number of strong and narrow crystalline features but at doses exceeding 3×10^{13} ions/cm^2 a broad, featureless band remains: the structure of the material has amorphized

Thus, while small grains can be amorphized when entrained in high temperature gas or when engulved by 500 km/s shocks, on a galactic scale, the amount of interstellar material processed this way is very small. Iron Cosmic Rays with energies of $\sim$10–100 MeV/nucleon, on the other hand, are very effective in transform-

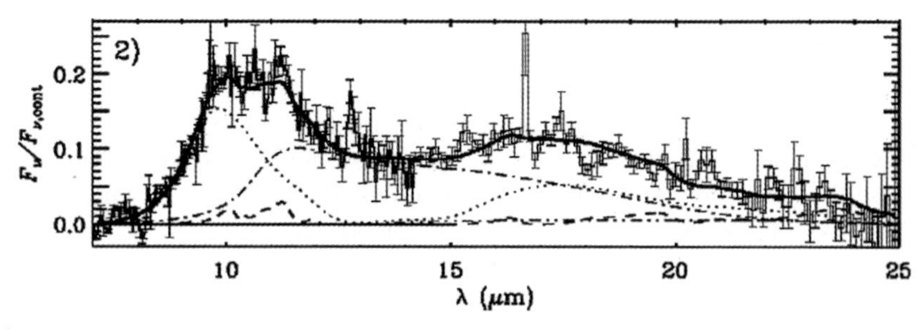

Fig. 11.17 The continuum subtracted infrared spectrum of the quasar, PG 2112+059, at a redshift of $z = 0.466$ measured by the IRS-Spitzer (Markwick-Kemper et al. 2007). The *thick solid line* is a fit to the data obtained by combining emission from amorphous olivine (*dotted*), forsterite (*dashed*), corundum (Al_2O_3, *dash-dotted*), and MgO (*dash-triple-dotted*)

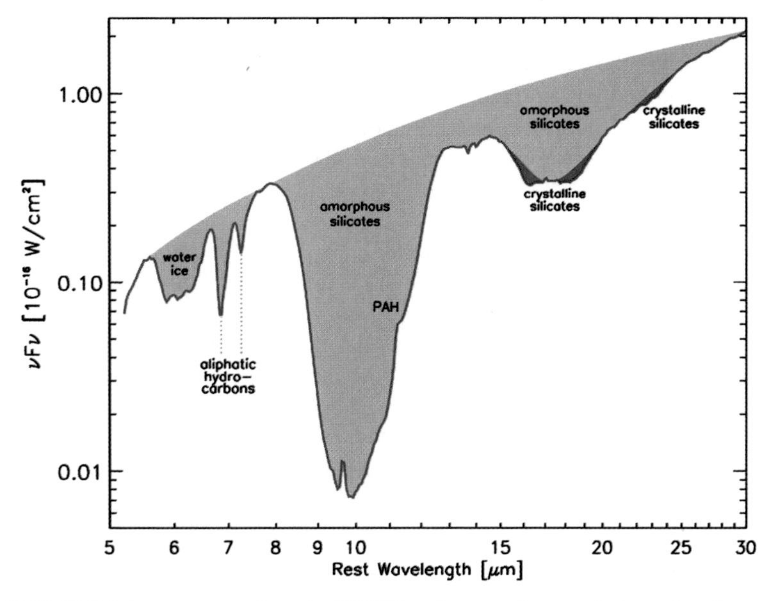

Fig. 11.18 IRS-Spitzer spectra of nearby ULIRGs often reveal a set of absorption features due to amorphous silicates, crystalline silicates, Hydrogenated Amorphous Carbon (HAC), and simple ices (Armus et al. 2007, Spoon, et al. 2006)

ing crystalline silicates into amorphous silicates. Scaling the energy deposited with the electronic interaction cross section and adopting the measured/extrapolated Cosmic Ray spectrum, the lifetime against amorphization of interstellar silicate grains is calculated to be only 70 million years (Bringa et al. 2007) which is very short compared to other relevant timescales. Conversely, the presence of the features due to crystalline silicates in the IR spectrum of an interstellar environment (cf., Fig. 11.17 and 11.18) provides a signature of recent addition of freshly synthesized stardust to the interstellar dust mix.

11.5 Summary

The evolution of the interstellar medium is driven by a number of complex processes which are deeply interwoven, including mass accretion, stellar mass injection, star formation, mechanical energy input by stellar winds and supernova explosions, and radiative energy input. Large molecules and dust play an important role in these processes but these links are only partially understood. The resulting ISM is highly structured where different phases interact and interchange dynamically on a rapid timescale. The properties of the ISM are expected to vary systematically reflecting the local (stellar/ISM) conditions.

In this review, I have illustrated the interwoven aspects of the interstellar medium with a few examples selected based upon personal interest: Specifically, the molecular characteristics of the interstellar medium and the energy balance and phase structure of the gas; the processing of dust by SN shocks and by cosmic rays; the evaporation and entrainment of cold gas and dust in hot gas. While we have made much progress in our understanding of the ISM over the last decades – largely driven by ever better observational opportunities from space – many questions remain. These include:

- What is the role of the Halo and the IGM in mass and energy budget of the ISM?
- What are the characteristics of interstellar dust and how does that depend on metallicity, star formation activity, ISM conditions (e.g., density, UV field, turbulence) ?
- What is the role of PAHs and very small grains in the energy & ionization balance of the ISM ? And how does that influence the structure of the ISM and the star formation activity ?
- What is the role of mechanical energy – SNR and turbulence – in the energy balance and phase structure of the ISM ? How do supernova shocks process dust and PAHs and how does this couple back to the structure of the ISM and the star formation activity ?
- What is the relative role of adiabatic expansion and evaporation/entrainment in the energy balance of SNR ? How does this depend on and couple back to the structure of the ISM ?
- What are the best tracers of star formation rate on a galactic scale ?

11.6 Future

We live in the best of times, we live in the worst of times (Dickens 1859). We are building ever more sophisticated instruments to peer ever farther in the universe and back of time to study processes which are really not understood. In particular, it is clear that in order to relate JWST and ALMA observations of the early universe to the characteristics of the emitting regions and objects, we need to study these processes in the local universe and determine their dependencies on the physical conditions such as density, temperature, metallicity, radiation field, etc. Fortunately, SOFIA and Herschel, in combination with JWST and ALMA, as well as a further mining of the Spitzer data archive provide us with just the right tools to accomplish this. In this section, I outline a few programs that I feel could be instrumental in this respect. These focus on methods that will allow us to analyze and interpret IR spectral data to arrive at the physical characteristics of the emitting regions as well as the star formation rate driving the activity because that is JWST's forte and represents my own interest.

11.6.1 What Processes Play a Role in Dust Formation and Evolution in the ISM ?

Significant amounts of dust are observed in high redshift galaxies and quasars (Isaak et al. 2002, Bertoldi & Cox 2002) and these are generally ascribed to injection by massive stars, because low mass stars evolve too slowly. Among the sources considered are Red Supergiants, Luminous Blue Variables, Wolf-Rayet stars, and Supernova (Barlow 2008). In addition, dust may grow oxide mantles in the diffuse ISM and icy mantles in the dense ISM. It is clear that dust in unique environments will differ in many characteristics from dust observed in the local ISM. Figures 11.17 and 11.18 show two examples taken from the Spitzer data. These two objects may actually be much more relevant for the interpretation of high-z observations than the local ISM. In order to interpret the data on the early universe, we will need to address a number of questions:

- How important are high mass stars in the dust budget of galaxies ?
- What are the spectroscopic and physical differences between 'high-mass' and low-mass' dust ?
- How important is dust injection by YSOs ?
- How important is dust growth in the ISM relative to dust injection by stars ?
- Is carbon involved in dust mantle formation ?
- How does this depend on metalicity ? On galaxy characteristics ?

The Spitzer-SAGE legacy program – and its 'offspring' SAGE-SMC, SAGE-Spec – will be very instrumental in determining the dust budget on a galaxy-wide scale and the IR characteristics of different source types in a low metallicity, dwarf galaxy environment (Meixner et al. 2006). As emphasized by Barlow (2008), this type of studies can be continued using JWST on normal local galaxies such as M31, while in the galaxy – where sources will be too bright for JWST – SOFIA can complete this census. For the latter, interpretation will require, though, accurate distances which GAIA can provide (Barlow 2008). The issue of carbon depletion and grain mantle growth can be addressed by the Cosmic Origin Spectrograph on HST by measuring elemental carbon depletion in translucent sight-lines to determine whether carbon mantles grow at higher density or in more shielded environments than normal diffuse clouds provide. JWST can study this problem from the other end by studying the IR characteristics of dust in dense clouds through absorption studies against faint background stars. In particular, the signatures of HAC grains at 3.4 μm and the XCN-compound at 4.62 μm could be a key indicator of mixing and processing of dust in these environments since both are thought to reflect a balance between accretion and UV photolysis. Likewise, MIRI on JWST could search for the IR features of crystalline silicates in field stars behind molecular clouds; a signature of high temperature processing in YSO environments and subsequent dispersal by YSO winds into the molecular cloud environment.

11.6.2 Heating and Cooling of the ISM

The physical characteristics of the neutral phases in the ISM are directly related to the photo-electric heating by interstellar PAHs and very small dust grains (Section 11.4.1). Likewise, the observational characteristics of regions of massive star formation (e.g., PDRs) are also directly related to the photo-electric heating (Section 11.4.2). Hence, it is important to quantify the role of PAHs in the heating of the neutral ISM and the dependence of the heating rate on the density and UV field intensity, metalicity, and galaxy characteristics.

SOFIA is uniquely suited to address these issues through observations of PDRs. SOFIA's discovery space – exemplified by its wide instrument complement with a wide wavelength coverage, wide-field imaging capabilities, and wide range in spectral resolution – will allow it measure the characteristics of the emitting PAH family, the heating efficiency, and the physical conditions in the emitting region. In this way, the various components involved in the energy balance of neutral interstellar gas can be directly interrelated to each other and to the local physical conditions. This is illustrated in Fig. 11.19 which 'calibrates' the observed 6.2/11.2 μm PAH band ratio as a function of the local physical conditions. SOFIA will be able to spectrally map very extended, bright PDRs efficiently in the IR emission features

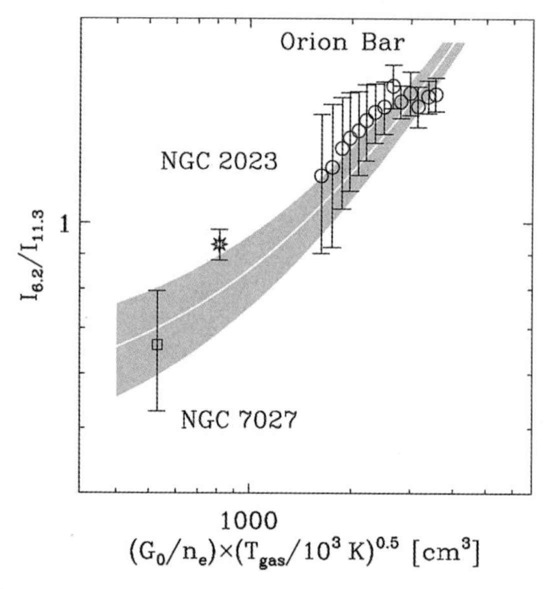

Fig. 11.19 The observed ratio of the 6.2 to 11.2 μm bands – a measure of the degree of ionization of PAHs – is related to the ionization parameter, $(\gamma=)G_oT^{1/2}/n_e$ for a few well-studied PDRs, where the physical conditions – the strength of the UV radiation field, G_o, the electron density, n_e, and, the temperature, T – have been well determined from a multitude of atomic fine-structure lines and molecular rotational lines. The degree of ionization of PAHs increases to the right as either G_o increases or n_e, decreases. The temperature, T, enters through the velocity dependence of the Coulomb focussing factor. Figure taken from Galliano et al. (2008b)

and to determine the physical conditions using atomic fine-structure lines ([OI], [CII], [SiII]) and molecular rotational lines (CO, H_2) and can really determine these relationships semi-empirically. We note that such observations will also directly determine the heating efficiency by comparing the intensity of the cooling lines with the intensity of the local radiation field (e.g., the dust continuum; cf., Fig. 11.11).

11.6.3 The CII Emission from the Galaxy

COBE has shown that the [CII] 158 μm line is the dominant emission line of the Milky Way with a luminosity of 5×10^7 $L_\odot$; e.g., about 0.003 of the total IR luminosity of the Milky Way is emitted in this one line. This is a very general result: in a sample of 60 normal, star forming galaxies, the [CII] line is in general the dominant IR cooling line. The [OI] 63 μm line is a close second and in a handful of galaxies even takes over (Fig. 11.20) (Malhotra et al. 2001). The origin of the [CII] line is controversial (cf., Hollenbach & Tielens 1999). Theoretically, because of their density and temperature, it is expected that HI clouds (e.g., the CNM) will radiate most of their energy through the [CII] line (Section 11.3.1; Dalgarno and McCray 1972). This is supported by measurements of the [CII] emission from high lattitude clouds using sounding rockets (Bock et al. 1993). However, COBE demonstrated that the intensity of the NII line – which must originate from ionized gas

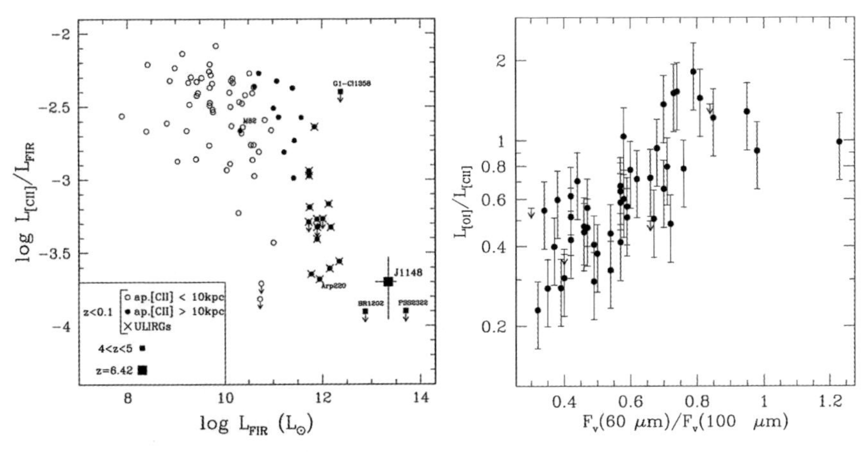

Fig. 11.20 *Left*: The ratio of the [CII] 158 μm line to the far-IR continuum as a function of the IR luminosity. Open (*closed*) circles are galaxies for which the aperture encompasses less (more) than 10 kpc. Crosses mark ULIRGS. The filled squares are high redshift objects. The large filled square represents the most distant known quasar, SDSS J114816.64+525150.3 (Maiolino et al. 2005). Right: The ratio of the [CII] 158 μm to the [OI] 63 μm line flux as a function of the 60–100 μm continuum flux ratio (this dust color temperature is a measure of the radiation field. Figures adapted from Malhotra et al. (2001)

since the N ionization potential is 14.5 eV – correlates with the intensity of the CII line (to the 1.5 power), suggesting that the low density ionized gas (the WIM) contributes a portion of the observed CII emission (Heiles 1994). In addition, given the high observed [OI] intensity (which has a critical density of $2 \times 10^5\,\mathrm{cm}^{-3}$) in the sample of star forming galaxies studied by ISO-LWS, a substantial fraction of the observed [CII] emission likely originates from dense, bright PhotoDissociation Regions associated with molecular cloud surfaces near regions of massive star formation (Section 11.4.2). Theoretical models have been developed based upon each of these three premisses (WNM, WIM, and PDR origin) and all are in reasonable agreement with the COBE observations of the Milky Way.

Understanding the origin of the [CII] line on a galactic scale has recently received additional impetus with the detection of this line in the spectrum of the most distant quasar, J1148, at a redshift of 6.42 using IRAM and Plateau de Buren (Fig. 11.21; Maiolino et al. 2005). The observed flux of this line – in conjunction with other PDR tracers (e.g., CO $J = 7 - 6$; Bertoldi et al. (2003)) – has been interpreted as evidence for vigorous star formation (3000 $M_{\odot}$/yr) in the host galaxy. Because of the high luminosity in this one spectral line, the CII line has often been considered as one key tracer of star formation in the early universe, particularly for heavily obscured galaxies. Thus, one of the three key scientific goals of the ALMA project is to use the CII line to probe star formation in the high redshift universe. However, we can only use the CII line as a quantitative measure of the star formation rate at high z if we properly understand its dependence on the local physical conditions. Again, this is a project that has SOFIA written all over it. The program outlined in Section 11.6.2 will be very instrumental in analyzing and

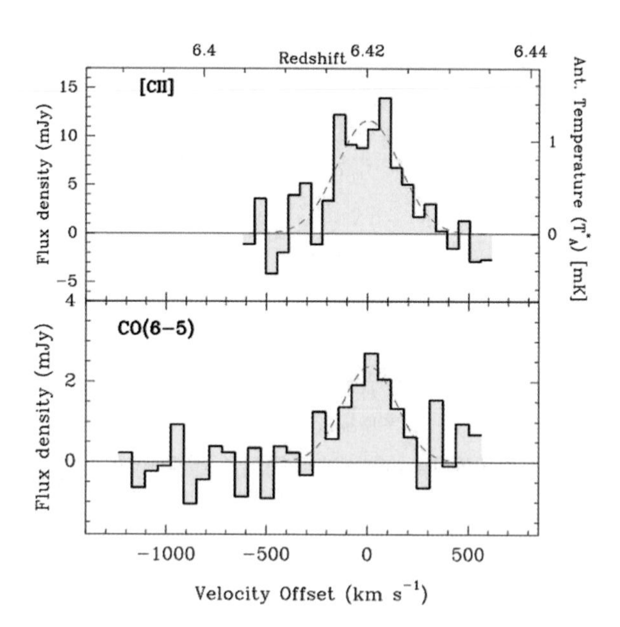

Fig. 11.21 The CII 157 μm line (*top*) and CO $J = 6 - 5$ line (*bottom*) in the most distant known quasar, SDSS J114816.64+525150.3. Figure taken from Maiolino et al. (2005)

interpreting [CII] and other PDR line observations of the high-z universe in terms of the physical conditions in the emitting regions and in translating these into the star formation rate.

11.6.4 PAHs as Tracers of Star Formation

The ubiquity of the PAH emission features and their predominance in the mid-IR spectra of regions of massive star forming regions make them potentially powerful tools for the study of star formation throughout the universe. Essentially, the PAHs act as a 'dye' for the presence of pumping FUV photons and hence trace the presence of massive stars. This aspect of the IR emission features has already been employed by Genzel et al. (1998) to conclude that ULIRGs are largely powered by star formation rather than AGN activity. While ISO was able to probe these emission features in the local universe (out to redshifts of $\sim$0.1), with Spitzer, the use of the IR emission features as probes of star formation has been extended to $z \sim 3$ for lensed galaxies and luminous submillimeter galaxies (Fig. 11.22) (Rigby et al. 2007, Pope et al. 2007, Lutz et al. 2007). Ultimately, the James Webb Space Telescope and SPICA can be expected to open up the whole universe to studies of the infrared emission features as tracers of star formation. Of course, this all rests on a validation and quantification of the relationship between the IR emission features and the star formation rate. One such calibration – based upon local starburst galaxies – is shown in Fig. 11.22.

Calzetti et al. (2007) have addressed this relationship based upon Spitzer-IRAC 8 μm (PAH) and HST-NICMOS (Pα) images of (part of) the SINGS sample of galaxies as well as low-metallicity starburst galaxies and Luminous InfraRed Galaxies (Fig. 11.23). They find an almost linear correlation between the PAH emission and the ionizing photon rate for galaxies with metallicities close to the Solar value. However, in the wider sample, this relationship shows a strong dependence on metallicity which likely reflects a decreased PAH abundance in low metallicity environments (Draine et al. 2007, Madden et al. 2006, Wu et al. 2006, Galliano et al. 2008a). In addition, the starburst galaxies also show a decreased 8 μm-to-IR ratio (see also below). A combination of 24 μm Spitzer-MIPS and Hα data provides a good handle on the star formation rate in these types of galaxies: Extinction in the inner galaxy hampers the use of Hα while the outer galaxy is optically thin in the dust and the IR underestimates the star formation rate.

In their study, Calzetti et al. (2007) focused on the 'quiescent' mode of star formation characteristic for spiral disks and the regions probed can be best typified as exposed PDRs: Orion-like regions of star formation characterized by a compact HII region with well developed PDRs. For more deeply embedded regions of massive star formation, the ratio of the 6.2 μm PAH feature to the far-IR continuum is known to be less by one to two orders of magnitude (Peeters et al. 2004). Possibly, the increased FUV field incident on PDRs associated with more compact HII regions destroys small PAHs. Alternatively, more of the FUV is absorbed by dust inside

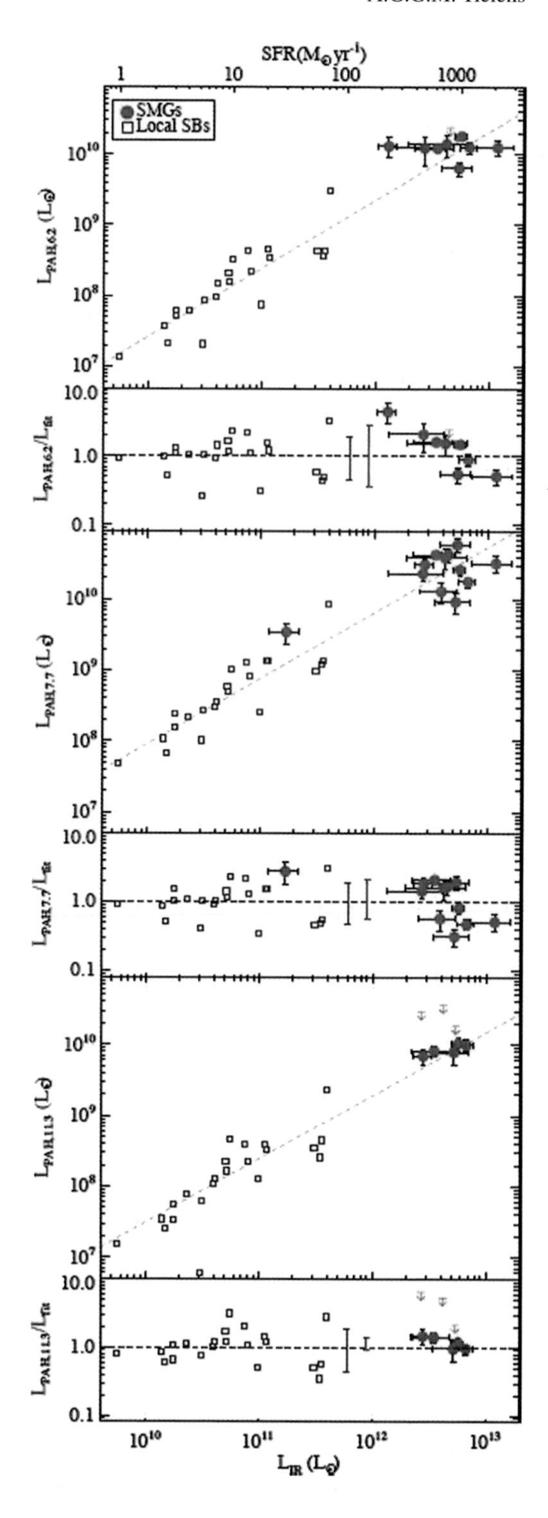

Fig. 11.22 The luminosity in various PAH features versus the IR luminosity for a sample of local starburst galaxies (*open blue squares*) and sub-millimeter galaxies (*red circles*). The dashed line provides the best fit to the starburst galaxies only. The observations of the sub-millimeter galaxies are consistent with them being more distant brothers of the starburst galaxies (Pope et al. 2007)

the HII region and never reaches the PAHs in the PDRs. Likely, this dependence on the character of the HII region probed causes the decreased 8 μm-to-IR ratio in Fig. 11.23. It is clear that the character of the star formation process has direct influence on the PAH tracer. Therefore, quantifying a PAH flux – measured, say, for local ULIRGs or more distant sub-millimeter or lensed galaxies with Spitzer or, in the future, with JWST in the far universe – into a star formation rate will require knowledge of the properties of their regions of star formation and identification and characterization of appropriate local templates (Peeters et al. 2004). We have only started to address these issues (cf., Fig. 11.22). Fortunately, the PAH spectrum itself contains information on the local physical conditions experienced by the emitting species. A concerted effort directly calibrating the PAH feature strength and their ratios as a function of the local physical conditions in nearby objects – as outlined in Section 11.6.2 – will be required to fully appreciate the implications of future JWST observations of the far-universe. Again, the synergy of SOFIA and Herschel with JWST comes in good stead. In particular, the large format mid-IR and far-IR spectroscopic capabilities of SOFIA make it ultimately suited for the study of the characteristics of the IR emission features and their relationship to the local physical conditions in regions of massive star formation (cf., Section 11.6.2).

11.6.5 Mass and Energy Flows on Galactic Scales

The mass flow in and out of galaxies is a key driver of galaxy evolution. Generally, these mass exchanges have associated cold molecular gas. The galactic winds associated with extreme star formation are a case in point where bright PAH

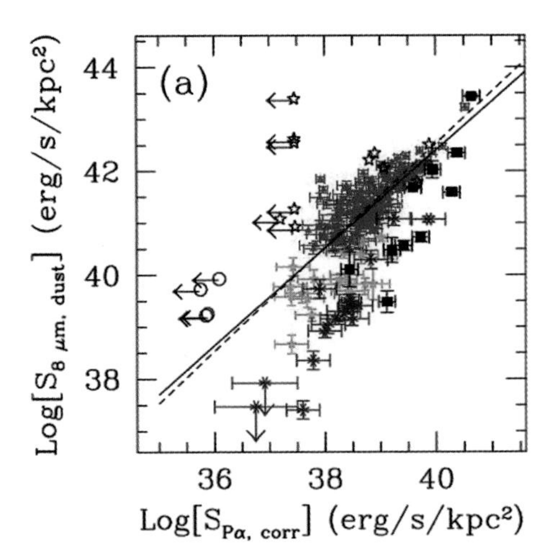

Fig. 11.23 The surface brightness in the 8 μm PAH feature compared to the surface brightness of the Pα HI recombination line. The colors indicate metallicity: red = high, green = intermediate, blue = low. Filled squares are local star burst galaxies. Open stars are Seyfert 2's or liners. Open black circles are extended background sources. Solid line is best fit through red data. Dashed line is best fit with a slope of 1. Figure taken from Calzetti et al. (2007)

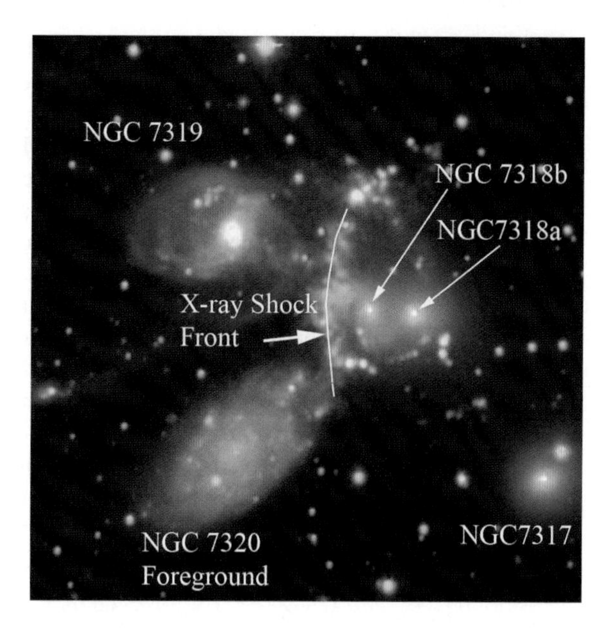

Fig. 11.24 False color image of the interaction galaxies in Stephan's quintet. Visible light is shown as blue. Emission from Hα (*green*) traces the shock front. The Spitzer-IRAC 8 μm emission traces the ISM of the galaxies (Appleton et al. 2006)

emission and pure rotational H_2 line emission has been detected (cf., Fig. 11.2) (Engelbracht et al. 2005). Cluster cooling flows provide another example. Gas at a wide range of temperatures is present embedded in the hot cluster gas down to cool, dense molecular gas. Spitzer-IRS observations suggest that some 10% of the cooling

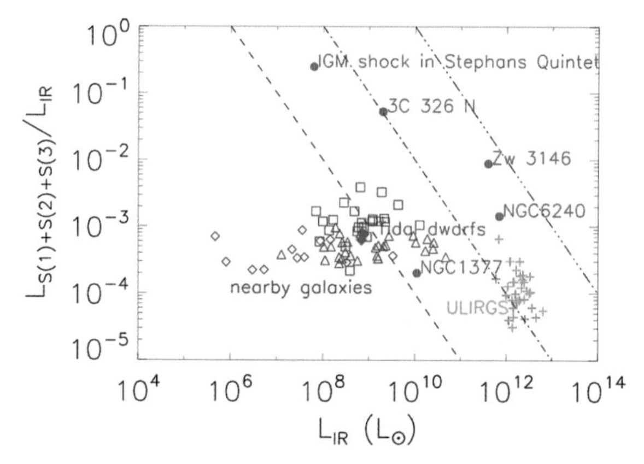

Fig. 11.25 Compilation of H_2 measurements for normal galaxies, ULIRGs, and a variety of other objects. The ratio of the luminosity in the sum of the S(1), S(2) and S(3) lines (of the ground vibration state) of H_2 to the infrared luminosity of the system is plotted against the infrared luminosity. Nearby galaxies are shown as open symbols, diamonds represent dwarf galaxies, triangles represent star-forming galaxies with no other type of activity, squares represent galaxies with Seyfert or LINER nuclei. ULIRGs are shown as crosses. Lines of constant H_2 luminosity (10^6, 10^8, and 10^{10} $L_\odot$) show that, in spite of the variations in LH2/Lir, ULIRGs are still more luminous in H_2 rotational lines, while intergalactic shocks can easily outshine whole galaxies in these lines Figure taken from Soifer et al. (2008)

luminosity in these flows may be emitted in the pure rotational lines of molecular hydrogen (Johnstone et al. 2007). The 11.3 μm PAH feature is also present in these spectra. Another example of bright molecular emission on galaxy wide scales is provided by Stephan's quintet (cf., Fig. 11.24) where the rapid motion of galaxy NGC 7318B has shock-heated gas to 6×10^6 K (Trinchieri et al. 2003). Spitzer-IRS observations have revealed pure rotational H_2 emission over a 24 kpc scale which is about 10 times the luminosity of the soft X-rays (Appleton et al. 2006).

Spitzer has started to reveal the importance of pure rotational H_2 emission on a galactic scale from local and more distant galaxies (Fig. 11.25; Soifer et al. (2008)). For the nearby galaxies and ULIRGS, the H_2 likely traces massive star formation through PDR activity. However, for the galaxies with much excess, the H_2 emission probably results from shocks associated with mass infall or outflow activity. It is clear that the role of molecular gas in the mass and energy flow on galactic scales is only now beginning to be appreciated. JWST will have the sensitivity to study this gas in detail. However, its small spectroscopic footprint will hamper large scale observations. A wide-field, mid-IR, cooled spectroscopic mapping mission may be called for to address this issue. As for the PAHs and [CII], SOFIA might be instrumental in clarifying and calibrating the relationship between this tracer of activity and the local physical conditions, particular if a future generation instrument were dedicated to wide-field imaging in H_2 lines.

Acknowledgments I am grateful for insightful discussions with Francois Boulanger, Pierre Cox, Frederic Galliano, George Helou, Ciska Marwick-Kemper, Els Peeters, Henrik Spoon, and Leisa Townsley which helped focus this review. I like to thank the moderator, Alberto Noriega-Crespo, for his comments on the presentation and his careful reading of an earlier version of this manuscript.

References

Abergel, A., et al. 2002, A & A, 389, 239
Acke, B., & van den Ancker, M. E. 2004, A & A, 426, 151
Allamandola, L. J., Tielens, G. G. M., & Barker, J. R. 1989, ApJS, 71, 733
Allamandola, L. J., Hudgins, D. M., & Sandford, S. A. 1999, ApJ, 511, L115
Appleton, P. N., et al. 2006, ApJ, 639, L51
Armus, L., et al. 2007, ApJ, 656, 148
Bakes, E. L. O., & Tielens, A. G. G. M. 1994, ApJ, 427, 822
Barlow, M., 2008, Astrophysics in the Next Decade (Springer)
Bertoldi, F., et al. 2003, A & A, 409, L47
Bertoldi, F., & Cox, P. 2002, A & A, 384, L11
Bock, J.J., et al., 1993, ApJ, 410, L115
Brandl, B. R., et al. 2006, ApJ, 653, 1129
Bregman, J. N. 1980, ApJ, 236, 577
Bringa, E. M., et al. 2007, ApJ, 662, 372
Brucato, J. R., Strazzulla, G., Baratta, G., & Colangeli, L. 2004, A & A, 413, 395
Burgdorf, M., Ashby, M. L. N., & Williams, R. 2007, ApJ, 668, 918
Burton, M. G., et al. 2000, ApJ, 542, 359
Calzetti, D., et al. 2007, ApJ, 666, 870
Cartledge, S. I. B., et al. 2005, ApJ, 630, 355

Cartledge, S. I. B., Lauroesch, J. T., Meyer, D. M., & Sofia, U. J. 2006, ApJ, 641, 327
Cowie, L. L., 1978, ApJ, 225, 887
de Avillez, M. A., & Mac Low, M.-M. 2002, ApJ, 581, 1047
Dalgarno, A., McCray, R.A., 1972, Annu Rev Astron Astrophys, 10, 375
Demyk, K., d'Hendecourt, L., Leroux, H., Jones, A. P., & Borg, J. 2004, A & A, 420, 233
Dickens, C., 1859, A tale of two cities (London, Penguin Books)
Draine, B. T., et al. 2007, ApJ, 663, 866
Draine, B. T., Salpeter, E. E. 1979, ApJ, 231, 438
Dwek, E., Scalo, J. M, 1980, ApJ, 239, 193
Ellison, D. C., Drury, L. O'C., Meyer, J.-P., 1997, ApJ, 487, 197
Engelbracht, C. W., Gordon, K. D., Rieke, G. H., Werner, M. W., Dale, D. A., & Latter, W. B. 2005, ApJ, 628, L29
Field, G. B. 1965, ApJ, 142, 531
Field, G. B. 1974, ApJ, 187, 453
Field, G. B., Goldsmith, D. W., & Habing, H. J. 1969, ApJ, 155, L149
Galliano, F., Dwek, E., & Chanial, P. 2008a, ApJ, 672, 214
Galliano, F. et al., 2008b, ApJ, 679, 310
Geers, V. C., et al. 2006, A & A, 459, 545
Genzel, R., et al. 1998, ApJ, 498, 579
Greenberg, J. M., & Li, A. 1999, Advances in Space Research, 24, 497
Grün et al. 1993, Nature, 362, 428
Gry, C., Lequeux, J., & Boulanger, F. 1992, A & A, 266, 457
Heckman, T., et al. 2002, ApJ, 577, 691
Heiles, C, 1994, ApJ, 436, 720
Heiles, C., & Troland, T. H. 2003, ApJ, 586, 1067
Hollenbach, D. J., & Tielens, A. G. G. M. 1999, Reviews of Modern Physics, 71, 173
Hony, S., Van Kerckhoven, C., Peeters, E., Tielens, A. G. G. M., Hudgins, D. M., & Allamandola, L. J. 2001, A & A, 370, 1030
Howk, J. C., & Savage, B. D. 1997, AJ, 114, 2463
Irwin, J. A., Kennedy, H., Parkin, T., & Madden, S. 2007, A & A, 474, 461
Isaak, K. G., Priddey, R. S., McMahon, R. G., Omont, A., Peroux, C., Sharp, R. G., & Withington, S. 2002, MNRAS, 329, 149
Jäger, C., Fabian, D., Schrempel, F., Dorschner, J., Henning, T., & Wesch, W. 2003, A & A, 401, 57
Jones A. P., Tielens, A. G. G. M., Hollenbach, D. J., McKee, C. F. 1994, ApJ, 433, 797
Jones A. P., Tielens, A. G. G. M., Hollenbach, D. J. 1996, ApJ, 469, 740
Johnstone, R. M., Hatch, N. A., Ferland, G. J., Fabian, A. C., Crawford, C. S., & Wilman, R. J. 2007, MNRAS, 382, 1246
Kaufman, M. J., Wolfire, M. G., & Hollenbach, D. J. 2006, ApJ, 644, 283
Kemper, F., Vriend, W. J., & Tielens, A. G. G. M., 2004, ApJ, 609, 826
Langhoff, S. 1996, J Phys Chem, 100, 2819
Lutz, D., et al. 2007, ApJ, 661, L25
Madden, S. C., Galliano, F., Jones, A. P., & Sauvage, M. 2006, A & A, 446, 877
Maiolino, R., et al. 2005, A & A, 440, L51
Malfait, K., et al. 1998, A & A, 332, L25
Malhotra, S., et al. 2001, ApJ, 561, 766
Markwick-Kemper, F., Gallagher, S. C., Hines, D. C., & Bouwman, J. 2007, ApJ, 668, L107
Mattila, K., Lemke, D., Haikala, L. K., Laureijs, R. J., Leger, A., Lehtinen, K., Leinert, C., & Mezger, P. G. 1996, A & A, 315, L353
McKee, C. F., & Ostriker, J. P. 1977, ApJ, 218, 148
Meixner, M., et al. 2006, AJ, 132, 2268
Mennella, V., Baratta, G. A., Esposito, A., Ferini, G., & Pendleton, Y. J. 2003, ApJ, 587, 727
Moutou, C., Sellgren, K., Verstraete, L., & Léger, A. 1999, A & A, 347, 949
Norman, C.A., Ikeuchi, S. 1989, ApJ, 345, 372

Peeters, E., Hony, S., Van Kerckhoven, C., Tielens, A. G. G. M., Allamandola, L. J., Hudgins, D. M., & Bauschlicher, C. W. 2002, A & A, 390, 1089
Peeters, E., Spoon, H. W. W., & Tielens, A. G. G. M. 2004, ApJ, 613, 986
Pendleton, Y. J., Sandford, S. A., Allamandola, L. J., Tielens, A. G. G. M., & Sellgren, K. 1994, ApJ, 437, 683
Pope, A., et al. 2007, ArXiv e-prints, 711, arXiv:0711.1553
Pottasch, S. R., Wesselius, P. R., & van Duinen, R. J. 1979, A & A, 74, L15
Povich, M. S., et al., 2007, ApJ, 660, 346
Puget, J. L., & Leger, A. 1989, Annu Rev Astron Astrophys, 27, 161
Rapacioli, M., Joblin, C., & Boissel, P. 2005, A & A, 429, 193
Regan, M. W., et al. 2004, ApJS, 154, 204
Rigby, J. R., et al. 2007, ArXiv e-prints, 711, arXiv:0711.1902
Roche, P., Aitken, 1985, MNRAS, 215, 425
Routly, P. M., Spitzer, L., Jr, 1952, ApJ, 115, 227
Savage, B. D., Sembach, K.R. 1996, Annu Rev Astron Astrophys, 34, 279
Savage, B. D., et al., 2003, ApJS, 146, 125
Seab, C. G., Shull, J. M. 1983, ApJ, 275, 652
Sellgren, K. 1984, ApJ, 277, 623
Shull, J. M. 1987, Interstellar Processes, 134, 225
Slavin, J. D., Jones, A. P., Tielens, A. G. G. M., 2004, ApJ, 614, 796
Smith, J. D. T., et al. 2007, ApJ, 656, 770
Soifer, B. T., Helou, G., Werner, M. 2008, Annu Rev Astron Astrophys, 46, 201
Spoon, H., et al. 2006, ApJ, 638, 759
Sylvester, R. J., Kemper, F., Barlow, M. J., de Jong, T., Waters, L. B. F. M., Tielens, A. G. G. M., & Omont, A. 1999, A & A, 352, 587
Tenorio-Tagle, G. 1979, A & A, 71, 59
Thompson, T. W. J., Howk, J. C., & Savage, B. D. 2004, AJ, 128, 662
Tielens, A. G. G. M., 1998, ApJ, 499, 267
Tielens, A. G. G. M., 2001, in Tetons 4: Galactic Structure, Stars and the Interstellar Medium, eds. C. E. Woodward, M. D. Bicay, and J. M. Shull (San Francisco, ASP) 231, 92
Tielens, A. G. G. M. ed. 2005, The Physics and Chemistry of the Interstellar Medium, (Cambridge, UK: Cambridge University Press)
Tielens, A. G. G. M., 2008, Ann Rev Astron Astrophys, 46, 289
Townsley, L. K., Feigelson, E. D., Montmerle, T., Broos, P. S., Chu, Y.-H., & Garmire, G. P. 2003, ApJ, 593, 874
Trinchieri, G., Sulentic, J., Breitschwerdt, D., & Pietsch, W. 2003, A & A, 401, 173
van Dishoeck, E. F., & Black, J. H. 1986, ApJS, 62, 109
Verstraete, L., et al. 2001, A & A, 372, 981
Watson, W. D. 1973, Interstellar Dust and Related Topics, 52, 335
Weaver, R., McCray, R., Castor, J., Shapiro, P., & Moore, R. 1977, ApJ, 218, 377
Weingartner, J. C., & Draine, B. T. 2001, ApJS, 134, 263
Wolfire, M. G., McKee, C. F., Hollenbach, D., & Tielens, A. G. G. M. 2003, ApJ, 587, 278
Wolfire, M. G., Hollenbach, D., McKee, C. F., Tielens, A. G. G. M., & Bakes, E. L. O. 1995, ApJ, 443, 152
Wu, Y., Charmandaris, V., Hao, L., Brandl, B. R., Bernard-Salas, J., Spoon, H. W. W., & Houck, J. R. 2006, ApJ, 639, 157
Zinner, E. K., 2003, In Treatise on Geochemistry Vol. 1, eds. K. K. Turekian, H. D. Holland, A. M. Davis (Amsterdam: Elsevier), p. 17

Jon Gardner, Dale Cruikshank, and Yvonne Pendleton are obviously enjoying having their picture taken

Chapter 12
Astrophysics in the Next Decade: The Evolution of Galaxies

Alice Shapley

Abstract One of the primary goals of observational cosmology is to understand how galaxies form and evolve into the current population of objects we observe around us today. The redshift range, $1.5 \leq z \leq 3.5$ represents a crucial epoch for observing the assembly of the modern galaxy population, which is largely in place by $z \sim 1$. There has been tremendous progress assembling observations at $z > 1.5$ over the past decade. Yet, many challenges remain. The next generation of large telescopes on the ground and in space will enable us to place unprecedented contraints on the connection between distant galaxies in the early universe and their local counterparts; the detailed stellar populations of distant galaxies while the bulk of their stellar mass is being formed; the structure and dynamics of galaxies as they settle onto the Hubble sequence; and the physics of "feedback", a crucial, yet poorly understood ingredient in models of galaxy formation.

12.1 Introduction

Understanding the formation and evolution of different types of galaxies represents one of the great challenges of modern cosmology. We know the universe began in a nearly uniform state shortly after the Big Bang. There is a detailed theoretical framework that accurately describes how tiny initial fluctuations evolved under the influence of gravity into the large-scale spatial distribution of matter in the current universe. However, there are many critical unanswered questions about how the interplay among gas, stars, radiation and heavy elements led to the rich variety of galaxy masses, luminosities, colors, and shapes that inhabit this large-scale structure.

Along with ab initio simulations of galaxy formation, and the study of fossil record of our own and other nearby galaxies, a powerful method of studying galaxy formation consists of direct observations of distant galaxies (Nagamine et al. 2006).

A. Shapley (✉)
Princeton University, Peyton Hall – Ivy Lane, Princeton, NJ 08544, USA
e-mail: aes@astro.princeton.edu

H.A. Thronson et al. (eds.), *Astrophysics in the Next Decade,* Astrophysics and Space Science Proceedings, DOI 10.1007/978-1-4020-9457-6_12,

Such observations indicate that the redshift range between $z \sim 1.5$ and $z \sim 3.5$ may hold many important keys to the puzzle of galaxy formation.

Indeed, efforts to characterize the global evolution of "activity" in the universe indicate that, once the effects of dust extinction are taken into account, the level of star-formation density was roughly an order of magnitude higher from $z \sim 1$–4 than it is at the current epoch. Figure 12.1 from Reddy et al. (2007) shows recent determinations of the star-formation rate density as a function of redshift, based on both rest-frame ultraviolet (UV) and far-infrared light. It is also worth keeping in mind that, at $z \sim 2$, even a large fraction of the progenitors of today's massive early-type galaxies were actively forming stars. Daddi et al. (2007) show that roughly 40% of galaxies with stellar masses $\geq 10^{11} M_{\odot}$ at $z \sim 2$ would be classified as ULIRGs, with $L_{IR} > 10^{12} L_{\odot}$. At the same time, it has long been known that the space density of optically luminous QSOs peaks at $z \sim 2.5$ (Schmidt et al. 1995).

The star-formation rate density is an instantaneous quantity. There have also been attempts to characterize the integral of this quantity: the evolution of the global stellar mass density (Bundy et al. 2006; Drory et al. 2005; Fontana et al. 2006). It appears that roughly 50% of the local stellar mass density is in place by $z \sim 1$, while perhaps only as much as 10–15% is in place by $z \sim 3$ (Drory et al. 2005, Fig. 12.2; Fontana et al. 2006). Clearly, there is a steep increase in the stellar mass content of the universe prior to $z \sim 1$. It is also possible to characterize the differential number counts of galaxies as a function of mass. In such studies, there is evidence for differential evolution in the galaxy stellar mass function between $z \sim 1$ and 0, in the sense that the evolution of the mass density of objects with $M_{*} > 10^{11} M_{\odot}$ is milder than that of less massive galaxies (Fontana et al. 2006; Bundy et al. 2006).

Along with the formation and assembly of the stellar mass in galaxies, we also want to understand how galaxy structural and morphological properties evolve with redshift and as a function of large-scale environment. Locally, the bimodal distribution of galaxies in color-luminosity space (Strateva et al. 2001; Baldry et al. 2004) extends into the realm of morphologies, with massive, quiescent red galaxies characterized typically by early-type elliptical morphologies and dispersion-supported

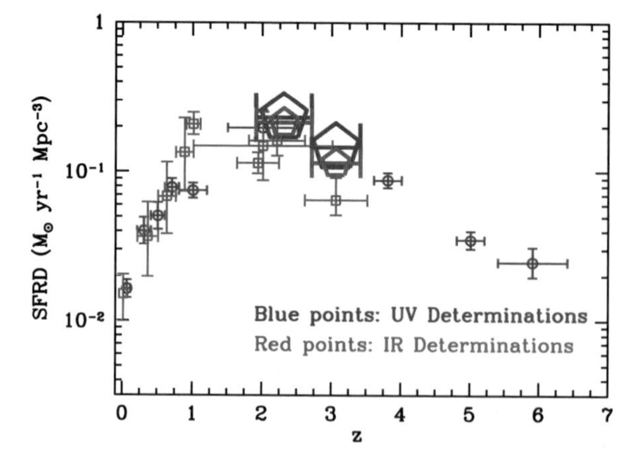

Fig. 12.1 From Reddy et al. (2007). This figure summarizes UV (*blue open circles*) and IR (*red open squares*) estimates of the global star-formation rate density as a function of redshift. Star-formation rate estimates are based on the integrated luminosity density at each wavelength, with a redshift-dependent correction for dust extinction. At $z \sim 2$ and $z \sim 3$, the average correction factor to the UV luminosity density is 4.5

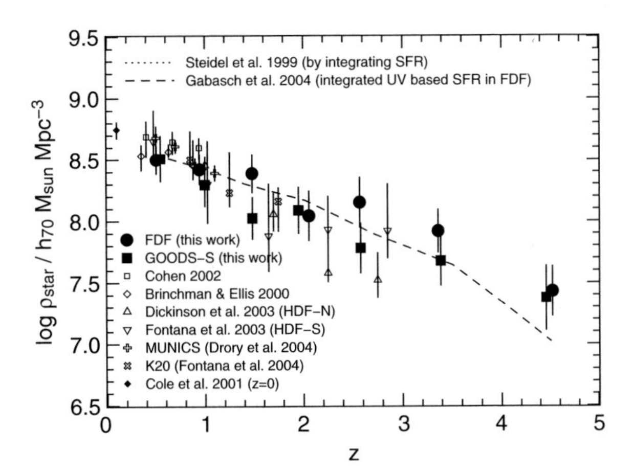

Fig. 12.2 From (Drory et al. 2005). The global stellar mass density as a function of redshift. The integral of the star-formation rate density is indicated as a *dotted line*

kinematics, and less massive, star-forming, blue galaxies described by disk-like morphologies, and rotationally-supported dynamics. The well-defined local morphological Hubble sequence of disks and spheroids appears to break down by $z \sim 1.5$ (Dickinson 2000; Lotz et al. 2006; Law et al. 2007a). Furthermore, the typically irregular morphologies of distant star-forming galaxies are found not only at the rest-frame UV wavelengths probed by the most commonly used optical imagers on the *Hubble Space Telescope* (HST) and shown in Fig. 12.3, but also at the rest-frame optical wavelengths probed by near-IR images (Dickinson 2000). Understanding the significance of these disturbed morphologies, and how they are connected in detail with galaxy assembly, is another crucial goal for studies of galaxy formation.

While we have recently made huge strides identifying and characterizing galaxies in this critical redshift range of $1.5 \leq z \leq 3.5$, in terms of their global and detailed properties, there are still many respects in which our knowledge is highly incomplete. A new generation of multi-wavelength facilities will be coming on-line within the next decade including the James Webb Space Telescope (JWST), the Thirty-Meter Telescope (TMT), the Giant Magellan Telescope (GMT), the European Extremely Large Telescope (ELT), and the Atacama Large Millimeter Array (ALMA). This next generation of large telescopes on the ground and in space will help us address these large gaps in our understanding of galaxy formation. In what follows, I will briefly review several techniques for identifying distant galaxies, and then highlight the areas of galaxy formation in which significant progress will be enabled by future facilities.

12.2 High-Redshift Galaxy Selection

Over the last several years, there has been a veritable explosion of surveys for galaxies at $z > 1.5$. Unlike traditional, apparent-magnitude-limited surveys, these new

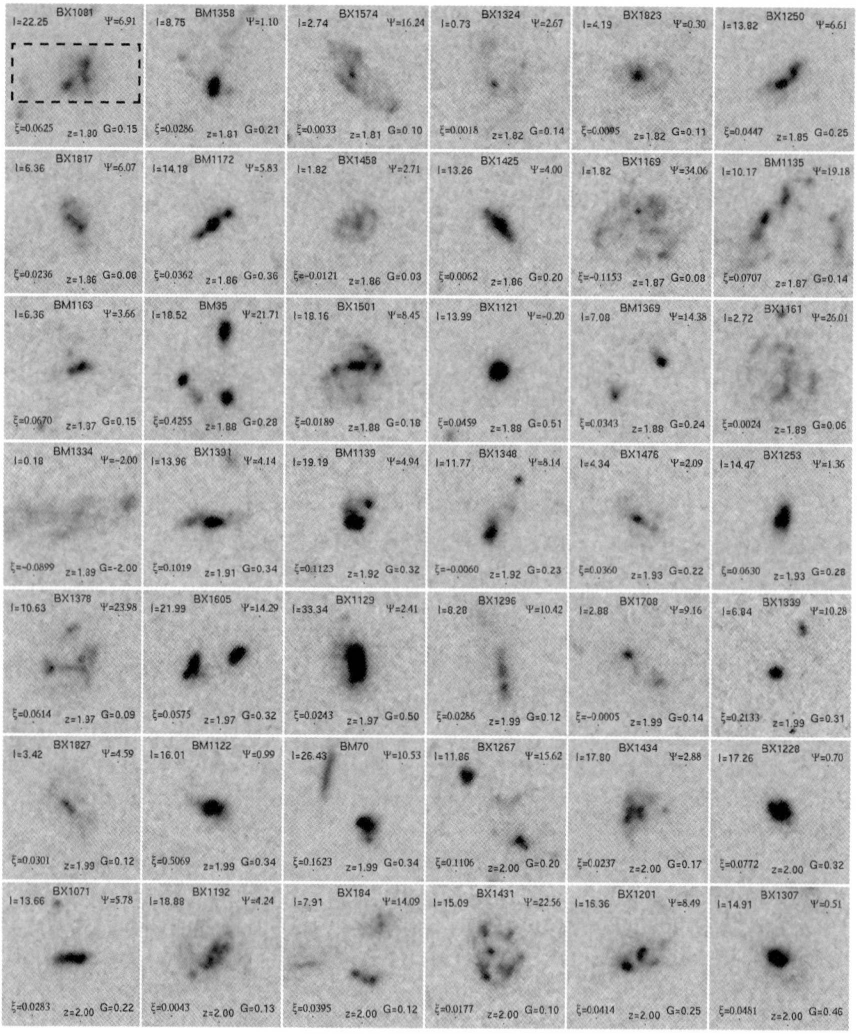

I=22.25 BX1081 Ψ=6.91
ξ=0.0625 z=1.80 G=0.15
I=8.75 BM1358 Ψ=1.10
ξ=0.0286 z=1.81 G=0.21
I=2.74 BX1574 Ψ=16.24
ξ=0.0033 z=1.81 G=0.10
I=0.73 BX1324 Ψ=2.67
ξ=0.0018 z=1.82 G=0.14
I=4.19 BX1823 Ψ=0.30
ξ=0.0095 z=1.82 G=0.11
I=13.82 BX1250 Ψ=6.61
ξ=0.0447 z=1.85 G=0.25
I=6.36 BX1817 Ψ=6.07
ξ=0.0236 z=1.86 G=0.08
I=14.18 BM1172 Ψ=5.83
ξ=0.0362 z=1.86 G=0.36
I=1.82 BX1458 Ψ=2.71
ξ=-0.0121 z=1.86 G=0.03
I=13.26 BX1425 Ψ=4.00
ξ=0.0062 z=1.86 G=0.20
I=1.82 BX1169 Ψ=34.06
ξ=-0.1153 z=1.87 G=0.08
I=10.17 BM1135 Ψ=19.18
ξ=0.0707 z=1.87 G=0.14
I=6.36 BM1163 Ψ=3.66
ξ=0.0670 z=1.87 G=0.15
I=18.52 BM35 Ψ=21.71
ξ=0.4255 z=1.88 G=0.28
I=18.16 BX1501 Ψ=8.45
ξ=0.0189 z=1.88 G=0.18
I=13.99 BX1121 Ψ=-0.20
ξ=0.0459 z=1.88 G=0.51
I=7.08 BM1369 Ψ=14.38
ξ=0.0343 z=1.88 G=0.24
I=2.72 BX1161 Ψ=26.01
ξ=0.0024 z=1.89 G=0.06
I=0.18 BM1334 Ψ=-2.00
ξ=-0.0899 z=1.89 G=-2.00
I=13.96 BX1391 Ψ=4.14
ξ=0.1019 z=1.91 G=0.34
I=19.19 BM1139 Ψ=4.94
ξ=0.1123 z=1.92 G=0.32
I=11.77 BX1348 Ψ=8.14
ξ=-0.0060 z=1.92 G=0.23
I=4.34 BX1476 Ψ=2.09
ξ=0.0360 z=1.93 G=0.22
I=14.47 BX1253 Ψ=1.36
ξ=0.0630 z=1.93 G=0.28
I=10.63 BX1378 Ψ=23.98
ξ=0.0614 z=1.97 G=0.09
I=21.99 BX1605 Ψ=14.29
ξ=0.0575 z=1.97 G=0.32
I=33.34 BX1129 Ψ=2.41
ξ=0.0243 z=1.97 G=0.50
I=8.28 BX1296 Ψ=10.42
ξ=0.0286 z=1.99 G=0.12
I=2.88 BX1708 Ψ=9.16
ξ=-0.0005 z=1.99 G=0.14
I=6.84 BX1339 Ψ=10.28
ξ=0.2133 z=1.99 G=0.31
I=3.42 BX1827 Ψ=4.59
ξ=0.0301 z=1.99 G=0.12
I=16.01 BM1122 Ψ=0.99
ξ=0.5069 z=1.99 G=0.34
I=26.43 BM70 Ψ=10.53
ξ=0.1623 z=1.99 G=0.34
I=11.86 BX1267 Ψ=15.62
ξ=0.1106 z=2.00 G=0.20
I=17.80 BX1434 Ψ=2.88
ξ=0.0237 z=2.00 G=0.17
I=17.26 BX1228 Ψ=0.70
ξ=0.0772 z=2.00 G=0.32
I=13.66 BX1071 Ψ=5.78
ξ=0.0283 z=2.00 G=0.22
I=18.88 BX1192 Ψ=4.24
ξ=0.0043 z=2.00 G=0.13
I=7.91 BX184 Ψ=14.09
ξ=0.0395 z=2.00 G=0.12
I=15.09 BX1431 Ψ=22.56
ξ=0.0177 z=2.00 G=0.10
I=16.36 BX1201 Ψ=8.49
ξ=0.0414 z=2.00 G=0.25
I=14.91 BX1307 Ψ=0.51
ξ=0.0481 z=2.00 G=0.46

Fig. 12.3 From Law et al. (2007a). HST Advanced Camera for Surveys (ACS) rest-UV morphologies of star-forming galaxies at $z \sim 2$. All panels include the galaxy name, redshift z, and morphological parameters size (I), Gini (G), multiplicity (Ψ), and color dispersion (ξ) (See Law et al. (2007a) for details). Images are 3 on a side, oriented with north up and east to the left

results utilize several complementary selection techniques for finding high-redshift galaxies. Several of these techniques are displayed in Fig. 12.4. It has been demonstrated that, while efficient in identifying distant galaxies, all of these selection methods suffer from incompleteness with respect to a sample defined in terms of physical quantities such as stellar mass, or star-formation rate (Reddy et al. 2005; van Dokkum et al. 2006). Future surveys of high-redshift galaxies need to address this issue of sample definition and completeness.

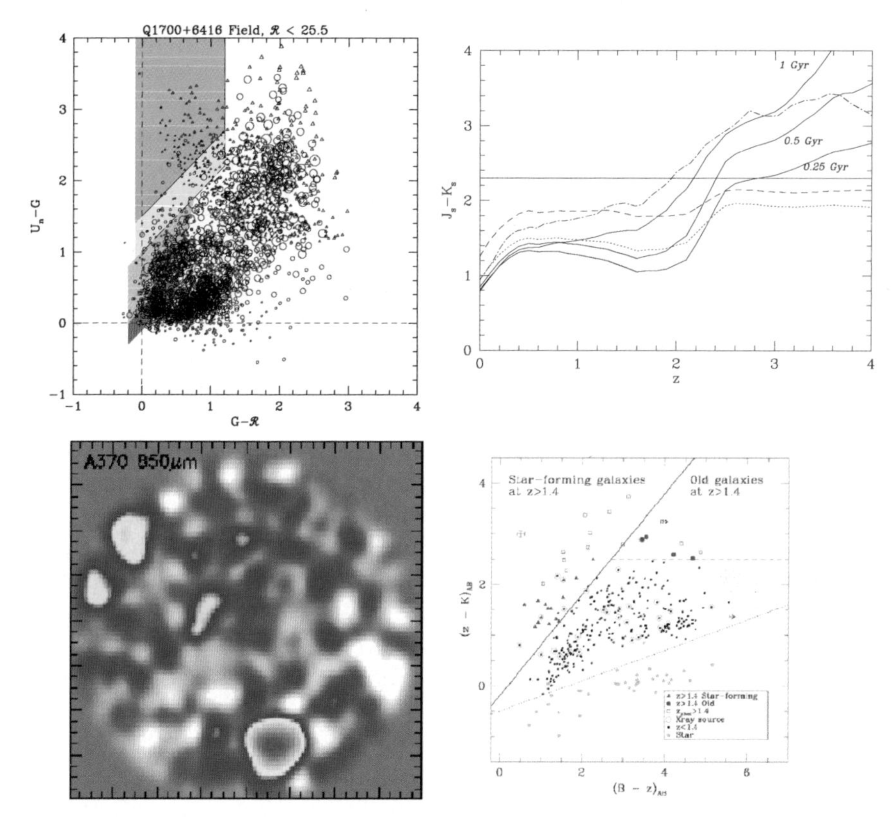

Fig. 12.4 Various high-z selection techniques. (*Top Left*) From Steidel et al. (2004). Rest-frame UV-selection, based on $U_n - G$ and $G - R$ colors. The green and yellow regions correspond the colors of galaxies at typical redshifts of $z \sim 3.0$. The cyan region corresponds to galaxies typicall at $z \sim 2.3$, while the magenta region corresponds to $z \sim 1.7$. (*Top Right*) From (Franx et al. 2003). $J-K$ selection of DRGs. The expected $J-K$ color is shown for different stellar population synthesis models as a function of redshift. (*Bottom Left*) From (Smail et al. 1997). Deep submillimeter map of the rich cluster Abell 370, showing the submillimeter detection of luminous background galaxies. (*Bottom Right*) From (Daddi et al. 2004). BzK selection. Here, different regions of the $z - K$ vs. $B - z$ color space correspond to the expected colors of star-forming or passive galaxies at $z \sim 1.4$–2.5

Rest-frame UV color selection (Steidel et al. 2003, 2004) has been used to identify galaxies over a broad range of redshifts. First applied at $z{\sim}3$ and known as the "Lyman Break" technique, this selection method has since been applied at both lower and higher redshift, and is tuned to find star-forming galaxies with moderate amounts of dust extinction (typically a factor of ~5 in the rest-frame UV). The top left panel of the figure from Steidel et al. (2004) shows how U_n-G and $G-\mathcal{R}$ colors are used to identify star-forming galaxies (LBGs) at $z{\sim}3$ (green/yellow region), $z{\sim}2.3$ (cyan region), and $z{\sim}1.7$ (magenta region). It is also possible to select galaxies on the basis of rest-frame optical colors. As shown in the top right panel of Fig. 12.4, a large observed $J-K$ color may be indicative of evolved "Distant

Red Galaxies" (DRGs) at $z>2$ with pronounced Balmer or 4000 Å breaks (Franx et al. 2003). In practice, a large fraction of high-redshift galaxies so identified are characterized by dusty star formation, which can also produce a red $J-K$ color (Papovich et al. 2006). A combination of optical and infrared filters have been used to formulate the "BzK" criteria (Daddi et al. 2004), based on which $B-z$ and $z-K$ colors of galaxies isolate both star-forming and passive galaxies at $1.4 \leq z \leq 2.5$ (bottom right of Fig. 12.4). There appears to be considerable overlap between the population of star-forming galaxies selected with this method and those identified through $U_nG\mathcal{R}$ colors (Reddy et al. 2005). Shown in the lower left-hand panel of Fig. 12.4, dusty galaxies with large bolometric luminosities ($L > 10^{12}$) have been identified by their fluxes at 850 µm, measured with the SCUBA instrument on the JCMT. The redshift distribution of bright submillimeter galaxies with 1.4 GHz radio counterparts above 30 μJy peaks at roughly $z \sim 2.3$ (Smail et al. 1997; Chapman et al. 2003), therefore probing similar redshifts to those of galaxies identified with the other multi-wavelength techniques described here. Though not shown in Fig. 12.4, narrow-band imaging techniques have also been widely used to identify high-redshift galaxies based on the presence of strong Lyα emission (Cowie & Hu 1998; Kodaira et al. 2003), ranging in redshift from $z \sim 2$ to epochs beyond the redshift of reionization.

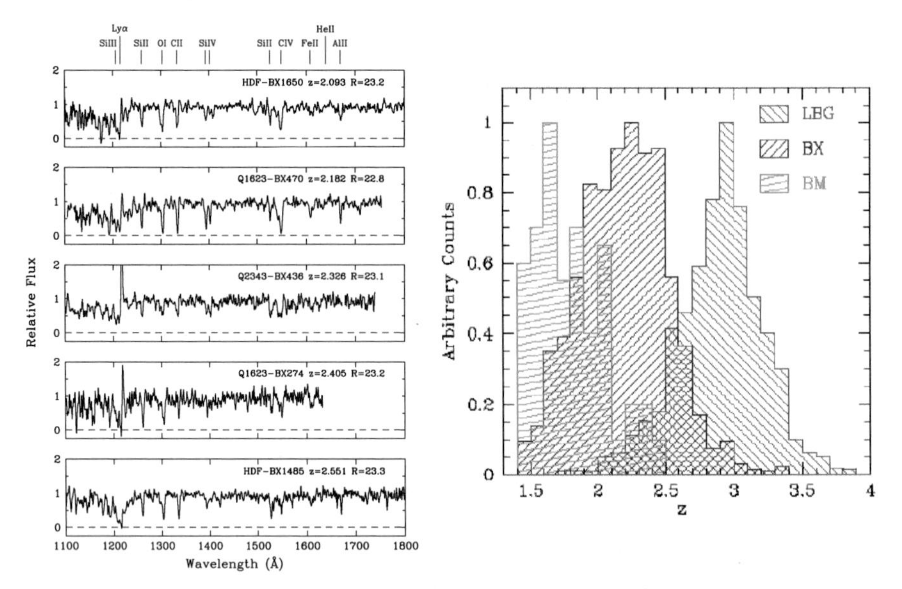

Fig. 12.5 (*Left*) From Steidel et al. (2004). Examples of spectra for UV-selected galaxies at $z{\sim}2$. The spectra have been shifted into the rest frame and normalized to unity in the continuum. The strongest spectral features are indicated at the top of the panel. (*Right*) Normalized redshift distributions of spectroscopically confirmed UV-selected galaxies. "BM" objects (*green histogram*) have $\langle z \rangle = 1.7$, "BX" objects (*blue histogram*) have $\langle z \rangle = 2.3$, while LBGs (*red histogram*) have $\langle z \rangle = 3.0$. From Reddy (2007, private communication)

One critical limiting factor in interpreting observations of high redshift galaxies is spectroscopic completeness. Spectroscopic data is crucial for constraining both the large-scale spatial distribution of galaxies, and the detailed properties of individual objects. Currently, UV-selected samples are characterized by the highest level of spectroscopic completeness. As shown in Fig. 12.5, there are more than 2000 spectroscopically-confirmed UV-selected galaxies at $1.5 \leq z \leq 3.5$. Spectra are typically obtained for galaxies with $\mathcal{R}$-band magnitudes brighter than 25.5 (AB). At fainter magnitudes, obtaining spectroscopic redshifts in reasonable exposure times becomes unfeasible with even 8–10-meter class telescopes. The problem is that a significant fraction of galaxies down to an interesting limit in stellar mass are fainter than $R = 25.5$. van Dokkum et al. (2006) show that, at $M_* \sim 10^{11} M_{\odot}$, the median $\mathcal{R}$-band magnitude is $\sim$ 26 (AB). Obtaining optical (rest-frame UV) spectroscopic redshifts for such faint objects will be straightforward with the next generation of large ground-based telescopes. Figure 12.6 shows a simulated 1-hour spectrum of $R = 26$ mag galaxies with either strong Lyα emission or absorption, using the planned Wide-Field Optical Spectrograph (WFOS) for TMT [1]. In both cases, spectroscopic redshifts would be easily obtained. Deep near-IR spectroscopy with the next generation of ground- and space-based telescopes, described in Section 12.4, will also provide redshifts for optically-faint sources.

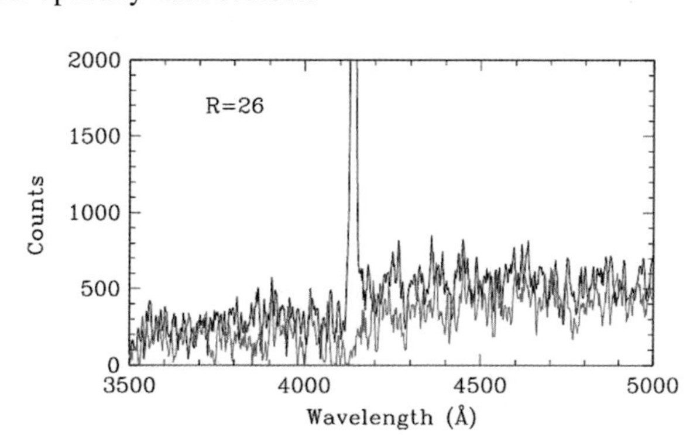

Fig. 12.6 Simulated TMT/WFOS spectrum for $z = 2.4$ galaxies with $R = 26$ at a resolution of $R = 1000$. From (Abraham et al. 2006)

12.3 Clustering

The study of galaxy clustering is one of the applications for which the spectroscopic identification of large and complete samples of objects is critically important. Spectroscopic redshifts are required to obtain accurate space densities and measures of spatial clustering. Reproduced from Benson et al. (2001), Figure 12.7 shows a z=0 slice through an cosmological dark matter simulation combined with a semi-analytic model for galaxy formation. The large-scale distribution of galaxies within the web of dark matter is related to properties of the individual galaxies. Accordingly, in concert with cosmological dark matter simulations, the large-scale spatial distribution of a population of galaxies at a specific redshift can be used to connect this population

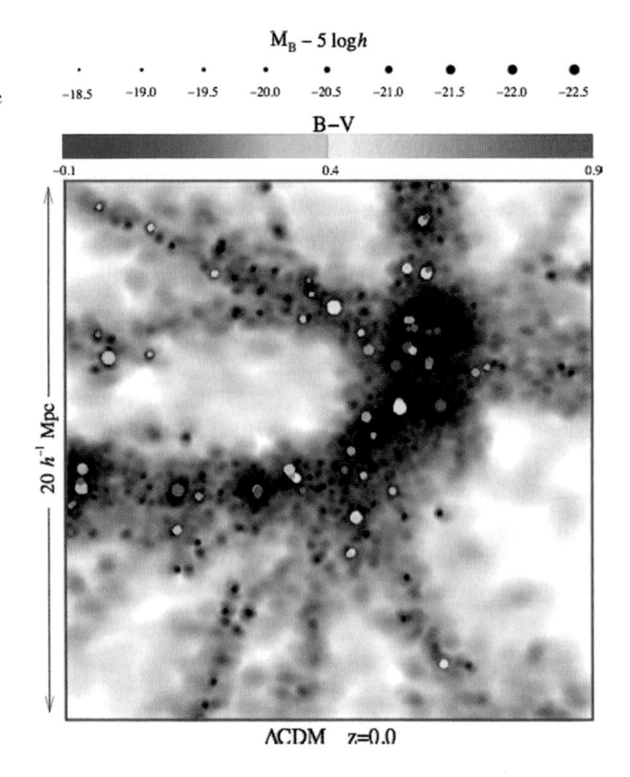

Fig. 12.7 From Benson et al. (2001). Slices through an N-body simulation volume at $z = 0$, in comoving coordinates. Positions of galaxies from a corresponding semi-analytic model are indicated by colored circles, with sizes indicating B-band absolute magnitude, and color indicating rest-frame $B - V$ color

to its host dark matter halos at that redshift. Indeed, the clustering strength of a given sample of galaxies can be used to estimate the minimum dark matter halo mass, M_{min}, hosting such galaxies, for a specified cold-dark matter (CDM) cosmology (see e.g. Wechsler et al. 1998; Adelberger et al. 2005a). This approach assumes that a sample of galaxies above a given luminosity threshold corresponds, at least approximately, to a sample of halos above a given mass threshold. Once a correspondence between galaxies and dark matter halos at high redshift is established, the evolution of these host dark matter halos to later epochs is easily extracted from cosmological simulations. The clustering of the lower-redshift descendant halos can then be compared with the clustering of known lower-redshift galaxy populations, to identify the galaxy descendants of the high-redshift objects. The simulations can also be used in an analogous way to identify the progenitors of a given galaxy population.

Building on previous work by Adelberger et al. (2005a), Conroy et al. show that UV-selected galaxies at $z{\sim}2$ are hosted by dark matter halos with $M \geq 10^{11.4} h^{-1} M_{\odot}$ (see Fig. 12.8, left). Based on a comparison between the space densities, clustering strengths, and satellite fractions of the descendant dark matter halos, and known galaxy populations at $z{\sim}1$ and $z{\sim}0$ (see Fig. 12.8, right), Conroy et al. conclude that the majority of UV-selected galaxies at $z{\sim}2$ evolve into "typical" L^* galaxies by $z{\sim}0$, with a mixture of red and blue colors. The same detailed analysis of the descendants of red galaxies at $z{\sim}2$ (e.g., DRGs, passive BzKs, and SCUBA

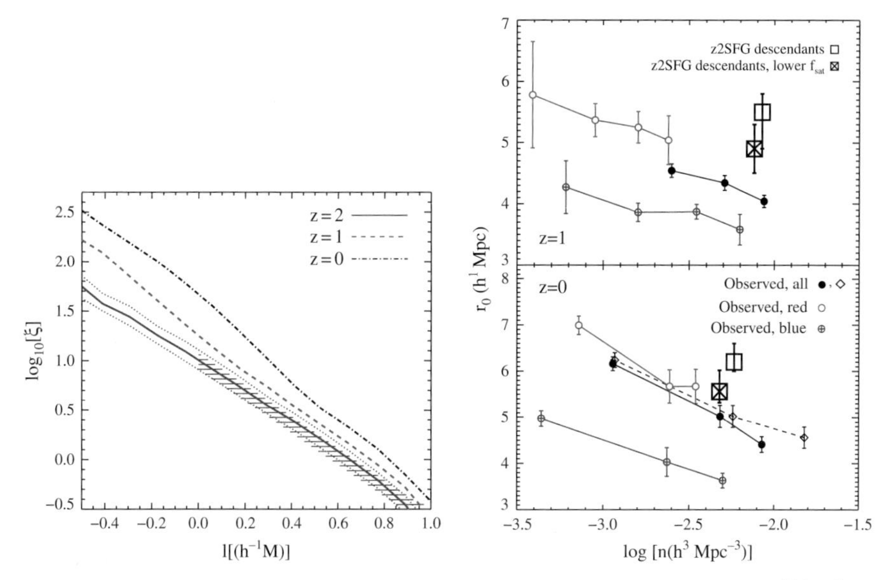

Fig. 12.8 From Conroy et al. (*Left*) The correlation function for halos with $M \geq 10^{11.4}h^{-1}M_\odot$ at $z{\sim}2$ (*solid line*) that have clustering properties similar to observed UV-selected star-forming galaxies at $z{\sim}2$ (*hatched region* (Adelberger et al. 2005a)). Also plotted are the correlation functions of those halos evolved to $z{\sim}1$ and $z{\sim}0$. (*Right*) Relationship between clustering strength (r_0) and sample number density (n) for observed galaxies and the descendants of the halos hosting $z{\sim}2$ UV-selected star-forming galaxies at $z{\sim}1$ (*top panel*) and $z{\sim}0$ (*bottom panel*). The data at $z{\sim}1$ are for samples defined above various magnitude thresholds (from -19.5 to -20.5 in half magnitude steps for the overall sample, and from -19 to -21 for the color-defined samples; (Coil et al. 2006, 2007), while at $z \sim 0$ they are defined for magnitude bins (in one magnitude intervals from -19 to -22 (Zehavi et al. 2005))

sources) has not been possible thus far, due to a limited number of spectroscopic redshifts for these complementary populations. Such fundamental studies await deeper optical and near-IR spectroscopic surveys that will be possible with the next generation of ground- and space-based telescopes.

12.4 Galaxy Stellar Populations

Another important constraint on the past star-formation histories and assembly of galaxies at high redshift consists of stellar population synthesis modeling of the observed spectral energy distributions (SEDs). In the local universe, with samples such as the Sloan Digital Sky Survey (SDSS), optical spectral diagnostics including the strength of the 4000 Å break, and the strength and profiles of hydrogen Balmer and metal absorption lines, have been used to infer the stellar mass-to-light ratio, the fraction of stellar mass formed in a burst of star formation, the stellar velocity dispersion, metallicity and abundance pattern (see Fig. 12.9) (Kauffmann et al. 2003; Gallazzi et al. 2005). In contrast, most constraints on

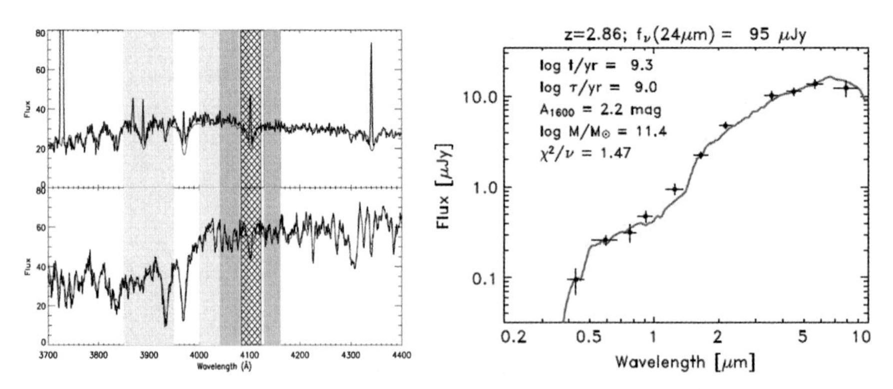

Fig. 12.9 (*Left*) From Kauffmann et al. (2003). SDSS galaxy spectra plotted from 3700–4400 Å. The shaded regions indicate the bandpasses over which the 4000Å break and $H\delta_A$ index are measured. The combination of these two spectroscopic parameters provide constraints on galaxy stellar populations (*Right*) From Papovich et al. (2006). The best-fit stellar population model for a massive galaxy at $z = 2.86$. Broad-band data modeled in this fit include HST/ACS $BViz$, VLT/ISAAC JHK_s, and Spitzer/IRAC 3.6–8.0 μm photometry

high-redshift stellar populations are derived from fitting stellar population synthesis models (e.g., Bruzual & Charlot 2003) to broad-band photometry alone (e.g., (Papovich et al. 2006; Erb et al. 2006b)). Such modeling constitutes a rather blunt tool, where simple exponentially-declining models are typically assumed for the past star-formation history, and an attempt is made to account for how much stellar mass might be hidden under the glare of the current episode of star formation. Furthermore, there is no direct spectroscopic information gained about the burst mass fractions, velocity dispersions, and metallicities and abundance patterns of stars forming the bulk of the stellar mass.

A sampling of near-IR spectra of $z\sim2$ galaxies demonstrates the source of these limitations. The largest Hα survey at $z\sim2$ has been presented by (Erb et al. 2006b), containing 114 objects with near-IR spectra. This survey has been used to study the Hα kinematics, fluxes, H II-region metallicities and physical conditions, and to establish systemic (center-of-mass) redshifts for objects undergoing large-scale outflows of gas, for which rest-frame UV redshift indicators are biased either blue or red. As shown in Figure 12.10, with typical Keck/NIRSPEC long-slit exposure times (1–2 objects along the slit in ~1 hour), and median near-IR magnitudes of $K_s\sim20.6$ (Vega), significant rest-frame optical continuum is not detected. Therefore, unlike in the local universe, rest-frame optical stellar population studies based on the shape of the high-redshift galaxy continuum or the strength of stellar or interstellar absorption lines, cannot be pursued with these spectra. Currently, coarse continuum spectroscopy has been performed using 8–10-meter-class telescopes for the rest-frame optically brightest ($K\sim19$) objects at $z\sim2$ (Kriek et al. 2006). However, since the surface density of such objects is 0.05 arcmin^{-2}, they must be observed one at a time, which results in a low survey efficiency. Ultimately, we will need to assemble similar continuum and absorption spectra for more typical $z\sim2$ galaxies, with

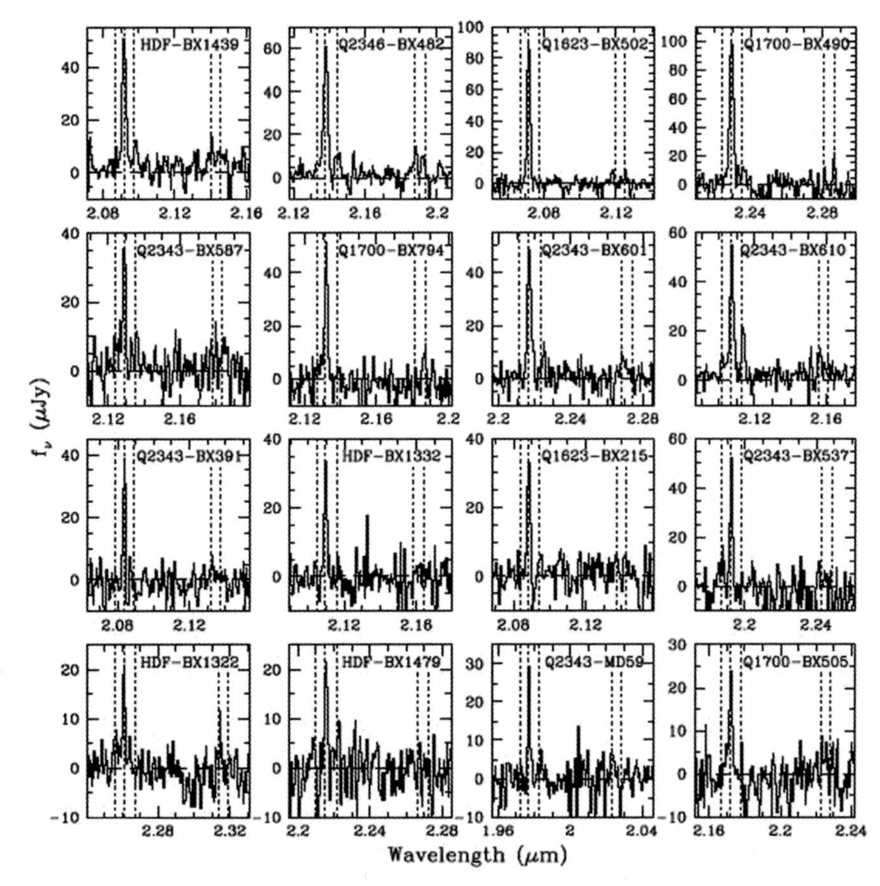

Fig. 12.10 From Erb et al. (2006b). Representative examples of $H\alpha$ spectra from the Keck/NIRSPEC survey of 114 star-forming galaxies at $z \sim 2$. Each row shows objects drawn from each quartile of Hα flux: the top row contains galaxies drawn from the highest quartile in flux, the second row from the second highest quartile, the third row from the third highest quartile, and the bottom row from the lowest quartile. The dotted lines in each panel show, from left to right, the locations of [N II]λ6548, Hα , [N II]λ6583, [S II]λ, λ6716, 6732

$K\sim$21.5. Planned near-IR spectrographs on future facilities, including NIRSpec on JWST and IRMS on TMT, will allow for multiplexed surveying of the rest-frame optical spectra of high-redshift galaxies.

High quality rest-frame UV spectra of star-forming galaxies also provide detailed insight into the nature of stellar populations. For example, as presented by Steidel et al. (2004), high signal-to-noise rest-frame UV spectra have been obtained for Q1307-BM1163, one of the brightest unlensed UV-selected star-forming galaxy with $z = 1.41$ and $\mathcal{R} = 21.8$ (Fig. 12.11). Analysis of the P-Cygni stellar wind lines of massive stars in BM1163 indicates that this galaxy is best described by a standard Salpeter initial mass function (IMF) at the high-mass end, and solar metallicity. To date, the detailed analysis of stellar wind lines has only been possible for BM1163,

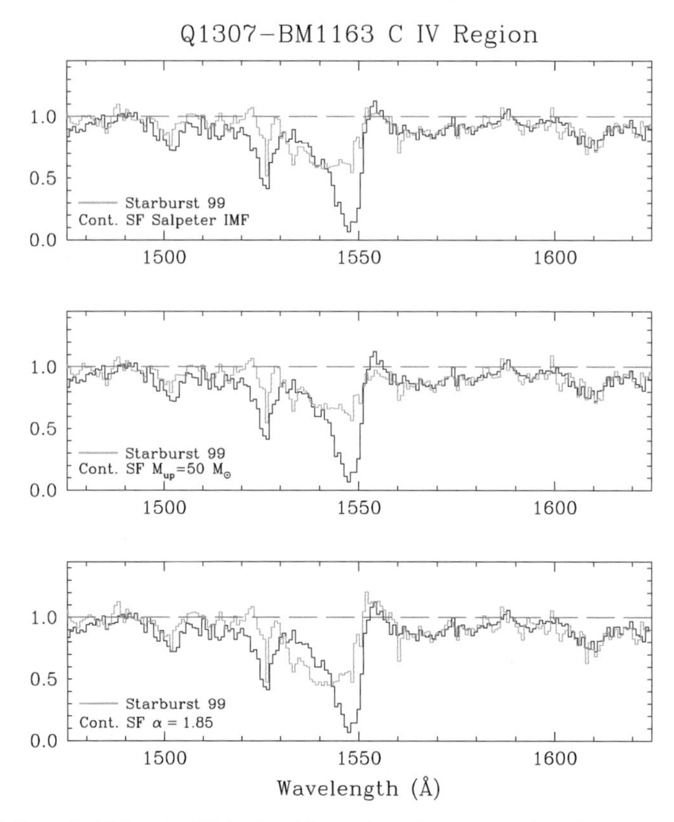

Fig. 12.11 From Steidel et al. (2004). Sensitivity of the C IV P-Cygni profile to the upper end of the IMF. Black histogram: Portion of the observed spectrum of Q1307-BM1163 ($z = 1.41$). *Green histogram*: Model spectra produced assuming a standard Salpeter IMF (*top*), an IMF lacking stars more massive than 50 $M_{\odot}$ (*middle*), and an IMF flatter than Salpeter (*bottom*). Overall, there is no evidence for a departure from the standard Salpeter IMF at the upper end of the mass distribution of stars in Q1307-BM1163. The P-Cygni lines are also consistent with solar metallicity

and gravitationally-lensed high-redshift galaxies (e.g., Pettini et al. 2002). However, with future optical spectrographs on the next generation of large ground-based facilities (e.g., WFOS on TMT), and moderate (≤ 5 hour) exposure times, it will be possible to carry out the same type of study for more typical star-forming galaxies at $z > 2$, with $R \sim 24$ mag.

12.5 Galaxy Assembly and Dynamics

While the luminosity of a galaxy is straightforward to measure, its mass is the more fundamental quantity. Indeed, one of the most important goals in the study of galaxy formation is to understand how mass is assembled in galaxies as a function of cosmic time. There are several methods for inferring the masses of distant galaxies. Stellar masses can be estimated by fitting population synthesis models to broad-band spectral energy distributions, while the clustering strength and number density of a galaxy population can be compared with cosmological simulations to determine the mass of the typical dark matter halo host (Adelberger et al. 2005a). In principle, the most direct estimate of galaxy mass is a dynamical one. In practice, kinematics for distant galaxies are measured from observations of rest-frame optical emission lines from ionized gas or millimeter-wave CO transitions tracing molecular gas. Furthermore, in order to detect velocity shear independent of long-slit position angle and sample the full gravitational potential, emission-line kinematics must be traced in two dimensions on the sky and to large radii. Robust estimates of dynamical masses in high-redshift galaxies therefore require deep, spatially-resolved integral-field spectroscopic maps of line emission out to > 10 kpc radii.

Instruments at both the Keck (OSIRIS) and VLT (SINFONI) observatories have been used to perform such observations at $z\sim1-3$, with results presented in Law et al. (2007a) and Förster Schreiber et al. (2006) (see Fig. 12.12). The objects surveyed thus far typically have integrated Hα fluxes of $>10^{-16}$ erg s^{-1} cm^{-2}, and are therefore drawn from among the most luminous in Hα at these redshifts. The resulting kinematic maps indicate a range of properties, from rotating disks, to dispersion-dominated objects, to complex merging systems. Luminous, star-forming galaxies in the early universe do not fit neatly into a simple picture of regular thin disks, nor can they be described as dominated by major merger events. Furthermore, a full understanding of the assembly process of these systems requires spatial information at scales below the typical seeing limit in the near-IR. Therefore, ground-based integral field unit observations enhanced by adaptive optics are crucial. As shown in Figure 12.13. With the next generation of integral field unit spectrographs on the ground and in space, it will be possible to extend the IFU observations to more galaxies with more typical Hα fluxes, and out to larger radii. These observations will place vital constraints on the galaxy assembly process at high redshift.

12.6 Feedback: Outflows and the Escape of Ionizing Radiation

The process described as "feedback" is considered a crucial ingredient in models of galaxy formation. Feedback commonly refers to large-scale outflows of mass, metals, energy, and momentum from galaxies, which therefore regulate the amount of gas available to form stars, as well as the thermodynamics and chemical enrichment of the surrounding intergalactic medium (IGM), and, in dense environments the intracluster medium (ICM). Examples of feedback in nearby starburst galaxies with extended emission from warm and hot gas are shown in Fig. 12.14. There are many puzzles and trends in galaxy formation for which feedback has been offered as a solution: the fact that the faint end of the $z\sim0$ optical galaxy luminosity function is significantly flatter than the low-mass end of the dark-matter halo mass function; the so-called "overcooling" and "angular momentum" problems, and the form of the stellar mass – metallicity relation. Energy and momentum input from

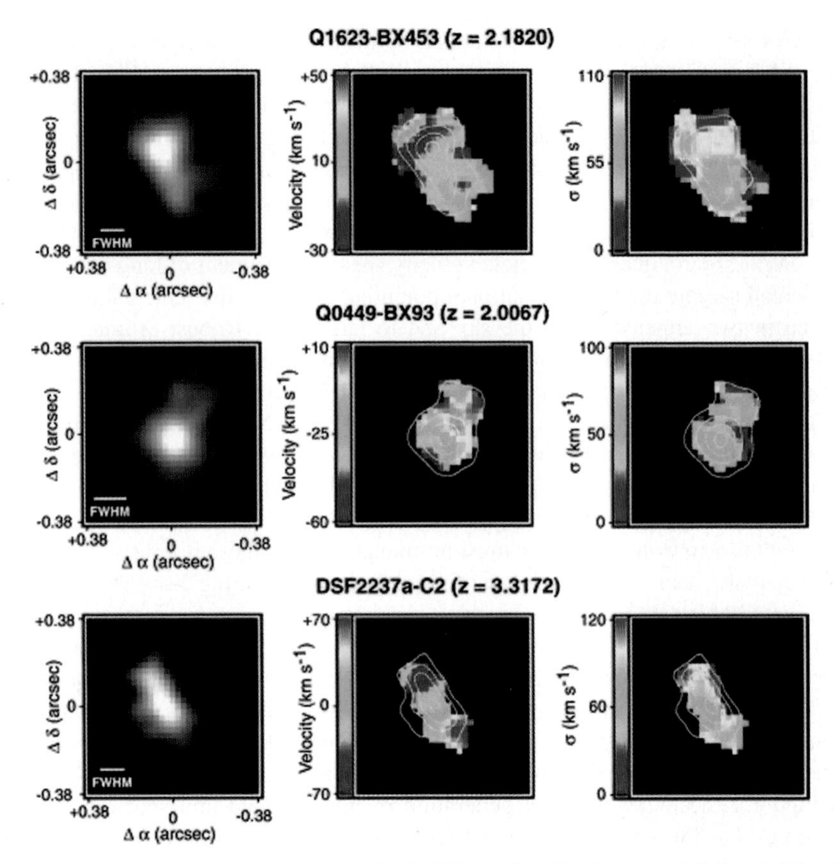

Fig. 12.12 From Law et al. (2007b). Keck/OSIRIS integral-field unit maps of (left to right) nebular emission line flux density, velocity, and velocity dispersion for the three target galaxies. The FWHM of the PSF after smoothing is 110 mas in Q1623-BX453 and DSF2237a-C2 and 150 mas in Q0449-BX93, indicated by solid lines in the left-hand panels. The total field of view (0.75×0.75 arcseconds) corresponds to 6.3, 6.4, and 5.7 physical kpc at the redshift of Q1623-BX453, Q0449-BX93, and DSF2237a-C2 respectively

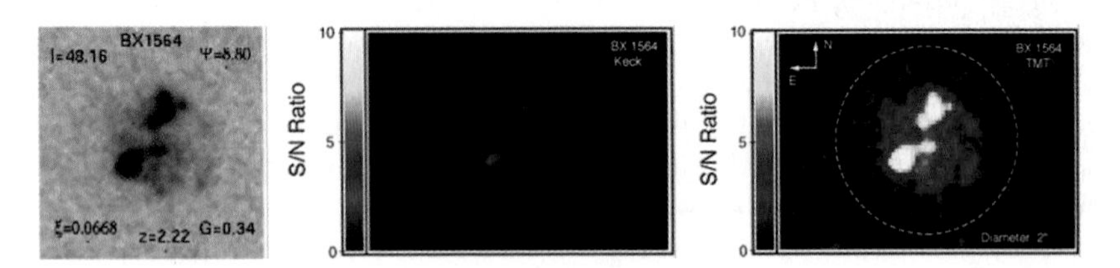

Fig. 12.13 From Law (2007, Private communication). (*Left*) HST/ACS image of the UV-selected $z = 2.2$ galaxy HDF-BX1564 (*Center*) Simulated Keck/OSIRIS integral field unit map of the Hα S/N (5 hour integration time). (*Right*) Simulated TMT integral field unit map of the Hα S/N in the same exposure time

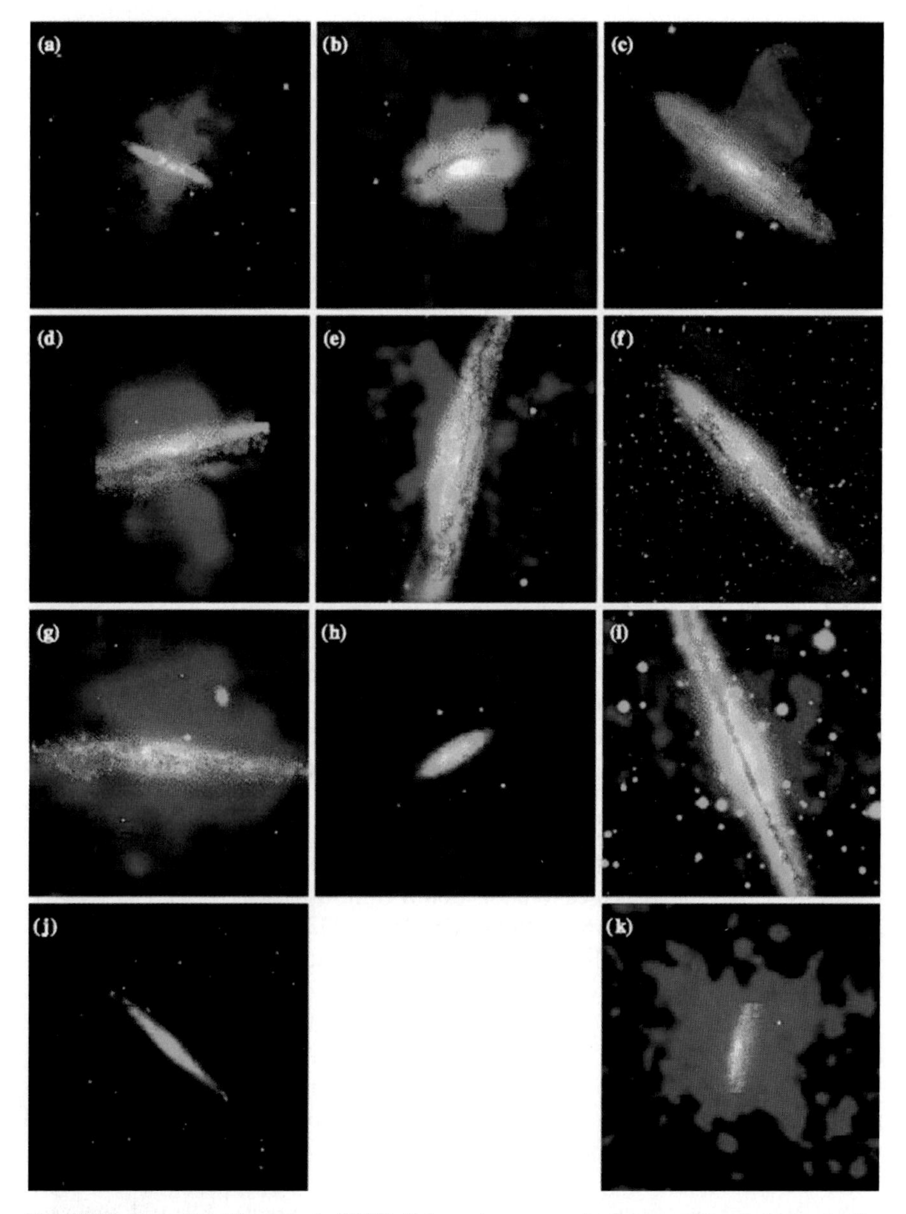

Fig. 12.14 From Strickland et al. (2004). False-color composite images of local starbursts. Hα emission is in red, optical R-band light is in green, and $0.3 - 2.0$keV soft X-ray emission is in blue. Each postage stamp shows a 20 kpc$\times$20 kpc region

processes related to both star-formation and nuclear (AGN) activity may contribute to resolving these issues, and determining the relative importance of both types of activity is an important yet challenging goal. While there have been many exciting observational developments in studies of feedback at both low and high redshift over the past several years, reviewed below, the next generation of large telescopes will enable still greater progress in describing a crucial element of the galaxy formation problem.

Here I will focus mainly on new results related to feedback as observed in star-forming galaxies identified at $z \sim 2 - 3$ (described in Section 12.2), when the universal star-formation rate density was at its peak level. In, for example, UV-selected systems (Steidel et al. 2003, 2004), high star-formation rate surface densities are observed, which far exceed the critical threshold value of $\Sigma_{SFR} = 0.1 M_{\odot} \mathrm{yr}^{-1} \mathrm{kpc}^{-2}$ above which large-scale outflows appear to be important in local star-forming galaxies (Heckman 2001). The evidence for feedback in high-redshift star-forming galaxies comes in many forms. Blue-shifts hundreds of $\mathrm{km\,s}^{-1}$ in interstellar absorption lines relative to galaxy systemic redshifts are one indication of large-scale motions (Pettini et al. 2001). The nature of the IGM environments of vigorously star-forming galaxies, in terms of the optical depth and kinematics of the surrounding H I and heavy elements provides further evidence (Adelberger et al. 2003, 2005b). The form of the mass-metallicity relation provides independent evidence for the importance of feedback (Erb et al. 2006a). We will start with this galaxy scaling relation.

In the local universe, the mass-metallicity relationship among star-forming galaxies has been exquisitely traced out by a sample of $\sim$53, 000 objects drawn from the SDSS (Tremonti et al. 2004). At $z > 1$, the largest sample of analogous measurements is presented by Erb et al. (2006a). In this work, a sample of 87 star-forming galaxies at $z \sim 2$ with both M_* and oxygen abundance (O/H) measurements is analyzed. The O/H values are estimated from measured [NII]/Hα line ratios, using the abundance calibration from Pettini & Pagel (2004). Not only do Erb et al. (2006a, Fig. 12.15) demonstrate a monotonic increase in O/H with increasing stellar mass, but these authors also investigate the change in metallicity with gas fraction. Such analysis requires an estimate of the gas content of these $z{\sim}2$ galaxies. (Erb et al. 2006a) estimate the gas masses and associated gas fractions by assuming that the Schmidt Law (Kennicutt 1998) relating gas and star-formation rate surface densities holds at high redshift. Accordingly, the measured Hα star-formation rate surface densities can be converted into gas surface densities, and then into gas masses and fractions, based on the Hα-emitting areas. The resulting relation between metallicity and gas fraction – i.e., a shallow increase in metallicity with decreasing gas fraction – can be compared with simple chemical evolution models with different values for the yield of heavy elements, and rate of infall and outflow of gas. The data are best described by models in which the gas outflow rate is greater than or equal to the star-formation rate (Erb et al. 2006a; Erb 2007), which provides indirect evidence for feedback.

The current sample of Keck/NIRSPEC long-slit observations, the largest of its kind, probe down to $\sim 10^{-17}$ erg s^{-1} cm^{-2} ($\sim 4 M_{\odot}$ yr^{-1}), in $\sim$1 hour exposures. By 2009-2010, for example, the planned Keck/MOSFIRE multi-object spectrograph

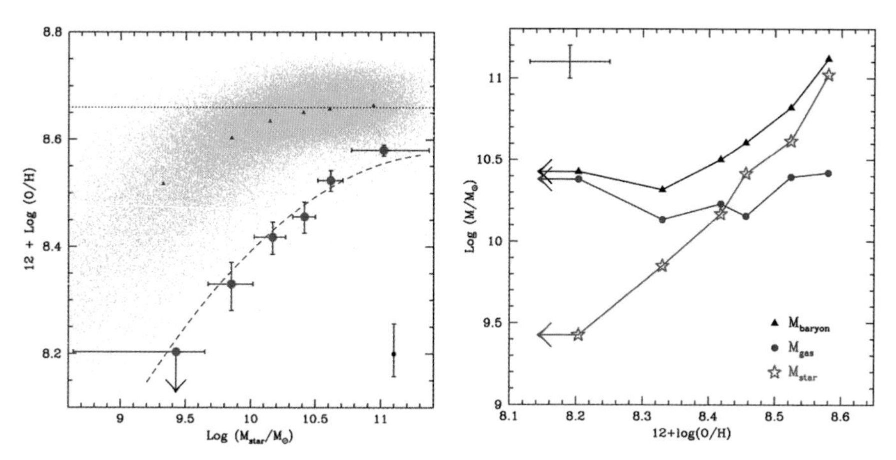

Fig. 12.15 From Erb et al. (2006a) (*Left*) Observed relation between stellar mass and oxygen abundance at $z \sim 2$, shown by large red circles. Oxygen abundances are based on the [NII]/Hα ratio in composite spectra of 14 or 15 galaxies. The small grey dots are metallicities measured from individual local star-forming galaxies in the SDSS (Tremonti et al. 2004), using the same [NII]/Hα indicator, while small blue triangles indicate mean values as a function of stellar mass. The blue dashed line is the local best-fit mass-metallicity relation from Tremonti et al. (2004), shifted down by 0.56 dex. The dotted line indicates solar metallicity. The vertical error bar shows the uncertainty in the [NII]/Hα abundance calibration from Pettini & Pagel (2004). (*Right*) Stellar, gas, and baryonic mass as a function of oxygen abundance. While stellar mass increases strongly with metallicity, gas mass is roughly constant, and, therefore baryonic mass increases only weakly. Finally, gas fraction is therefore a decreasing function of stellar mass

will add the capability for ~50-object multiplexing, with increased sensitivity down to star-formation rates of $\sim 1 M_{\odot}$ yr^{-1} in similar exposure times. Next decade instruments, including NIRSpec on JWST, and planned near-IR spectrographs for large ground-based telescopes, will probe an order of magnitude fainter in emission-line flux, down to lower stellar masses and metallicities. Additionally, JWST, free from the constraints of atmospheric absorption, will remove restrictions on the redshift ranges over which rest-frame optical emission lines can be measured. Finally, ALMA will offer more direct probes of the molecular gas content of typical star-forming galaxies with measurements of CO.

The study of mass-metallicity relationship provides indirect evidence for large-scale outflows from high-redshift galaxies. We can also try to address the question with more direct evidence about the characteristics of outflows at high redshift. Indeed, there are many properties of outflows which are important to measure, in order to assess their impact on the star-formation histories of galaxies. These include the mass, energy, and momentum outflow rates, a constraint on the multi-phase nature of the outflows, and the relative amounts of mass in each phase. It is also of interest to determine whether energy or momentum is being conserved in the flows, and what the ultimate fate of outflowing material is (i.e. bound or unbound). Finally, an essential ingredient to understanding feedback is constraining the relative importance of star-formation or AGN activity in driving material from galaxies.

Perhaps the most fundamental quantity to measure is the mass outflow rate. This quantity, $\dot{M}_w$, can be expressed as:

$$\dot{M}_w = \Omega r^2 \mu(r) n(r) v(r) \tag{12.1}$$

where Ω is the solid angle of the gas covering fraction, r is the radius at which the outflow rate is being evaluated, μ is the mean mass per particle, n is the number-density of particles, and v is the speed at which material is flowing. This can be re-expressed in more empirical quantities, including gas column-density N_{HI}, as follows:

$$\dot{M}_w \approx \Omega R_0 m_p N_{HI} v_\infty \tag{12.2}$$

In order to evaluate this expression, estimates must be obtained for the gas covering factor, Ω; the radius at which the wind is being measured, R_0; the column-density of hydrogen, N_{HI}; and the wind terminal velocity, v_∞. While we have established the widespread existence of outflows at high redshift (Pettini et al. 2001; Steidel et al. 2004), it is very difficult to measure the required parameters to estimate mass outflow rate (i.e. Ω, R_0, N_{HI}, v_∞). Indeed, the low S/N and spectral resolution of typical high-redshift galaxy spectra ($\sim$400 km s^{-1}) precludes us from measuring N_{HI} (or N_X from metal species, X), Ω, v_∞ (from th blue edge of interstellar absorption lines). Indeed, most of the lines we observe are strongly saturated, so we only observe their equivalent widths rather than column densities. Furthermore, it is not trivial to obtain the zeropoint of the galaxy velocity field in these distant objects, against which the redshifts of interstellar absorption lines can be compared to derive outflow speeds. Such a determination requires measurements of weak stellar photospheric absorption lines, or rest-frame optical nebular emission lines shifted into the near-IR. Finally, the small size of distant galaxies and the limited spatial resolution provided by seeing-limited ground-based observations preclude us from estimating R_0, the radius at which we are probing the winds. There are only limited indications of this size scale based on fortuitous spaced galaxy pairs, in which the higher-redshift pair member is *bright* (Adelberger et al. 2005b).

There are very special exceptions to this rather grim rule about our knowledge of outflows at high redshift. Gravitational lensing of typical L^* star-forming galaxies at $z>2$ boosts them into the realm where detailed study becomes possible. For example, MS1512-cB58 (or cB58), a star-forming galaxy at $z = 2.73$, has been magnified by a factor of $\sim$30 in brightness, such that it has a measured R-band magnitude of $\sim$20.5 (AB) (Yee et al. 1996). The extreme brightness of cB58 has enabled exquisitely detailed Keck/ESI rest-frame UV spectroscopic studies of its outflowing interstellar medium (ISM), interstellar abundance pattern, stellar population properties, and IMF (Pettini et al. 2000, 2002) (see Fig. 12.16). For 10 years, this object has been the only LBG with an estimated mass outflow rate, where $\dot{M}_w \sim \dot{M}_*$. At the same time, it is worth mentioning that cB58 has a distinctly non-average rest-frame UV spectrum, compared to the bulk of the LBG population (Shapley et al. 2003). Finally, within the last year, multiple additional lensed LBGs with very high surface

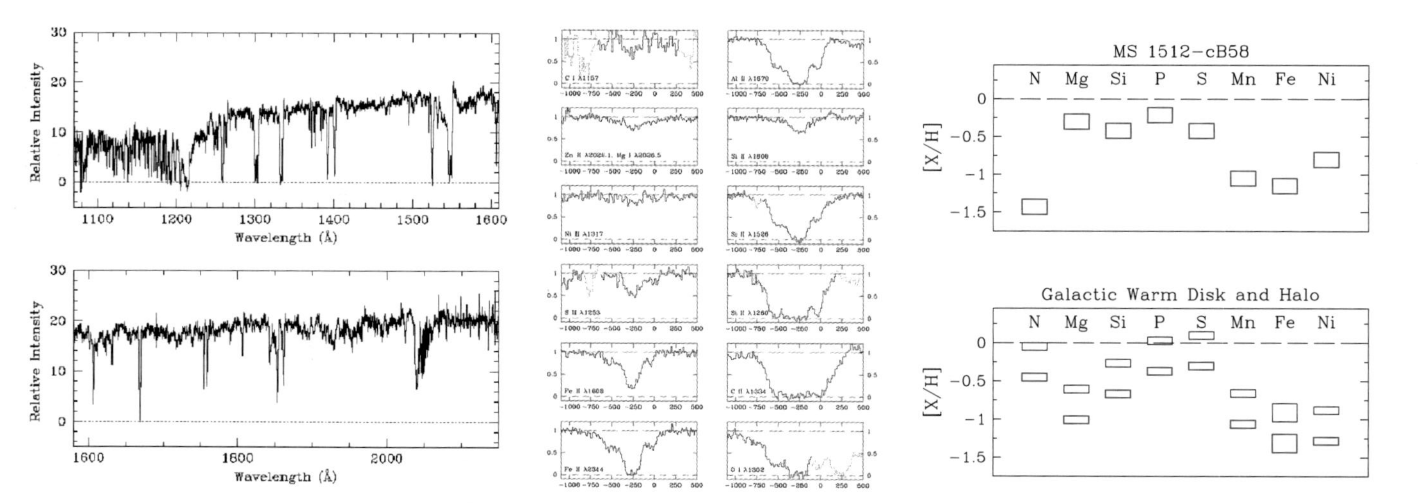

Fig. 12.16 From Pettini et al. (2002). Observations of the gravitationally-lensed LBG, MS1512-cB58 ($z = 2.73$. (*Left*) Keck/ESI spectrum of cB58, shifted into the rest frame. The resolution of this spectrum is 60 km s^{-1}, significantly higher than typical unlensed high-redshift galaxy spectra. The S/N per resolution element ranges between ~30 and ~55, also significantly better than in typical spectra. (*Center*) Velocity profiles of low-ionization interstellar absorption lines in the spectrum of cB58, relative to the stellar systemic redshift. The largest optical depth occurs at a blueshift of ~ -255 km s^{-1}, while absorption extends as far to the blue as -775 km s^{-1}. (*Right*) The abundance pattern in the interstellar medium of cB58 (*top*), compared with that in the diffuse gas in the Milky Way. The abundance pattern in cB58 indicate evidence for enhancement of α elements relative to Fe-peak elements such as Mn, Fe, and Ni

brightness have been identified, which will allow our sample of objects with detailed outflow measurements to grow by a significant factor (well, from one to several!).

However, in order to constrain the properties of outflows for a population rather than a handful, we must assemble these detailed observations for a statistical sample of high-redshift galaxies. With optical spectrographs on TMT, GMT, and ELT (e.g. TMT/WFOS), ~5 hour integration times, and resolution of ~60 km s^{-1}, spectra of similar quality to that of cB58 will be obtained for relatively typical z~2 – 3 objects with $R = 24$ mag (AB), which have a surface density of >1 arcmin^{-2} (see Fig. 12.17). With such spectra, we will obtain unprecedented constraints on outflow kinematics, mass outflow rate, chemical enrichment, stellar populations for star-forming galaxies at high redshift. Furthermore, near-IR integral field unit observations may be used to constrain the physical scale of warm and cool outflowing gas through the detection of extended Hα emission and Na I absorption, respectively.

Feedback is important not only for regulating the gas available for star formation in galaxies, but also for its impact on the surrounding intergalactic medium. Galaxies affect their intergalactic environments not only through the exchange of matter, kinetic energy and momentum, but also through radiative feedback, and the transfer of ionizing radiation. Indeed, a fundamental goal for observational cosmology is a proper understanding of the process through which the universe transformed from having a neutral to highly ionized IGM. The steep drop-off in the number density of luminous QSOs at high redshift (z>3), compared with the more gradual decline in the level of ionizing background, implies QSOs cannot dominate the ionizing background by z~5 (Fan et al. 2001; MacDonald & Miralda-Escudé 2001). To test whether galaxies can serve as the dominant source of ionizing radiation at these redshifts, it is crucial to determine the value of f_{esc}, i.e. the escape fraction of ionizing Lyman continuum radiation.

Based on ~10-hour integration times in the Lyman continuum region, detections of ionizing radiation have been achieved at z~3 for 2 out of 14 bright LBGs (Fig. 12.18). Both objects have very high inferred values of f_{esc}, i.e. >15%. While we've achieved the first individual detections of ionizing radiation escaping from

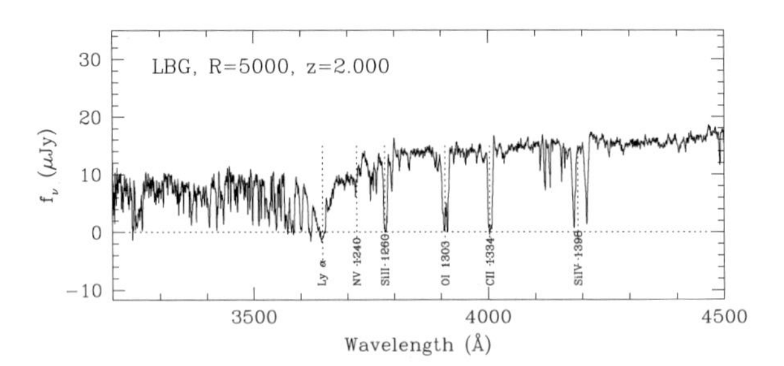

Fig. 12.17 From Steidel (2007, private communication). Simulated 5-hour spectrum of an $R = 24$ (AB) galaxy at $z = 2$ (based on actual observations of MS1512-cB58), illustrating the quality expected with the WFOS spectrograph on TMT. Therefore, with future facilities, it will be possible to place constraints on the chemical abundance pattern, the form of the high-mass end of the stellar IMF, the outflow rate of interstellar material for *unlensed* high-redshift star-forming galaxies

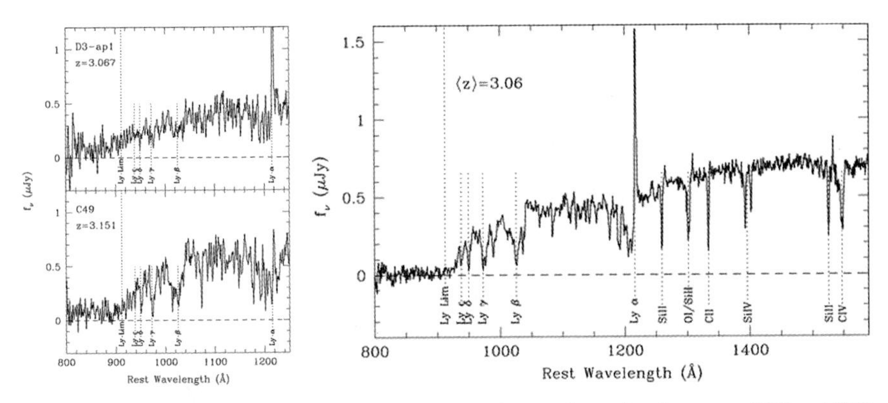

Fig. 12.18 From Shapley et al. (2006). (*Left*) Zoomed-in one-dimensional spectra of D3 and C49, $z \sim 3$ galaxies with Lyman-continuum detections. The region extending down from Lyα through the Lyman-continuum region is featured. (*Right*) Composite spectrum of deep LBG spectra. This plot represents the average of a sample of 14 objects, and shows flux in the Lyman-continuum region that is formally significant at greater than a 3σ level. The sample average ratio of 1500 Å to Lyman-continuum flux density, uncorrected for IGM opacity, is $\langle f_{1500}/f_{900}\rangle = 58 \pm 18_{stat} \pm 17_{sys}$

$z{\sim}3$ galaxies using Keck/LRIS-B, these measurements are difficult, long, and only apply to the brightest $z{\sim}3$ LBGs. These measurements will become impossible by $z{\sim}5$, where the IGM optical depth simply too high to detect ionizing radiation from galaxies. In contrast, it will be more valuable to observe at more typical $z{\sim}3$ and $z \sim 4$ objects, where the IGM optical depth will not prevent measurement of flux. If optical spectrographs on TMT, GMT, and ELT have good sensitivity in the near UV (throughput $>20\%$ at $\lambda = 3400 - 4500$Å), it will be possible to perform this fundamental measurement over an large dynamic range in f_{esc} for typical star-forming galaxies at $z{\sim}3$–4 – not simply the brightest.

12.7 IGM Tomography

Many important questions in the study of galaxy formation address the physics of baryons. These include a description of gas cooling, star-formation, metal enrichment and feedback. The IGM is a reservoir for a significant fraction of these baryons at high redshift. Historically, QSO sightlines have been probes of IGM baryons, including spectra of hydrogen and heavy elements such as C IV and O VI, which allow for description of the thermal properties and metal-enrichment distributions in intergalactic gas. Bright QSOs are rare, however, so it is difficult to use them to infer the three-dimensional distribution of H I and metals relative to galaxies. As shown in Fig. 12.19, at $R{\sim}24$ mag (AB), galaxies at $1.5 \leq z \leq 3.5$ outnumber QSOs by factor of $\sim$30. With such background targets, it will be possible to sample the IGM on scales $\leq$ 500 kpc, which constitutes the sphere of influence of the outflows of individual galaxies (Adelberger et al. 2003).

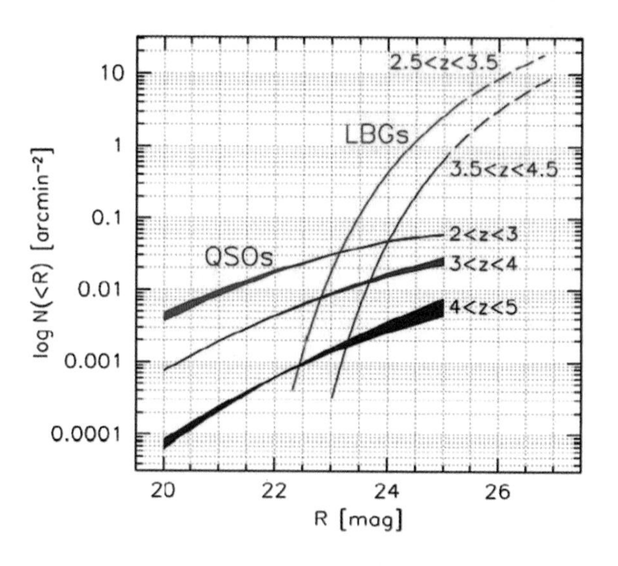

Fig. 12.19 From Thirty Meter Telescope Detailed Science Case (2007) Surface density of QSOs and UV-bright galaxies as a function of R-mag. By $R \sim 24$, the combined surface density of galaxies and QSOs exceeds $1\ \text{arcmin}^{-2}$, sufficient for tomographic mapping of the IGM

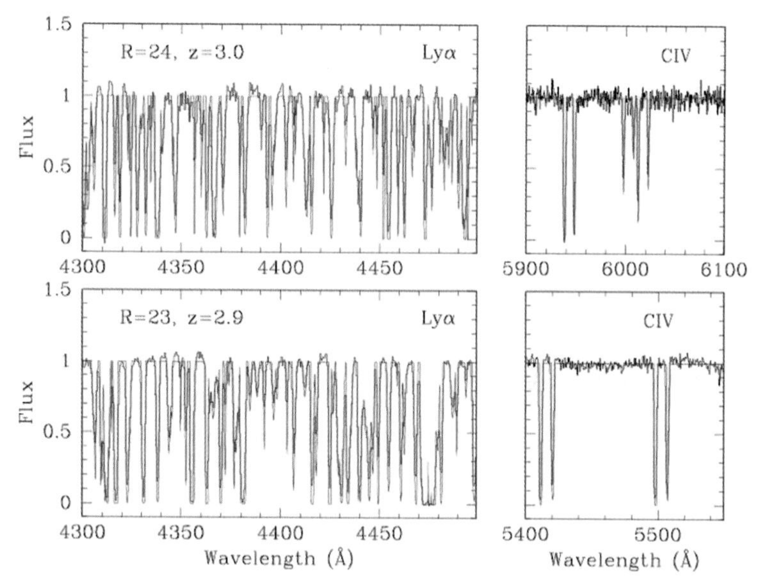

Fig. 12.20 From Abraham et al. (2006). Examples of simulated 3-hour TMT/WFOS spectra for star-forming galaxies at $z = 3$, with $R = 23$ and 24, assuming a resolution of $60\,\text{km}\,\text{s}^{-1}$. With such spectra, it will be possible to study the both hydrogen (*left*) and heavy elements (*right*) in the intergalactic medium

As shown in Fig. 12.20, with optical spectroscopy on the next generation large, ground-based telescopes (e.g. TMT/WFOS), ~5 hour integration times, resolution $\sim 60\,\mathrm{km\,s^{-1}}$ spectra for $1.5 \leq z \leq 3.5$ objects with $R = 24$ mag (AB) will provide a three-dimensional map of H I and metals, to be compared with the galaxy distribution in same volumes. Along with these high-resolution spectra for relatively bright sightline galaxies, one potential survey design consists of resolution ~1000 spectra for much larger (factor of 10) sample of foreground galaxies with a range of properties, including both actively star-forming and passive galaxies. With background galaxies as tomographic probes, such a survey will provide an unprecedented view of the interaction between galaxies and the IGM – an important component of the galaxy-formation problem.

12.8 Summary

We have now identified galaxies continuously in redshift space out to $z \sim 7$. The epoch at $1.5 \leq z \leq 3.5$ is an important one for understanding galaxy formation, and features many recently-identified galaxy populations. Extensive optical and near-IR spectroscopy down to faint magnitudes ($R > 26$ mag, SFR$< 1 M_\odot\,\mathrm{yr}^{-1}$) is critical for characterizing the spatial distribution of the progenitors of today's galaxies. Deep near-IR and optical spectroscopy with next-decade facilities will allow the spectroscopic characterization of high-z galaxy stellar populations, H II region properties, dynamics, and assembly. In particular the use of deep, high-resolution optical spectroscopy will provide new insight into "feedback," both in terms of its effect on the ISM of individual galaxies, and, more globally, on the properties of the IGM.

References

Abraham, R. G. et al. (2006), "TMT Wide-Field Optical Spectrograph Operational Concepts Definition Document", TMT Internal Report

Adelberger, K. L., Steidel, C. C., Shapley, A. E., and Pettini, M. (2003), ApJ, **584**, 45

Adelberger, K. L., Steidel, C. C., Pettini, M., Shapley, A. E., Reddy, N. A., and Erb, D. K. (2005a), ApJ, **619**, 697

Adelberger, K. L., Shapley, A. E., Steidel, C. C., Pettini, M., Erb, D. K., and Reddy, N. A. (2005b), ApJ, **629**, 636

Baldry, I. K., Glazebrook, K., Brinkmann, J., Ivezić, Ž., Lupton, R. H., Nichol, R. C., and Szalay, A. S. (2004), ApJ, **600**, 681

Benson, A. J., Frenk, C. S., Baugh, C. M., Cole, S., and Lacey, C. G. (2001), MNRAS, **327**, 1041

Bruzual, G. and Charlot, S. (2003), MNRAS, **344**, 1000

Bundy, K., Ellis, R. S., Conselice, C. J., Taylor, J. E., Cooper, M. C., Willmer, C. N. A., Weiner, B. J., Coil, A. L., Noeske, K. G., and Eisenhardt, P. R. M. (2006), ApJ, **651**, 120

Chapman, S. C., Blain, A. W., Ivison, R. J., and Smail, I. R. (2003), Nature, **422**, 695

Coil, A. L., Newman, J. A., Cooper, M. C., Davis, M., Faber, S. M., Koo, D. C., and Willmer, C. N. A. (2006), ApJ, **644**, 671

Coil, A. L. et al. (2007), ArXiv Astrophysics e-prints, ArXiv:0708.0004

Conroy, C., Shapley, A. E., Tinker, J. L., Santos, M. R., and Lemson, G. ArXiv Astrophysics e-prints, arXiv:astro-ph/0711.0001

Cowie, L. L., and Hu, E. M. (1998), AJ, **115**, 1319
Daddi, E., Cimatti, A., Renzini, A., Fontana, A., Mignoli, M., Pozzetti, L., Tozzi, P., and Zamorani, G. (2004), ApJ, **617**, 746
Daddi, E. et al. (2007), ApJ, **670**, 156
Dickinson, M. (2000), RSPTA, **358**, 2001
Drory, N., Salvato, M., Gabasch, A., Bender, R., Hopp, U., Feulner, G., and Pannella, M. (2005), ApJL, **619**, 131
Erb, D. K., Shapley, A. E. Pettini, M. Steidel, C. C. Reddy, N. A., and Adelberger, K. L. (2006a), ApJ, **644**, 813
Erb, D. K., Steidel, C. C., Shapley, A. E., Pettini, M., Reddy, N. A., and Adelberger, K. L. (2006b), ApJ, **646**, 107
Erb, D. K. (2007), ArXiv Astrophysics e-prints, arXiv:astro-ph/0710.4146
Fan X. et al. (2001), AJ, **122**, 2833
Fontana, A., Salimbeni, S., Grazian, A., Giallongo, E., Pentericci, L., Nonino, M., Fontanot, F., Menci, N., Monaco, P., Cristiani, S., Vanzella, E., de Santis, C., and Gallozzi, S. (2006), A&A, **459**, 745
Förster Schreiber, N. M. et al. (2006), ApJ, **645**, 1062
Franx, M., Labbé, I., Rudnick, G., van Dokkum, P. G., Daddi, E., Förster Schreiber, N. M., Moorwood, A., Rix, H., Röttgering, H., van de Wel, A., van der Werf, P., and van Starkenburg, L. (2003), ApJL, **587**, 79
Gallazzi, A., Charlot, S., Brinchmann, J., White, S. D. M., and Tremonti, C. A. (2005), MNRAS, **362**, 41
Heckman, T. M. (2001) ASPCS, **240**, 345
Kodaira K. et al. (2003), PASJ, **55**, L17
Kauffmann, G. et al. (2003), MNRAS, **341**, 33
Kennicutt, R. C. (1998), ApJ, **498**, 541
Kriek, M. et al. (2006), ApJ, **645**, 44
Law, D. R., Steidel, C. C., Erb, D. K., Pettini, M., Reddy, N. A., Shapley, A. E., Adelberger, K. L., and Simenc, D. J. (2007a), ApJ, **656**, 1
Law, D. R., Steidel, C. C., Erb, D. K., Larkin, J. E., Pettini, M., Shapley, A. E., and Wright, S. A. (2007b), ApJ, **669**, 929
Lotz, J. M., Madau, P., Giavalisco, M., Primack, J., and Ferguson, H. C. (2006), ApJ, **636**, 592
McDonald, P., and Miralda-Escudé, J. (2001), ApJL, **549**, 11
Nagamine, K., Ostriker, J. P., Fukugita, M., and Cen, R. (2006), ApJ, **653**, 881
Papovich, C. et al. (2006), ApJ, **640**, 92
Pettini, M., Shapley, A. E., Steidel, C. C., Cuby, J., Dickinson, M., Moorwood, A. F. M., Adelberger, K. L., and Giavalisco, M. (2001), ApJ, **554**, 981
Pettini, M., Steidel, C. C., Adelberger, K. L., Dickinson, M., and Giavalisco, M. (2000), ApJ, **528** 96
Pettini, M., Rix, S. A., Steidel, C. C., Adelberger, K. L., Hunt, M. P., and Shapley, A. E. (2002), ApJ, **569**, 742
Pettini, M., and Pagel, B. E. J. (2004), MNRAS, **348**, L59
Reddy, N. A., Erb, D. K., Steidel, C. C., Shapley, A. E., Adelberger, K. L., and Pettini, M. (2005), **633**, 748
Reddy, N. A., Steidel, C. C., Pettini, M., Adelberger, K. L., Shapley, A. E., Erb, D. K., and Dickinson, M. (2007), ArXiv Astrophysics e-prints, arXiv:astro-ph/0706.4091
Schmidt, M., Schneider, D. P., and Gunn, J. E. (1995), AJ, **110**, 68
Strateva, I. et al. (2001), AJ, **122**, 1861
Shapley, A. E., Steidel, C. C., Pettini, M., and Adelberger, K. L. (2003), ApJ, **588**, 65
Shapley, A. E., Steidel, C. C., Pettini, M., Adelberger, K. L., and Erb, D. K. (2006), **651**, 688
Smail, I., Ivison, R. J., and Blain, A. W. (1997), ApJL, **490**, 5
Steidel, C. C., Adelberger, K. L., Shapley, A. E., Pettini, M., Dickinson, M., and Giavalisco, M. (2003), ApJ, **592**, 728

Steidel, C. C., Shapley, A. E., Pettini, M., Adelberger, K. L., Erb, D. K., Reddy, N. A., and Hunt, M. P. (2004), ApJ, **604**, 534
Strickland, D. K., Heckman, T. M., Colbert, E. J. M., Hoopes, C. G., and Weaver, K. A. (2004), ApJ, **606**, 829
Thirty Meter Telescope Detailed Science Case (2007), http://www.tmt.org
Tremonti, C. A. et al. (2004), ApJ, **613**, 898
van Dokkum et al. (2006), ApJL, **638**, 59
Wechsler, R. H., Gross, M. A. K., Primack, J. R., Blumenthal, G. R., and Dekel, A. (1998), ApJ, **506**, 19
Yee, H. K. C., Ellingson, E., Bechtold, J., Carlberg, R. G., and Cuillandre, J.-C. (1996), AJ, **111**, 1783
Zehavi, I. et al. (2005), ApJ, **630**, 1

Brenae Bailey, still smiling after the Peace Corps and the University of Wyoming

Chapter 13
The Co-Evolution of Galaxies and Black Holes: Current Status and Future Prospects

Timothy M. Heckman

Abstract I begin by summarizing the evidence that there is a close relationship between the evolution of galaxies and supermassive black holes. They evidently share a common fuel source, and feedback from the black hole may be needed to suppress over-cooling in massive galaxies. I then review what we know about the co-evolution of galaxies and black holes in the modern universe ($z < 1$). We now have a good documentation of which black holes are growing (the lower mass ones), where they are growing (in the less massive early-type galaxies), and how this growth is related in a statistical sense to star formation in the central region of the galaxy. The opportunity in the next decade will be to use the new observatories to undertake ambitious programs of 3-D imaging spectroscopy of the stars and gas in order to understand the actual astrophysical processes that produce the demographics we observe. At high redshift ($z > 2$) the most massive black holes and the progenitors of the most massive galaxies are forming. Here, we currently have a tantalizing but fragmented view of their co-evolution. In the next decade the huge increase in sensitivity and discovery power of our observatories will enable us to analyze the large, complete samples we need to achieve robust and clear results.

13.1 Introduction

As we look forward to the powerful suite of space- and ground-based observatories that will dominate astronomy in the coming decade, it is worthwhile to reflect upon the extraordinary state of astronomy here and now.

Capped by the spectacular results from NASA's WMAP mission, we now have precision measurements of the geometry, age, composition, and density-fluctuation power spectrum of the universe (Spergel et al. 2007). The predictions of inflationary cosmology and the ΛCDM paradigm have been confirmed, so that we also have a robust understanding of the development and evolution of the structure of the dark-matter backbone of the universe (e.g. Tegmark et al. 2004). This might suggest that

T.M. Heckman (✉)
Center for Astrophysical Sciences, Department of Physics and Astronomy, Johns Hopkins University, Baltimore, MD, USA
e-mail: heckman@pha.jhu.edu

H.A. Thronson et al. (eds.), *Astrophysics in the Next Decade,* Astrophysics and Space Science Proceedings, DOI 10.1007/978-1-4020-9457-6_13,

our work is largely done. Paradoxically however, it is that minority-component of the mass-energy content of the universe with which we have the most experience by far – ordinary baryonic matter – whose cosmic evolution has proven to be most difficult to understand. Ordinary matter can interact dissipatively, radiate, cool, respond to pressure gradients, generate and respond to magnetic fields, undergo nuclear reactions, etc. This richness of physical processes leads directly to a corresponding richness of astrophysical phenomena that have inspired and challenged astronomers for centuries.

So, with the basic cosmological framework firmly in place, the frontier in the study of the formation and evolution of galaxies is to understand the complex "gastro-physics" of the gas-star-black hole cosmic ecosystem. While the crucial role of star formation and resulting feedback in the evolution of galaxies has been understood for decades, it has only been in the last few years that the role of supermassive black holes has been appreciated. The tight correlation between the mass of the black hole and the velocity dispersion and mass of the galactic bulge within which it resides (Ferrarese & Merritt 2000, Gebhardt et al. 2000, Marconi & Hunt 2003, Häring & Rix 2004) is compelling evidence for a close connection between the formation of the black hole and that of its host galaxy (e.g. Kauffmann & Haehnelt 2000, Granato et al. 2001).

Feedback from supermassive black holes is believed to be an essential process in the formation and evolution of galaxies. Galaxy formation models, which attempt to understand how galaxies form through gas cooling and condensation within a merging hierarchy of dark matter halos predict that massive galaxies should in general be surrounded by a substantial reservoir of gas. This gas should be cooling and forming stars. Over a Hubble time, these processes should have produced far more present-day galaxies with large stellar masses than are actually observed. Simply put, the mass spectrum of dark matter halos is a power-law, while the galaxy stellar-mass spectrum has an exponential cut-off (the standard Schechter function). Similarly, the cooling rate of hot gas in the haloes of massive galaxies today should be high enough to produce an easily-observable population of young stars (e.g. Kauffmann et al. 1993). A great deal of theoretical and observational effort is now being expended on understanding why this simple expectation does not appear to be borne out by observations. Matter accreting onto central supermassive black holes can, in principle, provide a vast source of energy, and jets or outflows can provide mechanisms for transporting the energy to large enough radii to regulate cooling and star formation in the massive galaxies where such black holes live (e.g. Churazov et al. 2001, Croton et al. 2006, Bower et al. 2006, Hopkins et al. 2007, Di Matteo et al. 2007).

Thus, while we do have a rough conceptual framework for the co-evolution of galaxies and supermassive black holes, there is much that we do not know. How does the gas get into the galaxy bulge in the first place (mergers, secular processes, cold and/or hot accretion)? Once in the galaxy, how is gas transported from the bulge (radii of a few kpc) to the black hole's accretion disk (radii of tens of astronomical units)? What astrophysics sets the mass ratio of the gas turned into stars compared to the gas accreted by the black hole at the observed value of $\sim 10^3$ (Marconi & Hunt 2003, Häring & Rix 2004)? What about the magnitude and specific form(s) of the feedback from the supermassive black hole? What is the sequencing? Does the black hole or galaxy form first? Is this a once-in-a-lifetime transformative event in the life of a galaxy, or a more gradual, intermittent process?

My plan for the rest of the paper is as follows. I will begin on the firmest ground by discussing the modern universe ($z < 1$). We know quite a lot already about the overall landscape, but not much about the actual astrophysical processes that connect the fueling of the black hole to star formation in the galaxy bulge. Then I will discuss the early universe ($z > 2$), the epoch when the most massive black holes and galaxies were forming. There have been remarkable developments over the past few years, but the picture is still fragmented. Throughout, I will try to highlight how the next generation of observatories can be used to answer the questions posed above.

13.2 The Modern Universe – What Do We Know Now?

13.2.1 Overview

In the present-day universe there are two nearly independent modes of black hole activity. The first is associated with Seyfert galaxies and QSO's. Here, the black hole is radiating strongly (typically at greater than a few percent of the Eddington limit). This mode is the one primarily responsible for the growth in mass of the population of black holes today. Only the population of lower mass black holes ($<10^8$ solar masses) are growing at a significant rate today, and these black holes live in the lower mass galaxies ($<10^{11}$ solar masses). For this AGN mode, there is a strong link (at least in a statistical sense) between the growth of the black hole and star formation. Since star formation requires cold gas, this suggests that the fuel source for the black hole is also cold gas. The associated star formation can provide a significant source of feedback through supernova-driven winds.

The second mode (radio galaxies) is dominated by the most massive black holes living in the most massive galaxies (giant ellipticals). Here, the accretion rates are low and the accretion process is radiatively inefficient. Instead, energy is being extracted from the black hole in the form of highly collimated relativistic jets that produce nonthermal radio emission. The strong dependence of the radio luminosity function on the mass of the galaxy and the enhancement in the luminosity function for galaxies living at or near the centers of clusters can both be understood if the black hole is being fueled by the accretion of slowly cooling hot gas. Little star formation accompanies this mode, and this may be in part because the radio jets are heating the surrounding hot gas, decreasing its cooling rate, and suppressing star formation.

13.2.2 Seyfert Galaxies: The Population of Rapidly Growing Black Holes

The relatively rapid growth of black holes that takes place in Seyfert nuclei and QSOs is accompanied by strong emission from the accretion disk and its environment.

In principle this emission can swamp that of the host galaxy, making it extraordinarily difficult to determine the properties of the host galaxy and relate these to the growth of the black hole. Fortunately, nature has provided us with its own "coronagraph". The central black hole and its accretion disk are encircled by a dusty torus that is opaque to the ultraviolet and optical emission from the accretion disk (Krolik & Begelman 1988). When viewed nearer the polar axis of the torus the accretion disk is seen directly and the object is called a Type 1 Seyfert galaxy or (if sufficiently luminous) a QSO. When viewed from nearer the equatorial plane of the torus, the accretion disk can not be seen directly and the object is called a Type 2 Seyfert galaxy (or Type 2 QSO). In the Type 2 Seyferts the observed ultraviolet through near-infrared continuum is dominated by the host galaxy (e.g. Kauffmann et al. 2003c). The presence of the obscured AGN can be determined by the strong mid-infrared emission from the dusty torus, by the high-ionization narrow UV, optical, and IR emission-lines produced as the ionizing radiation from the accretion disk escapes along the polar axis of the torus and photoionizes gas in the surrounding kpc-scale Narrow-Line Region (NLR), and by hard X-rays that have passed directly through the torus (if the torus is Compton-thin).

Our largest and most homogeneous sample of 'local' ($z \sim 0.1$) Type 2 AGN and their host galaxies has come from the Sloan Digital Sky Survey (SDSS). These AGN were recognized by the emission of the NLR in the SDSS spectra (Kauffmann et al. 2003c, Hao et al. 2005). The AGN luminosity can be estimated from the strong [OIII]λ5007 emission-line and an empirically-derived bolometric correction (Heckman et al. 2004), while the black hole mass can be determined from the bulge velocity dispersion (σ) and the M_{BH} vs. σ relation in Tremaine et al. (2002). The SDSS spectra plus multicolor optical images and GALEX near- and far-ultraviolet images can be used to determine the fundamental properties of the host galaxies (star formation rate, stellar mass, velocity dispersion, structure, and morphology) and of the local clustering environment (Kauffmann et al. 2003a, 2004, 2007a, Brinchmann et al. 2004, Li et al. 2007, Wild et al. 2007, Reichard et al. 2008, Martin et al. 2007).

The first question we can ask is: Which black holes are growing? The SDSS sample of Type 2 AGN shows that most present-day accretion occurs onto lower mass black holes ($<10^8$ solar masses). In fact, the volume-averaged accretion rates of low mass black holes imply that this population is currently growing (doubling its mass) on a timescale that is comparable to the age of the Universe (Heckman et al. 2004). In contrast, the mass-doubling timescale is more than two orders of magnitude longer for the population of the most massive black holes ($>10^9$ solar masses). These dormant black holes evidently formed at early times, and once blazed as powerful QSOs. Thus, the strong cosmological evolution of the AGN luminosity function documented in recent X-ray surveys (e.g. Barger et al. 2005, Hasinger et al. 2005) is primarily driven by a decrease in the characteristic mass scale of actively accreting black holes.

The next question we can ask is: In which galaxies are black holes currently growing? It is now well-known that the present-day galaxy population consists of two families (e.g. Kauffmann et al. 2003b, Baldry et al. 2004, Brinchmann et al. 2004, Schiminovich et al. 2007). One family is disk-dominated, rich in cold gas, and experiencing significant rates of star-formation (the "blue" galaxies). The other family is bulge-dominated, depleted in cold gas, and experiencing little if any star formation (the "red" galaxies). The transition between these two families is remarkably abrupt: "red" galaxies dominate at stellar masses above $\sim 10^{10.5}$ solar masses and at stellar surface mass densities above $\sim 10^{8.5}$ solar masses/kpc^2 while "blue" galaxies dominate below.

It is very intriguing that a majority of galaxies that lie in "no-man's land" between the red and blue galaxy population (in a region in parameter space sometimes called the "green valley") are AGN (Martin et al. 2007, Kauffmann et al. 2007a). There are also systematic trends in host properties with AGN luminosity (Cid Fernandes et al. 2001, Kauffmann et al. 2003c, Kauffmann et al. 2007a, Wild et al. 2007). The higher the luminosity of the AGN, the younger is the age of the stellar population in the central-most few kpc, the higher is the likelihood that the galaxy has experienced a burst of star formation within the last Giga-year, and the greater is the amount of dust extinction towards the bulge (a proxy for the presence of cold gas).

In a volume-averaged sense, most present-day accretion onto black holes is taking place in galaxies with relatively young stellar populations, intermediate stellar masses, and high surface mass densities (Heckman et al. 2004, Wild et al. 2007). These are hybrid galaxies with the structures of the red galaxies, the young stellar populations of the blue galaxies, and masses near the boundary between the two families. These results at $z \sim 0.1$ are at least qualitatively consistent with what is known about AGN host galaxies at $z \sim 1$ (Nandra et al. 2007, Rovilos et al. 2007, Rovilos & Georgantopoulos 2007).

The assessment above can be done in a more quantitative way. If we integrate over the SDSS volume, we find that the average ratio of star formation to black hole accretion in bulge-dominated galaxies is $\sim 10^3$. This is in remarkable agreement with the observed ratio of stellar mass to black hole mass in nearby galaxy bulges. For the population of bulge-dominated galaxies, the volume-averaged mean growth time of the black hole (due to accretion) and the mean growth time of the surrounding bulge (due to star formation) are very similar. After volume averaging, the growth of black holes through accretion and the growth of bulges through star formation are thus related at the present time in the same way that they have been related, on average, throughout cosmic history.

Thus, it seems likely that the processes that established the tight correlation between bulge mass and black hole mass are still operating in low redshift AGN. These results also suggest a picture in which star formation and black hole growth have been moving steadily and in parallel to lower and lower mass scales since $z \sim 2$ ("downsizing"). This is consistent with the cosmic evolution in the AGN luminosity function and inferred black hole mass function (Ueda et al. 2003, Barger et al. 2005, Hasinger et al. 2005, Fine et al. 2006, Vestergaard et al. 2008).

But by what process is the evolution of the black hole and galaxy linked? Some interesting clues emerged from Kauffmann et al. (2007a), who examined the relationship between the growth of the black hole and star-formation in the both the inner (bulge) and outer (disk) regions using SDSS and GALEX. They showed that galaxies with red (old) disks almost never have a young bulge or a strong AGN. Galaxies with blue (young) disks have bulges and black holes that span a wide range in age and accretion rate, while galaxies with young bulges and strongly accreting black holes almost always have blue outer disks. Their suggested scenario is one in which the source of gas that builds the bulge and black hole is a reservoir of cold gas in the disk. The presence of this gas is a necessary, but not sufficient condition for bulge and black hole growth. Some mechanism must transport this gas inwards in a time variable way.

What is this mechanism? In most of the popular models for the co-evolution of black holes and galaxies, the dominant phase of black hole growth occurs in the aftermath of a major merger between two gas-rich disk galaxies (e.g. Di Matteo et al. 2007, Hopkins et al. 2007). This growth coincides with or closely follows a central burst of star-formation. However, at least in the modern universe ($z < 1$), a strong connection between major mergers and black hole growth has not been established. Simple by-eye examination of images of the host galaxies of the most rapidly-growing black holes in the SDSS sample of Type 2 AGN shows that most are normal or mildly perturbed systems (Kauffmann et al. 2003c). Similar results have been reported for Type 2 AGN at $z \sim 1$ (Grogin et al. 2005, Pierce et al. 2007).

This result has been quantified by Reichard et al. (2008) who measured the "lopsidedness" of the stellar distribution in the host galaxies of low-redshift Type 2 AGN in the SDSS. They find that the more rapidly growing black holes live in galaxies that are more lopsided on-average. However, the typical lopsidedness of the hosts of even the most powerful AGN is mild (~10%). Moreover, Li et al. (2007), find no evidence for an excess of close companion galaxies near powerful Type 2 AGN in the SDSS. Li et al. (2007) and Reichard et al. (2008) suggest that one way to reconcile their two results is by a two-stage process associated with a minor merger (e.g. Mihos & Hernquist 1994). During the first passage of the companion, an initial inflow of gas occurs that fuels star-formation. The growth of the black hole occurs only later in the end stages when the merger is complete (and during or just after a second and more intense episode of star formation).

This would be consistent with observed link between black hole growth and star formation described above, but there would be an offset in time. This scenario would also be qualitatively consistent with the results of Kauffmann et al. (2003c) and Wild et al. (2007) who found an unusually large fraction of powerful Type 2 AGN in "post-starburst" systems. However, these post-starbursts are still only a minority (~20%) of powerful Type 2 AGN. Wild et al. (2007) concluded that the merger-starburst-AGN scenario was a significant – but not the dominant – channel for the growth of black holes today.

As noted in the introduction, feedback from the AGN is believed to be an important (some would say essential) process in the evolution of massive galaxies. Some of the most successful models assume that a significant amount of the rest-mass energy of the matter accreted by the black hole is available to drive a powerful galactic wind (e.g. Hopkins et al. 2007, Di Matteo et al. 2007). To date, there is not clear evidence in the local universe that such processes typically operate in the generic radio-quiet AGN that are the majority population (I will discuss the radio-loud AGN below).

While outflows in radio-quiet AGN are commonly detected in the form of blueshifted absorption-lines (e.g. Weymann et al. 1981), their energetics are very poorly constrained because we do not know their size-scale. In the one case where the dimensions of the outflow have been directly determined, the estimated outflow rate in kinetic energy is only of-order 10^{-6} L_{Bol} (Krongold et al. 2007). On the other hand, Tremonti et al. (2007) have discovered outflows in extreme post-starburst galaxies at intermediate redshifts. The high outflow velocites ($\sim 10^3$ km/sec) strongly suggest these flows were powered by a past AGN episode. If this AGN episode happened as long-ago as the starburst (a few hundred Myr), the implied size scale of the outflow is several hundred kpc, and the associated outflow rates are very high. Finally, small (kpc-scale) and relatively low-power radio jets are seen in many Seyfert galaxies, but they do not appear to be having a significant effect on the bulk of the host galaxy's ISM (Veilleux et al. 2005).

It is also important here to emphasize that since there is strong link between the growth of black holes and star formation, feedback in the form of the kinetic energy supplied by supernovae and stellar winds from massive stars will be available even if the black hole doesn't help at all. As an illustration, the formation of a 10^8 solar mass black hole would be (eventually) accompanied by the formation of $\sim 10^{11}$ solar masses in stars, resulting in $\sim 10^9$ supernovae and $\sim 10^{60}$ ergs in kinetic energy. This is equivalent to $\sim 1\%$ of the rest-mass energy of the black hole, and is sufficient (in principle) to expel the entire ISM of such a galaxy.

13.2.3 Radio Galaxies: Homes of the Most Massive Black Holes

Unlike the case of the Seyfert galaxies and QSOs, typical radio galaxies in the present-day universe appear to be accreting at a highly sub-Eddington rate, and in a radiatively inefficient mode (e.g. Allen et al. 2006). Most of the energy extracted from the black hole appears in the form of highly collimated relativistic outflows of radio-emitting plasma.

For low-power radio galaxies, the large-scale radio emission takes the form of oppositely-directed twin jets whose surface-brightness steadily declines with increasing distance from the nucleus (Fanaroff & Riley 1974 – "FR"). These "FR I" sources totally dominate the local radio galaxy population by number. At very high radio power the morphology of radio sources is strikingly different: it is dominated by bright "hot spots" located on opposite sides of the galaxy at the two outer edges of the radio source. These "FR II" radio galaxies are so rare in the local universe that there are few of them in the SDSS main galaxy sample. However, their comoving density evolves strongly with redshift (e.g. Dunlop & Peacock 1990). Accordingly, I will discuss the FR I radio galaxies here, and their more powerful kin in the section below on the early universe.

A cross-match of the SDSS main galaxy sample and the FIRST and NVSS radio catalogs, yields a sample of several thousand local ($z \sim 0.1$) low-power radio galaxies (Best et al. 2005a). Best et al. (2005b) and Kauffmann et al. (2007b) showed that this class of AGN is almost totally disjoint from the Seyfert/QSO population described above. The properties of the radio galaxies differ in several strong and systematic ways from the Type 2 Seyfert galaxies described in the preceding section. The radio galaxies are selectively drawn from the population of the most massive galaxies (quantifying the well-known result that they are typically giant elliptical galaxies). More precisely, Best et al. found that the probability that a galaxy was a radio-loud AGN (at fixed radio power) increased as the total stellar mass of the galaxy to the $\sim$2.5 power and as the galaxy velocity dispersion to the $\sim$6.5 power. The radio galaxies also had the old stellar populations and highly concentrated structures of normal giant elliptical galaxies.

The strong mass-dependence of the radio luminosity function can be understood as reflecting the mass-dependence of the cooling rate of hot gas in elliptical galaxies (Best et al. 2005b). In fact, Chandra high- resolution X-ray imaging spectroscopy of the hot gas in the centers of the nearest such radio galaxies shows that the radio jet power scales directly with the estimated rate of Bondi accretion in these systems (Allen et al. 2006). Further support for this fueling scenario comes from the enhanced probability of radio emission from the brightest galaxies at the centers of groups and clusters of galaxies – just the locations where the cooling rates will be unusually high (Best et al. 2007).

While – as discussed above – the nature of feedback operating in the Seyfert/QSO mode is still unclear, there is now quite compelling evidence for feedback provided by radio sources. The most direct evidence comes from observations of the cavities evacuated in the hot gas in the cores of galaxy clusters by radio sources (e.g. Fabian et al. 2006, Bîrzan et al. 2004, McNamara & Nulsen 2007). The sizes of the cavities and the measured gas pressure allows the amount of $P\Delta V$ work done by the radio source to be calculated, while the sound crossing time of the cavity gives a characteristic time scale. This then allows a rough estimate to be made of the time-averaged rate of energy transported by the radio jets.

Best et al. (2006) used this approach to derive the scaling between the observed radio luminosity of the jet and its rate of energy transport. Combining this scaling relation with the mass-dependent radio luminosity function in Best et al. (2005b) they were able to show that the time-averaged mass-dependent heating rate due to radio jets roughly matches the measured average mass-dependent radiative cooling rates (X-ray luminosities) in elliptical galaxies.

While these low-power FR I sources do not evolve with look-back time as strongly as the FR II sources discussed below, work by Sadler et al. (2007) shows that their co-moving density roughly doubles between $z \sim 0$ and $z \sim 0.7$. The associated heating rate per co-moving volume would likewise have been larger then, while the total amount of stellar mass in elliptical galaxies was smaller (Bell et al. 2004, Faber et al. 2007). Thus, this form of feedback would have more important (more ergs per gram) than at present.

13.3 The Modern Universe: Prospects for the Next Decade

From the summary above, it appears that we have a pretty clear overall map of the basic landscape in the local universe: we know which black holes are growing and in what kinds of galaxies. Having said that, it is also clear that we have a very incomplete understanding of the actual astrophysical processes at work (we have good cartoons and slogans!). In what follows, I will highlight a few issues where I think the combination of major new facilities will allow us to make dramatic discoveries in the next decade.

13.3.1 The Universe at Redshift One

At a redshift of one, the global rates of both star-formation and black hole growth were about an order-of-magnitude larger than today (e.g. Marconi et al. 2004). This is very intriguing because (from what we have seen so far) the universe at $z \sim 1$ does not look radically different from that at $z \sim 0$. The familiar galaxies that define the Hubble sequence are in place, and the scaling relations and building blocks that define the basic structures of galaxies have not evolved strongly (e.g. Barden et al. 2005, Jogee et al. 2004). As summarized above, the growth of black holes seems to occur in galaxies that are at least qualitatively similar to those in which black holes grow today. The relationship between black hole mass and galaxy velocity dispersion has apparently evolved only weakly if at all (Salviander et al. 2007). So the $z \sim 1$ universe looks sort of like the $z \sim 0$ universe on steroids. The most important difference presumably is the higher overall amount of cold gas in (and/or accretion rate onto) galaxies at $z \sim 1$.

With the advent of major new wide field multi-object spectrographs on 8 and 10 meter-class telescopes it would be feasible in the next decade to undertake the rough equivalent of the SDSS at $z \sim 0.5$ to 1. I'd like to highlight WFMOS – the Wide Field Multi-Object Spectrograph which is planned as a collaboration between the Gemini and Subaru Observatories. It would consist of 4500 fibers that could be deployed over a field of view of 1.5 degrees. These fibers would feed optical spectrographs with resolutions of $R \sim 4000$ ($\sigma \sim 30$ km/s).

13.3.2 The Fueling of Black Holes

We now know that there is a strong link (at least in a statistical sense) between star formation in the innermost several kpc and the growth of black holes. Perhaps the most important unanswered question about AGN is how some small fraction of the cold gas supply being used to form stars is transported over many orders-of-magnitude in radius to the accretion disk around the black hole.

It is generally supposed that at least the early stages of this is driven by the angular momentum transport provided by a non-axisymmetric gravitational potential together with the loss of binding energy through dissipation and radiative cooling of the gas orbiting in such a system. Bars are the most well-known examples of such non-axisymmetric structures, but unfortunately, observational evidence linking the growth of black holes to the presence of either large-scale or nuclear bars has been inconclusive at best (e.g. Mulchaey & Regan 1997, Knapen et al. 2000, Laine et al. 2002, Erwin & Sparke 2002). Oval distortions and even global spiral arms can also drive inflows (Kormendy & Kennicutt 2004 and references therein). Simões Lopes et al. (2007) used HST imaging to show that all 65 AGN host galaxies in their sample have dust (and hence gas) in the central-most few hundred parsecs. Kinematic mapping of several such systems containing low-luminosity AGN shows the streaming of gas inward along the spiral arms seen in the dust (Fathi et al. 2006, Storchi-Bergmann et al. 2007, Riffel et al. 2008).

The phase of gas that will dominate by mass on such circum-nuclear scales will be the molecular gas traced at mm-wavelengths. Moreover, this dense and cool gas will be less subject to the sorts of AGN-driven flows often seen in the hotter and less dense gas (Veilleux, Cecil et al. 2005). Exciting steps in this direction are being taken with current facilities (e.g. Lindt-Krieg et al. 2007). The capabilities of ALMA will be superbly matched to this problem for nearby AGN. The AGN lifetime may well be shorter than the characteristic flow times on these scales (e.g. Martini et al. 2003). If so, it will be essential to investigate the amount of potential fuel (molecular gas), its structure, and its dynamics for a large and complete sample of galaxies (e.g. without regard to the presence of an AGN). The huge increase in sensitivity provided by ALMA would make this feasible.

A related problem is that we do not know the astrophysics that sets the ratio of star-formation to black hole accretion to a time-averaged value of $\sim 10^3$. To tackle this problem we need to combine the information about the rate of inward gas flows as a function of radius (above) with a mapping of the radial distribution and time-dependence of star-formation. This can be provided by mapping of the stellar population and ionized gas in these regions (e.g. González Delgado et al. 2004, Davies et al. 2007) using the new capabilities provided by JWST, GSMT, and ELT.

These capabilities will also be essential to probe what's going on at even smaller (few pc) scales. Here we enter a new regime where the gravitational potential of the black hole itself starts to dominates the dynamics. This transition radius is given by $R = GM_{BH}/\sigma^2$ where σ is the galaxy velocity dispersion. Using the observed relation between σ and M_{BH} (Tremaine et al. 2002, Ferrarese & Merritt 2000), this can be written as $R \sim 12\ (M_{BH}/10^8)^{1/2}$ pc. For the nearest AGN this has an angular scale of-order 0.1 arcsec. These scales are also where we approach the domain of the obscuring torus.

Of course, we have a lot of information about this region in the Galactic Center (e.g. Genzel & Karas 2007), but our own supermassive black hole there is currently quiescent. Observations of water masers using VLBI techniques provide our highest resolution maps of this region in nearby AGN (e.g. Moran et al. 2007), but only a minority of AGN have such masing regions (Braatz et al. 2004). With existing mid-IR interferometers and AO-fed ground-based near-IR spectrographs it is now just possible to resolve the hot dust emission from this region (Tristram et al. 2007) and map the kinematics of the hot molecular hydrogen (Davies et al. 2006, Hicks & Malkan 2008). In the next decade, JWST, ELT, and GSMT will make it possible to make detailed maps of the gas and dust on these small scales for complete samples of the nearest AGN.

13.3.3 AGN Feedback

The powerful capabilities for X-ray spectroscopy provided by the proposed International X-ray Observatory (IXO) will give us revolutionary insights into AGN-driven feedback. With an imaging X-ray calorimeter it will be possible for the first time to actually make spatially-resolved maps of the detailed kinematics of the hot gas that is being accelerated and heated by radio sources in clusters and giant elliptical galaxies. The combination of high spectral resolution and sensitivity will also make it possible to conduct a major campaign of time-domain studies of the AGN-driven outflows traced through their blue-shifted X-ray absorption-lines. Such studies are the best way to determine the size-scales of these outflows, and hence the outflow rates of mass and energy carried by them.

13.4 The Early Universe

13.4.1 Overview

In summarizing above what we know about the modern universe, I emphasized results from large, homogeneously selected, and complete samples. I also emphasized the importance of spectroscopy. Such an approach has allowed us to decisively resolve some long-standing controversies, quantify long-known or long-suspected qualitative results, and discover the unexpected. Unfortunately, at high-redshift this approach is essentially impossible with our current capabilities.

It's obvious that at high redshift the host galaxies of AGN are faint, and so it is not currently possible to observe very large samples. A less obvious but equally important problem is that it is very difficult with the present data to robustly characterize the basic properties of both the AGN (e.g. bolometric luminosity, black hole mass, Eddington ratio) and those of its host galaxy (e.g. stellar mass, star formation rate, velocity dispersion, structure/morphology). We have a reasonable handle on the AGN properties in high-z QSOs, but then have very limited information about their host galaxies. As at low redshift, it is easier to study the properties of the host galaxies in Type 2 AGN. The main techniques for finding Type 2 AGN at high-redshift are through observations in the rest-frame mid-IR or in the hard X-ray band, supplemented by radio continuum observations. Unfortunately, with only this limited coverage of the full spectral energy distribution, it is often difficult to cleanly separate out AGN from powerful starbursts.

Here's one way of thinking about the problem. We know that star formation and black hole growth are coupled, and that in a time-averaged sense the ratio of these two rates is of-order 10^3 (e.g. Marconi et al. 2004). Models suggest that this ratio is more like 10^2 during the phase when the black hole growth rate is maximized (e.g. Hopkins et al. 2007). The ratio of bolometric luminosity (stars/black hole) corresponding to the two ratios above are $\sim$10 and $\sim$1 respectively (assuming a radiative efficiency of 10% for black hole accretion and assuming a Kroupa IMF for the star formation). Thus, we expect both phenomena to be energetically significant in high-redshift AGN. Sorting out the luminosity attributable to the AGN (measuring the black hole accretion rate) and to the young stars (determining the star-formation rate) is not straightforward with the limited available data. Even assuming we can determine the AGN bolometric luminosity, we usually have only a lower limit to the black hole mass in Type 2 AGN (by assuming that the Eddington limit is obeyed).

13.4.2 Which Came First?

The tight relationship at $z \sim 0$ between the mass of the black hole and the mass and velocity dispersion of the galaxy bulge within which it lives is the result of a time integral over the history of the universe. It does not however tell us that the two formation processes must occur simultaneously in any given galaxy. If not, which comes first: the black hole or the galaxy?

There have been a number of attempts to determine the relationship between the galaxy mass (or velocity dispersion) and black hole mass at high-redshift. Peng et al. (2006) used HST images of eleven QSO host galaxies at $z \sim 2$ to estimate luminosities of the bulges. They estimated the black hole masses using the "photoionization method". This is based on the scaling relation seen in local Seyfert nuclei between black hole mass (determined via reverberation mapping), AGN ionizing luminosity, and the width of the emission-lines in the Broad Line Region (Kaspi et al. 2005). Peng et al. find that the ratio of black hole to bulge mass is about four times larger at $z \sim 2$ than at present (e.g. the black holes form first). McLure et al. (2006) find a similar result using a matched sample of radio galaxies and radio-loud QSOs.

In contrast, Shields et al. (2003) examined a sample of QSOs at redshifts up to $\sim$3. They used the velocity dispersion of the emission-lines in the NLR as a proxy for the stellar velocity dispersion (Nelson & Whittle 1996, Greene & Ho 2005). The black hole masses were estimated using the above photoionization method. They found no evidence that the relation between black hole mass and galaxy velocity dispersion had evolved strongly between $z \sim 0$ to 3. Borys et al. (2005) point out that if the black holes detected in sub-mm galaxies via their hard X-ray emission are accreting at the Eddington rate, their luminosites would imply that they have masses more than an order-of-magnitude below that of black holes today in galaxies with similar mass (e.g. the galaxy would come before the black hole).

Overall, the situation regarding the phasing of black hole formation at high-z is thus unclear. Hopkins et al. (2006) use integral constraints on the co-moving mass density of black holes to argue that there can not be a strong evolution in the ratio of stellar and black hole mass in galaxies. Small changes will be difficult to detect, especially when the ferocious systematic effects discussed by Lauer et al. (2007) are considered. Lauer et al. argue that a robust attack on this problem requires both an accurate measurement of the "scatter function" in the M_{BH} vs. M_{gal} or vs. σ relations, and samples at high and low redshift that have been selected (and investigated) in exactly the same way using precisely defined and objective criteria. This task will be far easier using the capabilities of JWST, GSMT, and ELT to characterize the salient properties of large samples of galaxies at high-redshift.

13.4.3 Where are Black Holes and Galaxies Growing?

I'd like now to briefly discuss observations of the connections between black hole growth and star formation at high redshift by looking for signs of both in a given population of galaxies. Because of the multitude of ways in which both high-z AGN and high-z galaxies are selected and investigated, I am strongly reminded of the fable of the blind men and the elephant (whom the blind men variously describe as a tree, a large leaf, a rope, and a wall depending upon their particular set of "observations").

The largest sample of high-z AGN by far are QSOs. Observations of small samples of high-z QSOs that were selected based on their relative brightness in the rest-frame far-IR (indicating strong dust emission) show that these have very large masses of molecular gas (Solomon & Vanden Bout 2005) and high star formation rates (Lutz et al. 2007). However, observations of more typical high-z QSOs (selected without regard to their far-IR properties) do not reveal similarly large gas masses or star formation rates (Maiolino et al. 2007a,b).

Investigations of high-redshift FR II radio galaxies gave us our first view of galaxies at high redshift (e.g. Chambers et al. 1996). As a class, they appear to be very massive systems – the progenitors of present-day giant ellipticals (e.g. Lilly 1989, Willott et al. 2003). Their properties imply a high redshift of formation. The fraction of these systems that are detected in rest-frame far-IR rises steeply above redshifts of 2.5 to 3 (Archibald et al. 2001, Reuland et al. 2004). If this emission is powered by star-formation, this may signal the epoch of formation of these systems (Willott et al. 2002, 2003).

The largest and best-studied sample of star forming galaxies at $z > 2$ are the (rest)-UV-selected Lyman Break Galaxies (e.g. Giavalisco 2002), and the related BX and BM samples (Steidel et al. 2004). These have the advantage that the amount of dust-obscuration is modest so that we have a relatively clean view. Based on the general lack of either hard X-ray emission (Laird et al. 2006) or of an AGN signature in their rest-UV spectra (Steidel et al. 2002) only a small minority have bright AGN (although see Groves et al. 2006 for a possible caveat). Certainly in these galaxies it appears that the star formation rate is generally much greater than a thousand times the black hole accretion rate.

Another UV-selected population of high-z galaxies are those detected by their Lyα emission-lines (e.g. Rhoads et al. 2000). These galaxies appear to be undergoing significant star formation (with Lyα emission due to photoionization by O stars). They are not detected even in deep stacked X-ray (Wang et al. 2004) or radio continuum (Carilli et al. 2007b) data, and so do not seem to be harboring rapidly growing black holes.

There are several multicolor techniques used to select high-z galaxies in the rest-frame optical. Kriek et al. (2007) have obtained near-IR spectra of a complete sample of 20 K-band-selected galaxies at $z \sim 2.3$. They find that these are massive galaxies (few times 10^{11} solar masses). Of the sample, 45% are quiescent (no star formation or AGN), 35% are star-forming galaxies, and 25% are Type 2 AGN. Comparing these demographics to the properties of lower mass systems at the same redshift, they find that AGN activity is preferentially occurring in the most massive galaxies (consistent with a picture of cosmic downsizing). However, in this small sample they see no correlation between the presence of an AGN and star formation.

Moving to longer wavelengths, Spitzer observations of the high-z universe in the rest-frame mid-IR have led to the discovery of a population of highly luminous and dusty systems. The high level of obscuration makes it difficult to robustly separate out the energetic contribution of an AGN and starburst. Mid-IR spectroscopy with Spitzer shows that this is a heterogeneous population, ranging from objects dominated by AGN to objects dominated by starbursts to objects that are clearly starburst-AGN composites (Brand et al. 2007, Sajina et al. 2007, Yan et al. 2007). Daddi et al. (2007) find that roughly a quarter of these mid-IR selected galaxies show both a mid-IR excess attributable to an AGN and have X-ray properties consistent with a heavily obscured (Compton-thick) AGN. Their high space density would make them a very significant population of high-z AGN.

At yet longer wavelengths, the sub-mm galaxies at high-z are extremely dusty systems selected on the basis of their rest-frame far-IR emission. Radio continuum observations (Chapman et al. 2001) suggest that the bulk of the far-IR emission is powered by star formation (since they roughly obey the local relation between the radio and far-IR emission defined by local star forming galaxies). This is consistent with inferences based on mid-IR spectroscopy (Valiante et al. 2007, Pope et al. 2007). Most are detected in the hard X-ray band, but the X-ray luminosities imply that the AGN contributes only about 10% of the bolometric luminosity (Alexander et al. 2005). This ratio agrees expectations for an object in which the star formation rate is $\sim 10^3$ times the black hole accretion rate (see above). Thus, these are excellent candidates for witnessing the co-formation of high-mass galaxies and their black holes.

In the next decade, I believe an ambitious program with our new facilities will enable us to make enormous progress in determining the relationship between black hole growth in the early universe. The key will be to undertake panchromatic spectroscopic investigations of large and complete samples of galaxies and AGN selected in a careful and complementary way.

Spectroscopy is the best way to robustly determine the black hole accretion rates and star formation rates. X-ray spectroscopy (IXO) can provide clear evidence of an AGN even when the AGN itself is hidden behind Compton-thick material along our line-of-sight (e.g. Fabian et al. 2000, Levenson et al. 2006). For relatively unobscured objects, spectra in the rest-frame ultraviolet with GSMT and ELT can directly detect the spectroscopic signature of young stars and reflected light from a hidden AGN. Rest-frame optical spectra with JWST can utilize the same diagnostics of star formation and black hole accretion at high-redshift that have been effectively exploited at low redshift. The mid-IR region (JWST with MIRI) is also rich in spectroscopic diagnostics (e.g. Genzel et al. 1998). Heating by the hard radiation field of an AGN has characteristic effects on the molecular gas that can be investigated spectroscopically with ALMA (Carilli et al. 2007a).

The power of large and complete samples to address these issues could be realized by using WFMOS to undertake a spectroscopic survey in the rest-frame UV of a million high-redshift galaxies. This would be a nice side benefit from the core WFMOS science program that aims to constrain Dark Energy through the measurement of baryon acoustic oscillations in the galaxy power spectrum at high redshift.

13.4.4 How are Black Holes Fueled?

At low redshift AGN activity seems to be triggered by both the inflow of relatively cold gas in medium-mass systems (the Seyfert galaxies) and of hot gas in the most massive galaxies (the radio galaxies). Models imply that the relative importance of the "cold" mode will increase at high redshifts and dominate in both the low and high mass galaxies (e.g. Kereš et al. 2005, Croton et al. 2006). In such models, cold accretion occurs episodically, and many models link the formation of black holes to major merger events (e.g. Hopkins et al. 2007, Di Matteo et al. 2007). Can we see clear evidence at high-z that the growth of black holes is indeed driven by mergers or major accretion events?

Imaging the host galaxies of high-z QSO's is very challenging, even with HST. The investigations to date (Ridgway et al. 2001, Kukula et al. 2001, Peng et al. 2006) have determined only the most basic properties (luminosity and size) of the host galaxies for relatively small samples. The images are not sensitive enough to detect large-scale low-surface brightness tidal features while the inner regions are strongly contaminated by the QSO.

In the absence of a bright central QSO, it is easier to look for morphological evidence for mergers in Type 2 AGN. Imaging of high-redshift FR II radio galaxies do show spectacular examples of what appear to be the coalescence of a massive galaxy (e.g. Miley et al. 2006), but these are exceedingly rare systems compared to typical high-z AGN. Based on HST imaging, Pope et al. (2005) and Chapman et al. (2003) conclude that high-z sub-mm galaxies are larger and more asymmetric than other galaxies at these redshifts, and have the complex irregular morphologies suggestive of mergers.

Progress in the next decade will come through two complementary approaches. First, the high angular resolution, stable point spread function, and superb sensitivity of NIRCAM on JWST will enable us to accurately characterized the morphologies of large samples of the host galaxies of high-redshift AGN, including QSOs. This will allow us to quantify the incidence rates of recent mergers in these objects. We may be able to learn about the relative sequencing of the star formation and black hole growth by comparing samples of systems that appear to be in the early vs. late stage of the merger.

Second, the IFUs on JWST, GSMT, and ELT will make it possible to map the ionized gas in the AGN host galaxies, while ALMA will do the same for the molecular gas. These data could provide direct kinematic evidence that we are witnessing the aftermath of a major accretion/merging event and would pinpoint the location and kinematics of the molecular gas that is fueling the star-formation traced by the ionized gas. Programs with the current generation of facilities give us tantalizing examples of the power of such an approach (e.g. Forster Schreiber et al. 2006, Bouche et al. 2007, Nesvadba et al. 2007, Law et al. 2007).

13.4.5 AGN Feedback

The co-moving rate of accretion onto black holes at $z \sim 2$ to 3 was about a factor of ~ 30 higher than today. Feedback related to this black hole building could have a dramatic effect on the formation and evolution of massive galaxies. Do we see it in action? Here, the situation is somewhat similar to what we see in the modern universe: the most direct and convincing evidence of feedback that is being driven by the AGN and is having a dramatic impact on gas on galactic-scales is found in the form of radio jets.

The presence of galaxy-scale emission-line nebulae around radio-loud QSOs and powerful FR II radio galaxies at high-z has long been known (e.g. McCarthy et al. 1996, Heckman et al. 1991a,b). Long-slit spectroscopy of these systems shows the presence of high gas velocities (of-order 10^3 km/s), strongly suggesting the outflow of substantial amounts of gas. The morphological connection between the ionized gas and the radio sources supports the idea that the outflow is driven by the radio source itself (not by radiation pressure or a spherical wind blown by the black hole). This is consistent with the much smaller, fainter, and more quiescent emission-line nebulae seen around radio-quiet QSOs at the same high redshifts (Christensen et al. 2006).

Recently, very detailed maps in the rest-frame optical have been made for several of the high-z FR II radio galaxies (Nesvadba et al. 2006 and private communication) using near-IR IFU spectroscopy. These data provide for the first time detailed maps of the kinematics and physical conditions in the regions, and provide convincing confirmation of the idea that the radio sources are likely to be blasting away the entire gaseous halo around the galaxy. This is very exciting, but we must keep in mind that black holes with radio sources this powerful constitute only a small minority ($\sim 0.1\%$) of the rapidly growing supermassive black holes at these redshifts.

With the enormous increase in capability provided by the IFUs on JWST, GSMT, and ELT it would be relatively easy to undertake detailed investigations like this of large and complete samples of all the important classes of AGN at high redshift. The capabilities for high resolution X-ray spectroscopy provided by IXO can be used to investigate the physics of AGN feedback traced by the hot gas. With GSMT and ELT it will possible to undertake high S/N spectroscopy with high spectral resolution for moderately large, complete samples of QSOs to try to directly measure the outflow rates implied by their blueshifted absorption-lines (Korista et al. 2008).

13.5 Final Thoughts

I began this paper by listing all the unanswered questions we have about how the co-evolution of black holes and galaxies actually works. How does gas get into galaxies in the first place? Once inside, how is it transported all the way to the black hole's accretion disk? What astrophysics sets the mass ratio of the gas turned into stars compared to the gas accreted by the black hole (why is this ratio $\sim 10^3$)? What is the real astrophysics of the feedback from the supermassive black hole that seems to be a crucial ingredient in galaxy evolution? What is the sequencing, both in the sense of overall cosmic history and in the life of an individual galaxy? Does the black hole or galaxy form first? Is the rapid growth of a black hole a once-in-a-lifetime transformative event in the life of a galaxy (e.g. a major merger followed by catastrophic feedback), or is it a more gradual, intermittent process? Does the answer depend on redshift and/or black hole mass?

I have tried to describe our current state of knowledge and ignorance, and have also tried to summarize how I think we can use the amazing new observatories of the next decade to best answer these questions. Given that we are trying to understand a complex cosmic eco-system consisting of hot gas, cold dusty gas, stars, and black holes, it seems clear to me that a panchromatic approach is essential. JWST, ALMA, GSMT/ELT, and IXO will all play crucial roles. I have also tried to emphasize the importance of spectroscopy and of 3-D imaging-spectroscopy in particular. Again, the capabilities of these observatories are a superb match to the problem at hand. Finally, I have emphasized the importance of large surveys of complete and carefully selected samples. The increase in raw sensitivity provided by the new observatories will help make this approach feasible. The huge gain in discovery power provided by of the next generation of multi-object optical and near-IR spectrographs (such as WFMOS) on existing 8 and 10-meter class telescopes will allow us to undertake SDSS-scale surveys at redshifts of one and beyond.

References

Alexander, D.M., Bauer, F.E., Chapman, S.C., Smail, I., Blain, A.W., Brandt, W.N., & Ivison, R.J. 2005, ApJ, 632, 736

Allen, S.W., Dunn, R.J.H., Fabian, A.C., Taylor, G.B., & Reynolds, C.S. 2006, MNRAS, 372, 21

Archibald, E.N., Dunlop, J.S., Hughes, D.H., Rawlings, S., Eales, S.A., & Ivison, R.J. 2001, MNRAS, 323, 417
Baldry, I.K., Glazebrook, K., Brinkman, J., Ivezic, Z., Lupton, R., Nichol, R., & Szalay, A. 2004, ApJ, 600, 681
Barden, M., et al. 2005, ApJ, 635, 959
Barger, A.J., Cowie, L.L., Mushotzky, R.F., Yang, Y., Wang, W.-H., Steffen, A.T., & Capak, P. 2005, AJ, 129, 578
Bell, E.F., Wolf, C., Meisenheimer, K., Rix, H.-W., Borch, A., Dye, S., Kleinheinrich, M., Wisotzki, L., & McIntosh, D.H. 2004, ApJ, 608, 752
Best, P.N., Kauffmann, G., Heckman, T.M., & Ivezić, Ž. 2005a, MNRAS, 362, 9
Best, P.N., Kauffmann, G., Heckman, T.M., Brinchmann, J., Charlot, S., Ivezić, Ž., & White, S.D.M. 2005b, MNRAS, 362, 52
Best, P.N., Kaiser, C.R., Heckman, T.M., & Kauffmann, G. 2006, MNRAS, 368, L67
Best, P.N., von der Linden, A., Kauffmann, G., Heckman, T.M., & Kaiser, C.R. 2007, MNRAS, 379, 894
Bîrzan, L., Rafferty, D.A., McNamara, B.R., Wise, M.W., & Nulsen, P.E.J. 2004, ApJ, 607, 800
Borys, C., Smail, I., Chapman, S.C., Blain, A.W., Alexander, D.M., & Ivison, R.J. 2005, ApJ, 635, 853
Bouche, N., et al. 2007, ApJ, 671, 303
Bower, R., Benson, A., Malbon, R., Helly, J., Frenk, C., Baugh, C., & Lacey, C. 2006, MNRAS, 370, 645
Brand, K., Weedman, D., Desai, V., Le Floc'h, E., Armus, L., Dey, A., Houck, J.R., Jannuzi, B.T., Smith, H.A., & Soifer, B.T. 2007, astro-ph:07093119
Braatz, J.A., Henkel, C., Greenhill, L.J., Moran, J.M., & Wilson, A.S. 2004, ApJ, 617, L29
Brinchmann, J., Charlot, S., White, S.D.M., Tremonti, C., Kauffmann, G., Heckman, T., & Brinkmann, J. 2004, MNRAS, 351, 1151
Carilli, C.L., Walter, F., Wang, R., Wootten, A., Menten, K., Bertoldi, F., Schinnerer, E., Cox, P., Beelen, A., & Omont, A. 2007a, astro-ph:0703799
Carilli, C., et al. 2007b, ApJS, 172, 518
Chambers, K.C., Miley, G.K., van Breugel, W.J.M., Bremer, M.A.R., Huang, J.-S., & Trentham, N.A. 1996, ApJS, 106, 247
Chapman, S.C., Richards, E.A., Lewis, G.F., Wilson, G., Barger, A.J. 2001, ApJ, 548, 147
Chapman, S.C., Windhorst, R., Odewahn, S., Yan, H., & Conselice, C. 2003, ApJ, 599, 92
Christensen, L., Jahnke, K., Wisotzki, L., & Sánchez, S.F. 2006, A&A, 459, 717
Churazov, E., Brüggen, M., Kaiser, C.R., Böhringer, H., & Forman, W. 2001, ApJ, 554, 261
Cid Fernandes, R., Heckman, T., Schmitt, H., Delgado González. R, & Storchi-Bergmann, T., 2001, ApJ, 558, 81
Croton, D.J., Springel, V., White, S.D.M., De Lucia, G., Frenk, C.S., Gao, L., Jenkins, A., Kauffmann, G., Navarro, J.F., & Yoshida, N. 2006, MNRAS, 365, 11
Daddi, E., et al. 2007, ApJ, 670, 173
Davies, R.I., Genzel, R., Tacconi, L., Mueller Sánchez, F., & Sternberg, A. 2006, astroph: 0612009
Davies, R.I., Mueller Sánchez, F., Genzel, R., Tacconi, L.J., Hicks, E.K.S., Friedrich, S., & Sternberg, A. 2007 ApJ, 671, 1388
Dunlop, J.S., & Peacock, J.A. 1990, MNRAS, 247, 19
Di Matteo, T., Colberg, J., Springel, V., Hernquist, L., & Sijacki, D. 2007, astroph: 07052269
Erwin, P., & Sparke, L.S. 2002, AJ, 124, 65
Faber, S., et al. 2007, ApJ, 665, 265
Fabian, A., Iwasawa, K, Reynolds, C.S., & Young, A.J. 2000, PASP, 112, 1145
Fabian, A., Sanders, J., Taylor, G., Allen, S., Crawford, C., Johnstone, R., & Iwasawa, K. 2006, MNRAS, 366, 417
Fanaroff, B.L., & Riley, J.M. 1974, MNRAS, 167, 31
Fathi, K., Storchi-Bergmann, T., Riffel, Rogemar, A., Winge, C., Axon, D.J., Robinson, A., Capetti, A., & Marconi, A. 2006, ApJ, 641, L25

Ferrarese, L. & Merritt, D. 2000, ApJ, 539, L9
Fine, S., et al. 2006, MNRAS, 373, 613
Forster Schreiber, N. et al. 2006, AJ, 131 1891
Gebhardt, K. et al. 2000, ApJ, 539, L13
Genzel, R. et al. 1998, ApJ, 498, 579
Genzel, R., & Karas, V. 2007, astro-ph: 07041281
Giavalisco, M. 2002, ARA&A, 40, 579
González Delgado, R., Cid Fernandes, R., Pérez, E., Martins, L.P., Storchi-Bergmann, T., Schmitt, H., Heckman, T., & Leitherer, C. 2004, ApJ, 605, 127
Granato, G.L., Silva, L., Monaco, P., Panuzzo, P., Salucci, P., De Zotti, G., & Danese, L. 2001, MNRAS, 324, 757
Greene, J., & Ho, L. 2005, ApJ, 627, 721
Grogin, N.A., et al. 2005, ApJ, 627, L97
Groves, B., Heckman, T., & Kauffmann, G. 2006, MNRAS, 371, 1559
Hao. L. et al. 2005, AJ, 129, 1783
Häring, N., & Rix, H.-W. 2004, ApJ, 604, L89
Hasinger, G., Miyaji, T., & Schmidt, M. 2005, A&A, 441, 417
Heckman, T.M., Miley, G.K., Lehnert, M.D., & van Breugel, W. 1991a, ApJ, 370, 78
Heckman, T.M., Lehnert, M.D., Miley, G.K., & van Breugel, W. 1991b, ApJ, 381, 373
Heckman, T., Kauffmann, G., Brinchmann, J., Charlot, S., Tremonti, C., & White, S. 2004, ApJ, 613, 109
Hicks, E., & Malkan, M. 2008, ApJS, 174, 31
Hopkins, P.F., Robertson, B., Krause, E., Hernquist, L., & Cox, T.J. 2006, ApJ, 652, 107
Hopkins, P.F., Hernquist, L., Cox, T.J., & Keres, D. 2007, astro-ph: 07061243
Jogee, S., et al. 2004, ApJ, 615, L105
Kauffmann, G., & Haehnelt, M. 2000, MNRAS, 311, 576
Kauffmann, G., White, S.D.M., & Guiderdoni, B. 1993, MNRAS, 264, 201
Kauffmann, G., et al. 2003a, MNRAS, 341, 33
Kauffmann, G., Heckman, T.M., White, S.D.M., Charlot, S., Tremonti, C., Peng, E.W., Seibert, M., Brinkmann, J., Nichol, R.C., SubbaRao, M., & York, D. 2003b, MNRAS, 341, 54
Kauffmann, G., et al. 2003c, MNRAS, 346, 1055
Kauffmann, G., White, S.D.M., Heckman, T.M., Ménard, B., Brinchmann, J., Charlot, S., Tremonti, C., & Brinkmann, J. 2004, MNRAS, 353, 713
Kauffmann, G., et al. 2007a, ApJS, 173, 357
Kauffmann, G., Heckman, T., & Best, P. 2007b, astro-ph: 07092911
Kaspi, S., Maoz, D., Netzer, H., Peterson, B.M., Vestergaard, M., & Jannuzi, B.T. 2005, 629, 61
Kereš, D., Katz, N., Weinberg, D.H., & Davé, R. 2005, MNRAS, 363, 2
Korista, K., Bautista, M., Arav, N., Moe, M., Constantini, E., & Benn, C. 2008, astro-ph: 0807230
Kormendy, J. & Kennicutt, R. 2004, ARA&A, 42, 603
Knapen, J., Schlosman, I., & Peletier, R. 2000, ApJ, 529, 97
Kriek, M. et al. 2007, ApJ, 669, 776
Krolik, J.H., & Begelman, M. 1988, ApJ, 329, 702
Krongold, Y., Nicastro, F., Elvis, M., Brickhouse, N., Binette, L., Mathur, S., & Jiménez-Bailón, E. 2007, ApJ, 659, 1022
Kukula, M.J., Dunlop, J.S., McLure, R.J., Miller, L., Percival, W.J., Baum, S.A., & O'Dea, C.P. 2001, MNRAS, 326, 1533
Laine, S., Shlosman, I., Knapen, J.H., & Peletier, R.F. 2002, ApJ, 567, 97
Laird, E.S., Nandra, K., Hobbs, A., & Steidel, C.C. 2006, MNRAS, 373, 217
Lauer, T.R., Tremaine, S., Richstone, D., & Faber, S.M. 2007, ApJ, 670, 249
Law, D.R., Steidel, C.C., Erb, D.K., Larkin, J., Pettini, M., Shapley, A., & Wright, S.A. 2007, ApJ, 669, 929
Levenson, N.A., Heckman, T.M., Krolik, J.H., Weaver, K.A., & Życki, P.T. 2006, ApJ, 648, 111
Li, C., Kauffmann, G., Heckman, T.M., White, S.D.M., & Jing, Y.P. 2007, astro-ph: 07120383
Lilly, S. 1989, ApJ, 340, 77

Lindt-Krieg, E., Eckart, A., Neri, R., Krips, M., Pott, J.-U., Garcia-Burillo, S., & Combes, F. 2007, astro-ph: 07123133
Lutz, D., Sturm, E., Tacconi, L.J., Valiante, E., Schweitzer, M., Netzer, H., Maiolino, R., Andreani, P., Shemmer, O., & Veilleux, S. 2007, ApJ, 661, L25
Maiolino, R., Shemmer, O., Imanishi, M., Netzer, H., Oliva, E., Lutz, D., & Sturm, E., 2007a, A&A, 468, 979
Maiolino, R., et al., 2007b, A&A, 472, L33
Marconi, A., & Hunt, L.K. 2003, ApJ, 589, L21
Marconi, A., Risaliti, G., Gilli, R., Hunt, L.K., Maiolino, R., & Salvati, M. 2004, MNRAS, 351, 169
Martin, D.C., et al. 2007, ApJS, 173, 342
Martini, P., Regan, M.W., Mulchaey, J.S., & Pogge, R.W. 2003, ApJ, 589, 774
McCarthy, P., Baum, S., & Spinrad, H. 1996, ApJS, 106, 281
McLure, R.J., Jarvis, M.J., Targett, T.A., Dunlop, J.S., Best, P.N. 2006, MNRAS, 368, 1395
McNamara, B.R., & Nulsen, P.E.J. 2007, ARA&A, 45, 117
Mihos, J.C., & Hernquist, L. 1994, 425, L13
Miley, G.K., et al. 2006, ApJ, 650, L29
Moran, J., Humphreys, L., Greenhill, L., Reid, M., & Argon, A. 2007, astro-ph: 07071032
Mulchaey, J., & Regan, M. 1997, ApJ, 482, L135
Nandra, K., Georgakakis, A., Willmer, C.N.A., Cooper, M.C., Croton, D.J., Davis, M., Faber, S.M., Koo, D.C., Laird, E.S., & Newman, J.A. 2007, ApJ, 660, L11
Nelson, C., & Whittle, M. 1996, ApJ, 465, 96
Nesvadba, N.P.H., Lehnert, M.D., Eisenhauer, F., Gilbert, A., Tecza, M., & Abuter, R. 2006, ApJ, 650, 693
Nesvadba, N.P.H., Lehnert, M.D., Davies, R.I., Verma, A., & Eisenhauer, F. 2007, astro-ph: 07111491
Peng, C.Y., Impey, C.D., Rix, H.-W., Kochanek, C.S., Keeton, C.R., Falco, E.E., Lehár, J., & McLeod, B.A. 2006, ApJ, 649, 616
Pierce, C.M., et al. 2007, ApJ, 660, L19
Pope, A., Borys, C., Scott, D., Conselice, C., Dickinson, M., & Mobasher, B. 2005, MNRAS, 358, 149
Pope, A., Chary, R.-R., Alexander, D.M., Armus, L., Dickinson, M., Elbaz, D., Frayer, D., Scott, D., & Teplitz, H. 2007, astro-ph: 70111553
Reichard, T., Heckman, T., Rudnick, G., Brinchmann, J., Kauffmann, G., & Wild, V. 2008, arXiv: 0809.3310
Reuland, M., Röttgering, H., van Breugel, W., & De Breuck, C. 2004, MNRAS, 353, 377
Rhoads, J.E., Malhotra, S., Dey, A., Stern, D., Spinrad, H., & Jannuzi, B.T. 2000, ApJ, 545, L85
Ridgway, S.E., Heckman, T.M., Calzetti, D., & Lehnert, M. 2001, ApJ, 550, 122
Riffel, R., Storchi-Bergmann, T., Winge, C., Beck, T., & Schmitt, H. 2008, MNRAS, 385, 1129
Rovilos, E., & Georgantopoulos, I. 2007, A&A, 475, 115
Rovilos, E., Georgakakis, A., Georgantopoulos, I., Afonso, J., Koekemoer, A.M., Mobasher, B., & Goudis, C. 2007, A&A, 466, 119
Sadler, E., et al. 2007, MNRAS, 381, 211
Sajina, A., Yan, L., Armus, L., Choi, P., Fadda, D., Helou, G., & Spoon, H. 2007, ApJ, 664, 713
Salviander, S., Shields, G.A., Gebhardt, K., & Bonning, E.W. 2007, ApJ, 662, 131
2007 Schiminovich, D., et al. 2007, ApJS, 173, 315
Shields, G.A., Gebhardt, K., Salviander, S., Wills, B.J., Xie, B., Brotherton, M.S., Yuan, J., Dietrich, M. 2003, ApJ, 583, 124
Simões Lopes, R.D., Storchi-Bergmann, T., de Fátima Saraiva, M., & Martini, P. 2007, ApJ, 665, 718
Solomon, P.M., & Vanden Bout, P.A. 2005, ARA&A, 43, 677
Spergel, D., et al. 2007, ApJS, 170, 377
Steidel, C.C., Hunt, M.P., Shapley, A.E., Adelberger, K.L., Pettini, M., Dickinson, M., & Giavalisco, M. 2002, ApJ, 576, 653

Steidel, C.C., Shapley, A.E., Pettini, M., Adelberger, K.L, ,Erb, D.K., Reddy, N.A., Hunt, M.P. 2004, ApJ, 604, 534
Storchi-Bergmann, T., Dors, O.L., Jr., Riffel, R.A., Fathi, K., Axon, D.J., Robinson, A., Marconi, A., & Östlin, G. 2007, ApJ, 670, 959
Tegmark, M., et al. 2004, Phys. Rev. D 69, 3051
Tremaine, S., et al. 2002, ApJ, 574, 740
Tremonti, C.A., Moustakas, J., & Diamond-Stanic, A.M. 2007, ApJ, 663, L77
Tristram, K.R.W., et al. 2007, A&A, 474, 837
Ueda, Y., Akiyama, M., Ohta, K., & Miyaji, T. 2003, ApJ, 598, 886
Valiante, E., Lutz, D., Sturm, E., Genzel, R., Tacconi, L.J., Lehnert, M.D., & Baker, A.J. 2007, ApJ, 660, 1060
Veilleux, S., Cecil, G., & Bland-Hawthorn, J. 2005, ARA&A, 43, 769
Vestergaard, M., Fan, X., Tremonti, C.A., Osmer, P.S., & Richards, G.T. 2008, astro-ph: 0801243
Wang, J.X., Rhoads, J.E., Malhotra, S., Dawson, S., Stern, D., Dey, A., Heckman, T.M., Norman, C.A., & Spinrad, H. 2004, ApJ, 608, L21
Weymann, R., Carswell, R., & Smith, M. 1981, ARA&A, 19, 41
Wild, V., Kauffmann, G., Heckman, T., Charlot, S., Lemson, G., Brinchmann, J., Reichard, T., & Pasquali, A. 2007, MNRAS, 381, 543
Willott, C.J., Rawlings, S., Archibald, E.N., & Dunlop, J.S. 2002, MNRAS, 331, 435
Willott, C.J., Rawlings, S., Jarvis, M.J., & Blundell, K.M. 2003, MNRAS, 339, 173
Yan, L., Sajina, A., Fadda, D., Choi, P., Armus, L., Helou, G., Teplitz, H., Frayer, D., & Surace, J. 2007, ApJ, 658, 778

Steve Unwin was prepared to have his photo taken. Charles Lada was not

Chapter 14
The Intergalactic Medium at High Redshifts

Steven R. Furlanetto

Abstract The intergalactic medium (IGM) contains $>95\%$ of the mass in the Universe at high redshifts, and its properties control the earliest phases of structure formation and the reionization process. Although its evolution may seem straightforward, a number of feedback mechanisms can dramatically affect it. Radiative feedback, through a Lyman-Werner background, an X-ray background, and photoionization, affect halo collapse and the clumping of the IGM. We describe how the redshifted 21 cm background can be used to study these effects. Chemical feedback, primarily through supernova winds, changes the modes of star formation and halo cooling; it can be studied through metal absorption lines with the JWST, as well as metal lines in the cosmic microwave background, direct observations of cooling radiation, and fossil evidence in the nearby Universe. Finally, we describe how uncertainties in our modeling of the IGM structure affect reionization models and observations. Detailed studies of helium reionization, which occurs at the much more accessible $z \sim 3$, will significantly improve these models over the next few years.

14.1 Introduction

Although the luminous universe is dominated by matter inside galaxies, the material between these objects – the *intergalactic medium* (IGM) – actually contains significantly more mass. It is also, of course, the source from which halos and galaxies collapse, so its properties are crucial for understanding structure formation.

During much of cosmic history, the IGM can be modeled in a straightforward fashion. Before the first sources appear, it evolves linearly through gravitational instability, and long after they appear it can be modeled rather well with quasilinear perturbation theory and photoionization equilibrium. However, once the first sources do appear, and until the IGM is fully reionized, its evolution is quite complex. In this contribution, we will examine several outstanding issues in IGM physics. These

S.R. Furlanetto (✉)
University of California – Los Angeles, CA, USA
e-mail: sfurlane@astro.ucla.edu

H.A. Thronson et al. (eds.), *Astrophysics in the Next Decade,* Astrophysics and Space Science Proceedings, DOI 10.1007/978-1-4020-9457-6_14,

focus, for the most part, on feedback from luminous sources. Radiation backgrounds can have dramatic effects: ionizing the gas, heating it, and modifying its chemistry. Chemical enrichment affects the thermal balance, especially once the gas is incorporated into galaxies. In Sections 14.2 and 14.3, we will describe the physics behind these mechanisms as well as methods to constrain them.

The detailed properties of the IGM can have a profound effect on models of reionization, as well as interpretations of observations of the high-redshift Universe. This is because its precise density distribution remains unknown – so we cannot effectively model recombinations during reionization – and because our existing observations (especially quasar Lyα forest spectra) are only sensitive to special locations in the IGM. Thus better model-building should be a major goal of theorists in the coming years, as we describe in Secion 14.4. Fortunately, although IGM properties are difficult to measure during hydrogen reionization, they are already very well-known at $z \sim 3$ when helium is reionized. In Section 14.4.4, we suggest that this era provides a useful testbed for sharpening models of the IGM and reionization.

14.2 Radiative Feedback and the IGM

We will first examine radiative feedback mechanisms and describe how they can best be constrained.

14.2.1 The Lyman-Werner Background

One of the most important feedback mechanisms for the first proto-galaxies is the photo-dissociation of H_2. Rotational transitions of H_2 provide a cooling channel that operates in halos with $T_{\rm vir} > 200$ K (Haiman et al. 1996, Tegmark et al. 1997), well below the threshold at which atomic cooling becomes efficient. The first halos to form stars in any hierarchical model have small masses and therefore rely on H_2 for their cooling (Abel & Bryan 2002, Bromm et al. 2002). However, H_2 is fragile and is easily dissociated by soft UV photons in the Lyman-Werner band (11.26–13.6 eV) (Haiman et al. 1997a,b, 2000, Machacek et al. 2001, Mesinger et al. 2006). Once the first stars build up a sufficient UV background, this cooling channel terminates and the minimum halo mass to form stars increases. Indeed, such molecules are fragile and a weak soft UV background (less than one-thousandth the magnitude expected during reionization) would suffice to dissociate them (Haiman et al. 2000, Machacek et al. 2001). Because the IGM is optically thin to these photons (until they redshift into a Lyman-series transition) any individual halo sees Lyman-Werner sources out to cosmological distances (Haiman et al. 1997a). Thus the UV background redward of the Lyman limit builds up quickly as stars form.

It is easy to show that the photodissociation threshold *must* be reached well before reionization is complete (Haiman et al. 2000). We wish to estimate J_ν in the Lyman-Werner bands; neglecting the line opacity of the IGM, we have

$$J_\nu = \frac{c}{4\pi} \int_z^{z_{\rm max}} {\rm d}z' \, \frac{{\rm d}t}{{\rm d}z'} \, \varepsilon_{\rm LW}(\nu') \, \frac{1+z}{1+z'}, \tag{14.1}$$

where $\varepsilon_{\rm LW}$ is the emissivity per unit frequency, $\nu' = \nu(1+z')/(1+z)$, and the last factor accounts for the cosmological redshifting of the photon energy. The upper limit $z_{\rm max}$ enters because ionizing photons cannot propagate through the neutral IGM. The Lyman series introduces a "sawtooth modulation" of the background (Haiman et al. 1997a), but this does not affect our estimate. To connect to the ionized fraction, we write $\varepsilon_{\rm LW}(\nu) = h\nu \times (1/\nu) \times \chi \dot{n}_{\rm ion}$, where $\dot{n}_{\rm ion}$ is the rate at which ionizing photons are produced per unit volume and $\chi \sim 10$ is the number of Lyman-Werner photons per frequency decade divided by the number of ionizing photons per frequency decade (Ciardi & Madau 2003, Furlanetto 2006). Neglecting recombinations, the ionized fraction $\bar{x}_m$ when we reach the photodissociation threshold $J_{m,21}$ is (Haiman et al. 2000, Furlanetto & Loeb 2005)

$$\bar{x}_m \approx 0.03 \left(\frac{f_{\rm esc}}{\chi}\right) \left(\frac{J_{m,21}}{0.1}\right) \left(\frac{1+z}{10}\right)^3, \tag{14.2}$$

where $f_{\rm esc}$ is the escape fraction of UV photons. Even with $f_{\rm esc} = 1$, $\chi = 1$, and $f_{\rm coll} \approx 3\Delta f_{\rm coll}$, the ionized fraction is at most several percent when we reach the photodissociation threshold. Thus this feedback mechanism is likely to come into play long before reionization – so we need probes that reach significantly higher redshifts in order to study the transition from exotic very massive Population III star formation to more normal generations.

14.2.2 The X-Ray Background

A second radiative feedback mechanism is X-ray heating. Because they have relatively long mean free paths, X-rays from galaxies and quasars are likely to be the most important heating agent for the low-density IGM. X-rays heat the IGM gas by first photoionizing a hydrogen or helium atom. The hot "primary" electron then distributes its energy through three main channels: (1) collisional ionizations, producing more secondary electrons, (2) collisional excitations of HeI (which produce photons capable of ionizing HI) and HI (which produces a Lyα background), and (3) Coulomb collisions with free electrons. The relative cross sections of these processes determine what fraction of the X-ray energy goes to heating ($f_{X,h}$); Monte Carlo simulations find that $f_{X,h} \sim (1+2\bar{x}_i)/3$ (Shull & van Steenberg 1985, Chen & Kamionkowski 2004).

The properties of X-ray sources at high redshifts are currently more or less unknown. X-ray emission will inevitably accompany star formation through two channels: inverse-Compton scattering off of relativistic electrons accelerated in supernovae (which probably plays an increasingly important role at high redshifts because the photon energy density is $\propto (1+z)^4$, Oh 2001) and high-mass

X-ray binaries, in which material from a massive main sequence star accretes onto a compact neighbor. Such systems are born as soon as the first massive stars die, only a few million years after star formation commences. So they certainly ought to exist in high-redshift galaxies (Glover & Brand 2003), although their abundance depends on the metallicity and stellar initial mass function. In addition, any quasars (or "mini-quasars") will produce copious amounts of X-rays.

What observational constraints can we place on high-redshift X-ray emission? The most obvious limit is the present-day soft X-ray background (SXRB), part of which could originate from a high-redshift hard X-ray background (≥ 10 keV) that free streams until today (Ricotti & Ostriker 2004, Dijkstra et al. 2004, Salvaterra et al. 2005). Approximately $94^{+6}_{-7}\%$ of the SXRB has been resolved (Moretti et al. 2003a). Unfortunately, constraining X-ray reionization from the unresolved X-ray background (XRB) requires an uncertain extrapolation of the spectral energy distribution to hard energies, but we can perform a simple order-of-magnitude estimate that roughly matches the more detailed constraints of Dijkstra et al. (2004) and Salvaterra et al. (2005).

Suppose the high-redshift XRB is emitted at a median redshift z by a population of black holes with a fraction $f_{\rm HXR}$ of their radiation emerging in the [0.5–2] $(1+z)$ keV range and a fraction $f_{\rm UV}$ emerging above 1 Rydberg. Then, the SXRB observed at the present day is

$$J_X \approx 3.2 N_{\rm ion} \times 10^{-13} \left(\frac{f_{\rm HXR}/f_{\rm UV}}{0.2} \right) \left(\frac{\langle E \rangle}{5\,{\rm Ry}} \right) \times \left(\frac{10}{1+z} \right) \,{\rm erg\,s^{-1}\,cm^{-2}\,deg^{-2}}, \quad (14.3)$$

where $\langle E \rangle$ is the average energy per ionization, $N_{\rm ion}$ is the number of ionizing photons produced per baryon, and $f_{\rm HXR}/f_{\rm UV}$ and $\langle E \rangle$ are appropriate for a spectrum with $L_\nu \propto \nu^{-1}$ (significantly harder than lower-redshift quasars) ranging from 13.6 eV to 10 keV. The prefactor is comparable to the currently unresolved component, so while the observed SXRB makes reionization purely through X-rays difficult, there could be significant heating: raising the IGM temperature from $\sim$30 K to $\sim$1000 K requires $N_{ion} \sim 1$.

Finally, we note that X-rays also affect the chemistry of H_2. They could catalyze its formation by increasing the free electron fraction (Haiman et al. 2000, Machacek 2003b); however, Oh & Haiman (2003) argued that the heating that inevitably accompanies X-rays impedes collapse, ensuring a net suppression of the formation of small halos. Others have argued for more complicated positive feedback mechanisms if the free electron fraction grows large (Ricotti et al. 2002, Cen 2003).

14.2.3 The 21 cm Transition

The best way to constrain these feedback mechanisms on the IGM *directly* is to observe them through the hyperfine transition of HI. The brightness temperature of the IGM, relative to the CMB, in this transition is (see Furlanetto et al. (2006) for a detailed discussion)

$$\delta T_b(\nu) = \frac{T_S - T_\gamma(z)}{1+z}(1 - e^{-\tau}) \approx \frac{T_S - T_\gamma(z)}{1+z}\,\tau \tag{14.4}$$

$$\approx 9\, x_{\rm HI}(1+\delta)\,(1+z)^{1/2}\left[1 - \frac{T_\gamma(z)}{T_S}\right]\left[\frac{H(z)/(1+z)}{\mathrm{d}v_\parallel/\mathrm{d}r_\parallel}\right]\ \mathrm{mK}, \tag{14.5}$$

where T_γ is the CMB temperature, τ is the 21 cm optical depth, T_S is the spin temperature (or the excitation temperature of this transition), $1+\delta = \rho/\bar{\rho}$ is the fractional overdensity of the observed patch, and $\mathrm{d}v_\parallel/\mathrm{d}r_\parallel$ is the gradient of the peculiar velocity along the line of sight. On the first line, we have assumed $\tau \ll 1$ (typically it is ~1%). Note that δT_b saturates if $T_S \gg T_\gamma$, but it can become arbitrarily large (and negative) if $T_S \ll T_\gamma$. The observability of the 21 cm transition therefore hinges on the spin temperature, which as we will see is driven by precisely these feedback mechanisms.

As an example, Fig. 14.1 shows a series of snapshots of 21 cm emission during the reionization epoch. The 21 cm transition has four key advantages for IGM studies at high redshift. First, because it uses the CMB as a background source it allows us to image the entire sky (instead of isolated lines of sight, as with the Lyα forest). Second, because it is a spectral line each observed frequency corresponds to a different distance, and hence we can trace the entire history. Third, it does not suffer from saturation (which we will see is crucial for the Lyα forest). Finally, it is a direct measurement of the ~99% of mass that sits in the IGM – instead of requiring inferences about the IGM from observations of the galaxies.

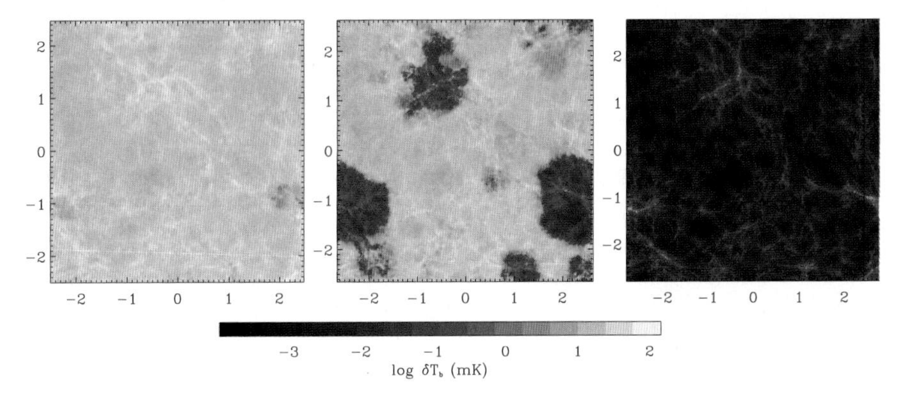

Fig. 14.1 The brightness temperature of the 21 cm transition at several redshifts, as predicted by the "late reionization" simulation analyzed in (Furlanetto et al. 2004). Each panel shows the same slice of the simulation box (with width $10h^{-1}$ comoving Mpc and depth $\Delta\nu = 0.1$ MHz), at $z = 12.1$, 9.2, and 7.6, from *left* to *right*. The three epochs shown correspond to the early, *middle*, and late stages of reionization in this simulation

To measure feedback, we are primarily interested in T_S. Three competing processes determine it: (1) absorption of CMB photons (as well as stimulated emission); (2) collisions with other hydrogen atoms, free electrons, and protons; and (3) scattering of UV photons. Then (Field 1958)

$$T_S^{-1} = \frac{T_\gamma^{-1} + x_c T_K^{-1} + x_\alpha T_c^{-1}}{1 + x_c + x_\alpha}, \tag{14.6}$$

where x_c and x_α are coupling coefficients for collisions and UV scattering, respectively, T_K is the gas kinetic temperature, and T_C is the color temperature of the UV radiation field. In the limit in which $T_c \to T_K$ (a reasonable approximation in most situations of interest), and at $z < 30$ when $x_c \approx 0$, equation (14.6) may be written

$$1 - \frac{T_\gamma}{T_S} = \frac{x_\alpha}{1 + x_\alpha}\left(1 - \frac{T_\gamma}{T_K}\right). \tag{14.7}$$

The UV radiation field affects T_S through the Wouthuysen-Field mechanism (Wouthuysen 1952, Field 1958). It is illustrated in Fig. 14.2, where we have drawn the hyperfine sublevels of the $1S$ and $2P$ states of HI. Suppose a hydrogen atom in the hyperfine singlet state absorbs a Lyα photon. The electric dipole selection rules allow $\Delta F = 0, 1$ except that $F = 0 \to 0$ is prohibited (here F is the total angular momentum of the atom). Thus the atom will jump to either of the central $2P$ states. However, these rules allow this state to decay to the ${}_1S_{1/2}$ triplet level.[1] Thus atoms can change hyperfine states through the absorption and spontaneous re-emission of a Lyα photon (or indeed any Lyman-series photon; Pritchard & Furlanetto (2006), Hirata (2006)).

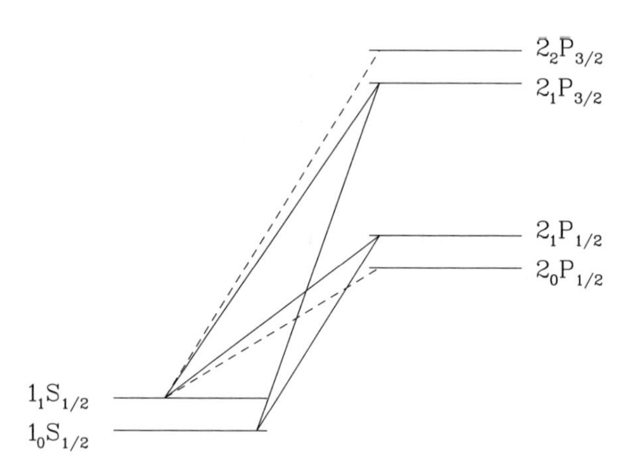

Fig. 14.2 Level diagram illustrating the Wouthuysen-Field effect. We show the hyperfine splittings of the $1S$ and $2P$ levels. The solid lines label transitions that mix the ground state hyperfine levels, while the dashed lines label complementary transitions that do not participate in mixing. From Pritchard & Furlanetto (2006)

[1] Here we use the notation ${}_F L_J$, where L and J are the orbital and total angular momentum of the electron.

We will give a relatively simple and intuitive treatment of this process here. Reality is considerably more complicated; see Hirata (2006), Furlanetto et al. (2006) for more precise calculations. The Wouthuysen-Field coupling must depend on the total rate (per atom) at which Lyα photons are scattered within the gas,

$$P_\alpha = 4\pi \chi_\alpha \int \mathrm{d}\nu \, J_\nu(\nu)\phi_\alpha(\nu), \tag{14.8}$$

where $\sigma_\nu \equiv \chi_\alpha \phi_\alpha(\nu)$ is the local absorption cross section, $\chi_\alpha \equiv (\pi e^2/m_e c) f_\alpha$, $f_\alpha = 0.4162$ is the oscillator strength of the Lyα transition, $\phi_\alpha(\nu)$ is the Lyα absorption profile (typically a Voigt profile, with thermal broadening $\Delta\nu_D$), and J_ν is the angle-averaged specific intensity of the background radiation field (by number, not energy). In the simplest approximation, we let J_ν be constant across the line; in reality it can change by factors of order unity over that span.

Our goal here is to relate this total scattering rate P_α to the indirect de-excitation rate P_{10} between hyperfine levels (Field 1958, 1959a, Meiksin 2000). We first label the $1S$ and $2P$ hyperfine levels a–f, in order of increasing energy, and let A_{ij} and B_{ij} be the spontaneous emission and absorption coefficients for transitions between these levels. We write the background flux at the frequency corresponding to the $i \to j$ transition as J_{ij}. Then

$$P_{01} \propto B_{\mathrm{ad}} J_{\mathrm{ad}} \frac{A_{\mathrm{db}}}{A_{\mathrm{da}} + A_{\mathrm{db}}} + B_{\mathrm{ae}} J_{\mathrm{ae}} \frac{A_{\mathrm{eb}}}{A_{\mathrm{ea}} + A_{\mathrm{eb}}}. \tag{14.9}$$

The first term contains the probability for an a→d transition ($B_{\mathrm{ad}} J_{\mathrm{ad}}$), together with the probability for the subsequent decay to terminate in state b; the second term is the same for transitions to and from state e. We can relate the individual A_{ij} to $A_\alpha = 6.25 \times 10^8$ Hz, the total Lyα spontaneous emission rate (averaged over all the hyperfine sublevels), with a sum rule stating that the sum of decay intensities ($g_i A_{ij}$) for transitions from a given nFJ to all the $n'J'$ levels (summed over F') is proportional to $2F+1$ (e.g., Bethe & Salpeter 1957); the relative strengths of the permitted transitions are then $(1, 1, 2, 2, 1, 5)$, where we have ordered the lines (bc, ad, bd, ae, be, bf) and the two letters represent the initial and final states. With our assumption that the background radiation field is constant across the individual hyperfine lines, we then find $P_{10} = (4/27)P_\alpha$ (Field 1958, 1959a) (see Meiksin (2000) for a detailed derivation).

The coupling coefficient x_α may then be written

$$x_\alpha = \frac{4P_\alpha}{27A_{10}} \frac{T_\star}{T_\gamma} = S_\alpha \frac{J_\alpha}{J_\nu^c}, \tag{14.10}$$

where in the second equality we evaluate $J_\nu = J_\alpha$ neglecting radiative transfer effects and set $J_\nu^c \equiv 1.165 \times 10^{-10}[(1+z)/20]$ cm^{-2} s^{-1} Hz^{-1} sr^{-1}, and S_α is a correction factor accounting for radiative transfer. The important point for us is that this Lyα photon background is produced by photons that redshift directly into Lyα

from farther in the UV and photons that redshift into higher Lyman-series transitions and then radiatively decay to Lyα. This is the same radiation background that dissociates H_2 – and hence measuring the evolution of the spin temperature constrains this feedback effect.

The Lyα coupling also depends on the effective temperature T_c of the UV radiation field, because the energy defect between the different hyperfine splittings of the Lyα transition implies that the mixing process is sensitive to the gradient of the background spectrum near the Lyα resonance. Simple arguments show that $T_c \approx T_K$: all boil down to the observation that, so long as the medium is extremely optically thick, the enormous number of Lyα scatterings must bring the Lyα profile to a blackbody of temperature T_K near the line center (Wouthuysen 1952). This condition is easily fulfilled in the high-redshift IGM, where in our cosmology the mean Lyα optical depth experienced by a photon that redshifts across the entire resonance is (Gunn & Peterson 1965)

$$\tau_{\rm GP} = \frac{\chi_\alpha \, n_{\rm HI}(z)\, c}{H(z)\nu_\alpha} \approx 3 \times 10^5 \, \bar{x}_{\rm HI} \left(\frac{1+z}{7} \right)^{3/2}. \tag{14.11}$$

Field (1959b) showed that atomic recoil during scattering, which tilts the spectrum to the red, is primarily responsible for establishing this equilibrium (though see (Hirata 2006)). This is the reason why the 21 cm field also measures the X-ray background: T_K itself is determined primarily by X-ray heating.

We will now use some representative models chosen from (Furlanetto 2006) to illustrate the expected features of the *average* 21 cm background (see also (Chen & Miralda-Escudé 2004, Sethi 2005, Hirata 2006)). We will use a fiducial set of Population II star formation parameters, with their X-ray and UV properties calibrated to starbursts in the nearby Universe. Figure 14.3a shows the resulting temperature history. The dotted curve is T_γ, the thin solid curve is T_K, and the thick solid curve is T_S. We note three major features. First, T_S is forced away from T_γ – into absorption – at quite high redshifts. This absorption epoch strengthens until $z \sim 15$, when X-ray heating increases T_K above T_γ. From that point, the IGM continues to be heated until $T_K \gg T_\gamma$.

Clearly Lyα coupling is extremely efficient for normal stars and precedes substantial heating; is this behavior generic? For Population II stars, the net X-ray heat input ΔT_c at z_c, the redshift where $x_\alpha = 1$, is (Furlanetto 2006)

$$\frac{\Delta T_c}{T_\gamma} \sim 0.08 f_X \left(\frac{f_{X,h}}{0.2} \frac{1}{S_\alpha} \right) \left(\frac{20}{1+z} \right)^3. \tag{14.12}$$

This creates an absorption epoch whose duration, timing, and amplitude offer a meaningful probe of UV feedback during the first sources. For example, very massive Pop III stars have a smaller fraction of soft-UV photons compared to X-ray photons, shortening the absorption era, and an early miniquasar population could completely eliminate it.

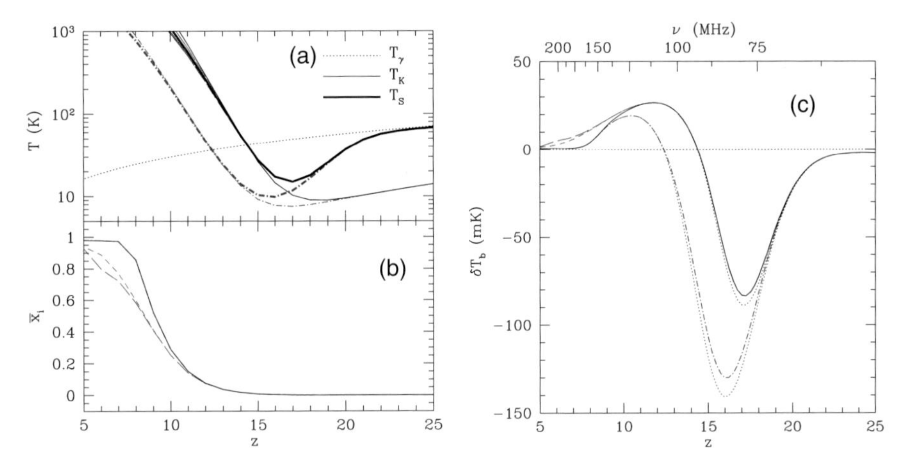

Fig. 14.3 Global IGM histories for Pop II stars. The solid curves take our fiducial parameters; the dot-dashed curve assumes weak X-ray heating. The short- and long-dashed curves include strong photoheating feedback. (**a**) Thermal properties. (**b**) Ionized fraction. (**c**) Differential brightness temperature against the CMB. In this panel, the two dotted lines show δT_b without including shock heating. From Furlanetto (2006)

Together with the ionization history from this model (shown in Fig. 14.3b), this allows us to compute $\delta \bar{T}_b(z)$ averaged across the entire Universe. This brightness temperature increment is shown in Fig. 14.3c. Here we have also labeled the corresponding (observed) frequency ν for convenience. The signal clearly has interesting structure. At the highest frequencies, reionization causes a steady decline in the signal, with the rate of decline depending on feedback processes in galaxy formation (especially photoheating, which we will discuss below).

For our current purposes, Fig. 14.3c contains an even more striking feature at higher redshifts. At $z \sim 30$, the IGM is nearly invisible even though $T_K \ll T_\gamma$. However, as the first galaxies form, the Wouthuysen-Field effect drives $T_S \to T_K$. Because Lyα coupling precedes heating, this produces a relatively strong absorption signal ($\delta T_b \approx -80$ mK) over the range $z \sim 21$–14 (or $\nu \sim 70$–95 MHz). At that point, the IGM heats up – and, crucially, this occurs well before reionization begins in earnest, requiring a push to quite high redshifts (and hence low frequencies) to observe it.

This temporal structure is also generic. The ionization fraction $\bar{x}_{i,c}$ at z_c, assuming Population II star formation, is (Furlanetto 2006)

$$\bar{x}_{i,c} \sim 0.02 \left(\frac{f_{\rm esc}}{1 + \bar{n}_{\rm rec}} \frac{1}{S_\alpha} \right) \left(\frac{20}{1+z} \right)^2, \qquad (14.13)$$

where $\bar{n}_{\rm rec}$ is the mean number of recombinations per ionized hydrogen atom. Even in the worst case of $f_{\rm esc} = 1$ and $\bar{n}_{\rm rec} = 0$, coupling would become efficient during the initial stages of reionization. However, very massive Population III stars have much harder spectra, with about fifteen times as many ionizing photons per coupling photon. In principle, it is therefore possible for Population III stars to reionize the universe *before* z_c, although that would require *extremely* unusual parameters (cf. Ciardi & Madau (2003)).

Finally, we ask whether the IGM will appear in absorption or emission during reionization. Again for Population II stars, we have (Furlanetto 2006)

$$\frac{\Delta T}{T_\gamma} \sim \left(\frac{\bar{x}_i}{0.025}\right)\left(f_X \frac{f_{X,h}}{f_{\rm esc}} \frac{10}{1+z}\right)(1+\bar{n}_{\rm rec}) \tag{14.14}$$

for the heat input ΔT as a function of $\bar{x}_i$. Here f_X is the X-ray luminosity per unit star formation, relative to its value in local starbursts. Provided $f_X > 1$, the IGM will be much warmer than the CMB during the bulk of reionization. This is significant in that δT_b becomes independent of T_S when $T_S \gg T_\gamma$, so it is easier to isolate the effects of the ionization field – but on the other hand, probing thermal feedback requires pushing to higher redshifts.

14.2.4 The Fluctuating 21 cm Signal

So far, we have described the globally-averaged spin and kinetic temperatures (as well as $\delta\bar{T}_b$). Because of their long mean free paths, X-ray photons are often portrayed as providing a uniform background. This is of course not strictly true. The comoving mean free path of an X-ray photon with energy E is:

$$\lambda_X \approx 4.9\,\bar{x}_{\rm HI}^{-1/3}\left(\frac{1+z}{15}\right)^{-2}\left(\frac{E}{300\text{ eV}}\right)^3 \text{ Mpc}; \tag{14.15}$$

thus, the universe will be optically thick to all photons below $\sim 1.8[(1+z)/15]^{1/2}\bar{x}_{\rm HI}^{1/3}$ keV. By comparison, a photon produced just redward of Lyβ can travel a comoving distance

$$\lambda_\alpha \approx 330\left(\frac{1+z}{15}\right)^{-1/2} \text{ Mpc} \tag{14.16}$$

before redshifting into the Lyα resonance. Although both are relatively large, several factors combine to render fluctuations in the backgrounds significant (Barkana & Loeb 2005, Pritchard & Furlanetto 2007): (i) the flux is weighted by r^{-2} and is hence more sensitive to nearby sources; (ii) for Lyα-coupling, the higher Lyman series transitions are more closely spaced in frequency and so have much smaller effective horizons, while for X-rays, it is the low-energy photons (with the shortest mean free paths) that provide most of the heating; (iii) the first sources of light are highly clustered; and (iv) the finite speed of light implies that more distant sources are sampled earlier in their evolution (when they were presumably less luminous). As a result, both backgrounds are much brighter near clumps of high-redshift sources.

Observations of this patchwork of emission or absorption would measure when these sources first appeared, how they affected the IGM and other sources, and their properties.

The fluctuations resulting from the Wouthuysen-Field effect have been calculated in the limiting case of $\bar{x}_i = 0$ and uniform (or zero) X-ray heating (Barkana & Loeb 2005, Pritchard & Furlanetto 2006). They have two parts. The first traces the underlying density field: large-scale overdensities contain extra sources and hence have large x_α. The stochastic distribution of galaxies provides a second set of fluctuations (Zuo 1992a,b): two nearby points sample nearly the same distribution of sources, inducing strong intrinsic correlations in the flux field. As with any Poisson process, the fractional fluctuations are proportional to $N^{-1/2}$, where N is the effective number of sources inside the sampled volume. Thus this term is important on scales comparable to the mean separation of sources. Because these two components relate to the density field in different ways, they are in principle separable using redshift space distortions.

Figure 14.4 shows some example power spectra at $z = 20$ calculated following Pritchard & Furlanetto (2006) and assuming standard Population II star formation parameters. The solid curves in the upper panel shows the μ^2 component of the

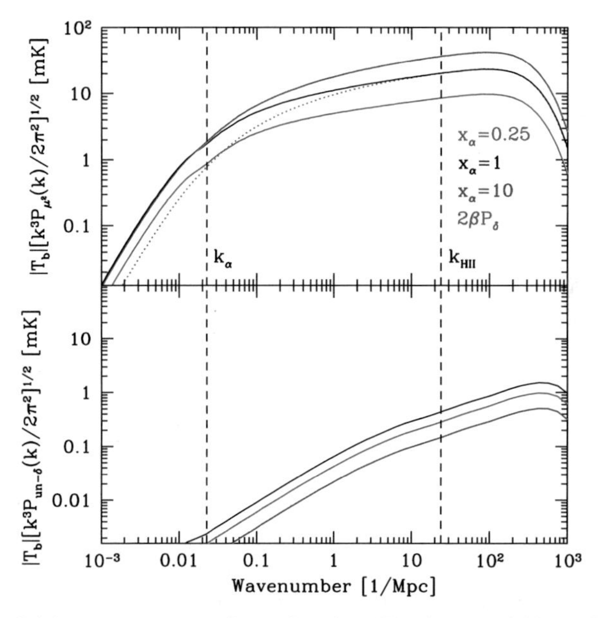

Fig. 14.4 21 cm brightness temperature fluctuations from Wouthuysen-Field coupling [39]. *Upper panel*: The μ^2 component of the power spectrum. *Lower panel*: Power sourced by stochastic variations in the galaxy density. In each panel, we assume $z = 20$, ignore X-ray heating, and assume Population II star formation. We show $x_\alpha = 0.25$, 1, and 10 (*solid curves, bottom to top*). The dotted curves in the upper panel show the component from density fluctuations when $x_\alpha = 1$. The *vertical dashed lines* show the scales corresponding to the "Lyα horizon" k_α and the minimum HII region size $k_{\rm HII}$. From Pritchard & Furlanetto (2006)

brightness temperature power spectrum (where μ is the cosine of the angle to the line of sight), which is sensitive to the density-dependent component only (Barkana & Loeb 2005). The lower panel shows the Poisson component (uncorrelated with the density field).

Of course, the fluctuation amplitude depends on the source properties: the signal increases for halos that are rarer and more biased (Barkana & Loeb 2005). The boost is much larger for the Poisson fluctuations, because those depend directly on the source density. On scales much smaller than the effective horizon of these photons, the power vanishes since the radiation field is smooth. On intermediate scales, the horizons of the various Lyn transition modulate the effective bias (Pritchard & Furlanetto 2006). The stellar spectral shapes matter on intermediate scales where the distribution of photons between the different Lyn transitions has a weak effect on the k-dependence. On large scales, the fluctuations are several times larger than the density fluctuations (because of the source bias), but the amplification vanishes at smaller scales. In general, on observable scales the Poisson fluctuations are much weaker than the density-induced component because the corresponding volumes are rather large.

The fluctuations peak when $x_\alpha \sim 1$, because the coupling saturates when $x_\alpha \gg 1$. This implies that fluctuations in the Lyα background are only relevant over a limited redshift range, so identifying that epoch, and studying the power spectrum during it, constrains the parameters of the first stars and Lyman-Werner feedback.

As we have seen, soft X-ray photons will fluctuate on relatively small scales and provide non-uniform heating, but, because of the steep energy dependence, there *will* be a uniform component to the X-ray background (unlike in the UV), plausibly providing $\sim$60% of the total heating. Thus we expect fluctuations from X-rays to peak on smaller scales than those from Lyα, but also to taper off to much larger scales. These fluctuations sourced by T_K have one unique property: the cross-correlation between temperature and density can be *negative* if $T_K < T_\gamma$. Physically, dense gas (which tends to sit near luminous sources) is warmer than average and has a *smaller* δT_b than average. Detection of such a feature would provide a clear indication of a cold IGM with substantial temperature fluctuations.

Figure 14.5 shows the net power spectra (including both the spherically-averaged component in the top panels and the angular component, which includes the correlation with density, in the bottom panels) in a Population II reionization model (Pritchard & Furlanetto 2006). At high redshifts, $z > 18$, Wouthuysen-Field fluctuations dominate and the power spectra appear similar to those in Fig. 14.4. X-ray heating begins to kick in at $z \sim 17$ and immediately has a dramatic effect. The angular component, which includes the density-temperature cross-correlation, becomes negative on intermediate scales, where temperature fluctuations are strong. The spherically-averaged power remains positive (as it must), but it contains two distinctive troughs separating the regimes where density, temperature, and Lyα fluctuations dominate (from small to large scales). The peak on the largest scales eventually disappears (once $x_\alpha \gg 1$), and the troughs disappear entirely once $T_K > T_\gamma$ (at $z \sim 14$ here). Tracing this shape will show how these radiation backgrounds interact and reveal details about both the source populations and the IGM evolution.

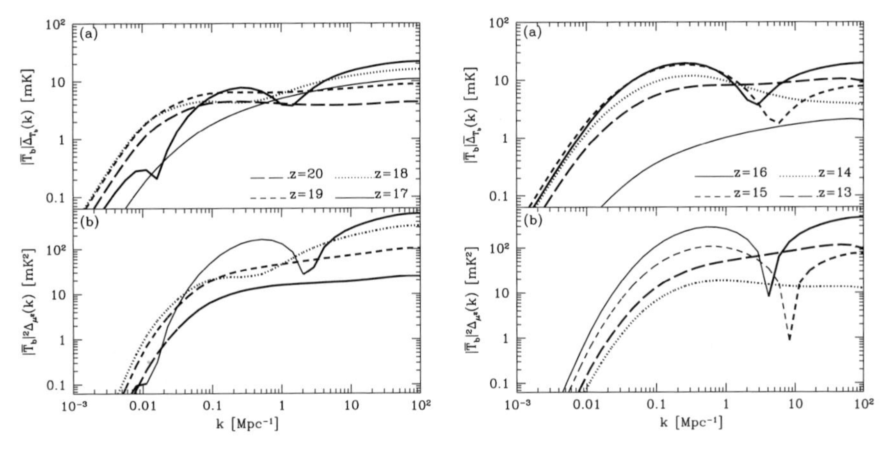

Fig. 14.5 Power spectra of δT_b including fluctuations in both T_K (from X-ray heating) and the Wouthuysen-Field coupling. The *upper panels* show the spherically-averaged power, while the *lower panels* show the μ^2 component. We show several different redshifts in a reionization history similar to our fiducial Pop II model of Fig. 14.3. In the *lower panels*, thick curves denote a positive signal and thin curves a negative signal. The thin curves in the *top panels* also show the fluctuations with uniform heating at $z = 19$ and 14 (*left and right*, respectively) for reference. From Pritchard & Furlanetto (2006)

14.3 Metal Enrichment

Of course, neutral hydrogen (through the 21 cm and Lyα transitions) is not the only way to study the IGM. Metal transitions redward of Lyα are still visible, and the tiny abundance of metals in the IGM means that these transitions do not saturate. But their absorption is also not necessarily negligible: the equivalent Gunn-Peterson optical depth at $z = 6$ for a metal transition is typically $\tau \sim 0.1$ for a metallicity $Z = 10^{-2.5}$ Z$_\odot$ (see Table 1 of Oh 2002). Here we will consider what we can learn by studying these metals.

14.3.1 Supernova Winds

Most importantly, such lines will allow us to probe the dispersal of metals throughout the early universe. This chemical feedback is crucial for understanding the history of galaxy and star formation. The most likely agent for enriching the IGM is supernova winds from starburst galaxies. These may be especially effective at high redshifts because of the small gravitational binding energies of the host halos (Madau & Rees 2000, Dekel & Silk 1986, Scannapieco et al. 2002). The effective transition between (very massive) Population III and Population II star formation appears to be sudden, occurring at a critical metallicity $Z_t \sim 10^{-3.5}$ Z$_\odot$ in the gas phase (Bromm et al. 2001, Bromm & Loeb 2003) or at $Z_t \sim 10^{-5}$ Z$_\odot$ if dust is included (Schneider et al. 2003). Because the transition could be accompanied by a large drop in the ionizing efficiency, it may have had an enormous effect on the reionization history.

The simplest prescription for chemical feedback is to track the mean cosmic metallicity $\bar{Z}$ and to switch the mode of star formation once $\bar{Z} > Z_t$. Because this imposes a global drop in the ionizing emissivity, it led to several predictions of "double reionization" in which $\bar{x}_i(z)$ actually decreased over some finite time interval (Wyithe & Loeb 2003ab, Cen 2003, Wyithe & Cen 2006). However, this approximation is not a good one, because metal enrichment (like reionization) is highly inhomogeneous and *must* be modeled physically (rather than with an arbitrary prescription) (Haiman & Holder 2003, Furlanetto & Loeb 2005). For example, once an individual galaxy forms even a tiny number of stars, any subsequent star formation in that galaxy will be metal-rich (provided small-scale mixing is efficient). On the other hand, newly-formed halos can produce Population III stars only if they collapse from pristine material. As galactic winds expand into the IGM, more and more halos form out of pre-enriched gas, eventually choking off the supply of Population III stars. Wind enrichment has been studied extensively from a theoretical perspective (Ferrara et al. 2000, Madau et al. 2001, Scannapieco et al. 2002, Mori et al. 2002, Furlanetto & Loeb 2003), though firm predictions are even more difficult than for reionization because the physics is so complex. The crucial point, however, is obvious: winds expand relatively slowly compared to ionizing photons, broadening the transition from Population III to Population II (Scannapieco et al. 2003, Furlanetto & Loeb 2005). For the same reason, it is difficult to arrange for complete enrichment to precede reionization.

Figure 14.6 shows the sizes achieved by supernova winds as a function of the source galaxy mass, assuming a Monte Carlo implementation of their star forma-

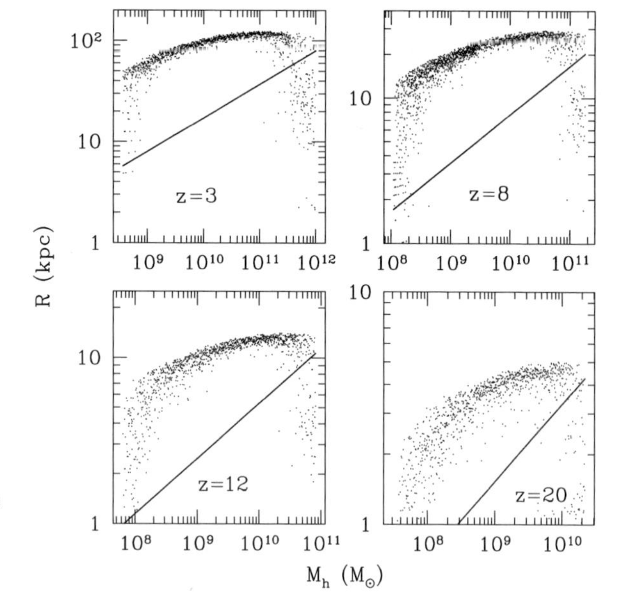

Fig. 14.6 Physical radius of wind bubbles at various redshifts. The points are Monte Carlo realizations of a simple wind expansion model. The solid lines show the halo virial radii; note that small galaxies are much more efficient polluters than massive galaxies. From Furlanetto & Loeb (2003)

tion histories in the extended Press-Schechter model (Furlanetto & Loeb 2003) (see Furlanetto & Loeb (2005) for an even simpler, but much more approximate, model). As expected, small galaxies are by far the most efficient polluters: this is because gravity confines winds around massive halos. Given the steep mass function at high redshifts, dwarf galaxies thus dominate the enrichment. These models, as well as numerical simulations of enrichment, have many uncertainties, but most agree that we can expect ~1–10% of the IGM to be enriched by $z \sim 6$, with a mean metallicity of $Z \sim 0.01\ Z_{\odot}$ inside these bubbles (Oppenheimer & Davé 2006). This is high enough both to trigger Population II star formation and to provide significant absorption along lines of sight that intersect such regions.

14.3.2 Metal Line Absorption

To estimate the actual absorption, we must first identify the most important line transitions. If the enriched medium is highly ionized, states such as CIV and SiIV are likely targets. However, earlier in reionization metals may be confined to compact, dense shells surrounding wind hosts, or they may be distributed through a mostly neutral universe. Then low ionization states, such as OI and CII, may be more common.

Figure 14.7 shows the resulting absorption at a range of redshifts in a simple model, assuming that all of the oxygen is neutral (but the absorption statistics are similar for SiIV, CII, and CIV). Note that these curves correspond to a filling factor of enriched material $Q \sim 10^{-4},\ 10^{-3},\ 10^{-2}$, and 0.1, from bottom to top. Obviously,

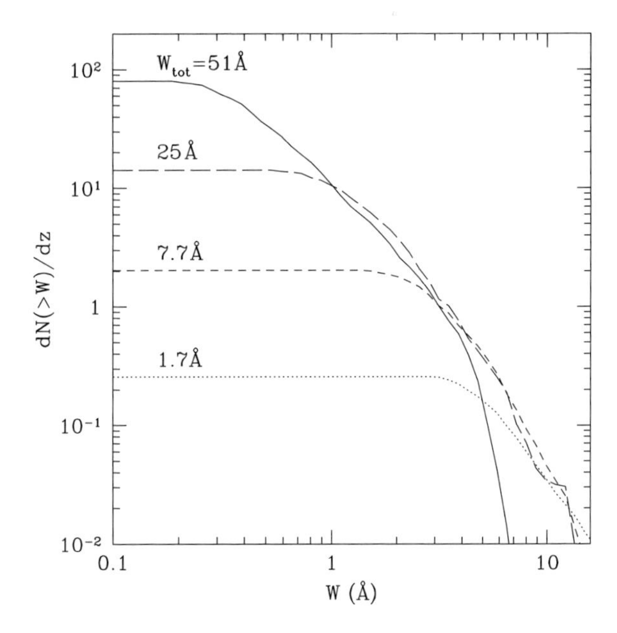

Fig. 14.7 Number of OI $\lambda 1302$ systems intersected along a random line of sight per unit redshift, as a function of (observed) equivalent width W, in a simple enrichment model. From top to bottom, the curves are for $z = 8,\ 12,\ 16$, and 20. Each curve is also marked with the total integrated absorption per unit redshift. See Furlanetto & Loeb (2003) for details

the absorption is never tremendously strong, but it is substantial even when $Q \sim 1$–10%. For example, we expect $\sim$10 absorbers per unit redshift with equivalent widths $W > 1$ Å once $Q > 1\%$.

The model suggests a number of observable trends. First, the total absorption depends on the total mass of metals ejected into the IGM, but the *distribution* of absorber strengths depends on the source parameters. Powerful winds from small halos produce widely distributed weak absorption, while less efficient winds from massive galaxies create only a few features (albeit strong ones). Thus we can directly study the metal ejection process. Second, the strongest absorbers will appear at high redshift, because the winds are most compact then. Third, comparing the strengths of different metal lines constrains the metal yield, allowing us to distinguish enrichment by Population II and Population III stars. Finally, extrapolating Figure 14.7 to filling factors near unity, metal absorption offers a powerful probe of the most extreme enrichment scenarios, in which an early generation of Pop III stars pollutes the entire universe, even if only to a low level.

14.3.3 The "OI Forest"

Oh (2002) pointed out another extremely interesting possibility. OI has an ionization potential of 13.62 eV; as a result it remains in tight charge-exchange equilibrium with HI in most astrophysical environments. Therefore we will see OI λ1302 absorption wherever the universe is neutral (and enriched). For a fully neutral medium at the mean density, the optical depth will be near unity for $Z \approx 10^{-1.5}\ Z_{\odot}$; thus, we might expect to see an "OI forest" marking regions of dense neutral material. Unlike the Lyα forest, these features will *not* be saturated, making them useful measures when $x_{\rm HI} \approx 1$ (although there is a degeneracy with the metallicity, of course). We would expect observable features to occur wherever the line-of-sight passes through an optically thick Lyman-limit cloud, which are obviously common before reionization (Oh 2002).

In fact, six $z > 5$ OI absorbers have already been observed by Becker et al. (2006), including four such systems along the line of sight to a $z = 6.42$ quasar – many, many more than would be expected from lower redshift lines of sight. However, their interpretation is ambiguous: other arguments show that this line of sight is actually highly ionized, so it is certainly not an "OI forest" in the above sense. Their implications for metal enrichment and/or reionization remain a mystery. CIV has also been found at high redshifts (Ryan-Weber et al. 2006).

14.3.4 Other Methods to Observe Metal Enrichment

Metal absorption lines viewed toward quasars are not the only way to observe chemical enrichment in the early IGM. A number of groups have studied how fine-structure absorption viewed against the CMB can constrain the overall metallicity evolution as well as its spatial distribution (Basu et al. 2004, Hernández-Monteagudo et al. 2007, 2008). These lines also provide most of the radiative cooling in enriched gas, and they may be directly visible. Finally, measurements of the metal distribution at lower redshifts also provide constraints on the history of enrichment. These include the standard Lyα forest studies, "near-field" cosmology of the metal distribution in Milky Way stars, and even the possibility of ongoing Population III star formation at moderate redshifts ($z \sim 3$) due to inefficient micro-mixing (Jimenez & Haiman 2006) or incomplete IGM enrichment (Scannapieco et al. 2003, Tornatore et al. 2007).

14.4 Reionization and the Intergalactic Medium

At least at present, most of the theoretical and observational work on the high-redshift Universe is focused on understanding the reionization epoch. Obviously, deep knowledge of the IGM is crucial to understanding this epoch in detail. Here we will identify several areas in which improved IGM modeling is vital and suggest that helium reionization (at $z \sim 3$) may provide clues.

14.4.1 Interpreting the Lyα Forest

Just as with the 21 cm line, Lyman series absorption in the spectra of high-redshift quasars traces neutral hydrogen in the early Universe. However, since the cross sections for such permitted transitions are $\sim 10^7$ times larger, optical/UV absorption spectra saturate much more rapidly than the 21 cm line. Quasar spectra are therefore primarily useful for studying the tail end of reionization, when the neutral fraction is small – but even then, they pose significant problems because making inferences about the ionization state of the universe from a saturated absorber is fraught with uncertainty.

Figure 14.8 illustrates this point. The solid curves show the logarithmic contribution to the transmission in the Lyα, Lyβ, and Lyγ transitions, as a function of the overdensity $\Delta = \rho/\bar{\rho}$, while the dashed curves show the actual volume and mass-weighted neutral hydrogen distributions. The small overlap between the two sets illustrates that inferring $x_{\rm HI}$ from measurements of the extreme tail of the distribution is dangerous – but unfortunately, that is all that is possible. Higher-order transitions do help, but they suffer from other contamination problems. As a result, quasar constraints on reionization are model dependent. According to (Fan et al. 2006) (who use the IGM model of (Miralda-Escudé et al. 2000)), the effective optical depth increases rapidly at high redshift, which they interpret as sudden evolution in the neutral fraction and radiation field, as might be expected when the ionized bubbles percolate (e.g., Cen & McDonald 2002, Lidz et al. 2002, Fan et al. 2002). On the other hand, an empirically motivated model of the mean optical depth evolution does not show such a break and implies that no such dramatic event is occurring

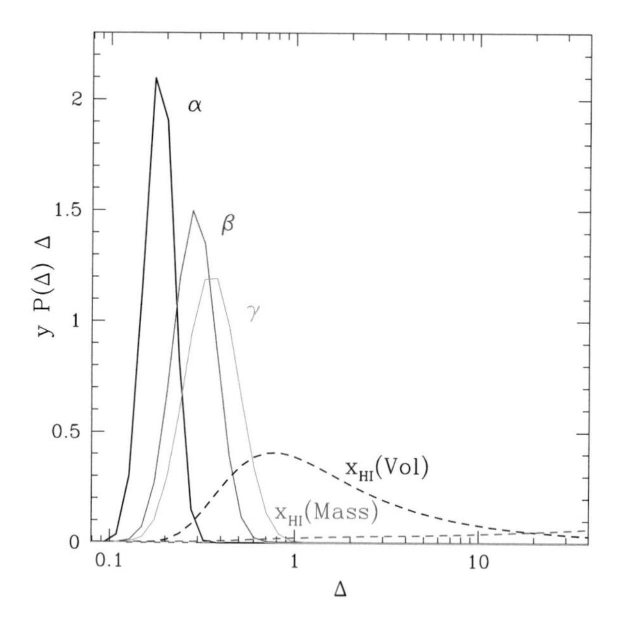

Fig. 14.8 The logarithmic integrands of the Lyα, Lyβ, and Lyγ transmission, as a function of IGM density (solid curves, left to right), and of the volume and mass-weighted neutral fractions (dashed curves). As the overlap between the different Lyman series lines and the neutral fraction drops, the uncertainty in inferring the latter increases. From Oh (2005)

(Becker et al. 2007) (see also Songaila & Cowie 2002). Better data and modeling will clearly be needed to settle this debate.

A complementary approach uses the increasing variance in the Lyα absorption (even smoothed over large scales), which has been interpreted as evidence for patchy reionization (Wyithe & Cen 2006, Fan et al. 2006). However, such claims should be viewed with caution: numerical simulations show that the observed sightline-to-sightline variance is in fact consistent with density fluctuations in a uniform radiation field (Lidz et al. 2006, see also Liu et al. 2006). Transmission fluctuations can be of order unity on large ($\sim 50\,h^{-1}$ Mpc) scales for two reasons: (i) transmission spectra are highly biased tracers of the underlying density fluctuations, because they are mainly sensitive to rare voids, and (ii) projected power from small-scale transverse modes is aliased to long wavelength line-of-sight modes. Thus, although it is quite likely that the ionizing background does contain substantial fluctuations at these epochs, it is extremely difficult to detect them. Again, much better modeling of the IGM density structure is needed to draw firm conclusions from the existing data.

14.4.2 Photoheating Feedback

A third radiative feedback mechanism, not discussed above, is photoionization itself, which heats the gas because the liberated electrons are typically left with energies >1 eV. The increased thermal pressure suppresses accretion onto small halos and hence decreases the rate of star formation. This is usually quantified by raising the Jeans mass in heated regions (Rees 1986, Efstathiou 1992, Thoul

& Weinberg 1996, Kitayama & Ikeuchi 2000). Unfortunately, the degree of suppression is not entirely clear, both because the final post-ionization temperatures are not well-constrained and because the feedback interacts with collapsing, not static, halos (Dijkstra et al. 2004). By suppressing star formation, photoheating provides a "self-regulation" mechanism for reionization and can significantly extend the process (Furlanetto & Loeb 2005, Illiev et al. 2007, McQuinn et al. 2007). But we must understand the heating and suppression processes in much more detail to quantify this important effect on the star formation (and reionization) histories.

14.4.3 IGM Recombinations

Although everyone agrees that IGM recombinations are a necessary, and probably significant, aspect of reionization models, there is little agreement on how to treat them. Differences in existing models range from the mundane (the assumed gas temperature, for example) to the profound. For example, one must choose whether case-A or case-B is more appropriate.[2] If ionizations (and hence recombinations) are distributed uniformly throughout the IGM, one would want case-B (e.g., Wyithe & Loeb 2003a, Cen 2003). On the other hand, in the highly-ionized low-redshift universe, most recombinations actually take place in dense, partially neutral gas (so-called Lyman-limit systems) because high-energy photons can penetrate inside these high-column density systems. However, the ionizing photons produced after recombinations to the ground state usually lie near the Lyman-limit (where the mean free path is small) so are consumed inside the systems. Thus these photons would not help ionize the IGM, and case-A would be more appropriate (Miralda-Escudé 2003).

Even more problematic is the clumping factor $C(z) = \langle n_e^2 \rangle / \langle n_e \rangle^2$, which describes the density dependence of the recombination process. In principle, of course, this can be computed through numerical simulations, but an accurate calculation requires enormous dynamic range as well as a self-consistent treatment of photoheating during reionization. Even in simulations, clumping is usually accounted for through a "subgrid" model built from semi-analytic techniques or smaller simulations (Sokasian et al. 2003, Illiev et al. 2006, Ciardi et al. 2006, Kohler et al. 2007). One recent estimate, from a $3.5h^{-1}$ Mpc N-body simulation resolving the Jeans mass in the IGM, is well-fit by (Mellema et al. 2006)

$$C(z) = 27.466 \exp(-0.114z + 0.001328z^2). \qquad (14.17)$$

But such subgrid models do not account for another problem: how do the sources and absorbers relate to each other, and how does ionization affect the small-scale

[2] Here "case-A" allows recombinations directly to the ground state while "case-B" does not; in the latter, ionizing photons produced from recombinations into the ground state are assumed to be immediately absorbed again, so that they do not affect the net ionization rate.

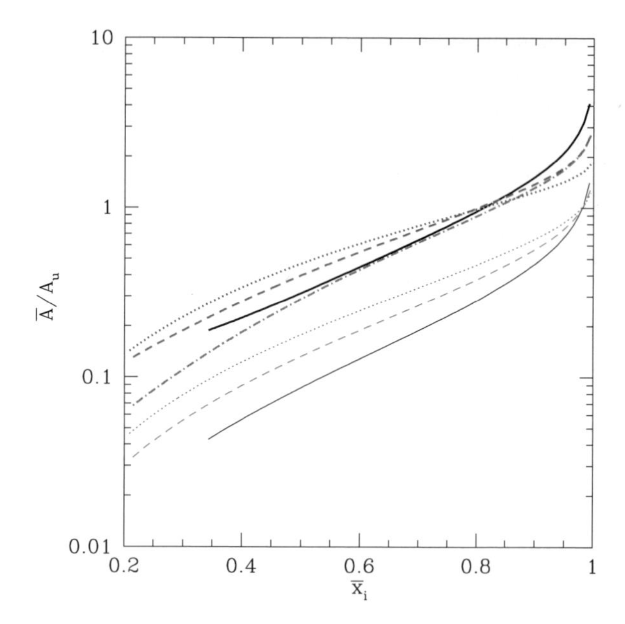

Fig. 14.9 Clumping factor during inhomogeneous reionization ($C = \bar{A}/A_u$). The lower thin set of curves assumes that the lowest-density gas in the IGM is ionized first. The upper thick set of curves assumes that HII regions begin near overdensities of sources, but that within each bubble the lowest-density gas is ionized first. Solid, dashed, and dotted curves are for $z = 6$, 9, and 12, respectively. From Furlanetto & Oh (2005)

clumping? For example, if low-density gas is ionized first, $C < 1$ throughout most of reionization (Miralda-Escudé et al. 2000), because all the dense gas would remain locked up in self-shielded systems. The thin curves in Fig. 14.9 show how the clumping factor evolves in such a model: note that it remains small throughout reionization, significantly less than the model of equation (14.17). However, such a model is obviously too naive, because the lowest-density gas is always far from the ionizing sources. The thick curves in Fig. 14.9 make the next-order correction, by placing the ionized regions around dense clumps of galaxies according to the model of Furlanetto et al. (2004, 2005). The increase from even this relatively modest change is a factor of four or so – showing that systematic assumptions limit the precision of our models.

The wide range of clumping factor predictions in the literature presents a substantial problem for interpretations of reionization that can only be resolved by improved modeling (or empirical data from high redshifts, which seems difficult at this point). As with the Lyα forest measurements, the first requirement is an accurate model of the IGM structure and density distribution. But more challenging – and probably more important – is to decide between case-A and case-B and to model the source-absorber relation better.

14.4.4 A Path Forward: Helium Reionization

Although better theoretical models will clearly help resolve these issues, as in all astrophysical questions observational constraints are also desperately needed. But, given the saturation in high-redshift Lyα forest spectra, our principle existing tool for detailed IGM studies, and the poor sensitivity of 21 cm experiments, our great hope for the future, to the final phases of reionization, such progress seems difficult. Fortunately, there is one powerful analog to this epoch that *can* be studied in more detail: helium reionization. Crucially, we understand the $z \sim 3$ Universe much better than the $z \sim 6$ Universe, in terms of the ionizing sources, the IGM, and other galaxies. It can thus provide much sharper tests of reionization models.

Observations indicate that this occurred at $z \sim 3$. The strongest evidence comes from far-ultraviolet spectra of the HeII Lyα forest along the lines of sight to several bright quasars at $z \sim 3$: the apparent HeII optical depth decreases rapidly at $z \approx 2.9$, with a spread of $\Delta z \approx 0.1$ along different lines of sight (Jakobsen et al. 1994, Davidsen et al. 1996, Anderson et al. 1999, Heap et al. 2000, Smette et al. 2002, Zheng et al. 2004, Shull et al. 2004, Reimers et al. 2005, 2006, Fechner et al. 2006), analogously to the rapidly-increasing HI optical depth observed at $z \sim 6$ and usually attributed to reionization (Fan et al. 2001, 2006).

There are several other indirect lines of evidence for helium reionization at $z \sim 3$, although in each case they are controversial. Helium reionization should at least double the IGM temperature, which has been detected at $z \sim 3.3$ by examining the Doppler widths of HI Lyα forest lines (Schaye et al. 2000, 1999, Theuns et al. 2002); at about the same time, the equation of state of the IGM also appears to become nearly isothermal, another indication of recent reionization (Schaye et al. 2000, Ricotti et al. 2000). However, temperature measurements via the Lyα forest flux power spectrum show no evidence for any sudden change, although the errors are rather large (Zaldarriaga et al. 2001, Viel et al. 2004, McDonald et al. 2006). Such a temperature jump should also decrease the recombination rate of hydrogen, thereby decreasing the HI opacity; such a feature does indeed appear in some Lyα forest data sets (Theuns et al. 2002, Bernardi et al. 2003, Faucher-Giguére et al. 2008), but not others (McDonald et al. 2006). Finally, the hardening of the radiation background that should accompany helium reionization may have been observed directly through the evolving HeII/HI ratio (Heap et al. 2000) and through metal line ratios (Songaila 1998, 2005, Vladilo et al. 2003, Agafonova et al. 2005, 2007), although again other data of comparable quality show no evidence for rapid evolution (Kim et al. 2002, Aguirre et al. 2004).

Despite this wealth of (often controversial) data, helium reionization has received relatively little theoretical attention. Here we point out how new observations, and especially better modeling of this event, can improve our understanding of the role of the IGM during reionization events. Recent efforts are beginning to address this possibility (Sokasian et al. 2002, Gleser et al. 2005, Furlanetto & Oh 2008ab, Paschos et al. 2007). Figure 14.10 illustrates one recent model of helium reionization, a nearly direct carry-over of a hydrogen reionization model. It shows the characteristic radius of HeIII regions throughout reionization. Because quasar are rare and bright (with the photon budget dominated by quasars with luminosities $L > L_\star$), random fluctuations in the quasar population determine the morphology of ionized gas when the global ionized fraction $\bar{x}_i$ is small, with the typical radius R_c of a HeIII bubble ~15–20 comoving Mpc – this is shown by the long-dashed curve in Fig. 14.10. This era provides a relatively poor analog for hydrogen reionization,

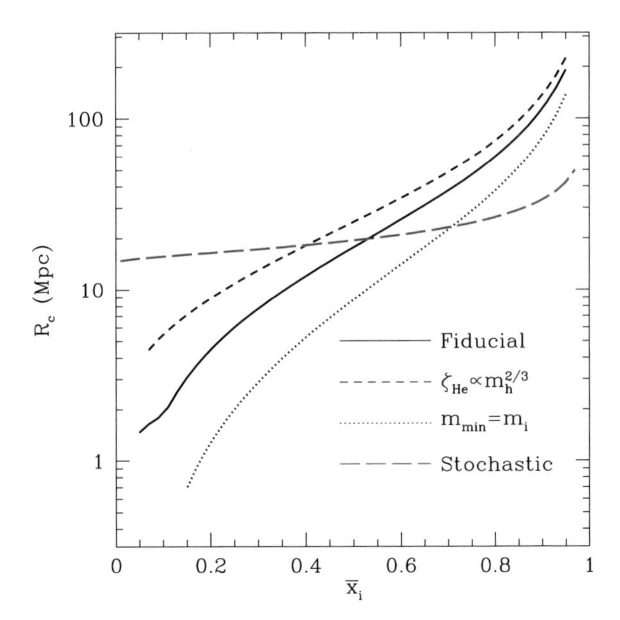

Fig. 14.10 Characteristic HeIII bubble sizes as a function of the HeIII fraction ($\bar{x}_i$) at $z = 3$. The solid curve assumes the quasar fluence is proportional to the host halo mass and that they live inside massive galaxies. The dotted curve allows them to live in smaller galaxies. The short-dashed curve assumes that the fluence increases rapidly with host mass. The long-dashed curve is the characteristic size in a stochastic model. From Furlanetto & Oh (2008a)

because then the sources were so common that the "stochastic" phase was short. Only when $\bar{x}_i > 0.5$ did the large-scale clustering of the quasars drive the characteristic size of ionized regions above this value; at this point, helium reionization does become a good analog for hydrogen reionization. The solid, short-dashed, and dotted curves show that the ionized bubbles then depend on the source population – which is how we can use the morphology of reionization to learn about the distribution of sources in the IGM.

In the late stages of reionization, when $\bar{x}_i > 0.75$, most ionizing photons are consumed by dense, recombining clumps before they reach the edge of their ionized region, halting the bubble growth. During hydrogen reionization, this phase is quite important but extremely sensitive to the properties of the IGM and hence difficult to model (Furlanetto & Oh 2005). Fortunately, the well-known characteristics of the $z = 3$ intergalactic medium allow a much more robust description of the recombination-dominated phase. We therefore hope to study it in detail to understand how sources and the IGM interact.

Figure 14.11 shows how we can use helium reionization to understand the photoheating process. Reionization of helium, just like hydrogen, heats the IGM because the liberated electrons typically have one or a few extra eV of energy (which is quickly shared with the rest of the IGM). This may suppress galaxy formation at moderate redshifts (Wyithe & Loeb 2007), which would help to disentangle the similar process occurring during hydrogen reionization. The figure shows that the median temperature increases quite rapidly, but that (at least during the latter phase of reionization, when large-scale structure drives the morphology of the ionized bubbles), the temperature depends on the local environment. The evolving equation of state can be used to understand the photoheating process in detail. The model also

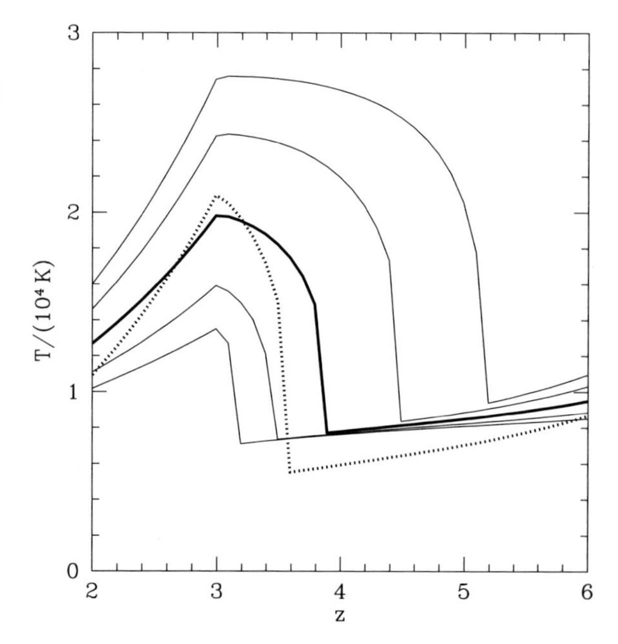

Fig. 14.11 Temperature evolution near helium reionization (assumed to be at $z_{\mathrm{He}} = 3$). The thick solid curve shows the median temperature for gas elements at the mean density; the thin solid curves show the 10th, 25th, 75th, and 90th percentiles of the temperature distribution. The dotted curve shows the median temperature for an underdense gas element. From Furlanetto & Oh (2008b)

shows a great deal of scatter in the temperature at each density, which can be used to measure the stochasticity of the process (Gleser et al. 2005, Furlanetto & Oh 2008b) and to diagnose the properties of reionization (Hui & Haiman 2003, Furlanetto & Oh 2008b).

14.5 Discussion

The physics of the high-redshift IGM is quite rich, because feedback from the first sources dramatically changes the properties of the material between the galaxies – radiatively, through ionization (and heating), X-ray heating, and the Lyman-Werner background, and chemically, through supernova (or quasar) winds. These processes strongly affect subsequent generations of structure, so they are important to understand in detail.

Radiative feedback can best be measured through the 21 cm transition, which is directly sensitive to the thermal history of the IGM. Experiments to detect this signal, such as the Mileura Widefield array, LOFAR, and the Precision Array to Probe the Epoch of Reionization, are now under construction and hope to begin collecting data before JWST is launched.

It should be possible to study chemical feedback with JWST through metal absorption lines in quasar spectra. These can probe the abundance, spatial distribution, and correlation of the enriched material with galaxies; in some circumstances, they should also teach us about the radiation backgrounds at this epoch through the "OI forest."

The detailed properties of the IGM are also crucial for understanding the reionization process (and hence subsequent structure formation). Its density structure has huge implications for interpreting quasar spectra, and the process by which reionization proceeds through the IGM strongly affects the significance of photoheating and clumping. This requires improved theoretical models of the high-redshift IGM, but we can also learn more by leveraging the well-studied IGM at $z \sim 3$ by studying the detailed progress of helium reionization, both directly (through UV instruments such as the Cosmic Origins Spectrograph on the Hubble Space Telescope) and indirectly (through its effect on the Lyα forest or even galaxy formation). Models of helium reionization are improving rapidly, and we hope that the physics of the IGM and reionization will be much better understood by the launch of JWST.

References

Abel, T., Bryan, G.L., Norman, M.L., Science **295**, 93 (2002)
Agafonova, I.I., Centurión, M., Levshakov, S.A., Molaro, P., Astron. Astrophys. **441**, 9 (2005). DOI 10.1051/0004-6361:20052960
Agafonova, I.I., Levshakov, S.A., Reimers, D., Fechner, C., Tytler, D., Simcoe, R.A., A. Songaila, Astron. Astrophys. **461**, 893 (2007). DOI 10.1051/0004-6361:20065721
Aguirre, A., Schaye, J., Kim, T.S., Theuns, T., Rauch, M., Sargent, W.L.W., Astrophys. J. **602**, 38 (2004). DOI 10.1086/380961
Anderson, S.F., Hogan, C.J., Williams, B.F., Carswell, R.F., Astron. J. **117**, 56 (1999). DOI 10.1086/300698
Barkana, R., Loeb, A., Astrophys. J. **626**, 1 (2005). DOI 10.1086/429954
Basu, K., Hernández-Monteagudo, C., Sunyaev, R.A., Astron. Astrophys. **416**, 447 (2004). DOI 10.1051/0004-6361:20034298
Becker, G.D., Rauch, M., Sargent, W.L.W., Astrophys. J. **662**, 72 (2007). DOI 10.1086/517866
Becker, G.D., Sargent, W.L.W., Rauch, M., Simcoe, R.A., Astrophys. J. **640**, 69 (2006). DOI 10.1086/500079
Bernardi, M., Sheth, R.K., SubbaRao, M., Richards, G.T., Burles, S., Connolly, A.J., Frieman, J., Nichol, R., Schaye, J., Schneider, D.P., Vanden Berk, D.E., York, D.G., Brinkmann, J., Lamb, D.Q., Astron. J. **125**, 32 (2003). DOI 10.1086/344945
Bethe, H.A., Salpeter, E.E., *Quantum Mechanics of One- and Two-Electron Atoms* (New Bromm, V., A. Loeb, Nature **425**, 812 (2003)
Bromm, V., Coppi, P.S., Larson, R.B., Astrophys. J. **564**, 23 (2002)
Bromm, V., Ferrara, A., Coppi, P.S., Larson, R.B., Mon. Not. R. Astron. Soc. **328**, 969 (2001)
Cen, R., Astrophys. J. **591**, 12 (2003)
Cen, R., Astrophys. J. **591**, L5 (2003)
Cen, R., McDonald, P., Astrophys. J. **570**, 457 (2002). DOI 10.1086/339723
Chen, X., Kamionkowski, M., Phys. Rev. D **70**(4), 043502 (2004). DOI 10.1103/PhysRevD.70.043502
Chen, X., Miralda-Escudé, J., Astrophys. J. **602**, 1 (2004). DOI 10.1086/380829
Ciardi, B., Madau, P., Astrophys. J. **596**, 1 (2003)
Ciardi, B., Scannapieco, E., Stoehr, F., Ferrara, A., Iliev, I.T., Shapiro, P.R., Mon. Not. R. Astron. Soc. **366**, 689 (2006). DOI 10.1111/j.1365-2966.2005.09908.x
Davidsen, A.F., Kriss, G.A., Wei, Z., Nature **380**, 47 (1996)
Dekel, A., Silk, J., Astrophys. J. **303**, 39 (1986). DOI 10.1086/164050
Dijkstra, M., Haiman, Z., Loeb, A., Astrophys. J. **613**, 646 (2004). DOI 10.1086/422167

Dijkstra, M., Haiman, Z., Rees, M.J., Weinberg, D.H., Astrophys. J. **601**, 666 (2004). DOI 10.1086/380603
Efstathiou, G., Mon. Not. R. Astron. Soc. **256**, 43 (1992)
Fan, X., et al., Astron. J. **122**, 2833 (2001)
Fan, X., et al., Astron. J. **123**, 1247 (2002)
Fan, X., Strauss, M.A., Becker, R.H., White, R.L., Gunn, J.E., Knapp, G.R., Richards, G.T., Schneider, D.P., Brinkmann, J., Fukugita, M., Astron. J. **132**, 117 (2006). DOI 10.1086/504836
Faucher-Giguére, C.A., Prochaska, J.X., Lidz, A., Hernquist, L., Zaldarriaga, M., ApJ, **681**, 831 (2008)
Fechner, C., et al., Astron. Astrophys. **455**, 91 (2006). DOI 10.1051/0004-6361:20064950
Ferrara, A., Pettini, M., Shchekinov, Y., Mon. Not. R. Astron. Soc. **319**, 539 (2000)
Field, G.B., Astrophys. J. **129**, 551 (1959b)
Field, G.B., Proceedings of the Institute of Radio Engineers **46**, 240 (1958)
Furlanetto, S.R., A. Loeb, Astrophys. J. **634**, 1 (2005)
Furlanetto, S.R., Loeb, A., Astrophys. J. **588**, 18 (2003)
Furlanetto, S.R., Mon. Not. R. Astron. Soc. **371**, 867 (2006). DOI 10.1111/j.1365-2966.2006.10725.x
Furlanetto, S.R., Oh, S.P., Astrophys. J., **681**, 1 (2008a)
Furlanetto, S.R., Oh, S.P., Astrophys. J., **682**, 14 (2008b)
Furlanetto, S.R., Oh, S.P., Briggs, F.H., Phys. Rep. **433**, 181 (2006). DOI 10.1016/j.physrep.2006.08.002
Furlanetto, S.R., Oh, S.P., Mon. Not. R. Astron. Soc. **363**, 1031 (2005). DOI 10.1111/j.1365-2966.2005.09505.x
Furlanetto, S.R., Sokasian, A., Hernquist, L., Mon. Not. R. Astron. Soc. **347**, 187 (2004)
Furlanetto, S.R., Zaldarriaga, M., Hernquist, L., Astrophys. J. **613**, 1 (2004)
G.B. Field, Astrophys. J. **129**, 536 (1959a)
Gleser, L., Nusser, A., Benson, A.J., Ohno, H., Sugiyama, N., Mon. Not. R. Astron. Soc. **361**, 1399 (2005). DOI 10.1111/j.1365-2966.2005.09276.x
Glover, S.C.O., Brand, P.W.J.L., Mon. Not. R. Astron. Soc. **340**, 210 (2003). DOI 10.1046/j.1365-8711.2003.06311.x
Gunn, J.E., Peterson, B.A., Astrophys. J. **142**, 1633 (1965)
Haiman, Z., Abel, T., Rees, M.J., Astrophys. J. **534**, 11 (2000). DOI 10.1086/308723
Haiman, Z., Holder, G.P., Astrophys. J. **595**, 1 (2003)
Haiman, Z., Rees, M.J., Loeb, A., Astrophys. J. **467**, 522 (1996). DOI 10.1086/177628
Haiman, Z., Rees, M.J., Loeb, A., Astrophys. J. **476**, 458 (1997a). DOI 10.1086/303647
Haiman, Z., Rees, M.J., Loeb, A., Astrophys. J. **484**, 985 (1997b). DOI 10.1086/304386
Heap, S.R., Williger, G.M., Smette, A., Hubeny, I., Sahu, M.S., Jenkins, E.B., Tripp, T.M., Winkler, J.N., Astrophys. J. **534**, 69 (2000). DOI 10.1086/308719
Hernández-Monteagudo, C., Haiman, Z., Jimenez, R., Verde, L., Astrophys. J. **660**, L85 (2007). DOI 10.1086/518090
Hernández-Monteagudo, C., Haiman, Z., Verde, L., Jimenez, R., Astrophys. J., **672**, 33 (2008)
Hirata, C.M., Mon. Not. R. Astron. Soc. **367**, 259 (2006)
Hui, L., Haiman, Z., Astrophys. J. **596**, 9 (2003)
Iliev, I.T., Mellema, G., Pen, U.L., Merz, H., Shapiro, P.R., Alvarez, M.A., Mon. Not. R. Astron. Soc. **369**, 1625 (2006). DOI 10.1111/j.1365-2966.2006.10502.x
Iliev, I.T., Mellema, G., Shapiro, P.R., Pen, U.L., Mon. Not. R. Astron. Soc. **376**, 534 (2007). DOI 10.1111/j.1365-2966.2007.11482.x
Jakobsen, P., Boksenberg, A., Deharveng, J.M., Greenfield, P., Jedrzejewski, R., Paresce, F., Nature **370**, 35 (1994)
Jimenez, R., Haiman, Z., Nature **440**, 501 (2006). DOI 10.1038/nature04580
Kim, T.S., Cristiani, S., D'Odorico, S., Astron. Astrophys. **383**, 747 (2002). DOI 10.1051/0004-6361:20011812
Kitayama, T., Ikeuchi, S., Astrophys. J. **529**, 615 (2000)

Kohler, K., Gnedin, N.Y., Hamilton, A.J.S., Astrophys. J. **657**, 15 (2007). DOI 10.1086/509907
Lidz, A., Hui, L., Zaldarriaga, M., Scoccimarro, R., Astrophys. J. **579**, 491 (2002). DOI 10.1086/342983
Lidz, A., Oh, S.P., Furlanetto, S.R., Astrophys. J. **639**, L47 (2006). DOI 10.1086/502678
Liu, J., Bi, H., Feng, L.L., Fang, L.Z., Astrophys. J. **645**, L1 (2006). DOI 10.1086/506149
M. P. van Haarlem (2000), p. 37
Machacek, M.E., Bryan, G.L., Abel, T., Astrophys. J. **548**, 509 (2001). DOI 10.1086/319014
Machacek, M.E., Bryan, G.L., Abel, T., Mon. Not. R. Astron. Soc. **338**, 273 (2003b). DOI 10.1046/j.1365-8711.2003.06054.x
Madau, P., Ferrara, A., Rees, M.J., Astrophys. J. **555**, 92 (2001). DOI 10.1086/321474
Madau, P., Rees, M.J., Astrophys. J. **542**, L69 (2000). DOI 10.1086/312934
McDonald, P., et al., Astrophys. J. Supp. **163**, 80 (2006). DOI 10.1086/444361
McQuinn, M., Lidz, A., Zahn, O., Dutta, S., Hernquist, L., Zaldarriaga, M., Mon. Not. R. Astron. Soc. **377**, 1043 (2007). DOI 10.1111/j.1365-2966.2007.11489.x
Meiksin, A., in *Perspectives on Radio Astronomy: Science with Large Antenna Arrays*, ed. Mellema, G., Iliev, I.T., Pen, U.L., Shapiro, P.R., Mon. Not. R. Astron. Soc. p. 1014 (2006). DOI 10.1111/j.1365-2966.2006.10919.x
Mesinger, A., Bryan, G.L., Haiman, Z., Astrophys. J. **648**, 835 (2006). DOI 10.1086/506173
Miralda-Escudé, J., Astrophys. J. **597**, 66 (2003). DOI 10.1086/378286
Miralda-Escudé, J., Haehnelt, M., Rees, M.J., Astrophys. J. **530**, 1 (2000)
Moretti, A., Campana, S., Lazzati, D., Tagliaferri, G., Astrophys. J. **588**, 696 (2003a). DOI 10.1086/374335
Mori, M., Ferrara, A., Madau, P., Astrophys. J. **571**, 40 (2002). DOI 10.1086/339913
Oh, S.P., Astrophys. J. **553**, 499 (2001). DOI 10.1086/320957
Oh, S.P., Furlanetto, S.R., Astrophys. J. **620**, L9 (2005)
Oh, S.P., Haiman, Z., Mon. Not. R. Astron. Soc. **346**, 456 (2003)
Oh, S.P., Mon. Not. R. Astron. Soc. **336**, 1021 (2002). DOI 10.1046/j.1365-8711.2002.05859.x
Oppenheimer, B.D., Davé, R., Mon. Not. R. Astron. Soc. **373**, 1265 (2006). DOI 10.1111/j.1365-2966.2006.10989.x
Paschos, P., Norman, M.L., Bordner, J.O., Harkness, R., Astrophys. J. submitted (arXiv.org/0711.1904 [astro-ph]) **711** (2007)
Pritchard, J.R., Furlanetto, S.R., Mon. Not. R. Astron. Soc. **367**, 1057 (2006). DOI 10.1111/j.1365-2966.2006.10028.x
Pritchard, J.R., Furlanetto, S.R., Mon. Not. R. Astron. Soc. **376**, 1680 (2007). DOI 10.1111/j.1365-2966.2007.11519.x
Rees, M.J., Mon. Not. R. Astron. Soc. **222**, 27 (1986)
Reimers, D., et al., in *Astrophysics in the Far Ultraviolet: Five Years of Discovery with FUSE, ASP Conf. Series*, eds. G. Sonneborn, W. Moos & B.-G. Andersson, **348**, 41 (2006)
Reimers, D., Fechner, C., Hagen, H.J., Jakobsen, P., Tytler, D., Kirkman, D., Astron. Astrophys. **442**, 63 (2005). DOI 10.1051/0004-6361:20053365
Ricotti, M., Gnedin, N.Y., Shull, J.M., Astrophys. J. **534**, 41 (2000). DOI 10.1086/308733
Ricotti, M., Gnedin, N.Y., Shull, J.M., Astrophys. J. **575**, 49 (2002). DOI 10.1086/341256
Ricotti, M., Ostriker, J.P., Mon. Not. R. Astron. Soc. **352**, 547 (2004). DOI 10.1111/j.1365-966.2004.07942.x
Ryan-Weber, E.V., Pettini, M., Madau P., Mon. Not. R. Astron. Soc. **371**, L78 (2006). DOI 10.1111/j.1745-3933.2006.00212.x
Salvaterra, R., Haardt, F., Ferrara, A., Mon. Not. R. Astron. Soc. **362**, L50 (2005). DOI 10.1111/j.1745-3933.2005.00074.x
Scannapieco, E., Ferrara, A., Madau, P., Astrophys. J. **574**, 590 (2002)
Scannapieco, E., Schneider, R., Ferrara, A., Astrophys. J. **589**, 35 (2003)
Schaye, J., Theuns, T., Leonard, A., Efstathiou, G., Mon. Not. R. Astron. Soc. **310**, 57 (1999)
Schaye, J., Theuns, T., Rauch, M., Efstathiou, G., Sargent, W.L.W., Mon. Not. R. Astron. Soc. **318**, 817 (2000)

Schneider, R., Ferrara, A., Salvaterra, R., Omukai, K., Bromm, V., Nature **422**, 869 (2003)
Sethi, S.K., Mon. Not. R. Astron. Soc. **363**, 818 (2005). DOI 10.1111/j.1365-2966.2005.09485.x
Shull, J.M., Tumlinson, J., Giroux, M.L., Kriss, G.A., Reimers, D., Astrophys. J. **600**, 570 (2004). DOI 10.1086/379924
Shull, J.M., van Steenberg, M.E., Astrophys. J. **298**, 268 (1985). DOI 10.1086/163605
Smette, A., Heap, S.R., Williger, G.M., Tripp, T.M., Jenkins, E.B., Songaila, A., Astrophys. J. **564**, 542 (2002). DOI 10.1086/324397
Sokasian, A., Abel, T., Hernquist, L., Mon. Not. R. Astron. Soc. **332**, 601 (2002). DOI 10.1046/j.1365-8711.2002.05291.x
Sokasian, A., Abel, T., Hernquist, L., Springel, V., Mon. Not. R. Astron. Soc. **344**, 607 (2003)
Songaila, A., Astron. J. **115**, 2184 (1998). DOI 10.1086/300387
Songaila, A., Astron. J. **130**, 1996 (2005). DOI 10.1086/491704
Songaila, A., Cowie, L.L., Astron. J. **123**, 2183 (2002). DOI 10.1086/340079
Tegmark, M., Silk, J., Rees, M.J., Blanchard, A., Abel, T., Palla, F., Astrophys. J. **474**, 1 (1997). DOI 10.1086/303434
Theuns, T., Bernardi, M., Frieman, J., Hewett, P., Schaye, J., Sheth, R.K., Subbarao, M., Astrophys. J. **574**, L111 (2002). DOI 10.1086/342531
Theuns, T., et al., Astrophys. J. **567**, L103 (2002)
Thoul, A.A., Weinberg, D.H., Astrophys. J. **465**, 608 (1996)
Tornatore, L., Ferrara, A., Schneider, R., Mon. Not. R. Astron. Soc. 1013 (2007). DOI 10.1111/j.1365-2966.2007.12215.x
Viel, M., Haehnelt, M.G., Springel, V., Mon. Not. R. Astron. Soc. **354**, 684 (2004). DOI 10.1111/j.1365-2966.2004.08224.x
Vladilo, G., Centurión, M., D'Odorico, V., Péroux, C., Astron. Astrophys. **402**, 487 (2003). DOI 10.1051/0004-6361:20030294
Wouthuysen, S.A., Astron. J. **57**, 31 (1952). DOI 10.1086/106661
Wyithe, J.S.B., Loeb, A., Astrophys. J. **586**, 693 (2003a)
Wyithe, J.S.B., Loeb, A., Astrophys. J. **588**, L69 (2003b)
Wyithe, J.S.B., Loeb, A., Astrophys. J. **646**, 696 (2006). DOI 10.1086/502620
Wyithe, J.S.B., Loeb, A., Mon. Not. R. Astron. Soc. **382**, 921 (2007). DOI 10.1111/j.1365-2966.2007.12447.x
Wyithe, S., Cen, R., submitted to Astrophys. J. (astro-ph0602503) (2006)
York: Academic Press), (1957)
Zaldarriaga, M., Hui, L., Tegmark, M., Astrophys. J. **557**, 519 (2001). DOI 10.1086/321652
Zheng, W., et al., Astrophys. J. **605**, 631 (2004). DOI 10.1086/382498
Zuo, L., Mon. Not. R. Astron. Soc. **258**, 36 (1992)
Zuo, L., Mon. Not. R. Astron. Soc. **258**, 45 (1992)

Klaus Hodapp and Yvonne Pendleton have finished debating whose shirt is most likely to attract one of the conference parrots

Chapter 15
Observing the First Stars and Black Holes

Zoltán Haiman

Abstract The high sensitivity of *JWST* will open a new window on the end of the cosmological dark ages. Small stellar clusters, with a stellar mass of several $\times 10^6 M_\odot$, and low-mass black holes (BHs), with a mass of several $\times 10^5 M_\odot$ should be directly detectable out to redshift $z = 10$, and individual supernovae (SNe) and gamma ray burst GRB afterglows are bright enough to be visible beyond this redshift. Dense primordial gas, in the process of collapsing from large scales to form protogalaxies, may also be possible to image through diffuse recombination line emission, possibly even before stars or BHs are formed. In this article, I discuss the key physical processes that are expected to have determined the sizes of the first star–clusters and black holes, and the prospect of studying these objects by direct detections with *JWST* and with other instruments. The direct light emitted by the very first stellar clusters and intermediate-mass black holes at $z > 10$ will likely fall below *JWST*'s detection threshold. However, *JWST* could reveal a decline at the faint-end of the high-redshift luminosity function, and thereby shed light on radiative and other feedback effects that operate at these early epochs. *JWST* will also have the sensitivity to detect individual SNe from beyond $z = 10$. In a dedicated survey lasting for several weeks, thousands of SNe could be detected at $z > 6$, with a redshift distribution extending to the formation of the very first stars at $z \gtrsim 15$. Using these SNe as tracers may be the only method to map out the earliest stages of the cosmic star–formation history. Finally, we point out that studying the earliest objects at high redshift will also offer a new window on the primordial power spectrum, on $\sim$100 times smaller scales than probed by current large-scale structure data.

15.1 Introduction

The formation of the first astrophysical objects and the subsequent epoch of reionization are at the frontiers of research in Astronomy. Recent years have seen significant progress both in our theoretical understanding and in observational probes

---

Z. Haiman (✉)
Department of Astronomy, Columbia University, New York, NY 10027
e-mail: zoltan@astro.columbia.edu

H.A. Thronson et al. (eds.), *Astrophysics in the Next Decade,* Astrophysics and Space Science Proceedings, DOI 10.1007/978-1-4020-9457-6_15,

of this transition epoch in the early universe. Through a combination of methods, including measurements of the cosmic microwave background (CMB) anisotropies, culminating in the precise determination of the temperature and polarization power spectra from the *Wilkinson Microwave Anisotropy Probe (WMAP)* experiment (Bennett et al. 2003, Spergel et al. 2003, Spergel et al. 2007, Page et al. 2007, Dunkley et al. 2008, Komatsu et al. 2008), Hubble diagrams of distant Supernovae (Riess et al. 1998, Perlmutter et al. 1999), and various probes of large scale structure (see, e.g., references in Bahcall et al. 1999, Tegmark et al. 2004, Slosar et al. 2007) the key cosmological parameters have been determined to high accuracy.

Because of the emergence of a concordance (ΛCDM) cosmology, we can securely predict the collapse redshifts of the first non–linear dark matter condensations: 2–3σ peaks of the primordial density field on mass scales of 10^{5-6} $M_{\odot}$, corresponding to the cosmological Jeans mass, collapse at redshifts z = 15–20. Of course, in addition to the cosmological parameters describing the average background universe (ΛCDM), one also needs a description of the seed fluctuations, to make such predictions. The nearly scale–invariant nature of the initial fluctuation power spectrum has also been empirically confirmed by the above–mentioned large–scale structure data. However, the predictions for the earliest halos rely on extrapolating the primordial power spectrum to a wavenumber of $k \sim 10\,h\mathrm{Mpc}^{-1}$, a scale that is 2–3 orders of magnitude smaller than the smallest scale directly probed by current data (e.g. (Viel et al. 2007) and references therein). It is possible that the small–scale power differs substantially from the extrapolated value. For example, in the case of warm dark matter (WDM), the power can be reduced by many orders of magnitude on the relevant scales, and, to zeroth order, the first generation of halos predicted in ΛCDM would not exist (Barkana et al. 2001). Recent simulations indicate that the formation of the first halos would indeed be delayed (O'Shea and Norman 2006, Yoshida et al. 2006), and the details of the collapse dynamics modified, possibly ultimately affecting the properties of the first stars (Gao and Theuns 2007). Within the ΛCDM paradigm, however, robust predictions can be made, using three–dimensional simulations (Yoshida et al. 2004, Springel et al. 2005), rather than semi–analytical methods (Press and Schechter 1974, Sheth et al. 2001). Such predictions are now limited mainly by the $\approx$5% uncertainty in the normalization of the primordial power spectrum, $\sigma_{8h^{-1}}$ (e.g., Slosar et al. 2007), which translates essentially to a $\approx$5% uncertainty in the collapse redshift of the first structures.

On the observational side, the recent measurement of the optical depth to electron scattering ($\tau_e = 0.09 \pm 0.03$) by *WMAP* suggests that the first sources of light significantly ionized the intergalactic medium (IGM) at redshift $z \sim 11 \pm 3$ (Page et al. 2007, Spergel et al. 2007). The Sloan Digital Sky Survey (SDSS) has uncovered a handful of bright quasars at redshifts as high as $z = 6.41$ (see Becker et al. 2001, Fan et al. 2001, 2003, 2004, 2006 or the recent review by Fan 2006). The luminosity of these sources rivals those of the most powerful quasars at the peak of their activity at $z \simeq 2.5$. A sizable population of $z \sim 6$ galaxies have also been identified in the past several years (see, e.g., Ellis 2008 for a recent comprehensive review), many of them in broad–band photometry in deep, wide-area *Hubble Space Telescope* fields (Dickinson et al. 2004, Malhotra et al. 2005, Bouwens et al. 2006, Stanway et al. 2004,

2007). Narrow–band searches have also been successful in discovering numerous $z > 5$ Lyman-α emitting galaxies. Starting with the initial success of (Weymann et al. 1998), 10 m class telescopes have now identified about two dozen such objects out to $z \approx 6.7$ (see Malhotra and Rhoads 2004 for a compilation of datasets; and the more recent Subaru survey results by, e.g., Taniguchi et al. 2005, Kashikawa et al. 2006). In summary, the above observational data leave little doubt that the first galaxies and quasars were indeed well in place by redshift $z \simeq 6$, but not in significant numbers prior to redshift $z \simeq 15$.

These developments have been complemented by progress in theoretical work focusing on the cooling and collapse of gas in primordial low–mass halos (see reviews by Haiman 2004, Abel and Haiman 2000 and references therein). A broad picture has emerged, identifying several processes that are important in the formation of the first stars and black holes (BHs) out of the baryonic gas in the earliest halos. The key ingredient in this picture is the abundance of H_2 molecules that form via gas–phase reactions in the early universe (this beautifully simple fact was recognized as early as in 1967 by Saslaw and Zipoy 1967). The first objects form out of gas that cools via H_2 molecules, and condenses at the centers of virialized dark matter "minihalos" with virial temperatures of $T_{\rm vir} \sim 200$ K (Haiman et al. 1996, Tegmark et al. 1997). Detailed numerical simulations have shown convergence toward a gas temperature $T \sim 300$ K and density $n \sim 10^4\,{\rm cm}^{-3}$, dictated by the thermodynamic properties of H_2 (Abel et al. 2002, Bromm et al. 2002, Yoshida et al. 2003, 2006), which allows the collapse of a single clump of mass 10^2–$10^3\,M_\odot$ at the center of the halo. The soft UV and X-ray radiation emitted by the stars and black holes formed in the first handful of such clumps provide prompt and significant global feedback on the chemical and thermal state of the IGM. A relatively feeble early background radiation field can already have strong effects on gas cooling and H_2 chemistry, and can produce significant global impact, affecting the formation of the bulk of the first astrophysical structures, and therefore the reionization history of the IGM (Haiman et al. 1997, 2000, Machacek et al. 2001, 2003, Oh and Haiman 2003, Ricotti et al. 2002a, Kuhlen and Madau 2005, Furlanetto and Loeb 2005, Mesinger et al. 2006, Ahn and Shapiro 2007, Johnson et al. 2007). The observational data summarized above suggests that such feedback processes indeed shaped the reionization history, and hints that star–formation in early minihalos was suppressed (Haiman and Bryan 2006).

15.2 Summary of Relevant Physics at High Redshifts

In the context of ΛCDM cosmologies, it is natural to identify the first dark matter halos as the hosts of the first sources of light. A simplified picture for the emergence of the first sources of light, and consequent reionization is as follows: the gas in dark matter halos cools and turns into ionizing sources (stars and/or black holes), which produce UV (and possibly X-ray) radiation, and drive expanding ionized regions into the IGM. The volume filling factor of ionized regions then, at least initially, roughly tracks the global formation rate of dark halos. Eventually, the ionized regions percolate (when the filling factor F_{HII} reaches unity), and the remaining neutral hydrogen in the IGM is rapidly cleared away as the ionizing background builds up. However, a soft UV radiation background at photon energies of <13.6 eV, as well as possibly a soft X-ray background at $\gtrsim 1$ keV, will build up before reionization, since the early IGM is optically thin at these energies.

We now briefly describe the various effects that should determine the evolution of the first sources and of reionization at high redshifts. For in–depth discussion, we refer the reader to extended reviews (e.g. Barkana and Loeb 2001), and to reviews focusing on the roles of H_2 molecules in reionization (Abel and Haiman 2000), on the effect of reionization on CMB anisotropies (Haiman and Knox 1999), and on progress in the last three years in studying reionization (Haiman 2004, Ferrara 2007).

In the discussions that follow, it will be useful to distinguish three different types of dark matter halos, which can be roughly divided into three different ranges of virial temperatures, as follows:

$300\,\mathrm{K} \lesssim T_{\mathrm{vir}} \lesssim 10^4\,\mathrm{K}$	(Type II – susceptible to H_2–feedback)
$10^4\,\mathrm{K} \lesssim T_{\mathrm{vir}} \lesssim 2\times10^5\,\mathrm{K}$	(Type Ia – susceptible to photo–heating feedback)
$T_{\mathrm{vir}} \gtrsim 2\times10^5\,\mathrm{K}$	(Type Ib – gas can fall in, even in the face of photo –heating)

We will hereafter refer to these three different types of halos as Type II, Type Ia, and Type Ib halos. The motivation in distinguishing Type II and Type I halos is based on H_2 molecular vs atomic H cooling, whereas types "a" and "b" reflect the ability of halos to allow infall and cooling of photoionized gas. While the actual values of the critical temperatures, and the sharpness of the transition from one category to the next is uncertain (see discussion below), as we argue below, star formation in each type of halo is likely governed by different physics, and each type of halo likely plays a different role in the reionization history. In short, Type II halos can host ionizing sources only in the neutral regions of the IGM, and only if H_2 molecules are present; Type Ia halos can form new ionizing sources only in the neutral IGM regions, but can do so by atomic cooling, irrespective of the H_2 abundance, and Type Ib halos can form ionizing sources regardless of the H_2 abundance, and whether they are in the ionized or neutral phase of the IGM. It is useful to next provide a summary of the various important physical effects that govern star–formation and feedback in these halos.

H_2 *Molecule Formation:* In ΛCDM cosmologies, structure formation is bottom–up: the earliest nonlinear dark matter halos form at low masses. Gas contracts together with the dark matter only in dark halos above the cosmological Jeans mass, $M_{\mathrm{J}} \approx 10^4$. However, this gas can only cool and contract to high densities in somewhat more massive halos, with $M \gtrsim M_{\mathrm{H2}} \equiv 10^5 M_{\odot}[(1+z)/11]^{-3/2}$ (i.e. Type II halos), and only provided that there is a sufficient abundance of H_2 molecules, with a relative number fraction at least $n_{\mathrm{H2}}/n_{\mathrm{H}} \sim 10^{-3}$ (Haiman et al. 1996, Tegmark

Table 15.1 Collapse redshifts and masses of 1, 2 and 3σ dark matter halos with virial temperatures of $T_{\rm vir} = 100$ and 10^4K, reflecting the minimum gas temperatures required for cooling by H_2 and neutral atomic H, respectively

$T_{\rm vir}$(K)	ν	$z_{\rm coll}$	$M_{\rm halo}(M_\odot)$
100	1	7	2×10^5
	2	16	5×10^4
	3	26	3×10^4
10^4	1	3.5	4×10^8
	2	9	1×10^8
	3	15	6×10^7

et al. 1997). Because the typical collapse redshift of halos is a strong function of their size, the abundance of H_2 molecules is potentially the most important parameter in determining the onset of reionization. For example, 2σ halos with virial temperatures of 100K appear at $z = 16$, while 2σ halos with virial temperatures of 10^4K (Type Ia halos, in which gas cooling is enabled by atomic hydrogen lines) appear only at $z = 9$. Table 15.1 summarizes the collapse redshifts and masses of dark halos at these two virial temperatures in the concordance cosmology. As a result, the presence or absence of H_2 in Type II halos makes a factor of $\sim$2 difference in the redshift for the onset of structure formation, and thus potentially effects the reionization redshift and the electron scattering opacity τ by a similar factor. In the absence of any feedback processes, the gas collecting in Type II halos is expected to be able to form the requisite amount of H_2 (Haiman et al. 1996, Tegmark et al. 1997). However, both internal and external feedback processes can alter the typical H_2 abundance (see discussion below).

The Nature of the Light Sources in the First Halos: A significant uncertainty is the nature of the ionizing sources turning on inside halos collapsing at the highest redshifts. Three dimensional simulations using adaptive mesh refinement (AMR; Abel et al. 2002) and smooth particle hydrodynamics (SPH; Bromm et al. 2002, Yoshida et al. 2008) techniques have followed the contraction of gas in Type II halos at high redshifts to high densities. These works have shown convergence toward a temperature/density regime of T $\sim$ 200 K, n $\sim$ 10^4 cm^{-3}, dictated by the critical density at which the excited states of H_2 reach equilibrium population levels. These results have suggested that the first Type II halos can only form unusually massive stars, with masses of at least $\sim$100 $M_\odot$, and only from a small fraction ($\lesssim$0.01) of the available gas. Such massive stars are effective producers of ionizing radiation, making reionization easier to achieve. In addition, it is expected that the first massive stars, forming out of metal–free gas, have unusually hard spectra (Tumlinson and Shull 2000, Bromm et al. 2001, Schaerer 2002), which is important for possibly ionizing helium, in addition to hydrogen. An alternative possibility is that a similar fraction of the gas in the first Type II halos forms massive black holes (Haiman and Loeb 1997, Madau et al. 2004, Ricotti et al. 2005). In fact, when massive, non–rotating metal–free stars end their life, they are expected to leave behind stellar–mass seed BHs, unless their mass is in the range of 140–260 $M_\odot$ (Heger et al. 2003). These early black holes can then accrete gas (possibly only after a delay – if the progenitor star clears the host halo of gas, the BH will have to await until the host halo merges with another, gas–rich halo), acting as "miniquasars". The miniquasars will produce a hard spectrum extending to the soft X-rays, which could be important in catalyzing H_2 formation globally (see discussion below).

Efficiencies and the Transition from Metal Free to "Normal" Stars: Another fundamental question of interest is the efficiency at which the first sources inject ionizing photons into the IGM. This can be parameterized by the product $\epsilon_* \equiv N_\gamma f_* f_{esc}$, where $f_* \equiv M_*/(\Omega_b M_{halo}/\Omega_m)$ is the fraction of baryons in the halo that turns into stars; N_γ is the mean number of ionizing photons produced by an atom cycled through stars, averaged over the initial mass function (IMF) of the stars; and f_{esc} is the fraction of these ionizing photons that escapes into the IGM. It is difficult to estimate these quantities at high–redshifts from first principles, but the discussion below can serve as a useful guide.

Although the majority of the baryonic mass in the local universe has been turned into stars (Fukugita et al. 1998), the global star formation efficiency at high redshifts was likely lower. To explain a universal carbon enrichment of the IGM to a level of $10^{-2} - 10^{-3}$ $Z_\odot$, the required efficiency, averaged over all halos at $z \gtrsim 4$, is $f_* = 2-20\%$ (Haiman 1997). However, the numerical simulations mentioned in § 2.2 above suggest that the fraction of gas turned into massive stars in Type II halos is $f_* \lesssim 1\%$.

The escape fraction of ionizing radiation in local starburst galaxies is of order $\sim$10%. The higher characteristic densities at higher redshifts could decrease this value (Dove et al. 2000, Wood and Loeb 1999), although there are empirical indications that the escape fraction in $z \sim 3$ galaxies may instead be higher (Steidel et al. 2001), at least in some galaxies (Shapley et al. 2006). Radiation can also more readily ionize the local gas, and escape from the small Type II halos which have relatively low total hydrogen column densities ($\lesssim 10^{17}$ cm^{-2}), effectively with $f_{esc} = 1$ in the smallest halos (Whalen et al. 2004).

The ionizing photon yield per proton for a normal Salpeter IMF is $N_\gamma \approx 4000$. However, if the IMF consists exclusively of massive $M \gtrsim 200$ $M_\odot$ metal–free stars, then N_γ can be up to a factor of $\sim$20 higher (Bromm et al. 2001, Schaerer 2002). The transition from metal–free to a "normal" stellar population is thought to occur at a critical metallicity of $Z_{cr} \sim 5 \times 10^{-4} Z_\odot$, above which cooling and fragmentation becomes efficient and stops the IMF from being biased toward massive stars (Bromm et al. 2001). It is natural to associate this transition with that of the assembly of halos with virial temperatures of $>10^4$K (Type Ia halos). Type II halos are fragile, and likely blow away their gas and "shut themselves off" after a single episode of (metal–free) star–formation. They are therefore unlikely to allow continued formation of stars with metallicities above Z_{crit}. Subsequent star–formation will then occur only when the deeper potential wells of Type Ia halos are assembled and cool their gas via atomic hydrogen lines. The material that collects in these halos will then have already gone through a Type II halo phase and contain traces of metals.

As we argue in Section 15.3 below, there exists an alternative, equally plausible scenario. Most Type II halos may not have formed any stars, due to global H_2 photodissociation by an early cosmic soft–UV background. In this case, the first generation of metal–free stars must appear in Type Ia halos. Halos above this threshold can eject most of their self–produced metals into the IGM, but, in difference from Type II halos, can retain most of their gas (MacLow and Ferrara 1999), and can have significant episodes of metal–free star formation. These halos will also start the process of reionization by driving expanding ionization fronts into the IGM. The metals that are ejected from Type Ia halos will reside in these photoionized regions of the IGM. As discussed in Section 15.3 below, photoionization heating in these regions may suppress gas infall and cooling, causing a pause in the formation of new structures, until larger dark matter halos, with virial temperatures of $T_{\rm vir} \gtrsim 2 \times 10^5$K (Type Ib halos) are assembled. The material that collects in Type Ib halos will then have already gone through a previous phase of metal–enrichment by Type Ia halos, and it is unlikely that Type Ib halos can form significant numbers of metal–free stars.

15.3 Feedback Effects

Several feedback effects are likely to be important for modulating the evolution of the early cosmic star formation history and reionization. There can be significant *internal* feedback in or near each ionizing source, due to the presence of supernovae (Ferrara 1998), or of the radiation field (Omukai and Nishi 1999, Ricotti et al. 2002a), on the local H_2 chemistry. The net sign of these effects is difficult compute, as it depends on the source properties and spectra, and on the detailed density distribution internal and near to the sources. For practical purposes of computing the feedback that modulates the global star–formation or reionization, we may, however, think of any internal feedback effect as regulating the efficiency parameter ε_* defined above.

Since the universe is optically thin at soft UV (below 13.6 eV), and soft X-ray ($\gtrsim$1 keV) photon energies, radiation from the earliest Type II halos can build up global backgrounds at these energies, and provide prompt *external*, global feedback on the formation of subsequent structures, which are easier to follow. A very large number of studies over the past several years have assessed the feedback from both the LW and X-ray backgrounds quantitatively, including many works that employed three–dimensional simulations (Haiman et al. 2000, Ciardi et al. 2000, Machacek et al. 2001, 2003, Ricotti et al. 2002a,b, Glover and Brandt 2003, Kuhlen and Madau 2005, Furlanetto and Loeb 2005, Mesinger et al. 2006, Ahn and Shapiro 2007, Johnson et al. 2007). In isolation, Type II halos with virial temperatures as low as a few 100 K could form enough H_2, via gas–phase chemistry, for efficient cooling and gas contraction (Haiman et al. 1996, Tegmark et al. 1997). However, H_2 molecules are fragile, and can be dissociated by soft UV radiation absorbed in their Lyman-Werner (LW) bands (Haiman et al. 1997, 2000, Ciardi et al. 2000, Ricotti et al. 2001). In patches of the IGM corresponding to fossil HII regions that have recombined (after the death of short–lived ionizing source, such as a massive star), the gas retains excess entropy for a Hubble time. This "entropy floor" can reduce gas densities in the cores of collapsing halos (analogously to the case of "preheating" of nearby galaxy clusters), and decrease the critical LW background flux that will photodissociate H_2 (Oh and Haiman 2003). On the other hand, positive feedback effects, such as the presence of extra free electrons (beyond the residual electrons from the recombination epoch) from protogalactic shocks (Shapiro and Kang 1987, Ferrara 1998), from a previous ionization epoch (Oh and Haiman 2003, Susa et al. 1998), or from X-rays (Haiman et al. 1997, Oh 2001, Ricotti et al. 2002a,b), can enhance the H_2 abundance.

The extent to which star–formation in Type II halos was quenched globally has remained unclear, with numerical simulations generally favoring less quenching (Machacek et al. 2001, 2003, Ricotti et al. 2002a,b, Wise and Abel 2007) than predicted in semi–analytical models. Likewise, simulations find a smaller, if any, effect from X-rays (Kuhlen and Madau 2005). In a recent study (Mesinger et al. 2006), we used three-dimensional hydrodynamic simulations to investigate the effects of a transient ultraviolet (UV) flux and a LW background on the collapse and cooling of pregalactic clouds, with masses in the range 10^5 – 10^7 $M_\odot$, at high redshifts ($z \gtrsim 18$). In the absence of a LW background, we found that a critical specific intensity of $J_{UV} \sim 0.1$ (in units of 10^{-21}ergs s^{-1} cm^{-2} Hz^{-1} sr^{-1}) demarcates the transition from net negative to positive feedback for the halo population. Note that this flux is $\sim$2 orders of magnitude below the level required for reionization (defined by the requirement of producing a few photons per H atom). A weaker UV flux stimulates subsequent star formation inside the fossil HII regions, by enhancing the H_2 molecule abundance. A stronger UV flux significantly delays star–formation by reducing the gas density, and increasing the cooling time at the centers of collapsing halos. At a fixed J_{UV}, the sign of the feedback also depends strongly on the density of the gas at the time of UV illumination. In either case, once the UV flux is turned off, its impact starts to diminish after $\sim$30% of the Hubble time. In the more realistic case when a permanent LW background is present (in addition to a short–lived ionizing neighbor), with $J_{LW} \gtrsim 0.01 \times 10^{-21}$ergs s^{-1} cm^{-2} Hz^{-1} sr^{-1}, strong suppression persists down to the lowest redshift ($z = 18$) in these simulations. The feedback was also found to depend strongly on the mass of the Type II halo, with the smaller halos strongly suppressed, but the larger halos ($\gtrsim 10^7$ $M_\odot$) nearly immune to feedback. In recent works, (Wise and Abel 2007) and (O'Shea and Norman 2008) found that the largest Type II halos, with $T_{vir} \gtrsim 4000$K, can eventually cool their gas even in the face of a LW background with $J_{LW} > 0.1$, due to the increased electron abundance and elevated temperature and cooling rate in the inner regions of these halos.

A second type of important feedback is that photo–ionized regions are photo-heated to a temperature of $\gtrsim 10^4$K, with a corresponding increase in the Jeans mass in these regions, and possible suppression of gas accretion onto low-mass halos (e.g. Efstathiou 1992, Thoul and Weinberg 1996, Gnedin 2000, Shapiro et al. 1994). Reionization is then expected to be accompanied by a drop in the global SFR, corresponding to a suppression of star formation in small halos (i.e. those with virial temperatures below $T_{vir} \lesssim 10^4$–10^5 K). The size of such a drop is uncertain, since the ability of halos to self–shield against the ionizing radiation is poorly constrained at high redshifts. Early work on this subject, in the context of dwarf galaxies at lower redshift (Thoul and Weinberg 1996), suggested that an ionizing background would completely suppress star formation in "dwarf galaxy" halos with circular velocities $\upsilon_{\rm circ} \lesssim 35$ km s^{-1}, and partially suppress star–formation in halos with 35 km s^{-1} $\lesssim$ v$_{\rm circ}$ $\lesssim$ 100 km s^{-1}. However, more recent studies (Kitayama and Ikeuchi 2000, Dijkstra et al. 2004) find that at high–redshifts ($z \gtrsim 3$), self-shielding and increased cooling efficiency could be strong countering effects. These calculations, however, assume spherical symmetry, leaving open the possibility of strong feedback for a halo with non-isotropic gas profile, illuminated along a low-column density line of sight (see (Shapiro et al. 2004) for a detailed treatment of three-dimensional gas dynamics in photo-heated low-mass halos).

Because the earliest ionizing sources formed at the locations of the rare density peaks, their spatial distribution was strongly clustered. Since most feedback mechanisms operate over a limited length scale, their effects will depend strongly on the spatial distribution of halos hosting ionizing sources. Numerical simulations are a promising way to address feedback among clustered sources, since they capture the full, three-dimensional relationships among the host halos (Iliev et al. 2007). However, the dynamic range required to resolve the small minihalos, within a large enough cosmic volume to be representative (Barkana & Loeb 2004b), remains a challenge, especially in simulations that include radiative transfer (Iliev et al. 2006). Semi-analytical models avoid the problems associated with the large dynamical range; they are also an efficient way to explore parameter space and serve as important sanity checks for more complicated simulations. Semi-analytical studies to date have included *either* various feedback effects (e.g., Haiman and Loeb 1998, Haiman and Holder 2003, Wyithe and Loeb 2003, Cen 2003, Greif and Bromm 2006, Furlanetto and Loeb 2005, Wyithe and Cen 2007, Johnson et al. 2007) *or* the effect of source clustering on the HII bubble–size distribution (e.g. Furlanetto et al. 2004), but *not both*. In a recent study (Kramer et al. 2006), we have incorporated photo-ionization feedback, in a simplified way, into a model that partially captures the source clustering (i.e., only in the radial direction away from sources). Source clustering was found to increase the mean HII bubble size by a factor of several, and to dramatically increase the fraction of minihalos that are suppressed, by a factor of up to ~60 relative to a randomly distributed population. We argue that source clustering is likely to similarly boost the importance of a variety of other feedback mechanisms. (This enhanced suppression can also help reduce the electron scattering optical depth τ_e, as required by the three-year data from WMAP Haiman and Bryan 2006.)

15.4 How Can We Detect this Feedback?

There are several ways, at least in principle, to discover the presence of global feedback mechanisms that modulate the early star–formation rate and reionization. Here I simply list several possibilities, in order to give an (admittedly crude) overview. In Section 15.6, I will discuss in more detail the one most likely to be relevant to

JWST – tracing the cosmic star–formation history with SNe, and searching for a feature in this ultra-high redshift version of the "Lilly–Madau" diagram, caused by the feedback.

In general, the feedback processes discussed above can produce an extended and complex (possibly even non-monotonic) reionization history, especially at the earliest epochs. The earliest stages of the global reionization history (ionized fraction versus cosmic time) can be probed in 21 cm studies (see (Furlanetto et al. 2006) and the contributions by Steve Furlanetto and Avi Loeb in these proceedings). Additionally, the measurement of polarization anisotropies by *Planck* can go beyond a constraint on the total electron scattering optical depth τ_e that is measured by *WMAP*, and at least distinguish reionization histories that differ significantly (Kaplinghat et al. 2003, Haiman and Holder 2003, Mortonson and Hu 2007) from each other. Reionization may modify the small–angle CMB anisotropies, as well, through the kinetic SZ effect, at a level that may be detectable in the future (Santos et al. 2003, 2007, Salvaterra et al. 2005, McQuinn et al. 2005, Johnson et al. 2007).

A recent study (Hernández-Monteagudo et al. 2007a) considered the pumping of the 63.2 μm fine-structure line of neutral oxygen in the high-redshift intergalactic medium (IGM), in analogy with the Wouthuysen-Field effect for the 21 cm line of cosmic HI. This showed that the soft UV background at 1300 Å can affect the fine–structure population levels in the ground state of OI. If a significant fraction of the IGM volume is filled with "fossil H II regions" that have recombined and contain neutral OI, then this can produce a non-negligible spectral distortion in the cosmic microwave background (CMB). A measurement of this signature can trace the global metallicity at the end of the dark ages, prior to the completion of cosmic reionization, and is complementary to the cosmological 21 cm studies. In addition to the mean spectrum, fluctuations in the OI pumping signal may be detectable, provided the background fluctuations on arcminute scales, around 650GHz, can be measured to the nJy level (Hernández-Monteagudo et al. 2007b). If the IGM is polluted with metals at high-redshift, then CMB angular fluctuations due to density fluctuations alone (without pumping, just due to the geometrical effect of scattering) could also be detectable for several metal and molecular species (Basu et al. 2004).

At lower redshift, the details of the later stages of reionization can be probed in more detail, through studying the statistics of Lyman line absorption in the spectra of quasars (e.g. (Mesinger and Haiman 2004, 2007) and references therein), gamma-ray burst afterglows, and galaxies (see, e.g., Fan et al. 2006 and references therein). In particular, the drop in the cosmic star–formation history near the end–stages of reionization at $z \sim 6$–7 due to photo–heating could be detected in the high–redshift extension of the "Lilly – Madau" diagram (Lilly et al. 1996, Madau et al. 1996), by directly counting faint galaxies (Barkana and Loeb 2000b). In practice, the low-mass galaxies susceptible to the reionization suppression are faint and may fall below *JWST*'s detection limit. Whether or not these galaxies will be detected (and in sufficient numbers so that they can reveal the effect of reionization), depends crucially on the feedback effects discussed above, the redshift of reionization, the size of the affected galaxies and their typical star–formation efficiencies, as well as the amount of dust obscuration. Alternatively, by analyzing the Lyman α absorption spectra of SDSS quasars at $z \sim 6$, (Cen and McDonald 2002) suggested, from the non-monotonic evolution of the mean IGM opacity, that we may already have detected a drop in the SFR at $z \sim 6$. In order to improve on this current, low signal-to-noise result, deep, high-resolution spectra of bright quasars would be required from beyond the epoch of reionization at $z \gtrsim 6$ (this could be possible with *JWST*, see (Haiman and Loeb 1999)). A suppression of low-mass galaxies would also increase the effective clustering of the reionizing sources (since higher-mass halos are more strongly clustered), and increase the fluctuations in the Lyman α forest opacity at somewhat lower redshifts. This effect may already have been observed (Wyithe and Loeb 2006), although the modeling details still matter, and the presently measured scatter in opacity could perhaps still be consistent with density fluctuations alone (Lidz et al. 2006).

In Section 15.6 below, I will return to the issue of feedback, and discuss tracing the cosmic star–formation history with distant SNe.

15.5 Direct Detections of the First Sources

The most basic question to ask is whether *JWST* could directly detect the first stars or black holes. As discussed above, the very first star may have formed in isolation in a 10^6 $M_\odot$ dark matter halo at $z \gtrsim 15$, and it will then be beyond the reach of direct imaging even by *JWST*. The critical mass for detection with *JWST* depends on the IMF and star–formation efficiency. At $z = 10$, assuming $\sim$10% of the gas in a halo turns into stars, with a normal Salpeter IMF, a 1nJy broad-band threshold at near–IR wavelengths would allow the detection of a stellar cluster whose mass is a several $\times 10^6$ $M_\odot$ (Haiman and Loeb 1997). It will help if the IMF is biased toward more massive, and more luminous stars. If the stars were all metal–free, on the other hand, they would have a lower flux than metal–enriched stars at *JWST*'s wavelengths, due to their high effective temperatures which shifts their flux to (observed) UV wavelengths (Tumlinson and Shull 2000, Bromm et al. 2001, Schaerer 2002). The conclusion is that while *JWST* is very unlikely to detect the first individual stars directly, it will most likely directly measure the luminosity function of faint galaxies, extending down to sufficiently small sizes, corresponding to the halo masses $M = 10^{8-10}$ $M_\odot$. This will directly probe the feedback effects discussed above. For example, a clear turn–over in the LF at the luminosity corresponding to the atomic cooling threshold (or lack of it) would be compelling evidence for H_2–feedback (or lack of it).

Likewise, one can ask whether *JWST* can see the first BHs directly? The relevant threshold – i.e. the lowest BH mass – will again depend on the spectrum and luminosity (in terms of, say, the Eddington value) of the BHs. With the average spectrum of quasars at lower redshift (Elvis et al. 1994), and assuming Eddington luminosity, at the 1nJy threshold, *JWST* could detect a BH whose mass is a several $\times 10^5$ $M_\odot$ (Haiman and Loeb 1998). If the first BHs are the remnants of massive, metal–free stars, then their initial masses will be below this threshold. Furthermore, it is not clear when such stellar–seed BHs would start shining at a significant fraction of the Eddington limit (since the progenitor star may clear their host halo of gas, as mentioned above). On the other hand, if H_2–suppression prevents most Type II halos from forming stars, then the first stars and BHs would appear in large numbers only inside more massive dark halos, with virial temperatures exceeding $T_{\rm vir} \gtrsim 10^4$K. How gas cools, condenses, and fragments in such halos is presently not well understood. Using semi-analytical toy models, (Oh and Haiman 2002) argued that the initial cooling by atomic H allows the gas to begin to collapse – even in the face of a significant LW background. If the gas remained at $\sim 10^4$K, the high Jeans mass $\sim 10^6$ $M_\odot$ in these halos would suggest that a supermassive black hole (SMBH) of a similar mass may form at the nucleus (Oh and Haiman 2002, Bromm and Loeb 2003, Volonteri and Rees 2005, Begelman et al. 2006, Lodato and Natarajan 2006). Such BHs could be directly detectable by *JWST* at $z \sim 10$, provided they shine at $\gtrsim$10% of their Eddington luminosity.

Predictions for the number counts of high redshift galaxies and quasars at *JWST*'s thresholds at near-infrared wavelengths have been made using simple semi-analytic models (Haiman and Loeb 1997, 1998). Surface densities as high as several sources per square arcminute are predicted from $z \gtrsim 5$, with most of these sources at $z \gtrsim 10$. These predictions, obtained from the simplest models, represent the DM halo abundance, multiplied with optimistic star/BH formation efficiencies, and corrected for stellar/quasar duty cycles. As such, they could be considered as upper bounds – feedback effects will certainly reduce the counts. If the galaxies occupy a fair fraction ($\sim$5%) of the virial radius of their host halos, then a large fraction ($\gtrsim$50%) of them can potentially be resolved with *JWST*'s planned angular resolution of $\sim 0.06''$ (Haiman and Loeb 1998a, Barkana and Loeb 2000a).

In addition to broad-band searches, one may look for the earliest light–sources in emission lines. The strongest recombination lines of H and He from $5 < z < 20$ will fall in the near-infrared bands of *JWST* and could be bright enough to be detectable. Specific predictions have been made for the source counts in the Hα emission line (Nemiroff 2003) and for the three strongest HeII lines that could be powered either by BHs or metal–free stars with a hard spectrum (Oh et al. 2001, Tumlinson et al. 2001). The key assumption is that most of the ionizing radiation produced by the miniquasars is processed into such recombination lines (rather than escaping into the IGM). Under this optimistic assumption (which is likely violated at least in the smallest minihalos (Whalen et al. 2004)), the lines are detectable for a fiducial 10^5 $M_\odot$ miniquasar at $z = 10$, or for a "microgalaxy" with a star-formation rate of 1$M_\odot$ yr^{-1}. The simultaneous detection of H and He lines would be especially significant. As already argued above, the hardness of the ionizing continuum from the first sources of ultraviolet radiation plays a crucial role in the reionization of the IGM. It would therefore be very interesting to directly measure the ionizing continuum of any $z > 6$ source. While this may be feasible at X-ray energies for exceptionally bright sources, the absorption by neutral gas within the source and in the intervening IGM will render the ionizing continuum of high redshift sources inaccessible to direct observation out to $\sim$1keV. The comparison of Hα and HeII line strengths can be used to infer the ratio of HeII to HI ionizing photons, $Q =$

$\dot{N}_{\rm ion}^{\rm HeII}/\dot{N}_{\rm ion}^{\rm HI}$. A measurement of this ratio would shed light on the nature of the first luminous sources, and, in particular, it could reveal if the source has a soft (stellar) or hard (AGN-like) spectrum. Note that this technique has already been successfully applied to constrain the spectra of sources in several nearby extragalactic HII regions (Garnett et al. 1991). An alternative method to probe the hardness of the spectrum of the first sources would be to study the thickness of their ionization fronts (Shapiro et al. 2004, Zaroubi and Silk 2005, Kramer and Haiman 2008); this may be possible for bright quasars in Lyman line absorption (Kramer and Haiman 2008) or in 21 cm studies (Zaroubi and Silk 2005, Thomas and Zaroubi 2008, Kramer and Haiman 2008).

Provided the gas in the high redshift halos is enriched to near–solar levels, several molecular lines may be visible. In fact, CO has already been detected in the most distant $z = 6.41$ quasar (Walter et al. 2003), and, in fact, spatially resolved (Walter et al. 2004). The detectability of CO for high redshift sources in general has been considered by (Silk and Spaans 1997) and by (Gnedin et al. 2001). For a star formation rate of $\gtrsim$30 $M_{\odot}/$ yr, the CO lines are detectable at all redshifts $z = 5$–30 by the Millimeter Array (the redshift independent sensitivity is due to the increasing CMB temperature with redshift), while the Atacama Large Millimeter Array (ALMA) could reveal fainter CO emission. The detection of these molecular lines will provide valuable information on the stellar content and gas kinematics in the earliest halos.

Finally, the baryons inside high – redshift halos with virial temperatures $T \gtrsim 10^4$K need to cool radiatively, in order to condense inside the dark matter potential wells, even before any stars or black holes can form. The release of the gravitational binding energy, over the halo assembly time-scale, can result in a significant and detectable Lyα flux (Haiman et al. 2000, Fardal et al. 2001). At the limiting line flux $\approx 10^{-19}$ erg s^{-1} cm^{-2} asec^{-2} of *JWST*, several sufficiently massive halos, with velocity dispersions $\sigma \gtrsim 120$ km s^{-1}, would be visible per $4' \times 4'$ field. The halos would have characteristic angular sizes of $\approx 10''$, would be detectable in a broad–band survey out to $z \approx 6$–8, and would provide a direct probe of galaxies in the process of forming. They may be accompanied by He$^+$ Lyα emission at the $\approx 10\%$ level (Yang et al. 2006), but remain undetectable at other wavelengths. The main challenge, if such blobs are detected, without any continuum source, will likely be the interpretation – as is the case for the currently detected extended Lyman α sources (there are currently $\sim$ three dozen extended Lyα blobs known, with 1/3rd of such objects in the largest sample (Matsuda et al. 2004) consistent with being powered by cooling radiation).

Monte Carlo calculations of Lyα radiative transfer through optically thick, spherically symmetric, collapsing gas clouds were presented in (Dijkstra et al. 2006a,b), with the aim of identifying a clear diagnostic of gas infall (a similar effort in 3D is being carried out (Tasitsiomi 2006)). These represent simplified models of proto–galaxies that are caught in the process of their assembly. Such galaxies can produce Lyα flux over an extended solid angle, either from a spatially extended Lyα emissivity, or from scattering effects, or both. We presented a detailed study of the effect of the gas distribution and kinematics, and of the Lyα emissivity profile, on the emergent spectrum and surface brightness distribution. The emergent Lyα spectrum is typically double–peaked and asymmetric. The surface brightness distribution is typically flat, and the detection of a strong wavelength dependence of its slope (with preferential flattening at the red side of the line) would be a robust indication that Lyα photons are being generated (rather than just scattered) in a spatially extended region around the galaxy, as in the case of a cooling flow. An alternative, robust diagnostic for scattering is provided by the polarization of an extended Lyman α source (Dijkstra and Loeb 2008). Spectral polarimetry (in particular, the wavelength-dependence of the polarization) can differentiate between Lya scattering off infalling gas and outflowing gas.

15.6 Tracing the Beginning of the Cosmic Star–Formation History with Supernovae

As argued above, the very first stellar clusters may fall below the direct detection of *JWST*. A promising way (and probably the only way) to directly observe the first stars individually is through their explosions, either as Supernovae or gamma ray bursts (GRBs). Before we present the expectations for SNe in detail, we first briefly discuss an alternative, GRB afterglows. In the years leading up to the launch of the *Swift* satellite,[1] it has been increasingly recognized that distant gamma ray bursts (GRBs) offer a unique probe of the high redshift universe. In particular, GRBs are the brightest known electromagnetic phenomena in the universe, and can be detected up to very high redshifts (e.g. Wijers et al. 1998, Lamb and Reichart 2000, Ciardi and Loeb 2000), well beyond the redshift $z \approx 6.5$ of the most distant currently known quasars (Fan 2006) and galaxies (Ellis 2008). There is increasing evidence that GRBs are associated with the collapse of short–lived, massive stars, including the association of bursts with star–forming regions (e.g. Fruchter et al. 1999), a contribution of supernova light to the optical afterglow (e.g. Bloom et al. 1999, Garnavich et al. 2003), and most directly, association with a supernova (Stanek et al. 2003, Hjörth et al. 2003).

As a result, the redshift distribution of bursts should follow the mean cosmic star–formation rate (SFR). Several studies have computed the evolution of the expected GRB rate under this assumption, based on empirical models of the global SFR (Totani 1997, 1999, Wijers et al. 1998, Lamb and Reichart 2000, Ciardi and Loeb 2000). Determinations of the cosmic SFR out to redshift $z \sim 5$ (e.g. Bunker et al. 2004, Gabasch et al. 2004, Giavalisco et al. 2004) have shown that star–formation is already significant at the upper end of the measured redshift range, with $\gtrsim$10% of all stars forming prior to $z = 5$, which would result in a significant population of GRBs at these redshifts. Further associating star-formation with the formation rate of non-linear dark matter halos, and using theoretical models based on the Press–Schechter formalism (Press and Schechter 1974), it is possible to extrapolate the

[1] See http://swift.gsfc.nasa.gov

SFR and obtain the GRB rates expected at still higher redshifts (Bromm and Loeb 2002, Choudhury and Srianand 2002, Mesinger et al. 2005). These studies have concluded that a significant fraction (exceeding several percent) of GRBs detected at *Swift*'s sensitivity should originate at redshifts as high as $z > 10$. The spectra of bright optical/IR afterglows of such distant GRBs can provide information both on the progenitor, and also reveal absorption features by neutral hydrogen in the intergalactic medium (IGM), and can serve as an especially clean probe of the reionization history of the universe (Miralda-Escudé 1998, Lamb and Reichart 2000, Choudhury and Srianand 2002, Lamb and Haiman 2003, Barkana and Loeb 2004a).

In summary, the main advantage of GRBs is that they are bright and can be seen essentially from arbitrary far away. The main drawback, however, is that they are rare. Even the most optimistic among the above-listed models predict only a handful of detectable GRBs per year from $z > 6$ – indeed, in the past two years, only one GRB has been discovered at $z > 6$ (GRB050904 at $z = 6.3$; Kawai et al. 2006).

15.6.1 Supernovae from the First Stars

Although SNe are not as bright as GRB afterglows, they are bright compared to galaxies at the very faint end of the luminosity function, and individual core-collapse SNe would still be visible from beyond $z = 10$. Furthermore, since they occur much more frequently than GRBs, and they remain visible for a longer time than GRB afterglows, they will likely offer a better (and possibly only) chance to trace out the cosmic star formation history accurately, with a statistically significant sample extending beyond $z > 10$. The above is true even if the bulk of early supernovae are similar to the core–collapse SNe in the local universe (note that Type Ia supernovae will not have time to form at $z \gg 6$). In addition, the pair-instability supernovae from massive, metal-free stars are expected to be much brighter than Type II supernovae from normal (metal-enriched) stars (Heger and Woosley 2002).

There have been several studies of the expected early supernova rate (SNR), calibrated to the observed metallicity of the Lyman α forest (Miralda-Escudé and Rees 1997) or the observed SFR at lower redshifts (Dahlén and Fransson 1999). The expected rate of pair-instability SNe from very high-z from the first generation of metal–free stars was studied by (Wise and Abel 2005). In a recent work (Mesinger et al. 2006), we constructed the expected detection rate of high-z SNe in SNe surveys for *JWST*. We also quantified the prospects of detecting a drop in the SN rate due to photo–heating feedback at reionization. Given that SNe may be our only hope to directly map out the beginning of the cosmic star-formation history, most of the rest of this article is devoted to discussing this possibility in detail.

15.6.2 The Global Star Formation and Supernova Rates

The global SFR density can be obtained and extrapolated by a common approach, based on the dark matter halo formation rate, calibrating the star–formation efficiency to the SFR at redshift $z \lesssim 6$. From this extrapolated SFR density, we can then obtain the intrinsic supernova rate by using the properties of core–collapse SNe in the local universe as a guide.

In particular, we can estimate the global SFR density at redshift z as

$$\dot{\rho}_*(z) = \varepsilon_* \frac{\Omega_b}{\Omega_M} \int_{M_{\min}(z)}^{\infty} dM \int_{\infty}^{z} dz' M \frac{\partial^2 n(>M, z')}{\partial M \partial z'} P(\tau) , \tag{15.1}$$

where ε_* is the efficiency (by mass) for the conversion of gas into stars, $M dM(\partial n(>M, z)/\partial M)$ is the mass density contributed by halos with total (dark matter + baryonic) masses between M and $M + dM$, $t(z)$ is the age of the universe at redshift z, and $P(\tau)$ is the probability per unit time that new stars form in a mass element of age $\tau \equiv t(z) - t(z')$ (normalized to $\int_0^\infty d\tau P(\tau) = 1$). We adopt the fiducial value of $\varepsilon_* = 0.1$ (see, e.g., (Cen 2003)). Note that the star formation efficiency in minihalos (i.e. halos with virial temperatures below 10^4K) that contain pristine metal–free gas could be significantly lower than 0.1, as suggested by numerical simulations of the first generation of stars (Abel et al. 2002, Bromm et al. 2002). The expected pre-reionization SNe rates would then lie closer to our $T_{\rm vir} \gtrsim 10^4$ K curves prior to reionization (see discussion below), making the detection of the reionization feature significantly more difficult. However, the efficiency is likely to be very sensitive to even trace amounts of metallicity (Bromm et al. 2001), and conditions for star–formation may result in a standard initial mass function (IMF) in gas that has been enriched to metallicities above a fraction 10^{-4} of the solar value. Indeed, it is unlikely that metal–free star–formation in minihalos can produce enough ionizing photons to dominate the ionizing background at reionization (e.g. Haiman et al. 2000, Haiman and Holder 2003, Sokasian et al. 2004). We assume further that star–formation occurs on an extended time-scale, corresponding to the dynamical time, $t_{\rm dyn} \sim [G\rho(z)]^{-1/2}$ (Cen and Ostriker 1992, Gnedin 1996):

$$P(\tau) = \frac{\tau}{t_{\rm dyn}^2} \exp\left[-\frac{\tau}{t_{\rm dyn}}\right] , \tag{15.2}$$

where $\rho(z) \approx \Delta_c \rho_{\rm crit}(z)$ is the mean mass density interior to collapsed spherical halos (e.g. (Barkana and Loeb 2001)), and Δ_c is obtained from the fitting formula in (Bryan and Norman 1998), with $\Delta_c = 18\pi^2 \approx 178$ in the Einstein–de Sitter model. The minimum mass, $M_{\min}(z)$ in Eq. (15.1), depends on the efficiency with which gas can cool and collapse into a dark matter halo. Prior to reionization and without molecular hydrogen, $M_{\min}(z)$ corresponds to a halo with virial temperature, $T_{\rm vir} \sim 10^4$ K; with a significant H_2 abundance, the threshold decreases to $T_{\rm vir} \sim 300$ K (Haiman et al. 2000); we use the conversion between halo mass and virial temperature as given in (Barkana and Loeb 2001). Post reionization, the Jeans mass is raised, so $M_{\min}(z)$ could increase. The degree of self-shielding, the ability of the halo gas to cool, as well as the amount of H_2 present in the high–redshift low–mass halos is uncertain (Dijkstra et al. 2004), and so below we present results for several values of $M_{\min}(z)$, which we will henceforth express in terms of $T_{\rm vir}(z)$.

Next, from $\dot{\rho}_*(z)$ we obtain the intrinsic differential SNR (number of core collapse SNe per unit redshift per year) with

$$\frac{d\dot{N}}{dz} = \eta_{\rm SN} \frac{1}{1+z} \frac{dV(z)}{dz} \dot{\rho}_*(z) \,, \tag{15.3}$$

where the factor $1/(1+z)$ accounts for time dilation, $dV(z)/dz$ is the comoving volume in our past light cone per unit redshift, and $\eta_{\rm SN}$ is the number of SNe per solar mass in stars. For a fiducial Salpeter initial mass function (IMF), we obtain $\eta_{\rm SN} \sim 1/180\ {\rm M}_\odot^{-1}$, assuming all stars with masses $9\ M_\odot \lesssim M \lesssim 40\ M_\odot$ become core collapse SNe (Heger et al. 2003). We neglect the lifetime of these high mass stars in determining our SNR; this is a reasonable assumption as the lifetimes (as well as the spread in the lifetimes) are shorter than a unit redshift interval for redshifts of interest. Note that an alternative extreme shape for the IMF, consisting entirely of 100–200 $M_\odot$ stars (Abel et al. 2002, Bromm et al. 2002) would yield a similar value for $\eta_{\rm SN}$.

We present our SFR densities (*top panel*) and SNRs (*bottom panel*) in Fig. 15.1. The curves correspond to redshift–independent virial temperature cutoffs of $T_{\rm vir} = 300,\ 10^4,\ 4.5 \times 10^4$, and 1.1×10^5 K (or circular velocities of $v_{\rm circ} = 3,\ 17,\ 35$, and 55 km s^{-1}, respectively), top to bottom, spanning the expected range (Thoul and Weinberg 1996, Dijkstra et al. 2004). Also shown are results from GOODS (Giavalisco et al. 2004): the bottom points assume no dust correction and the top points are dust corrected according to (Adelberger and Steidel 2000); the statistical error bars lie within the points (Adelberger and Steidel 2000). As there are large uncertainties associated with dust correction, each pair of points (top and bottom) serves to encompass the expected SFR densities.

Our SFRs are consistent with other theoretical predictions (e.g. (Somerville et al. 2001, Barkana and Loeb 2000b, Bromm and Loeb 2002)), as well as other estimates from the Hubble Ultra Deep Field (Bunker et al. 2004) and the FORS Deep Field on the VLT (Gabasch et al. 2004), after they incorporate a factor of 5–10 increase in the SFR (Adelberger and Steidel 2000) due to dust obscuration. Furthermore, we note that our SNRs, which at $z \gtrsim 5$ yield 0.3–2 SNe per square arcminute per year, are in good agreement with the ~1 SN per square arcminute per year estimated by (Miralda-Escudé and Rees 1997) by requiring that high-redshift SNe produce a mean metallicity of ~0.01 $Z_\odot$ by $z \sim 5$. Note, however, that the rates we obtain are significantly higher (by a factor of ~60 – 2000 at $z \sim 20$) than those recently found by (Wise and Abel 2005). The reason for this large difference is that Wise & Abel consider star–formation only in minihalos, and they assume a very low star–formation efficiency of a single star per minihalo, as may be appropriate for star–formation out of pristine (metal-free) gas in the first generation of minihalos (Abel et al. 2002, Bromm et al. 2002). In contrast, we assume an efficiency of $\varepsilon_* = 0.1$, which may be more appropriate for star-formation in pre-enriched gas that dominates the SFR just prior to reionization (including star-formation in minihalos).

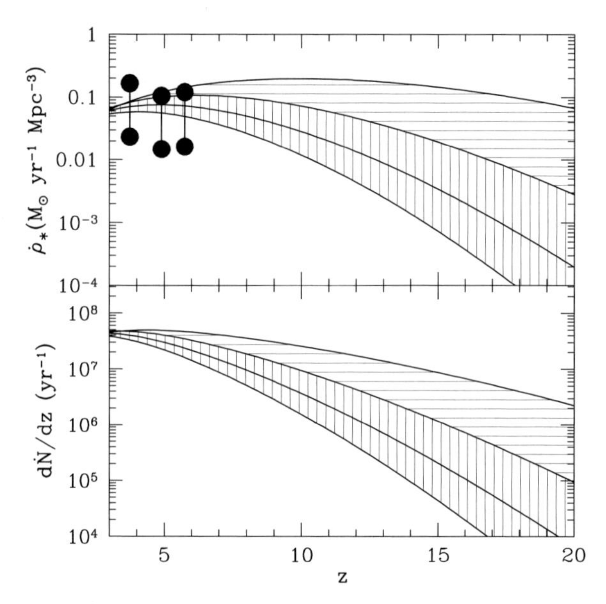

Fig. 15.1 *Upper Panel*: SFR densities obtained in a model based on dark matter halo abundances. The curves (*top to bottom*) correspond to different lower cutoffs on the virial temperatures of star–forming halos, $T_{vir} \gtrsim 300,\ 10^4,\ 4.5 \times 10^4$, and 1.1×10^5 K (corresponding to circular velocity thresholds of $v_{circ} \gtrsim 3,\ 17,\ 35$, and 55 km s^{-1}). Dots indicate results from GOODS (Giavalisco et al. 2004): the lower set of points assume no dust correction, while the upper set of points are dust corrected; statistical 1-σ error bars lie within the points. The lines connecting each pair of points span the expected range of SFR densities. *Lower Panel*: the total global supernova rate accompanying the SFR densities in the top panel (adapted from Mesinger et al. 2006)

As mentioned above, by increasing the cosmological Jeans mass, reionization is expected to cause a drop in the SFR (and hence the SNR), with the rates going from the horizontally striped region in Fig. 15.1 at $z > z_{re}$ to the vertically striped region at $z < z_{re}$.

The redshift width of this transition is set by a combination of large-scale cosmic variance, radiative transfer, and feedback effects. For the majority of the paper, we use $\Delta z_{re} \sim 1$ as a rough indicator of the width of the transition we are analyzing. We distinguish between "reionization" and a "reionization feature", and use Δz_{re} as an indicator of the width of the later. Even with an extended reionization history ($\Delta z \sim 10$), fairly sharp ($\Delta z_{re} \lesssim 3$) features are likely, as discussed in detail in (Mesinger et al. 2006).

The other important factor determining the usefulness of the method proposed here is the factor by which the SFR drops during the reionization epoch. The size of this drop is mediated by the effectiveness of self-shielding and gas cooling during photo–heating feedback: i.e. on whether or not the star-formation efficiency is significantly suppressed in those halos that dominate the SFR and SNR immediately preceding the reionization epoch. Given the uncertainties about this feedback discussed above, we will consider a range of possibilities below, parameterized by the modulation in the virial temperature threshold for star-formation during reionization.

15.6.3 The Rate of High-Redshift SNe Detectable in a Future Survey

Given the intrinsic star-formation and SN rates, our next task is to estimate the number of SNe that could actually be detected in a future search with *JWST*. In general, the number of SNe per unit redshift, $dN_{\rm exp}/dz$, that are bright enough to be detectable in an exposure of duration $t_{\rm exp}$ can be expressed as

$$\frac{dN_{\rm exp}}{dz} = \frac{d\dot{N}}{dz}\int_0^\infty f_{\rm SN}(>t_{\rm obs})\,dt_{\rm obs}\;, \tag{15.4}$$

where $(d\dot{N}/dz)dt_{\rm obs}$ is the number of SNe which occurred between $t_{\rm obs}$ and $t_{\rm obs} + dt_{\rm obs}$ ago (per unit redshift; note that the global mean SNR will evolve only on the Hubble expansion time-scale, and can be considered constant over several years), and $f_{\rm SN}(>t_{\rm obs})$ is the fraction of SNe which remains visible for at least $t_{\rm obs}$ in the observed frame. Then the total number of SNe detected in a survey of duration $t_{\rm surv}$ is

$$N_{\rm surv} = \frac{t_{\rm surv}}{2\,t_{\rm exp}}\frac{\Delta\Omega_{\rm FOV}}{4\pi}N_{\rm exp}\;, \tag{15.5}$$

where $\Delta\Omega_{\rm FOV}$ is the instrument's field of view, and $t_{\rm surv}/(2t_{\rm exp})$ is the number of fields which can be tiled in the survey time, $t_{\rm surv}$ (we add a factor of 1/2 to allow for a second pair of filters to aid in the photometric redshift determination; note that this provides for imaging in 4 different *JWST* bands; see discussion below). Note also that Eq. (15.5) is somewhat idealized, in that it assumes continuous integration for a year, and e.g., does not account for time required to slew the instrument to observe different fields. In principle, each field has to have repeated observations (to detect SNe by their variability), and therefore any dedicated survey should target fields that have already been observed. Furthermore, a dedicated year–long program may not be necessary, because the effect could be detected with relatively few fields (at least under optimistic assumptions; see discussion below), and several fields with repeated imaging (separated by $>$1–2 years) may already be available from other projects; these fields can then be used for the SN search.

In general, $f_{\rm SN}(>t_{\rm obs})$ in Eq. (15.4), i.e. the fraction of SNe which remain visible for at least $t_{\rm obs}$, depends on (i) the properties of the SN, in particular their peak magnitude and lightcurve, and the distribution of these properties among SNe, and (ii) on the properties of the telescope, such as sensitivity, spectral coverage, and field of view. In the next two subsections, we discuss our assumptions and modeling of both of these in turn.

15.6.3.1 Empirical Calibration of SN Properties

At each redshift, we run Monte-Carlo simulations to determine $f_{\rm SN}(>t_{\rm obs})$ in equation (15.4). We use the observed properties of local core–collapse SNe (CCSNe) in estimating $f_{\rm SN}(>t_{\rm obs})$. For the high redshifts of interest here, we only consider core collapse SNe of Type II. SNe resulting from the collapse of Chandrasekhar–mass white dwarfs (Type Ia) are expected to be extremely rare at high redshifts ($z \gtrsim 6$), as the delay between the formation of the progenitor and the SN event ($\gtrsim$1 Gyr; (Strolger et al. 2004)) is longer than the age of the universe at these redshifts. Local core CCSNe come in two important varieties, types IIP and IIL, differentiated by their lightcurve shapes. We ignore the extremely rare additional CCSN types, e.g. Type IIn and IIb, which appear to have significant interaction with circumstellar material and constitute less than 10% of all CCSNe. Type Ib/c, which may or may not also result from core collapse, have luminosities and light-curves that are similar to Type IIL and occur less frequently. While the relative numbers of Type IIP and Type IIL SNe are not known even for nearby SNe, estimates imply that they are approximately equal in frequency (Cappellaro et al. 1997). We therefore assume that 50% of the high-redshift SNe are Type IIP and 50% are Type IIL.

CCSNe result from the collapse of the degenerate cores of high-mass stars. The luminosity of CCSNe is derived from the initial shock caused by the core-collapse which ionizes material and fuses unstable metal isotopes (see (Leibundgut and Suntzeff 2003) and references therein for a more detailed description of SN lightcurves). In the early stages of the SN, the shock caused by the core collapse breaks out from the surface of the progenitor (typically high mass red giants), resulting in a bright initial peak in the light curve that lasts less than a few days in the rest-frame of the SN. As the shock front cools, the SN dims. However, the SN may then reach a plateau of constant luminosity in the light curve, believed to be caused by a wave of recombining material (ionized in the shock) receding through the envelope. The duration and strength of this plateau depends on the depth and mass of the progenitor envelope, as well as the explosion energy, with those SNe exhibiting a strong plateau classified as Type IIP. A typical plateau duration is $\lesssim$100 rest-frame days (Patat et al. 1994). In Type IIL SNe, this plateau is nearly non-existent, and the lightcurve smoothly transitions from the rapid decline of the cooling shock to a slower decline where the luminosity is powered by the radioactive decay of metals in the SN nebula. After the plateau, Type IIP SNe also enter this slowly declining "nebular" phase.

These observationally determined behaviors have been summarized in a useful form as lightcurve templates in (Doggett and Branch 1985). We use these template lightcurves in determining $f_{\rm SN}(>t_{\rm obs})$, and normalize the lightcurves using Gaussian–distributed peak magnitudes (i.e. log–normally distributed in peak flux) determined by (Richardson et al. 2002) from a large sample of local Type IIP and Type IIL SNe (see Table 15.2). We perform the Monte-Carlo simulations with both

Table 15.2 Means and standard deviations of the adopted peak absolute magnitudes of core collapse SNe. Values are taken from (Richardson et al. 2002). Note: an $M_B = -17$ SN would be detectable out to $z \approx 8.2$ at the flux threshold of 3 nJy in the 4.5 μm band with *JWST*

	Corrected for Dust		Not Corrected for Dust	
SN Type	$\langle M_B \rangle$	σ	$\langle M_B \rangle$	σ
IIP	−17.00	1.12	−16.61	1.23
IIL	−18.03	0.9	−17.80	0.88

the dust corrected, and dust uncorrected values in (Richardson et al. 2002), since the dust production history of the early universe is poorly understood and is essentially unconstrained empirically.

We use a combined high-resolution HST STIS + ground–based spectrum of the Type IIP supernova SN1999em (the "November 5th" spectrum of (Baron et al. 2000)) as the template SN spectrum in order to obtain K-corrections (with the Doggett and Branch 1985 lightcurves given in the restframe B filter). This spectrum was obtained within 10 days of maximum light, i.e. during the initial decline of the SN brightness, and has been dereddened by $A_V = 0.3$ mag (Baron et al. 2000, Hamuy et al. 2001). While the spectrum, and hence the K-corrections, of SNe evolve during the lightcurve, this effect is not strong for the wavelengths of interest (Patat et al. 1994), especially since the lightcurve template we use is well matched to the wavelengths being probed by the observations we consider below (leading to small K-corrections). We have also used this Type IIP SN template spectrum to calculate K-corrections for Type IIL SNe. This is necessary due to the lack of restframe UV spectra of Type IIL SNe that can be combined with optical spectra, and justifiable because the K-corrections are relatively small and the broadband colors of both Type IIP and IIL SNe (a measure of the spectral shape that determines the K-corrections) are similar, at least in the optical (Patat et al. 1994). Note that we assume that very-high redshift core-collapse SNe are similar to local SNe in their spectra, peak luminosities, and temporal evolution. However, these assumption do not significantly impact our conclusions below, as long as (i) the average properties of the SNe do not change *rapidly* at high redshift (which could mimic the reionization drop), (ii) they do not become preferentially under-luminous (which would make high-z SNe less detectable, lowering the statistical confidence at which the reionization drop is measured), and (iii) the detection efficiency does not change rapidly with redshift (e.g. due to instrument parameters or spectral lines).

In order to use the method outlined above to probe reionization, the SN redshifts must also be known to an accuracy of $\Delta z \lesssim 1$. SN redshifts can be determined via spectroscopy of either the SN itself, or of the host galaxy. However, as we have already noted, the host galaxies may only be marginally detectable even in imaging, and the SNe may be too faint for anything other than extremely low resolution spectroscopy. We present here only a very brief example of the possibility of obtaining redshifts from the extremely low resolution ($\lambda/\Delta\lambda \sim 5$) spectra provided by multi-band imaging: a complete investigation of this possibility is warranted, but is beyond the scope of this paper.

To the extent that Type II SNe spectra can be represented as a sequence of blackbodies of different temperatures (e.g. Dahlén and Fransson 1999) photometric redshifts will be impossible to obtain without information about the SN epoch, since temperature and redshift would be degenerate. However, local Type IIP SNe show significant deviations from a blackbody in the UV ($\lambda < 3500$Å) due to metal-line blanketing in the SN photosphere, providing spectral signatures that could be used as redshift indicators, depending on their strength. In Fig. 15.2, we show the evolution with redshift of the infrared colors (in bands accessible with *JWST*; see below) of our template SN spectrum, compared with the color evolution of a blackbody. The figure shows that the template spectrum deviates significantly from a blackbody. If the spectrum of the SN is always the same as the template spectrum, then there are good prospects for obtaining photometric redshifts for these SNe, at least in the redshift range $z = 7$–13. While the figure shows that multiple redshifts may be possible at fixed observed colors, the degenerate solutions would correspond to $z > 16$ SNe; contamination from such high redshift will be mitigated by the fact that these SNe are likely to be too faint to be detected. Of course, more detailed studies of the UV behavior of local SNe, especially their variety and spectral evolution, will be necessary to confirm the possible use of photometric redshifts.

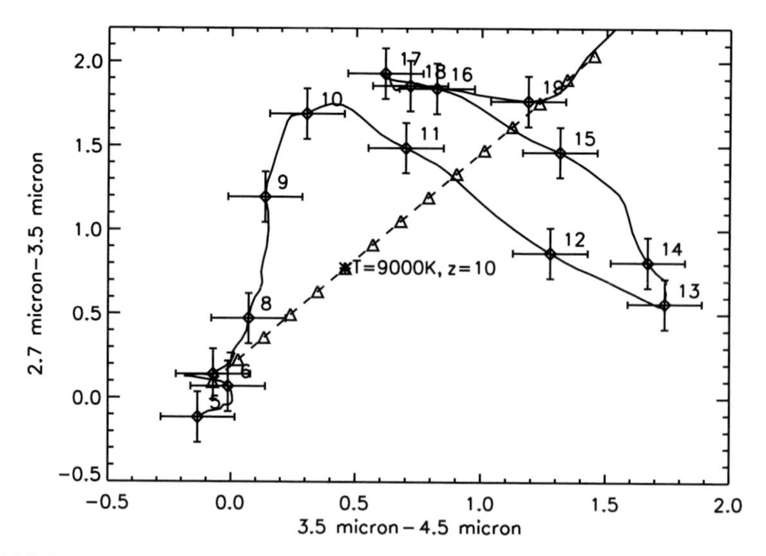

Fig. 15.2 Infrared colors of a Type II SN as a function of its redshift. *Solid curve*: The template spectrum of the text, from the Type IIP SN 1999em. Diamonds are placed at intervals of $\Delta z = 1$, labeled with the redshift. Error bars of 0.15 mag are shown to give a sense of the photometric errors that may be expected, excluding the many possible systematic effects. *Dashed curve:* The colors of a blackbody at different redshifts (temperatures) are shown for reference. The asterisk marks the color of a 9000K blackbody at $z = 10$, triangles are at intervals of $\Delta z = 1$; or equivalently, at a constant redshift, but with correspondingly different temperatures (adapted from Mesinger et al. 2006)

15.6.3.2 SNe Detectability and Survey Parameters

As a specific example for the number of detectable SNe in a future sample, we consider observations by *JWST*. The relevant instrument on *JWST* is NIRcam,[2] a near–infrared imaging detector with a FOV of $2.3' \times 4.6'$. A field can be observed in two filters simultaneously. NIRcam will have five broadband filters (with resolution $\lambda/\Delta\lambda \sim 5$). We model the filter response as tophat functions with central wavelengths of 1.5, 2.0, 2.7, 3.5, and 4.5 μm. For concreteness, below we will present results only for the 4.5 and 3.5 μm filters, since they are the longest–wavelength *JWST* bands; however, we allow time for imaging in two other bands, if needed for photometric redshift determinations. The current estimate of the *JWST* detection threshold at 4.5 μm is $\gtrsim$3 nJy for a 10 σ detection and an exposure time of 10^5 s. The 3.5 μm band is more sensitive, with a detection threshold of $\gtrsim$1 nJy for a 10 σ detection and an exposure time of 10^5 s.

We show our results for $f_{\rm SN}(>t_{\rm obs})$ in Fig. 15.3. The solid curves correspond to $z = 7$, the dashed curves correspond to $z = 10$, and the dotted curves correspond to $z = 13$. In each panel, the top set of curves assumes a flux threshold of 3 nJy (or an exposure time of $t_{\rm exp} = 10^5$ seconds in the 4.5 μm band), and the bottom set assumes 9.5 nJy ($t_{\rm exp} = 10^4$ seconds in the 4.5 μm band, background dominated). The top panel further assumes no dust extinction, while in the bottom panel, we adopt the same dust extinction as in the low redshift sample (Richardson et al. 2002). Understandably, the distributions get wider as redshift increases (due to time dilation), but the total visible fraction, $f_{\rm SN}(>0)$, gets smaller (due to the increase in luminosity distance). The double bump feature in some of the curves corresponds to the plateau of Type IIP SNe lightcurves discussed above.

As seen in Fig. 15.3, most of the supernovae that are bright enough to be visible at all, will remain visible for up to ~1–2 years. Hence, in order to catch the most SNe, it will be necessary to have repeat observations of the SN survey fields a few years apart, to insure that most of the observed SNe will be identified as new sources or sources that have disappeared. A time between observations that is comparable to or larger than the SN duration will be the optimal strategy for detecting the most SNe (Nemiroff 2003). Note that we ignore the time required to collect reference images, since observations conducted for other programs will likely provide a sufficient set of such reference images. However, as shown in Eq. (15.5), we allow time for the field to be imaged in four different bands, to aid in photometric redshift determination.

To be more explicit, we find that in order to obtain the largest number of high–redshift SNe, in general it is a more efficient use of *JWST*'s time to "tile" multiple fields rather than 'stare' for extended periods of time ($\gtrsim 10^4$ s) at the same field (Nemiroff 2003). This is because of strong time dilation at these redshifts. As can be seen from Fig. 15.3, most detectable SNe will remain above the detection threshold for several months, even assuming 10^4 s integration times. Also evident from Fig. 15.3 is that the increase in the total visible fraction of SNe going from $t_{\rm exp} = 10^4$ s to $t_{\rm exp} = 10^5$ s, is less than the factor of 10 increase in exposure time. As a result, a

[2] See http://ircamera.as.arizona.edu/nircam for further details.

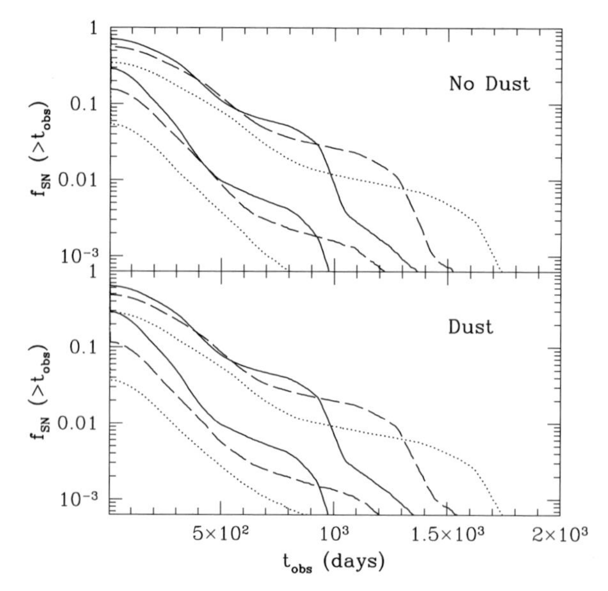

Fig. 15.3 The fraction of SNe which remain visible for an observed duration of t_{obs} or longer. The curves correspond to SN redshifts of $z = 7$ (*solid curve*), $z = 10$ (*dashed curve*), and $z = 13$ (*dotted curve*). In each panel, the top set of curves assumes a flux threshold of 3 nJy (exposure time of $t_{exp} = 10^5$s with the 4.5 μm *JWST* band), and the bottom set assumes 9.5 nJy ($t_{exp} = 10^4$ s). The top panel assumes no dust extinction, while the bottom panel assumes the same dust extinction as observed within the low–redshift sample (Richardson et al. 2002) (adapted from (Mesinger et al. 2006))

fiducial 1–yr *JWST* survey would therefore detect more SNe using $t_{exp} = 10^4$ s than using $t_{exp} = 10^5$ s (see Figs. 15.4 and 15.5). Understandably, this conclusion does not hold for very high-redshifts, $z \gtrsim 14$, where SNe are extremely faint, and require very long exposure times to be detectable. However, even with such long exposure times, very few SNe will be detectable at these large redshifts, rendering the use of longer exposure times unnecessary.

15.6.3.3 SNe Detection Rates

The number of SNe that could be detectable in putative future surveys are shown in Figs. 15.4 and 15.5. The curves correspond to the same virial temperature cutoffs for star-forming halos as in Fig. 15.1. Solid lines assume no dust obscuration; dashed lines include a correction for dust obscuration as discussed above. Figure 15.4 shows results assuming flux density thresholds of 9.5 nJy (or $t_{exp} = 10^4$ s with the 4.5 μm *JWST* filter) (*top panel*) and 3 nJy ($t_{exp} = 10^5$ s) (*bottom panel*). Figure 15.5 shows results with the 3.5 μm filter assuming equivalent exposure times: flux density thresholds of 3.2 nJy ($t_{exp} = 10^4$ s with the 3.5 μm *JWST* filter) (*top panel*) and 1 nJy ($t_{exp} = 10^5$ s) (*bottom panel*). The right vertical axis displays the number of SNe per unit redshift per FOV (2.3 × 4.6); the left vertical axis shows the number of SNe per unit redshift in a fiducial 1-year survey. As mentioned above, reionization

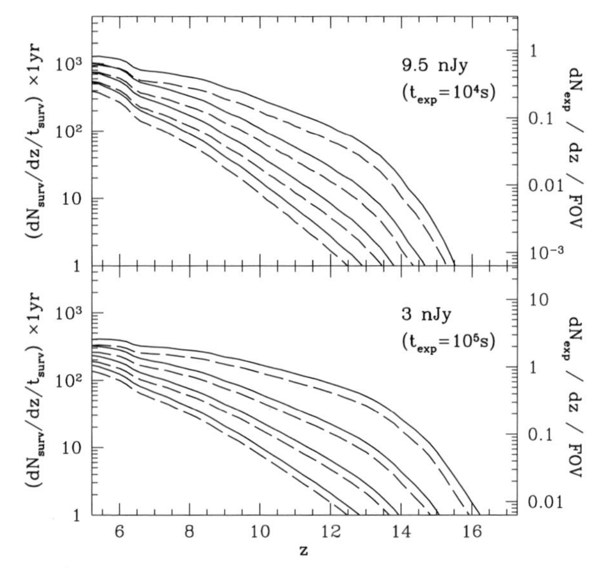

Fig. 15.4 The number of high-redshift SNe detectable with the 4.5 μm *JWST* filter. The curves correspond to the same virial temperature cutoffs as in Fig. 15.1. *Solid curves* assume no dust obscuration; *dashed curves* adopt dust obscuration in the same amount as observed in the low redshift SNe sample. The figure shows results assuming flux density thresholds of 9.5 nJy (or $t_{exp} = 10^4$ s with *JWST*) (*top panel*) and 3 nJy (or $t_{exp} = 10^5$ s with *JWST*) (*bottom panel*). The right vertical axis displays the number of SNe per unit redshift per field; the left vertical axis shows the number of SNe per unit redshift found in $t_{surv}/(2t_{exp})$ such fields (i.e. the differential version of Eq. (15.5) with $t_{surv} = 1$ yr). Reionization should be marked by a transition from the region bounded by the top two solid curves to the region enclosed by the bottom three solid curves (or the analog with the dashed curves if dust is present at the time of reionization). Adapted from (Mesinger et al. 2006)

should be marked by a transition from the region bounded by the top two solid curves to the region enclosed by the bottom three solid curves (or the analog with the dashed curves if dust is present at the time of reionization).

We note that our expected rates are somewhat higher than those in (Dahlén and Fransson 1999), a previous study which included SNe lightcurves and spectra in the analysis. For example, we find 4–24 SNe per field at $z \gtrsim 5$ in the 4.5 μm filter with $t_{exp} = 10^5$ s, compared to ∼0.7 SNe per field at $z \gtrsim 5$ obtained by (Dahlén and Fransson 1999) (after updating their *JWST* specifications to the current version). However, they use SFRs extrapolated from the low-redshift data available at the time, which are not a good fit to recent high-z SFR estimates (Giavalisco et al. 2004, Gabasch et al. 2004, Bunker et al. 2004), and are lower than our $z \gtrsim 5$ SFRs by a factor of 6 – 40.[3] Taking this factor into account, their procedure yields 4–27

[3] The possibility that the SFR in a "Lilly-Madau diagram" remains flat, or even increases, at redshifts $z \gtrsim 5$, owing to star–formation in early, low–mass halos, is also expected theoretically (see, e.g., Fig. 15.1 in (Bromm and Loeb 2002) and associated discussion).

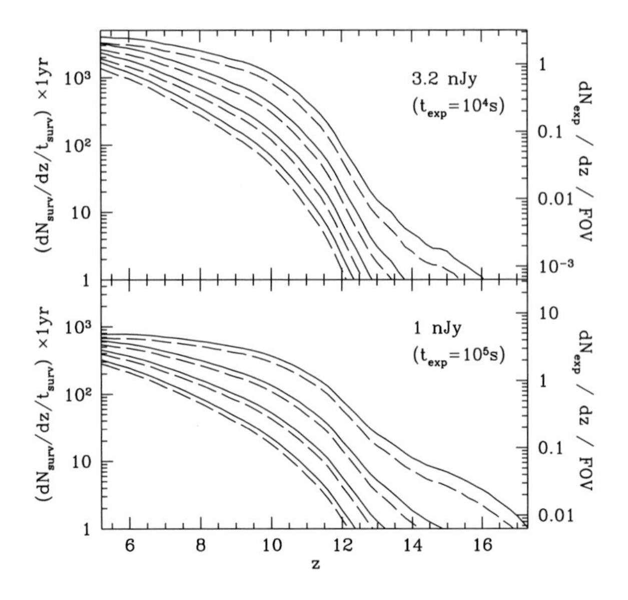

Fig. 15.5 Same as Fig. 15.4, but with the 3.5 μm filter, instead of the 4.5 μm one. Note that we present results for comparable exposure times, hence the sensitivity thresholds are three times lower than in Fig. 15.4, due to the disparate sensitivities of the 3.5 μm and 4.5 μm filters (adapted from (Mesinger et al. 2006))

SNe per field at $z \gtrsim 5$, which is in excellent agreement with our estimate of 4–24 SNe per field at $z \gtrsim 5$.

15.6.3.4 Pop–III SNe

As discussed above, high-redshift SNe whose progenitor stars are formed from metal–free gas within minihalos could be intrinsically very different from the low-redshift SNe, due to differences in the progenitor environments (e.g. very low metallicities; (Abel et al. 2002, Bromm et al. 2002)). If such differences could be identified and detected, then these "pop–III" SNe could provide valuable information about primordial stars and their environments. Indeed, (Wise and Abel 2005) studied the redshift–distribution of such primordial SNe, but only briefly addressed the issue of their detectability. In order to directly assess the number of such SNe among the hypothetical SNe samples we obtained here, in Fig. 15.6, we plot the fraction of SNe whose progenitor stars are located in minihalos. The solid curve assumes our fiducial model with a Salpeter IMF and $\varepsilon_{*\rm minihalo} = 0.1$; the dotted curve assumes that each minihalo produces only a single star, and hence a single SN, over a dynamical time (assuming that strong feedback from this star disrupts any future star formation; as in (Wise and Abel 2005)). The figure shows that with an unevolving star formation efficiency, progenitor stars in minihalos would account for $\gtrsim$ half of the SNe at $z \gtrsim 9$. On the other hand, in the extreme case of a single SNe per minihalo, progenitor stars in minihalos would account for less than $\sim$1% of the $z \sim 10$ SNe (although at the earliest epochs, $z \gtrsim 22$, they would still constitute $\gtrsim$ half of all SNe). Given the overall detection rate of several hundred $z \gtrsim 10$ SNe with a 1 year *JWST* survey we have found above, even in this extreme case, Fig. 15.6 implies that several of these SNe could be caused by pop–III stars.

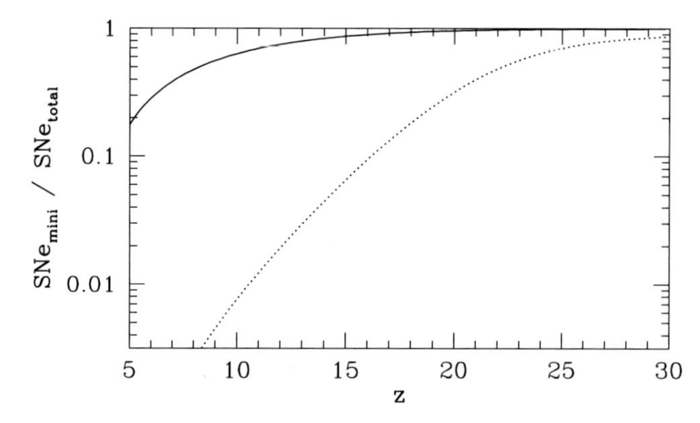

Fig. 15.6 The fraction of SNe whose progenitor stars are located in minihalos (with virial temperatures of $T_{\rm vir} < 10^4$K). These SNe may be pair instability SNe, which would be much brighter than the normal core-collapse SNe used in the computation of SN rates in the previous figures. The solid curve assumes our fiducial model with a Salpeter IMF and $\varepsilon_{*\rm minihalo} = 0.1$; the dotted curve assumes that each minihalo produces only a single star (and therefore a single SN) over a dynamical time (adapted from (Mesinger et al. 2006))

15.6.4 Detecting Features due to Feedback

In summary, we find that 4–24 SNe may be detectable from $z \gtrsim 5$ at the sensitivity of 3 nJy (requiring 10^5 s exposures in a 4.5 μm band) in each ~10 arcmin^2 *JWST* field. In a hypothetical one year survey, we expect to detect up to thousands of SNe per unit redshift at $z \sim 6$. These rates are high, and, if reionization produces a fairly sharp features in the reionization history (with a drop in the SFR by a factor of a few, spread at most over $\Delta z_{\rm re} \sim 1$–3), we have also shown (Mesinger et al. 2006) that the number of SNe is sufficient to detect the feature out to $z \sim 13$, as well as set constraints on the photo-ionization heating feedback on low–mass halos at the reionization epoch. Specifically, for a wide range of scenarios at $z_{\rm re} \lesssim 13$, the drop in the SNR due to reionization can be detected at S/N $\gtrsim 3$ with only tens of deep *JWST* exposures. These results therefore suggest that future searches for high–z SNe could be a valuable new tool, complementing other techniques, to study the process of reionization, as well as the feedback mechanism that regulates it.

15.7 The Dark Ages as a Probe of the Small Scale Power Spectrum

Since, as discussed above, GRBs are a rare tracer of the cosmic SFR, a mere presence of a GRB at, say, $z > 10$ will indicate that non-linear structures exist already at this redshift: the stars that give birth to the GRBs must form out of gas that collected inside dense dark matter potential wells. Structure formation in a cold dark matter (CDM) dominated universe is "bottom–up", with low–mass halos condensing first.

In the current "best–fit" cosmology, with densities in CDM and "dark energy" of $(\Omega_\mathrm{M}, \Omega_\Lambda) \approx (0.3, 0.7)$ that has emerged from *WMAP* and other recent experiments, DM halos with the masses of globular clusters, $10^{5-6} M_\odot$, condense from $\sim 3\sigma$ peaks of the initial primordial density field as early as $z \sim 25$. It is natural to identify these condensations as the sites where the first astrophysical objects, including the first massive stars, were born. As a result, one expects to possibly find GRBs out to this limiting redshift, but not beyond.

With a scale-invariant initial fluctuation power spectrum, the CDM theory has been remarkably successful, and matched many observed properties of large-scale structures in the universe, and of the cosmic microwave background radiation. However, the power spectrum on scales corresponding to masses of $M \lesssim 10^9$ $\mathrm{M}_\odot$ remains poorly tested. Some observations have suggested that the standard model predicts too much power on small scales: it predicts steep cusps at the centers of dark matter halos, whereas the rotation curves of dwarf galaxies suggest a flat core; it also predicts more small satellites than appear to be present in the Local Group (reviewed by, e.g. Sellwood and Kosowsky 2001). Although astrophysical explanations of these observations are possible, several proposals to solve the problem have been put forward that involve the properties of dark matter. These include self-interacting dark matter (Spergel and Steinhardt 2000), adding a repulsive interaction to gravity (Goodman 2000, Peebles 2000), the quantum-mechanical wave properties of ultra-light dark matter particles (Hu et al. 2000), and a resurrection of warm dark matter (WDM) models (Bode et al. 2001).

By design, a common feature of models that attempt to solve the apparent small-scale problems of CDM is the reduction of fluctuation power on small scales. The loss of small-scale power modifies structure formation most severely at the highest redshifts, where the number of self-gravitating objects is reduced. *In each model, there exists a redshift beyond which the number of GRBs (or any other object) is exponentially suppressed; a detection of a GRB beyond this redshift can be used to rule out such models.* As an example, (Barkana et al. 2001) showed that in the case of WDM models, invoking a WDM particle mass of $\sim$1keV (the mass needed to solve the problems mentioned above), it becomes difficult to account for the reionization of the universe by redshift $z \sim 6$, due to the paucity of ionizing sources beyond $z = 6$.

GRBs, if discovered in sufficient numbers at $z > 6$, have the potential to improve this constraints significantly, since increasingly higher redshifts probe increasingly smaller scales. The constraints, in particular, that the detection of distant GRBs would place on structure formation models with reduced small–scale power was quantified in (Mesinger et al. 2005). In this work, we computed the number of GRBs that could be detectable by the *Swift* satellite at high redshifts ($z \gtrsim 6$), assuming that the GRBs trace the cosmic star formation history, which itself traces the formation of non–linear structures. Simple models of the intrinsic luminosity function of the bursts were calibrated to the number and flux distribution of GRBs observed by the *Burst And Transient Source Experiment (BATSE)*. Under these assumptions, the discovery of high–z GRBs would imply strong constraints on models with reduced small-scale power. For example, a single GRB at $z \gtrsim 10$, or, alternatively, 10 GRBs at $z \gtrsim 5$, discovered within a two–year period, would rule out an exponential suppression of the power spectrum on scales below $R_c = 0.09$ Mpc (exemplified by warm dark matter models with a particle mass of $m_x = 2$ keV). Models with a less sharp suppression of small–scale power, such as those with a red tilt or a running scalar index, n_s, are more difficult to constrain, because they are more degenerate with an increase in the power spectrum normalization, σ_8, and with models in which star–formation is allowed in low–mass minihalos. We find that a tilt of $\delta n_s \approx 0.1$ is difficult to detect; however, an observed rate of 1 GRB/yr at $z \gtrsim 12$ would yield an upper limit on the running of the spectral index, $\alpha \equiv dn_s/d\ln k > -0.05$.

Acknowledgments This article draws on joint work with numerous colleagues over the past several years. I would like to thank all my collaborators, but especially Andrei Mesinger, Ben Johnson, Greg Bryan, Mark Dijkstra, and Roban Kramer, whose recent works were particularly emphasized here. The work described in this review was supported by NASA, the NSF, and by the Polányi Program of the Hungarian National Office for Research and Technology (NKTH).

References

Abel, T., Bryan, G. L., & Norman, M. L. 2002, Science, 295, 93
Abel, T., & Haiman, Z. 2000, in *Molecular Hydrogen in Astrophysics*, eds. F. Combes and G. P. des Forets, Cambridge University Press, pp. 237–246
Adelberger, K. L., & Steidel, C. C. 2000, ApJ, 544, 218
Ahn, K., & Shapiro, P. R. 2007, MNRAS, 375, 881
Bahcall, N. A., Ostriker, J. P., Perlmutter, S., & Steinhardt, P. J. 1999, Science, 284, 1481
Barkana, R., Haiman, Z., & Ostriker, J. P. 2001, ApJ 539, 20
Barkana, R., & Loeb, A. 2000a, ApJ, 531, 613
Barkana, R., & Loeb, A. 2000b, ApJ, 539, 20
Barkana, R., & Loeb, A. 2001, Phys. Reports, 349, 125
Barkana, R., & Loeb, A. 2004a, ApJ, 601, 64
Barkana, R., & Loeb, A. 2004b, ApJ, 609, 474
Baron, E., et al. 2000, ApJ, 545, 444
Basu, K., Hernandez-Monteagudo, C., & Sunyaev, R. 2004, A&A, 416, 447
Becker, R. H., et al. 2001, AJ, 122, 2850
Begelman, M., Volonteri, M., & Rees, M. J. 2006, MNRAS, 370, 289
Bennett, C. L., et al. 2003, ApJ, 583, 1
Bloom, J. S., et al. 1999, Nature, 401, 453
Bode, P., Ostriker, J. P., & Turok, N. 2001, ApJ, 556, 903
Bouwens, R. J., Illingworth, G. D., Blakeslee, J. P., Franx, M. 2006, ApJ, 653, 53
Bromm, V., Coppi, P. S., & Larson, R. B. 2002, ApJ, 564, 23
Bromm, V., Ferrara, A., Coppi, P. S., & Larson, R. B. 2001, MNRAS, 328, 969
Bromm, V., Kudritzki, R. P., & Loeb, A. 2001, ApJ, 552, 464
Bromm, V., & Loeb, A. 2002, ApJ, 575, 111
Bromm, V., & Loeb, A. 2003, ApJ, 596, 34
Bryan, G. L., & Norman, M. L. 1998, ApJ, 495, 80
Bunker, A. J., Stanway, E. R., Ellis, R. S., & McMahon, R. G. 2004, MNRAS, 355, 374
Cappellaro, E., Turatto, M., Tsvetkov, D. Y., Bartunov, O. S., Pollas, C., Evans, R., & Hamuy, M. 1997, A&A, 322, 431
Cen, R. 2003, ApJ, 591, 12
Cen, R., & McDonald, P. 2002, ApJ, 570, 457

Cen, R., & Ostriker, J. P. 1992, ApJl, 399, L113
Choudhury, T. R., & Srianand, R. 2002, MNRAS, 336, L27
Ciardi, B., Ferrara, A., & Abel, T. 2000, ApJ, 533, 594
Ciardi, B. & Loeb, A. 2000, ApJ, 540, 687
Dahlén, T., & Fransson, C. 1999, A&A, 350, 349
Dickinson, M., et al. 2004, ApJ, 600, L49
Dijkstra, M., Haiman, Z., Rees, M. J., & Weinberg, D. H. 2004, ApJ, 601, 666
Dijkstra, M., Haiman, Z., & Spaans, M. 2006a, ApJ, 649, 14
Dijkstra, M., Haiman, Z., & Spaans, M. 2006b, ApJ, 649, 37
Dijkstra, M., & Loeb, A 2008, MNRAS, **386**, 492
Doggett, J. B., & Branch, D. 1985, AJ, 90, 2303
Dove, J. B., Shull, J. M., & Ferrara, A. 2000, ApJ, 531, 846
Dunkley, J., et al. 2008, ApJ, submitted, arXiv.org:0803.0586
Efstathiou, G. 1992, MNRAS, 256, 43P
Ellis, R. S. 2008, in *First Light in Universe*, Saas-Fee Advanced Courses, No. 36, Swiss Soc. Astrophys. Astron., p. 259
Elvis, M., Wilkes, B. J., McDowell, J. C., Green, R. F., Bechtold, J., Willner, S. P., Oey, M. S., Polomski, E., & Cutri, R. 1994, ApJS, 95, 1
Fan, X., et al. 2001, AJ, 122, 2833
Fan, X., et al. 2003, AJ, 125, 1649
Fan, X., et al. 2004, AJ, 128, 515
Fan, X., et al. 2006, AJ, 131, 1203
Fan, X. 2006, New Astronomy Reviews, 50, 665
Fan, X., Carilli, C. L.., & Keating, B. 2006, ARA&A, 44, 415
Fardal, M. A., Katz, N., Gardner, J. P., Hernquist, L., Weinberg, D. H., & Davé, R. 2001, ApJ, 562, 605
Ferrara, A. 1998, ApJ, 499, L17
Ferrara, A. 2007, in *Chemodynamics: From First Stars to Local Galaxies*, eds. E. Emsellem, H. Wozniak, G. Massacrier, J.-F. Gonzalez, J. Devriendt & N. Champavert, EAS Publications Series, 24, pp. 229–243
Fruchter, A. S., et al. 1999, ApJ, 519, L13
Fukugita, M., Hogan, C. J., & Peebles, P. J. E. 1998, ApJ, 503, 518
Furlanetto, S. R., Oh, S.-P., & Briggs, F. 2006, Phys Rep, 433, 181
Furlanetto, S., & Loeb, 2005, ApJ, 634, 1
Furlanetto, S. R., Zaldarriaga, M., & Hernquist, L. 2004, ApJ, 613, 1
Gabasch, A., et al. 2004, ApJl, 616, L83
Gao, L., & Theuns, T. 2007, Science, 317, 1527
Garnavich, P., et al. 2003, ApJ, 582, 2003
Garnett, D. R., Kennicutt, R. C., Chu, Y.-H., & Skillman, E. D. 1991, ApJ, 373, 458
Giavalisco, M., et al. 2004, ApJl, 600, L103
Glover, S. C. O., & Brandt, P. W. J. L. 2003, MNRAS, 304, 210
Gnedin, N. Y. 1996, ApJ, 456, 1
Gnedin, N. Y. 2000, ApJ, 542, 535
Gnedin, N. Y., Silk, J., & Spaans, M. 2001, astro-ph/0106110
Goodman, J. 2000, New Astronomy, 5, 103
Greif, T. H., & Bromm, V. 2006, MNRAS, 373, 128
Haiman, Z. 2004, in *Carnegie Observatories Astrophysics Series, Vol. 1: Coevolution of Black Holes and Galaxies*, ed. L. C. Ho, Cambridge; Cambridge University Press, pp. 67–87
Haiman, Z., Abel, T., & Rees, M. J. 2000, ApJ, 534, 11
Haiman, Z., & Bryan, G. L. 2006, ApJ, 650, 7
Haiman, Z., & Holder, G. P. 2003, ApJ, 595, 1
Haiman, Z., & Knox, L. 1999, in *Microwave Foregrounds*, ASP Conference Series #181, eds. A. de Oliveira-Costa and M. Tegmark, p. 227

Haiman, Z., & Loeb, A. 1997, ApJ, 483, 21
Haiman, Z., & Loeb, A. 1998, ApJ, 503, 505
Haiman, Z., & Loeb, A. 1998a, in *Science with the NGST*, eds. E. P. Smith & A. Koratkar. San Francisco, ASP Conference Series, 133, p. 251
Haiman, Z., & Loeb, A. 1999, ApJ, 519, 479
Haiman, Z., Rees, M. J., & Loeb, A. 1997, ApJ, 476, 458
Haiman, Z., Spaans, M., & Quataert, E. 2000, ApJL, 537, L5
Haiman, Z., Thoul, A. A., & Loeb, A. 1996, ApJ, 464, 523
Hamuy, M., et al. 2001, ApJ, 558, 615
Heger, A., Fryer, C. L., Woosley, S. E., Langer, N., & Hartmann, D. H. 2003, ApJ, 591, 288
Heger, A., & Woosley, S. E. 2002, ApJ, 567, 532
Hernández-Monteagudo, C., Haiman, Z., Jimenez, R., & Verde, L. 2007, ApJ, 660, L85
Hernández-Monteagudo, C., Haiman, Z., Verde, L. & Jimenez, R. 2008, ApJ, **672**, 33
Hjörth, J., et al. 2003, Nature, 423, 847
Hu, W., Barkana, R., & Gruzinov, A. 2000, PRL, 85, 1158
Iliev, I. T., et al. 2006, MNRAS, 371, 1057
Iliev, I. T., Mellema, G., Shapiro, P. R., Pen, U.-L. 2007, MNRAS, 376, 534
Iliev, I. T., Pen, U.-L., Bond, J. R., Mellema, G., & Shapiro, P. R. 2007, ApJ, 660, 933
Johnson, J. L., Greif, T. H., & Bromm, V. 2007, ApJ, 665, 85
Kaplinghat, M., Chu, M., Haiman, Z., Holder, G. P., Knox, L., & Skordis, C. 2003, ApJ, 2003, 583, 25
Kashikawa, N., et al. 2006, ApJ, 648, 7
Kawai, N., et al. 2006, Nature, 440, 184
Kitayama, T., & Ikeuchi, S. 2000, ApJ, 529, 615
Komatsu, E., et al. 2008, ApJS, submitted, arXiv.org:0803.0547
Kuhlen, M., & Madau, P. 2005, MNRAS, 363, 1069
Kramer, R., Haiman. Z., & Oh, S. P. 2006, ApJ, 649, 570
Kramer, R., & Haiman. Z., 2008, MNRAS, **385**, 1561
Lamb, D., & Haiman, Z. 2003, in the proceedings of the 3rd Rome Workshop on Gamma-Ray Bursts in the Afterglow Era, held in Rome, September 2002, eds. L. Piro, F. Frontera, N. Masetti & M. Feroci, ASP Conf. Series, astro-ph/0312502
Lamb, D. Q., & Reichart, D. E. 2000, ApJ, 536, L1
Leibundgut, B., & Suntzeff, N. B. 2003, LNP Vol. 598: Supernovae and Gamma-Ray Bursters, 598, 77
Lidz, A., Oh, S. P., & Furlanetto, S. R. 2006, ApJL, 639, 47
Lilly, S. J., Le Fevre, O., Hammer, F., & Crampton, D. 1996, ApJl, 460, L1
Lodato, G., & Natarajan, P. 2006, MNRAS, 371, 1813
Machacek, M. E., Bryan, G. L., & Abel, T. 2001, ApJ, 548, 509
Machacek, M. E., Bryan, G. L., & Abel, T. 2003, MNRAS, 338, 273
MacLow, M.-M. & Ferrara, A. 1999, ApJ, 513, 142
Madau, P., Ferguson, H. C., Dickinson, M. E., Giavalisco, M., Steidel, C. C., & Fruchter, A. 1996, MNRAS, 283, 1388
Madau, P., Rees, M. J., Volonteri, M., Haardt, F. & Oh, S. P. 2004, ApJ, 604, 484
Malhotra, S., & Rhoads, J. E. 2004, ApJ, 617, L5
Malhotra, S., et al. 2005, ApJ, 626, 666
Matsuda, Y., Yamada, T., Hayashino, T., Tamura, H., Yamauchi, R., Ajiki, M., Fujita, S. S., Murayama, T., Nagao, T., Ohta, K., Okamura, S., Ouchi, M., Shimasaku, K., Shioya, Y., & Taniguchi, Y. 2004, AJ, 128, 569
McQuinn, M., Furlanetto, S. R.; Hernquist, L., Zahn, O., & Zaldarriaga, M. 2005, ApJ, 630, 643
Mesinger, A., Bryan, G. L., & Haiman, Z. 2006, ApJ, 648, 835
Mesinger, A., & Haiman, Z. 2004, ApJL, 611, L69
Mesinger, A., & Haiman, Z. 2007, ApJ, 660, 923
Mesinger, A., Johnson, B., & Haiman, Z. 2006, ApJ, 637, 80

Mesinger, A., Perna, R., & Haiman, Z. 2005, ApJ, 623, 1
Miralda-Escudé, J. 1998, ApJ, 501, 15
Miralda-Escudé, J., & Rees, M. J. 1997, ApJ, 478, L57
Mortonson, M. J., & Hu, W. 2007, ApJ, submitted, arxiv preprint astro-ph:0705.1132
Nemiroff, R. J. 2003, AJ, 125, 2740
Oh, S. P. 2001, ApJ, 553, 499
Oh, S. P., & Haiman, Z. 2002, ApJ, 569, 558
Oh, S. P., & Haiman, Z. 2003, MNRAS, 346, 456
Oh, S. P., Haiman, Z., & Rees, M. J. 2001, ApJ, 553, 73
Omukai, K., & Nishi, R. 1999, ApJ, 518, 64
O'Shea, B. W., & Norman, M. L. 2006, ApJ, 648, 31
O'Shea, B. W., & Norman, M. L. 2008, ApJ, **673**, 14
Page, L., et al. 2007, ApJS, 170, 335
Patat, F., Barbon, R., Cappellaro, E., & Turatto, M. 1994, A&A, 282, 731
Peebles, P. J. E. 2000, ApJ, 534, 127
Press, W. H., & Schechter, P. L. 1974, ApJ, 181, 425
Perlmutter, S., et al. 1999, ApJ, 517, 565
Riess, A. G., et al. 1998, AJ, 116, 1009
Richardson, D., Branch, D., Casebeer, D., Millard, J., Thomas, R. C., & Baron, E. 2002, AJ, 123, 745
Ricotti, M., Gnedin, N. Y., & Shull, J. M. 2001, ApJ, 560, 580
Ricotti, M., Gnedin, N. Y., & Shull, J. M. 2002a, ApJ, 575, 33
Ricotti, M., Gnedin, N. Y., & Shull, J. M. 2002b, ApJ, 575, 49
Ricotti, M., Ostriker, J. P., & Gnedin, N. Y. 2005, MNRAS, 357, 207
Salvaterra, R., Ciardi, B., Ferrara, A., & Baccigalupi, C. 2005, MNRAS, 360, 1063
Santos, M. G., Cooray, A., Haiman, Z., Knox, L., & Ma, C.-P. 2003, ApJ, 598, 756
Santos, M. G., Amblard, A., Pritchard, J., Trac, H., Cen, R., & Cooray, A. 2007, ApJ, submitted, arxiv preprint astro-ph:0708.2424
Saslaw, W. C., & Zipoy, D. 1967, Nature, 216, 976
Schaerer, D. 2002, A&A, 382, 28
Sellwood, J., & Kosowsky, A. 2001, in *Gas & Galaxy Evolution*, eds. J. E. Hibbard, M. P. Rupen, & J. van Gorkom, astro-ph/0009074
Shapiro, P. R., Giroux, M. L., & Babul, A. 1994, ApJ, 427, 25
Shapiro, P. R., Iliev, I. T., & Raga, A. C. 2004, MNRAS, 348, 753
Shapiro, P. R., & Kang, H. 1987, ApJ, 318, 32
Shapley, A. E., Steidel, C. C., Pettini, M., Adelberger, K. L., & Erb, D. K. 2006, ApJ, 651, 688
Sheth, R. K., Mo, H. J., & Tormen, G. 2001, MNRAS, 323, 1
Silk, J., & Spaans, M. 1997, ApJ, 488, L79
Slosar, A., McDonald, P., & Seljak, U. 2007, New Ast Rev, 51, 327
Sokasian, A., Yoshida, N., Abel, T., Hernquist, L., & Springel, V. 2004, MNRAS, 350, 47
Somerville, R. S., Primack, J. R., & Faber, S. M. 2001, MNRAS, 320, 504
Spergel, D. N., et al. 2003, ApJS, 148, 175
Spergel, D. N., et al. 2007, ApJS, 170, 377
Spergel, D. N., & Steinhardt, P. J. 2000, PRL, 84, 3760
Springel, V., et al. 2005, Nature, 435, 629
Stanek, K. Z., et al. 2003, ApJ, 591, L17
Stanway, E. R., et al. 2004, ApJ, 607, 704
Stanway, E. R., et al. 2007, MNRAS, 376, 727
Steidel, C. C., Pettini, M., & Adelberger, K. L. 2001, 546, 665
Strolger, L., et al. 2004, ApJ, 613, 200
Susa, H., Uehara, H., Nishi, R., & Yamada, M. 1998, Prog Theor Phys 100, 63
Taniguchi, Y., et al. 2005, PASJ, 57, 165
Tasitsiomi, A. 2006, ApJ, 645, 792

Tegmark, M., Silk, J., Rees, M. J., Abel, T., & Blanchard, A., 1997, ApJ, 474, L1
Tegmark, M., et al. 2004, Phys Rev D, id. 693501
Thomas, R. M. & Zaroubi, S. 2008, MNRAS, **384**, 1080
Thoul, A. A., & Weinberg, D. H. 1996, ApJ, 465,608
Totani, T. 1997, ApJ, 486, L71
Totani, T. 1999, ApJ, 511, 41
Tumlinson J., Shull J. M. 2000, ApJ, 528, L65
Tumlinson, J., Giroux, M. L., & Shull, M. J. 2001, ApJ, 550, L1
Viel, M., Becker, G. D., Bolton, J. S., Haehnelt, M. G., Rauch, M., & Sargent, W. L. W. 2007, Phys Rev Lett., **100**, 1304
Volonteri, M., & Rees, M. J. 2005, ApJ, 633, 624
Walter, F., et al. 2003, Nature, 424, 406
Walter, F., Carilli, C., Bertoldi, F., Menten, K., Cox, P., Lo, K. Y., Fan, X., & Strauss, M. A. 2004 ApJ, 615, L17
Weymann, R. J., Stern, D., Bunker, A., Spinrad, H., Chaffee, F. H., Thompson, R. I., & Storrie-Lombardi, L. J. 1998, ApJ, 505, L95
Whalen, D., Abel, T., & Norman, M.L. 2004, ApJ, 610, 14
Wijers, R. A. M. J. et al. 1998, MNRAS, 294, L13
Wise, J. H., & Abel, T. 2005, ApJ, 629, 615
Wise, J. H., & Abel, T. 2007, ApJ, 671, 1559
Wood, K. & Loeb, A. 1999, ApJ, 545, 86
Wyithe, J. S. B., & Cen, R. 2007, ApJ, 659, 890
Wyithe, J. S. B., & Loeb, A. 2003, ApJ, 588, L69
Wyithe, J. S. B., & Loeb A. 2006, Nature, 441, 322
Yang, Y., Zabludoff, A. I., Davé, R., Eisenstein, D. J., Pinto, P. A., Katz, N., Weinberg, D. H., & Barton, E. J. 2006, ApJ, 640, 539
Yoshida, N., Abel, T., Hernquist, L., & Sugiyama, N. 2003, ApJ, 592, 645
Yoshida, N., Omukai, K., Hernquist, L., & Abel, T. 2006, ApJ, 652, 1
Yoshida, N., Omukai, K., & Hernquist, L. 2008, Science, 321, 669
Yoshida, N., Sokasian, A., Hernquist, L., & Springel, V. 2004, ApJ, 591, 1
Zaroubi, S. & Silk, J. 2005, MNRAS, 360, L64

David Jewitt, Massimo Stiavelli, and a balloon

Chapter 16
Baryons: What, When and Where?

Jason X. Prochaska and Jason Tumlinson

Abstract We review the current state of empirical knowledge of the total budget of baryonic matter in the Universe as observed since the epoch of reionization. Our summary examines on three milestone redshifts since the reionization of H in the IGM, z = 3, 1, and 0, with emphasis on the endpoints. We review the observational techniques used to discover and characterize the phases of baryons. In the spirit of the meeting, the level is aimed at a diverse and non-expert audience and additional attention is given to describe how space missions expected to launch within the next decade will impact this scientific field.

16.1 Introduction

Although baryons are believed to be a minor constituent of the mass-energy budget of our universe, they have played a dominant role in astronomy because they are the only component that interacts directly and frequently with light. Indeed, much of modern astrophysics focuses on the production and destruction of heavenly bodies comprised of baryons. Current observations of baryons extend from our solar system to the very early universe ($z > 7$), i.e. spanning over 95% of the lifetime of our known universe. In terms of cosmology, the principal areas of baryonic research include measuring their total mass density, identifying the various elements and phases that comprise them, and resolving their distribution throughout the universe. In turn, astronomers are increasingly interested in probing the baryonic processes that feed galaxy formation and the feedback processes of galaxies that transform and return baryons to the intergalactic medium (IGM). Our mandate from the organizers was to review our knowledge of baryons to a broad astronomical audience from the epoch of reionization ($z \approx 6$–10) to the present day. Granted limited time, we focus on our empirical knowledge and limit the discussion of theoretical inquiry.

J.X. Prochaska (✉)
University of California Observatories - Lick Observatory, University of California, Santa Cruz, CA 95064, USA
e-mail: xavier@ucolick.org

H.A. Thronson et al. (eds.), *Astrophysics in the Next Decade,* Astrophysics and Space Science Proceedings, DOI 10.1007/978-1-4020-9457-6_16,

In the past two decades, advances in telescopes and instrumentation have significantly advanced our understanding of the distribution of baryonic matter in the cosmos. Modern surveys of the local universe yield an increasingly complete census of galaxies and the large-scale structure in which they reside, the distribution of galaxy clusters and mass estimates of the hot gas within them, and a view of the diffuse gas that lies between galaxies. In this proceeding, we review this work, placing particular emphasis on techniques related to observing diffuse baryonic phases. Presently these techniques are most efficiently pursued at $z \sim 3$ and $z \sim 0$–0.5, and observational constraints on the properties and distribution of baryons are most precise at these epochs. It is somewhat embarrassing that there lies a nearly 10 Gyr gap in our knowledge of the majority of baryons between these two epochs. Future missions need to address this hole in addition to the primary uncertainties remaining for the $z \sim 0$ and $z \sim 3$ epochs.

This proceeding is organized as follows: In Section 16.2, we provide a basic introduction to the observational techniques used to trace baryons throughout the universe. The current best estimates for the total baryonic matter density is reviewed in Section 16.3. We review the observational constraints on the main phases of baryons at $z = 3$ in Section 16.4, briefly comment on our general absence of knowledge at $z = 1$ (Section 16.5), and then review the $z \sim 0$ missing baryons problem in Section 16.6. We finish with a brief review of planned, proposed, and desired mission and telescopes that would significantly impact this science.

16.2 A Primer on Tracing Baryons

A complete baryon census poses a diverse set of observational challenges to astronomy, where we most often study baryons by observing the radiation that they emit, absorb or reflect. This includes stars, HII regions, cluster gas, planets, debris disks, etc. With these observations, one can assess the luminosity, temperature, chemical composition, and size of the object under study. In terms of performing a complete census and analysis of baryons in the present and past universe, however, this approach is currently limited by several factors. First, there is the simple fact that more distant objects are fainter. Although recent advances in telescopes, instrumentation, and search strategies now provide samples of galaxies at high redshift, these samples are sparse and (at $z > 1$) are comprised of only the brightest objects. There is no *a priori* reason why the budget of baryons would be dominated by the bright luminous objects, so these samples do not necessarily account for a large fraction of the budget. Second, it is generally difficult to estimate precisely the mass of an emitting object. With stars and galaxies, for example, mass estimates generally rely on stellar evolution modeling which is subject to many uncertainties (e.g. the initial mass function) that are difficult to resolve even in the local universe. Third, it turns out that only a small fraction of the universe's baryons are in the collapsed, luminous objects that are most easily detected. This point is especially true in the young universe but it even holds in our modern universe. And finally, although diffuse gas emits line and continuum radiation, it is very difficult to detect this radiation even from gas in the nearby universe.

It is the last reason, in particular, that has driven observers to absorption-line techniques for characterizing the mass density, temperature, and distribution of the bulk of the baryons in the universe. The absorption-line experiment is standard observational astronomy in reverse: one observes a bright, background source (e.g. a quasar or gamma-ray burst) with the aim to study the absence of light due to absorption by intervening baryons. These basic analytic techniques were developed primarily to study the interstellar medium (ISM) of our Galaxy (e.g. Spitzer (1978)) via UV spectroscopy of bright O and B stars. Baryons with at least one electron will exhibit both resonant line absorption (e.g. Lyα) and continuum opacity (e.g. the Lyman limit feature), primarily at UV and X-ray frequencies. The principal observable of an absorption line is its equivalent width W_λ, the fraction of light over a spectral interval that is absorbed by the gas. The principal physical parameter is the column density N, the number of atoms per unit area[1] along the sightline. This is the number density equivalence of a surface density. For weak absorption, the column density scales linearly with the equivalent width. For stronger (i.e. saturated) lines, N depends weakly on W_λ and may be difficult to estimate. To assess these issues, it is highly desirable to obtain data at spectral resolution and to measure directly the optical depth profile of an absorption feature. In this manner, one can integrate the column density provided accurate knowledge of the atomic data of the transition.

Table 16.1 presents the principal absorption line features used to study H, He, and metals in diverse phases. Because most of these spectral features have rest wavelengths at ultraviolet or higher energies, one must utilize UV and X-ray telescopes to probe the past ≈7 Gyr of the universe (i.e. $z < 1$). These satellite missions have limited aperture and (often) low-throughput instrumentation which requires very bright background sources. Most of the gas, at least at $z > 1$, is relatively cool ($T \sim 10^4$ K) and the observed absorption lines have widths less than ≈40 km s^{-1}. To spectrally resolve such features, one is driven toward echelle spectrometers and generally the largest telescopes on Earth (or in space). In these respects, the $z \sim 0$ absorption-line experiment is much more challenging than studies of the high z universe, where several of the key features are redshifted into optical pass-bands and accessible to large ground-based telescopes. This has led, in a few ways, to a greater understanding of baryons in the $z \sim 3$ universe than our recent past.

A tremendous advantage of the absorption-line experiment is that it achieves extremely sensitive limits for studying diffuse gas. For example, the signal-to-noise and resolution easily afforded by current telescopes and instrumentation allows detection of the HI Lyα transition to column densities $N_{\rm HI} \sim 10^{12}$ cm^{-2}. This is ten *orders* of magnitude lower than the surface density of a typical molecular cloud in a star-forming galaxy. With coverage of the full Lyman series, one can measure the surface density of atomic hydrogen across these ten orders of magnitude and probe the densest material to very diffuse environments. By the same token, one can study cold H_2 ($T < 100$ K) molecules through Lyman/Werner absorption bands while also probing diffuse, hot ($T > 10^6$ K) gas via metal-line transitions of O, N, and Ne.

[1] Although we generally view stars and quasars as point sources, they have finite area. One may envision the concept of column density like a core sample of the Earth, where spectra resolve the location of gas in velocity not depth.

Table 16.1 Key baryonic diagnostic lines and features

Line	Phase	T (K)	λ_{rest} (Å)	$\lambda_{z=1}$ (Å)	$\lambda_{z=3}$ (Å)	$\lambda_{z=9}$ (μm)
Lyman-Werner	Molecular gas	10–100	~1000	2000	4000	1
21 cm	Atomic gas	100–1000	21cm	0.7 Ghz	0.4 Ghz	140 MHz
Lyα	Atomic+Ionized gas	100–40000	1216	2400	4800	1.2
Hα	Ionized gas	10000–40000	6560	13000	26000	65000
Lyman limit	Ionized gas	10000–40000	912	1800	3600	0.9
HeII	Ionized gas	10000–40000	304	450	912	0.2
CIV	Ionized Gas	20000–40000	1550	3000	6000	1.5
OVI	Warm/Hot Gas	20000–10^6	1030	2000	4000	1
OVII,OVIII	Hot Gas	10^6–10^8	21.6,18.9	40	8	200
NeVIII	Hot Gas	10^7	775	1550	3100	7750

One thereby traces structures in the universe with a wide dynamic range of sizes: molecular clouds with diameters less than 100 pc in the centers of galaxies to 'voids' spanning tens of Mpc.

There are several disadvantages, however, to absorption-line studies: (1) the majority of key diagnostics have rest-frame UV or X-ray frequencies. At low z, one requires a space mission with large collecting area and efficient spectrometers to perform the experiment. This is expensive and often technologically infeasible; (2) One requires a bright, background UV/X-ray source[2] subjecting the final analysis to biases in the discovery and selection of these sources; (3) Bright UV/X-ray sources are rare and sparsely distributed across the sky. Therefore, multiplexing has limited value and observational surveys are relatively expensive. Furthermore, with only a limited number of sightlines to probe the universe, the densest (i.e. smallest) structure are at best sparsely sampled; (4) The technique is limited to the most distant luminous objects detected, currently $z \approx 6$; (5) dust intrinsic to the gas under study may extinguish the background source and preclude the analysis altogether.

Several key aspects of absorption-line research are illustrated in Fig. 16.1 which presents the echelle spectrum of quasar J133941.95+054822.1 acquired with the Magellan/MIKE spectrometer (Bernstein et al. 2003). This optical spectrum shows the rest-frame UV emission of the $z_{QSO} = 2.980$ quasar. One notes the broad Lyα+NV emission line superimposed on a underlying, approximately power-law spectrum. All of the absorption features, meanwhile, are associated with gas foreground to the quasar. The most prominent is the continuous opacity at $\lambda < 3600$Å which marks the Lyman limit of a gas 'cloud' at $z_{LLS} = 2.97$. This Lyman limit system (LLS) must have an HI column density $N_{\rm HI} > 10^{17}$ cm^{-2} to exhibit a Lyman limit opacity $\tau_{LL} > 1$. The thicket of narrow absorption lines at $\lambda = 3600$ to 4850Å is dominated by Lyα absorption from gas with $z < z_{QSO}$. These lines are termed the Lyα forest and they trace the intergalactic medium (IGM) of the high z universe. Their HI column densities are generally less than 10^{15} atoms per cm^2. Within the Lyα forest one also notes a very broad absorption line at $\lambda \approx 4360$Å whose observed

[2] Traditionally, researchers have used quasars although increasingly transient gamma-ray burst afterglows are analyzed (e.g. Chen et al. 2005, Vreeswijk et al. 2004).

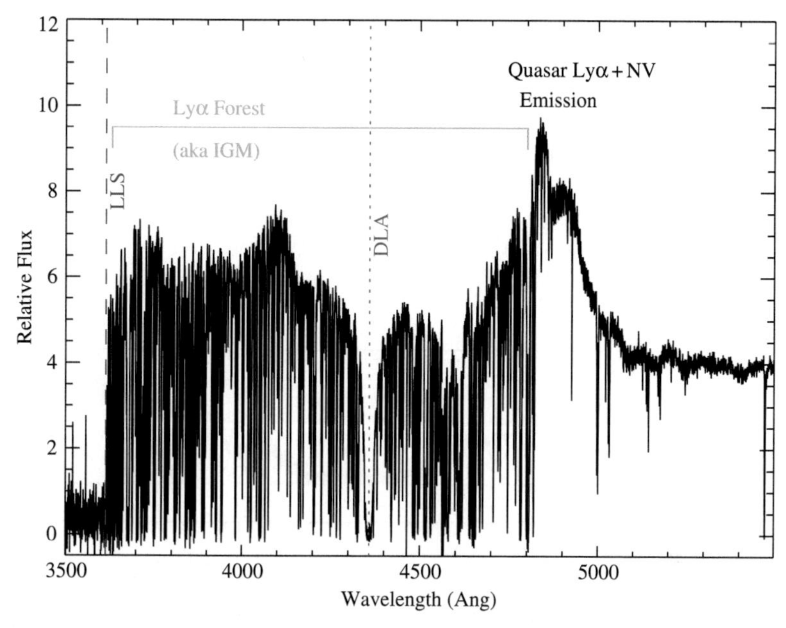

Fig. 16.1 Magellan/MIKE echelle spectrum of the $z_{QSO} = 2.980$ quasar J133941.95+054822.1. The data reveal the broad Lyα+NV emission line of the quasar imposed on its otherwise nearly power-law spectrum. Imprinted on this spectrum is a plethora of absorption features due to gas foreground to the quasar. In particular, one identifies continuum opacity at $\lambda < 3600$Å due to a Lyman limit system at $z_{LLS} = 2.97$ and a thick of absorption features related to the Lyα forest. We also mark the damped Lyα profile at $\lambda \approx 4360$Å of a foreground galaxy ($z_{DLA} = 2.59$)

equivalent width is $W_\lambda \approx 50$Å. This 'redwood' of the Lyα forest is a damped Lyα system (DLA); its very large HI column density, $N_{\rm HI} > 2 \times 10^{20}\,{\rm cm}^{-2}$, allow the Lorentzian damping wings to be resolved in the Lyα transition. This feature marks the presence of a foreground galaxy and has a larger HI column density than total of all other lines in the Lyα forest. Finally, one observes a more sparse set of absorption features at $\lambda > 4900$Å which are metal-line transitions from ions and atoms of O^0, Si^+, C^+, C^{+3}, Fe^+, etc. These lines enable studies of the metal abundance of our universe (Prochaska et al. 2003, Schaye et al. 2003) and also probe gas phases that are difficult to examine with Lyα alone.

16.3 The Baryonic Mass Density

Our first problem is to assess the total budget of baryonic matter in the Universe, preferably by independent measurement. The baryonic mass density ρ_b, generally defined relative to the critical density $\rho_c = 9.20 \times 10^{-30} h_{70}^2\,{\rm g\,cm}^{-3}$, $\Omega_b \equiv \rho_b/\rho_c$ and h_{70}, the Hubble constant in units of $70\,{\rm km\,s}^{-1}\,{\rm Mpc}^{-1}$, are fundamental cosmological parameters. The first successful method for estimating Ω_b is to compare measurements of isotopes for the light elements (H, D, ^{3}He, ^{4}He, Li) against predictions from Big Bang Nucleosynthesis (BBN) theory. The relative abundances of these isotopes are sensitive to the entropy of the universe during BBN, i.e. the ratio of photons to baryons. At first, observers focused on He even though Li and D are much more sensitive 'baryometers'. This was because He/H was inferred from nearby emission-line galaxies where one could obtain exquisite S/N observations. It is now realized, however, that the precision of the inferred He/H ratio is limited by systematic uncertainties in the modeling of HII regions and the precision of the atomic data (Peimbert et al. 2007).

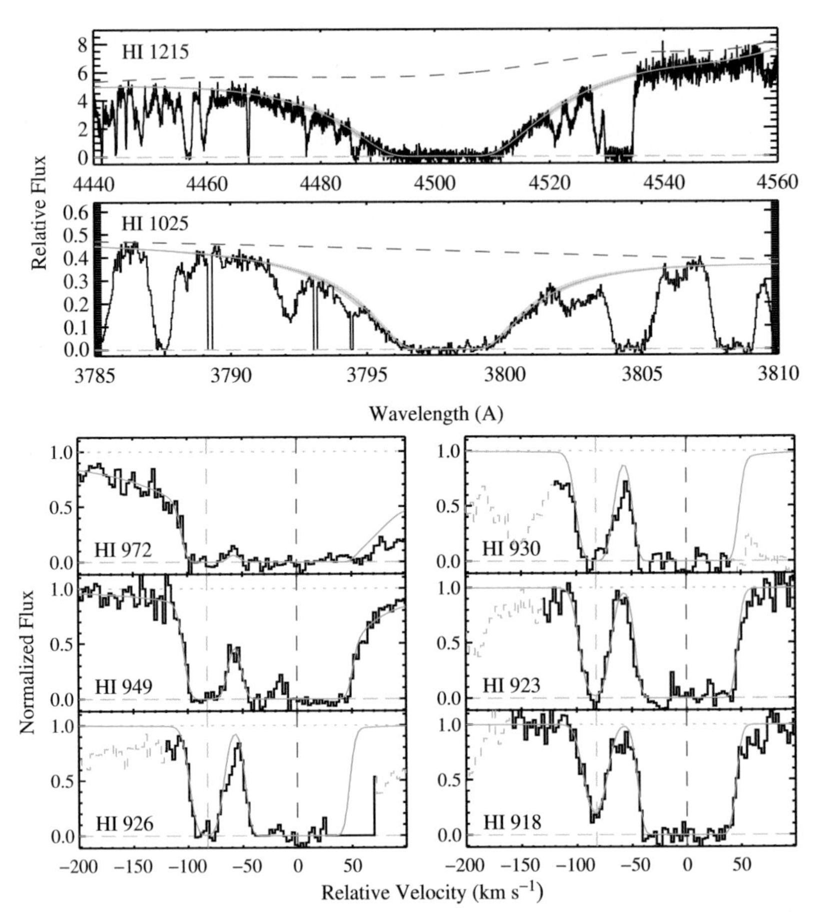

Fig. 16.2 H I and D I Lyman series absorption in the $z = 2.70262$ DLA towards SDSS1558-0031 (O'Meara et al. 2006). For the Lyα and Lyβ transitions, the data is unnormalized and the dashed line traces the estimate for the local continuum level. The remaining Lyman series transitions are shown continuum normalized. The solid green line shows the best single-component fit to the D I and H I absorption. The estimate of the HI column density comes from analysis of the damping wings present in the Lyman α–δ lines whereas constraints on the DI column density come from the unsaturated D I Lyman-11 transition ($\sim$918 Å rest wavelength)

The tightest constraints on Ω_b using BBN theory come from direct measurements of the D/H ratio. Ideally this ratio is measured in a metal-poor gas to minimize the likelihood that D has been astrated by stars. The first detections of D, however, were from absorption-line studies in the metal-rich Galactic ISM (Rogerson and York 1973). These observations establish a lower limit to the primordial D/H value of $\approx 1 \times 10^{-5}$ (e.g. Moos et al. (2002)). In the past decade, echelle spectrometers on 10m-class telescopes have extended the experiment to high redshift (Burles and Tytler 1998). Researchers have searched for the very rare 'clouds' which are metal-poor and therefore minimally astrated, have large HI column density, and are sufficiently quiescent that the higher order lines DI and HI Lyman series are resolved (the velocity separation is 82 km s^{-1}). Figure 16.2 shows a recent example (O'Meara et al. 2006). The first few transitions of the HI Lyman series are broadened by the damping wings of their 'natural' Lorentzian profiles and a fit to the profile establishes $N_{\mathrm{HI}} = 10^{20.67\pm0.05}$ cm^{-2}. The DI transitions are lost within the cores of these strong absorption lines but higher order lines show resolved and unsaturated DI absorption. Analysis of these lines provide a precise estimation of the DI column density.

The D/H ratio is estimated from the $N(\mathrm{DI})/N(\mathrm{HI})$ ratio under the reasonable assumption that these isotopes have identical ion configurations[3] and differential depletion. Allowing for these assumptions, the estimation of D/H is direct: abundances measurements from absorption-line studies are simply a matter of counting atoms. Unlike most other methods to measure abundances, the results are essentially independent of the local physical conditions in the gas. Figure 16.3 presents the 'gold-standard' set of D/H measurements from quasar absorption line surveys of high z, metal-poor absorbers. The observations are scattered about log D/H $= -4.55$ with a boot-strap dispersion of 0.04 dex. Adopting standard BBN theory, this constrains $\eta \equiv n_b/n_\gamma = 5.8 \pm 0.7 \times 10^{-10}$ where n_b and n_γ are the number densities

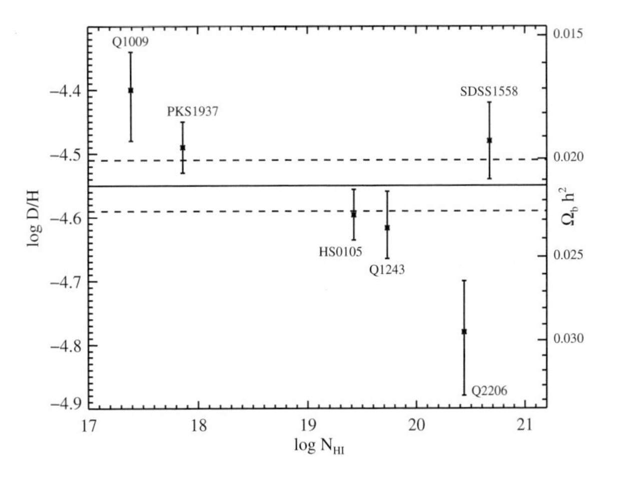

Fig. 16.3 Values of the D/H ratio vs. N_{HI}(O'Meara et al. 2006). The horizontal lines represent the weighted mean and jackknife errors of the 6 measurements, $\log \mathrm{D/H} = -4.55 \pm 0.04$. The right hand axis shows how the values of D/H translate into values for $\Omega_b h^2$ using BBN

[3] The gas is self-shielding and likely has a high neutral fraction.

of baryons and photons respectively. Direct measurements of the cosmic microwave background give n_γ, and one derives a baryonic matter density relative to the critical density ρ_c, $\Omega_b = \rho_b/\rho_c = (0.043 \pm 0.003)h_{70}^{-2}$. It is a triumph of the BBN theory and observational effort that this value has been confirmed by observations of the CMB power spectrum (Dunkley et al. 2008, Netterfield et al. 2002).

Nevertheless, a few outstanding issues remain. The dispersion in D/H at high z exceeds[4] that predicted from the reported errors on the individual measurements (Kirkman et al. 2003). Although this most likely reflects over-optimistic error estimates by the observers, it is worth further exploring models of inhomogeneous BBN (Jedamzik 2004) and systematic error associated with the astrophysics of H and D (Draine 2006). The other outstanding issue concerns an inconsistency between the BBN-predicted Li abundance based on the η value from D/H and observation (Boesgaard et al. 2005, Spite et al. 1984). It is now generally expected that this observed Li underabundance will be explained by non-primordial stellar astrophysics (e.g. Korn et al. (2006)), but a comprehensive model of these processes has not yet been developed.

16.4 Baryons at $z \sim 3$

This section reviews constraints on the baryon census during the few Gyr following reionization, starting with the densest phases and proceeding to the least dense. Perhaps not coincidentally (for hierarchical cosmology), this organization also proceeds roughly from the lowest cosmological mass density to the largest. Along the way, we highlight the key uncertainties in these values and comment on the impact of future missions.

16.4.1 The Stellar Component

The majority (but not all) of galaxy samples identified at $z > 2$ are discovered through spectral or photometric signatures associated with ongoing star-formation. These include the Lyman break galaxies (see Shapley's presentation), the Lyα emitters (e.g. Gawiser et al. 2007), sub-mm galaxies (Chapman et al. 2005), and gamma-ray burst host galaxies (e.g. Le Floc'h et al. 2003). These populations dominate star formation activity during the young universe. It is likely that they also dominate the stellar mass density in this epoch because 'red and dead' galaxies are very rare at this time (e.g. Kriek et al. 2006). It is difficult, however, to estimate the stellar mass density of these star-forming galaxies. This is especially true when one is limited to optical and near-IR photometry which corresponds to rest-frame UV and blue light,

[4] Interestingly, the same is true of the Galactic ISM measurements (Jenkins et al. 1999). This result is unlikely to be due to observational error and the nature of the dispersion remains an open question (Linsky et al. 2006, Prochaska et al. 2005).

i.e. photons from massive star formation that provide little, if any, information on the low-mass end of the IMF.

To infer the stellar mass, one must adopt a mass to light ratio (M_*/L). Standard practice is to model the stellar population that gives rise to the photometry and spectroscopy of the galaxy. This modeling is sensitive to the presumed star-formation history, reddening by dust intrinsic to the galaxy, and uncertainties in the initial mass function. Altogether, these lead to large systematic uncertainties in the stellar mass estimate. These issues aside, current surveys do not probe the fainter galaxies of what appears to be a very steep luminosity function (e.g. Reddy et al. 2007). The contribution by Dickinson provides a deeper discussion of these issues. Current estimates of the stellar mass density at $z = 3$ range from $\Omega_* = 2$ to $4 \times 10^{-7}\ h_{70} M_\odot\ \mathrm{Mpc}^{-3}$ with a systematic uncertainty of $\approx 50\%$ (Fontana et al. 2006, Rudnick et al. 2006). This corresponds to $0.005 \pm 0.002 \Omega_b h_{70}$. The launch of new satellites with greater sensitivity in near and mid-IR pass-bands (Herschel, JWST) will reduce the uncertainties by observing the stellar component at redder wavelengths and also by extending surveys to much fainter levels. Even with the large current uncertainty, however, it is evident that stars make a nearly negligible contribution to the total baryon budget at these redshifts.

16.4.2 Molecular Gas

In the local universe, star formation is associated with molecular gas in the form of molecular clouds. Although the link between molecular gas and star formation has not been extensively established at $z \sim 3$, it is reasonable to expect that the large SFR density observed at this epoch implies a large reservoir of molecular gas within high z galaxies (e.g. Hopkins et al. 2005). Indeed, molecular gas is detected in CO emission from a number of $z > 2$ galaxies (Tacconi et al. 2006, Walter et al. 2003) with facilities such as CARMA, IRAM, and CSO. The observations focus on CO and other trace molecules because the H_2 molecule has no dipole moment and therefore very weak emission lines. Unfortunately, current millimeter-wave observatories do not afford extensive surveys for CO gas in high z galaxies. The Large Millimeter Telescope (LMT) and, ultimately, the Atacama Large Millimeter Array (ALMA) will drive this field wide open. These facilities will determine the CO masses for a large and more representative sample of high z galaxies and (for a subset) map out the spatial distribution and kinematics of this gas. Similar to the local universe, to infer the total molecular gas mass from these CO measurements one must apply a CO to H_2 conversion factor, the so-called ‘X-factor’. The X-factor is rather poorly constrained even in the local universe and nearly unconstrained at high z. Therefore, a precise estimate of the molecular component at high z will remain a challenging venture.

Although observations of H_2 in emission are possible at cosmological redshifts (Appleton et al. 2006, Ogle et al. 2007), these measurements are not yet common enough to assess the total molecular gas budget at $z \sim 3$. The H_2 molecule does exhibit Lyman/Werner absorption bands in the rest-frame UV that are strong enough to be detectable in many astrophysical environments with UV and optical spectrographs. These ro-vibrational transitions have large oscillator strengths and permit very sensitive searches for the H_2 molecule, i.e. H_2 surface densities of less than 10^{15} particles per cm^2. To date (Ge and Bechtold 1997, Petitjean et al. 2000), the search for H_2 has been targeted toward gas with a known, large HI column density ($N_{HI} > 10^{19}\,cm^{-2}$), i.e. the Lyman limit and damped Lyα systems. This is a sensible starting point; any sightline intersecting molecular gas in a high z galaxy would also be likely to intersect the ambient ISM of that galaxy and exhibit a large HI column density. Although present surveys are small, it is evident that the molecular fraction in these HI absorbers is very low. The frequency of positive detections is $\approx$15% (Noterdaeme et al. 2008) and the majority of these have molecular fractions $f(H_2) < 10^{-3}$ (Ledoux et al. 2003). Only a handful of these galaxies show molecular fractions approaching the typical Galactic values of even diffuse molecular clouds, $f(H_2) \sim 0.1$, along sightlines with comparable HI column density (Cui et al. 2005).

In hindsight, the paucity of H_2 detections, especially ones with large $f(H_2)$, is not surprising. The formation of H_2 clouds represents a transition from the atomic phase, one that leads to gas with much lower T and much higher density. Consequently, cold molecular clouds phase exhibit much lower cross-section than when the gas is atomic. In turn, these clouds will have significantly lower probability of being detected by absorption line surveys. Analysis of the distribution of molecular gas in the local universe suggests that fewer than 1 in 1000 quasar sightlines should pierce a cold, dense molecular cloud (Zwaan and Prochaska 2006). In addition to this probalistic argument, these clouds generally have very high dust-to-gas ratios and correspondingly large extinction. Therefore, even in the rare cases that a quasar lies behind the molecular cloud of a high z galaxy, it may be reddened and extinguished out of the optical, magnitude-limited surveys that dominate quasar discovery. It is somewhat curious, however, that GRB afterglow spectra also show very low molecular fractions in gas that is believed to lie very near star-forming regions (Tumlinson et al. 2007).

In summary, there are very limited current constraints on the mass density and distribution of molecules in the young universe. Furthermore, future studies will focus primarily on tracers of H_2 and will also be challenged to precisely assess this very important phase. At present, our best estimate for the molecular mass density may be inferred from the atomic and stellar phases. Allowing for an order of magnitude uncertainty, we can estimate $\Omega_{H_2} = (0.001\text{–}0.1)\Omega_b$. Therefore, we suggest that the molecular phase contributes less than 10% of the mass density budget at $z \sim 3$.

16.4.3 Neutral Hydrogen Gas

In the local universe, galaxies have roughly equal measures of molecular and atomic hydrogen gas. The latter phase is primarily studied through 21-cm emission line surveys using large radio telescopes. Unfortunately, these radio telescopes do not have enough collecting area to extend 21-cm emission-line observations beyond $z \approx 0.2$ (Lah et al. 2007). Although one can trace the 21-cm line in absorption to much greater redshifts (e.g. Briggs & Wolfe 1983, Kanekar and Chengalur 2003), detections are exceedingly rare. This is due in part to the paucity of bright background radio sources and in part because galaxies have a relatively low cross-section of cold, neutral gas. For these reasons, neutral hydrogen gas at high z is principally surveyed via Lyα absorption. A total HI column density of $\approx 10^{20}\ \mathrm{cm}^{-2}$ is required for self-shielding to maintain neutral gas with a standard gas-to-dust ratio against the extragalactic UV background and local UV sources (Viegas 1995). At these column densities, the damping wings of the Lyα transition are resolved even with moderate spectral resolution (FWHM $\approx$2Å; Fig. 16.1). This spectral feature lends the name 'damped Lyα system' to the absorbers with $N_{\mathrm{HI}} \geq 2 \times 10^{20}\ \mathrm{cm}^{-2}$.

Wolfe and collaborators pioneered damped Lyα research beginning with the Kast spectrometer on the 3m Shane telescope at Lick Observatory (Wolfe et al. 2005, 1986). These optical observations cover redshifts $z > 1.7$ which shifts Lyα above the atmospheric cutoff. The most recent DLA surveys leverage the tremendous public database of quasar spectroscopy from the Sloan Digital Sky Survey (SDSS; (Adelman-McCarthy et al. 2007)). Figure 16.4 presents the frequency distribution $f(N_{\mathrm{HI}}, X)$ of N_{HI} values per dN_{HI} interval per absorption pathlength dX for 719 damped Lyα systems identified in the SDSS-DR5 database (see http://www.ucolick.org/~xavier/SDSSDLA; (Prochaska et al. 2005)). This distribution function, akin to the luminosity function of galaxies, is reasonably well described by a Gamma function with a 'faint-end slope' $\alpha = -2$ and a characteristic column density $N_* \approx 10^{21.5}\ \mathrm{cm}^{-2}$. The first moment of $f(N_{\mathrm{HI}}, X)$ yields the mass density of neutral gas $\Omega_{neut}(X)dX \equiv \frac{\mu m_H H_0}{c\rho_c} \int_{N_{min}}^{\infty} N_{\mathrm{HI}} f(N_{\mathrm{HI}}, X)\, dX$ where μ is the mean molecular mass of the gas (taken to be 1.3), N_{min} is the column density which marks the transition from neutral to ionized gas (taken here as $10^{20.3}\ \mathrm{cm}^{-2}$), H_0 is Hubble's

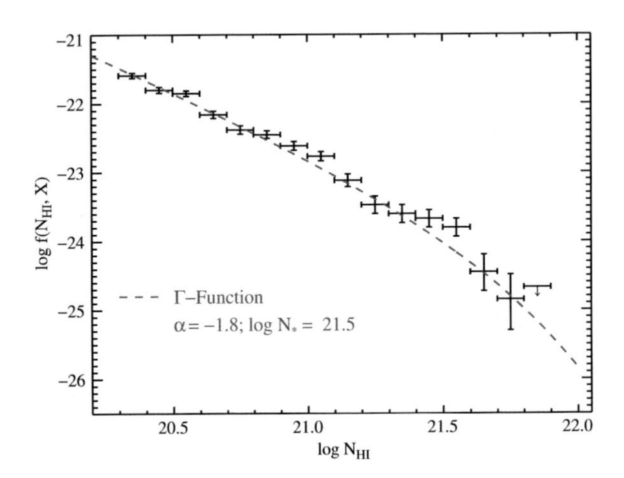

Fig. 16.4 The HI frequency distribution $f(N_{\mathrm{HI}}, X)$ for the damped Lyα systems identified in the SDSS-DR5 quasar database (mean redshift $z = 3.1$, (Prochaska et al. 2005)). Overplotted on the discrete evaluation of $f(N_{\mathrm{HI}}, X)$ is the best fitting Γ-function which is a reasonable (albeit unphysically motivated) description of the observations

constant, and ρ_c is the critical mass density. The faint-end slope of $f(N_{\rm HI}, X)$ is logarithmically divergent, but the break at N_* gives a finite density.

Current estimates for Ω_{neut} from the SDSS-DR5 database give $\Omega_{neut} = 0.016 \pm 0.002 \Omega_b$. There is a modest decline by a factor of $\approx$2 from $z = 4$ to $z = 2$ where the latter value is (remarkably) in agreement with the neutral gas mass density observed today (Section 16.6.3). One may speculate, therefore, that the neutral gas mass density has been nearly invariant for the past $\approx$10 Gyr (but see (Rao et al. 2006)). This mild evolution aside, we conclude that the neutral, atomic gas phase at $z = 3$ comprises only $\simeq 1-2\%$ percent of the baryon budget at high z. Cosmological simulations suggest that this gas comprises the ISM of galaxies in formation, an assertion supported by measurements of DLA clustering with bright star-forming galaxies (Bouché et al. 2005, Cooke et al. 2006) and the detection of heavy elements in all DLAs (Prochaska et al. 2003). We can conclude that the sum of galactic baryonic components is a minor fraction ($<$10%) of the total baryonic mass density at $z \sim 3$. This is, of course, a natural consequence of hierarchical cosmology (e.g., Nagamine et al. 2007).

With the DLA samples provided by the SDSS survey, the statistical error on Ω_{neut} approaches 10% in redshift bins $\Delta z = 0.5$. At this level, systematic errors become important, especially biases associated with the selection criteria and magnitude limit of optically bright quasars. The bias which has received the most attention is dust obscuration (Fall and Pei 1993, Ostriker and Heisler 1984), i.e. dust within DLAs redden and extinguish background quasars removing them from optical surveys. The dust-to-gas ratio in the ISM of these high z galaxies is small (Murphy and Liske 2004, Pettini et al. 1994, Vladilo et al. 2008), however, and the predicted impact on Ω_{neut} is only of the order of 10%. This expectation has been tested through DLA surveys toward radio-selected quasars (Ellison et al. 2001, Jorgenson et al. 2006), which show similar results, albeit with poorer statistical power.

Although we would like to extend estimates of Ω_{neut} to $z > 5$, the experiment is challenged by the paucity of bright quasars (although this may be overcome with GRB afterglows) and because the collective, blended opacity of gas outside of galaxies (i.e., the IGM) begins to mimic damped Lyα absorption. By $z = 6$, the mean opacity of the universe approaches unity at Lyα (White et al. 2003) and the experiment is entirely compromised. Of course, current expectation is that the universe is predominantly neutral not long before $z = 6$, i.e. $\Omega_{neut} = \Omega_b$. These expectations and the evolution of neutral gas at $z > 6$ will be tested through 21cm studies of the young universe (see contributions by Furlanetto and Loeb for discussions of this reionization epoch).

16.4.4 Ionized Gas

As Fig. 16.6 and Table 16.2 indicate, the mass densities of stars, molecular gas, and neutral gas are unlikely to contribute significantly to Ω_b at $z \sim 3$. Unless one invokes an exotic form of dense baryonic matter (e.g. compact objects), the

Table 16.2 Empirical summary of baryons in the universe

Phase	Temperature (K)	$\Omega(z=3)$ Location	Estimate[a]	$\Omega(z=0)$ Location	Estimate[a]
Stars	–	Galaxies	0.005 ± 0.002	Galaxies	0.06 ± 0.03
Molecular Gas	10^2	Galaxies	>0.001	Galaxies	0.0029 ± 0.0015
Neutral Gas	10^3	Galaxies	0.016 ± 0.002	Galaxies	0.011 ± 0.001
Ionized Gas	10^4	IGM	>0.80	IGM	$0.17^{+0.2}_{-0.05}$
Warm/Hot Gas	10^6	Galaxies?	>0.01	Filaments?	??
Hot Gas	10^7	??	??	Clusters	0.027 ± 0.009

[a] Relative to an assumed Ω_b value of 0.043.

remainder of baryons in the early universe must lie in a diffuse component outside of the ISM of galaxies (Petitjean et al. 1993). Indeed, the absence of a complete Gunn-Peterson trough in $z < 6$ quasars demonstrates that the majority of baryons are highly ionized at these redshifts (Fan et al. 2006). Because it is impossible to directly trace H^+ for the vast majority of the mass in the diffuse IGM, we must probe this phase through the remaining trace amounts of HI gas or, in principle, HeI and HeII. Unfortunately, the resonance-line transitions of He have wavelengths that preclude easy detection (Table 16.1; though the Gunn-Peterson effect has been observed in HeII at $z > 2.1$, see (Kriss et al. 2001, Shull et al. 2004)). Therefore, the majority of research has focused on the frequency distribution of HI column densities for gas associated with the IGM (a.k.a. the Lyα forest).

The modern description for the Lyα forest is an undulating density fields with minor, but important peculiar velocities and temperature variations. In this paradigm, it may be more fruitful to assess distribution functions of the smoothly varying optical depth of the IGM (Becker et al. 2007, Faucher-Giguere et al. 2007, Kim et al. 2007, McDonald et al. 2006) and compare these against theoretical models (Bolton et al. 2005, Miralda-Escudé et al. 1996). Nevertheless, the classical approach of discretizing the Lyα forest into individual absorption lines does yield valuable insight into the distribution and mass density of baryons in the intergalactic medium. This analysis is generally summarized through the distribution function of HI column densities, $f(N_{\rm HI}, X)$. Figure 16.5 summarizes our current knowledge of the $f(N_{\rm HI}, X)$ distribution for both the ionized and neutral gas. Quasar absorption line analysis to date has provided an assessment of $f(N_{\rm HI}, X)$ from $N_{\rm HI} = 10^{12}$ to $10^{22}\,{\rm cm}^{-2}$ with an important gap spanning from $N_{\rm HI} \approx 10^{15-19}\,{\rm cm}^{-2}$. At column densities larger than $10^{22}\,{\rm cm}^{-2}$, the systems are too rare to be measured with existing quasar samples. Detections below $10^{12}\,{\rm cm}^{-2}$ are limited by S/N and by the difficulty of distinguishing absorption from single 'clouds' from unresolved blends of weaker and stronger lines (e.g., Kirkman & Tytler 1997), while at $10^{11}\,{\rm cm}^{-2}$, the clouds are unphysically large for typical IGM density and ionization conditions.

Let us focus first on the systems exhibiting the largest $N_{\rm HI}$ values, the super Lyman limit systems (SLLS; also referred to as sub-DLAs). These absorbers are defined to have $10^{19}\,{\rm cm}^{-2} < N_{\rm HI} < 10^{20.3}\,{\rm cm}^{-2}$. Similar to the DLAs, one can measure the HI column densities of SLLS through fits to the damping wings of the Lyα profile although higher resolution spectroscopy is required (O'Meara

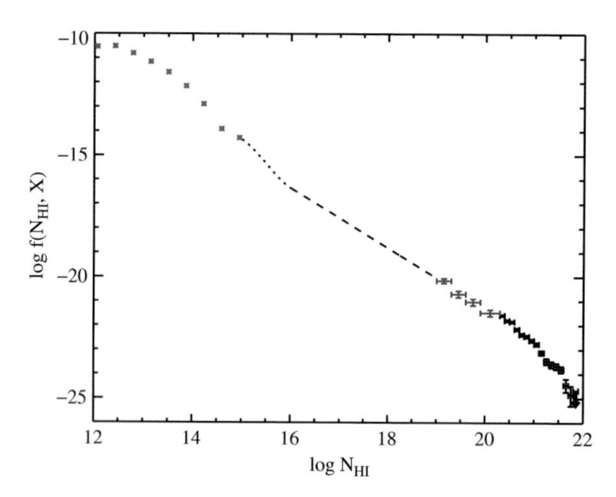

Fig. 16.5 The $f(N_{\rm HI}, X)$ distribution function of ionized $N_{\rm HI} < 10^{20}\,{\rm cm}^{-2}$ and neutral $N_{\rm HI} > 10^{20}\,{\rm cm}^{-2}$ at $z = 3$. The data points at $N_{\rm HI} \geq 2 \times 10^{20}\,{\rm cm}^{-2}$ correspond to the damped Lyα systems and are drawn from the SDSS-DR5 analysis of (Prochaska et al. 2005). The data at $19 < \log(N_{\rm HI}/\,{\rm cm}^{-2}) < 20.3$ correspond to the super Lyman limit systems (SLLS) and were drawn from the analysis of (O'Meara et al. 2007). The dashed line traces their best guess at the frequency distribution for the Lyman limit systems with $16 < \log(N_{\rm HI}/\,{\rm cm}^{-2}) < 19$. For the Ly$\alpha$ forest ($N_{\rm HI} < 10^{15}\,{\rm cm}^{-2}$) we adopt the results from (Kirkman & Tytler 1997) normalized to our assumed cosmology (Spergel et al. 2007). Finally, the dotted line is a spline fit to the Lyα forest data and the functional form for the LLS given by O'Meara et al. (2007). Although the turnover in $f(N_{\rm HI}, X)$ at $N_{\rm HI} < 10^{13.5}\,{\rm cm}^{-2}$ may be partially due to incompleteness, it is likely a real effect (see also (Hu et al. 1995, Kim et al. 2002))

et al. 2007, Péroux et al. 2003). Current surveys indicate that $f(N_{\rm HI}, X)$ flattens below $N_{\rm HI} \approx 10^{20}\,{\rm cm}^{-2}$ with $d \log f / d \log N \approx -1.4$ (O'Meara et al. 2007). If the SLLS are predominantly neutral, then they would contribute only a few percent to Ω_{neut} and a negligible fraction of Ω_b (Fig. 16.6). Most of these absorbers, however, are partially ionized with mean ionization fraction $<x> \approx 0.9$ (Péroux et al. 2007, Prochaska 1999). In Figure 16.6 we display the baryonic mass density of the SLLS inferred from the HI frequency distribution and assuming an $N_{\rm HI}$-weighted, average ionization fraction $<x> = 0.9$. We infer that the SLLS contribute on the order of 2% of the baryonic mass density at $z = 3$. The majority of the baryons must reside in yet more diffuse and highly ionized gas.

At column densities $N_{\rm HI} = 10^{16}$ to $10^{19}\,{\rm cm}^{-2}$ (the Lyman limit systems; LLS), the majority of the Lyman series lines are saturated and the damping wings of Lyα are too weak to measure. For these reasons, it is difficult to estimate $N_{\rm HI}$ values in this interval. In principle, one can precisely measure $N_{\rm HI}$ values for the partial Lyman limit absorbers with $N_{\rm HI} < 10^{17}\,{\rm cm}^{-2}$ whose opacity at the Lyman limit is $\tau_{912} \approx 1$. To date, however, only small samples from heterogeneous surveys exist in this range of $N_{\rm HI}$ (e.g., Petitjean et al. 1993). Because the Lyman limit feature is easily identified, one more easily derives an integral constraint on $f(N_{\rm HI}, X)$

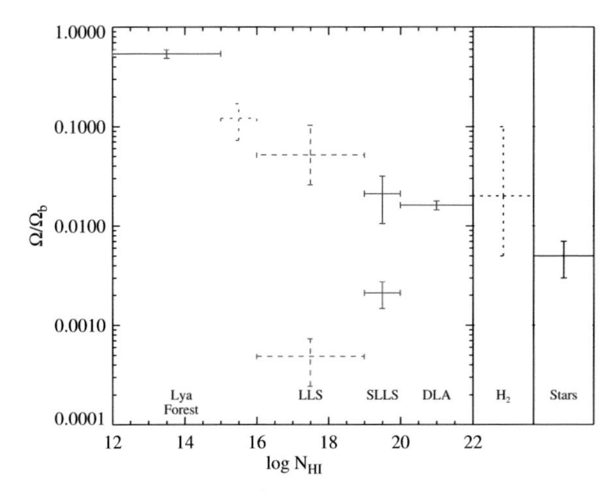

Fig. 16.6 Estimates of the baryonic mass density (relative to Ω_b) for various phases of baryons in the $z = 3$ universe. On the right side of the figure, we show the stellar mass density estimate from several recent works (Fontana et al. 2006, Rudnick et al. 2006). The uncertainty in this measurement is dominated by uncertainties in stellar population modeling and the presumed IMF. The phases corresponding to log $N_{\rm HI} < 22$ refer to absorption line systems traced by HI gas. The molecular gas content is very crudely estimated to lie near the stellar and atomic mass densities. The neutral atomic phase is traced by the DLAs where one estimates a mass density on the order of 1 to $2\%\Omega_b$ (Prochaska et al. 2005). Gas with $N_{\rm HI} < 10^{20}\,{\rm cm}^{-2}$ is predominantly ionized and one must make significant ionization corrections to estimate the mass density from the observed HI atoms (O'Meara et al. 2007). The red data points show the mass density from HI atoms alone for the LLS whereas the blue points represent estimates of the total mass density (see the text for details). The dashed symbols have weak empirical constraints and the dotted symbols have very weak empirical constraints

by surveying the incidence of LLS in quasar spectra $\ell(X) = \int f(N_{\rm HI}, X) dN_{\rm HI}$ (e.g., Storrie-Lombardi et al. 1994). This integral constraint includes gas with $N_{\rm HI} > 10^{19}\,{\rm cm}^{-2}$, but is dominated by absorbers with low $N_{\rm HI}$ values. Current estimates indicate $\ell(X) = 7.5$ at $z = 3$ (Péroux et al. 2003). Analysis of the SDSS database will greatly improve this estimate for $z > 3.2$ (O'Meara et al., in prep). Furthermore, complete analysis of the Lyα series for LLS will yield tighter constraints on $f(N_{\rm HI}, X)$ within the interval $N_{\rm HI} = 10^{17}$ to $10^{19}\,{\rm cm}^{-2}$ (Prochter et al., in prep.). (O'Meara et al. 2007) combined the current observational constraints to predict $f(N_{\rm HI}, X)$ across the LLS regime. These are presented as the dashed curve in Figure 16.5 and the uncertainty in this estimation is $\approx$50%.

Not only is $f(N_{\rm HI}, X)$ very uncertain for the LLS, so too is the ionization state of this gas. For a few systems, analysis of metal-line transitions have confirmed the gas is predominantly ionized with ionization fractions $x > 0.95$ (e.g., Prochaska & Burles 1999). The samples are too small, however, to reveal the distribution or trend of ionization fraction, x, with $N_{\rm HI}$. Unfortunately, there is little insight from theory. Previous work with cosmological simulations has severely underestimated the incidence of Lyman limit systems (e.g., Gardner et al. 2001) and it is likely that a careful treatment of radiative transfer is required to assess these systems that are near the transition from neutral to ionized (Kohler & Gnedin 2007). We can set a very conservative lower limit to their contribution by taking the $f(N_{\rm HI}, X)$ function of O'Meara et al. and assume $x = 0$, to obtain $\Omega_{LLS} > 0.0004\Omega_b$. Ionization corrections will undoubtedly increase this limit by one or more orders of magnitude. Even in the case of a 100 times increase, it is very unlikely that LLS dominate the baryonic budget at $z \sim 3$. Figure 16.6 presents an estimate of the baryonic mass density of the LLS where we have adopted the HI frequency distribution presented in Fig. 16.5 and estimated the ionization correction from simple Cloudy calculations assuming an ionization parameter $\log U = -2.5$, inferred from metal-line analysis of several LLS (Prochaska & Burles 1999). This implies a mass density estimate of $\approx 0.05\Omega_b$.

For column densities $N_{\rm HI} \approx 10^{12}$ to $10^{14.2}\,{\rm cm}^{-2}$, where Lyβ and/or Lyα are optically thin, the frequency distribution is reasonably well characterized by a power-law $f(N_{\rm HI}, X) \propto N_{\rm HI}^{-1.5}$ (e.g., Kim et al. 2002, Kirkman & Tytler 1997). Observers do report, however, a departure from this power-law at $N_{\rm HI} < 10^{13}\,{\rm cm}^{-2}$ which is likely not simply a consequence of incompleteness in the analysis (Kirkman & Tytler 1997). Although the Lyα forest absorbers dominate the spectrum of any quasar by number, the total number of HI atoms that they contribute is negligible ($\approx 4 \times 10^{-6}\Omega_b$). From cosmological simulations of the high-z universe, we expect that the Lyα forest arises in undulating density fields of highly ionized gas filling large regions ($\approx$Mpc) with modest overdensity $\delta\rho/\rho \sim 3$. These same simulations argue that the majority of the baryonic mass density resides in the Lyα forest, with a contribution possibly exceeding $0.9\,\Omega_b$ (e.g., Bryan et al. 1999, Miralda-Escudé et al. 1996, Rauch et al. 1997). Empirically, the gas is observed to only exhibit high-ion states of elements like C, Si, and O which demonstrate a high degree of ionization.

Because this gas is optically thin, if one can estimate its volume density and the intensity of the ionizing radiation field it is relatively straightforward to estimate its ionization fraction and thereby the total baryonic mass density of the Lyα forest. (Schaye 2001a) has presented a simple but intuitive prescription for modeling the Lyα forest as gravitationally bound gas clouds whose size is of order the Jeans length. Within this prescription, which matches scaling laws derived from cosmological simulations, the ionization fraction x of the gas can be related to the HI column density:

$$x = 1 - 4.6 \times 10^{-6} \left(\frac{N_{\rm HI}}{2 \times 10^{13}\,{\rm cm}^{-2}} \right)^{2/3} T_4^{-0.6}\, \Gamma_{12}^{-1/3} \tag{16.1}$$

where T_4 is the gas temperature in units of 10^4 K and Γ_{12} is the photoionization rate in units of $10^{-12}\,{\rm s}^{-1}$.

Photoionization should maintain the gas at a temperature near 2×10^4K, as predicted by the cosmological simulations. Within this prescription, therefore, the only major uncertainty is the ionization rate of the extragalactic UV background (EUVB) radiation field, Γ_{EUVB}. At $z = 3$, the EUVB will have contributions from both quasars and star-forming galaxies. The contribution of the former is directly measured from the luminosity function of $z = 3$ quasars. The contribution of galaxies, however, is far more uncertain. Although recent studies offer relatively well constrained UV luminosity functions (e.g., Reddy et al. 2007), these are measured at $\lambda \approx 1500$Å in the rest frame, and so must be extrapolated to estimate the flux of ionizing photons. Furthermore, one must adopt an escape fraction f_{esc} to estimate the flux emitted by these star-forming galaxies. Current empirical constraints are extremely limited (Chen et al. 2007, Shapley et al. 2006, Siana et al. 2007); the estimates range from $f_{esc} = 0$ to $\approx$10%. Because the comoving number density of star-forming galaxies greatly exceeds that of quasars, it is possible that galaxies contribute as much or more of the ionizing flux to the EUVB. Thus this ionizing background is uncertain at high redshift, with a corresponding uncertainty in the ionization correction for diffuse ionized gas.

The most conservative approach toward estimating the baryonic mass density of the Lyα forest is to assume $\Gamma_{EUVB} = \Gamma_{QSO}$. Hopkins et al. (2007) provide an estimate of $\Gamma_{QSO} = 4.6 \times 10^{-13}$ s^{-1} at $z = 3$ with an $\approx$30% uncertainty. Convolving Equation 16.1 (assuming $\Gamma_{12} = 0.46$, $T_4 = 2$) with $f(N_{\rm HI}, X)$, we infer a baryonic mass density $\Omega_{\rm Ly\alpha} = 0.65\Omega_b$. Again, this should be considered a lower limit to $\Omega_{\rm Ly\alpha}$ subject to the uncertainties of Equation (16.1). More detailed comparisons of the opacity of the Lyα forest with cosmological simulations reach a similar conclusion: the EUVB ionization rate exceeds that from quasars alone and $\Omega_{\rm Ly\alpha} = 0.95\Omega_b$ (e.g., Jena et al. 2005, Meiksin 2007, Miralda-Escudé et al. 1996). From a purely empirical standpoint, however, we must allow that as many as 20% of the baryons at $z = 3$ are still unaccounted for. It is very unlikely that the difference lies in photoionized gas with $N_{\rm HI} < 10^{12}$ cm^{-2} because the mass contribution of the Lyα forest peaks near $N_{\rm HI} = 10^{13.5}$ cm^{-2} (Schaye 2001a). Instead, one would have to invoke yet another phase of gas, e.g. hot diffuse baryons.

Ongoing surveys will fill in the current gap in $f(N_{\rm HI}, X)$ for $N_{\rm HI} = 10^{19}$ cm^{-2} down to at least $N_{\rm HI} = 10^{16.5}$ cm^{-2}. This gas, which likely is the interface between galaxies and the IGM, is very unlikely to contribute significantly to the baryonic mass density. It may be critical, however, to the metal mass density at $z \sim 3$ (Bouché et al. 2006, Prochaska et al. 2006). Furthermore, the partial Lyman limits ($N_{\rm HI} \sim 10^{17}$ cm^{-2}) dominate the opacity of the universe to HI ionizing radiation and therefore set the "attenuation length"(Fardal et al. 1998). This quantity is necessary to convert the observed luminosity function of ionizing radiation sources into an estimate of Γ_{EUVB} (Haardt & Madau 1996, Meiksin & White 2003). These observations may also provide insight into the processes of reionization at $z > 6$ and the escape fraction of star-forming galaxies.

16.4.5 Warm-Hot Diffuse Baryons

In the previous section, we demonstrated that the photoionized IGM may contain all the baryons in the $z \sim 3$ universe that are not associated with dense gas and stars. We also noted, however, that this conclusion hinges on the relatively uncertain intensity of the EUVB radiation field. As a consequence even as much as $30\%\Omega_b$ may be unaccounted for in the IGM. One possible reservoir in this category is a hot ($T > 10^5$ K) diffuse phase. Indeed, there are several processes already active at $z \sim 3$ that can produce a reservoir of hot and diffuse gas. The hard, intense radiation field from quasars and other energetic AGN processes (e.g. radio jets) may heat and ionize their surroundings out to several hundred kpc (e.g., Chelouche et al. 2007). Similarly, the supernovae that follow star formation may also drive a hot diffuse medium into the galactic halo and perhaps out into the surrounding IGM (e.g., Aguirre et al. 2001, Kawata & Rauch 2007). Outflows of ionized gas are observed in the spectra of quasars and star-forming galaxies at these redshifts. On the other hand, gas accreting onto galactic halos is expected to shock heat to the virial temperature ($T > 10^6$K), especially in halos with $M > 10^{11} M_\odot$ (Dekel & Birnboim 2006). Altogether these processes may heat a large mass of gas in and around high z galaxies.

Presently, there are very weak empirical constraints on the mass density and distribution of a hot, diffuse medium at $z \sim 3$. Detecting this gas in emission is not technically feasible and there are very few observable absorption-line diagnostics. The most promising approach uses the OVI doublet to probe highly ionized gas associated with the IGM (e.g., Bergeron & Herbert-Fort 2005, Simcoe et al. 2002) and galaxies (Fox et al. 2007, Simcoe et al. 2006). These studies suggest a hotter component ($T >\sim 10^5$K) contributes on the order of a few percent Ω_b at $z \approx 2$.

The mass density of an even hotter phase ($\simeq 10^{6-7}$ K), meanwhile, is unconstrained by observation. Ideally we would assess this component with soft X-ray absorption and/or emission but this will require a major new facility such as Con-X. We will take up these issues again in Section 16.6.6 on the $z \sim 0$ universe, where the warm/hot component is predicted to be substantial.

16.5 Baryons in the $z \sim 1$ Universe

In contrast to the success that astronomers have had in probing baryons at $z \sim 3$, the knowledge of the distribution of baryons at $z \sim 1$ is quite limited. This is the result of several factors. One aspect is psychological: astronomers have pushed harder to study the universe at earlier times. This emphasis is shifting, however, with advances in IR spectroscopy (where the rest-frame optical diagnostics fall for $z \sim 1$), the impact of the Spitzer mission (e.g. Dickinson's contribution), and the generation of large spectroscopic samples of $z \sim 1$ galaxies (GDDS, DEEP, VIRMOS). The other aspect is technical. As with the $z \sim 3$ universe, it is difficult or impossible with current facilities to build large samples of clusters to assess the host phase, or to survey the molecular phase with CO, or to measure HI gas with 21-cm techniques. Furthermore, the majority of key absorption-line diagnostics (e.g. Lyα, CIV; Table 16.1) fall below the atmospheric cutoff and UV spectroscopy on space-based telescopes is necessary and expensive. Key emission-line diagnostics (Hα, [OIII]), meanwhile, shift into the near-IR making it more difficult to pursue large galaxy surveys.

Surprisingly, the prospects are also limited for major advances in exploring baryons at $z \sim 1$ during the next decade. The notable positive examples are (i) Sunyaev-Zeldovich experiments, which should establish the mass function of $z \sim 1$ clusters, (ii) multi-object, near-IR spectrometers which will improve detections of star-forming galaxies and the luminosity function of early-type galaxies and (iii) the construction of mm telescopes (LMT, ALMA) that will enable searches for molecular gas. These advances will allow astronomers to characterize the dense, baryonic component at $z \sim 1$. Similar to $z \sim 3$ and $z \sim 0$, however, we expect that these components will be minor constituents. It is a shame, therefore, that there is currently no funded path to empirically constrain the properties of diffuse gas, neutral or ionized. Upcoming 'pathfinders' for the proposed Square Kilometer Array (SKA) will extend 21-cm observations beyond the local universe but are unlikely to reach $z = 1$. This science awaits an experiment on the scale of SKA. The classical absorption-line experiments must access hundreds of faint QSOs at $z \simeq 1 - 2$ and so require a UV space mission probably on the scale of JWST (Sembach et al. 2005). At present, however, there is no development funding path toward such a mission. In this respect, our ignorance on the distribution and characteristics of the majority of baryons in the $z \sim 1$ universe may extend beyond the next decade.

16.6 Baryons in the $z \sim 0$ Universe

In this section, we follow closely the footsteps of (Fukugita et al. 1998), who presented a comprehensive census of baryons in the local universe (see (Fukugita 2004) for an update). With this exercise in accounting, the authors stressed a startling problem: roughly 50% of the baryons in the local universe are missing! They further suggested that these missing baryons are likely hidden in a warm/hot ($T \approx 10^5$ to 10^7K) diffuse medium that precluded easy detection (Table 16.2). These claims were soon supported by cosmological simulations that predict a warm/hot intergalactic medium (WHIM) comprising 30 to 60% of today's baryons (Cen & Ostriker 1999, Davé et al. 2001). Motivated by these semi-empirical and theoretical studies, the search for missing baryons gained great attention and continues today. We review the current observational constraints on the distribution and phases of baryons in the local universe and stress paths toward future progress. As in previous sections, we roughly order the discussion from dense phases to the diffuse.

16.6.1 Stars

Large-area imaging and spectroscopic surveys (e.g., 2MASS, SDSS, 2dF) have precisely measured the luminosity function of galaxies in the local universe to faint levels (Blanton et al. 2003, Cole et al. 2001). The light emitted by galaxies in optical and IR bands, therefore, is well characterized. Establishing the mass budget of baryons in these stars, however, requires translating the luminosity function into a mass function (Salucci & Persic 1999). The typical procedure is to estimate a stellar mass-to-light ratio (M/L) for each galaxy based on stellar population modeling of the colors of the integrated light. These estimates are sensitive to the metallicity, star-formation history, and (most importantly) the initial mass function (IMF) of star formation. Because the stellar mass of a galaxy is dominated by lower mass stars even at early times, the analysis is most accurately performed in red passbands.

The SDSS and 2dF galaxy surveys have led the way in this line of research; SDSS possesses the advantage of z-band imaging ($\lambda \approx 9000$Å) which reduces the uncertainty in the M/L estimate. Kauffmann et al. (2003) analyzed the mass-to-light ratio for a large sample of SDSS galaxies, for which they report a luminosity function-weighted ratio $<M/L_z> = 1.5$ with an uncertainty of $\approx$20% and assuming a Kroupa IMF. Taking the Kauffmann et al. (2003) estimation of M/L with the SDSS luminosity function (Blanton et al. 2003, Fukugita 2004) estimates a stellar mass density $\Omega_*(z = 0) = 0.0025$, which includes an estimate of stellar remnants. Assuming a 50% uncertainty, we find $\Omega_*(z = 0) = (0.06 \pm 0.03)\Omega_b$.

Further progress in estimating the stellar contribution to Ω_b is not limited by neither statistical error nor depth of the luminosity function, but rather by systematic uncertainty, especially in the IMF. Kauffmann et al. (2003) did not vary this assumption when estimating the error budget. Because sub-solar mass stars contribute most of the mass in stellar systems, uncertainties in the IMF can easily dominate the error budget. A conservative estimate of this uncertainty is $\approx$50%, especially if we allow for a Salpeter IMF.

The subject of stellar masses and their sensitivity to IMF is worthy of a review article in its own right, and so it is well beyond the scope of our mandate. Here we only pause to note three recent developments. First, the uncertainty in *total* stellar masses is probably only a factor of 1.5–2, even though the error may be larger for individual cases. Second, this topic is receiving more attention in stellar population synthesis models (e.g. Maraston, Bruzual and Charlot), which can now include IMF uncertainties in their budgets of systematic error. Finally, there have been recent, speculative suggestions that the IMF at high redshift may depart significantly from the Galactic form. At early times, generally hotter environments within galaxies and an elevated CMB may raise the typical stellar mass (Davé 2008, Tumlinson 2007, van Dokkum 2008). Though much theoretical work and experimental testing remains to be done in this area, it is possible that these ideas will ultimately lead to a more self-consistent picture of the evolution of stellar mass with redshift, with a consequent improvement in our understanding of the baryon budget. This is one of the key science goals of the JWST and ALMA facilities. These uncertainties aside, it is evident that stars are a sub-dominant baryonic component in our current universe. As an editorial aside, we note the irony that even the baryonic component that is easiest to see – the luminous stars – is difficult to weigh. This uncomfortable fact should be kept in mind later as we attempt to assess the unseen, hot diffuse phases of the IGM.

16.6.2 Molecular Gas

With the exception of the Galaxy (Savage et al. 1977), the Magellanic Clouds (Tumlinson et al. 2002), and some nearby starburst galaxies (Hoopes et al. 2004), there are very few direct detections of H_2 in the local universe (Appleton et al. 2006, Ogle et al. 2007). Commonly these detections are of absorption in the FUV Lyman-Werner bands, and so are accessible only in optically thin, low-dust ($A_V < 1$) environments. H_2 can also be detected in UV fluorescence lines from interstellar gas, but these are intrinsically very faint. Neither of these H_2 reservoirs are likely to contribute much to the total cosmic mass budget of molecular gas (judging from the Milky Way). Because it is a homonuclear molecule with no dipole moment, H_2 has only very weak quadrupole emission even if the density is high. The usual technique, then, is to track molecular gas with trace optically-thin species and infer the total molecular content through scaling factors calibrated against our Galaxy and others in the local universe. The most common indirect tracer is CO and its conversion factor is termed the 'X-factor'. In contrast with HI gas (see below), 'blind' surveys for CO have not yet been performed. Instead, researchers have estimated the molecular mass density by correlating CO mass with another galaxy observable (e.g. HI mass, infrared flux density) that has a well defined, volume-limited distribution function. The convolution of the two functions gives an estimate of the CO mass function (Keres et al. 2003, Young & Scoville 1991, Zwaan & Prochaska 2006). Finally, this is converted to the H_2 mass function with the X-factor which is generally assumed to be independent of any galaxy properties. This latter assumption is known to be invalid for faint metal-poor galaxies and luminous starbursts, but variations are believed to be small for the $L \approx L_*$ galaxies that likely dominate the molecular mass density (Devereux & Young 1990).

Taking $X \equiv N(\mathrm{H_2})/I(CO) = 3 \times 10^{20}\,\mathrm{cm^{-2}(K\,km\,s)^{-1}}$ Young & Scoville (1991), Keres et al. (2003) report $\rho_{\mathrm{H_2}} = 2.2 \times 10^7 h_{70} M_\odot\,\mathrm{Mpc^{-3}}$ for an IR sample of galaxies and (Zwaan & Prochaska 2006) report $\rho_{\mathrm{H_2}} = 1.2 \times 10^7 h_{70} M_\odot\,\mathrm{Mpc^{-3}}$ for the optically selected BIMA sample. The factor of two difference in these estimates reflects the systematic error associated with sample selection. We adopt the mean value of the two estimates and assume a 30% uncertainty: $\Omega_{\mathrm{H_2}} = 0.0029 \pm 0.0009\Omega_b$. The next major advance for surveying molecular gas in the local universe awaits a wide-field, deep, and blind survey of CO gas with depth and area comparable to current 21-cm surveys. This is a difficult technical challenge and we know of no such survey that is currently planned.

16.6.3 Neutral Hydrogen Gas

Over the past decade, radio astronomers have performed blind surveys for 21-cm emission to derive the HI mass function of galaxies (e.g., Rosenberg and Schneider 2002, Zwaan et al. 1997, 2005a). The modern version is the HIPASS survey (Meyer et al. 2004), an all Southern-sky survey for HI with the Parkes radio telescope. Zwaan et al. (2005a) present the HI mass function from HIPASS and derive an integrated mass density $\Omega_{\rm HI} = 3.8 \times 10^{-4} h_{70}^{-1}$. Including He, this implies a neutral gas mass density $\Omega_{neut} = (0.011 \pm 0.001)\Omega_b$. Although some uncertainty remains regarding the shape of the low-column ('faint-end') slope for the HI mass function, the total mass density is dominated by $L \approx L_*$ galaxies. Another important result of these surveys is the demonstration that very little HI mass, if any, is associated with star-free systems. Ultimately, the ongoing ALFALFA survey (Giovanelli et al. 2005) will surpass the HIPASS sample in sensitivity and provide the definitive measure of the HI mass function at $z \sim 0$. In the future, facilities like the EVLA or Square Kilometer Array will be able to push such observations to $z \sim 1$ or beyond.

Interferometric observations can map the HI surface density profiles of low-redshift galaxies and thereby infer the HI frequency distribution at $N_{\rm HI} > 10^{20}\,{\rm cm}^{-2}$ (Ryan-Weber et al. 2003, Zwaan et al. 2005b). This distribution function is illustrated in Fig. 16.7. Remarkably, it has nearly identical shape to the $f(N_{\rm HI}, X)$ function derived for $z \sim 3$ galaxies from damped Lyα observations (Fig. 16.5). Both of these $f(N_{\rm HI}, X)$ functions follow a roughly $N_{\rm HI}^{-2}$ power-law at lower col-

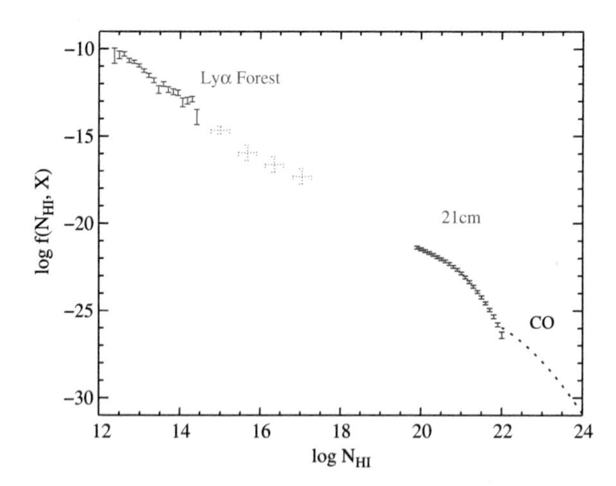

Fig. 16.7 The $f(N_{\rm HI}, X)$ distribution function of ionized $N_{\rm HI} < 10^{20}\,{\rm cm}^{-2}$, neutral $N_{\rm HI} = 10^{20-22}\,{\rm cm}^{-2}$, and molecular $N_{\rm HI} > 10^{22}\,{\rm cm}^{-2}$ gas at $z \sim 0$. The data at $N_{\rm HI} < 10^{17}\,{\rm cm}^{-2}$ which correspond to HST/STIS observations of the low redshift Lyα forest are taken from Penton et al. (2004), the distribution function neutral, atomic hydrogen gas is taken from 21cm WSRT observations (Zwaan et al. 2005b), and the distribution function of molecular gas is estimated from BIMA CO observations (Zwaan & Prochaska 2006)

umn densities and exhibit a break at $\log N_{\rm HI} \approx 21.7$. This suggests that HI gas is distributed in a similar manner within galaxies at $z \sim 0$ and $z \sim 3$. Perhaps even more remarkable is the fact that the integrated mass density is also very comparable. The Ω_{neut} value at $z \sim 0$ is at most three times smaller than the largest Ω_{neut} values derived from the $z \sim 3$ damped Lyα systems (0.0045 vs. 0.0014) and actually coincides with the Ω_{neut} value at $z = 2.2$ (Prochaska et al. 2005). One might infer that Ω_{neut} has been constant for the past $\approx$10 Gyr, which suggests that all of the gas accreted into the neutral phase within galaxies is converted into stars (e.g., Kereš et al. 2005).

16.6.4 Intracluster Medium (ICM)

The majority of galaxies in clusters are early-type with little or no neutral hydrogen gas. It may have been thought at first, therefore, that the baryonic content of clusters was almost entirely stars. X-ray observations, however, have revealed that clusters are filled with a hot plasma that extends to at least their virial radii ($\sim$1 Mpc). This intracluster medium (ICM) contains the baryons that have collapsed within the cluster structure and have been shock-heated to the virial temperature ($T > 10^7$ K) of the cluster. Analysis of the X-ray surface brightness indicate that the ICM has an electron density profile $n_e(r) = n_0[1 + (r/r_c)^2]^{3\beta/2}$ with $\beta \approx 0.6$ and n_0 ranging from 10^{-1} to 10^{-2} at r_c of a few hundred kpc (e.g., Reiprich & Böhringer 2002). In fact, the ICM mass significantly exceeds the stellar mass inferred for the galaxies within the cluster. From a cosmological standpoint, therefore, the ICM may represent a major baryonic component of our current universe.

Detailed analysis of the X-ray observations show that the gas fraction, f_{gas}, or the ratio of the gas mass to the dynamical mass, is nearly constant with cluster mass, $f_{gas} = 0.11 \pm 0.01$ (Allen et al. 2008). Therefore, one can estimate the total baryonic mass density of the ICM in all clusters simply by taking the product of f_{gas} with the cosmological mass density of clusters. A recent estimation of the latter comes from an X-ray selected sample of clusters with overlapping SDSS observations (Rines et al. 2008). Fitting these data to the (Jenkins et al. 2001) dark-matter halo mass function for a WMAP5 cosmology (Dunkley et al. 2008), and integrating from high mass down to $M_{min} = 10^{14} M_\odot$, we derive a mass density of clusters, $\Omega_{cluster} = 0.010$. Adopting $f_{gas} = 0.11$, the ICM has a baryonic mass density $\Omega_{ICM} = 0.027\Omega_b$. Uncertainties in this estimate are dominated by the uncertainty in $\Omega_{cluster}$, which are in turn driven by uncertainty in the power spectrum normalization specified by σ_8. These errors translate into a roughly 30% uncertainty in $\Omega_{cluster}$. This diffuse, hot component represents a non-negligible mass density in our modern universe. It raises the possibility that a similar medium associated with collapsed structures having $M < M_{min}$ (i.e. groups and isolated galaxies) could contribute significantly to the mass density of baryons at $z \sim 0$ (Fukugita et al. 1998, Mulchaey et al. 1996).

16.6.5 Photoionized Gas

Summing the mass densities for the gas and stars of the preceding sub-sections, we estimate a total mass density of only $\approx 0.10\Omega_b$. Similar to the high-z universe, therefore, we conclude that the majority of baryons at $z \sim 0$ lie in diffuse gas outside of virialized structures. It is reasonable to speculate first that the photoionized component, i.e. the intergalactic medium, is also the major baryonic reservoir at $z \sim 0$. As with the $z \sim 3$ universe, this component is best traced by Lyα absorption using UV absorption-line spectroscopy. Even a visual inspection of low-z quasar spectra shows that the incidence of Lyα absorption $\ell(X)$ is quite low relative to the $z \sim 3$ universe. This sharp decline in $\ell(X)$ is a natural consequence, however, of the expanding universe. Quantitatively, the decline in $\ell(X)$ is actually less than predicted from expansion alone (Weymann et al. 1998) because of a coincident decrease in the EUVB intensity which implies a higher neutral fraction and correspondingly higher incidence of HI absorption (Davé et al. 1999).

The HI column density distribution function of the $z \sim 0$ IGM has been characterized by several groups using UV spectrometers onboard the Hubble Space Telescope (Davé & Tripp 2001, Penton et al. 2004). In Fig. 16.7, we show the Penton et al. $f(N_{\rm HI}, X)$ results for a sample with mean redshift $z = 0.03$. These results assume a Doppler parameter $b_{\rm IGM} = 25$ km s^{-1} for all absorption lines identified in their survey (Danforth & Shull 2007) have obtained $b_{\rm IGM} = 27$ km s^{-1} for their larger sample of 650 Lyα absorbers). The results are insensitive to this assumption for low column densities ($N_{\rm HI} < 10^{14}$ cm^{-2}) because the Lyα profiles are unsaturated but measurements at higher $N_{\rm HI}$ are sensitive to $b_{\rm IGM}$ and have much greater uncertainty. Similar to the $z \sim 3$ universe, the $f(N_{\rm HI}, X)$ function is sufficiently steep that one expects the mass density is dominated by clouds with $N_{\rm HI} < 10^{15}$ cm^{-2}. Danforth & Shull (2005) have fitted the distribution function at low $N_{\rm HI}$ with a power-law function $f(N_{\rm HI}, X) = AN_{\rm HI}^{\beta}$ and report $A = 10^{10.3\pm1.0}$ and $\beta = -1.73 \pm 0.04$ for their recent sample with $\langle z \rangle = 0.14$, and there is significant degeneracy between the two parameters.

We may estimate the IGM baryonic mass density at $z \sim 0$ using the formalism described in 16.4.4 following Schaye (2001b). In addition to $f(N_{\rm HI}, X)$, this analysis requires an estimate of the HI photoionization rate Γ and the gas temperature $T_{\rm IGM}$. The dependences of the mass density on these quantities are weak ($\Omega_{{\rm Ly}\alpha} \propto \Gamma^{1/3} T^{3/5}$) but so too are the empirical constraints. Estimates of Γ range from 0.3 to 3 $\times$ 10^{-13} s^{-1} (Scott et al. 2000, Shull et al. 1999, Weymann et al. 2001) and we adopt $\Gamma = 10^{-13}$ s^{-1} based on calculations of the composite quasar and galaxy background estimated by Haardt & Madau (CUBA; in prep). The characteristic temperature of the IGM is constrained empirically only through the observed distribution of line widths, $b_{\rm IGM}$. Several studies report a characteristic line width at $z \sim 0$ of $b_{\rm IGM} \approx 25$ km s^{-1} which implies $T_{\rm IGM} < 38,000$K (Davé & Tripp 2001, Paschos et al. 2008, Penton et al. 2002). This is an upper limit because it presumes the line widths are thermally and not turbulently, broadened. Indeed, (Davé & Tripp 2001) have argued from analysis of numerical simulations that the observed $b_{\rm IGM}$ distribution is in fact dominated by non-thermal motions. These same simulations suggest a density-temperature relation $T_{\rm IGM} \approx 5000(\rho/\bar{\rho})^{0.6}$K (the so-called "IGM equation of state").

Combining $\Gamma = 10^{-13}$ s^{-1} and the (Davé & Tripp 2001) $\rho - T$ "equation-of-state" with the (Penton et al. 2004) estimation for $f(N_{\rm HI}, X)$ and the (Schaye 2001b) formalism, we estimate the baryonic mass density of the $z \sim 0$ Lyα forest ($N_{\rm HI} = 10^{12.5-14.5}$ cm^{-2}) to be $\Omega_{\rm Ly\alpha} = 0.0075 = 0.17\Omega_b$. From their sample of 650 Lyα absorbers at $\langle z \rangle = 0.14$, (Danforth et al. 2007) obtain $\Omega_{\rm Ly\alpha} = 0.0131 \pm 0.0017 = 0.29\Omega_b$. This range is still several times smaller than the mass density estimate for the IGM at $z \sim 3$ (Fig. 16.6). The immediate implication is that the IGM may not be the majority reservoir of baryons at $z \sim 0$. Indeed, the missing baryons problem hinges on this result. It is crucial, therefore, to critically assess the uncertainty in this estimate, in particular whether current observational constraints allow for a significantly larger value.

First, consider the photoionization rate Γ which is constrained to no better than a factor of a few (at any redshift). As noted above, $\Omega_{\rm Ly\alpha}$ is largely insensitive to the photoionization rate ($\Omega_{\rm Ly\alpha} \propto \Gamma^{1/3}$). Furthermore, our adopted value lies toward the upper end of current estimations which leads us to conclude that a more precise measure of Γ will not markedly increase the baryonic mass density inferred for the IGM at $z \sim 0$.

Another consideration is the temperature of the IGM, although again the mass density is not especially sensitive to this quantity. Even if the Lyα forest were uniformly at $T_{\rm IGM} = 2 \times 10^4$K, we would derive only a 50% higher baryonic mass density. It is evident that modifications to Γ and $T_{\rm IGM}$ will not make a qualitative difference in $\Omega_{\rm Ly\alpha}$, though these effects in combination could increase it by a factor of 2. Turning to the HI frequency distribution, (Penton et al. 2004, Davé & Tripp 2001) emphasize that $f(N_{\rm HI}, X)$ shows no break from a power-law down to the sensitivity limit of the HST/STIS observations, $N_{lim} = 10^{12.5}$ cm^{-2}. It is reasonable, therefore, to extend the distribution function to lower column densities. Taking $N_{lim} = 10^{11.5}$ cm^{-2} with the power-law exponent fixed at $\beta = -1.65$, leads to a $2\times$ increase in $\Omega_{\rm Ly\alpha}$. Finally, both cosmological simulations and some observational analysis suggest a steeper power-law ($\beta \approx -2$) (Davé & Tripp 2001, Lehner et al. 2007, Paschos et al. 2008) which, in turn, may lead to an enhancement in $\Omega_{\rm Ly\alpha}$. If we allow for a combination of these effects, it is possible to recover $\Omega_{\rm Ly\alpha}$ values exceeding $0.5\Omega_b$. For example, the following parameter set – $\Gamma = 10^{-13}$ s^{-1}, $T_{\rm IGM} = 2 \times 10^4$K, $\beta = -1.8$, $N_{lim} = 10^{11.3}$ cm^{-2} – yields $\Omega_{\rm Ly\alpha} = 0.6\Omega_b$.

On empirical grounds alone, then, the current observations do not rule out a scenario where the photoionized Lyα forest contains most of the baryons not locked in denser phases, which we estimate at $\simeq 0.1\Omega_b$. This would require, however, a warmer IGM than predicted by numerical simulations and a $f(N_{\rm HI}, X)$ function that remains steep to $N_{\rm HI} \approx 10^{11}$ cm^{-2}. The latter point may be tested with new observations of the IGM using the planned HST/COS instrument, especially if it achieves S/N >50 per pixel. Given that the uncertainties must be pressed in concert to their limits for the budget to be complete, it is also reasonable to conclude that the IGM does not contain all of the missing baryons at $z \sim 0$. In the next section we will take up the question of what could fill this gap, assuming it exists.

16.6.6 The WHIM

In Fig. 16.8, we present the $z \sim 0$ baryonic mass density estimates for the phases described in the previous sub-section. A simple summation of the central value estimate of each component[5] yields a total $\Omega_{stars} + \Omega_{H_2} + \Omega_{neut} + \Omega_{\rm ICM} + \Omega_{\rm Ly\alpha} = 0.27\Omega_b$. Therefore, having followed the footsteps of FHP98, we reach a similar conclusion: current observations have not revealed roughly 70% of the baryons in the $z \sim 0$ universe. The observational techniques described in the previous section are sensitive to baryons in luminous matter, hot ($T > 10^7$K) and diffuse gas, molecular and atomic neutral gas, and diffuse photoionized gas. From a practical standpoint, this leaves one obvious phase: warm/hot (10^5 K $< T < 10^7$ K), diffuse gas.

Indeed, FHP98 identified this phase as the likely reservoir for the remainder of baryons and suggested it would be associated with galaxy groups and galactic halos (see also (Mulchaey et al. 1996)). Cosmological simulations of the low redshift universe, meanwhile, argue that a warm/hot phase exists yet external to galaxies, i.e. within the intergalactic medium (Cen & Ostriker 1999, Davé et al. 2001). This warm/hot intergalactic medium (WHIM), believed to be heated by shocks during large-scale gravitational collapse, is predicted by the simulations to comprise roughly 50% of baryons in the present universe. Bregman (2007) has recently

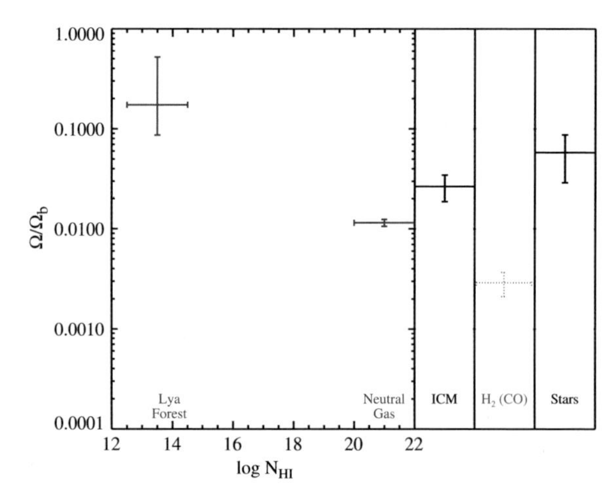

Fig. 16.8 Estimates of the baryonic mass density (relative to Ω_b) for various phases of baryons in the $z \sim 0$ universe. On the left side of the figure, we show the mass densities estimated from the photoionized Lyα forest and the neutral gas in low z galaxies. One notes that the sum of the central values for the various components is significantly less than unity. This suggests that a significant mass component is missing from this census, e.g. the warm/hot intergalactic medium (WHIM). Even adopting the maximum values for each component, one may require an additional baryonic phase to contribute at $z = 0$

---

[5] Although there lies a gap in our knowledge or gas with $N_{\rm HI} \approx 10^{17}$ cm^{-2}, it is unlikely that this gas contributes significantly to Ω_b.

reviewed the observational and theoretical evidence for the WHIM at $z \sim 0$. We refer the reader to that manuscript for a broader discussion. We highlight here only a few points with emphasis on absorption-line observations.

The WHIM gas, as conceived in numerical simulations or as gas reservoirs of galaxy groups, has too low density and temperature to be detected in emission by current X-ray instrumentation and even the next generation of X-ray facilities would be unlikely to detect the most diffuse gas. The neutral fraction, meanwhile, is far too low to permit detections via 21-cm observations. For these reasons, searches for WHIM gas have essentially been limited to absorption-line techniques. The current approaches include (1) a search for the wisps of HI gas associated with the WHIM and (2) surveys for ions of O (and Ne) which trace warm/hot gas.

The first (and most popular) approach has been to survey intergalactic O^{+5} along quasar sightlines. This is primarily driven by observational efficiency; oxygen exhibits a strong OVI doublet at ultraviolet wavelengths (1031, 1037Å) that can be surveyed with high resolution, moderate signal-to-noise observations. To date, these have been acquired by spectrometers on the HST and FUSE satellites. In collisional ionization equilibrium (CIE; (Gnat & Sternberg 2007, Sutherland & Dopita 1993)), the O^{+5} ion is abundant in gas with $T = 10^5$ to 10^6 K and therefore may trace the cooler WHIM. A hard radiation field, however, may also produce substantial O^{+5} ions in a diffuse gas. Therefore, the detection of OVI is not definitive proof of WHIM gas. Tripp and collaborators were the first to demonstrate that intergalactic OVI may be detected along quasar sightlines (Tripp & Savage 2000, Tripp et al. 2000). Surveys along tens of sightlines have now been comprised with HST/STIS and FUSE observations (Danforth & Shull 2005, 2008, Thom and Chen 2008, Tripp et al. 2007) which reveal a line density $\ell(z)$ of absorbers $\ell(z) \approx 10$ per unit redshift path down to $W_{1031} \geq 50$mÅ and $\ell(z) \approx 40$ down to $W_{1031} \geq 10$mÅ.

Intriguingly this incidence and the observed equivalent width distribution are in broad agreement with the predictions from $z \sim 0$ cosmological simulations (Cen 2006). At the least, this should be considered circumstantial evidence for WHIM gas. On the other hand, at least 30% of OVI systems showing OVI absorption also exhibit coincident CIII absorption (Danforth & Shull 2007, Danforth et al. 2006, Prochaska et al. 2004, Thom and Chen 2008). The O^{+5} and C^{++} ions cannot coexist in CIE; their coincident detection either indicates a multi-phase medium or a photoionized gas. Tripp et al. (2007) have also emphasized that ≈40% of OVI absorbers show HI absorption which tracks the OVI very closely and have line-widths indiciative of gas with $T < 10^5$K (Fig. 16.9, see also (Thom & Chen 2008)). In their analysis, Danforth & Shull (2007) have placed this fraction at <10%, perhaps owing to their choice to not divide the absorbers into velocity subcomponents. Thus there is agreement that at least some, if not all, of these absorbers have temperatures of $T < 10^5$ K, i.e. photoionized material.[6] In other cases, the OVI absorption is offset in velocity from the strongest HI absorption and there is little diagnostic

[6] Although one could invoke undetected, broad Lyα absorption in these cases, this would not predict the observed identical line-profiles for OVI and Lyα absorption.

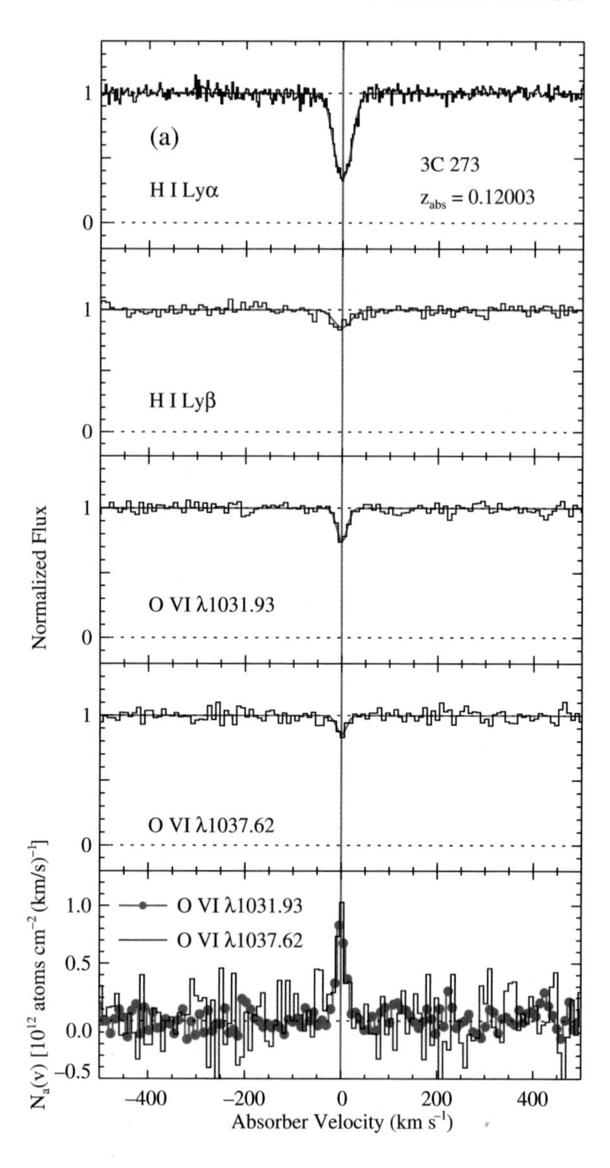

Fig. 16.9 Plot of the Lyα, Lyβ, and OVI line-profiles for an OVI absorber at $z = 0.12$ toward 3C273 (taken from (Tripp et al. 2007)). Note that the profiles have identical velocity centroid. Furthermore, an analysis of the line-widths suggests a temperature $T \approx 2 \times 10^4$K. In turn, this implies the material is predominantly photoionized, not collisionally ionized and is therefore unlikely to correspond to warm/hot intergalactic gas. Roughly 40% of OVI absorbers share these properties. It remains an open question as to whether the majority of observed OVI gas corresponds to the WHIM

constraint on the gas temperature. In only a few cases (e.g. the $z = 0.1212$ absorber toward HS1821+643; (Tripp et al. 2001)) does one detect broad Lyα absorption aligned with the OVI absorption that clearly favors collisionally ionized gas with $T > 10^5$ K. Nevertheless, the exact ionization origin of the majority of OVI absorbers remains an open question.

In support of the view that the OVI absorbers are predominantly hot and collisionally ionized, Danforth & Shull (2007) offer the arguments that (1) the column-density and linewidth distributions of the two high ions OVI and NV differ significantly from those of commonly photoionized species, such as CIII/IV and

HI, and (2) the observed systems imply clouds with an unrealistically wide range of ionization parameters in close proximity to yield OVI and lower ions. Of the known OVI systems, the majority have not been confirmed by multiple investigators as truly arising from collisional ionization, so there is presently no consensus that OVI in fact measures the $T > 10^5$ WHIM. The upshot of these ongoing controversies is that while OVI can be a tracer of WHIM, its origins are ambiguous and further work, including detailed cross-checking by the interested groups, is needed to assess the fraction of OVI absorbers that are truly hot and collisionally ionized, and from there to calculate their contributions to the baryon budget.

It may, in fact, be premature to identify this OVI-traced gas with the intergalactic medium. Although the incidence of OVI absorbers is too high for all of them to be associated with the halos of $L \approx L^*$ galaxies, there are sufficient numbers of dwarf galaxies to locate the gas within their halos or "zones of metal enrichment" if these extend to ≈150–250 kpc (Danforth & Shull 2007, Stocke et al. 2006, Tumlinson and Fang 2005). Indeed, searches for galaxies linked with OVI absorbers have indicated a range of associations from individual galactic halos to galaxy groups to intergalactic gas (Cooksey et al. 2008, Prochaska et al. 2006, Stocke et al. 2006, Tripp et al. 2006, Tumlinson et al. 2005). We await a larger statistical sample and detailed comparison to numerical simulations to resolve this issue (Ganguly et al. 2008). The upcoming Cosmic Origins Spectrograph on HST will make a large and important contribution to solving this problem.

Setting aside the controversial aspects of OVI, one may speculate on the mass density of the OVI-bearing gas $\Omega_{\rm OVI}$ using similar techniques as for the Lyα forest. Because OVI is simply a tracer of the material, one must assume both the metallicity and an ionization correction for the gas. Current estimations give $\Omega_{\rm OVI} \approx 0.1\Omega_b$, assuming that all the OVI detections arise in collisionally ionized gas and that the gas has 1/10 solar metallicity (Danforth & Shull 2007). Therefore, even in the case that the OVI-bearing gas traces only WHIM material, it is evident that it comprises a relatively small fraction of the baryon census.

Ideally, one would assess the mass density and spatial distribution of WHIM gas through observations of hydrogen which dominates the mass. This is, however, a difficult observational challenge. Assuming CIE, a WHIM 'absorber' with (an optimistic) total hydrogen column of $N_H = 10^{20}\,{\rm cm}^{-2}$ and $T = 10^6$K will have an HI column density of $N_{\rm HI} = 10^{13.4}\,{\rm cm}^{-2}$ and Doppler parameter $b_{HI} = 129\,{\rm km\,s}^{-1}(T/10^6{\rm K})^{1/2}$. This gives a peak optical depth $\tau_0 = 1.5 \times 10^{-2}(N\lambda_{{\rm Ly}\alpha} f_{{\rm Ly}\alpha})/b_{HI} = 0.25$ and a total equivalent width $W_{{\rm Ly}\alpha} = 137$mÅ. The detection of such a line-profile requires at least moderate resolution UV spectra at relatively high S/N and also a precise knowledge of the intrinsic continuum of the background source.

Detections of line profiles consistent with broad Lyα absorption have now been reported in the few HST/STIS echelle spectra with high S/N (Lehner et al. 2007, Richter et al. 2006). Because these detections lie near the sensitivity limit of the spectra, it is not possible to exclude the possibility that these broad features arise from the blends of several weak, narrow absorbers. Furthermore, independent analysis of these same datasets have questioned most of the putative detections

(Stocke, priv. comm.; Danforth & Shull, in preparation). These issues can only be resolved by higher S/N observations and, ideally, analysis of the corresponding Lyβ profiles. We look forward to the installation of HST/COS which should enable such observations. Even if one can conclusively demonstrate the presence of warm/hot gas with broad Lyα absorbers, there are limitations to directly assessing the mass density of this gas from empirical observation. First, studies of numerical simulations reveal that broad Lyα lines arise from gas whose kinematics are a mixture of thermally and turbulently broadened motions making it difficult to derive the temperature and obtain an ionization correction (Richter et al. 2006). Second, these same models predict that photoionization processes are significant for some gas with $T < 10^6$K and this contribution to the ionization correction cannot be assessed from HI observations alone. Finally, the detection of broad Lyα features becomes prohibitive for gas with $T > 10^{6.5}$K (i.e. $\tau_0 < 0.05$ for $N_H = 10^{20}\,\mathrm{cm}^{-2}$), which may account for the majority of WHIM material by mass (e.g. (Davé et al. 2001)). In short, one is driven to statistical comparisons between numerical simulations and observations of broad Lyα absorption and, ultimately, may be limited to a minor fraction of the WHIM.

At $T > 10^{6.3}$K, the ion fraction of O^{+5} is negligible and broad Lyα lines are generally too weak to detect with previous or planned UV spectrometers. To more efficiently trace this gas, we must turn to higher ionization states of oxygen and other elements. One option is the NeVIII doublet which probes gas to $T \approx 10^7$K and is accessible with UV spectrometers for $z > 0.25$ (Savage et al. 2005). The most sensitive tracers of $T > 10^6$K gas, however, are resonance lines of OVII (21.602 Å), OVIII (18.969 Å), and the NeVIII doublet (770,780 Å). While NeVIII may be detectable at $z > 0.5$ with COS, for current X-ray satellites the oxygen lines are a demanding observational challenge; one is limited by both spectral resolution and instrument collecting area. Nicastro and collaborators have addressed the latter issue (in part) by monitoring X-ray variable sources that show occasional, bright flares. During a flare event, one obtains target-of-opportunity observations to yield a relatively high S/N dataset (Fang et al. 2002, Nicastro et al. 2005, Williams et al. 2006). While this program has been successful observationally, the purported detections of intergalactic OVII and OVIII remain controversial (Kaastra et al. 2006, Rasmussen et al. 2007). While we are optimistic that these programs will reveal a few indisputable detections before the demise of the Chandra and XMM observatories, it is evident that a statistical survey for OVII and OVIII awaits future instrumentation (e.g. Con-X, XEUS). Because OVI may indicate cooling gas within a hot medium (or hot-cold interfaces), the numerous detections of OVI can serve as signposts to the 'true' WHIM (Shull et al. 2003).

Ultimately, however, the challenge will remain to convert these tracers of hot gas into constraints on the mass density and spatial distribution of WHIM gas. At a minimum, this requires assumptions about the metal-enrichment of the intergalactic medium.

In summary, we find the current empirical evidence for the WHIM to be intriguing but far from definitive. One must consider additional means for exploring the WHIM that would be enabled by new, proposed or planned instrumentation. One of these is to explore the line *emission* (e.g. Lyα, OVI, OVII) from the denser WHIM gas (Bertone et al. 2008, Cen & Fang 2006), which could be accomplished with UV and X-ray missions currently in design (e.g. *IGM,XENIA*). If combined with absorption-line studies, one may map out the mass density and spatial distribution of this gas at low z. Finally, it may be possible to cross-correlate the Sunyaev-Zeldovich (SZ) signal of the CMB with WHIM structures to infer the density and spatial distribution of this ionized gas (Ho et al. 2008). This might be accomplished with currently planned SZ experiments.

16.7 Future Facilities and the Baryon Census

In the spirit of the meeting 'Astrophysics in the Next Decade: JWST and Concurrent Facilities', we consider how the baryon census will be advanced by the facilities in development now for operations in the next decade. To stimulate discussion at the conference, we composed a chart showing the applicability of each mission or facility to the important reservoirs of baryonic matter. The list of missions under consideration was obtained from the conference website. This chart appears in Fig. 16.10, with the contributions of each mission color coded for major (green), supporting (yellow), minor (red), or no (grey) role. These assignments were made in the idiosyncratic judgment of one of us (J. T.) and should not be taken too literally. Each mission has its own, presumably compelling, scientific justification that in

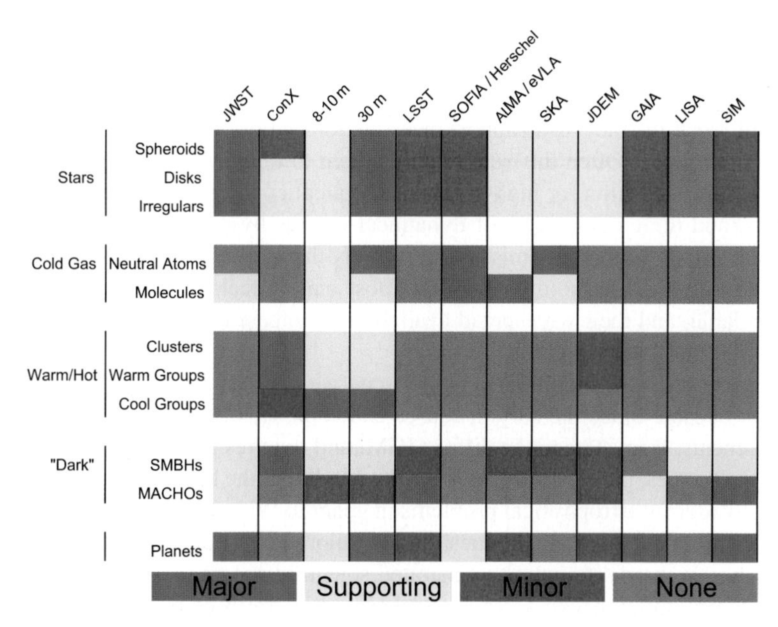

Fig. 16.10 For specific missions or categories of facility, the colors represent the role they are expected to play in measuring the budget of baryons in each phase discussed in the text

most cases does not explicitly include the baryon census (though JWST and Con-X are notable exceptions). The assignments were based primarily on wavelength coverage, spectral and spectroscopic resolution, field of view, and a review of the mission's public science case where available.

The overlap between missions and baryonic components raise some interesting points. First, the future looks relatively bright for the baryon census as a whole. JWST, Con-X, and ALMA are the three most important missions for resolving uncertainties in the census of baryons in stars, hot gas, and cold gas respectively. These are the areas of greatest uncertainty today, so this is somewhat encouraging for the future. However, only two of these facilities are in development phases: Con-X is still under study and could not reasonably be expected to fly until late in the next decade or after 2020.

The 'Assembly of Galaxies' is one of the four top-level science drivers of JWST, which will have the sensitivity and wavelength coverage to improve on stellar mass estimates at high redshift and the ability to measure IMFs by star counts over much larger local volumes than has been possible to date. As this is the most uncertain step in the derivation of the cosmic stellar mass budget, assessing the variation of the IMF from place to place and over time should be a primary goal of galaxy formation and evolution studies with JWST.

It looks as though the next decade will bring major advances in the study of the cold neutral and molecular phases of baryons over a wide redshift range, thanks to ALMA and eVLA. As mentioned above, the greatest need in this area is for a more thorough testing of the conversion factor from CO mass to H_2 mass in a broad range of astrophysical environments.

One pleasant surprise from this exercise is the notable versatility of large, ground-based optical telescopes (Keck and VLT today, and the 20–30-m ELTs in the future). Though they can lead the way for direct studies of stellar masses over a wide range of redshifts, they also support investigations whose primary use is in another wavelength range. Often this role requires them to obtain supporting data such as spectroscopic redshifts for high-z galaxies, optical counterparts for X-ray or radio sources, and the measurement of dynamical masses by high-resolution studies of linewidths in absorption or emission. It is likely that this versatility reflects the long history of optical astronomy, where the most mature technologies and techniques are available, and their widespread availability compared with the scarce resource of space-based observing time.

This exercise also highlights a broad trend we may expect from astrophysics in the next decade. Since the baryon census encompasses many diverse astrophysical environments, from stars to the diffuse IGM, and requires a wide range of observational techniques, the capability of a facility to address the baryon census reflects its ability to address astrophysical problems in general. With this in mind, we note that many of the space missions in question are tailored to specific scientific questions rather than a broad, community-generated scientific program. While it is difficult to see how a mission like LISA could have collateral uses beyond the detection of gravity waves from a few astrophysical sources, even missions carrying more conventional instruments like JDEM, GAIA, and SIM are fine-tuned to specific scientific cases. This trend toward specialization, even among flagship-scale missions, will evidently be a hallmark of astrophysics in the next decade.

Finally, we note that perhaps the greatest need for new capabilities to address the "missing baryons" lies in the X-ray, and that there is no realistic chance for Con-X or something like it to launch before the very end of the next decade. Thus it is possible that astronomers will, by 2015, see the 'first galaxies' (JWST), measure DE equation of state (LSST+JDEM), weigh molecules at $z \sim 6$ (ALMA), discover gravitational waves (LISA), find earthlike extrasolar planets (SIM+GAIA), and still not know where all the ordinary matter is or what phase it is in. This long interval would place the solution to the missing baryons problem at least 20 years away from when it was first recognized. This may be one of those cases where the urgency of performing an observation is tempered by the conventional wisdom that we already know the answer (the WHIM). But theory has been wrong before, and the only way to know is to do the experiment. If the last decade of astrophysics is a reliable guide, the next ten years will bring a few surprises. The what, when, and where of the baryons may well be among them.

Acknowledgments We acknowledge helpful criticism from M. Shull. J. X. P. is partially supported by NASA/Swift grant NNX07AE94G and an NSF CAREER grant (AST-0548180). J. T. gratefully acknowledges the generous support of Gilbert and Jaylee Mead for their namesake fellowship in the Yale Center for Astronomy and Astrophysics.

References

Adelman-McCarthy, J. K., et al. 2007, ApJS, 172, 634
Aguirre, A., Hernquist, L., Schaye, J., Weinberg, D. H., Katz, N., & Gardner, J. 2001, ApJ, 560, 599
Allen, S. W., Rapetti, D. A., Schmidt, R. W., Ebeling, H., Morris, R. G., & Fabian, A. C. 2008, MNRAS, 383, 879
Appleton, P. N., Xu, K. C., Reach, W., Dopita, M. A., Gao, Y., Lu, N., Popescu, C. C., Sulentic, J. W., Tuffs, R. J., & Yun, M. S. 2006, ApJL, 639, L51
Becker, G. D., Rauch, M., & Sargent, W. L. W. 2007, ApJ, 662, 72
Bergeron, J., & Herbert-Fort, S. 2005, in IAU Colloq. 199: Probing Galaxies through Quasar Absorption Lines, ed. P. Williams, C.-G. Shu, & B. Menard, 265–280
Bernstein, R., Shectman, S. A., Gunnels, S. M., Mochnacki, S., & Athey, A. E. 2003, in Instrument Design and Performance for Optical/Infrared Ground-based Telescopes, eds. M. Iye, A.F.M. Moorwood, Proceedings of the SPIE, Vol. 4841, 2003, 1694–1704
Bertone, S., Schaye, J., & Dolag, K. 2008, Space Science Reviews, 134, 295
Blanton, M. R., et al. 2003, ApJ, 592, 819
Boesgaard, A. M., Stephens, A., & Deliyannis, C. P. 2005, ApJ, 633, 398
Bolton, J. S., Haehnelt, M. G., Viel, M., & Springel, V. 2005, MNRAS, 357, 1178
Bouché, N., Gardner, J. P., Katz, N., Weinberg, D. H., Davé, R., & Lowenthal, J. D. 2005, ApJ, 628, 89
Bouché, N., Lehnert, M. D., & Péroux, C. 2006, MNRAS, 367, L16
Bregman, J. N. 2007, ARAA, 45, 221
Briggs, F. H., & Wolfe, A. M. 1983, ApJ, 268, 76
Bryan, G. L., Machacek, M., Anninos, P., & Norman, M. L. 1999, ApJ, 517, 13
Burles, S., & Tytler, D. 1998, ApJ, 507, 732
Cen, R., & Fang, T. 2006, ApJ, 650, 573

Cen, R., & Ostriker, J. P. 1999, ApJ, 514, 1
Chapman, S. C., Blain, A. W., Smail, I., & Ivison, R. J. 2005, ApJ, 622, 772
Chelouche, D., Ménard, B., Bowen, D. V., & Gnat, O. 2007, ArXiv e-prints, 706
Chen, H.-W., Prochaska, J. X., Bloom, J. S., & Thompson, I. B. 2005, ApJL, 634, L25
Chen, H.-W., Prochaska, J. X., & Gnedin, N. Y. 2007, ApJL, 667, L125
Cole, S., et al. 2001, MNRAS, 326, 255
Cooke, J., Wolfe, A. M., Gawiser, E., & Prochaska, J. X. 2006, ApJL, 636, L9
Cooksey, K. L., Prochaska, J. X., Chen, H.-W., Mulchaey, J. S., & Weiner, B. J. 2008, ApJ, 676, 262
Cui, J., Bechtold, J., Ge, J., & Meyer, D. M. 2005, ApJ, 633, 649
Danforth, C. W., & Shull, J. M. 2005, ApJ, 624, 555
Danforth, C. W., & Shull, J. M. 2007, ArXiv e-prints, 709
Danforth, C. W., Shull, J. M., Rosenberg, J. L., & Stocke, J. T. 2006, ApJ, 640, 716
Davé, R. 2008, MNRAS, 385, 147
Davé, R., Cen, R., Ostriker, J. P., Bryan, G. L., Hernquist, L., Katz, N., Weinberg, D. H., Norman, M. L., & O'Shea, B. 2001, ApJ, 552, 473
Davé, R., Hernquist, L., Katz, N., & Weinberg, D. H. 1999, ApJ, 511, 521
Davé, R., & Tripp, T. M. 2001, ApJ, 553, 528
Dekel, A., & Birnboim, Y. 2006, MNRAS, 368, 2
Devereux, N. A., & Young, J. S. 1990, ApJ, 359, 42
Draine, B. T. 2006, in Astronomical Society of the Pacific Conference Series, Vol. 348, Astrophysics in the Far Ultraviolet: Five Years of Discovery with FUSE, eds. G. Sonneborn, H. W. Moos, & B.-G. Andersson, 58–+
Dunkley, J., et al. 2008, ArXiv e-prints, 803
Ellison, S. L., Yan, L., Hook, I. M., Pettini, M., Wall, J. V., & Shaver, P. 2001, A & A, 379, 393
Fall, S. M., & Pei, Y. C. 1993, ApJ, 402, 479
Fan, X., Carilli, C. L., & Keating, B. 2006, ARAA, 44, 415
Fang, T., Marshall, H. L., Lee, J. C., Davis, D. S., & Canizares, C. R. 2002, ApJL, 572, L127
Fardal, M. A., Giroux, M. L., & Shull, J. M. 1998, AJ, 115, 2206
Faucher-Giguere, C., Prochaska, J. X., Lidz, A., Hernquist, L., & Zaldarriaga, M. 2007, ArXiv e-prints, 709
Fontana, A., et al. 2006, A & A, 459, 745
Fox, A. J., Petitjean, P., Ledoux, C., & Srianand, R. 2007, A & A, 465, 171
Fukugita, M. 2004, in IAU Symposium, Vol. 220, Dark Matter in Galaxies, eds. S. Ryder, D. Pisano, M. Walker, & K. Freeman, 227–+
Fukugita, M., Hogan, C. J., & Peebles, P. J. E. 1998, ApJ, 503, 518
Ganguly, R., Cen, R., Fang, T., & Sembach, K. 2008, ArXiv e-prints, 803
Gardner, J. P., Katz, N., Hernquist, L., & Weinberg, D. H. 2001, ApJ, 559, 131
Gawiser, E., et al. 2007, ApJ, 671, 278
Ge, J., & Bechtold, J. 1997, ApJL, 477, L73+
Giovanelli, R., et al. 2005, AJ, 130, 2598
Gnat, O., & Sternberg, A. 2007, ApJS, 168, 213
Haardt, F., & Madau, P. 1996, ApJ, 461, 20
Ho, S., Hirata, C. M., Padmanabhan, N., Seljak, U., & Bahcall, N. 2008, ArXiv e-prints, 801
Hoopes, C. G., Sembach, K. R., Heckman, T. M., Meurer, G. R., Aloisi, A., Calzetti, D., Leitherer, C., & Martin, C. L. 2004, ApJ, 612, 825
Hopkins, A. M., Rao, S. M., & Turnshek, D. A. 2005, ApJ, 630, 108
Hopkins, P. F., Richards, G. T., & Hernquist, L. 2007, ApJ, 654, 731
Hu, E. M., Kim, T.-S., Cowie, L. L., Songaila, A., & Rauch, M. 1995, AJ, 110, 1526
Jedamzik, K. 2004, PhRD, 70, 083510
Jena, T., Norman, M. L., Tytler, D., Kirkman, D., Suzuki, N., Chapman, A., Melis, C., Paschos, P., O'Shea, B., So, G., Lubin, D., Lin, W.-C., Reimers, D., Janknecht, E., & Fechner, C. 2005, MNRAS, 361, 70

Jenkins, A., Frenk, C. S., White, S. D. M., Colberg, J. M., Cole, S., Evrard, A. E., Couchman, H. M. P., & Yoshida, N. 2001, MNRAS, 321, 372
Jenkins, E. B., Tripp, T. M., Woźniak, P. R., Sofia, U. J., & Sonneborn, G. 1999, ApJ, 520, 182
Jorgenson, R. A., Wolfe, A. M., Prochaska, J. X., Lu, L., Howk, J. C., Cooke, J., Gawiser, E., & Gelino, D. M. 2006, ApJ, 646, 730
Kaastra, J. S., Werner, N., Herder, J. W. A. d., Paerels, F. B. S., de Plaa, J., Rasmussen, A. P., & de Vries, C. P. 2006, ApJ, 652, 189
Kanekar, N., & Chengalur, J. N. 2003, A & A, 399, 857
Kauffmann, G., et al. 2003, MNRAS, 341, 33
Kawata, D., & Rauch, M. 2007, ApJ, 663, 38
Keres, D., Yun, M. S., & Young, J. S. 2003, ApJ, 582, 659
Kereš, D., Katz, N., Weinberg, D. H., & Davé, R. 2005, MNRAS, 363, 2
Kim, T.-S., Bolton, J. S., Viel, M., Haehnelt, M. G., & Carswell, R. F. 2007, MNRAS, 382, 1657
Kim, T.-S., Carswell, R. F., Cristiani, S., D'Odorico, S., & Giallongo, E. 2002, MNRAS, 335, 555
Kirkman, D., & Tytler, D. 1997, ApJ, 484, 672
Kirkman, D., Tytler, D., Suzuki, N., O'Meara, J. M., & Lubin, D. 2003, ApJS, 149, 1
Kohler, K., & Gnedin, N. Y. 2007, ApJ, 655, 685
Korn, A. J., Grundahl, F., Richard, O., Barklem, P. S., Mashonkina, L., Collet, R., Piskunov, N., & Gustafsson, B. 2006, Nature, 442, 657
Kriek, M., et al. 2006, ApJL, 649, L71
Kriss, G. A., et al. 2001, Science, 293, 1112
Lah, P., et al. 2007, MNRAS, 376, 1357
Le Floc'h, E., et al. 2003, A & A, 400, 499
Ledoux, C., Petitjean, P., & Srianand, R. 2003, MNRAS, 346, 209
Lehner, N., Savage, B. D., Richter, P., Sembach, K. R., Tripp, T. M., & Wakker, B. P. 2007, ApJ, 658, 680
Linsky, J. L., et al. 2006, ApJ, 647, 1106
McDonald, P., et al. 2006, ApJS, 163, 80
Meiksin, A., & White, M. 2003, MNRAS, 342, 1205
Meiksin, A. A. 2007, ArXiv e-prints, 711
Meyer, M. J., et al. 2004, MNRAS, 350, 1195
Miralda-Escudé, J., Cen, R., Ostriker, J. P., & Rauch, M. 1996, ApJ, 471, 582
Moos, H. W., et al. 2002, ApJS, 140, 3
Mulchaey, J. S., Davis, D. S., Mushotzky, R. F., & Burstein, D. 1996, ApJ, 456, 80
Murphy, M. T., & Liske, J. 2004, MNRAS, 354, L31
Nagamine, K., Wolfe, A. M., Hernquist, L., & Springel, V. 2007, ApJ, 660, 945
Netterfield, C. B., et al. 2002, ApJ, 571, 604
Nicastro, F., Mathur, S., Elvis, M., Drake, J., Fang, T., Fruscione, A., Krongold, Y., Marshall, H., Williams, R., & Zezas, A. 2005, Nature, 433, 495
Noterdaeme, P., Ledoux, C., Petitjean, P., & Srianand, R. 2008, ArXiv e-prints, 801
Ogle, P., Antonucci, R., Appleton, P. N., & Whysong, D. 2007, ApJ, 668, 699
O'Meara, J. M., Burles, S., Prochaska, J. X., Prochter, G. E., Bernstein, R. A., & Burgess, K. M. 2006, ApJL, 649, L61
O'Meara, J. M., Prochaska, J. X., Burles, S., Prochter, G., Bernstein, R. A., & Burgess, K. M. 2007, ApJ, 656, 666
Ostriker, J. P., & Heisler, J. 1984, ApJ, 278, 1
Paschos, P., Jena, T., Tytler, D., Kirkman, D., & Norman, M. L. 2008, ArXiv e-prints, 802
Peimbert, M., Luridiana, V., & Peimbert, A. 2007, ApJ, 666, 636
Penton, S. V., Stocke, J. T., & Shull, J. M. 2002, ApJ, 565, 720
Penton, S.V., Stocke, J.T., & Shull, J.M. 2004, ApJS, 152, 29
Péroux, C., Dessauges-Zavadsky, M., D'Odorico, S., Kim, T.-S., & McMahon, R. G. 2003a, MNRAS, 345, 480
Peroux, C., Dessauges-Zavadsky, M., D'odorico, S., Kim, T.S., & McMahon, R.G. 2007, MNRAS, 382, 177

Péroux, C., McMahon, R. G., Storrie-Lombardi, L. J., & Irwin, M. J. 2003b, MNRAS, 346, 1103
Petitjean, P., Srianand, R., & Ledoux, C. 2000, A & A, 364, L26
Petitjean, P., Webb, J. K., Rauch, M., Carswell, R. F., & Lanzetta, K. 1993, MNRAS, 262, 499
Pettini, M., Smith, L. J., Hunstead, R. W., & King, D. L. 1994, ApJ, 426, 79
Prochaska, J. X. 1999, ApJL, 511, L71
Prochaska, J. X., & Burles, S. M. 1999, AJ, 117, 1957
Prochaska, J. X., Chen, H.-W., Howk, J. C., Weiner, B. J., & Mulchaey, J. 2004, ApJ, 617, 718
Prochaska, J. X., Gawiser, E., Wolfe, A. M., Castro, S., & Djorgovski, S. G. 2003, ApJL, 595, L9
Prochaska, J. X., Herbert-Fort, S., & Wolfe, A. M. 2005a, ApJ, 635, 123
Prochaska, J. X., O'Meara, J. M., Herbert-Fort, S., Burles, S., Prochter, G. E., & Bernstein, R. A. 2006a, ApJL, 648, L97
Prochaska, J. X., Tripp, T. M., & Howk, J. C. 2005b, ApJL, 620, L39
Prochaska, J. X., Weiner, B. J., Chen, H.-W., & Mulchaey, J. S. 2006b, ApJ, 643, 680
Rao, S. M., Turnshek, D. A., & Nestor, D. B. 2006, ApJ, 636, 610
Rasmussen, A. P., Kahn, S. M., Paerels, F., Herder, J. W. d., Kaastra, J., & de Vries, C. 2007, ApJ, 656, 129
Rauch, M., Miralda-Escude, J., Sargent, W. L. W., Barlow, T. A., Weinberg, D. H., Hernquist, L., Katz, N., Cen, R., & Ostriker, J. P. 1997, ApJ, 489, 7
Reddy, N. A., Steidel, C. C., Pettini, M., Adelberger, K. L., Shapley, A. E., Erb, D. K., & Dickinson, M. 2007, ArXiv e-prints, 706
Reiprich, T. H., & Böhringer, H. 2002, ApJ, 567, 716
Richter, P., Fang, T., & Bryan, G. L. 2006a, A & A, 451, 767
Richter, P., Savage, B. D., Sembach, K. R., & Tripp, T. M. 2006b, A & A, 445, 827
Rines, K., Diaferio, A., & Natarajan, P. 2008, ArXiv e-prints, 803
Rogerson, J. B., & York, D. G. 1973, ApJL, 186, L95+
Rosenberg, J. L., & Schneider, S. E. 2002, ApJ, 567, 247
Rudnick, G., Labbé, I., Förster Schreiber, N. M., Wuyts, S., Franx, M., Finlator, K., Kriek, M., Moorwood, A., Rix, H.-W., Röttgering, H., Trujillo, I., van der Wel, A., van der Werf, P., & van Dokkum, P. G. 2006, ApJ, 650, 624
Ryan-Weber, E. V., Webster, R. L., & Staveley-Smith, L. 2003, MNRAS, 343, 1195
Salucci, P., & Persic, M. 1999, MNRAS, 309, 923
Savage, B. D., Bohlin, R. C., Drake, J. F., & Budich, W. 1977, ApJ, 216, 291
Savage, B. D., Lehner, N., Wakker, B. P., Sembach, K. R., & Tripp, T. M. 2005, ApJ, 626, 776
Schaye, J. 2001a, ApJ, 559, 507
Schaye, J. 2001b, ApJL, 559, L1
Schaye, J., Aguirre, A., Kim, T.-S., Theuns, T., Rauch, M., & Sargent, W. L. W. 2003, ApJ, 596, 768
Scott, J., Bechtold, J., Dobrzycki, A., & Kulkarni, V. P. 2000, ApJS, 130, 67
Sembach, K. R., et al. 2005, in Bulletin of the American Astronomical Society, Vol. 37, Bulletin of the American Astronomical Society, 1197–+
Shapley, A. E., Steidel, C. C., Pettini, M., Adelberger, K. L., & Erb, D. K. 2006, ApJ, 651, 688
Shull, J. M., Roberts, D., Giroux, M. L., Penton, S. V., & Fardal, M. A. 1999, AJ, 118, 1450
Shull, J. M., Tumlinson, J., & Giroux, M. L. 2003, ApJL, 594, L107
Shull, J. M., Tumlinson, J., Giroux, M. L., Kriss, G. A., & Reimers, D. 2004, ApJ, 600, 570
Siana, B., Teplitz, H. I., Colbert, J., Ferguson, H. C., Dickinson, M., Brown, T. M., Conselice, C. J., de Mello, D. F., Gardner, J. P., Giavalisco, M., & Menanteau, F. 2007, ApJ, 668, 62
Simcoe, R. A., Sargent, W. L. W., & Rauch, M. 2002, ApJ, 578, 737
Simcoe, R. A., Sargent, W. L. W., Rauch, M., & Becker, G. 2006, ApJ, 637, 648
Spergel, D. N., et al. 2007, ApJS, 170, 377
Spite, M., Spite, F., & Maillard, J. P. 1984, A & A, 141, 56
Spitzer, L. 1978, Physical processes in the interstellar medium (New York Wiley-Interscience, 1978. 333.)
Stocke, J. T., Penton, S. V., Danforth, C. W., Shull, J. M., Tumlinson, J., & McLin, K. M. 2006, ApJ, 641, 217
Storrie-Lombardi, L. J., McMahon, R. G., Irwin, M. J., & Hazard, C. 1994, ApJL, 427, L13

Sutherland, R. S., & Dopita, M. A. 1993, ApJS, 88, 253
Tacconi, L. J., et al. 2006, ApJ, 640, 228
Thom, C., & Chen, H.-W. 2008a, ArXiv e-prints, 801
Tripp, T. M., Aracil, B., Bowen, D. V., & Jenkins, E. B. 2006, ApJL, 643, L77
Tripp, T. M., Giroux, M. L., Stocke, J. T., Tumlinson, J., & Oegerle, W. R. 2001, ApJ, 563, 724
Tripp, T. M., & Savage, B. D. 2000, ApJ, 542, 42
Tripp, T. M., Savage, B. D., & Jenkins, E. B. 2000, ApJL, 534, L1
Tripp, T. M., Sembach, K. R., Bowen, D. V., Savage, B. D., Jenkins, E. B., Lehner, N., & Richter, P. 2007, ArXiv e-prints, 706
Tumlinson, J. 2007, ApJL, 664, L63
Tumlinson, J., & Fang, T. 2005, ApJL, 623, L97
Tumlinson, J., Prochaska, J. X., Chen, H.-W., Dessauges-Zavadsky, M., & Bloom, J. S. 2007, ApJ, 668, 667
Tumlinson, J., Shull, J. M., Giroux, M. L., & Stocke, J. T. 2005, ApJ, 620, 95
Tumlinson, J., Shull, J. M., Rachford, B. L., Browning, M. K., Snow, T. P., Fullerton, A. W., Jenkins, E. B., Savage, B. D., Crowther, P. A., Moos, H. W., Sembach, K. R., Sonneborn, G., & York, D. G. 2002, ApJ, 566, 857
van Dokkum, P. G. 2008, ApJ, 674, 29
Viegas, S. M. 1995, MNRAS, 276, 268
Vladilo, G., Prochaska, J. X., & Wolfe, A. M. 2008, A & A, 478, 701
Vreeswijk, P. M., et al. 2004, A & A, 419, 927
Walter, F., Bertoldi, F., Carilli, C., Cox, P., Lo, K. Y., Neri, R., Fan, X., Omont, A., Strauss, M. A., & Menten, K. M. 2003, Nature, 424, 406
Weymann, R. J., et al. 1998, ApJ, 506, 1
Weymann, R. J., Vogel, S. N., Veilleux, S., & Epps, H. W. 2001, ApJ, 561, 559
White, R. L., Becker, R. H., Fan, X., & Strauss, M. A. 2003, AJ, 126, 1
Williams, R. J., Mathur, S., & Nicastro, F. 2006, ApJ, 645, 179
Wolfe, A. M., Gawiser, E., & Prochaska, J. X. 2005, ARAA, 43, 861
Wolfe, A. M., Turnshek, D. A., Smith, H. E., & Cohen, R. D. 1986, ApJS, 61, 249
Young, J. S., & Scoville, N. Z. 1991, ARAA, 29, 581
Zwaan, M. A., Briggs, F. H., Sprayberry, D., & Sorar, E. 1997, ApJ, 490, 173
Zwaan, M. A., Meyer, M. J., Staveley-Smith, L., & Webster, R. L. 2005a, MNRAS, 359, L30
Zwaan, M. A., & Prochaska, J. X. 2006, ApJ, 643, 675
Zwaan, M. A., van der Hulst, J. M., Briggs, F. H., Verheijen, M. A. W., & Ryan-Weber, E. V. 2005b, MNRAS, 364, 1467

Matt Mountain and James Bullock obviously enjoy having their photos included in this book

Chapter 17
Observational Constraints of Reionization History in the JWST Era

Xiaohui Fan

Abstract The epoch of reionization is a crucial phase of cosmic evolution, when the UV photons from the first generation of galaxies and quasars ionized the neutral hydrogen in the IGM, ending the cosmic dark ages. In this Chapter, I will first review the techniques and current results on constraining reionization history. Measurement of CMB polarization indicates the peak of reionization activity at $z \sim 10$. Observations of IGM transmission in high-redshift quasar spectra show that the Universe was ionized by $z \sim 6$, while rapid increase of Gunn-Peterson optical depth towards high redshift suggests that reionization is ending at that epoch. Reionization is likely a prolonged and complex process, requiring detailed observations of high-redshift sources to completely unveal its history. JWST and concurrent ground-based facilities will allow dramatic progress in high-redshift observations. New high-redshift quasar surveys will provide JWST ideal luminous sources to map IGM evolution at $z > 7$; galaxy surveys using Lyman Break galaxies and Lyα emitters as tracers will not only detect the dwarf galaxies that are likely the primary sources of reionization, but also allow detailed measurements of IGM topology during reionization era. Aided by further development in theoretical modeling and better understanding of high-redshift galaxy population, JWST will play a central goal in understanding the history of reionization.

17.1 Introduction

After the recombination epoch at $z \sim 1100$, the universe became mostly neutral, until the first generation of stars and quasars reionized the intergalactic medium (IGM) and ended the cosmic "dark ages". Cosmological models predict reionization at redshifts between 6 and 20. When and how the universe reionized remains one of the fundamental questions of modern cosmology (see reviews of Barkana and Loeb 2001, Loeb and Barkana 2001, Ciardi and Ferrara 2005, Fan et al. 2006).

X. Fan (✉)
Steward Observatory, University of Arizona, USA
e-mail: fan@as.arizona.edu

H.A. Thronson et al. (eds.), *Astrophysics in the Next Decade,* Astrophysics and Space Science Proceedings, DOI 10.1007/978-1-4020-9457-6_17,

The last few years have witnessed the first direct observational constraints on the history of reionization. Lack of complete Gunn and Peterson (1965) absorption at $z < 6$ indicates that the IGM is highly ionized by that epoch (e.g., Fan et al. 2000, Becker et al. 2001, Djorgovski et al. 2001, Songaila and Cowie 2002), while the GP optical depth appear to be increasing dramatically at $z > 6$, suggesting that we are closing in on the end of reionization epoch at those redshifts. At the other end of the reionization history, polarization measurements of the cosmic microwave background (CMB) from WMAP (Spergel et al. 2007, Komatsu et al. 2008) show a large optical depth due to Thompson scattering of electrons in the early Universe, suggesting that the IGM was largely ionized by $z \sim 10.8 \pm 1.4$. However, detailed reionization history remain highly uncertain beyond these two crude constraints.

Some of the key questions regarding reionization include:

- **When** did reionization happen? Whether it is early ($z \sim 15$) or late ($z \sim 6$–8), whether it has an extended history or resembles a phase transition.
- **How** did reionization proceed? Whether it is homogeneous or with large scatter? How did HII regions grow during reionization and how did overlap happen?
- **What** did it? Whether the main sources of reionization photons come from dwarf galaxies, AGNs, or even more exotic sources such as decay particles.

To answer these questions, the main observations goals are: (1) to map the evolution and spatial distribution of IGM ionization state; and (2) to find the highest redshift galaxies and quasars which are the main sources of reionization. To probe early galaxy/quasar formation and reionization is one of the primary goals of JWST, and is closely related to all JWST instruments, as well as the missions of a number of future generation ground-based and space telescopes. In this Chapter, we first review the current observational constraints of reionization (Section 17.2), then discuss the future reionization probes using high-redshift quasars (Section 17.3) and galaxies (Section 17.4). We summarize main issues concerning reionization probes in Section 17.5. We will focus on probes using luminous sources (quasars and galaxies), and refer the reader to other Chapters of these proceeding to future 21cm and CMB probes of the reionization.

17.2 Current Observational Constraints

17.2.1 CMB Polarization

Thomson scattering produces CMB polarization when free-electron scatterers are illuminated by an anisotropic photon distribution. Large scale CMB polarization therefore probes the ionization history by measuring the total optical depths to be CMB produced by free electrons generated from reionization process. The strongest polarization constraint comes from the new WMAP 5-yr results (Komatsu et al. 2008), with the best-fit values of $\tau = 0.084 \pm 0.016$, and $z_{\rm reion} = 10.8 \pm 1.4$, assuming an instantaneous reionization at $z_{\rm reion}$. However, the CMB polarization

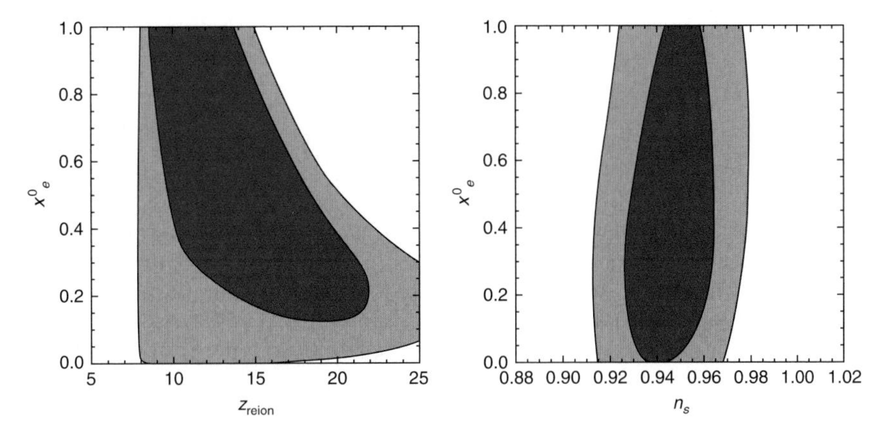

Fig. 17.1 WMAP 3-yr constraints on the reionization history. (*Left*) The 68% and 95% joint 2D marginalized confidence level contours for $x_e^0 - z_{\text{reion}}$ for a power-law cold dark matter (CDM) model with the reionization history that assuming the universe to be partially reionized at to an ionization fraction of x_e^0 and then became fully ionized at $z = 7$. (*Right*) The 68% and 95% joint 2D marginalized confidence level contours for $x_e^0 - n_s$, showing that reionization history and CDM power-law index are nearly independent for a given value of τ, indicating that WMAP determinations of cosmological parameters are not affected by details of the reionization history. Adapted from Spergel et al. (2007)

only measures an integrated reionization signal. Figure 17.1 illustrates the degeneracy of reionization redshift and detailed reionization history based on CMB data alone, using a simple reionization model assuming that the universe was partically reionization at z_{reion} to an ionization fraction of x_e^0 and then became fully ionizated at $z = 7$.

Detailed measurements of reionization history, however, requires observations of high-redshift sources during the reionization era. These observations will not only map the evolution of IGM ionization state, $x_{HI}(z)$, but also reveal the topology and spatial distribution of reionization process, and identify the sources that are responsible to reionization.

17.2.2 Gunn-Peterson Tests

Gunn and Peterson (GP, 1965) first proposed using Lyα resonance absorption in the spectrum of distant quasars as a direct probe to the neutral hydrogen density in the IGM at high-redshift. The Gunn-Peterson optical depth to Lyα photons is

$$\tau_{\text{GP}} = \frac{\pi e^2}{m_e c} f_\alpha \lambda_\alpha H^{-1}(z) n_{\text{HI}}, \tag{17.1}$$

where f_α is the oscillator strength of the Lyα transition, $\lambda_\alpha = 1216\text{Å}$, $H(z)$ is the Hubble constant at redshift z, and n_{HI} is the density of neutral hydrogen in the IGM.

At high redshifts:

$$\tau_{GP}(z) = 4.9 \times 10^5 \left(\frac{\Omega_m h^2}{0.13}\right)^{-1/2} \left(\frac{\Omega_b h^2}{0.02}\right) \left(\frac{1+z}{7}\right)^{3/2} \left(\frac{n_{HI}}{n_H}\right), \tag{17.2}$$

for a uniform IGM. Even a tiny neutral fraction, $x_{HI} \sim 10^{-4}$, gives rise to complete Gunn-Peterson absorption. Note that this test is only sensitive at the end of the reionization when the IGM is already mostly ionized, and saturates for the higher neutral fraction in the earlier stage.

Over the last seven years, the Sloan Digital Sky Survey (SDSS) and other wide-field imaging surveys have discovered $\sim$30 quasars at $z > 5.5$, with the highest redshift at $z = 6.43$ (Willott et al. 2007). Figure 17.2 presents the spectra of 27 quasars at $z > 5.7$ from the SDSS. It shows strong redshift evolution of the transmission of the IGM: transmitted flux is clearly detected in the spectra of quasars at $z < 6$ and blueward of the Lyα emission line; the absorption troughs deepen for the high-redshift quasars, and complete Gunn-Peterson absorption begins to appear along lines of sight at $z > 6.1$.

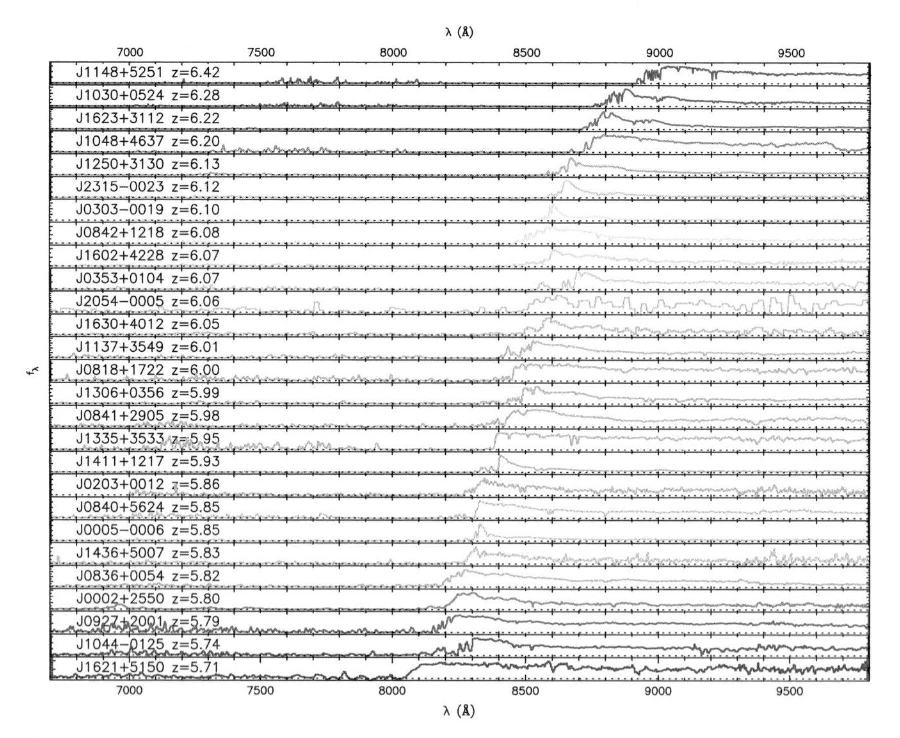

Fig. 17.2 Moderate resolution spectra of twenty-seven lumouns SDSS quasars at $5.74 < z < 6.42$. Note the strong Gunn-Peterson absorption on the blue side of Lyα emission at $z > 6$

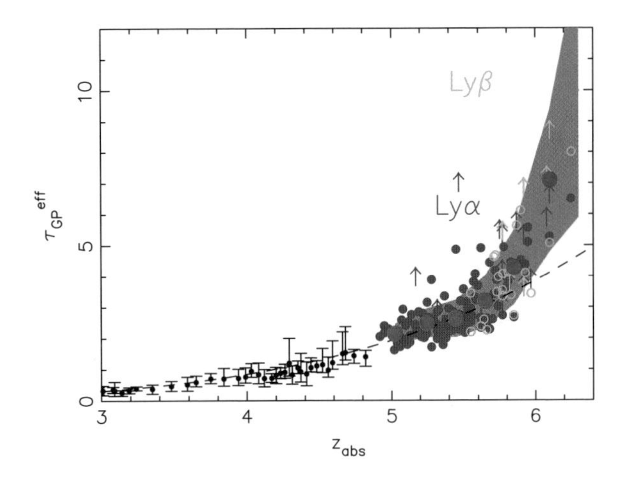

Fig. 17.3 Evolution of Optical depth with combined Lyα and Lyβ results. The dash line is for a redshift evolution of $\tau_{\mathrm{GP}} \propto (1+z)^{4.3}$. At $z > 5.5$, the best fit evolution has $\tau_{\mathrm{GP}} \propto (1+z)^{>10.9}$, indicating an accelerated evolution. The large open symbols with error bars are the average and standard deviation of optical depth at each redshift. The sample variance increases also increases rapidly with redshift. Adapted from Fan et al. (2006)

Fan et al. (2006) measured the evolution of Gunn-Peterson optical depths along the line of sight of the nineteen $z > 5.7$ quasars from the SDSS (Fig. 17.3). We found that at $z_{\mathrm{abs}} < 5.5$, the optical depth can be best fit as $\tau \propto (1+z)^{4.3}$, while at $z_{\mathrm{abs}} > 5.5$, the evolution of optical depth accelerates: $\tau \propto (1+z)^{>10}$. There is also a rapid increase in the variation of optical depth along different lines of sight: $\sigma(\tau)/\tau$ increases from $\sim$15% at $z \sim 5$, to >30% at $z > 6$, in which τ is averaged over a scale of $\sim$60 comoving Mpc. Assuming photoionization equilibrium and a model of IGM density distribution, one can convert the measured effective optical depth in Fig. 17.3 to IGM properties, such as the level of UV ionizing background and average neutral fraction. We find that at $z > 6$ the volume-averaged neutral fraction of the IGM has increased to $>10^{-3.5}$, with both ionizing background and neutral fraction experiencing about one order of magnitude change over a narrow redshift range, and the mean-free-path of UV photons is shown to be <1 physical Mpc at $z > 6$ (Fan et al. 2006).

However, with the emergence of complete Gunn-Peterson troughs at $z > 6$, it becomes increasingly difficult to place stringent limits on the optical depth and neutral fraction of the IGM. More sensitive tests are required to probe the ionization state of the IGM towards neutral era. Songaila and Cowie (2002) suggested using the distribution of optically thick, dark gaps in the spectrum as an alternative statistic. Fan et al. (2006) examined the distribution of dark gaps with $\tau > 2.5$ among their sample of SDSS quasars at $z > 5.7$, and showed a dramatic increase in the average length of dark gaps at $z > 6$ (Fig. 17.4), similar to the model prediction of Paschos and Norman (2005). Gap statistics provides a powerful new tool to characterize the IGM ionization at the end of reionization (Gallerani et al. 2008), and can be sensitive to larger neutral fractions, as they carry higher-order information than optical depth alone. Songaila and Cowie (2002), Pentericci et al. (2002), Fan et al. (2002, 2006) also studied using the distribution function of transmitted fluxes or statistics of threshold crossing. On the other hand, the finite length of dark gaps at $z > 6$ and the existence of transmission peaks indicate that the GP absorption trough is

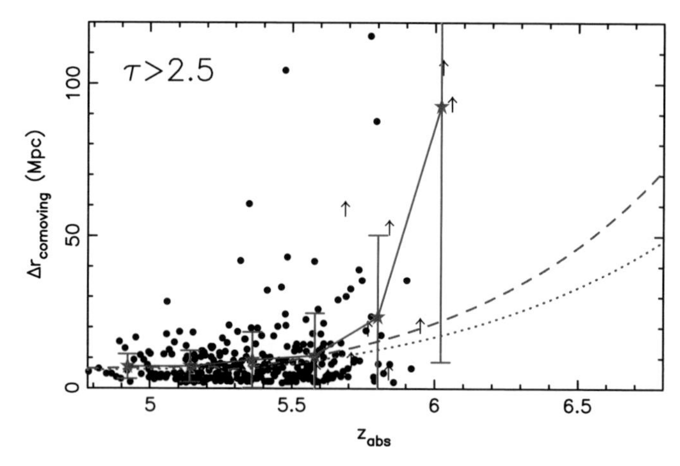

Fig. 17.4 Distributions of dark gaps, defined as regions in the spectra where all pixel having observed optical depth larger than 2.5 for Lyα transition. Upward arrows are gaps immediately blueward of quasar proximity zone, therefore the length is only a lower limit. Solid lines with error bars are average depth lengths with 1-σ dispersion at each redshift bin. Long dark gaps start to appear at $z \sim 5.6$, with the average gap length increases rapidly at $z > 6$, marking the end of reionization. It is compared with simulation of Paschos and Norman (2005), in which dashed and dotted lines are for moderate and high spectral resolutions. The simulation has an overlapping redshift at $z \sim 7$. Adapted from Fan et al. (2006)

not yet completely saturated, in the sense that isolated transmissions have not been suppressed by the GP damping wing. Fan et al. (2006) placed an upper limit of $x_{HI} < 30\%$ based on the length of dark gaps and the observed galaxy luminosity function at $z \sim 6$. Qualitatively, these statistics reveal very similar trends to the GP optical depth measurements. But they also contain high order information, related to the clustering of transmission spikes and dark gaps. The evolution of dark gaps appears to be the most sensitive quantity tracing the reionization process: of the statistics we examine, it shows the most dramatic evolution at $z > 6$. The distribution of dark gap length is readily computable from both observations and cosmological simulations (Gallerani et al. 2008). Simulations show that even when the average neutral fraction is high, the dark gap length is still finite due to the presence of regions that were ionized by star-forming galaxies.

Luminous quasars produce a highly ionized HII region around them even when the IGM is still mostly neutral otherwise. The presence of (time bounded) cosmic Strömgren spheres around the highest redshift SDSS QSOs has been deduced from the observed difference between the redshift of the onset of the GP effect and the systemic redshift of the host galaxy (White et al. 2005, Wyithe and Loeb 2004). The physical size of these spheres is typically ~5 Mpc at $z > 6$. The size of the Strömgren spheres is determined by the UV luminosity of the QSO, the HI density of the IGM, and the age of the QSO. However, the size of this HII region and the observed size of proximity zone around quasars are strongly affected by a number of factors, in particular, details of radiative transfer inside the quasar Strömgren sphere.

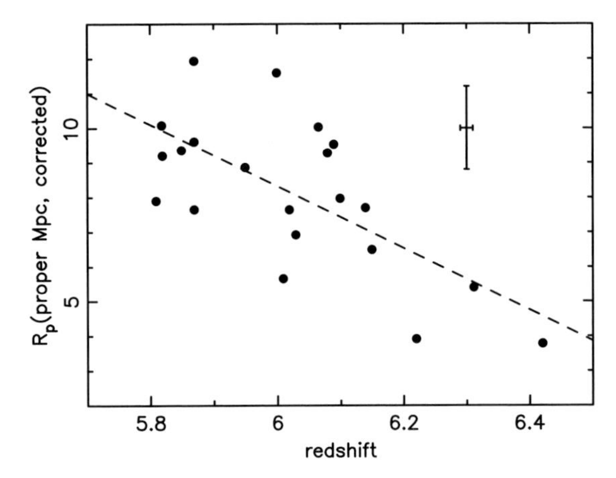

Fig. 17.5 Size of quasar proximity zone as a function of the redshift of the quasar. The size R_p is the line of sight distance from the quasar to the point at which the transmitted flux ratio falls to 0.1 of the continuum level. The radius also has been scaled to a common absolute magnitude ($M_{1450} = -27$). The linear fit shows a factor of 2.8 decrease in the HII region size around quasars from $z = 5.7$ to 6.4. Adapted from Fan et al. (2006)

In practice, it is also non-trivial to define the proximity zone size in the presence of patchy Lyα forest absorption.

Instead of using the Strömgren sphere radius, we define the size of the proximity zone R_p as the region where the transmitted flux ratio is above 0.1, when smoothed to a resolution of 20Å. Figure 17.5 shows the evolution of proximity zone sizes among the SDSS quasars at $z > 5.7$, after correcting for the luminosity dependence. The significant evolution of the size of quasar HII regions validates the physical picture that quasar HII regions are expanding into an increasingly neutral IGM at higher redshift. Since $R_p \propto [(1 + z)x_{\rm HI}]^{-1/3}$ (after scaling out the dependence on quasar luminosity), a decrease of HII region sizes by a factor of 2.8 from $z = 5.7$ to 6.4 suggests an increase of the neutral fraction by a factor of 14, consistent with results from GP optical depth measurements.

17.2.3 Lyα Emitters

Surveys of galaxies with strong Lyα emission lines through narrow-band imaging in selected dark windows of the night sky OH emission forest have proven to be a powerful technique for discovering the highest redshift galaxies (e.g., Hu et al. 2002, Rhoads et al. 2003, Taniguchi et al. 2005, Iye et al. 2006). Figure 17.6 shows the spectra of currently the highest redshift galaxy with confirmed spectroscopic redshift at $z = 6.96$, discovered using narrow-band imaging technique (Iye et al. 2006).

Lyα galaxies represent a significant fraction of star forming galaxies at high redshift. Furthermore, properties of Lyα galaxies can be used to directly probe

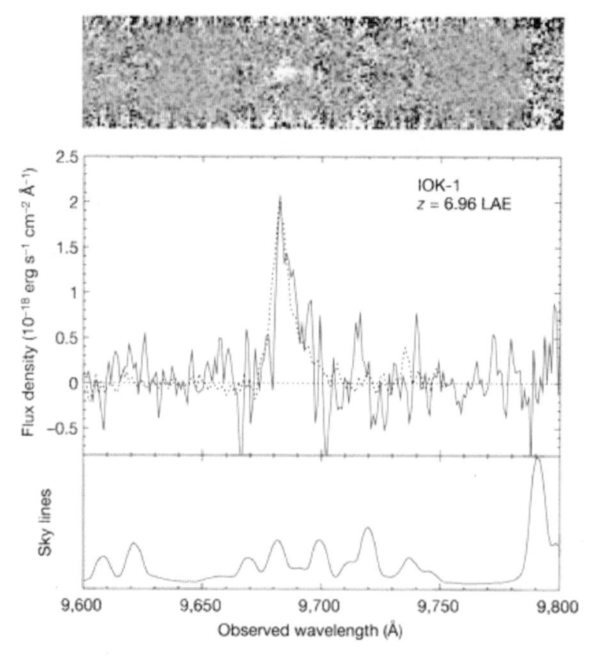

Fig. 17.6 Combined spectrum of the most distant Lyα emitter with confirmed spectroscopic redshift. Adapted from Iye et al. (2006)

the IGM neutral fraction. For a largely neutral IGM, $\tau \sim 10^5$, the damping wing of the GP trough arising from the large GP optical depth of the neutral medium will extend into the red side of the Lyα emission line (Miralda-Escudé 1998). For $z \sim 6$, at $\sim$10Å redward of Lyα of the host galaxy, the optical depth is of order unity for a neutral IGM. However, this GP damping wing test cannot be applied to luminous quasars, due to the proximity effect from the quasar itself, as shown by Madau and Rees (2000) and Cen and Haiman (2000) and discussed in the last section.

Without a large Strömgren sphere, the intrinsic Lyα emission will be considerably attenuated. In the simplest picture, one predicts: (1) the Lyα galaxy luminosity function will decrease sharply in an increasingly neutral IGM, even if the total star formation rate in the Universe remains roughly constant, and (2) the Lyα profiles will have a stronger red wing and a smaller average equivalent width before the onset of reionization (Haiman 2002).

Malhotra and Rhoads (2004) and Stern et al. (2005) combined the LALA survey of Lyα galaxies (Rhoads et al. 2003) with other Lyα surveys in the literature to determine the luminosity function of Lyα galaxies at $z = 6.5$ and 5.7. They found no evolution between these two redshift bins, consistent with the IGM being largely ionized by $z \sim 6.5$. However, using deep Subaru data, Ota et al. (2008) show a rapid decline of Lyα galaxy density at $z \sim 6.6$–7.0, while only mild evolution in

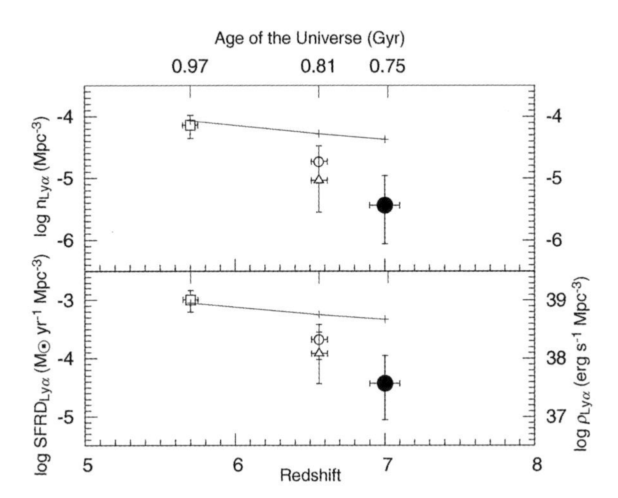

Fig. 17.7 Decline of Lyα emitter number density and star formation rate at high-redshift, comparing with the expected intrisic evolution. At $z > 6$, the densities clearly decrease with increasing redshifts and are smaller than the model-predicted values, implying that the Ly lines might be attenuated by the possibly increasing neutral IGM at the reionization epoch. Adapted from Ota et al. (2008)

the density of continuum-selected galaxies over the same redshift range (Fig. 17.7). The extra density evolution can be interpreted as the attenuation of due to a rapid evolution of neutral hydrogen fraction with $x_{HI} \sim 0.3$–0.6 at $z \sim 7$. The interpretations of these results, however, require more detailed modelling. Haiman (2002), Santos (2004), Cen (2005) showed that the local HII regions around Lyα galaxies reduce the attenuations of Lyα flux. Furthermore, the clustering of ionizing sources increases the HII region size and further reduces the attenuation. Including large scale clustering, the constraint on the IGM neutral fraction becomes less stringent (Furlanetto et al. 2006).

17.2.4 Sources of Reionization

Regardless of the detailed reionization history, the IGM has been almost fully ionized since at least $z \sim 6$. This places a minimum requirement on the emissivity of UV ionizing photons per unit comoving volume required to keep up with recombination and maintain reionization (Miralda-Escudé et al. 2000):

$$\dot{\mathcal{N}}_{ion}(z) = 10^{51.2}\,\mathrm{s}^{-1}\,\mathrm{Mpc}^{-3}\left(\frac{C}{30}\right) \times \left(\frac{1+z}{6}\right)^3 \left(\frac{\Omega_b h^2}{0.02}\right)^2, \tag{17.3}$$

where $C \equiv \frac{\langle n_H^2 \rangle}{\langle n_H \rangle^2}$ is the clumping factor of the IGM.

Quasars and AGN are effective emitters of UV photons. Luminous quasar density declines exponentially towards high-redshift: it is ~40 times lower at $z \sim 6$ than at its peak at $z \sim 2.5$. However, quasars have a steep luminosity function at the bright end – most of the UV photons come from the faint quasars that are currently below the detection limit at high-redshift. Jiang et al. (2008) recently determined the

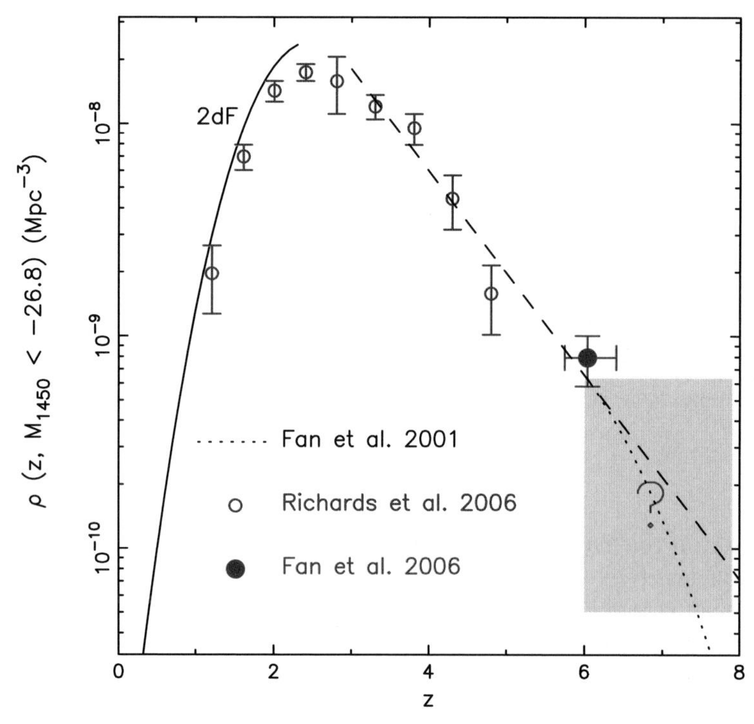

Fig. 17.8 Evolution of the density of luminous quasars based on the SDSS and 2dF surveys. Quasar density peaks at $z \sim 2$–3 and declines rapidly towards higher redshift

bright-end luminosity function of quasars at $z \sim 6$ using a sample of faint quasars. Based on the derived luminosity function, we find that the quasar/AGN population cannot provide enough photons to ionize the intergalactic medium (IGM) at $z \sim 6$ unless the IGM is very homogeneous and the luminosity at which the QLF power law breaks is very low (Fig. 17.8).

Due to the rapid decline in the AGN populations at very high z, most theoretical models assume stellar sources reionized the universe. Figure 17.9 shows the present constraints on the evolution of UV luminosity and star formation rate of Lyman Break galaxies (Bouwens and Illingworth 2006). Comparing to AGNs, galaxy population show only mild evolution at $z > 5$. However, despite rapid progress, there is still considerable uncertainty in estimating the total UV photon emissivity of star-forming galaxies at high-redshift, especially the IMF and the UV escape fraction from dwarf galaxies (Section 17.4.3). Given these uncertainties, the current data are consistent with star forming galaxies, in particular, relatively low luminosity galaxies, as being the dominant sources of reionizing photons, although more exotic sources, such as high-redshift mini-quasars, can not yet be ruled out as minor contributors of reionization budget.

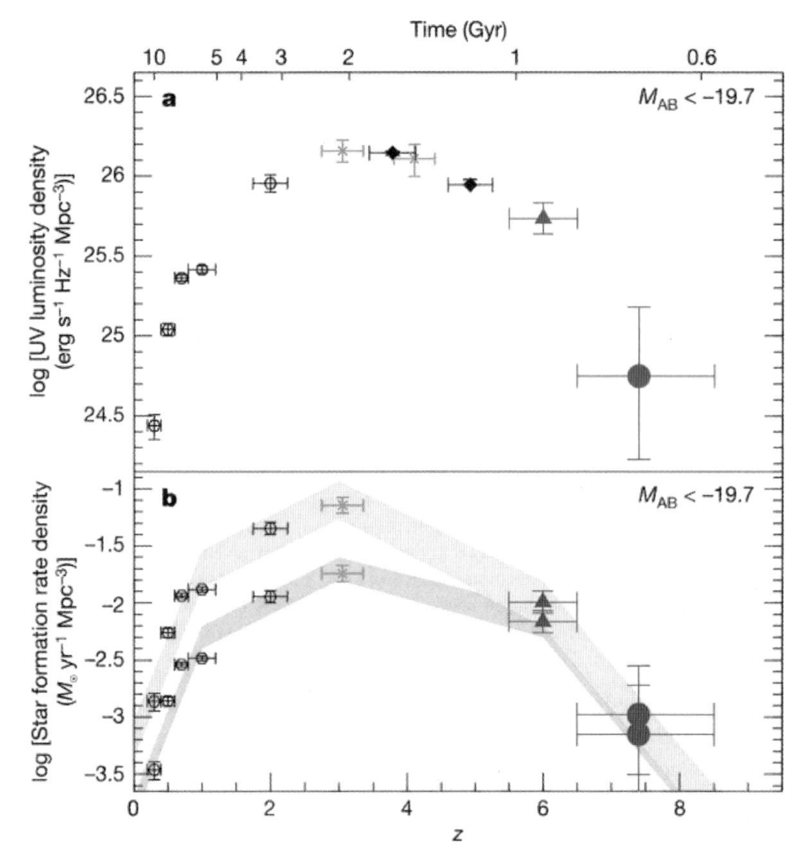

Fig. 17.9 The luminosity density and star-formation rate density evolution of Lyman Break Galaxy population. Adapted from Bouwens and Illingworth (2006)

17.2.5 Summary on Current Status

Figure 17.10 shows the current limits on the cosmic neutral fraction versus redshift. The observations paint an interesting picture. On the one hand, studies of GP optical depths and variations, and the GP "gap" distribution, as well as of the thermal state of the IGM at high z, and of cosmic Strömgren spheres and surfaces around the highest redshift QSOs, suggest a qualitative change in the state of the IGM at $z \sim 6$. These data indicate a significant neutral fraction, $x_{HI} > 10^{-3}$, and perhaps as high as 0.1, at $z \geq 6$, as compared to $x_{HI} \leq 10^{-4}$ at $z < 5.5$. The IGM characteristics at this epoch are consistent with the end of the 'percolation' stage of reionization (Gnedin and Fan 2006). On the other hand, transmission spikes in the GP trough and study of the evolution of Lyα galaxy luminosity function indicate a neutral fraction smaller than 50% at $z \sim 6.5$. Moreover, the measurement of the large scale polarization of the CMB suggests a significant ionization fraction extending to higher redshifts, $z \sim 11 \pm 2$.

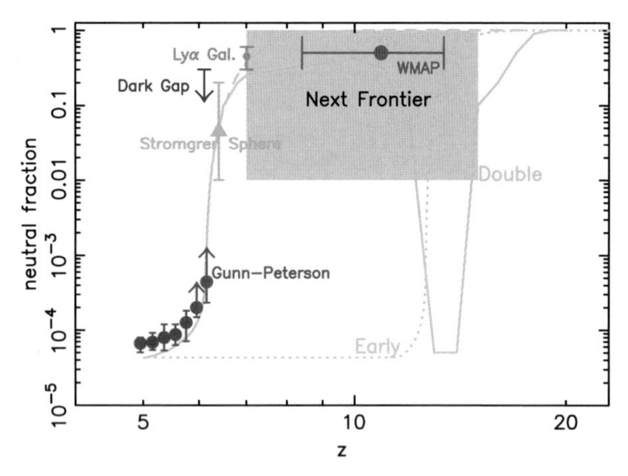

Fig. 17.10 The volume averaged neutral fraction of the IGM versus redshift using various techniques. The dashed line shows the model of Gnedin (2004) with late reionization at $z = 6$–7, the solid line shows an idealized model with double reionization as described in Cen (2003), and the dotted line illustrates the model with early reionization at $z \sim 14$. Current observational constraints indicate that the peak of the reionization activity is at $z \sim$ 7–13, but still cannot differentiate between various models of reionization history. By probing IGM evolution at $z > 7$, we will be able to determine whether the reionization is a phase transition or is a prolonged process that started at much higher redshifts

Note that all these measurements have implicit assumptions and uncertainties, as discussed throughout. Indeed, the GP effect and CMB large scale polarization studies can be considered complimentary probes of reionization, with optical depth effects limiting GP studies to the end of reionization, while CMB studies are weighted toward the higher redshifts, when the densities were higher. The data argue against a simple reionization history in which the IGM remains largely neutral from $z \sim 1100$ to $z \sim$ 6–7, with a single phase transition at $z \sim 6$ (the "late" model in Fig. 17.10), as well as against a model in which the Universe reaches complete ionization at $z \sim$ 15–20 and remained so ever since (the "early" model in Fig. 17.10). These facts, combined with the large line of sight variations at the end of reionization as indicated by Gunn-Peterson measurements, suggest a more extended reionization history. However, theoretical models with reionization caused by Population II star formation are consistent with both GP optical depth and WMAP CMB polarization measurement (e.g. Gnedin and Fan 2006). Current data do not present strong evidence for major contributions from metal-free Population III star formation at $z > 15$ to the reionization, and have weak constraints on models with multiple episodes of reionization (the "double" model in Fig. 17.10), which were suggested by a possible high CMB Thomspon optical depth based on early WMAP measurements.

17.3 Probing Reionization History with $z > 7$ Quasars

Current observational constraints show that we are already at the threshold of reionization, while the peak of reionization activity is likely at $7 < z < 15$, highlighting the need for higher redshift probes. Figure 17.11 summaries the galaxies and quasars

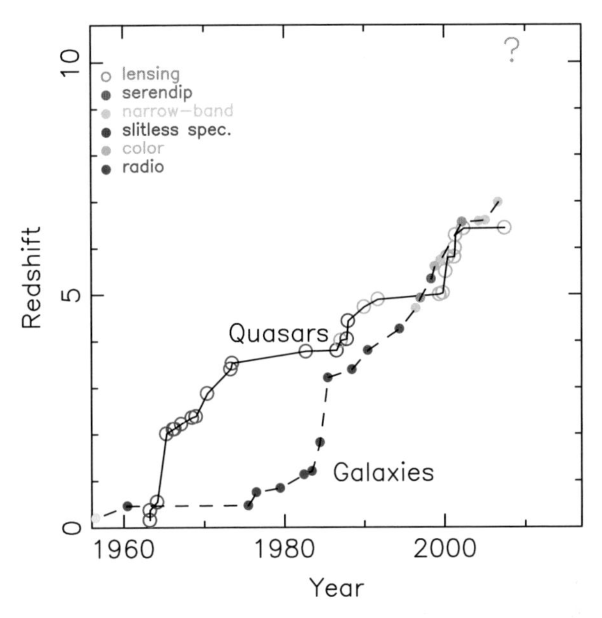

Fig. 17.11 The highest redshift known galaxies and quasars, as a function of discovery date. Different symbols represent different discovery techniques. The question mark at $z \sim 10$ refers to the $z \sim 10$ Lyα emitter candidates reported in Stark et al. (2007)

with the highest confirmed spectroscopic redshifts in the past half century. The ability of detecting high redshift sources is mainly driven by technilogical advances, as the advant of sensitivity radio observations, digital optical detectors, 10-m class telescopes, and wide-angle ground-based surveys are clearly identified with the breaking of new redshift records. In this area, the observational goals in the next decade include: (1) to discover sufficient number of luminous quasars and AGNs at $z > 7$ to be used for absorption line studies; (2) to carry out surveys of $z > 7$ star forming galaxies, in order to identify the population that reionized the Universe, and use them as probes of the topology of reionization.

17.3.1 Probing the Neutral Era with JWST/Nirspec

Detailed spectroscopy of luminous quasars and AGNs at $z > 7$ can provide powerful probes to the neutral era of the IGM. As shown in Section 17.2, the classic Gunn-Peterson test saturates at volume averaged neutral fraction $x_{HI} \sim 10^{-3}$. However, new approaches such as dark gap statistics and HII region sizes around luminous quasars offer tests that are sensitive to IGM with higher neutral fraction. Gallerani et al. (2008) show that the distribution functions of dark gaps are dramatically different for different reionization. In order to characterize dark gap distribution, spectra with moderate (R of a few thousand) and high S/N (>30) are needed. Even with the largest ground-based telescope and JWST, such data are only possible with the most luminous, and rarest quasars.

Sizes of HII region around quasars are sensitive to IGM neutral fraction of order unity. Figure 17.12 (Bolton and Haehnelt 2007) presents the Lyα near-zone sizes in luminous quasars as a function of x_{HI} based on synthetic spectra extracted from cosmological reionization simulations that include detailed radiative transfer models. It both shows the promise – HII regions are only few physical Mpcs for neutral IGM, comparing to tens of Mpcs for a mostly ionized IGM; and uncertainties – large scatter on individual HII region sizes due to density and radiative transfer effects. The study shows the need of samples of a few dozen quasars to average out these large scatter. The advantage of HII region size measurement is that it can be applied to fainter AGNs, since low S/N, low resolution spectra are all that are needed to measure the regions with flux transmission around quasars.

Observations of the evolution of heavy element in the IGM provide yet another powerful probe of reionization and early galaxy formation/feedback. Oh (2002) proposed using OI absorption in quasar spectra to probe reionization, since OI and HI have almost identical ionization potential, and are in a tight charge exchange equilibrium. OI has much lower abundance and therefore will not saturate even when the IGM is mostly neutral and HI optical depth $\gg$1. The fluctuation OI forest during the neutral era constrains both ionization topology and metal pollution of the early IGM. Becker et al. (2006) presented the first detection of IGM OI absorption at $z > 6$. A major drawback is that detection of weak OI lines require even higher quality data than those needed for the dark gap tests.

The Nirspec instrument on JWST will be the most powerful near-IR spectrograph for faint object spectroscopy. Figure 17.13 presents simulations of quasars at $z = 9.4$ with different flux level and different resolution, observed using Nirspec, after a long

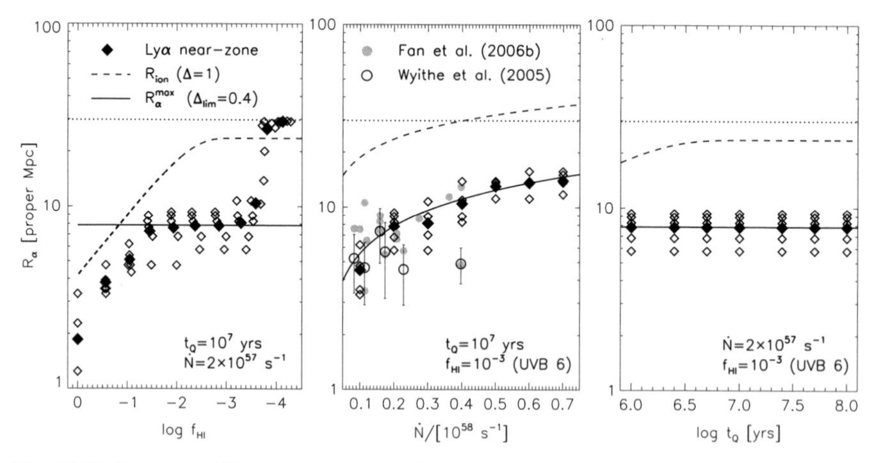

Fig. 17.12 The sizes of Ly near-zones around quasars at $z = 6$ computed from radiative transfer simulations in an inhomogeneous hydrogen and helium IGM. Adapted from Bolton and Haehnelt (2007)

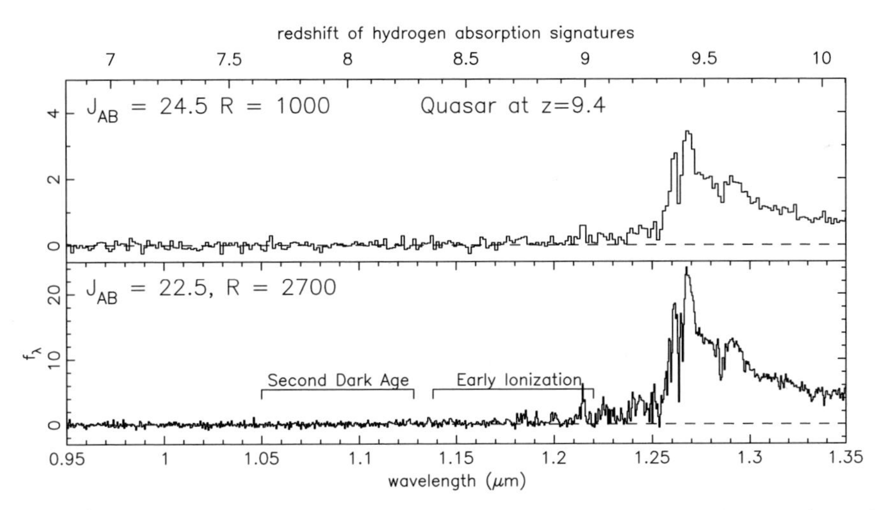

Fig. 17.13 Simulated JWST/NIRSpec spectra of quasars at $z \sim 9.4$, with total exposure times of 300k sec. The top panels shows a faint ($J_{AB} \sim 24.5$) quasar with $R \sim 1000$ resolution, allowing accurate measurements the HII region sizes and average Gunn-Peterson optical depth; the bottom panel shows a bright ($J_{AB} \sim 22.5$) quasars with $R \sim 2700$ resolution, allowing measurements of dark gap statistics. In both cases, we assume a complex reionization history in which the Universe was highly ionized at $z \sim 10$, and has a part recombination between $z \sim 7$–9 before it is completely ionized at $z \sim 7$

(300 k second) integration. Under the lowest resolution mode ($R \sim 100$), NIRSpec is sensitive to very faint AGNs; however, the spectra can only be used to measure average Gunn-Peterson optical depth which is insensitive to a neutral IGM. The intermediate resolution mode ($R \sim 1000$) provides sufficient resolution and S/N to measure HII region sizes for faint quasars ($J_{AB} \sim 24.5$), possible with the next generation near-IR surveys. Measurement of dark gap statistics and metal absorption requires the highest NIRSpec resolution ($R \sim 2700$), and can only be carried out on bright ($J_{AB} < 22.5$) quasars, discoverable with wide angle surveys. The key question is whether there will be quasars surveys capable to discovering a good number of $z > 7$ quasars by the time JWST is launched.

17.3.2 Quasar Surveys at $z > 7$

Luminous quasars are extremely rare at high redshift (Fig. 17.9), requiring wide-angeled, deep sky survey in the far optical and near-IR wavelength to discover even a small sample of them. Figure 17.14 shows the number counts of quasars at $z > 7$–10 in a 100 deg^2 survey as a function of survey depth, based on an extrapolation of the SDSS quasar luminosity function measured up to $z \sim 6$ and $M_{1450} \sim -25$. The two dashed lines show the approximate flux limit in order to carry out detailed spectroscopic studies to characterize dark gap distribution and to measure HII region sizes using the Nirspec instrument on JWST. Clearly, surveys of a few hundred deg^2

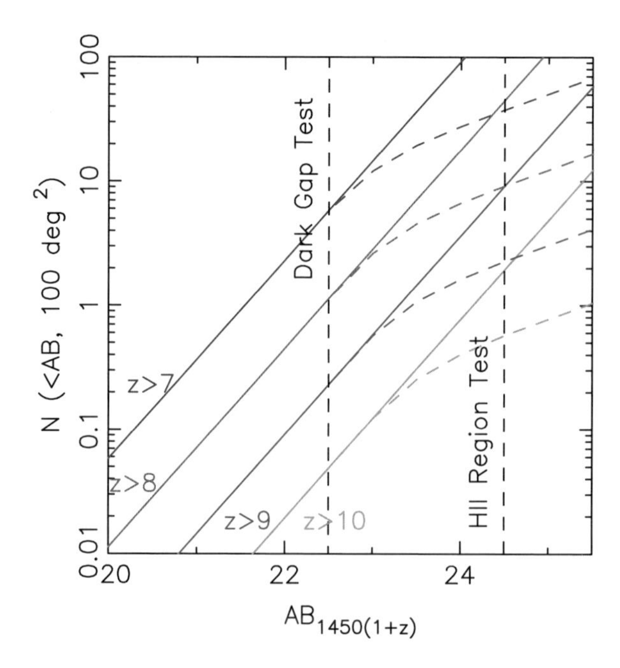

Fig. 17.14 Predicted number of quasars at high redshift over a survey of $\sim$100 deg^2, as a function of survey limiting magnitudes. The *solid line* shows an extrapolation from the power-law quasar luminosity function determined by the SDSS at $z \sim 6$; the *dashed line* assumes a break in the quasar luminosity function with L^* comparable to that in the 2dF quasar luminosity function at low-redshift

is needed at $z \sim 7$, while in order to establish reliable statistics at $z \sim 10$, surveys of $\sim$1000 deg^2 down to AB $\sim$ 24 is in order.

Optical surveys such as the SDSS are limited to detect sources at $z < 7$, due to the fact that Lyα emission line would redshift out of the sensitivity window of CCD detectors at $z \sim 7$. The new generations of red-sensitive CCD devices greatly improved the quantum efficiency of CCDs at $\sim$1 μm. A number of new surveys, such as the new Hyper-suprimecam survey on Subaru telescope, and the Pan-Starrs and LSST projects, all plan to ultilize the 1 μm sensitivity to search for quasars at $z \sim 7$, likely resulting in dozens of such objects discovered in the next five years.

At even higher redshift, wide-field near-IR surveys are required. Because of the rapidly increasing difficulty of working in the near-IR from the ground, the situation is less clear. Dedicated near-IR surveys using the UKIRT and VISTA telescopes are on-going or being planned. These surveys might recover a small sample of quasars at $z = 7$–9, but more ambitious programs might be needed in order to find higher-redshift/brighter objects. For example, a Spitzer survey during the warm mission over $\sim$500 deg^2 in areas in which deep optical data exist might uncover $\sim$20 quasars at $z > 8$. The new ESA mission e-ROSITA might discover similar number of quasars at these redshift, over a large fraction of the whole sky. Near-infrared wide-angle space-based surveys such as proposed JDEM and EUCLID dark energy surveys will have the sensitivity and sky coverage that allow discoveries of large number of quasars at $z > 7$.

17.4 Mapping Reionization History with Star-Forming Galaxies

17.4.1 Surveying Galaxies that Reionized the Universe

Currently, the highest redshift confirmed Lyα galaxy is at $z \sim 7$. In addition, Stark et al. (2007) carried out a spectroscopic survey of gravitational lensed high-redshift galaxies along critical lines of a number of galaxies cluster, and reported the discoveries of up to six highly promising Lyα emitter candidates at $z = 8.7$ to 10.2. If the detected lines are indeed high-redshift Lyα, these sources will already provide sufficient photons that reionize the Universe by $z \sim 10$. However, at these redshift, we are at the limit of ground-based spectroscopic identification.

A number of dedicated high-redshift Lyα survey for $z > 7$ will be carried out over the next few years on 8-10m class telescopes. They use the dark windows in the near-IR OH airglow forest in sky spectrum to archive high sensitivity. Combination of new OH suppression technical and adaptive optics would further improve survey efficiency. Ground-based Lyα surveys could probe to be the most effective tool of finding Lyα emitters up to $z \sim 10$; at even higher redshift, the JWST tunable filter will likely be the best survey tool.

Broad-band surveys using JWST/Nircam (and as a precuror, deep surveys using WFC3 on HST) will provide order of magnitude increase in survey speed and sensitivity for high-redshift continuum surveys, and will identify the reionization population. A JWST survey will reach an AB magnitude of 32, and detect dwarf galaxies at $z > 15$. A key project surveying over the HST/GOODS field will detect large number of galaxies even up to $z \sim 15$. In addition, JWST/Nirspec will provide spectroscopic confirmation to at least the brighter candidates. These observations will provide the most valuable samples of high-redshift star-forming galaxies to probe the reionization history.

17.4.2 Lyα Emitter Surveys and Reionization Topology

Observations using quasar absorption spectra only provide one-dimensional reionization probe, since quasars have low spatial density; it is also possible that at $z \gg 7$, quasar density declines very steeply with redshift as growth of supermassive black holes is severely limited by the black hole accretion timescale – the Salpeter time. Lyα Emitters provide an alternative, and likely more powerful, probe of the reionization history, especially the reionization topology during overlapping.

Figure 17.15 is a series of cosmological simulations of the distribution of Lyα emitters during different stages of overlapping stage of reionization (McQuinn et al. 2007). Both the *observed* density and spatial distribution of Lyα emitters are strikely different, even though the *intrinsic* density are comparable over this narrow time span. The reason, as discussed in Section 17.2.2, is the selective attenuation of Lyα emission line in more neutral regions due to Lyα damping wing absorption. The evolution of Lyα emitter luminosity function is sensitive to the average neutral

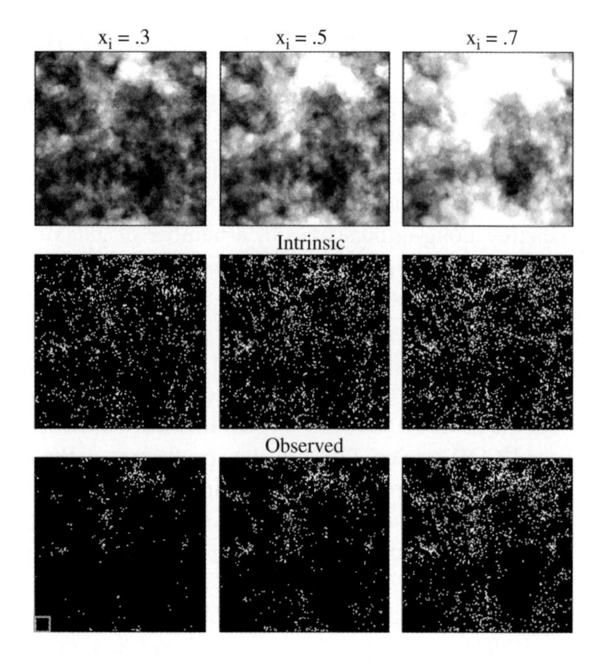

Fig. 17.15 Simulation of Lyα emitters over a half-degree field corresponding to a $94 \times 94 \times 35$ Mpc box. The top panel shows the distribution of IGM neutral fraction in three different stages of reionization at $x_i \sim 0.3,\ 0.5$ and 0.7. The middle panel shows the intrinsic distribution of Lyα emitters while the bottom panel shows the observed distribution as a result of IGM absorption. The observed distribution is highly sensitive to the reionization history. Adapted from McQuinn et al. (2007)

fraction, while the clustering properties and higher order statistics (such as genus numbers) provide sensitive tests to the overlapping topology and the growth of HII bubbles during reionization.

However, interpretation of Lyα emitter surveys is highly model-dependent, requiring understanding of the evolution of continuum luminosity function and clustering of star-forming galaxies and detailed models of Lyα radiative transfer in the galactic ISM. As shown in Fig. 17.16, a combination of continuum (dropout) selected sample and Lyα selected sample is needed to differentiate the intrinsic evolution of galaxy population and attenuation due to a neutral IGM. Ultimately, such probe requires a concerted, synergetic survey using different JWST instruments – NIRCam for dropout survey, NIRSpec for spectroscopy and FGS/TFI for Lyα survey at $z > 12$, as well as large ground-based telescopes for Lyα surveys at lower redshift, and ALMA to characterize star formation and ISM kinematics in the early galaxy population. This kind of survey should be a major focus for JWST mission.

17.4.3 Key Uncertainties

Although the next generation observing facilities, especially the JWST, will make order of magnitude improvement in our ability to find and characterize the high-redshift quasar and galaxy population, and make detailed measurements of IGM properties at $z > 7$ possible, two key uncertainties remain in interpretating these new observations in the context of reionization history.

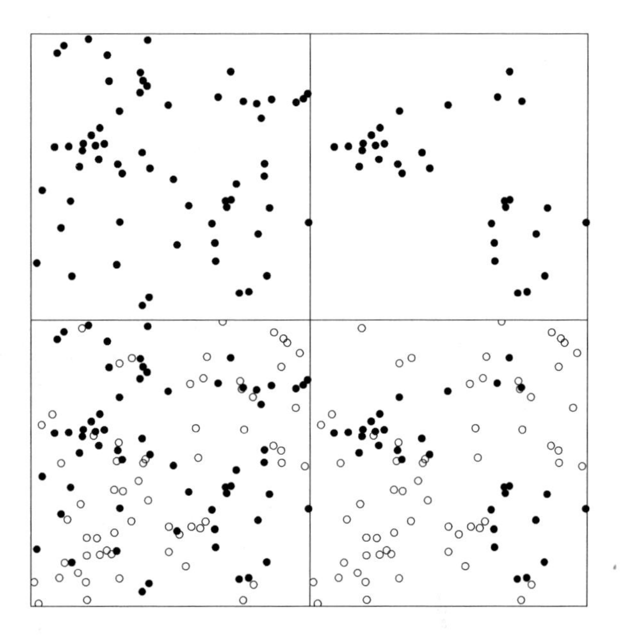

Fig. 17.16 Mapping ionized bubbles using Lyα emitters. *Top left* shows a hypothetical projected Lyα emitter distribution; *Top right* shows the observed distribution modified by IGM absorption during pathy reionization. *Bottom panels* and the same distributions, now include continnum-selected samples. Using Lyα emitter to map reionization requires the continuum-selected sample as a control population. Adapted from Rhoads (2008)

- Realistic reionization simulations. Measurement of IGM neutral fraction and its spatial distribution requires detailed comparisons between observations and theoretical models, since neutral fraction is not directly observable. Even converting the simpliest Gunn-Peterson optical depth measurement to neutral fraction is subject to the uncertainties in IGM density distribution models (e.g. Becker et al. 2007). Tests using HII region sizes, dark gaps and Lyα emitters require detailed modeling of high-redshift galaxy or quasar environments. Furthermore, many of the tracer populations are thought to be highly biased and clustered at high redshift: to fully understand their spatial distribution, large volume simulation with sufficient resolution is needed. Many of the tests also require accurate treatment of radiative transfer effect, a highly non-trivial task for cosmological simulations. As an illustration, Fig. 17.17 shows a simple convergence test by measuring the Thompson optical depth in reionization simulations with different mesh sizes (Gnedin and Fan 2006): these simulations are not yet converged in τ, and a simple fit shows even though the current simulation has much lower τ values, that the final τ could be significantly higher, consistent with the high WMAP values.
- Galaxy UV photon escape fraction. High-redshift dwarf galaxies are so-far the most likely candidates of the sources of reionization. However, a key uncertainty in this picture is the fraction of UV photons that are generated in galactic star forming regions actually escapting from the local ISM to the general IGM as sources of reionization. Observational constraints on escape fraction are highly uncertain. Steidel et al. (2001) inferred a relatively large escape fraction based on compositive spectrum of 29 Lyman Break Galaxies. Shapley et al. (2006)

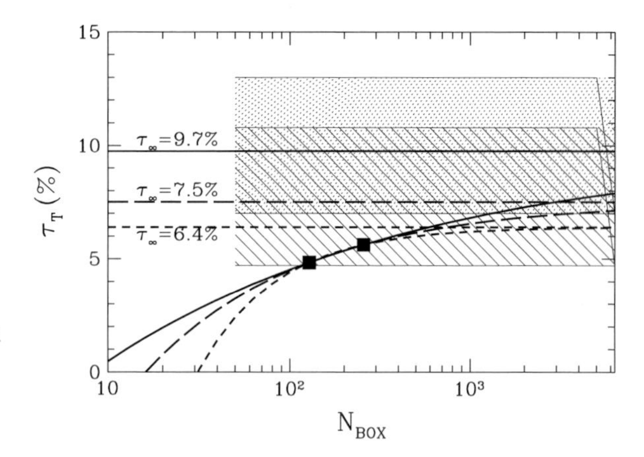

Fig. 17.17 Thomson optical depth as a function of resolution. The *gray* and *hatched bands* show WMAP constraints on τ. Different curves are fits to the simulation results assuming different convergence behaviors. Large simulation is clearly needed even to calculate simple global properties such as τ. Adapted from Gnedin and Fan (2006)

presented the first direct detection of Lyman continuum from Lyman Break Galaxies, indicating a lower escape fraction. The measurement, however, is based on only two detections (Fig. 17.18). Siana et al. (2007) presented deep far-UV imaging of sub-L^* galaxies at $z \sim 1.3$, showing no detection of Lyman continuum photons among 21 galaxies observed with HST. Theoretical models (e.g. Gnedin et al. 2008) also suggest a relatively low escape fraction in dwarf galaxies. If this is true, Gnedin (2008) argue that there might not be enough UV photons from dwarf galaxies to reionize the Universe by $z > 6$. This is a key uncertainty that affects almost all aspects of our understanding of reionization history. Detailed rest-frame far-UV observations of large sample of galaxies, spanning a wide range in luminosity, mass and metallicity are crucially needed to resolve this issue.

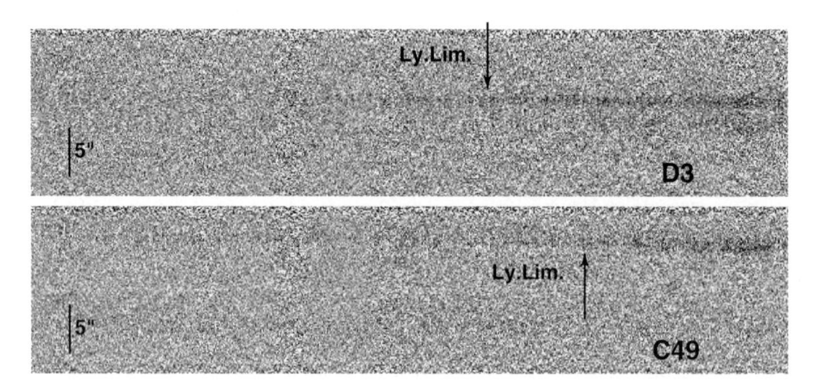

Fig. 17.18 Two dimensional spectra of two Lyman Break Galaxies with Lyman continuum detections from high S/N observations of 14 $z \sim 3$ galaxies. This is the first direct detection of Lyman continuum photons from high-redshift galaxies. Adapted from Shapley et al. (2006)

17.5 Summary

There have been tremendous progress in the observations of galaxy population and IGM at high-redshift in the last decade. However, our understanding of the history of cosmic reionization, a crucial event in the cosmic evolution, remain to be very limited. Strong evolution of Gunn-Peterson optical depths observed in high-redshift quasars indicates a rapid IGM evolution at $z > 6$, suggesting that reionization activities might be ending at $z \sim 6$–7; on the other hand, detections of large scale CMB polarization suggest abundant of free elections at high-redshift, and the peak of reionization activity at $z \sim 10$. Those two observations remain our main constraints on the reionzation thus far, and suggest a complex and extended reionization process. Although classic Gunn-Peterson test saturates at low neutral fraction, other tests, such as the sizes of quasar HII regions, distribution of gap Gunn-Peterson gaps and density of Lyα galaxies show promise in constraining IGM neutral fraction in the era in which the IGM might be substantially neutral.

JWST will represent order(s) of magnitude increase in the sensitivity of detecting and observing high-redshift galaxies and IGM. Large ground-based surveys will uncover luminous quasars at $z \sim 6$–10 over the next five years, providing ideal targets for JWST/Nirspec for high S/N spectroscopic observations of the IGM. JWST, combined with ground-based telescopes, will carry out large surveys of Lyα emitters and continuum-selected Lyman Break Galaxies at $7 < z < 15$. These new observations will not only identify galaxies that are responsible to the reionization, but also allow detailed mapping of reionization history, especially through Lyα galaxy surveys.

Acknowledgments I would like to that many colleagues that I have been working with on studies of high-redshift quasars and reionization, in particular, Bob Becker, Chris Carilli, Fabian Walter, Nick Gnedin, Jim Gunn, Linhua Jiang, Gordon Richards, Don Schneider, Michael Strauss and Rick White. My research has been supported by an Alfred P. Sloan Fellowship, a David and Lucile Packard Fellowship, and NSF Grant AST 03-07384.

References

Barkana, R., Loeb, A. 2001, Phys. Rep. 349, 125
Becker, R. H., et al. 2001, AJ, 122, 2850
Becker, G. D., et al. 2006, ApJ, 640, 69
Becker, G., D., et al. 2007, ApJ, 662, 72
Bolton, J. S., & Haehnelt, M. G., 2007, MNRAS, 374, 493
Bouwens, R., & Illingworth, G. 2006, Nature, 443, 189
Cen, R., & Haiman, Z. 2000, ApJ, 542, L75
Cen, R. 2003, ApJ, 591, 12
Cen, R., Haiman, Z., & Mesinger, A. 2005, ApJ, 621, 89
Ciardi, B., & Ferrara, A. 2005, Space Science Reviews, 116, 625
Djorgovski, S. G., Castro, S., Stern, D., & Mahabal, A. A. 2001, ApJ, 560, L5
Fan, X., et al. 2000, AJ, 120, 1167
Fan, X., et al. 2002, AJ, 123, 1247

Fan, X. et al. 2006, AJ, 132, 117
Fan, X., Carilli, C. L., & Keating, B. 2006, ARAA, 44, 415
Furlanetto, S. R., Zaldarriaga, M., & Hernquist, L. 2006, MNRAS, 365, 1012
Gallerani, S., Ferrara, A., Fan, X., & Choudhury, T. R., 2008, MNRAS, 386, 359
Gnedin, N. 2004, ApJ, 610, 9
Gnedin, N. Y.., & Fan, X. 2006, ApJ, 648, 1
Gnedin, N. Y. 2008, ApJ, 671, 1
Gnedin, N. Y., Kravstov, A. V., & Chen, H.-W. 2008, ApJ, 672, 765
Gunn, J. E., & Peterson, B. A. 1965,, ApJ, 142, 1633
Haiman, Z. 2002, ApJ, 576, L1
Hu, E. M., Cowie, L. L., McMahon, R. G., Capak, P., Iwamuro, F., Kneib, J.-P., Maihara, T., & Motohara, K. 2002, ApJ, 568, L75
Iye, M., et al. 2006, Nature, 443, 1861
Jiang, L., et al. 2008, AJ, 135, 1057
Loeb, A., Barkana, R. 2001, ARAA, 39, 19
Komatsu, E., et al. 2008, ApJS, submitted (astro-ph/0803.0547)
Madau, P., & Rees, M. J. 2000, ApJ, 542, L69
Malhotra, S., & Rhoads, J. E. 2004, ApJ, 617, L5
McQuinn, M., Hernquist, L., Zaldarriaga, M., & Suvendra, D. 2007, MNRAS, 381, 75
Miralda-Escudé, J. 1998, ApJ, 501, 15
Miralda-Escudé, J., Haehnelt, M., & Rees, M. J. 2000, ApJ, 530, 1
Oh, S. P. 2002, MNRAS, 336, 1021
Ota, K., et al. 2008, ApJ, 677, 12
Paschos, P., & Norman, M. L. 2005, ApJ, 631, 59
Pentericci, L., et al. 2002, AJ, 123, 2151
Rhoads, J. E., et al. 2003, AJ, 125, 1006
Rhoads, J. E. 2008, ApJ, submitted (astro-ph/0708.2909)
Santos, M. R. 2004, MNRAS, 349,1137
Shapley, A. E., et al. 2006, ApJ, 651, 688
Siana, B. D., et al. 2007, ApJ, 668, 62
Songaila, A., & Cowie, L. L. 2002, AJ, 123, 2183
Spergel, D., et al., 2007, ApJS, 130, 377
Stark, D. P., et al. 2007, ApJ, 663, 10
Steidel, C. C., et al. 2001, ApJ, 546, 665
Stern, D., et al. 2005, ApJ, 619, 12
Taniguchi, Y., et al. 2005, PASJ, 57, 165
White, R. L., Becker, R. H., Fan, X., & Strauss, M. A. 2005, AJ, 129, 2102
Willott et al. 2007, AJ, 134, 2435
Wyithe, J. S. B., & Loeb, A. 2004, Nature, 432, 194

Knox Long and Peter 'Tex' Stockman discuss the success of the conference

Chapter 18
The Frontier of Reionization: Theory and Forthcoming Observations

Abraham Loeb

Abstract The cosmic microwave background provides an image of the Universe 0.4 million years after the Big Bang, when atomic hydrogen formed out of free electrons and protons. One of the primary goals of observational cosmology is to obtain follow-up images of the Universe during the epoch of reionization, hundreds of millions of years later, when cosmic hydrogen was ionized once again by the UV photons emitted from the first galaxies. To achieve this goal, new observatories are being constructed, including low-frequency radio arrays capable of mapping cosmic hydrogen through its redshifted 21 cm emission, as well as imagers of the first galaxies such as the *James Webb Space Telescope (JWST)* and large aperture ground-based telescopes. The construction of these observatories is being motivated by a rapidly growing body of theoretical work. Numerical simulations of reionization are starting to achieve the dynamical range required to resolve galactic sources across the scale of hundreds of comoving Mpc, larger than the biggest ionized regions.

18.1 Preface

When we look at our image reflected off a mirror at a distance of 1 meter, we see the way we looked 6.7 nanoseconds ago, the light travel time to the mirror and back. If the mirror is spaced 10^{19} cm $\simeq$ 3 pc away, we will see the way we looked twenty-one years ago. Light propagates at a finite speed, and so by observing distant regions, we are able to see what the Universe looked like in the past, a light travel time ago. The statistical homogeneity of the Universe on large scales guarantees that what we see far away is a fair statistical representation of the conditions that were present in our region of the Universe a long time ago. This fortunate situation makes cosmology an empirical science. We do not need to guess how the Universe evolved. Using telescopes we can simply see how the Universe appeared at earlier cosmic times. In principle, this allows the entire 13.7 billion year cosmic history of

A. Loeb (✉)
Harvard University, CfA, MS 51, 60 Garden Street, Cambridge MA 02138, USA
e-mail: aloeb@cfa.harvard.edu

H.A. Thronson et al. (eds.), *Astrophysics in the Next Decade,* Astrophysics and Space Science Proceedings, DOI 10.1007/978-1-4020-9457-6_18,

our Universe to be reconstructed by surveying galaxies and other sources of light out to large distances. From these great distances, the wavelength of the emitted radiation is stretched by a large redshift factor $(1 + z)$ until it is observed, owing to the expansion of the Universe. Since a greater distance means a fainter flux from a source of a fixed luminosity, the observation of the earliest, highest-redshift sources of light requires the development of sensitive infrared telescopes such as the *James Webb Space Telescope (JWST)*.

Our cosmic photo album contains an early image of the Universe when it was 0.4 million years old in the form of the cosmic microwave background (CMB) (Bennett et al. 1996, Sperget et al. 2006), as well as many snapshots of galaxies more than a billion years later ($z < 6$; see overview in Ellis 2007). But we are still missing some crucial pages in this album. In between these two epochs was a period when the Universe was dark, stars had not yet formed, and the cosmic microwave background no longer traced the distribution of matter. And this is precisely the most interesting period, when the primordial soup evolved into the rich zoo of objects we now see. The situation that cosmologists face is similar to having a photo album of a person that contains the first ultrasound image of him or her as an unborn baby and some additional photos as a teenager and an adult. If you tried to guess from these pictures what happened in the interim, you could be seriously wrong. A child is not simply a scaled-up fetus or scaled-down adult. The same is true with galaxies. They did not follow a straightforward path of development from the incipient matter clumping evident in the microwave background.

18.2 Preliminaries

About 400,000 years after the Big Bang the temperature of the Universe dipped for the first time below a few thousand degrees Kelvin. The protons and electrons were then sufficiently cold to recombine into hydrogen atoms. It was just before the moment of cosmic recombination (when matter started to dominate in energy density over radiation) that gravity started to amplify the tiny fluctuations in temperature and density observed in the CMB data (Sperget et al. 2006). Regions that started out slightly denser than average began to contract because the gravitational forces were also slightly stronger than average in these regions. Eventually, after hundreds of millions of years of contraction, galaxies and the stars within them were able to form.

The detailed statistical properties of the CMB anisotropies (Sperget et al. 2006) indicate that indeed the structure apparent in the present-day Universe was seeded by small-amplitude inhomogeneities, mostly likely induced by quantum fluctuations during the early epoch of inflation. The growth of structure from these seeds was enhanced by the presence of dark matter – an unknown substance that makes up the vast majority (84%) of the cosmic density of matter. The motion of stars and gas around the centers of nearby galaxies indicates that each is surrounded by an extended mass of dark matter, and so dynamically-relaxed dark matter concentrations are generally referred to as "halos".

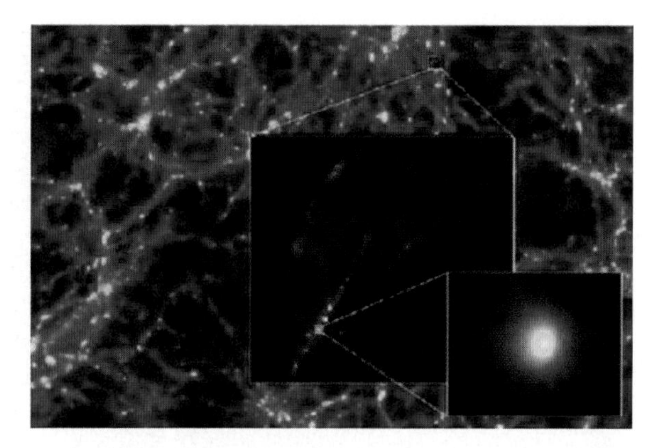

Fig. 18.1 A slice through a numerical simulation of the first dark matter condensations to form in the Universe (from Diemand et al. 2005). Colors represent the dark matter density at $z = 26$. The simulated volume is 60 comoving pc on a side, simulated with 64 million particles each weighing $1.2 \times 10^{-10} M_{\odot}$

Under the assumption that general relativity describes the evolution of the Universe, the measured CMB anisotropies indicate conclusively that most of the matter in the Universe must be very weakly coupled to electromagnetism and hence cannot be the matter that we are made of (baryons). This follows from the fact that prior to hydrogen recombination, the cosmic plasma was coupled to the radiation through Thomson scattering. Small-scale fluctuations were then damped in the radiation-baryon fluid by photon diffusion. The damping is apparent in the observed suppression of the CMB anisotropies on angular scales well below a degree on the sky, corresponding to spatial scales much smaller than 200 comoving Mpc. To put this scale in context, the matter that makes up galaxies was assembled from scales of $<$2Mpc. In order to preserve the primordial inhomogeneities that seeded the formation of galaxies, it is necessary to have a dominant matter component that does not couple to the radiation fluid. The most popular candidate for making up this component is a weakly interacting massive particle (WIMP). If this particle is the lightest supersymmetric particle, it might be discovered over the coming decade in the data stream from the *Large Hadron Collider*.

The natural temperature for the decoupling of WIMPs is expected to be high (tens of MeV), allowing them to cool to an extremely low temperature by the present epoch. The resulting *Cold Dark Matter (CDM)* is expected to fragment down to a Jupiter mass scale (Fig. 18.1, Diemand et al. 2005, Loeb and Zaldarriaga 2005, Bertschinger 2006). The baryons, however, cannot follow the CDM on small scales because of their higher thermal pressure. The minimum scale for the fragmentation of the baryons, the so-called “filtering scale” (which is a time-averaged Jeans mass), corresponds to $\sim 10^5 M_{\odot}$ prior to reionization (Loeb 2006).

18.2.1 The First Stars

For the scale-invariant Λ CDM power spectrum (Sperget et al. 2006), the first dark matter halos to contain gas have formed at a redshift of several tens. The assembly and cooling of gas in these halos resulted in the formation of the first stars

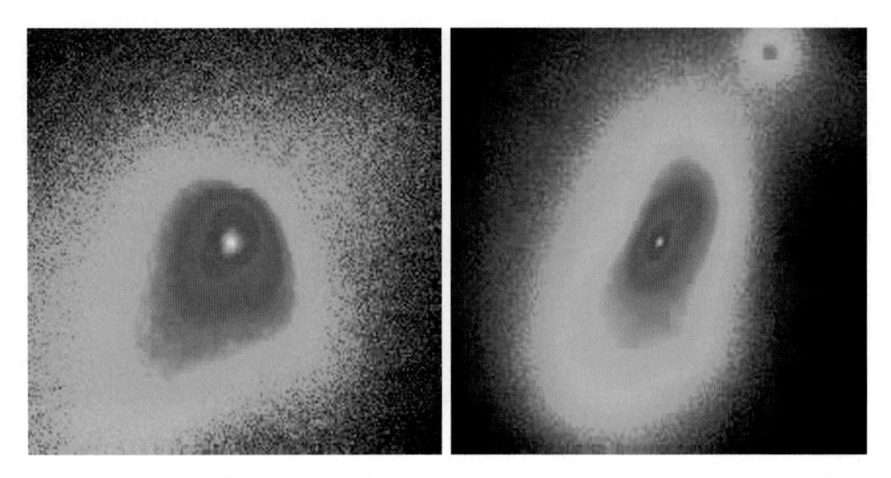

Fig. 18.2 Collapse and fragmentation of a primordial cloud (from Bromm and Loeb 2004). Shown is the projected gas density at a redshift $z \simeq 21.5$, briefly after gravitational runaway collapse has commenced in the center of the cloud. *Left:* The coarse-grained morphology in a box with linear physical size of 23.5 pc. At this time in the unrefined simulation, a high-density clump (sink particle) has formed with an initial mass of $\sim 10^3 M_{\odot}$. *Right:* The refined morphology in a box with linear physical size of 0.5 pc. The central density peak, vigorously gaining mass by accretion, is accompanied by a secondary clump

(Bromm and Larson 2004). Hydrodynamical simulations indicate that the primordial (metal-free) gas cooled via the radiative transitions of molecules such as H_2 and HD down to a temperature floor of a few hundred K, dictated by the energy levels of these molecules. At the characteristic density interior to the host clouds, the gas fragmented generically into massive ($>100 M_{\odot}$) clumps which served as the progenitors of the first stars. The relatively high sound speed (c_s) resulted in a high accretion rate ($\dot{M} \sim c_s^3/G$ over the stellar lifetime of a few million years) and a high characteristic mass for the first (so-called *Population III*) stars (Fig. 18.2, Bromm and Loeb 2004, Abel et al. 2002). The lowest-mass halos most likely hosted one star per halo.

Population III stars in the mass range of 140–$260 M_{\odot}$ led to pair-instability supernovae that enriched the surrounding gas with heavy elements (Heger et al. 2003). Enrichment of the gas to a carbon or oxygen abundance beyond $\sim 10^{-3.5} Z_{\odot}$ resulted in efficient cooling and fragmentation of the gas to lower-mass stars (Bromm and Loeb 2003a, Frebel et al. 2007, Schneider et al. 2006). The hierarchical growth in halo mass eventually led to the formation of halos with a virial temperature of $\sim 10^4$K in which cooling was mediated by atomic transitions. Fragmentation of gas in these halos could have led to the direct formation of the seeds for quasar black holes (Bromm and Loeb 2003b, Loeb and Rasio 1994).

A massive metal-free star is an efficient factory for the production of ionizing photons. Its surface temperature ($\sim 10^5$ K) and emission spectrum per unit mass are nearly independent of its mass above a few hundred $M_{\odot}$, as it radiates at the Eddington luminosity (i.e. its luminosity is proportional to its mass). Therefore the cumulative emissivity of the first massive stars was proportional to their total cumulative mass, independent of their initial mass function. These stars produced $\sim 10^5$ ionizing photons per baryon incorporated into them (Bromm et al. 2001, Tumlinson and Shull 2000). In comparison, low-mass stars produce ~4,000 ionizing photons per baryon (Loeb 2006). In both cases, it is clear that only a small fraction of the baryons in the Universe needs to be converted into stars in order for them to ionize the rest.

Given the formation rate of galaxy halos as a function of cosmic time, the course of reionization can be determined by counting photons from all sources of light (Arons and Wingert 1972, Shapiro and Giroux 1987, Tegmark et al. 1997, Kamionkowski et al. 1994, Furlanetto and Loeb 2003, Shapiro et al. 1994, Haiman and Loeb 1997). Both stars and black holes contribute ionizing photons, but the early Universe is dominated by small galaxies which in the local Universe have central black holes that are disproportionately small, and indeed quasars are rare above redshift 6 (Fan et al. 2003). Thus, stars most likely dominated the production of ionizing UV photons during the reionization epoch [although high-redshift galaxies should have also emitted X-rays from accreting black holes and accelerated particles in collisionless shocks (Oh 2001, Furlanetto and Loeb 2004)]. Since most stellar ionizing photons are only slightly more energetic than the 13.6 eV ionization threshold of hydrogen, they are absorbed efficiently once they reach a region with substantial neutral hydrogen. This makes the intergalactic medium (IGM) during reionization a two-phase medium characterized by highly ionized regions separated from neutral regions by sharp ionization fronts (see Fig. 18.4).

We can obtain a first estimate of the requirements of reionization by demanding one stellar ionizing photon for each hydrogen atom in the IGM. If we conservatively assume that stars within the ionizing galaxies were similar to those observed locally, then each star produced ~4000 ionizing photons per baryon. Star formation is observed today to be an inefficient process, but even if stars in galaxies formed out of only ~10% of the available gas, it was still sufficient to accumulate a small fraction (of order 0.1%) of the total baryonic mass in the Universe into galaxies in order to ionize the entire IGM. More accurate estimates of the actual required fraction account for the formation of some primordial stars (which were massive, efficient ionizers, as discussed above), and for recombinations of hydrogen atoms at high redshifts and in dense regions.

From studies of quasar absorption lines at $z \sim 6$ we know that the IGM is highly ionized a billion years after the Big Bang. There are hints, however, that some large neutral hydrogen regions persist at these early times (Wyithe and Loeb 2004a, Mesinger and Haiman 2004, Lidz et al. 2006) and so this suggests that we may not need to go to much higher redshifts to begin to see the epoch of reionization. We now know that the Universe could not have fully reionized earlier than an age of ~300 million years, since WMAP3 observed the effect of the freshly created plasma at reionization on the large-scale polarization anisotropies of the CMB and this limits the reionization redshift (Sperget et al. 2006); an earlier reionization, when the Universe was denser, would have created a stronger scattering signature that would be inconsistent with the WMAP3 observations. In any case, the redshift

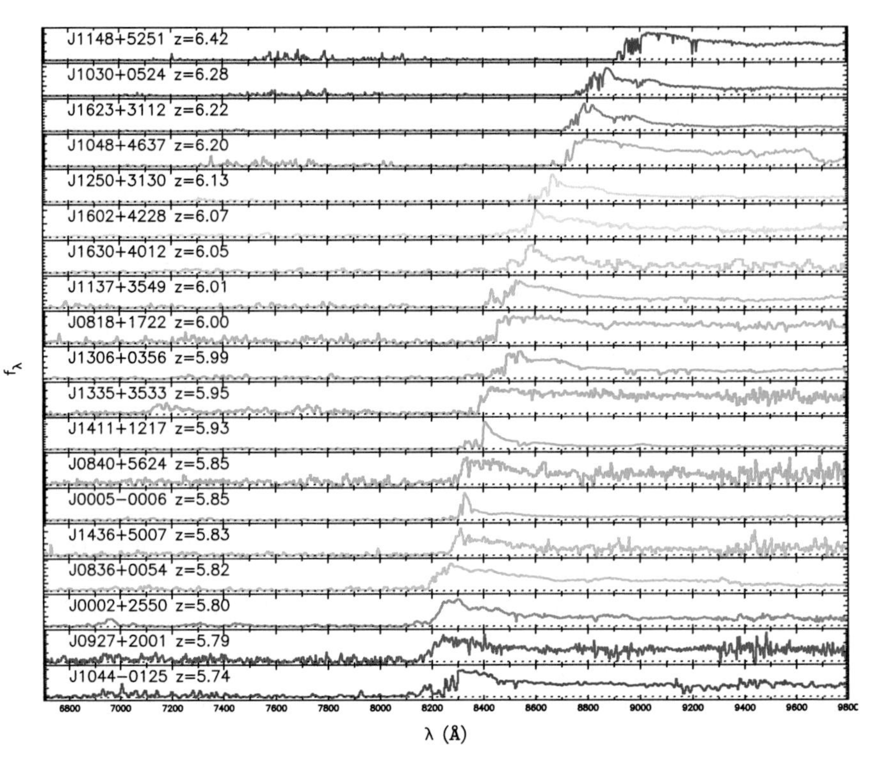

Fig. 18.3 Spectra of 19 quasars with redshifts $5.74 < z < 6.42$ from the *Sloan Digital Sky Survey*, taken from Fan et al. (2005). For some of the highest-redshift quasars, the spectrum shows no transmitted flux shortward of the Lyα wavelength at the quasar redshift (the so-called "Gunn-Peterson trough"), indicating a non-negligible neutral fraction in the IGM

at which reionization ended only constrains the overall cosmic efficiency of ionizing photon production. In comparison, a detailed picture of reionization as it happens will teach us a great deal about the population of young galaxies that produced this cosmic phase transition.

Several quasars beyond $z \sim 6$ show in their spectra a Gunn-Peterson trough, a blank spectral region at wavelengths shorter than Lyα at the quasar redshift (Fig. 18.3). The detection of Gunn-Peterson troughs indicates a rapid change (Fan et al. 2002, White et al. 2003, 2006) in the neutral content of the IGM at $z \sim 6$, and hence a rapid change in the intensity of the background ionizing flux. However, even a small atomic hydrogen fraction of $\sim 10^{-3}$ would still produce nearly complete Lyα absorption.

A key point is that the spatial distribution of ionized bubbles is determined by clustered groups of galaxies and not by individual galaxies. At such early times galaxies were strongly clustered even on very large scales (up to tens of Mpc), and these scales therefore dominated the structure of reionization (Barkana and Loeb 2004b). The basic idea is simple (Kaiser 1984). At high redshift, galactic halos are rare and correspond to high density peaks. As an analogy, imagine searching on Earth for mountain peaks above 5000 meters. The 200 such peaks are not at all distributed uniformly but instead are found in a few distinct clusters on top of large mountain ranges. Given the large-scale boost provided by a mountain range, a small-scale crest need only provide a small additional rise in order to become a 5000 meter peak. The same crest, if it formed within a valley, would not come anywhere near 5000 meters in total height. Similarly, in order to find the early galaxies, one must first locate a region with a large-scale density enhancement, and then galaxies will be found there in abundance.

The ionizing radiation emitted from the stars in each galaxy initially produces an isolated ionized bubble. However, in a region dense with galaxies the bubbles quickly overlap into one large bubble, completing reionization in this region while the rest of the Universe is still mostly neutral (Fig. 18.4). Most importantly, since the abundance of rare density peaks is very sensitive to small changes in the density threshold, even a large-scale region with a small enhanced density (say, 10% above the mean density of the Universe) can have a much larger concentration of galaxies than in other regions (e.g., a 50% enhancement). On the other hand, reionization is harder to achieve in dense regions, since the protons and electrons collide and recombine more often in such regions, and newly-formed hydrogen atoms need to be reionized again by additional ionizing photons. However, the overdense regions end up reionizing first since the number of ionizing sources in these regions is increased so strongly (Barkana and Loeb 2004b, McQuinn et al. 2007a). The large-scale topology of reionization is therefore inside out, with underdense voids reionizing only

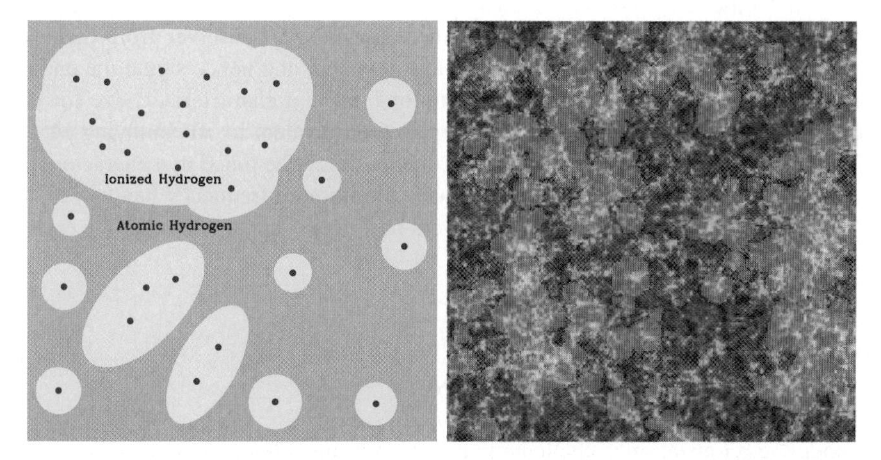

Fig. 18.4 The spatial structure of cosmic reionization. The illustration (*left panel*, based on Barkana and Loeb 2004b) shows how regions with large-scale overdensities form large concentrations of galaxies (*dots*) whose ionizing photons produce enormous joint ionized bubbles (*upper left*). At the same time, galaxies are rare within large-scale voids, in which the IGM is still mostly neutral (*lower right*). A numerical simulation of reionization (*right panel*, from Mellema et al. 2006) indeed displays such variation in the sizes of ionized bubbles (*orange*), shown overlayed on the density distribution (*green*)

at the very end of reionization, with the help of extra ionizing photons coming in from their surroundings (which have a higher density of galaxies than the voids themselves). This is a key prediction awaiting observational tests.

Detailed analytical models that account for large-scale variations in the abundance of galaxies (Furlanetto et al. 2004) confirm that the typical bubble size starts well below a Mpc early in reionization, as expected for an individual galaxy, rises to 5–10 comoving Mpc during the central phase (i.e., when the Universe is half ionized), and then by another factor of $\sim$5 towards the end of reionization. (These scales are given in comoving units that scale with the expansion of the Universe, so that the actual sizes at a redshift z were smaller than these numbers by a factor of $1+z$.) Numerical simulations have only recently begun to reach the enormous scales needed to capture this evolution (Ciardi et al. 2003, Mellema et al. 2006, Zahn et al. 2006). Accounting precisely for gravitational evolution over a wide range of scales but still crudely for gas dynamics, star formation, and the radiative transfer of ionizing photons, the simulations confirm that the large-scale topology of reionization is inside out, and that this topology can be used to study the abundance and clustering of the ionizing sources (Figs. 18.4 and 18.9).

Wyithe and Loeb (2004b) showed that the characteristic size of the ionized bubbles at the end reionization can be calculated based on simple considerations that only depend on the power-spectrum of density fluctuations and the redshift. As the size of an ionized bubble increases, the time it takes a 21-cm photon to traverse it gets longer. At the same time, the variation in the time at which different regions reionize becomes smaller as the regions grow larger. Thus, there is a maximum size above which the photon crossing time is longer than the cosmic variance in ionization time. Regions bigger than this size will be ionized at their near side by the time a 21-cm photon will cross them towards the observer from their far side. They would appear to the observer as one-sided, and hence signal the end of reionization. These "light cone" considerations imply a characteristic size for the ionized bubbles of $\sim$10 physical Mpc at $z \sim 6$ (equivalent to 70 comoving Mpc). This result implies that future radio experiments should be tuned to a characteristic angular scale of tens of arcminutes and have a minimum frequency band-width of 5-10 MHz for an optimal detection of 21-cm brightness fluctuations near the end of reionization.

18.2.2 Simulations of Reionization

Simulating reionization is challenging for two reasons. First, one needs to incorporate radiative transfer at multiple photon frequencies into a code that follows the dynamics of gas and dark matter. This implies that the sources of the radiation, i.e. galaxies, need to be resolved. Second, one needs to simulate a sufficiently large volume of the Universe for cosmic variance not to play a role (Barkana 2004b). Towards the end of reionization, the sizes of individual ionized regions grow up to a scale of $\sim$50–100 comoving Mpc (Wyithe and Loeb 2004b, Furlanetto et al. 2004) and the representative volume needs to include many such region in order for it to fully describe the large-scale topology of reionization. There is an obvious tension between the above two requirements for simulating small scales as well as large scales simultaneously.

Numerical simulations of reionization are starting to achieve the dynamic range required to resolve galaxy halos across the scale of hundreds of comoving Mpc, larger than the size of the ionized regions at the end of the reionization process (Zahn et al. 2006, Iliev et al. 2007, Trac and Cen 2006). These simulations cannot yet follow in detail the formation of individual stars within galaxies, or the feedback that stars produce on the surrounding gas, such as photoheating or the hydrodynamic and chemical impact of supernovae, which blow hot bubbles of gas enriched with the chemical products of stellar nucleosynthesis. Thus, the simulations cannot directly predict whether the stars that form during reionization are similar to the stars in the Milky Way and nearby galaxies or to the primordial $100M_{\odot}$ behemoths. They also cannot determine whether feedback prevents low-mass dark matter halos from forming stars. Thus, models are needed that make it possible to vary all these astrophysical parameters of the ionizing sources.

18.3 Imaging Cosmic Hydrogen

18.3.1 Basic Principles

The ground state of hydrogen exhibits hyperfine splitting involving the spins of the proton and the electron (Fig. 18.5). The state with parallel spins (the triplet state) has a slightly higher energy than the state with anti-parallel spins (the singlet state). The 21-cm line associated with the spin-flip transition from the triplet to the singlet state is often used to detect neutral hydrogen in the local Universe. At high redshift, the occurrence of a neutral pre-reionization intergalactic medium (IGM) offers the prospect of detecting the first sources of radiation and probing the reionization era by mapping the 21-cm absorption or emission from neutral regions. Regions where the gas is slightly denser than the mean would produce a stronger signal. Therefore, the 21-cm brightness will fluctuate across the sky as a result of the inhomogeneous

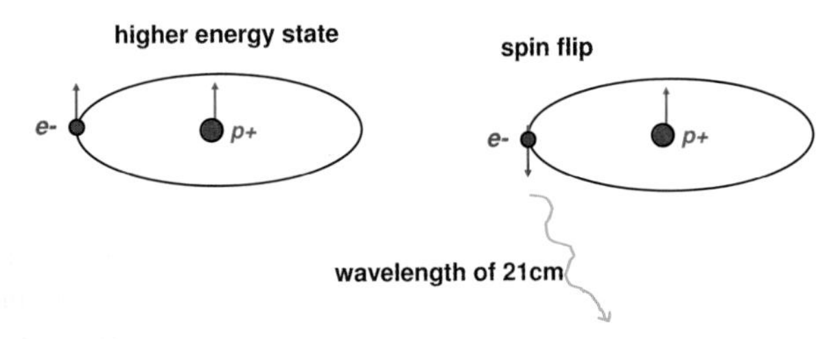

Fig. 18.5 The 21-cm transition of hydrogen. The higher energy level the spin of the electron (e-) is aligned with that of the proton (p+). A spin flip results in the emission of a photon with a wavelength of 21-cm (or a frequency of 1420MHz)

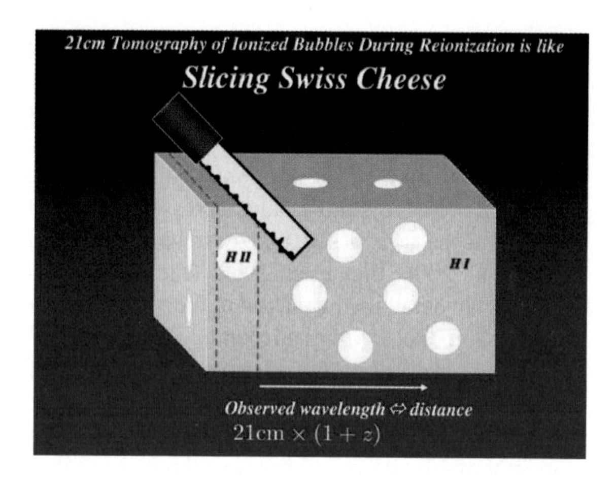

Fig. 18.6 21-cm imaging of ionized bubbles during the epoch of reionization is analogous to slicing swiss cheese. The technique of slicing at intervals separated by the typical dimension of a bubble is optimal for revealing different pattens in each slice

distribution of hydrogen. Moreover, this resonant line can be used to slice the Universe at different redshifts z by observing different wavelengths corresponding to 21-cm $(1 + z)$. Altogether, the 21-cm brightness fluctuation can be used to map the inhomogeneous hydrogen distribution in three dimensions (Fig. 18.6).

The atomic hydrogen gas formed soon after the big-bang, was affected by processes ranging from quantum fluctuations during the early epoch of inflation to irradiation by the first galaxies at late times. Mapping this gas through its resonant 21-cm line serves a dual role as a powerful probe of fundamental physics and of astrophysics. The facets of fundamental physics include the initial density fluctuations imprinted by inflation as well as the nature of the dark matter, which amplifies these fluctuations during the matter-dominated era. It is possible to avoid the contamination from astrophysical sources by observing the Universe before the first galaxies had formed. In the concordance Λ CDM cosmological model, the 21 cm brightness fluctuations of hydrogen were shaped by fundamental physics (inflation, dark matter, and atomic physics) at redshifts $z > 20$, and by the radiation from galaxies at lower redshifts.

Following cosmological recombination at $z \sim 10^3$, the residual fraction of free electrons coupled the gas thermally to the cosmic microwave background (CMB) for another 65 million years ($z \sim 200$), but afterwards the gas decoupled and cooled faster than the CMB through its cosmic expansion. In the redshift interval of the so-called *Dark Ages* before the first stars had formed, $30 < z < 200$, the *spin temperature* of hydrogen, T_s (defined through the level population of the spin-flip transition), was lower than the CMB temperature, T_γ, and the gas appeared in absorption. The primordial inhomogeneities of the gas produced varying levels of 21cm absorption and hence brightness fluctuations. Detection of this signal can be used to constrain models of inflation as well as the nature of dark matter (Hogan and Rees 1979, Scott and Rees 1990, Loeb and Zaldarriaga 2004, Lewis and Challinor 2007). Altogether, there are $\sim 10^{16}$ independent pixels on the 21cm sky from this epoch (instead of $\sim 10^7$ for the CMB). They make the richest data set on the sky, providing an unprecedented probe of non-Gaussianity and running of the spectral index of the power-spectrum of primordial density fluctuations from inflation. Detection of these small-scale fluctuations can also be used to infer the existence of massive neutrinos and other sub-dominant components in addition to the commonly inferred cold dark matter particles.

After the first galaxies formed and X-ray sources heated the gas above the CMB temperature, the gas appeared in 21-cm emission. The bubbles of ionized hydrogen around groups of galaxies were dark and dominated the 21 cm fluctuations (Furlanetto et al. 2006, Pritchard and Furlanetto 2007). After the first stars had formed, the 21-cm fluctuations were sourced mainly by the hydrogen ionized fraction, and spin temperature (Madau et al. 1997).

The basic physics of the hydrogen spin transition is determined as follows (for a more detailed treatment, see (Madau et al. 1997, Furlanetto et al. 2006)). The ground-state hyperfine levels of hydrogen tend to thermalize with the CMB bringing the IGM away from thermal equilibrium, then the gas becomes observable against the CMB in emission or in absorption. The relative occupancy of the spin levels is described in terms of the hydrogen spin temperature T_S, defined through the Boltzman factor,

$$\frac{n_1}{n_0} = 3 \exp\left\{-\frac{T_*}{T_S}\right\} , \tag{18.1}$$

where n_0 and n_1 refer respectively to the singlet and triplet hyperfine levels in the atomic ground state ($n = 1$), and $T_* = 0.068$ K is defined by $k_B T_* = E_{21}$, where the energy of the 21 cm transition is $E_{21} = 5.9 \times 10^{-6}$ eV, corresponding to a frequency of 1420 MHz. In the presence of the CMB alone, the spin states reach thermal equilibrium with the CMB temperature $T_S = T_\gamma = 2.725(1+z)$ K on a time-scale of $T_*/(T_\gamma A_{10}) \simeq 3 \times 10^5 (1+z)^{-1}$ yr, where $A_{10} = 2.87 \times 10^{-15}$ s^{-1} is the spontaneous decay rate of the hyperfine transition. This time-scale is much shorter than the age of the Universe at all redshifts after cosmological recombination.

The IGM is observable when the kinetic temperature T_k of the gas differs from the CMB temperature T_γ and an effective mechanism couples T_S to T_k (Fig. 18.7). Collisional de-excitation of the triplet level (Purcell and Field 1956) dominates at very high redshift, when the gas density (and thus the collision rate) is still high, making the gas observable in absorption. Once a significant galaxy population forms in the Universe, the X-rays they emit heat T_k above T_γ and the UV photons they emit couple T_s to T_k making the gas appear in 21cm emission. The latter coupling mechanism acts through the scattering of Lyα photons (Wouthuysen 1952, Field 1959). Continuum UV photons produced by early radiation sources redshift by the Hubble expansion into the local Lyα line at a lower redshift. These photons mix the spin states via the Wouthuysen-Field process whereby an atom initially in the $n = 1$ state absorbs a Lyα photon, and the spontaneous decay which returns it from $n = 2$ to $n = 1$ can result in a final spin state which is different from the initial one. Since the neutral IGM is highly opaque to resonant scattering, and the Lyα photons receive Doppler kicks in each scattering, the shape of the radiation spectrum near Lyα is determined by T_k (Field 1959), and the resulting spin temperature (assuming $T_S \gg T_*$) is then a weighted average of T_k and T_γ:

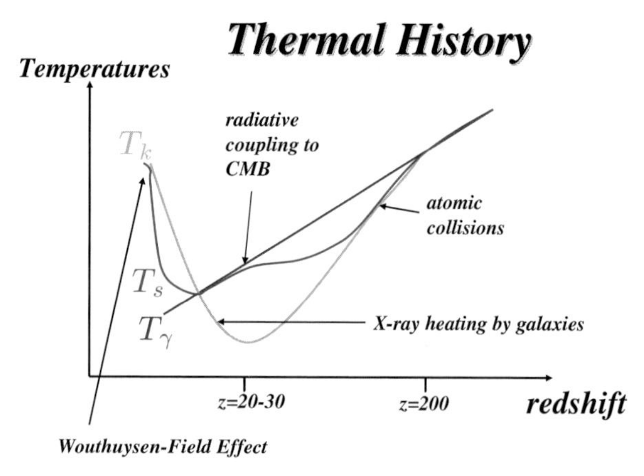

Fig. 18.7 Schematic sketch of the evolution of the kinetic temperature (T_k) and spin temperature (T_s) of cosmic hydrogen (from Loeb 2006). Following cosmological recombination at $z \sim 10^3$, the gas temperature (orange curve) tracks the CMB temperature (blue line; $T_\gamma \propto (1+z)$) down to $z \sim 200$ and then declines below it ($T_k \propto (1+z)^2$) until the first X-ray sources (accreting black holes or exploding supernovae) heat it up well above the CMB temperature. The spin temperature of the 21-cm transition (*red curve*) interpolates between the gas and CMB temperatures. Initially it tracks the gas temperature through collisional coupling; then it tracks the CMB through radiative coupling; and eventually it tracks the gas temperature once again after the production of a cosmic background of UV photons between the Lyα and the Lyman-limit frequencies that redshift or cascade into the Lyα resonance (through the Wouthuysen-Field effect (Wouthuysen 1952, Field 1959)). Parts of the curve are exaggerated for pedagogical purposes. The exact shape depends on astrophysical details about the first galaxies, such as the production of X-ray binaries, supernovae, nuclear accreting black holes, and the generation of relativistic electrons in collisionless shocks which produce UV and X-ray photons through inverse-Compton scattering of CMB photons

$$T_S = \frac{T_\gamma T_k (1 + x_{\rm tot})}{T_k + T_\gamma x_{\rm tot}}, \tag{18.2}$$

where $x_{\rm tot} = x_\alpha + x_c$ is the sum of the radiative and collisional threshold parameters. These parameters are $x_\alpha = \frac{P_{10} T_\star}{A_{10} T_\gamma}$, and $x_c = \frac{4\kappa_{1-0}(T_k) n_H T_\star}{3 A_{10} T_\gamma}$, where P_{10} is the indirect de-excitation rate of the triplet $n = 1$ state via the Wouthuysen-Field process, related to the total scattering rate P_α of Lyα photons by $P_{10} = 4P_\alpha/27$ (Field 1958). Also, the atomic coefficient $\kappa_{1-0}(T_k)$ is tabulated as a function of T_k (Allison 1969, Zygelman 2005). Note that we have adopted the modified notation (i.e., in terms of x_α and x_c) of Barkana and Loeb (2005b). The coupling of the spin temperature to the gas temperature becomes substantial when $x_{\rm tot} > 1$; in particular, $x_\alpha = 1$ defines the thermalization rate (Madau et al. 1997) of P_α: $P_{\rm th} \equiv \frac{27 A_{10} T_\gamma}{4 T_*} \simeq 7.6 \times 10^{-12} \left(\frac{1+z}{10}\right) \, {\rm s}^{-1}$.

A patch of neutral hydrogen at the mean density and with a uniform T_S produces (after correcting for stimulated emission) an optical depth at a present-day (observed) wavelength of $21(1+z)$ cm,

$$\tau(z) = 9.0 \times 10^{-3} \left(\frac{T_\gamma}{T_S}\right) \left(\frac{\Omega_b h}{0.03}\right) \left(\frac{\Omega_m}{0.3}\right)^{-1/2} \left(\frac{1+z}{10}\right)^{1/2}, \tag{18.3}$$

assuming a high redshift $z \gg 1$. The observed spectral intensity I_ν relative to the CMB at a frequency ν is measured by radio astronomers as an effective brightness temperature T_b of blackbody emission at this frequency, defined using the Rayleigh-Jeans limit of the Planck radiation formula: $I_\nu \equiv 2k_B T_b \nu^2/c^2$.

The brightness temperature through the IGM is $T_b = T_\gamma e^{-\tau} + T_S(1 - e^{-\tau})$, so the observed differential antenna temperature of this region relative to the CMB is (Madau et al. 1997)

$$T_b = (1+z)^{-1}(T_S - T_\gamma)(1 - e^{-\tau})$$
$$\simeq 28\,\text{mK}\left(\frac{\Omega_b h}{0.033}\right)\left(\frac{\Omega_m}{0.27}\right)^{-1/2}\left(\frac{1+z}{10}\right)^{1/2}\left(\frac{T_S - T_\gamma}{T_S}\right), \quad (18.4)$$

where $\tau \ll 1$ can be assumed and T_b has been redshifted to redshift zero. Note that the combination that appears in T_b is $\frac{T_S - T_\gamma}{T_S} = \frac{x_{\text{tot}}}{1+x_{\text{tot}}}\left(1 - \frac{T_\gamma}{T_k}\right)$. In overdense regions, the observed T_b is proportional to the overdensity, and in partially ionized regions T_b is proportional to the neutral fraction. Also, if $T_S \gg T_\gamma$ then the IGM is observed in emission at a level that is independent of T_S. On the other hand, if $T_S \ll T_\gamma$ then the IGM is observed in absorption at a level that is enhanced by a factor of T_γ/T_S.

To complement computationally-intensive simulations of reionization, various groups developed approximate schemes for simulating 21-cm maps in the regime where T_S is much larger than T_γ. For example, Furlanetto et al. (2004) developed an analytical model that allows the calculation of the probability distribution (at a given redshift) of the size of the ionizing bubble surrounding a random point in space. Zahn et al. (2006) have considered numerical schemes that apply the Furlanetto et al. (2004) model to either the initial conditions of their simulation or to part of its results (Fig. 18.10).

18.3.2 Predicted 21-cm Signal

In approaching redshifted 21-cm observations, although the first inkling might be to consider the mean emission signal, the signal is orders of magnitude fainter than foreground synchrotron emission from relativistic electrons in the magnetic field of our own Milky Way (Furlanetto et al. 2006) as well as other galaxies (Di Matteo et al. 2002). Thus cosmologists have focused on the expected characteristic variations in T_b, both with position on the sky and especially with frequency, which signifies redshift for the cosmic signal. The synchrotron foreground is expected to have a smooth frequency spectrum, and so it is possible to isolate the cosmological signal by taking the difference in the sky brightness fluctuations at slightly different frequencies (as long as the frequency separation corresponds to the characteristic size of ionized bubbles). The 21-cm brightness temperature depends on the density of neutral hydrogen. As explained in the previous subsection, large-scale patterns in the reionization are driven by spatial variations in the abundance of galaxies; the 21-cm fluctuations reach ~5 mK (root mean square) in brightness temperature (Fig. 18.9) on a scale of 10 comoving Mpc. While detailed maps will be difficult to extract due to the foreground emission, a statistical detection of these fluctuations (through the power spectrum) is expected to be well within the capabilities of the first-generation experiments now being built (Bowman et al. 2006, McQuinn et al. 2006). Current work suggests that the key information on the topology and timing of reionization can be extracted statistically (Fig. 18.8, Fig. 18.9).

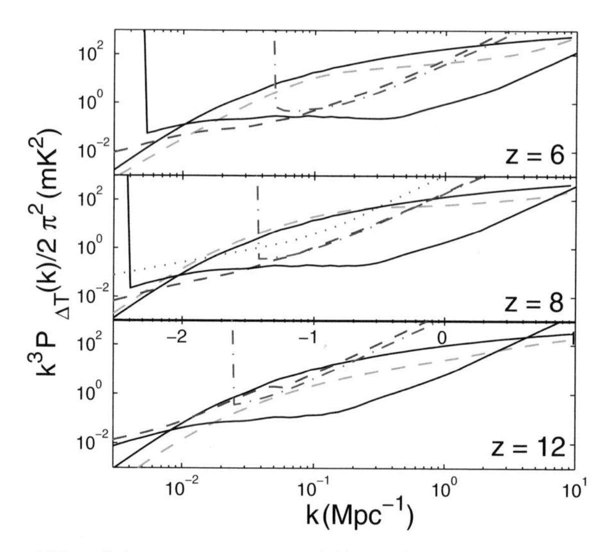

Fig. 18.8 Detectability of the power-spectrum of 21-cm fluctuations by different future observatories (from McQuinn et al. 2006). The detector noise plus sample variance errors is shown for a 1000 hr observation on a single field in the sky, assuming perfect foreground removal, for MWA (*thick dashed curve*), LOFAR (*thick dot-dashed curve*), and SKA (*thick solid curve*) for wavenumber bins of $\Delta k = 0.5\,k$. The thin solid curve represents the spherically averaged signal for a small ionization fraction and $T_s \gg T_\gamma$

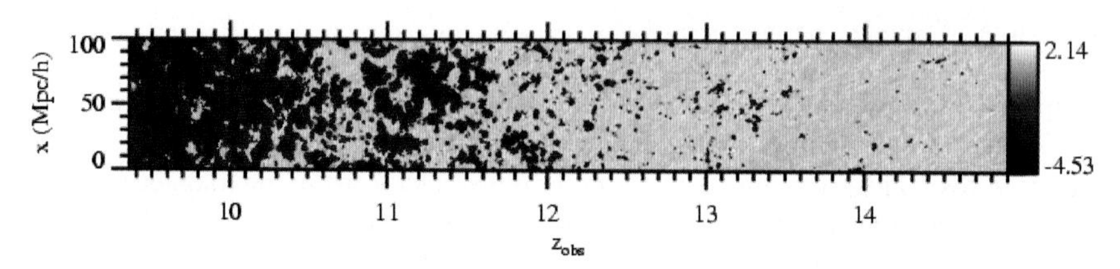

Fig. 18.9 Close-up of cosmic evolution during the epoch of reionization, as revealed in a predicted 21-cm map of the IGM based on a numerical simulation (from Mellema et al. 2006). This map is constructed from slices of the simulated cubic box of side 150 Mpc (in comoving units), taken at various times during reionization, which for the parameters of this particular simulation spans a period of 250 million years from redshift 15 down to 9.3. The vertical axis shows position χ in units of Mpc/h (where the Hubble constant in units of 100 km s^{-1} is $h = 0.7$). This two-dimensional slice of the sky (one linear direction on the sky versus the line-of-sight or redshift direction) shows $\log_{10}(T_b)$, where T_b (in mK) is the 21-cm brightness temperature relative to the CMB. Since neutral regions correspond to strong emission (i.e., a high T_b), this slice illustrates the global progress of reionization and the substantial large-scale spatial fluctuations in reionization history. Observationally it corresponds to a narrow strip half a degree in length on the sky observed with radio telescopes over a wavelength range of 2.2 to 3.4-m (with each wavelength corresponding to 21-cm emission at a specific line-of-sight distance and redshift)

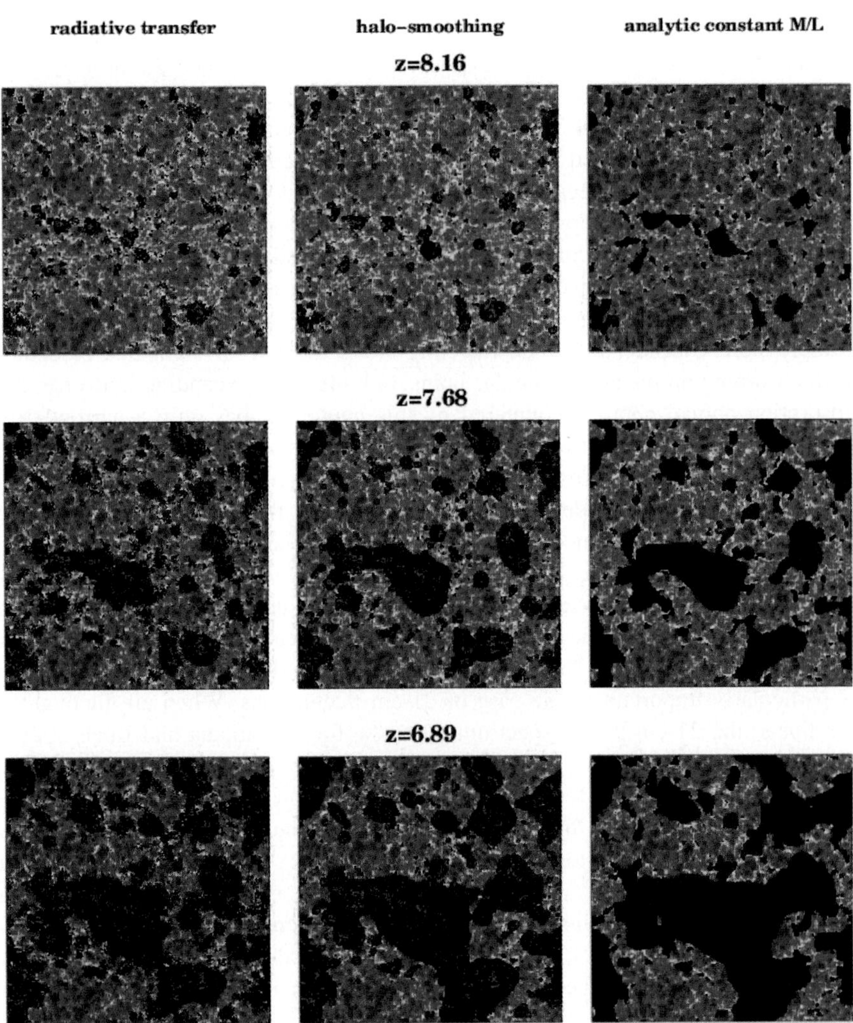

Fig. 18.10 Maps of the 21-cm brightness temperature comparing results of a numerical simulation and of two simpler numerical schemes, at three different redshifts (from Zahn et al. 2006). Each map is 65.6 Mpc/h on a side, with a depth (0.25 Mpc/h) that is comparable to the frequency resolution of planned experiments. The ionized fractions are $x_{\rm i} = 0.13$, 0.35 and 0.55 for $z = 8.16$, 7.26 and 6.89, respectively. All three maps show a very similar large-scale ionization topology. *Left column:* Numerical simulation, showing the ionized bubbles (black) produced by the ionizing sources (blue dots) that form in the simulation. *Middle column:* Numerical scheme that applies the Furlanetto et al. (2004) analytical model to the final distribution of ionizing sources that form in the simulation. *Right column:* Numerical scheme that applies the Furlanetto et al. (2004) analytical model to the linear density fluctuations that are the initial conditions of the simulation

The theoretical expectations for reionization and for the 21-cm signal are based on rather large extrapolations from observed galaxies to deduce the properties of much smaller galaxies that formed at an earlier cosmic epoch. Considerable surprises are thus possible, such as an early population of quasars or even unstable exotic particles that emitted ionizing radiation as they decayed.

An important cross-check on these measurements is possible by measuring the particular form of anisotropy, expected in the 21-cm fluctuations, that is caused by gas motions along the line of sight (Kaiser 1987, Bharadwaj and Ali 2004, Barkana and Loeb 2005a). This anisotropy, expected in any measurement of density that is based on a spectral resonance or on redshift measurements, results from velocity compression. Consider a photon traveling along the line of sight that resonates with absorbing atoms at a particular point. In a uniform, expanding Universe, the absorption optical depth encountered by this photon probes only a narrow strip of atoms, since the expansion of the Universe makes all other atoms move with a relative velocity that takes them outside the narrow frequency width of the resonance line. If there is a density peak, however, near the resonating position, the increased gravity will reduce the expansion velocities around this point and bring more gas into the resonating velocity width. This effect is sensitive only to the line-of-sight component of the velocity gradient of the gas, and thus causes an observed anisotropy in the power spectrum even when all physical causes of the fluctuations are statistically isotropic. Barkana and Loeb (2005a) showed that this anisotropy is particularly important in the case of 21-cm fluctuations. When all fluctuations are linear, the 21-cm power spectrum takes the form (Barkana and Loeb 2005a) $P_{21-\mathrm{cm}}(\mathbf{k}) = \mu^4 P_\rho(k) + 2\mu^2 P_{\rho-\mathrm{iso}}(k) + P_{\mathrm{iso}}$, where $\mu = \cos\theta$ in terms of the angle θ between the wavevector $\mathbf{k}$ of a given Fourier mode and the line of sight, P_{iso} is the isotropic power spectrum that would result from all sources of 21-cm fluctuations without velocity compression, $P_\rho(k)$ is the 21-cm power spectrum from gas density fluctuations alone, and $P_{\rho-\mathrm{iso}}(k)$ is the Fourier transform of the cross-correlation between the density and all sources of 21-cm fluctuations. The three power spectra can also be denoted $P_{\mu^4}(k)$, $P_{\mu^2}(k)$, and $P_{\mu^0}(k)$, according to the power of μ that multiplies each term. At these redshifts, the 21-cm fluctuations probe the infall of the baryons into the dark matter potential wells (Barkana and Loeb 2005c). The power spectrum shows remnants of the photon-baryon acoustic oscillations on large scales, and of the baryon pressure suppression on small scales (Naoz and Barkana 2005).

Once stellar radiation becomes significant, many processes can contribute to the 21-cm fluctuations. The contributions include fluctuations in gas density, temperature, ionized fraction, and Lyα flux. These processes can be divided into two broad categories: The first, related to "*physics*", consists of probes of fundamental, precision cosmology, and the second, related to "*astrophysics*", consists of probes of stars. Both categories are interesting – the first for precision measures of cosmological parameters and studies of processes in the early Universe, and the second for studies of the properties of the first galaxies. However, the astrophysics depends on complex non-linear processes (collapse of dark matter halos, star formation, supernova feedback), and must be cleanly separated from the physics contribution, in order to allow precision measurements of the latter. As long as all the fluctuations are linear, the anisotropy noted above allows precisely this separation of the physics from the astrophysics of the 21-cm fluctuations (Barkana and Loeb 2005a). In particular, the $P_{\mu^4}(k)$ is independent of the effects of stellar radiation, and is a clean probe of the gas density fluctuations. Once non-linear terms become important, there arises a significant mixing of the different terms; in particular, this occurs on the scale of the ionizing bubbles during reionization (McQuinn et al. 2006).

At early times, the 21-cm fluctuations are also affected by fluctuations in the Lyα flux from stars, a result that yields an indirect method to detect and study the early population of galaxies at $z \sim 20$ (Barkana and Loeb 2005b). The fluctuations are caused by biased inhomogeneities in the density of galaxies, along with Poisson fluctuations in the number of galaxies. Observing the power-spectra of these two sources would probe the number density of the earliest galaxies and the typical mass of their host dark matter halos. Furthermore, the enhanced amplitude of the 21 cm fluctuations from the era of Lyα coupling improves considerably the practical prospects for their detection. Precise predictions account for the detailed properties of all possible cascades of a hydrogen atom after it absorbs a photon (Hirata 2006, Pritchard and Furlanetto 2006). Around the same time, X-rays may also start to heat the cosmic gas, producing strong 21-cm fluctuations due to fluctuations in the X-ray flux (Pritchard and Furlanetto 2007).

18.3.3 Baryonic Acoustic Oscillation

As described above, the fluctuations in the emission of redshifted 21-cm photons from neutral intergalactic hydrogen will provide an unprecedented probe of the reionization era. Conventional wisdom assumes that this 21-cm signal disappears as soon as reionization is complete, when little atomic hydrogen is left through most of the volume of the IGM. However, the statistics of damped Lyα absorbers (DLAs) in quasar spectra indicate that a few percent of the baryonic mass reservoir remains in the form of atomic hydrogen at all redshifts. The residual hydrogen is self-shielded from UV radiation within dense regions in which the recombination rate is high. Wyithe and Loeb (2007) used a physically-motivated model to show that residual neutral gas would generate a significant post-reionization 21cm signal. Even though the signal is much weaker (see Fig. 18.11) than that of a fully-neutral IGM, the synchrotron foreground, whose brightness temperature scales as $\propto (1+z)^{2.6}$, is also much weaker at the corresponding low redshifts. Thus, the power-spectrum of fluctuations in this signal will be detectable by the first generation of low-frequency observatories at a signal-to-noise that is comparable to that achievable in observations of the reionization era. The statistics of 21-cm fluctuations will therefore probe not only the pre-reionization IGM, but rather the entire process of HII region overlap, as well as the appearance of the diffuse ionized IGM. With an angular resolution of an arcminute, the radio beam of future interferometers will contain many DLAs

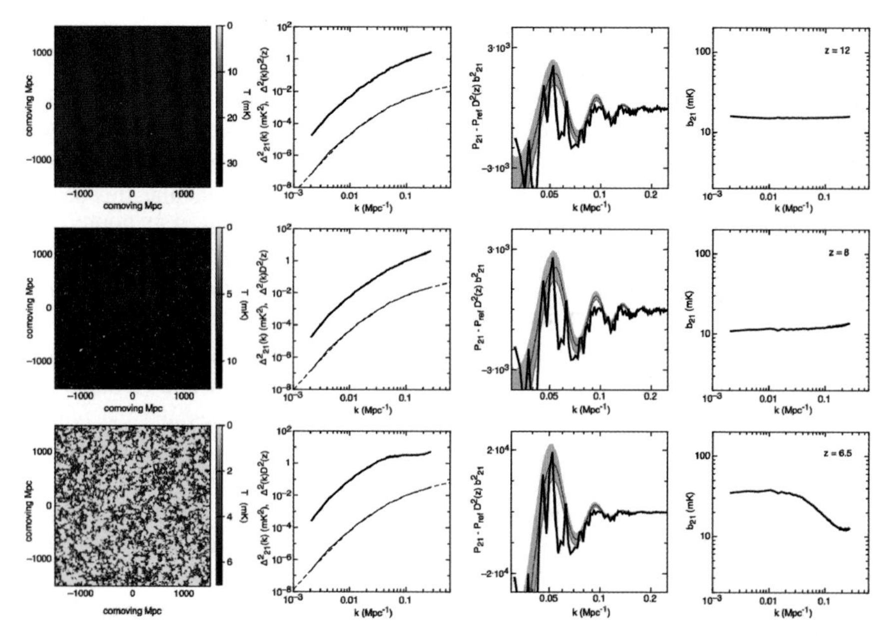

Fig. 18.11 Examples of the 21-cm power spectra during reionization (from Wyithe et al. 2007). *Left panels:* Maps of the 21cm emission from slices through the numerical simulation boxes, each 3000 co-moving Mpc on a side with a thickness of 12 co-moving Mpc. In these maps yellow designates the absence of redshifted 21-cm emission. *Central-left panels:* The corresponding matter power spectra multiplied by the growth factor squared (thin solid lines) as well as the 21 cm (*thick solid lines*) power spectra computed from the simulation box. The input co-moving power spectrum P (also multiplied by the growth factor squared) is shown for comparison (*short-dashed lines*). *Central-right panels:* The baryonic oscillations component of the simulated 21-cm power spectrum. The curves (*thick dark lines*) show the difference between the simulated 21-cm power spectrum, and a reference no wiggle 21-cm power spectrum computed from the theoretical no wiggle reference matter power spectrum multiplied by the square of the product between the bias and the growth factor [i.e. $P_{21} - P_{\rm ref} b_{21}^2 D^2$]. For comparison, the red lines show the difference between the input matter and the no-BAO reference matter power spectra, multiplied by the bias and growth factor squared [i.e. $(P - P_{\rm ref}) b_{21}^2 D^2$]. The grey band surrounding this curve shows the level of statistical scatter in realizations of the power spectrum due to the finite size of the simulation volume. *Right panels:* The scale dependent bias (b_{21}). The upper, central and lower panels show results at $z = 12$, $z = 8$ and $z = 6.5$, which have global neutral fractions of 98%, 48% and 11% respectively in the model shown

and will not resolve them individually. Rather, the observations will map the course-binned distribution of neutral hydrogen across scales of tens of comoving Mpc.

Wyithe et al. (2007) demonstrated that the power spectrum of the cumulative 21-cm emission during and after reionization will show baryonic acoustic oscillations (BAOs), whose comoving scale can be used as a standard ruler to infer the evolution of the equation of state for the dark energy. The BAO yardstick can be used to measure the dependence of both the angular diameter distance and Hubble parameter on redshift. The wavelength of the BAO is related to the size of the sound horizon at recombination, as this reflects the distance out to which different points were correlated in the radiation-baryon fluid. Its value depends on the Hubble constant and on the matter and baryon densities. However, it does not depend on the amount or nature of the dark energy. Thus, measurements of the angular diameter distance and Hubble parameter can in turn be used to constrain the possible evolution of the dark energy with cosmic time. This idea was originally proposed in relation to galaxy redshift surveys (Blake and Glazebrook 2003, Hu and Haiman 2003, Seo and Eisenstein 2003) and has since received significant theoretical attention (Glazebrook and Blake 2005, Seo and Eisenstein 2005, Angulo et al. 2007). Moreover, measurement of the BAO scale has been achieved within large surveys of galaxies at low redshift, illustrating its potential (Cole et al. 2005, Eisenstein et al. 2005). Galaxy redshift surveys are best suited to studies of the dark energy at relatively late times due to the difficulty of obtaining accurate redshifts for large numbers of high redshift galaxies. Wyithe et al. (2007) have found that the first generation of low-frequency experiments (such as MWA or LOFAR) will be able to constrain the acoustic scale to within a few percent in a redshift window just prior to the end of the reionization era. This sensitivity to the acoustic scale is comparable to the best current measurements from galaxy redshift surveys, but at much higher redshifts. Future extensions of the first generation experiments (involving an order of magnitude increase in the antennae number of the MWA) could reach sensitivities below one percent in several redshift windows and could be used to study the dark energy in the unexplored redshift regime of $3.5 < z < 12$. Moreover, new experiments with antennae designed to operate at higher frequencies would allow precision measurements ($< 1\%$) of the acoustic peak to be made at more moderate redshifts ($1.5 < z < 3.5$), where they would be competitive with ambitious spectroscopic galaxy surveys covering more than 1000 square degrees. Together with other data sets, observations of 21-cm fluctuations will allow full coverage of the acoustic scale from the present time out to $z \sim 12$ (Wyithe et al. 2007) and beyond (Barkana and Loeb 2005c).

The left hand panels of Fig. 18.11 show the 21-cm emission from 12 Mpc slices through a numerical simulation (Wyithe et al. 2007). The higher redshift example ($z = 12$) is early in the reionization era, and shows no HII regions forming at the resolution of the simulation (i.e. the IGM does not contain ionized bubbles with radii > 5 co-moving Mpc). The fluctuations in the 21-cm emission are dominated by the density field at this time. The central redshift ($z \sim 8$) shows the IGM midway through the reionization process, and includes a few HII regions above the simulation resolution. The lower redshift example is just prior to the overlap of the ionized regions (and hence the completion of reionization), when the IGM is dominated by large percolating HII regions.

18.3.4 Low-Frequency Arrays

The main obstacle towards detecting the 21-cm signal is the synchrotron foreground contamination from our Galaxy and extragalactic point sources, whose brightness temperature rises steeply towards lower frequencies ($\propto \nu^{-2.6}$) and makes the detection of the redshifted 21-cm line more challenging at higher redshifts. This fact directed most experimental and theoretical work so far towards the study of the astrophysics-dominated era at low redshifts ($z < 20$), during which the 21-cm fluctuations were sourced mainly by the growth of ionized bubbles around galaxies. For example, the Murchison Wide-Field Array (MWA; Fig. 18.12), which is currently funded by NSF and the Australian government, is designed to cover the redshift range of 6–17. The first generation MWA-demonstrator will have 512 antenna tiles of 4×4 dipole antennae each.

The prospect of studying reionization by mapping the distribution of atomic hydrogen across the Universe using its prominent 21-cm spectral line has motivated several teams to design and construct arrays of low-frequency radio telescopes; the Low Frequency Array,[1] MWA,[2] the 21CMA,[3] and ultimately the Square Kilometer Array[4] will search over the next decade for 21-cm emission or absorption from $z \sim 3.5$–15, redshifted and observed today at relatively low frequencies which correspond to wavelengths of 1 to 4-m. Producing resolved images even of large sources such as cosmological ionized bubbles requires telescopes which have a kilometer scale. It is much more cost-effective to use a large array of thousands of simple antennas distributed over several kilometers, and to use computers to cross-correlate

Fig. 18.12 Prototype of the tile design for the *Murchison Wide-Field Array* (MWA) in western Australia, aimed at detecting redshifted 21-cm from the epoch of reionization. Each 4×4-m tile contains 16 dipole antennas operating in the frequency range of 80–300MHz. Altogether the initial phase of MWA (the so-called "Low-Frequency Demonstrator") will include 500 antenna tiles with a total collecting area of 8000 m^2 at 150 MHz, scattered across a 1.5-km region and providing an angular resolution of a few arcminutes

[1] http://www.lofar.org/

[2] http://www.haystack.mit.edu/ast/arrays/mwa/index.html

[3] http://arxiv.org/abs/astro-ph/0502029

[4] http://www.skatelescope.org

the measurements of the individual antennas and combine them effectively into a single large telescope. The new experiments are being placed mostly in remote sites, because the cosmic wavelength region overlaps with more mundane terrestrial telecommunications.

18.4 Imaging Galaxies

18.4.1 Future Infrared Telescopes

Narrow-band searches for redshifted Lyα emission have discovered galaxies robustly out to redshifts $z = 6.96$ (Iye et al. 2006) and potentially out to $z = 10$ (Stark et al. 2007). Existing observations provide a first glimpse into the formation of the first galaxies (Malhotra and Rhoads 2002, Kashikawa et al. 2006, Taniguchi et al. 2005, Dijkstra et al. 2006) with potential theoretical implications for the epoch of reionization (Loeb and Rybicki 1999, Rybicki and Loeb 1999, Haiman and Spaans 1999, Malhotra and Rhoads 2004, Stark et al. 2007, McQuinn et al. 2007a, Mesinger and Furlanetto 2007). Future surveys intend to exploit this search strategy further by pushing to even higher redshifts and fainter flux levels (Horton et al. 2004, Willis et al. 2006, 2007, Cuby et al. 2007). The spectral break due to Lyα absorption by the IGM allows to identify high-redshift galaxies photometrically with even greater sensitivity (Bouwens et al. 2006, Bouwens and Illingworth 2006, 2007, Dow-Hygelund 2007) (see overview in Ellis 2007 and phenomenological model in Stark et al. 2007).

The construction of large infrared telescopes on the ground and in space will provide us with new photos of first generation of galaxies during the epoch of reionization. Current plans include the space telescope JWST (which will not be affected by the atmospheric background) as well as ground-based telescopes which are 24–42 meter in diameter (such as the GMT and Fig. 18.13,[5] TMT,[6] and EELT.[7]

18.4.2 Cross-Correlating Galaxies with 21-cm Maps

Given that the earliest galaxies created the ionized bubbles around them by their UV emission, the locations of galaxies should correlate with the cavities in the neutral hydrogen during reionization. Within a decade it should be possible to explore the environmental influence of individual galaxies by using large-aperture infrared telescopes in combination with 21-cm observatories of reionization (Wyithe and Loeb 2007, Wyithe et al. 2005, 2007a).

---

[5] http://www.gmto.org/
[6] http://celt.ucolick.org/
[7] http://www.eso.org/projects/e-elt/

Fig. 18.13 Artist's conception of the design for one of the future giant telescopes that could probe the first generation of galaxies from the ground. The *Giant Magellan Telescope* (*GMT*) will contain seven mirrors (each 8.4-m in diameter) and will have a resolving power equivalent to a 24.5-m (80 foot) primary mirror. For more details see http://www.gmto.org/

Wyithe and Loeb (2007a) calculated the expected anti-correlation between the distribution of galaxies and the intergalactic 21-cm emission at high redshifts. As already mentioned, overdense regions are expected to be ionized first as a result of their biased galaxy formation. This early phase leads to an anti-correlation between the 21-cm emission and the overdensity in galaxies, matter, or neutral hydrogen. Existing Lyα surveys probe galaxies that are highly clustered in overdense regions. By comparing 21-cm emission from regions near observed galaxies to those away from observed galaxies, future observations will be able to test this generic prediction and calibrate the ionizing luminosity of high-redshift galaxies. McQuinn et al. (2007b) showed that observations of high-redshift Ly-alpha emitters (LAEs) have the potential to provide definitive evidence for reionization because their Lyα line is damped by the neutral regions in the IGM. In particular, the clustering of the emitters is increased by incomplete reionization (Fig. 18.14). For stellar reionization scenarios, the angular correlation function of the 58 LAEs in the Subaru Deep Field $z = 6.6$ photometric sample (Taniguchi et al. 2005, Kashikawa et al. 2006) is already consistent with a mostly ionized IGM (McQuinn et al. 2007b). At higher redshifts near the beginning of reionization, when the ionized regions were small, this analysis needs to be combined with detailed radiative transfer calculations of the Lyα line, since the line shape is sensitive to the local infall/outflow profile of the gas around individual galaxies (Dijkstra et al. 2007).

18.4.3 Gamma-ray Bursts: Probing the First Stars One Star at a Time

Gamma-Ray Bursts (GRBs) are believed to originate in compact remnants (neutron stars or black holes) of massive stars. Their high luminosities make them detectable out to the edge of the visible Universe (Lamb and Reichart 2000, Ciardi et al. 2003). GRBs offer the opportunity to detect the most distant (and hence earliest) population of massive stars, the so-called Population III (or Pop III), one star at a time

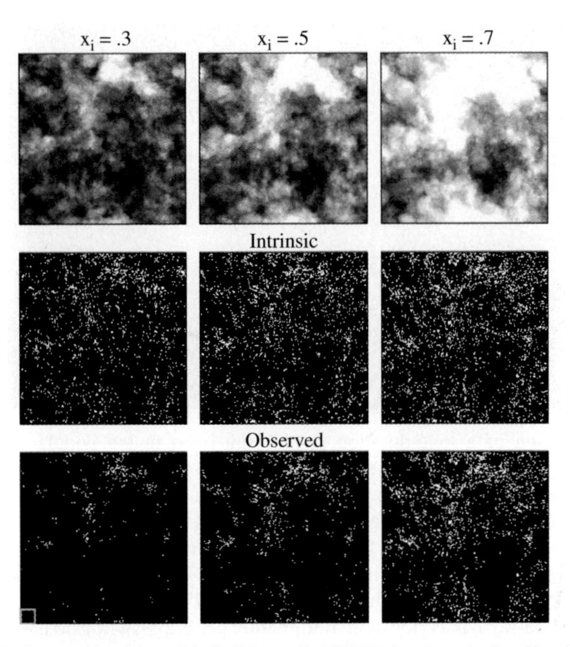

Fig. 18.14 Simulation results (from McQuinn et al. 2007b) for the relative distribution of neutral hydrogen and Lyα emitting galaxies (LAE). The top panels show the projection of the ionized hydrogen fraction x_i in the survey volume. In the white regions the projection is fully ionized and in black it is neutral. The left, middle, and right panels are for $z = 8.2$ (with an average ionized fraction $\bar{x}_i = 0.3$), $z = 7.7$ ($\bar{x}_i = 0.5$), and $z = 7.3$ ($\bar{x}_i = 0.7$). The middle and bottom rows are the intrinsic and observed LAE maps, respectively, for a futuristic LAE survey that can detect halos down to a mass $> 1 \times 10^{10} M_{\odot}$. The observed distribution of emitters is modulated by the location of the HII regions (compare bottom panels with corresponding top panels). Each panel is 94 comoving Mpc across (or 0.6 degrees on the sky), roughly the area of the current Subaru Deep Field (SDF) at $z = 6.6$ (Kashikawa et al. 2006). The depth of each panel is $\Delta\lambda = 130$Å, which matches the FWHM of the Subaru 9210Ånarrow-band filter. The number densities of LAEs for the panels in the middle row are few times larger than the number density in the SDF photometric sample of $z = 6.6$ LAEs. The large-scale modulation of LAE by the HII bubbles is clearly apparent in this survey. The square in the lower left-hand panel represents the $3' \times 3'$ field-of-view of *JWST* drawn to scale

(Fig. 18.15). In the hierarchical assembly process of halos that are dominated by cold dark matter, the first galaxies should have had lower masses (and lower stellar luminosities) than their more recent counterparts. Consequently, the characteristic luminosity of galaxies or quasars is expected to decline with increasing redshift. GRB afterglows, which already produce a peak flux comparable to that of quasars or starburst galaxies at $z \sim 1$–2, are therefore expected to outshine any competing source at the highest redshifts, when the first dwarf galaxies formed in the Universe.

GRBs, the electromagnetically-brightest explosions in the Universe, should be detectable out to redshifts $z > 10$. High-redshift GRBs can be identified through infrared photometry, based on the Lyα break induced by absorption of their spectrum at wavelengths below $1.216\,\mu\text{m}\,[(1+z)/10]$. Follow-up spectroscopy of

Fig. 18.15 Illustration of a long-duration gamma-ray burst in the popular "collapsar" model (Zhang et al. 2003). The collapse of the core of a massive star (which lost its hydrogen envelope) to a black hole generates two opposite jets moving out at a speed close to the speed of light. The jets drill a hole in the star and shine brightly towards an observer who happens to be located within with the collimation cones of the jets. The jets emanating from a single massive star are so bright that they can be seen across the Universe out to the epoch when the first stars formed. Upcoming observations by the *Swift* satellite will have the sensitivity to reveal whether Pop III stars served as progenitors of gamma-ray bursts (for more information see http://swift.gsfc.nasa.gov/)

high-redshift candidates can then be performed on a large aperture infrared telescope, such as *JWST*. GRB afterglows offer the opportunity to detect stars as well as to probe the metal enrichment level (Furlanetto and Loeb 2003) of the intervening IGM. Recently, the ongoing *Swift* mission (Gehrels et al. 2004) has detected GRB050904 originating at $z \simeq 6.3$ (Haislip et al. 2006), thus demonstrating the viability of GRBs as probes of the early Universe.

Another advantage of GRBs is that the GRB afterglow flux at a given observed time lag after the γ-ray trigger is not expected to fade significantly with increasing redshift, since higher redshifts translate to earlier times in the source frame, during which the afterglow is intrinsically brighter (Ciardi et al. 2003). For standard afterglow lightcurves and spectra, the increase in the luminosity distance with redshift is compensated by this *cosmological time-stretching* effect (Ciardi et al. 2003, Barkana and Loeb 2004a) as shown in Figure 18.16.

GRB afterglows have smooth (broken power-law) continuum spectra unlike quasars which show strong spectral features (such as broad emission lines or the so-called "blue bump") that complicate the extraction of IGM absorption features. In particular, the continuum extrapolation into the Lyα damping wing during the epoch of reionization is much more straightforward for the smooth UV spectra of GRB afterglows than for quasars with an underlying broad Lyα emission line (Barkana and Loeb 2004a). However, the interpretation regarding the neutral fraction of the IGM may be complicated by the presence of damped Lyα absorption by dense neutral hydrogen in the immediate environment of the GRB within its host galaxy (Barkana and Loeb 2004a, Totani et al. 2006) and by the patchiness of the neutral IGM during reionization (McQuinn et al. 2007c). Since long-duration GRBs originate from the

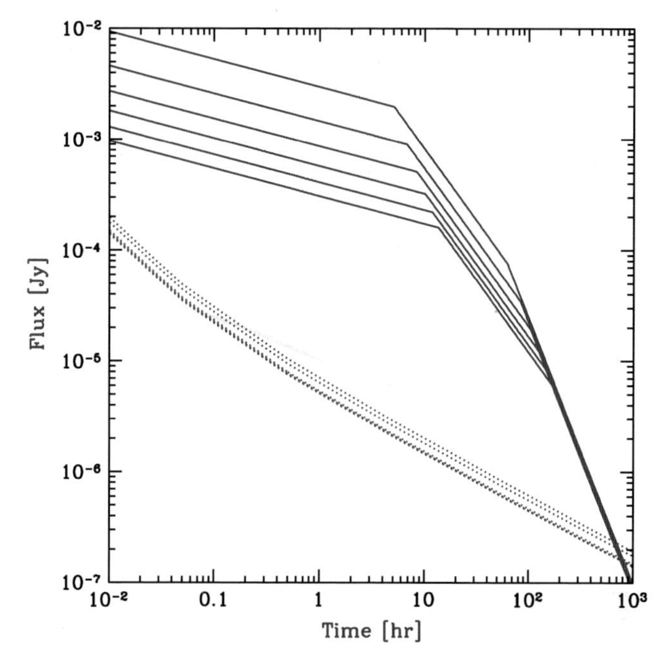

Fig. 18.16 GRB afterglow flux as a function of time since the γ-ray trigger in the observer frame (from Barkana and Loeb 2004a). The flux (*solid curves*) is calculated at the redshifted Lyα wavelength. The dotted curves show the planned detection threshold for the *James Webb Space Telescope* (*JWST*), assuming a spectral resolution $R = 5000$ with the near infrared spectrometer, a signal to noise ratio of 5 per spectral resolution element, and an exposure time equal to 20% of the time since the GRB explosion. Each set of curves shows a sequence of redshifts, namely $z = 5$, 7, 9, 11, 13, and 15, respectively, from top to bottom

dense environment of active star formation, the associated damped Lyα absorption from their host galaxy was so-far always observed (Prochaska et al. 2007a,b), including in the most distant GRB at $z = 6.3$ (Totani et al. 2006).

Acknowledgments I thank my collaborators on the work described in this review, Dan Babich, Rennan Barkana, Volker Bromm, Benedetta Ciardi, Mark Dijkstra, Richard Ellis, Steve Furlanetto, Zoltan Haiman, Jonathan Pritchard, George Rybicki, Dan Stark, Stuart Wyithe and Matias Zaldarriaga. I also thank Matt McQuinn for a careful reading of the manuscript.

References

Abel T, Bryan, G L, Norman M L (2002) Science **295**, 93
Allison A C, Dalgarno A (1969) ApJ **158**, 423
Angulo R, Baugh C M, Frenk C S, Lacey C G (2007) ArXiv Astrophysics e-prints, arXiv:astro-ph/0702543
Arons J, Wingert D W (1972) ApJ **177**, 1
Barkana R, Loeb A (2004a) ApJ **601**, 64
Barkana R, Loeb A (2004b) ApJ **609**, 474

Barkana R, Loeb A (2005a) ApJL **624**, L65
Barkana R, Loeb A (2005b) ApJ **626**, 1
Barkana R, Loeb A (2005c) MNRAS **363**, L36
Bennett C L et al. (1996) ApJL **464**, L1
Bertschinger E (2006) Phys Rev **D74**, 063509
Bharadwaj S and Ali S S (2004) MNRAS **352**, 142
Blake C, Glazebrook K (2003) ApJ **594**, 665
Bouwens R J, Illingworth G D, Blakeslee J P, Franx M (2006) ApJ **653**, 53
Bouwens R J, Illingworth G D (2006) Nature **443**, 189
Bouwens R J, Illingworth G D, Franx M, Ford H (2007) ArXiv e-prints, 707, arXiv:0707.2080
Bowman J D, Morales M F and Hewitt J N (2006) ApJ **638**, 20
Bromm V, Kudritzki R P, Loeb A (2001) ApJ **552**, 464
Bromm V, Larson R B (2004) Ann Rev Astron & Astrophys **42**, 79
Bromm V, Loeb A (2003a) Nature **425**, 812
Bromm V, Loeb A (2003b) ApJ **596**, 34
Bromm V, Loeb A (2004) New Astronomy **9**, 353
Ciardi B, Ferrara A, White S D M (2003) MNRAS **344** L7
Cole S, et al. (2005) **362**, 505
Cuby J-G, Hibon P, Lidman C, Le Fèvre O, Gilmozzi R, Moorwood A, van der Werf P (2007) A & A **461**, 911
Ciardi B, Ferrara A, White S D M (2003) MNRAS **344**, L7
Di Matteo T, Perna R, Abel T, Rees M J (2002) ApJ **564**, 576
Dijkstra M, Haiman Z, Spaans M (2006) ApJ **649**, 14
Dijkstra M, Lidz A, Wyithe J S B (2007) MNRAS **377**, 1175
Diemand J, Moore B, Stadel J (2005) Nature **433**, 389 (2 005)
Dow-Hygelund C C, et al. (2007) ApJ **660**, 47
Eisenstein D, et al. (2005) ApJ **633**, 560
Ellis R S (2007) ArXiv Astrophysics e-prints, arXiv:astro-ph/0701024
Fan X, et al (2002) AJ, **123**, 1247
Fan X, et al (2003) AJ **125**, 1649
Fan X, et al. (2005) AJ **132**, 117
Fan X, Carilli C L, Keating B (2006) Ann Rev Astron & Astrophys **44**, 415
Field G B (1958) Proc IRE **46**, 240
Field G B (1959) ApJ **129**, 551
Frebel A, Johnson J L, Bromm V (2007) MNRAS **380**, L40
Furlanetto S R, Loeb A (2004) ApJ **611**, 642
Furlanetto S R, Oh S P, Briggs F H (2006) Phys Rep **433**, 181
Furlanetto S R, Zaldarriaga M and Hernquist L (2004) ApJ **613** 1
Fukugita M, Kawasaki M (1994), MNRAS **269**, 563
Furlanetto S R, Loeb A (2003) ApJ **588**, 18
Gehrels N, et al. (2004) ApJ **611**, 1005
Glazebrook K, Blake C (2005) ApJ **631**, 1
Haiman Z, Loeb A (1997) ApJ **483**, 21
Haiman Z, Spaans M (1999) ApJ **518**, 138
Haislip J, et al. (2006) Nature **440**, 181
Heger A, Fryer C L, Woosley S E, Langer N, Hartmann D H (2003) ApJ **591**, 288
Hirata C M (2006) MNRAS **367**, 259
Hogan C J, Rees M J (1979) MNRAS **188**, 791
Horton A, Parry I, Bland-Hawthorn J, Cianci S, King D, McMahon R, Medlen S (2004) Proc SPIE **5492**, 1022
Hu W, Haiman Z (2003) Phys Rev **D68**, 063004
Iliev I T, Shapiro, P R, Mellema G, Pen U-L, McDonald P, Bond J R (2007) ArXiv e-prints, 708, arXiv:0708.3846

Iye M, et al. (2006) Nature **443**, 186
Kaiser N (1984) ApJL **284**, 9
Kaiser N (1987) MNRAS **227**, 1
Kamionkowski M, Spergel D N, Sugiyama N (1994), ApJL **426**, 57
Kashikawa N, et al. (2006) ApJ **648**, 7
Lamb D Q, Reichart D E (2000) ApJ **536**, 1
Lewis A, Challinor A (2007), ArXiv Astrophysics e-prints, arXiv:astro-ph/0702600
Lidz A, Oh S P, Furlanetto S R (2006), ApJL **639**, 47
Loeb A (2006), "First Light", extensive review for the SAAS-Fee 2006 Winter School, ArXiv Astrophysics e-prints, arXiv:astro-ph/0603360
Loeb A, Rasio F A (1994) ApJ **432**, 52
Loeb A, Rybicki G B (1999) ApJ **524**, 527
Loeb A, Zaldarriaga M (2004) Phys Rev Lett **92**, 211301
Loeb A, Zaldarriaga M (2005) Phys Rev **D71**, 103520
Madau P, Meiksin A, Rees M J (1997) ApJ **475**, 429
Malhotra S, Rhoads J E (2002) ApJL **565**, L71
Malhotra S, Rhoads J E (2004) ApJL **617**, L5
Mellema G, Iliev I T, Pen U-L, Shapiro P R (2006) MNRAS **372**, 679
Mesinger A, Haiman Z (2004) ApJL **611**, 69
Mesinger A, Furlanetto S (2007) ArXiv e-prints, 708, arXiv:0708.0006
McQuinn M, Zahn O, Zaldarriaga M, Hernquist L, Furlanetto S R (2006) ApJ **653**, 815
McQuinn M, Lidz A, Zahn O, Dutta S, Hernquist L, Zaldarriaga M (2007a) MNRAS **377**, 1043
McQuinn M, Hernquist L, Zaldarriaga M, Dutta S (2007b) MNRAS **381**, 75
McQuinn M, Lidz A, Zaldarriaga M, Hernquist L, Dutta S (2007c) ArXiv e-prints, 710, arXiv:0710.1018
Naoz S and Barkana R (2005) MNRAS **362**, 1047
Oh S P (2001) ApJ **553**, 499
Pritchard J R, Furlanetto S R (2006) MNRAS **367**, 1057
Pritchard J R, Furlanetto S R (2007) MNRAS **376**, 1680
Prochaska J X, et al. (2007a) ApJS **168**, 231
Prochaska, J X, Chen H-W, Dessauges-Zavadsky M, Bloom J S (2007b) ApJ **666**, 267
Purcell E M and Field G B (1956) ApJ **124**, 542
Rybicki G B, Loeb A (1999) ApJL **520**, L79
Schneider R, Omukai K, Inoue A K, Ferrara A (2006) MNRAS **369**, 1437
Scott D, Rees M J (1990) MNRAS **247**, 510
Seo H-J, Eisenstein D J (2003) ApJ **598**, 720
Seo H-J, Eisenstein, D J (2005) ApJ **633**, 575; **665**, 14
Shapiro P R, Giroux M L (1987) ApJL, **321**, 107
Shapiro P R, Giroux M L, Babul A (1994) ApJ **427**, 25
Spergel D N, et al. (2006) ApJS **170**, 377
Stark D P, Ellis R S, Richard J, Kneib J-P, Smith G P, Santos M R (2007) ApJ **663**, 10
Stark, D P, Loeb A, Ellis R S (2007) ApJ **668**, 627
Taniguchi Y, et al. (2005) PASJ **57**, 165
Tegmark M et al. (1997) ApJ **474**, 1
Totani T, Kawai N, Kosugi G, Aoki K, Yamada T, Iye M, Ohta K, Hattori T (2006) PASJ **58**, 485
Trac H, Cen R (2006) ArXiv Astrophysics e-prints, arXiv:astro-ph/0612406
Tumlinson J, Shull J M (2000) ApJL **528**, 65
Wouthuysen S A (1952) AJ **57**, 31
White R L, Becker R H, Fan X, Strauss M A (2003) AJ **126**, 1
Willis J P, Courbin F, Kneib J-P, Minniti D (2006) New Astronomy Review **50**, 70
Willis J P, Courbin F, Kneib J P, Minniti D (2007) ArXiv e-prints, 709, arXiv:0709.1761
Wyithe J S B, Loeb A (2004a) Nature **427**, 815
Wyithe J S B, Loeb A (2004b) Nature **432**, 194

Wyithe J S B, Loeb A, Barnes D G (2005) ApJ **634**, 715
Wyithe J S B, Loeb A (2007a) MNRAS **375**, 1034
Wyithe J S B, Loeb A (2007b) ArXiv e-prints, 708, arXiv:0708.3392
Wyithe J S B, Loeb A, Schmidt B (2007a) ArXiv e-prints, 705, arXiv:0705.1825
Wyithe J S B, Loeb A, Geil P (2007b) ArXiv e-prints, 709, arXiv:0709.2955
Zahn O, Lidz A, McQuinn M, Dutta S, Hernquist L, Zaldarriaga M, Furlanetto S R (2006) ApJ **654**, 12
Zhang W, Woosley S, MacFadyen A I (2003) ApJ **586**, 356
Zygelman B (2005) ApJ **622**, 1356

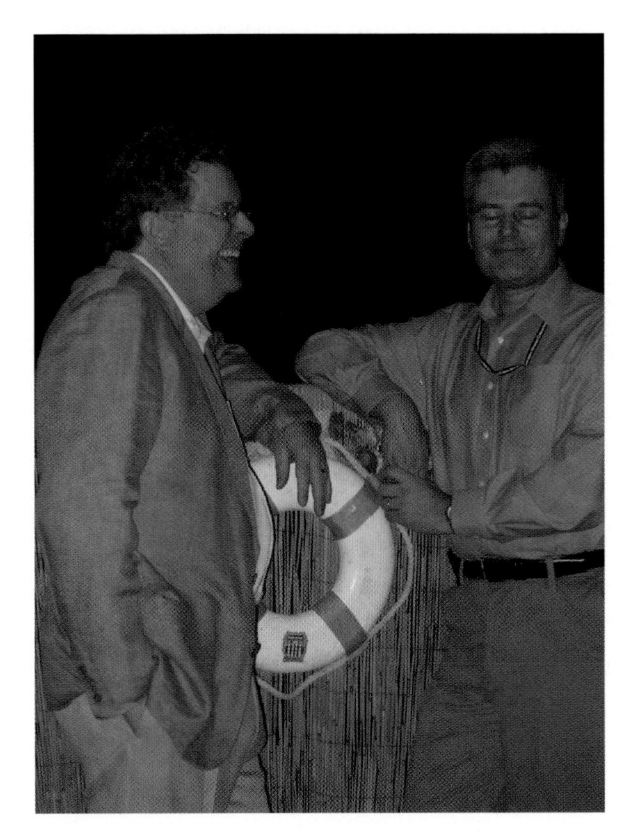

Matt Mountain enjoys a joke, which causes Mark Clampin to grimace

Index

A

Absorption lines, 421, 422
Accretion
  Bondi, 342
Accretion disk, 167–182, 338
  See also Disk
Active galactic nuclei (AGN), 324ff, 342ff
  feedback, 345, 350–351
  host, 344
  mass, 346
  obscured, 338
  radio-loud, 342
  radio-quiet, 341
  type I, 338, 342
  type II, 338ff, 345ff
AFGL 618, 253
AFGL 2199, 249
AFGL 2688, 253
AG Car, 261
Allen telescope array (ATA), 228
ALMA, 176, 178, 181, 206–208, 222, 230, 232, 238, 258, 262, 287, 427, 437, 450
Alpha Ori, 260
Amino acid, 200
Angular momentum
  transport, 344
Antennae galaxies, 226, 227ff
Arches cluster, 219, 223, 224
Arp 220, 222
Asteroids, 54–101
  See also by name; Kuiper Belt; Oort Cloud
Astronomy & Astrophysics Survey Committee, 42
  See also Bahcall Committee
Atacama pathfinder experiment (APEX), 237
Atacama submillimeter telescope experiment (ASTE), 237
Atmosphere
  chemistry, 104, 112, 113
  clouds, 108–110
  composition, 112–114
  giant planet, 136–138
  mixing, 108, 109, 112–114, 118
  temperature, 138
  terrestrial planet, 138
Atomic hydrogen, 227, 228, 281, 428–430, 440
AU Mic, 83
AU Pic, 82

B

B68, 196
Background
  ionizing, 328, 461
  ionizing emissivity, 465
  Lyman-Werner, 358–359, 391ff
  UV, 358
  X-ray, 359–360, 391
Bahcall Committee, 42–43, 45
  See also National Research Council
Bahcall, John, 41–42
Baryons, 419–451
  acoustic oscillations, 347, 349
  intracluster medium, 441
  ionized gas, 431–435, 442, 443
  mass density, 423ff
  molecular gas, 427, 428, 439
  neutral hydrogen, 428–430, 440
  stellar component, 426, 427, 438
  warm-hot diffuse medium, 435, 436, 444–449
Bely, Pierre, 34ff
Berkeley illinois maryland array (BIMA), 439
Big Bang nucleosynthesis, 424–426

Black holes, 336ff, 344, 351
accretion, 339
first, 395–398, 484
formation, 336, 347
fueling, 343–345, 349–350
growth, 338ff, 343ff
supermassive, 396
transition radius, 344
Bonner-Ebert spheres, 158–164, 195
Brightness temperature, 360, 494
Broad line region (BLR), 346
Brown dwarfs, 101, 102, 130
BX/BM selection, 347
BzK galaxy selection, 314, 316

C

CI line, 175, 198
CII line, 175, 237, 259, 299, 300
C_2, 195
C_3, 195
C_2H, 175, 237
C_2H_2, 176, 198, 199, 206
C_2H_6, 195
C_4H, 194
C_6H, 194
C_6H_6, 253
Caltech submillimeter observatory (CSO), 427
η Car, 261
CARMA, 230, 232, 237, 427
ρ Cas, 260
Cas A supernova remnant, 263
CH_2CHCH_3, 195
CH_2CHCN, 200
CH_3CH_2CN, 200
CH_3CHO, 202
CH_3CH_2OH, 202
CH_3CN, 200
CH_2CO, 202
CH_3CONH_2, 200
CH_3OCH_3, 200
CH_3OH, 196–198, 200, 202, 237
CH_2OHCHO, 200
CH_4, 102, 105–107, 118, 136, 138, 176, 195, 197, 198
Chandrasekhar limit, 258
Chandra X–ray observatory (Chandra), 448
Chemistry (interstellar), 187–209
Clouds, 102, 103, 108–110, 116
sedimentation, 111
silicates, 109, 110
Clumping factor, 375
21cm, 490–501
brightness fluctuations, 489ff, 494ff
fluctuation signal, 366–369
transition, 360–366
See also Hyperfine
CN, 175
CO, 108, 113, 118, 136, 170, 175, 176, 180ff, 228, 229, 237, 258
CO_2, 135, 136, 138–140, 176, 196–198, 206
Coagulation, 169, 188
COBE, 298, 299
Co-evolution, 339ff
Cold dark matter (CDM), 483
Collisional excitation, 359
Collisional ionization, 359
Color-luminosity, 310
Color-magnitude diagram, 115
Comets, 57ff
Halley family, 57–59
Kuiper Belt, 59
long period, 57ff
main-belt, 60–61
Compact HII region, 235
Concordance ΛCDM cosmology, 386
Constellation-X (Con-X), 450
Cosmic microwave background (CMB), 361
anisotropies, 386
polarization, 458–459, 485
Cosmic rays, 291–294
Cosmic star formation history, 394
with supernovae, 398–411
Cosmic variance, 402
Coulomb collisions, 359
Crab supernova remnant, 263
Critical density, 192
CS, 173, 175, 237
3C58 supernova remnant, 263

D

D_2CO, 196
D_2H^+, 176, 197, 209
Damped Lyα system (DLA), 423, 429–431, 441
Dark Ages (in the Universe), 2, 411, 457, 492
Dark energy, 412
Dark gaps, 461
Dark matter halo, 316
DCN, 175
DCO^+, 175
Debris disk, 82ff
See also, Disk, debris
Depletion, 290, 291
Deuterium abundance, 425, 426

Deuterium fractionation, 196, 197
Disk
accretion shock, 205
chemistry, 205, 252
debris, 82ff, 168
See also Grains, Dust, zodiacal
dynamics, 179, 180
evolution, 85, 188, 205–207
frequency, 85–88
gap, 180, 206
geometry, 173
infall, 173
mixing, 169
photodissociation Region, 171, 175, 205, 207
planet, 174
post-AGB object, 252
protoplanet, 180
protoplanetary, 167–182
rotation, 179, 180
snow line, 206
spectral Energy Distribution, 84, 168, 173, 206
surface density, 168
turbulence, 169, 180
viscosity, 168, 180
Distant red galaxies (DRG), 314ff
DM Tau, 175, 176
30 Dordus, 217
Dust, 193, 274–276
amorphous, 110, 118, 250, 260, 275, 292–295
asteroid, 72
budget, 274–276, 297
cometary, 72
crystalline, 110, 118, 250–252, 260, 275, 279, 292–295
destruction, 288–292
extinction, 310
formation, 260–267, 297
interplanetary, 73
Kuiper Belt, 73ff
solar system, 72
zodiacal, 72, 75
See also Grains
Dusty torus, 338
Dwarfs
atmosphere, 106–108
binary, 111, 116
brown, 101, 102, 130
color, 106
cooling, 102, 115
dataset, 106
effective temperature, 103, 106
spectra, 103
spectral indices, 104
spectral sequence, 101, 104–106
surface gravity, 116
ultracool, 101–119

E

Earth, 54, 125, 126
See also Terrestrial planets
Eddington rate, 346, 395ff
Edison (space telescope), 47
Elemental depletions, 290, 291
Embryo, 169
Emission line kinematics, 321
Escape fraction, 390, 475–476
Exoplanets, 102, 123–141
atmosphere, 134
Doppler technique, 123
effective temperature, 128, 129
formation, 168, 169
ε-εEri, 82ff
habitable, 138
microlensing, 124
migration, 124
moon, 134
ocean, 139
pulsar, 124
radii, 127
spectrum, 125
super-earths, 123, 135, 140
Extinction, 222, 279
Extragalactic UV background, 434, 435
Extremely large telescope (ELT), 208, 450
Extreme star formation, 215–238

F

Feedback, 321–329, 336ff, 358ff, 391–395, 411, 489
chemical, 369ff
photoheating, 374–375, 489
FeH, 102
FeII line, 176, 198
Field, George, 40
Filtering scale, 483
First black holes, 395–398
First star, 386, 389–390, 395–398, 483–489
Supernovae, 399
See also Population III
Fischer-Tropsch, 206
Fluctuation amplitude, 368
Fomalhaut, 7, 82
Fornax cluster, 255
Freeze out, 188, 196, 198, 199, 205

G
GAIA, 248, 262, 297, 450
Galactic
  center, 219, 224
  corona, 273, 282–284
  fountain, 272
  halo, 273, 283, 436, 444
  mass budget, 273
  mechanical energy, 273, 282
  winds, 272, 303, 436
Galaxies
  assembly, 320–321
  bars, 344
  blue, 339
  bulge-dominated, 339
  clusters, 315–317, 342
  dwarf, 216, 219, 222, 227, 231, 273, 297ff
  early-type, 310
  elliptical, 250, 337, 342, 347
  formation, 4, 336
  FR-I radio galaxy, 341
  FR-II radio galaxy, 341, 347
  gravitationally-lensed, 473
  green valley, 339
  irregular morphology, 311
  lopsided, 340
  luminous IR Galaxies, 216, 228, 230
  Lyman alpha emitters, 387, 463–465
  Lyman break, 347
  mass, 346
  morphology, 310
  post-starburst, 340, 341
  radio, 337, 341–342
  red, 339
  selection technique, 311–315
  Seyfert, 337–341
  starburst, 216, 217, 228, 230
  star forming, 430, 434, 435
  Ultraluminous IR galaxies (ULIRG), 216, 228, 231, 276, 295, 305, 310
  winds, 216, 232, 235, 272, 303
  Wolf-Rayet, 219, 225
Gamma-ray burst (GRB), 398, 412, 502–505
Gas
  cooling rate, 336, 342
Gas giant planets, 55
Gas phase reactions, 193
GG Tau, 175
Giacconi, Ricardo, 34ff
Giant HII region, 216
Giant planets, 55
GJ 436b, 128
GL 229B, 101, 113, 118
GL 570D, 112, 113
Globular cluster *see*, Star cluster, globular
Glycine, 200
Grains
  aggregates, 169
  carbonaceous, 253, 261, 275
  chemical equilibrium, 104
  chemistry, 197, 198, 202, 237, 291
  destruction, 288–292
  formation, 260–267, 297
  fragmentation, 169
  grain-grain collisions, 289
  growth, 169, 177, 188, 203, 206, 291
  ices, 172, 237, 275, 291
  iron, 102, 171, 275
  mineralogy, 178
  organic refractory mantle, 280
  oxide, 171, 193, 275, 297
  shattering, 289
  SiC, 255
  silicates, 102, 171, 193, 204, 275, 279
  size distribution, 109
  sputtering, 289
  thermal activation, 92
  very small grains, 285, 286, 296
  See also Dust
Gravitational instability, 169, 357
Gunn-Peterson
  absorption, trough, 458, 475, 486
  damping wing, 462ff
  optical depth evolution, 461, 467
  tests, 459–463

H
H_2, 102, 170, 176, 304, 305
H_3^+, 196, 197
H_2CO, 173, 175, 197, 237
H_2D^+, 196, 197, 209
H_2O, 105, 106, 118, 130ff, 170, 176, 180, 196, 197, 199, 206, 258
Hale-Bopp, 90
Halo, 282–284
HCN, 175, 176, 198, 199, 206, 228, 237, 258
HC_3N, 237
HC_5N, 194
HC_7N, 194
HC_9N, 194
HCO^+, 175, 176
HD 3651B, 112
HD 144432, 179
HD 189733, 128
HD 2094558, 128, 130
HD 32297, 82

HD 69830, 90
He 2–10, 235
Herbig star, 177, 178
Herschel, Space Observatory, 174, 181, 238, 252, 258, 259, 263, 287, 303
HH30, 178
High mass protostellar object (HMPO), 189, 197
High-redshift galaxy selection, 311–315
HII region, 276, 284, 462
  fossil, 391
  percolating, 499
  size, 470
HK Tau B, 177
HNC, 175, 237
Hot Core, 189, 200–202
  chemistry, 202
Hot Jupiters, 128, 130, 136–138, 169
HR 8752, 260
Hubble space telescope (HST), 31ff, 217, 227, 265, 405, 443, 447, 448
Hyper-compact HII region, 189
Hyperfine
  hydrogen levels, 362
  transition, 360
  See also 21 cm
Hypergiant, 260

I

IC 342, 237
IC 348, 162–163
Ice, 188, 196–199, 203–206, 236, 237
Ice giant planets, 56
  See also by name
Illingworth, Garth, 34ff
Inflation, 492
Initial mass function, 102, 117, 222–227, 233, 390
  See also Star formation
Institut de radio astronomie millimetrique (IRAM), 427
Interferometry, 192
Intergalactic medium (IGM), 329–331, 357ff, 375–376, 386, 422, 431–436, 442, 485, 493
Interstellar absorption, 325
Interstellar dust, 274–276
Interstellar gas, 272–274, 285–287
Interstellar medium (ISM), 271–305
  cooling, 280, 281
  corona, 282–284
  heating, 280, 281
  lifecycle, 279, 280
  mixing, 279
  phases, 280–282, 285, 291
Inverse Compton scattering, 359
Ionization history, 365
Ionized fraction, 359, 387
Ionizing
  background, 328
  continuum escape fraction, 390
  efficiency, 370
IRAM *see*, Institut de radio astronomie millimetrique (IRAM)
IRAS, 216
IRC 10011, 249
IRC 10420, 260
IR cirrus, 276
IR emission features, 276, 279, 301

J

James webb space telescope (JWST), 176, 177, 181, 208, 222, 225, 227, 253, 255, 297, 303ff, 437, 450ff
  fine guidance sensor (FGS), 22–23
  mid-infrared instrument (MIRI), 19–21, 108, 140
  near-infrared camera (NIRCam), 16–17, 134, 141, 407
  near-infrared spectrograph (NIRSpec), 18–19, 133ff, 469–471
  performance, 8ff
  tunable filter imager (TFI), 21–22
Jeans length, 161
Jeans mass, 374, 396, 402
Jupiter, 55, 103, 112, 125, 126
  See also Gas giant planets

K

Kepler supernova remnant, 263
Kuiper Belt, 62–67ff
  classical, 62
  detached, 64
  extrasolar, 81
  properties, 70ff
  resonant, 64
  scattered, 64
  See also Comets

L

Laboratory astrophysics, 193
Lane-Emden equation, 157–160
Large deployable reflector (LDR), 43–45
Large lunar telescope (LLT), 38
Large millimeter telescope (LMT), 427, 437
Large space telescope (LST), 42–43
Late heavy bombardment (LHB), 67ff

Light echo, 265
Lilly–Madau diagram, 394
Line trapping, 193
LIRG, *see* Galaxies, luminous IR galaxies
LMT, *see* Large millimeter telescope (LMT)
Longair, Malcolm, 34
Low Frequency Array (LOFAR), 499ff
Luminous Blue Variable (LBV), 232, 248, 260–262, 276, 297
Luminous IR Galaxies, *see* Galaxies, luminous IR galaxies
Lyman α
  absorption, 429–448
  forest, 373–374, 422, 431, 434, 435, 463
  selection, 348
Lyman break technique, 313
Lyman-limit system (LLS), 422, 432–434
Lyman-Werner background, 358–359, 391ff

M

M1-67, 261
M17, 284, 287
M31, 297
M74, 266
M81, 227, 228
M82, 217, 222, 223, 227, 228, 235, 273, 284
M83, 228
M87, 226
Machetto, Duccio, 41
Maffei 2, 219
Magellanic Clouds, 217, 248, 258, 273, 279, 297, 439
Magneto-rotational instability, 168
Mars, 55, 125, 126
  See also Terrestrial planets
2MASS 0415-0935, 114
2MASS J224-0158, 110
Mass loss, 247–267
Mass-metallicity relationship, 324ff
2MASS WJ 46425+2000321, 116
Mean-motion resonances (Solar System), 75
Mercury, 54
  See also Terrestrial planets
Merger, major, 340
Metal
  enrichment, 369
  line absorption, 371–372
  line blanketing, 406
MIDI-VLTI, 178, 179
Migration, 169
Minihalos, 387, 400, 410
Mira, 249
Molecular clouds, 228–230, 234, 427, 428, 439
  core mass function, 153
  linewidth-size relation, 156
  masses, 229, 230, 427
  pressure-confined cores, 156
  thermally dominated cores, 156
Molecular hydrogen
  cooling, 387
  feedback, 391
  formation, 388–389, 392
  photodissociation, 390
  suppression, 396
  X-ray catalysis, 360
Molecules
  chemical equilibrium, 104, 112, 113
Moon, transit, 134
MS1512-cB58, 326ff
MSX SMC, 255, 256
Murchison Wide-Field Array (MWA), 499ff

N

N_2, 113, 114, 135, 197, 198
N_2H^+, 197, 237
Narrow-Line Region (NLR), 338, 346
National Academy of Science (NAS, on NGST/JWST), 35, 42
National Research Council, 42
ND_3, 196
Negative ion, 194
NeII line, 176
NeIII line, 225
Neptune, 56, 76ff, 132
  See also Ice giant planets
Neutral fraction, 461, 467
  See also Ionized fraction
Next Generation Space Telescope (NGST), 31, 37–40, 42ff
NGC 253, 219
NGC 628, 266
NGC 1275, 217
NGC 1569, 217, 218, 222, 223
NGC 1705, 213, 217
NGC 1866, 218
NGC 2023, 287
NGC 2516, 258
NGC 3077, 227, 228
NGC 3603, 219, 223, 233, 235
NGC 5253, 222, 227ff
NGC 6302, 250, 258
NGC 6720, 259
NGC 6946, 219, 267
NGC 7023, 287
NGC 7027, 258
NH_2CH_2CN, 201

NH_2CH_2COOH, 201
NH_3, 102ff, 197, 198, 199
NH_4^+, 198
NHD_2, 196
Nice model, 67–68
NML Cyg, 260
Novae, 276
Nucleosynthesis, 248
Number counts, 310

O
O_2, 138
O_3, 140
OI line, 202–204, 209, 372, 470
OIV line, 225
OVI line, 285, 436, 445–447
Omicron Ceti, 249
Oort Cloud, 57ff
Optical depth
  Thompson (electron) scattering, 386, 458, 475
Origins of life, 6, 200
Orion bar, 287
Orion-KL, 200
Outflow, 188, 202, 249–253, 258–262, 321ff, 341
Oxygen budget, 203, 204

P
PAHs, *see* Polycyclic Aromatic Hydrocarbon
P-Cygni, 261, 319
PDR, *see* PhotoDissociation Region (PDR)
Perseus Cloud, 153
Perturbation theory, 357
Phoebe, 69
PhotoDissociation Region (PDR), 171, 175, 236, 237, 286–288, 298, 300
Photoelectric effect, 285–287, 298
Photoheating, 374–375, 378, 391
Photoionization equilibrium, 357
Photoionization rate, 434, 435, 442, 443
Photometric redshift, 406
Photospheric pollution (of stars by dust), 92
β-Pic, 82ff
Pipe Nebula, 153–156, 162–163
Planetary nebula, 252, 255, 258, 259, 276
Planets
  formation, 5–6, 168, 169
  migration, 169
  See also by name, Terrestrial planets, Giant planets
Plateau de Bure, 230, 232
Pleiades cluster, 111, 112, 232
Plutinos, 64
  See also Kuiper Belt
Pluto, 71
  See also Kuiper Belt
Polycyclic aromatic hydrocarbon, 171, 172, 190, 193, 206–208, 236, 252, 273, 276–279, 285–287, 298, 301–303, 305
Population II, 369, 468
Population III, 369, 468
  star, 359, 365, 369, 484, 503
  supernovae, 410, 484
Population synthesis models, 321
Porous aggregates, 169
Post-AGB object, 252, 253, 276
Power spectrum, 368
  Non-Gaussianity, 492
  Primordial, 386, 492
  small-scale, 411–413
Poynting–Robertson Effect, 75
Prebiotic species, 200, 201
Press-Schechter, 371, 398
Prestellar core, 194–197
Protoplanetary disk, 167–182
  solar, 57
Protostar, 188
Proximity zone, 463

Q
QSO, 328, 337ff, 341, 345ff, 350, 462, 467
  host galaxy, 337
  mini, 466
  survey, 471–472
Quasar, 260, 434
  HS 1821+643, 446
  J114816.64+5251503.3, 300
  J33941.95+054822.1, 422, 423
  PG 2112+059, 260, 279
  Q1307-BM1163, 319
Quintuplet Cluster, 219, 224

R
R136, 217, 218, 223, 224
Radiative transfer, 397, 475
Recombination epoch, 457, 492
Red Rectangle, 252
Reflection nebula, 276
Reionization
  double, 370
  helium, 376–379
  history, 2, 3, 387, 394, 402ff, 468–476, 485
  hydrogen, 358, 365, 370, 373ff, 402, 405

sources, 465–466
topology, 473–474, 487ff
X-ray, 360
Ring nebula, 259
Rotation diagram, 193

S

Salpeter time, 473
Saturn, *see* Gas giants
Schechter function, 336
Schmidt law, 228
SCUBA sources, 315
Serpens cloud, 152
Sgr B2, 200, 219
Shock, 202, 203, 205, 237, 262, 288–292
Sigma Orionis cluster, 111
SI line, 176, 198
SIV line, 225
SiII line, 176
Silicate
amorphous, 110, 118, 250, 260, 275, 292–295
crystalline, 110, 118, 250–252, 260, 275, 279, 292–295
enstatite, 109, 110, 118, 275
feature, 108–110, 118, 177, 196, 249
forsterite, 109, 110, 118, 250, 251, 275, 293
quartz, 118
SiO, 264
Sloan Digital Sky Survey (SDSS), 2, 386
SMA, *see* SubMillimeter Array (SMA)
SMP LMC 8, 255
SMP LMC 11, 253–255
SMP LMC 36, 255
Snow line, 206
SOFIA, *see* Stratospheric Observatory For IR Astronomy (SOFIA)
Solar corpuscular drag, 75
Space Studies Board (SSB) (on NGST/JWST), 35
Space Telescope Science Institute (STScI), 31ff
Spectroscopy
absorption, 190
completeness, 315
emission, 191
infrared, 189
sub-millimeter, 191
Spin temperature, 361
Spiral arm, 230
Spitzer, 237
Spontaneous emission, 363
Square Kilometer Array (SKA), 207, 228, 437, 440
Starburst, *see* Galaxies, starburst
Star cluster, 216, 217
evolution, 247
evolutionary sequence, 219–222
globular, 215, 217, 234, 236
masses, 222–225
mass function, 226, 227, 234
mass loss, 247–267
super, 215–217
Star dust, 262, 274
Star formation, 4–5, 147ff, 188, 205, 215–238, 336, 343, 346, 374, 392
density, 310, 401
efficiency, 228, 233
initial mass function (IMF), 148ff, 222–227, 233
outflow, 188, 202
rate, 273, 301
Stars
Asymptotic Giant Branch (AGB), 249–259, 273, 276, 279
binary, 261
carbon-rich, 255, 258
effective temperature, 103, 106, 110
evolution, 247, 248
intracluster, 255
multiplicity, 148–149
oxygen-rich, 255, 258
spectroscopy, 224
spectrum, 103
upper mass limit, 223, 224
winds, 234, 235
Stellar
mass density, 310
populations, 317–321
Stockman, Peter, 34
Stratospheric Observatory For IR Astronomy (SOFIA), 176, 181, 209, 222, 238, 252, 253, 287, 297, 298, 303
Stromgren sphere, 462ff
SubMillimeter Array (SMA), 230, 232, 237
Sugar, 200
Sun
formation, 69
Sunyaev-Zeldovich effect, 437, 449
SuperEarths, 123, 135, 140
Supergiant, 260, 276, 297
Super Lyman limit systems (SLLS), 431
Supernova, 272, 273, 276, 282–284, 288, 297, 359, 391, 489
lightcurve, 264

light echo, 265
pair-instability, 399, 484
properties, 404–406
rates, 399–411
remnant, 232, 234, 260, 262–267, 291
type II, 399
type IIP, 404ff
winds, 369–371
Supernovae
SN 1987A, 263
SN 2002h, 267
SN 2002hh, 264
SN 2003gd, 266
SN 2005af, 264
SN 2005df, 264
SN 2006jc, 267

T
Taurus cloud, 162–163
Terrestrial planets, 54
Thermal activation, 92
Thirty-Meter Telescope (TMT), 315
Thompson, Rodger, 46
Thompson scattering, 483
See also Optical depth
Thronson, Harley, 47
TiO, 102, 130
TMC1, 194, 195
Transits, 123–141
Trapezium (in Orion), 150–151
TrES, 136
Trojans, Trojan asteroids, 64, 67
T Tauri star, 177
Tulkoff, P. J., 38
Turbulent shear zone, 202
TW Hya, 174–176
Tycho supernova remnant, 263

U
ULAS J0031-00, 117
ULIRG, *see* Galaxies, Ultraluminous IR galaxies (ULIRG)
Ups And, 128, 130
Uranus, 56
See also Ice giant planets

V
Vega, 84
Venus, 54, 125, 126
See also Terrestrial planets
Very Large Array (VLA), 222, 237, 440, 450
Very small grains, 285, 286
Virgo cluster, 255
Virial temperature, 388
Voigt profile, 363
VX Sgr, 260
VY CMA, 258, 260

W
W49A, 219
Warm Dark Matter (WDM), 386, 412
Westerlund, 1, 219
Wide-field Infrared Survey Explorer (WISE), 222
Wilkinson Microwave Anisotropy Probe (WMAP), 2
Wolf-Rayet, 219, 225, 232, 248, 260–262, 267, 276, 277
Wouthuysen-Field mechanism, 362ff, 367ff, 394, 493ff
WR, 114, 261
WX Psc, 249

X
XMM Newton X–ray Observatory (XMM), 448
X-ray
background, 359–360
binaries, 360
X-ray Dissociation Region (XDR), 236

Z
Zodiacal cloud, 77
See also Dust, zodiacal